T0189048

Lecture Notes in Computer Science 1113

Edited by G. Goos, J. Hartmanis and J. van Leeuwen

Advisory Board: W. Brauer   D. Gries   J. Stoer

Springer
Berlin
Heidelberg
New York
Barcelona
Budapest
Hong Kong
London
Milan
Paris
Santa Clara
Singapore
Tokyo

Wojciech Penczek   Andrzej Szałas   (Eds.)

Mathematical Foundations of Computer Science 1996

21st International Symposium, MFCS '96
Cracow, Poland, September 2-6, 1996
Proceedings

 Springer

Series Editors

Gerhard Goos, Karlsruhe University, Germany

Juris Hartmanis, Cornell University, NY, USA

Jan van Leeuwen, Utrecht University, The Netherlands

Volume Editors

Wojciech Penczek
Polish Academy of Sciences, Institute of Computer Sciences
ul. Ordona 21, 01-237 Warsaw, Poland

Andrzej Szałas
University of Warsaw, Institute of Informatics
ul. Banacha 2, 02-097 Warsaw, Poland

Cataloging-in-Publication data applied for

Die Deutsche Bibliothek - CIP-Einheitsaufnahme

Mathematical foundations of computer science 1996 : 21th
international symposium ; proceedings / MFCS '96, Craków,
Poland, September 2 - 6, 1996 / Wojciech Penczek ; Andrzej
Szałas (ed.). - Berlin ; Heidelberg ; New York ; Barcelona ;
Budapest ; Hong Kong ; London ; Milan ; Paris ; Santa Clara ;
Singapore ; Tokyo : Springer, 1996
  (Lecture notes in computer science ; Vol. 1113)
  ISBN 3-540-61550-4
NE: Penczek, Wojciech [Hrsg.]; MFCS <21, 1996, Kraków>; GT

CR Subject Classification (1991): F.1-4, D.2-3, G.2

ISSN 0302-9743
ISBN 3-540-61550-4 Springer-Verlag Berlin Heidelberg New York

Typesetting: Camera-ready by author
SPIN 10513314   06/3142 – 5 4 3 2 1 0   Printed on acid-free paper

Foreword

The present volume contains the papers selected for presentation at the 21st Symposium on Mathematical Foundations of Computer Science (MFCS'96) held in Cracow, Poland, September 2–6, 1996. The symposium was organized in the Polonia Institute of the Jagiellonian University.

It was the 21st symposium in a series of international meetings organized alternately in the Czech Republic, Poland, and the Slovak Republic. The aim of these symposia is to bring together specialists in various theoretical fields of computer science and to stimulate mathematical research in all branches of theoretical computer science. The previous meetings took place in Jabłonna, 1972; Štrbské Pleso, 1973; Jadwisin, 1974; Mariánské Lázně, 1975; Gdańsk, 1976; Tatranská Lomnica, 1977; Zakopane, 1978; Olomouc, 1979; Rydzyna, 1980; Štrbské Pleso, 1981; Prague, 1984; Bratislava, 1986; Carlsbad, 1988; Porąbka-Kozubnik, 1989; Banská Bystrica, 1990; Kazimierz Dolny, 1991; Prague, 1992; Gdańsk, 1993; Košice, 1994; and Prague, 1995.

The program committee of MFCS'96 consisted of:

K.R. Apt (The Netherlands)	W. Marek (USA)
R. Back (Finland)	I. Németi (Hungary)
J.C.M. Baeten (The Netherlands)	R. De Nicola (Italy)
M. Crochemore (France)	M. Nielsen (Denmark)
D. Gabbay (UK)	E. R. Olderog (Germany)
M.C. Gaudel (France)	E. Orłowska (Poland)
U. Goltz (Germany)	Z. Pawlak (Poland)
W. Foryś (Poland)	W. Penczek (chair, Poland)
R. Kurshan (USA)	A. Pnueli (Israel)
G. Rozenberg (The Netherlands)	A. Szałas (co-chair, Poland)
P. Ružička (Slovak Republik)	A. Tarlecki (Poland)
W. Rytter (Poland)	I. Wegener (Germany)
C. Stirling (UK)	J. Zlatuška (Czech Republic)

The scientific program included invited lectures covering the areas of main interest and contributed papers. Altogether 35 contributed papers were selected out of 92 submissions. Thus the present volume contains 35 contributed papers, 8 invited papers, and 2 abstracts of the invited talks.

The selection process was organized in a distributed way via e-mail discussions. It took over one month to select the papers. This allowed Program Committee members to carefully discuss the evaluation of the submitted papers.

The following referees assisted the Program Committee in the extensive evaluation process: L. Aceto, S. Ambroszkiewicz, D. Ancona, R. Back, E. Bampis, T. Basten, M.A. Bednarczyk, J. Berstel, J. Blanco, R. Bloo, F. de Boer, J. Bohn,

M. Boreale, O. J. Boxma, T. Brauner, F. van Breugel, H. Buhrman, G. Castagna, J. Cassaigne, A. Cheng, B. Chlebus, P. Chrétienne, V. Dancik, S. Delaet, S. Demri, J. Desarmenien, H. Dierks, M. Dietzfelbinger, K. Diks, K. Doets, P. Duris, P. van Emde Boas, U. H. Engberg, E. Fachini, M. Fernandez, G. Ferrari, C. Fischer, P. Fischer, H. Fleischhack, W. Fokkink, L.C. van der Gaag, L. Gąsieniec, F. Gécseg, H. Geuvers, P. Godefroid, J. Goldsmith, B. Grahlmann, B. Graves, J. Grundy, A. Habel, J. Harrison, T. Hegedus, W. H. Hesselink, T. Hofmeister, M. Huhn, H. Huttel, M. Hühne, P. Idziak, W. Janssen, M. Jerrum, C. Johnen, B. Juurlink, S. Kahrs, L. Kari, M. Karpiński, C. Kenyon, J. Kleist, J. N. Kok, B. Konikowska, J.-C. König, M. Krause, H.-J. Kreowski, M. Kryszkiewicz, R. Kubiak, R. Kuiper, O. Kupferman, Á. Kurucz, A. Labella, K.G. Larsen, C. Lavault, J. van Leeuwen, H. Lefmann, C. Lynch, J. Madarász, A. Maggiolo-Schettini, M. Marchiori, A. Masini, A. Mateescu, G. Mirkowska, U. Montanari, M. Morvan, P. D. Mosses, J. Neraud, P. Niebert, D. Niwiński, E. Ochmański, D. Pardubska, G. Paun, W. Pawłowski, D.A. Peled, D. Perrin, A. Pierantonio, W. Plandowski, R. Pugliese, P. Quaglia, M. Raczunas, A. Ramer, M. Regnier, M.A. Reniers, A. Rensink, B. Reus, D. Revuz, G. Roussel, B. Rovan, J. Rutten, G. Sági, I. Sain, A. Salomaa, K. Salomaa, D. Sangiorgi, V. Sassone, M. Schenke, M. Schwartzbach, P. Seebold, D. Sieling, A. Simon, H. U. Simon, S. Sokołowski, N. Spyratos, M. Srebrny, P. Stevens, K. Sunesen, O. Sýkora, M. Ślusarek, R. Szelepcsenyi, A. Szepietowski, J.-Y. Thibon, J. Tiuryn, M. Truszczyński, P. Urzyczyn, J.J. Vereijken, L. Viennot, E. de Vink, I. Vrto, D. Walker, E. Waller, I. Walukiewicz, R. Wegner, H. Wehrheim, J. Winkowski, M. Wirsing, J. von Wright, S. Yu, M. Zawadowski, B. Zerrouk, W. Zielonka.

The Symposium was organized by the Institute of Computer Science of the Polish Academy of Sciences and the Institute of Informatics of Warsaw University. The Organizing Committee consisted of: S. Ambroszkiewicz, M. Białasik, W. Foryś, E. Gąsiorowska, B. Konikowska (chair), W. Penczek, M. Srebrny, A. Szałas, and A. Tarlecki.

The MFCS'96 Symposium was supported by the following institutions and companies: the European Association for Theoretical Computer Science, ICS PAS, Warsaw University, European MEDICIS project, Microsoft Poland, SUN Poland, the Batory Foundation, and the State Committee for Scientific Research (KBN).

We would like to thank the invited speakers, the authors of the papers, and the Program Committee members. We would also like to thank all the above institutions and companies for financial support. We are also indebted to Springer-Verlag for excellent co-operation in publishing this volume.

Warsaw, 1996   Wojciech Penczek and Andrzej Szałas

Contents

Word Level Model Checking

Edmund M. Clarke, Jr.
(joint work with Xudong Zhao)

School of Computer Science
Carnegie Mellon University
Pittsburgh, Pa. 15213

Abstract. Proving the correctness of arithmetic operations has always been an important problem. The importance of this problem has been recently underscored by the highly-publicized division error in the Pentium. Although symbolic model checking techniques based on Binary Decision Diagrams (BDDs) have been successful in verifying control logic, these techniques cannot be used directly for verifying arithmetic circuits. In order to verify arithmetic circuits in this manner, it is necessary to be able represent and manipulate functions that map boolean vectors to integer values.

We have extended the symbolic model checking system SMV so that it can handle properties involving relationships among data words. In the original SMV system, atomic formulas can only contain state variables. In the extended system, we allow atomic formulas to be equations or inequalities between expressions as well. These expressions are represented as "hybrid BDDs." We have developed efficient algorithms to compute the set of variable assignments that would make these equations and inequalities true. For the class of linear functions, which includes many of the functions that occur in practice, such operations are guaranteed to have complexity that is polynomial in the width of the data words.

By using the extended model checking system, we have successfully verified circuits for division and square root computation that are based on the SRT algorithm used by the Pentium. We are able to handle both the control logic and the data paths. The total number of state variables exceeds 600 (which is much larger than any circuit previously checked by SMV).

Code Problems on Traces

Volker Diekert   Anca Muscholl*

Universität Stuttgart, Institut für Informatik
Breitwiesenstr. 20-22, D-70565 Stuttgart

Abstract. The topic of codes in the framework of trace monoids leads to interesting and challenging decision problems of combinatorial flavour. We give an overview of the current state of some basic questions in this field. Among these, we consider the existence problem for strong codings, clique-preserving morphisms and the unique decipherability problem (code problem).

1   Introduction

Free partially commutative monoids [7] offer a mathematically sound framework for modelling and analyzing concurrent systems. This was made popular by the work of Mazurkiewicz. He investigated originally the behaviour of safe 1-labelled Petri nets [17] and the computer science community quickly recognized the importance of this work. The basic concept is to consider a system as a finite set of actions Σ, together with a fixed symmetric independence relation $I \subseteq \Sigma \times \Sigma$, denoting pairs of actions which can be scheduled in parallel. In the setting defined by a pair (Σ, I) we identify sequential observations (i.e., strings over Σ) modulo identities $ab = ba$ for $(a, b) \in I$. This yields the partial commutation and the resulting monoid has been called *trace monoid* by Mazurkiewicz. Trace monoids have been successfully considered in classical theories like formal languages, automata and logic. Some challenging decision problems for trace monoids are still open. One of the fundamental open problems is whether or not there is an algorithm to decide the solvability of equations involving constants. For the word case of free monoids the positive answer is known to due a famous result of Makanin [15]. The solution of an equation is a homomorphism. We think that a better understanding of homomorphisms will help to attack a possible extension of Makanin's result, and our contribution is restricted to some questions about homomorphisms. The surprising fact is that even basic questions on codings (injective homomorphisms) between trace monoids are still open. We focus on two problems. First, given two trace monoids $\mathbb{M}_1, \mathbb{M}_2$, does an injective homomorphism $h\colon \mathbb{M}_1 \to \mathbb{M}_2$ exist? Second, given a homomorphism $h\colon \mathbb{M}_1 \to \mathbb{M}_2$, is h injective, i.e., a coding? For both questions only partial answers are known. The first question has a positive answer for strong codings, see Sect. 3 below.

---

* This research has been supported in part by the French-German research programme PROCOPE.

The second question is known to be undecidable, even if the first monoid is free, as soon as (Σ_2, I_2) contains a cycle on four vertices as induced subgraph [14, 8]. It is known to be decidable when the independence relation (Σ_2, I_2) does not contain any induced subgraph isomorphic to the path P_4 or the cycle C_4 on four vertices, [1]. Recently it has been shown that for (Σ_2, I_2) being equal to the P_4, the code problem is decidable. However, the question of the precise borderline for decidability is still open. We present in Sect. 4 the solution to the code problem for P_4, which has been exhibited independently in [13, 16]. Another natural instance of the question whether a given homomorphism $h\colon \mathbb{M}_1 \to \mathbb{M}_2$ is a coding or not is given when h is described by morphisms between independence alphabets. This special case arises when actions in a system are refined in such a way that the refined action is a product of independent elements which therefore can be performed in parallel. However even for such a restricted class of homomorphisms the injectivity problem turns out again to be undecidable, see Sect. 5.

2 Notations and Preliminaries

A dependence alphabet is a pair (Σ, D), where Σ is a finite alphabet and $D \subseteq \Sigma \times \Sigma$ is a reflexive and symmetric relation, called dependence relation. The complement $I = (\Sigma \times \Sigma) \setminus D$ is called independence relation; it is irreflexive and symmetric. The pair (Σ, I) is denoted independence alphabet. We view both (Σ, D) and (Σ, I) as undirected graphs. The difference is that (Σ, D) has self-loops.

Given an independence alphabet (Σ, I) (or a dependence alphabet (Σ, D) resp.) we associate the trace monoid $\mathbb{M}(\Sigma, I)$. This is the quotient monoid $\Sigma^*/\equiv_I$, where $\equiv_I$ denotes the congruence generated by the set $\{uabv = ubav \mid (a, b) \in I, u, v \in \Sigma^*\}$; an element $t \in \mathbb{M}(\Sigma, I)$ is called a trace, the length $|t|$ of a trace t is given by the length of any representing word. By $\mathrm{alph}(t)$ we denote the alphabet of a trace t, which is the set of letters occurring in t. The initial alphabet of t is the set $in(t) = \{x \in \Sigma \mid \exists t' \in \mathbb{M}(\Sigma, I) : t = xt'\}$. By 1 we denote both the empty word and the empty trace. Traces $s, t \in \mathbb{M}(\Sigma, I)$ are called independent, if $\mathrm{alph}(s) \times \mathrm{alph}(t) \subseteq I$. We simply write $(s, t) \in I$ in this case. A trace $t \neq 1$ is called a *root*, if $t = u^n$ implies $n = 1$, for every $u \in \mathbb{M}(\Sigma, I)$. A trace t is called connected, if $\mathrm{alph}(t)$ induces a connected subgraph of the dependence alphabet (Σ, D).

3 Existence problem for codings

A homomorphism $h\colon \mathbb{M}(\Sigma, I) \to \mathbb{M}(\Sigma', I')$ is given by a mapping $h\colon \Sigma \to \mathbb{M}(\Sigma', I')$ such that $h(a)h(b) = h(b)h(a)$ for all $(a, b) \in I$. The homomorphism is called a *coding*, if it is injective. Representing each trace $h(a)$ by a word $\hat{h}(a) \in \Sigma'^*$ we obtain a homomorphism between free monoids $\hat{h}\colon \Sigma^* \to \Sigma'^*$. It may happen that the lifting $\hat{h}$ is injective, but h is not. (A trivial example is $\hat{h} = \mathrm{id}_{\Sigma'}$ with $I' \neq \emptyset$.) But if we know that $\hat{h}$ is not injective, then h

cannot be injective. This surprising fact has been recently announced in [5]: If $h\colon \mathrm{M}(\Sigma, I) \to \mathrm{M}(\Sigma', I')$ is a coding, then every lifting $\hat{h}\colon \Sigma^* \to \Sigma'^*$ is a word coding. The result is far from being trivial and gives some flavour about the unexpected behaviour of homomorphisms between trace monoids. It is completely open whether the existence problem for codings between trace monoids is decidable. Only some few results and conjectures have been established [4, 12]. The situation changes if we restrict our attention to a naturally arising subclass of homomorphisms. The existence problem of strong codings has a nice graph characterization being discussed in this section.

Definition 1. [6] A homomorphism $h\colon \mathrm{M}(\Sigma, I) \to \mathrm{M}(\Sigma', I')$ is called a *strong* homomorphism, if $(h(a), h(b)) \in I'$ holds for all $(a, b) \in I$. A *strong* coding is a strong homomorphism being injective.

Hence, strong homomorphisms map independent letters to independent traces. The perhaps most prominent example of a strong coding is given by the Projection Lemma [9, 10]. It can be rephrased as follows.

Example 1. Let $\Sigma' \subseteq \Sigma$ be a subalphabet such that $\Sigma' \times \Sigma' \subseteq D$. The canonical projection $\pi_{\Sigma'}\colon \mathrm{M}(\Sigma, I) \to \Sigma'^*$ is the homomorphism induced by setting $\pi_{\Sigma'}(a) = a$ for $a \in \Sigma'$ and $\pi_{\Sigma'}(a) = 1$ for $a \in \Sigma \setminus \Sigma'$. Consider (Σ, D) written as a union of cliques, i.e., $(\Sigma, D) = (\bigcup_{i=1}^{k} \Sigma_i, \bigcup_{i=1}^{k} \Sigma_i \times \Sigma_i)$. Then the canonical homomorphism

$$\pi\colon \mathrm{M}(\Sigma, I) \to \prod_{i=1}^{k} \Sigma_i^*, \qquad t \mapsto (\pi_{\Sigma_i}(t))_{1 \le i \le k}$$

is a strong coding.

The Projection Lemma leads to the upper bound in the next proposition. To derive the lower bound is left as an exercise to the reader.

Proposition 2. *Let (Σ, D) be a dependence alphabet and $|\Sigma_i| \ge 2$ for $i = 1, \ldots, k$. Then there exists a strong coding $h\colon \mathrm{M}(\Sigma, I) \to \prod_{i=1}^{k} \Sigma_i^*$ if and only if (Σ, D) allows a covering by k cliques. In particular, deciding the existence of strong codings into a k-fold direct product of free monoids is NP-complete.*

The following example suggests some of the differences between the existence problem for codings resp. strong codings.

Example 2. Let $(\Sigma, D) = C_n$ denote the cycle on $n > 3$ vertices, i.e., $\Sigma = \{a_1, \ldots, a_n\}$ and D is defined by the edges $\{a_m a_{m+1} \mid 1 \le m \le n\}$ (with addition modulo n). By the proposition above there exists a strong coding from $\mathrm{M}(\Sigma, I)$ into $\prod_{i=1}^{k} \Sigma_i^*$ (where $|\Sigma_i| \ge 2$ for all i) if and only if $k \ge n$. If we consider codings instead of strong codings we obtain $k = n - 1$ as an upper bound. Let $h\colon \mathrm{M}(\Sigma, I) \to \prod_{i=1}^{n-1} \Sigma_i^*$ be given by $h(a_m) = x_{m-1} y_m x_{m+1}$ for $m < n$ and $h(a_n) = x_{n-1} x_1$ (where by convention $x_0 = x_{n-1}$ and $x_n = x_1$). It is straightforward to check that h is a homomorphism. Suppose now $h(u) = h(v)$ for some u, v and let $|u|+|v|$ be minimal. The minimality implies $in(u) \cap in(v) = \emptyset$. Since $u \neq 1 \neq v$ we may assume by symmetry that for some $1 \leq m < n$ we have $a_m \in in(u) \setminus in(v)$. Then $y_m \in in(h(u))$. This implies $a_m \in alph(v)$ and even $a_m \in in(v)$ (otherwise $x_m \in in(h(v))$). Hence a contradiction.

It can be shown that $k \geq n - 1$ is also necessary for the existence of a coding for $n > 4$. On the other hand, $M(\Sigma, I)$ with $(\Sigma, D) = C_4$ can be already embedded in the direct product of two free monoids, [4, 12].

No difference between the existence of codings and strong codings occurs however if the left-hand side monoid is free commutative (see also [18]). It also yields a lower bound for the complexity in Cor. 4 and Cor. 7 below.

Proposition 3. *Let (Σ, D) be a dependence alphabet and $k \geq 1$. The following assertions are equivalent:*

i) The dependence alphabet (Σ, D) contains an independent set of size k.

ii) There exists a strong coding $h \colon \mathbb{N}^k \to M(\Sigma, I)$.

iii) There exists a coding $h \colon \mathbb{N}^k \to M(\Sigma, I)$.

Corollary 4. *Given (Σ, D) and k, it is NP-complete to decide whether there exists a (strong) coding of $\mathbb{N}^k$ into $M(\Sigma, I)$. Therefore, the problem whether there exists a (strong) coding between two given trace monoids is (at least) NP-hard.*

Strong homomorphisms are tightly connected to morphisms of dependence and independence alphabets, which we now define.

Definition 5. *Let (V', E'), (V, E) be undirected graphs (possibly with self-loops), $H \subseteq V' \times V$ be a relation between vertices, and for $a' \in V'$ let $H(a')$ denote the set $\{a \in V \mid (a', a) \in H\}$. The relation H is called a relational morphism, if $(a', b') \in E'$ implies $H(a') \times H(b') \subseteq E$ for all $a', b' \in V'$.*

Theorem 6 [12]. *Let (Σ, D) and (Σ', D') be dependence alphabets. The following assertions are equivalent:*

i) There exists a strong coding $h \colon M(\Sigma, I) \to M(\Sigma', I')$.

ii) There exists a relational morphism $H \colon (\Sigma', D') \to (\Sigma, D)$ of dependence alphabets such that for all $a \in \Sigma$ there exists $a' \in \Sigma'$ with $a \in H(a')$, and for all $(a, b) \in D$, $a \neq b$ there exists $(a', b') \in D'$, $a' \neq b'$, with $(a, b) \in H(a') \times H(b')$.

Furthermore there are effective constructions between h and H such that $H = \{(a', a) \in \Sigma' \times \Sigma \mid a' \in alph(h(a))\}$.

Corollary 7. *The following problem is NP-complete:*
Instance: Independence alphabets (Σ, I), (Σ', I').
Question: Does there exist a strong coding from $M(\Sigma, I)$ into $M(\Sigma', I')$?

4 Code problem

Rational languages in trace monoids (built over finite sets using concatenation, union and Kleene star) strictly contain the family of recognizable languages and do not form a Boolean algebra, in general. More precisely, the family of rational trace languages over $M(\Sigma, I)$ is a Boolean algebra if and only if the independence relation I is transitive [20]. This fact leads to nontrivial decision problems for rational trace languages, see [3, 1, 20]. It is known e.g. that the question whether the intersection of two rational trace languages is empty or not is undecidable, in general. More precisely, the intersection problem is undecidable if and only if the graph associated to the independence relation contains as induced subgraph either a cycle or a path on four vertices [1]. It is therefore decidable exactly for transitive forests [22].

We ask in this section a closely related question, namely whether a finite set X is a code, i.e., whether X freely generates X^+. More generally, we will consider the notion of trace code.

Definition 8. For a set $X \subseteq M(\Sigma, I)$ let Σ_X be an alphabet being in bijection ψ with the set X. Define $I_X \subseteq \Sigma_X \times \Sigma_X$ by $(x, y) \in I_X$ if $x \neq y$ and $\psi(x)\psi(y) = \psi(y)\psi(x)$.
Then X is called a *trace code*, if the induced homomorphism $\psi \colon M(\Sigma_X, I_X) \to M(\Sigma, I)$ is a coding.

Definition 9. The *(trace) code* problem for (Σ, I) is to decide on the instance of a finite set $X \subseteq M(\Sigma, I)$ whether or not it is a (trace) code.

The trace code problem can be reduced to the intersection problem of rational trace languages. To see this, assume that X is not a trace code, then we find two traces $u, v \in M(\Sigma_X, I_X), u \neq v$ such that $\psi(u) = \psi(v)$. For u, v with $|u| + |v|$ minimal we obtain some $x \in \Sigma_X$ with $x \in in(u)$ and either $alph(v) \subseteq I_X(x)$ or $v = v_1 y v_2$, for some $y \neq x$, $(x, y) \notin I_X$ and $alph(v_1) \subseteq I_X(x)$ ($I_X(x)$ denoting the set of letters from Σ_X independent of x). Hence, X is not a trace code if and only if for some $x \in \Sigma_X$ the following holds:

$$\psi(x)X^* \cap \left(\psi(I_X(x)^*) \cup \psi(I_X(x)^*)\psi(D_X(x) \setminus x)X^*\right) \neq \emptyset.$$

Since rational sets are closed under homomorphisms, the claim immediately follows. Hence, we obtain the following

Proposition 10. *i) If the trace code problem is decidable for (Σ, I), then the code problem is decidable for (Σ, I).*

ii) Both problems are decidable if the independence relation is a transitive forest [1].

iii) Both problems are undecidable as soon as the independence relation I contains C_4 as induced subgraph [14, 8] (equivalently, $M(\Sigma, I)$ contains a submonoid isomorphic to $\{a, b\}^* \times \{c, d\}^*$).

The precise borderline for the decidability of the (trace) code problem is currently open. A temptive idea would consist in conjecturing undecidability for independence alphabets containing P_4 as induced subgraph, since this would settle the problem. However, the decidability of the (trace) code problem for P_4 has been recently established, [13, 16]. Furthermore, not every independence alphabet which does not contain C_4 as induced subgraph has a decidable code problem, see [13, 16] for examples.

The code problem between free monoids is equivalent to the emptiness problem for finite automata. Given $X \subseteq A^*$, the states of the associated automaton $\mathcal{A}_X$ are suffixes of words in X. A transition $u \xrightarrow{1} v$ exists if $v \in X^{-1}u \cup u^{-1}X$. With $X^{-1}X \setminus \{1\}$ as initial states and 1 as unique final state one can verify that $1 \in L(\mathcal{A}_X)$ is equivalent to $L(\mathcal{A}_X) \neq \emptyset$, which holds if and only if X is not a code.

For the special case of free monoids, well-known techniques (see [2]) solve the code problem in polynomial time. More precisely, the code problem is complete for NL, the class of languages which can be recognized by nondeterministic Turing machines in logarithmic space (for the hardness see [19]). For the trace monoid $\mathbf{M} = \mathbb{M}(\Sigma, I)$, where the independence (or equivalently, the dependence) alphabet is P_4, the same complexity result holds, c.f. Thm. 13 below. However, we need a more complex device than finite automata.

We assume for the rest of this section that $\mathbf{M} = \mathbb{M}(\Sigma, I)$, where (Σ, I) is equal to P_4:

$$\Sigma = \{a, b, c, d\} \text{ and } I = \{(a, b), (b, a), (b, c), (c, b), (c, d), (d, c)\}.$$

We will use a one-counter automaton. Thus, the automaton can increment, decrement, and test the counter for zero. The value of the counter is an integer. By storing the sign in the finite control, we may also use a pushdown automaton where the pushdown alphabet contains besides the bottom symbol only one single symbol.

We consider below mainly the code problem, i.e., the question whether a finite set $X \subseteq \mathbf{M}$ freely generates X^+. The basic idea will be to guess two different factorizations of an element of X^+ by storing in the counter a certain number of b's resp. c's, while keeping a (finite) synchronization information in the finite control.

The technical lemma below explains how information can be stored in this way. We need some more notations. For traces $w_1, w_2, \ldots$ in $\mathbf{M}$ let $w[i]$ denote $w_1 \cdots w_i$. If u is a factor of v we also write $u \subseteq v$. (This means that $v \in \mathbf{M}u\mathbf{M}$.) For $u_1, \ldots, u_i, v_1, \ldots, v_j \in X$ we call the pair $(u[i], v[j])$ a partial solution, if $u[i]s = v[j]s'$ holds for some $s, s' \in \mathbf{M}$. Using Levi's Lemma [10] we can represent $u[i], v[j]$ as follows.

Fact 11 Let $(u[i], v[j])$ be a partial solution. Then unique traces $r, \alpha_0, \alpha, \beta_0, \beta \in \mathbf{M}$ exist such that

- $u[i] = r\alpha_0\alpha$, $v[j] = r\beta_0\beta$
- $(\alpha_0\alpha, \beta_0\beta) \in I$

- $\alpha_0 \subseteq u[i-1]$, $\alpha \subseteq u_i$, $\beta_0 \subseteq v[j-1]$, $\beta \subseteq v_j$.

Lemma 12. *Let $X \subseteq M$ be given and let $w = u_1 \cdots u_n = v_1 \cdots v_m$ be such that $u_i, v_j \in X$ with $(u_1, \ldots, u_n) \neq (v_1, \ldots, v_m)$. Let $(u[i], v[j])$ be a partial solution with $r, \alpha_0, \alpha, \beta_0, \beta$ defined as above, $(i, j) \neq (n, m)$. Moreover, suppose $|alph(\alpha_0\beta_0)| \leq 1$.*
Then there exist $s, t \geq 0$ with $s + t > 0$ satisfying the following properties:

i) *Either $s = 0$ or $t = 0$. Moreover, if $s = 0$ then $\mathrm{alph}(v_{j+1} \cdots v_{j+t-1})$ is a clique in (Σ, I). Symmetrically, if $t = 0$ then $\mathrm{alph}(u_{i+1} \cdots u_{i+s-1})$ is a clique in (Σ, I).*

ii) *$u[i+s] = r'\alpha_0'\alpha'$ and $v[j+t] = r'\beta_0'\beta'$ hold for uniquely determined traces $r', \alpha_0', \alpha', \beta_0', \beta'$ satisfying $\alpha_0' \subseteq u[i+s-1]$, $\alpha' \subseteq u_{i+s}$, $\beta_0' \subseteq v[j+t-1]$ and $\beta' \subseteq v_{j+t}$. Moreover, we have $|\mathrm{alph}(\alpha_0'\beta_0')| \leq 1$.*

Proof. First observe that $|\mathrm{alph}(\alpha_0\beta_0)| \leq 1$ together with $(\alpha_0, \beta_0) \in I$ implies that either $\alpha_0 = 1$ or $\beta_0 = 1$. Suppose therefore without loss of generality $\beta_0 = 1$. We distinguish the following cases:

i) Let $\alpha_0 = 1$:
Choose $(s, t) = (1, 0)$ and let $p, \alpha_0', \alpha', \beta'$ be such that $\alpha u_{i+1} = p\alpha_0'\alpha'$, $\beta = p\beta'$, $(\alpha', \beta') \in I$ and $\alpha_0' \subseteq \alpha$, $\alpha' \subseteq u_{i+1}$. Symmetrically we can choose $(s, t) = (0, 1)$ and define $p, \alpha', \beta_0', \beta'$ accordingly.

ii) Let $\beta = 1$:
Choose $(s, t) = (0, 1)$. Then we have $\alpha_0\alpha = p\alpha_0'\alpha'$, $v_{j+1} = p\beta'$ for uniquely determined $p, \alpha_0', \alpha', \beta'$ with $(\alpha_0'\alpha', \beta') \in I$, where $\alpha_0' \subseteq \alpha_0$, $\alpha' \subseteq \alpha$. Note that p, α', β' depend on α, v_{j+1} and on the comparison between $|\alpha_0|$ and a value bounded by α, v_{j+1} (thus bounded by X).

iii) Let $\alpha_0, \beta \neq 1$ and $|\mathrm{alph}(\alpha_0\alpha)| = 1$:
Choose $(s, t) = (1, 0)$. Then we have $\alpha_0\alpha u_{i+1} = p\alpha_0'\alpha'$ and $\beta = p\beta'$ for uniquely determined $p, \alpha_0', \alpha', \beta'$ with $(\alpha_0'\alpha', \beta') \in I$, where $\alpha_0' \subseteq \alpha_0\alpha$ and $\alpha' \subseteq u_{i+1}$. Due to $(\alpha_0\alpha, \beta) \in I$ we have $p \subseteq u_{i+1}$, hence $\alpha_0' = \alpha_0\alpha$. Moreover, α' and β' can be directly computed from u_{i+1}, β.

iv) Finally let $|\mathrm{alph}(\alpha_0\alpha)| > 1$ and $\alpha_0, \beta \neq 1$:
We know in this case that $\mathrm{alph}(\alpha_0\alpha)$ is a clique of the dependence relation D. In fact, either $\mathrm{alph}(\alpha_0\alpha) \subseteq \{a, c\}$ or $\mathrm{alph}(\alpha_0\alpha) \subseteq \{b, d\}$. Let e.g. $\alpha_0\alpha \in b^+d\{b, d\}^*$ and consider the least $t > 0$ such that $\mathrm{alph}(v_{j+t}) \cap (D(b) \setminus \{b\}) \neq \emptyset$, i.e., $d \in \mathrm{alph}(v_{j+t})$. Clearly, every $e \in \mathrm{alph}(\beta v_{j+1} \cdots v_{j+t-1})$ satisfies $(e, b) \in I$ or $e = b$. Moreover, if $e \neq b$ then $(e, d) \in I$. Otherwise, there would exist an edge from an e-labelled vertex x in $v_{j+1} \cdots v_{j+t-1}$ to a d-labelled vertex y in $v_{j+t} \cdots v_m$, whereas y precedes x in $\alpha_0\alpha u_{i+1} \cdots u_n$. Therefore, $e = c$ and $\mathrm{alph}(\beta v_{j+1} \cdots v_{j+t-1}) \subseteq \{b, c\}$.
Let $p, \alpha_0', \alpha', \beta_0', \beta'$ be the unique traces satisfying $(\alpha_0'\alpha', \beta_0'\beta') \in I$ and $\alpha_0\alpha = p\alpha_0'\alpha'$ and $\beta v_{j+1} \cdots v_{j+t} = p\beta_0'\beta'$. Moreover, $\alpha_0' \subseteq \alpha_0$, $\alpha' \subseteq \alpha$, $\beta_0' \subseteq \beta v_{j+1} \cdots v_{j+t-1}$ and $\beta' \subseteq v_{j+t}$. Since both $\mathrm{alph}(\alpha) \cap D(b) \neq \emptyset$ and $\mathrm{alph}(v_{j+t}) \cap D(b) \neq \emptyset$ hold, we obtain $\alpha_0 \subseteq p$ and $\pi_b(v_{j+1} \cdots v_{j+t-1}) \subseteq p$. Hence, $\alpha_0' = 1$ and $\beta_0' \subseteq \beta\pi_c(v_{j+1} \cdots v_{j+t-1})$. Clearly we have $\beta \in c^+$ due to $(\alpha_0\alpha, \beta) \in I$, and the claim is satisfied.

Remark. 1. Note that Lem. 12 still holds if (Σ, I) contains no triangle and no induced C_4. Moreover, for $(\Sigma, I) = P_4$ we note that the additional assumption $\alpha_0\beta_0 \in b^* \cup c^*$ yields in Lem. 12 $\alpha_0'\beta_0' \in b^* \cup c^*$, too. It suffices to choose in the case $\alpha_0 = 1$: $(s, t) = (1, 0)$, if $\alpha \in b^* \cup c^*$, resp. $(s, t) = (0, 1)$, if $\beta \in b^* \cup c^*$.

2. In order to be a code, X may contain at most one element of the form $b^k c^l$, since any two such traces commute.

3. Let us take a closer look at the last case in the previous proof. Assume $\alpha_0 = b^q$ and $\beta = c^r$ for some $q \geq 0$, $r > 0$.

Supposing that $X \cap b^* c^* = \{b^k c^l\}$ the following equations hold:

$$b^q \alpha = p\alpha',$$

(1)

$$c^r \binom{b^k}{c^l}^{t-1} v_{j+t} = p\beta_0'\beta'.$$

(2)

Note that q and $k(t - 1)$ differ by a value depending on α and v_{j+t}, only. Hence, $k(t - 1)$ is determined by q and a value depending on α and v_{j+t}, only (thus bounded by X). Moreover, due to $|\alpha|_c = 0$ we have $\beta_0' = c^{r+l(t-1)}$. Finally, with $p = b^q p'$ for some p' we have $\alpha = p'\alpha'$, $b^{k(t-1)}v_{j+t} = b^q p'\beta'$ and we see that p', α', β' can be computed using α and v_{j+t}, only.

Let $(u[i], v[j])$ be a partial solution with $u[i] = r\alpha_0\alpha$, $v[j] = r\beta_0\beta$ as in Lem. 12. A counter automaton can store the integer value $|\alpha_0| - |\beta_0|$ in the counter, whereas α, β are part of the finite control. The initial configurations (α, β), α_0, β_0 will be given by partial solutions $(u[i], v[j])$ of minimal length satisfying $x \in$ alph$(u_1) \cap$ alph$(v[j])$ for some $x \in \{a, d\}$ (thus, $\alpha_0\beta_0 \in b^* \cup c^*$). The automaton will accept if $\alpha = \beta = 1$ and the counter is empty. It remains to describe the transition relation corresponding to the one-step transition described in Lem. 12, i.e., from $(u[i], v[j])$ to $(u[i + s], v[j + t])$.

Updating α', β' according to the situations considered in Lem. 12 is not difficult. The only problem arises in the last case considered in Lem. 12, when $b^k c^l \in X$ with $kl \neq 0$ and the counter has to switch from b's to c's (or conversely). Roughly, the value of the counter has to be divided by k and multiplied by l (we may also have to increment/decrement the counter by a bounded value, i.e., a value depending on the finite state). Obviously, we cannot perform this combined operation using a single counter. The solution is to store instead of $|\alpha_0| - |\beta_0|$ the value $(|\alpha_0| - |\beta_0|)$ div k, if $\alpha_0\beta_0 \in b^*$. Of course, we keep $(|\alpha_0| - |\beta_0|)$ mod k in the finite control. (Analogously, we store $(|\alpha_0| - |\beta_0|)$ div l in the counter, if $\alpha_0\beta_0 \in c^*$.)

Remark. In order to be a trace code, X may contain at most two elements x_1, x_2 with alph(x_1), alph$(x_2) \subseteq \{b, c\}$. More precisely, if $x_i = b^{k_i} c^{l_i}$ then $k_1 l_2 \neq k_2 l_1$ should hold. The second equation in the previous remark has to be replaced then by

$$c^r \binom{b^{k_1}}{c^{l_1}}^{t_1} \binom{b^{k_2}}{c^{l_2}}^{t_2} v_{j+t} = p\beta_0'\beta',$$

(2)

for t_1, t_2 with $t_1 + t_2 = t - 1$. It can be shown in this case that if X is not a trace code, then we may suppose in the equation above that the difference $|t_1 - t_2|$ is bounded by X (i.e., there exists a suitable solution $u_1 \cdots u_n = v_1 \cdots v_m$ with this property). In this case, we use the same method as for the code problem, with $k_1 + k_2$ (resp. $l_1 + l_2$) replacing k (resp. l) and keeping $|t_1 - t_2|$ in the finite control.

Theorem 13 [13, 16]. *Let $\Sigma = \{a, b, c, d\}$ and $M(\Sigma, I)$ be defined by three equations $ab = ba$, $bc = cb$, and $cd = dc$. Then the (trace) code problem for the independence alphabet (Σ, I) is decidable in polynomial time, actually it is NL-complete.*

Proof sketch. For the complexity result in the theorem above recall that the hardness is already provided by the case of free monoids (over two letters alphabets) [19]. The code problem is shown to belong to NL by noting e.g. that one can test the existence of two different factorizations over a given $X \subseteq M$ by using a 2-way multi-head nondeterministic counter automaton. With the notations of Lem. 12 this automaton keeps track of α and β using two heads, which point at the corresponding elements of X in the input; the counter is used as in the proof of Thm. 13, whereas the modulo k, l values are handled by further heads on the input. Since the class of languages accepted by 2-way multi-head nondeterministic counter automata is known to coincide with NL [21], the result follows.

5 Clique-preserving morphisms

From the viewpoint of semantics codings may arise by refinement of actions. Assume we want to refine an action in such a way that it can be distributed to different parallel components. Then we may think of this as a homomorphism where a letter is mapped to a product of independent letters. This idea leads to the following definition, where (throughout this section) the notion of clique is meant w.r.t. independence alphabets.

Definition 14. A clique-preserving morphism of independence alphabets H, $H: (\Sigma, I) \to (\Sigma', I')$ is a relation $H \subseteq \Sigma \times \Sigma'$ such that $H(A) = \{\alpha \in \Sigma' \mid (a, \alpha) \in H, a \in A\}$ is a clique of (Σ', I') whenever $A \subseteq \Sigma$ is a clique of (Σ, I).

A clique-preserving morphism $H \subseteq \Sigma \times \Sigma'$ yields in a natural way a homomorphism $h: M(\Sigma, I) \to M(\Sigma', I')$ by letting $h(a) = \prod_{\alpha \in H(a)} \alpha$ for $a \in \Sigma$. Note that the product is well-defined since $H(a)$ is (by definition) a clique, i.e., a set of commuting elements. Moreover, for $(a, b) \in I$ we have

$$h(ab) = h(ba) = \prod_{\alpha \in H(a) \cup H(b)} \alpha \cdot \prod_{\alpha \in H(a) \cap H(b)} \alpha$$

Remark. A clique-preserving morphism is not necessarily a morphism of undirected graphs as defined in Sect. 3. The reason is that for $(a,b) \in I$ we may have $H(a) \cap H(b) \neq \emptyset$. Therefore the induced homomorphisms of trace monoids are not strong, in general. On the other hand, the strong coding defined in the Projection Lemma (Ex. 1) is clique-preserving, too.

Due to the next proposition we have a decidability result in the case where the left-hand side is free. This positive situation is in major contrast to Thm. 16 below.

Proposition 15. *Let $H \subseteq \Sigma \times \Sigma'$ be a relation such that $H(a)$ is a clique of (Σ', I') for all $a \in \Sigma$. Then the induced homomorphism $h \colon \Sigma^* \to M(\Sigma', I')$ with $h(a) = \prod_{\alpha \in H(a)} \alpha$ is injective if and only if for all $a, b \in \Sigma$, $a \neq b$ there exists some $(\alpha, \beta) \in D'$, $\alpha \neq \beta$ with $\alpha \in H(a), \beta \in H(b)$.*

The following result has been stated in [12] without proof:

Theorem 16. *Given a clique-preserving morphism of independence alphabets $H \colon (\Sigma, I) \to (\Sigma', I')$, it is undecidable whether the associated homomorphism $h \colon M(\Sigma, I) \to M(\Sigma', I')$, $h(a) = \prod_{\alpha \in H(a)} \alpha$ for $a \in \Sigma$, is a coding.*

The following proof uses the undecidability of Post's correspondence problem (PCP), stated as follows: given two homomorphisms $f, g \colon \Gamma^* \to \Gamma'^*$, does some $u \in \Gamma^+$ exist with $f(u) = g(u)$? For simplifying our reduction we impose following restrictions on the given PCP instances:

- $1 \leq |f(a)|, |g(a)| \leq 2$ for every $a \in \Gamma$
- There exist $x, y \in \Gamma$, $x \neq y$, such that $f(u) = g(u)$ with $u \in \Gamma^+$ implies $u \in x\Gamma^+y$; moreover, if the instance (f, g) has a solution, then also one in $x(\Gamma_0\Gamma_0)^*\Gamma_0 y$, where $\Gamma_0 = \Gamma \setminus \{x, y\}$. Finally, for some $\alpha, \beta \in \Gamma' \colon f(x), g(x) \in \alpha\Gamma'^*$, $f(y), g(y) \in \Gamma'^*\beta$ and α, β occur in no $f(a), g(a)$, for $a \notin \{x, y\}$.

One can show that PCP with these additional restrictions remains undecidable. This can be performed e.g. by slightly modifying the PCP pairs obtained in the reduction from the word problem for semi-Thue systems [11] (resp. imposing suitable restrictions on the semi-Thue systems).

For the alphabet Σ_0 defined below we will denote by Σ_0' the alphabet $\{a' \mid a \in \Sigma_0\}$. Let

$$\Sigma_0 = \{a_b, \bar{a}_b \mid a, b \in \Gamma_0\} \cup \{a, \bar{a} \mid a \in \Gamma_0\} \cup \{x_1, x_2, y_1, y_2\} \cup \{\bar{a}_x, y_b \mid a, b \in \Gamma_0\},$$

and $\Sigma_1 = \Sigma_0 \cup \Sigma_0'$. On Σ_1 we have the independence relation $I_1 \subseteq \Sigma_1 \times \Sigma_1$:

$$I_1 = \{(x_1, x_1'), (y_2, y_2'), (x_1', x_1), (y_2', y_2)\}.$$

For the second independence alphabet (Σ_2, I_2) let

$$\Sigma_2 = \{A_i, \bar{A}_i \mid a \in \Gamma_0, 1 \leq i \leq 6\} \cup \{X_i, Y_i \mid 1 \leq i \leq 6\} \cup \{F, G, \mathord{\mathrm{\notin}}, \$\} \cup \Gamma',$$

where we associate to every letter $a \in \Gamma_0$ the capital letters $A_i, \overline{A}_i$ (resp. X_i for x, resp. Y_i for y) for all $1 \leq i \leq 6$. On Σ_2 we define the dependence (independence, resp.) relation $D_2 \subseteq \Sigma_2 \times \Sigma_2$ ($I_2 \subseteq \Sigma_2 \times \Sigma_2$, resp.) as symmetric relations satisfying:

(1) For all $a \in \Gamma$ let $D_2 \cap \{A_i \mid 1 \leq i \leq 6\}^2$ be (with self-loops omitted)

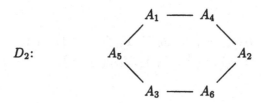

respectively, for $a \in \Gamma_0$ and $1 \leq i, j \leq 6$:

$$(\overline{A}_i, \overline{A}_j) \in D_2 \quad \text{iff} \quad (A_i, A_j) \in D_2 \,.$$

(2) For $a \neq b$ and i, j let $(A_i, B_j) \in D_2$ and $(\overline{A}_i, \overline{B}_j) \in D_2$.
(3) For all a, b, i, j we define $(A_i, \overline{B}_j) \in I_2$ if and only if

$$\{(i,j), (j,i)\} \cap (\{5,6\} \times \{1,2\} \cup \{4,6\} \times \{1,3\}) \neq \emptyset \,.$$

(4) Let $D_2(\gamma) = \Gamma'$ for all $\gamma \in \Gamma'$. Define further $D_2(F) = D_2(G) = \{F, G\}$, $I_2(\mathfrak{c}) = \{X_1, X_2, X_3, F, G\} \cup \Gamma'$ and $I_2(\$) = \{Y_4, Y_5, Y_6, F, G\} \cup \Gamma'$.

We define now a clique-preserving morphism $H \colon (\Sigma_1, I_1) \to (\Sigma_2, I_2)$ based on the given PCP instance $f, g \colon \Gamma^* \to \Gamma'^*$. The associated homomorphism h will simulate the given (word) homomorphisms f, g using the control alphabet $A_i, \overline{A}_i, F, G, \ldots$. More precisely, the control alphabet enforces by synchronization the presence of f- resp. g-images (since $f(a)$ or $g(a)$ may have length 2, we cannot generate them directly by a homomorphism associated to a clique-preserving morphism.) We denote in the following by $h(a)_i$, $i = 1, 2$, $a \in \Gamma$, the ith letter of $h(a)$ (resp. 1, if $|h(a)| < i$), with $h \in \{f, g\}$.
In the definition below we have by convention $k(a)_i \in H(c)$ only if $k(a)_i \neq 1$, where $c \in \Sigma_1$, $a \in \Gamma$, and $k \in \{f, g\}$. However, for simplifying notations we write $f(a)_i, g(a)_i$ throughout.

(1) For all $a, b \in \Gamma_0$ let

$$H(a_b) = \{\overline{B}_5, \overline{B}_6, A_1, A_2, f(a)_1, F\} \qquad H(a) = \{A_3, A_4, f(a)_2, G\},$$
$$H(a'_b) = \{\overline{B}_4, \overline{B}_6, A_1, A_3, g(a)_1, F\} \qquad H(a') = \{A_2, A_5, g(a)_2, G\},$$

and analogously

$$H(\overline{a}_b) = \{B_5, B_6, \overline{A}_1, \overline{A}_2, f(a)_1, F\} \qquad H(\overline{a}) = \{\overline{A}_3, \overline{A}_4, f(a)_2, G\},$$
$$H(\overline{a}'_b) = \{B_4, B_6, \overline{A}_1, \overline{A}_3, g(a)_1, F\} \qquad H(\overline{a}') = \{\overline{A}_2, \overline{A}_5, g(a)_2, G\}.$$

(2) Let for x_i, x_i' and $a \in \Gamma_0$:

$$H(x_1) = \{\varphi, X_1, X_2, f(x)_1, F\} \qquad H(x_2) = \{X_3, X_4, f(x)_2, G\},$$
$$H(x_1') = \{\varphi, X_1, X_3, g(x)_1\} \qquad H(x_2') = \{X_2, X_5, g(x)_2\},$$
$$H(\bar{a}_x) = \{X_5, X_6, \overline{A}_1, \overline{A}_2, f(a)_1, F\}$$
$$H(\bar{a}_x') = \{X_4, X_6, \overline{A}_1, \overline{A}_3, g(a)_1, F\}.$$

(3) For $b \in \Gamma_0$ let:

$$H(y_b) = \{\overline{B}_5, \overline{B}_6, Y_1, Y_2, f(y)_1, F\} \qquad \text{and}$$
$$H(y_b') = \{\overline{B}_4, \overline{B}_6, Y_1, Y_3, g(y)_1, F\},$$

together with

$$H(y_1) = \{Y_3, Y_4, f(y)_2\} \qquad H(y_2) = \{Y_5, Y_6, \$\},$$
$$H(y_1') = \{Y_2, Y_5, g(y)_2, G\} \qquad H(y_2') = \{Y_4, Y_6, F, \$\}.$$

With $I_1 = \{(x_1, x_1'), (y_2, y_2'), (x_1', x_1), (y_2', y_2)\}$ it is easily seen that H is a clique-preserving morphism of independence alphabets. Let $h\colon \mathbb{M}(\Sigma_1, I_1) \to \mathbb{M}(\Sigma_2, I_2)$ denote the associated homomorphism of trace monoids, $h(a) = \prod_{A \in H(a)} A$, $a \in \Sigma_1$. We defined H in such a way that every pair $(u, v) \in \mathbb{M}(\Sigma_1, I_1)^2$ of minimal length $|u| + |v|$ with $h(u) = h(v)$ and $u \neq v$ satisfies $\mathrm{alph}(u) \times \mathrm{alph}(v) \subseteq \Sigma_0 \times \Sigma_0' \cup \Sigma_0' \times \Sigma_0$.

Proposition 17. *The homomorphism $h\colon \mathbb{M}(\Sigma_1, I_1) \to \mathbb{M}(\Sigma_2, I_2)$ associated to the clique-preserving morphism H is not injective if and only if the PCP instance (f, g) has a solution.*

Proof. First, let us consider a solution $u \in x\Gamma_0^n y$ for (f, g) with n odd. We have $f(u) = g(u)$ with $u = x\, a(1) \cdots a(n)\, y$, $a(i) \in \Gamma_0$. For a trace $t = [a_1 \cdots a_n]$, $a_i \in \Sigma_0$, we will denote by t' the trace $[a_1' \cdots a_n']$. It is easy to see that with

$$z = [x_1 x_2\, \overline{a(1)}_x \overline{a(1)}\, a(2)_{a(1)} a(2) \cdots \overline{a(n)}_{a(n-1)} \overline{a(n)}\, y_{a(n)}\, y_1 y_2]$$

we obtain

$$h(z) = h(z') =$$

$$[\varphi\, X_1 \cdots X_6\, \overline{A(1)}_1 \cdots \overline{A(1)}_6 \ldots \ldots \overline{A(n)}_1 \cdots \overline{A(n)}_6\, Y_1 \cdots Y_6\, \$]$$

$$f(u)$$

$$F(GF)^{n+1}$$

(For the vector-like representation above we use the fact that $f(u)$, $F(GF)^{n+1}$ commute pairwise, resp. with the first component.)

For the converse, let us notice two properties of the clique-preserving morphism H used throughout the proof.

Fact 18 *1. $Zh(c) = h(c)Z$ for $c \in \Sigma_1$ and $Z \in \{A_i, \overline{A}_i \mid 1 \le i \le 6, a \in \Gamma\} \setminus \{X_2, X_3, Y_4, Y_5\}$ is equivalent with $(c, Z) \in H$, i.e., $Z \in \text{alph}(h(c))$.*
For $Z \in \{X_2, X_3, Y_4, Y_5\}$ we have $Zh(c) = h(c)Z$ if and only if $(c, Z) \in H$ or $(c, Z) \in \{(x_1', X_2), (x_1, X_3), (y_2, Y_4), (y_2', Y_5)\}$.

2. For all $Z \in \{A_i, \overline{A}_i \mid 2 \le i \le 5, a \in \Gamma_0\} \cup \{X_4, Y_2\}$ we have two distinct letters $c_1, c_2 \in \Sigma_1$ satisfying $(c_i, Z) \in H$. Additionally, either $F \in H(c_1)$ and $G \in H(c_2)$, or $F \in H(c_2)$ and $G \in H(c_1)$ holds. This property will allow to determine c_1 resp. c_2 in a unique way, using $(F, G) \in D_2$.

We denote in the following for $u \in \mathbb{M}(\Sigma_1, I_1)$ by $h^*(u)$ the projection of $h(u)$ on $\Sigma_2 \setminus \Gamma'$, i.e., we consider the control letters, only.

Assume now that $h: \mathbb{M}(\Sigma_1, I_1) \to \mathbb{M}(\Sigma_2, I_2)$ is not injective, and let $u, v \in \mathbb{M}(\Sigma_1, I_1)$, $u \ne v$, be of minimal length $|u| + |v|$ with $h(u) = h(v)$. Note that for every $(a, b) \in D_1 \setminus \text{id}_{\Sigma_1}$ one has $(H(a) \times H(b)) \cap (D_2 \setminus \text{id}_{\Sigma_2}) \ne \emptyset$. Moreover, by the minimality of $|u| + |v|$ the initial alphabets of u resp. v are pairwise independent, i.e., $in(u) \times in(v) \subseteq I_1$. Hence, two cases are possible: either $in(u) = \{x_1\}$ and $in(v) = \{x_1'\}$, or $in(u) = \{y_2\}$ and $in(v) = \{y_2'\}$.

We first consider the case where $in(u) = \{x_1\}$, $in(v) = \{x_1'\}$ and suppose $u = x_1^n u_1$, $v = x_1'^m v_1$ for some $u_1, v_1 \in \mathbb{M}(\Sigma_1, I_1) \setminus 1$ with $in(u_1) \ne \{x_1\}$, $in(v_1) \ne \{x_1'\}$, and $n, m \ge 1$. Hence, we have

$$h^*(u) = \mathfrak{c}^n \frac{X_1^n}{X_2^n} F^n h^*(u_1) = \mathfrak{c}^m \frac{X_1^m}{X_3^m} h^*(v_1) = h^*(v).$$

Due to the condition imposed on the initial alphabets of u_1, v_1 we immediately follow $n = m$. Moreover, recalling Fact 18(1) and the minimality of $|u| + |v|$, we obtain $in(u_1) = \{x_2\}$ and $in(v_1) = \{x_2'\}$. Actually, since $\{X_4, X_5\} \subseteq D_2(\mathfrak{c})$ we may observe that $u_1 = x_2^n u_2$ and $v_1 = x_2'^n v_2$ holds for some u_2, v_2 (neither $u_1 \in x_2^i x_1' \mathbb{M}(\Sigma_1, I_1)$ nor $v_1 \in x_2'^i x_1 \mathbb{M}(\Sigma_1, I_1)$ can lead to a solution, if $i < n$).

Therefore, we obtain the following situation (we omit in the representations below dependence edges from $\mathfrak{c}$):

$$h^*(u) = \mathfrak{c}^n \frac{X_1^n X_2^n X_3^n X_4^n}{F^n G^n} h^*(u_2) = \mathfrak{c}^n X_1^n X_3^n X_2^n X_5^n h^*(v_2) = h^*(v).$$

Fact 18(2) applied to X_4 yields now directly $v_2 = \overline{a}_x'^n v_3$, for some v_3 and $a \in \Gamma_0$. Since $X_5 \in in(h(u_2))$ we have $\{\overline{a}_x, x_2'\} \cap in(u_2) \ne \emptyset$. Due to $(\overline{a}_x', \overline{A}_1) \in H$ and $(\overline{A}_1, X_2) \in D_2$, we obtain $u_2 = \overline{a}_x'^n u_3$, for some u_3. This yields the partial solution:

$$h(u) = \begin{array}{ccc} \mathfrak{c}^n & X_1^n \cdots X_6^n & \overline{A}_1^n \overline{A}_2^n \\ F^n\, G^n & F^n & \\ (f(x)_1)^n\, (f(x)_2)^n & (f(a)_1)^n & \end{array} h(u_3) =$$

$$h(v) = \begin{array}{ccc} \mathfrak{c}^n & X_1^n \cdots X_6^n & \overline{A}_1^n \overline{A}_3^n \\ & F^n & \\ (g(x)_1)^n\, (g(x)_2)^n & (g(a)_1)^n & \end{array} h(v_3)\,.$$

More generally, let us assume

$$h^*(u) = W \frac{\overline{A}_1^n \overline{A}_2^n}{G^n\, F^n} h^*(u_1) = W \frac{\overline{A}_1^n \overline{A}_3^n}{} h^*(v_1) = h^*(v),$$

for $a \notin \{x, y\}$, $W \in M(\Sigma_2, I_2)$, $u_1, v_1 \in M(\Sigma_1, I_1)$.
Due to $\overline{A}_2 \in in(h(v_1))$, together with Fact 18(2), we immediately obtain $v_1 = \overline{a}'^n v_2$ for some v_2, hence $h^*(v) = W \frac{\overline{A}_1^n \overline{A}_3^n \overline{A}_2^n \overline{A}_5^n}{G^n} h^*(v_2)$. With $\overline{A}_3 \in in(h(u_1))$ we need a letter $c \in \Sigma$ satisfying $(c, \overline{A}_3) \in H$ and $alph(h(c)) \cap D(\overline{A}_5) \subseteq \{\overline{A}_3\}$; hence, $c = \overline{a}$ and $u_1 = \overline{a}^n u_2$, for some u_2. This yields $h^*(u) = W \frac{\overline{A}_1^n \overline{A}_2^n \overline{A}_3^n \overline{A}_4^n}{G^n F^n G^n} h^*(u_2)$.
Finally, $\overline{A}_4 \in in(h(v_2))$, together with Fact 18(2) yields $v_2 = b_a'^n v_3$ and

$$h^*(v) = W \frac{\overline{A}_1^n \cdots \overline{A}_6^n\ B_1^n B_3^n}{G^n F^n} h^*(v_3),$$

for some $b \neq x$ and $v_3 \in M(\Sigma_1, I_1)$; on the other hand, the condition $\overline{A}_5 \in in(h(u_2))$ requires a letter $c \in \Sigma_1$ with $(c, \overline{A}_5) \in H$ and $B_1 h(c) = h(c) B_1$, hence $c = b_a$ and $u_2 = b_a^n u_3$, with $h^*(u) = W \frac{\overline{A}_1^n \cdots \overline{A}_6^n\ B_1^n B_2^n}{G^n F^n G^n F^n} h^*(u_3)$, for some $u_3 \in M(\Sigma_1, I_1)$.
The case $h^*(u) = W \frac{A_1^n A_2^n}{G^n\ F^n} h^*(u_1) = h^*(v) = W A_1^n A_3^n h^*(v_1)$, $a \neq y$, is symmetric. Therefore, we suppose now $a = y$. We observe that the same arguments used above for $\overline{A}_2, \overline{A}_3$ hold (dually) for Y_2, Y_3 (in particular, by Fact 18(2)). Hence,

$$h^*(u) = W \frac{Y_1^n Y_2^n Y_3^n Y_4^n}{G^n\, F^n} h^*(u_2) = W \frac{Y_1^n Y_3^n Y_2^n Y_5^n}{G^n} h^*(v_2) = h^*(v)\,.$$

Thus we obtained $w = [x_1^n x_2^n\, \overline{a_1}^n \overline{a_1}^n\, a_2^{\,n}{}_{a_1} a_2^n \cdots \overline{a_m}^n{}_{a_{m-1}} \overline{a_m}^n\, y_{a_m}^n\, y_1^n]$, $a_i \in \Gamma_0$, $n, m \geq 1$, such that $u = wu_2$ and $v = w'v_2$ for some u_2, v_2 (recall that w' denotes the Σ_0'-copy of $w \in \Sigma_0^*$). Recall that $f(y), g(y) \in \Gamma'^* \beta$ and β occurs in no $f(a), g(a)$, for $a \notin \{x, y\}$. Hence, we obtained

$$(f(x)_1)^n (f(x)_2)^n (f(a_1)_1)^n (f(a_1)_2)^n \cdots (f(a_m)_1)^n (f(a_m)_2)^n (f(y)_1)^n (f(y)_2)^n =$$
$$(g(x)_1)^n (g(x)_2)^n (g(a_1)_1)^n (g(a_1)_2)^n \cdots (g(a_m)_1)^n (g(a_m)_2)^n (g(y)_1)^n (g(y)_2)^n\,.$$

It is now easy to check that $z = x\, a_1 \cdots a_m\, y$ satisfies $f(z) = g(z)$, hence z is a solution for the PCP instance (f, g).

It remains to consider a pair $(u, v) \in M(\Sigma, I)^2$, $u \neq v$, of minimal length $|u| + |v|$, with $h(u) = h(v)$, which satisfies $in(u) = \{y_2'\}$, $in(v) = \{y_2\}$. Assume $u = y_2'^{n} u_1$ and $v = y_2^m v_1$ with u_1, v_1 such that $in(u_1) \neq \{y_2'\}$, $in(v_1) \neq \{y_2\}$. Again, $m = n$ follows immediately. Hence, we obtain

$$h^*(u) = \$^n \frac{Y_6^n Y_4^n}{F^n} h^*(u_1) = \$^n {}^{Y_6^n Y_5^n} h^*(v_1) = h^*(v).$$

This case can now be handled as the first one, replacing $A_1, A_2, A_3, A_4, A_5, A_6$ by $A_6, A_4, A_5, A_2, A_3, A_1$ ($\overline{A}_i$ analogously) and interchanging x, y. This concludes the proof.

References

1. IJ. J. Aalbersberg and H. J. Hoogeboom. Characterizations of the decidability of some problems for regular trace languages. *Mathematical Systems Theory*, 22:1–19, 1989.
2. J. Berstel and D. Perrin. *Theory of Codes*. Pure and Applied Mathematics; 117. Academic Press, Orlando, Florida, 1985.
3. A. Bertoni, G. Mauri, and N. Sabadini. Equivalence and membership problems for regular trace languages. In *Proceedings of the 9th International Colloquium on Automata, Languages and Programming (ICALP'82)*, number 140 in Lecture Notes in Computer Science, pages 61–71, Berlin-Heidelberg-New York, 1982. Springer.
4. V. Bruyère and C. De Felice. Trace codings. In E. Mayr and C. Puech, editors, *Proceedings of the 12th Annual Symposium on Theoretical Aspects of Computer Science (STACS'95), 1995*, number 900 in Lecture Notes in Computer Science, pages 373–384, Berlin-Heidelberg-New York, 1995. Springer.
5. V. Bruyère and C. De Felice. Any lifting of a trace coding is a word coding. Submitted for publication, 1996.
6. V. Bruyère, C. De Felice, and G. Guaiana. Coding with traces. In P. Enjalbert, E. Mayr, and K. W. Wagner, editors, *Proceedings of the 11th Annual Symposium on Theoretical Aspects of Computer Science (STACS'94), 1994*, number 775 in Lecture Notes in Computer Science, pages 353–364, Berlin-Heidelberg-New York, 1994. Springer.
7. P. Cartier and D. Foata. *Problèmes combinatoires de commutation et réarrangements*. Number 85 in Lecture Notes in Mathematics. Springer, Berlin-Heidelberg-New York, 1969.
8. M. Chrobak and W. Rytter. Unique decipherability for partially commutative alphabets. *Fundamenta Informaticae*, X:323–336, 1987.
9. M. Clerbout and M. Latteux. Partial commutations and faithful rational transductions. *Theoretical Computer Science*, 34:241–254, 1984.
10. R. Cori and D. Perrin. Automates et commutations partielles. *R.A.I.R.O. — Informatique Théorique et Applications*, 19:21–32, 1985.
11. M. D. Davis and E. J. Weyuker. *Computability, complexity and languages*. Academic Press, New York, 1983.
12. V. Diekert, A. Muscholl, and K. Reinhardt. On codings of traces. In E. Mayr and C. Puech, editors, *Proceedings of the 12th Annual Symposium on Theoretical Aspects of Computer Science (STACS'95), 1995*, number 900 in Lecture Notes in Computer Science, pages 385–396, Berlin-Heidelberg-New York, 1995. Springer.

13. H. J. Hoogeboom and A. Muscholl. The code problem for traces – improving the boundaries. Submitted for publication.
14. G. Hotz and V. Claus. *Automatentheorie und Formale Sprachen, Band III.* Bibliographisches Institut, Mannheim, 1972.
15. G. S. Makanin. The problem of solvability of equations in free semigroups. *Math. USSR Izvestiya,* 21:483–546, 1983.
16. Yu. Matyiasevich. Cas décidables et indécidables du problème du codage pour les monoïdes partialement commutatifs. To appear in Quadrature.
17. A. Mazurkiewicz. Concurrent program schemes and their interpretations. DAIMI Rep. PB 78, Aarhus University, Aarhus, 1977.
18. E. Ochmański. On morphisms of trace monoids. In R. Cori and M. Wirsing, editors, *Proceedings of the 5th Annual Symposium on Theoretical Aspects of Computer Science (STACS'88),* number 294 in Lecture Notes in Computer Science, pages 346–355, Berlin-Heidelberg-New York, 1988. Springer.
19. W. Rytter. The space complexity of the unique decipherability problem. *Information Processing Letters,* 23:1–3, 1986.
20. J. Sakarovitch. The "last" decision problem for rational trace languages. In I. Simon, editor, *Proceedings of the 1st Latin American Symposium on Theoretical Informatics (LATIN'92),* number 583 in Lecture Notes in Computer Science, pages 460–473, Berlin-Heidelberg-New York, 1992. Springer.
21. K. Wagner and G. Wechsung. *Computational complexity.* VEB Deutscher Verlag der Wissenschaften, Berlin, 1986.
22. E. S. Wolk. A note on the comparability graph of a tree. *Proc. of the Amer. Math. Soc.,* (16):17–20, 1965.

Models of DNA Computation

Alan Gibbons[1] *, Martyn Amos[1], David Hodgson[2]

[1] Department of Computer Science, University of Liverpool, L69 3BX, UK
[2] Department of Biological Sciences, University of Warwick, Coventry, CV4 7AL, UK

Abstract. The idea that living cells and molecular complexes can be viewed as potential machinic components dates back to the late 1950s, when Richard Feynman delivered his famous paper describing sub-microscopic computers. Recently, several papers have advocated the realisation of massively parallel computation using the techniques and chemistry of molecular biology. Algorithms are not executed on a traditional, silicon-based computer, but instead employ the test-tube technology of genetic engineering. By representing information as sequences of bases in DNA molecules, existing DNA-manipulation techniques may be used to quickly detect and amplify desirable solutions to a given problem.

We review the recent spate of papers in this field and take a critical view of their implications for laboratory experimentation. We note that extant models of DNA computation are flawed in that they rely upon certain error-prone biological operations. The one laboratory experiment that is seminal for current interest and claims to provide an efficient solution for the Hamiltonian path problem has proved to be unrepeatable by other researchers. We introduce a new model of DNA computation whose implementation is likely to be far more error-resistant than extant proposals. We describe an abstraction of the model which lends itself to natural algorithmic description, particularly for problems in the complexity class NP. In addition we describe a number of linear-time parallel algorithms within our model, particularly for NP-complete problems. We describe an "in vitro" realisation of the model and conclude with a discussion of future work and outstanding problems.

1 Introduction

"Computers in the future may weigh no more than 1.5 tons.", so said *Popular Mechanics* in 1949. Today, in the age of smart cards and wearable PCs, this statement is striking because it falls so short of reality. In another fifty years, who would be prepared to predict how close to the levels of miniaturization described in Feynman's visionary paper [9] we will have come?

Huge advances in miniaturization have been made since the days of room-sized computers, yet the underlying computational model (the Von Neumann architecture) has remained the same. Today's supercomputers still employ the kind of sequential logic used by the mechanical dinosaurs of the 1930s. Some

---

* Email: amg@csc.liverpool.ac.uk

researchers are now looking beyond these boundaries and are investigating entirely new media and computational models. These include quantum, optical and DNA-based computers. It is the last of these developments that this paper concentrates on.

The idea that living cells and molecular complexes can be viewed as potential machinic components dates back to the late 1950s, when Richard Feynman delivered his famous paper describing "sub-microscopic" computers. More recently, several papers [1, 2, 18] (also see [13, 21]) have advocated the realisation of massively parallel computation using the techniques and chemistry of molecular biology. In [1], Adleman described how a computationally intractable problem, known as the *directed Hamiltonian Path Problem* (HPP) might be solved using molecular methods. Recall that the HPP involves finding a path through a graph that visits each vertex exactly once. Adleman's method employs a simple, massively parallel random search. The algorithm is not executed on a traditional, silicon-based computer, but instead employs the "test-tube" technology of genetic engineering. By representing information as sequences of bases in DNA molecules, Adleman shows how existing DNA-manipulation techniques may be used to quickly detect and amplify desirable solutions to a given problem.

How can we combine a flask of DNA with biological tools to solve a hard mathematical problem? The first stage is the creation of a flask of DNA molecules, each molecule encoding a potential solution to the problem. With reference to the HPP, for example, each strand encodes a path (not necessarily Hamiltonian) through the graph. Given every DNA molecule that encodes a path of length n, for a graph with n vertices, we can be sure that every possible solution is present, some legal, but most illegal. Once the entire solution space is present in a flask the DNA computer really comes into its own. We use a small set of biological tools to "sift" out DNA that we know encodes illegal solutions. These are those paths that don't visit every vertex, or paths that visit a particular vertex more than once. At the end of the sifting process, we are left only with strands that encode legal solutions.

Of course, for DNA computers, each individual operation, for example, extracting DNA strands can take minutes or even hours to perform. This cost of a computational step, when compared to that of supercomputers capable of executing a trillion operations a second, looks unimpressive. However, the real power of DNA computers lies in their inherent parallelism – each operation is performed not on one single DNA strand, but on every strand in the flask *simultaneously*. The fastest supercomputers in existence today are capable of executing around a trillion operations a second. DNA computers have the potential to execute more than a *thousand* trillion operations per second, as well as being a billion times more energy-efficient and requiring a trillionth of the space needed by existing storage media. Nature has information compression down to a fine art – over forty 1 Mb floppy discs are required to store the genome of a single fruit fly.

The rest of the paper is organized as follows. In Section 2 we review extant models of DNA computation, and explain why they are flawed. We also describe an alternative, *error-resistant* model of DNA computation, In Section 3 we provide a brief introduction to the structure and manipulation of DNA. In Section 4 we show how our model may be implemented in the laboratory and in Section 5 discuss the advantages of the model. In Section 6 we summarize the current state of DNA computation and some outstanding problems.

2 A review of DNA computation

This section briefly reviews the major extant models of DNA computation. Our descriptions are a mathematical abstraction in terms of manipulating multisets. We describe later how these models may be realised by the manipulation of DNA. In what follows we use the term *tube* to describe a multiset of strings over an alphabet α.

2.1 Models of computation

Adleman [1] provided the impetus for recent work through his experimental solution to the Hamiltonian Path Problem. In [18], Lipton generalised Adleman's model and showed how it can encompass solutions to other NP-complete problems. Here we summarize the main features of the Adleman/Lipton model.

Lipton employed the following operations:

- *separate*(T, S). Given a tube T and a substring S, create two new tubes $+(T, S)$ and $-(T, S)$, where $+(T, S)$ is all strings in T containing S, and $-(T, S)$ is all strings in T *not* containing S.
- *merge*$(T_1, T_2, \ldots, T_n)$. Given tubes $T_1, T_2, \ldots, T_n$, create $\cup(T_1, T_2, \ldots, T_n) = T_1 \cup T_2 \cup \ldots T_n$.
- *detect*(T). Given a tube T, return *true* if T is non-empty, otherwise return *false*.
- *amplify*(T, x, y). Given a tube T, create a new tube T', where T' contains an exponential replication of strings in T that start with x and terminate with y. If x and y are undefined, *all* strings in T are replicated.

In addition to these, Adleman used a fifth operation to sort strings on length:

- *length*(T, l). Given a tube T, create a new tube $\varphi(T, l)$ containing all strings in T of length l.

For example, given $\alpha = \{A, B, C\}$, the following algorithm only returns *true* if the initial multiset contains a string composed entirely of 'A's:

```
Input(T)
T ← −(T, B)
T ← −(T, C)
Output(detect(T))
```

Adleman's algorithm for the Hamiltonian Path problem is the only algorithm that to date has been both realised in the laboratory and published. For this reason we describe it in detail. A directed graph $G = (V, E)$ has a Hamiltonian path between the designated vertices v_{in} and v_{out} iff there exists a sequence of edges that begins at v_{in}, ends at v_{out} and enters every other vertex once and only once.

The problem instance solved by Adleman is depicted in Figure 1, with its single Hamiltonian Path outlined in bold ($v_{in} = v_0$, $v_{out} = v_6$).

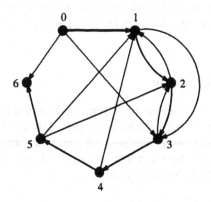

Fig. 1. Adleman's instance of the Hamiltonian Path Problem

We omit details of its construction, but the initial multiset T consists of strings each encoding a possible path (not necessarily a simple path) in the graph $G = (V, E)$, where $V = \{v_0, v_1, \ldots, v_{n-1}\}$. The following pseudo-code then describes Adleman's algorithm:

(1) Input(T)
(2) $T \leftarrow amplify(T, p_1, p_n)$
(3) $T \leftarrow \varphi(T, n)$
(4) **for** $i = 1$ to n **do begin** $T \leftarrow +(T, p_i)$ **end**
(5) Output(detect(T))

Step 1 generates random paths through the graph. Step 2 copies only those strings encoding paths that begin with v_{in} and end with v_{out}. Step 3 discards all strings encoding paths of length $\neq n$. Step 4 ensures that in remaining strings, each vertex of the graph appears. The detection stage at Step 5 reports whether or not a string encoding a Hamiltonian path is found.

We note that Adleman's experiment was performed on the single problem instance shown in Figure 1. This instance contains just a single Hamiltonian

Path. No control experiments were performed for cases *without* Hamiltonian Paths.

Lipton [18] described a solution to another NP-complete problem, namely the so-called *satisfiability* problem. Since any instance of any problem in the complexity class NP may be expressed in terms of an instance of any NP-complete problem, it follows that the multiset operations described earlier at least implicitly provide sufficient computational power to solve any problem in NP. They are unlikely to provide the full algorithmic computational power of a Turing Machine. However, as several authors (see, for example [10, 22]) have recently described, one further simple operation will provide full Turing computability.

Let S and T be two strings over the alphabet α. Then the *splicing* operation consists of cutting S and T at specific positions and concatenating the resulting prefix of S with the suffix of T and concatenating the prefix of T with the suffix of S. The name *splicing* is used in [15], although this type of operation has been used in various guises by several authors [10, 22]. In [10], the authors show that the generative power of *finite extended splicing systems* is equal to that of Turing Machines. In an earlier paper [22], Reif within his so-called *Parallel Associative Memory Model* describes a *Parallel Associative Matching* (PA-Match) operation. The essential constituent of the PA-Match operation is a restricted form of the splicing operation which we denote here by *Rsplice*, and describe as follows. If $S = S_1S_2$ and $T = T_1T_2$, then the result of $Rsplice(S, T)$ is the string S_1T_2 provided $S_2 = T_1$, but has no value if $S_2 \neq T_1$.

Leading results of Reif [22], made possible by his PA-Match operation, concern the simulation of nondeterministic Turing Machines and the simulation of Parallel Random Access Machines (specifically CREW PRAMs). We can capture the spirit of his Turing Machine simulation through the *Rsplice* operation as follows. The initial tube in the simulation consists of all strings of the form S_iS_j where S_i and S_j are encodings of *configurations* of the simulated nondeterministic Turing Machines and such that S_j follows from S_i after one (of possibly many) machine cycle. By a configuration here we mean an instantaneous description of the Turing Machine capturing the contents of the tape, the machine state and which tape square is being scanned. If the *Rsplice* operation is now performed between all pairs of initial strings, the tube will contain strings S_kS_l where S_l follows from S_k after two machine cycles. Similarly, after t repetitions of this operation the tube will contain strings S_mS_n where S_n follows from S_m after 2^t machine cycles. Clearly, if the simulated Turing Machine runs in time T, then after $O(\log T)$ operations the simulation will produce a tube containing strings S_oS_f where S_o encodes the initial configuration and S_f a final configuration.

Other papers have also addressed the problem of Turing Machine simulation (see [5, 23], for example).

2.2 Practical implementation problems

We describe later (Section 4) how the various multiset operations described in the previous section may be realised thorough standard DNA manipulation techniques. However, it is convenient at this point to emphasise two impediments to effective computation by this means. The first hampers the problem size that might be effectively handled, and the second casts doubt on the potential for biochemical success of the precise implementations that have been proposed.

Naturally, the strings making up the multisets are encoded in strands of DNA in all the proposed implementations. Consider for a moment what volume of DNA would be required for a typical NP-complete problem. The algorithms mentioned earlier and others described later in this paper require just a polynomial number of DNA manipulation steps. For the NP-complete problems there is an immediate implication that an exponential number of parallel operations would be required within the computation. This in turn implies that the tube of DNA must contain a number of strands which is exponential in the problem size. Despite the molecular dimensions of the strands, for only moderate problem sizes (say, n $\sim$ 20 for the Hamiltonian Path problem) the required volume of DNA would make the experiments impractical. As Hartmanis points out in [14], if Adleman's experiment were scaled up to 200 vertices the weight of DNA required would exceed that of the Earth.

We note that [3] have described DNA algorithms which reduce the problem just outlined, however, the "exponential curse" is inherent in the NP-complete problems. There is the hope, as yet unrealised (depsite the claims of [4]) that for problems in the complexity class P (i.e. those which can be solved in sequential polynomial-time) there may be DNA computations which only employ polynomial-sized volumes of DNA.

Now we consider the potential for biochemical success that was mentioned earlier. It is a common feature of *all the early proposed implementations* that the biological operations to be used are assumed to be error-free. An operation central to and frequently employed in most models is the *extraction* of DNA strands containing a certain sequence (known as *removal by DNA hybridization*). The most important problem with this method is that it is not 100% efficient[3], and may at times inadvertently remove strands that *do not* contain the specified sequence. Adleman did not encounter problems with extraction because only a few operations were required. However, for a large problem instance, the number of extractions required may run into hundreds, or even thousands. For example, a particular DNA-based algorithm may rely upon repeated "sifting" of a "soup" containing many strands, some encoding legal solutions to the given problem, but most encoding illegal ones. At each stage, we may wish to extract only strands that satisfy certain criteria (i.e., they contain a certain sequence). Only strands that satisfy the criteria at one stage go through to the next. At the end of the sifting process, we are hopefully left only with strands that encode legal solutions, since they satisfy all criteria. However, assuming 95% efficiency

---

[3] The actual efficiency depends on the concentration of the reactants.

of the extraction process, after 100 extractions the probability of us being left with a soup containing (a) a strand encoding a legal solution, and (b) no strands encoding illegal solutions is about 0.006. Repetitive extraction will not guarantee 100% efficiency, since it is impossible to achieve the conditions whereby only correct hybridization occurs. Furthermore, as the length of the DNA strands being used increases, so does the probability of incorrect hybridization.

These criticisms have been borne out by recent attempts [16] to repeat Adleman's experiment. The researchers performed Adleman's experiment twice; once on the original graph as a positive control, and again on a graph containing no Hamiltonian path as a negative control. The results obtained were inconclusive. The researchers state that *"At this time we have carried out every step of Adleman's experiment, but have not gotten an unambiguous final result."*

Although attempts [17] have been made to simulate highly reliable extraction using a sequence of imperfect operations, it is clear that for any non-trivial problem, reliance on this operation must be minimised, or, ideally, removed entirely. In the following section we describe a novel model of DNA computation that obviates the need for hybridization extraction within the main body of the computation.

3 An error-resistant model of DNA computation

Here we describe an abstract model of computation which is clearly inspired by much of the work described earlier. However, as we show in Section 4, this model has a clean implementation in DNA chemistry, essentially because the implementation avoids repetitive use of hybridization extraction.

Like previous models, our model is particularly effective for algorithmic description. It is sufficiently strong to solve any of the problems in the class NC which includes, of course, the notoriously intractable NP-complete problems [11]. As we shall see, these problems naturally have polynomial-time (often linear-time) parallel solutions within the model. This usually comes with the expense of exponentially large data sets. Note that the addition of a *splice* operation (the details of which we omit for space reasons) would make our model Turing-complete.

Like other models, a computation consists of a sequence of operations on finite sets of strings. It is normally the case that a computation begins and terminates with a single set. Within the computation, by applying legal operations of a computation, several sets may exist at the same time. We define legal operations on sets shortly but first consider the nature of an *initial set*.

An initial set consists of strings which are typically of length $O(n)$ where n is the problem size. As a subset, the initial set should include all possible solutions (each encoded by a string) to the problem to be solved. The point here is that the superset is supposed, in any implementation of the model, to be relatively easy to generate as a starting point for a computation. The computation then proceeds by filtering out strings which cannot be a solution.

For example, if the problem is to generate a permutation of the integers
$1 \ldots n$ then the initial set might include all strings of the form $p_1 i_1 p_2 i_2 \ldots p_n i_n$
where each i_k may be any of the integers in the range $[1 \ldots n]$ and p_k encodes the
information "position k". Here, as will be typical for many computations, the
set has cardinality which is exponential in the problem size. For our example of
finding a permutation, we should filter out all strings in which the same integer
appears in at least two locations p_k. Any of the remaining strings is then a legal
solution to the problem. We return to this problem (whose solution, incidentally,
may be regarded as a very useful standard operation within our model for the
solution of other problems) after defining some basic legal operations on strings.

3.1 Basic set operations

Here we define the basic legal operations on sets within the model. Our choice is
determined by what we know can be effectively implemented by very precise and
complete chemical reactions within the DNA implementation. The operation set
defined here provides the power we claim for the model but, of course, it might
be augmented by additional operations in the future to allow greater conciseness
of computation.

- $remove(U, \{S_i\})$. This operation removes from the set U, in parallel, any
  string which contains at least one occurrence of any of the substrings S_i.
- $union(\{U_i\}, U)$. This operation, in parallel, creates the set U whic h is the
  set union of the sets U_i.
- $copy(U, \{U_i\})$. In parallel, this operation produces a number of copie s, U_i,
  of the set U.
- $select(U)$. This operation selects an element of U uniformly at random, if U
  is the empty set then $empty$ is returned.

From the point of view of establishing the parallel time-complexities of algorithms within the model, these basic set operations will be assumed to take
constant-time. This is certainly what the DNA implementation described in Section 3 provides.

3.2 A First Algorithm

We now provide our first algorithmic description within the model. The problem
solved is that of generating the set of all permutations of the integers 1 to n.
The initial set and the filtering out of strings which are not permutations were
essentially described earlier. Although not NP-complete, the problem does of
course have exponential-sized input and output.

The algorithmic description below introduces a format that we utilise elsewhere. The particular device of copying a set (as in copy($U, \{U_1, U_2, \ldots, U_n\}$))
followed by parallel *remove* operations (as in remove($U_i, \{p_j \neg i, p_k i\}$)) is a very
useful compound operation as we shall see in several later algorithmic descriptions. Indeed, it is precisely this use of *Parallel Filtering* that is at the core of

most algorithms with in the model. The only non-selfevident notation employed below is $\neg i$ to mean (in this context) any integer in the range $i=1, 2, \ldots, n$ which is *not* equal to i.

- **Problem: Permutations**

  Generate the set P_n of all permutations of the integers $\{1, 2, \ldots, n\}$.
- **Solution**
  - *Input*: The input set U consists of all strings of the form $p_1 i_1 p_2 i_2 \ldots p_n i_n$ where, for all j, p_j uniquely encodes "positio n j" and each i_j is in $\{1, 2, \ldots, n\}$. Thus each string consists of n in tegers with (possibly) many occurences of the same integer.
  - *Algorithm*

    for $j = 1$ to n **do**

        begin

        copy$(U, \{U_1, U_2, \ldots, U_n\})$

        for $i=1, 2, \ldots, n$ and all $k > j$

                      in parallel do remove$(U_i, \{p_j \neg i, p_k i\})$

        union$(\{U_1, U_2, \ldots, U_n\}, U)$

        end

    $P_n \leftarrow U$
  - *Complexity*: $O(n)$ parallel-time.

After the jth iteration of the **for** loop, the computation ensures that in the surviving strings the integer i_j is not duplicated at positions $k > j$ in the string. The integer i_j may be any in the set $\{1, 2, \ldots, n\}$ (which one it is depends in which of the sets U_i the containing string survived). At the end of the computation each of the surviving strings contains exactly one occurence of each integer in the set $\{1, 2, \ldots, n\}$ and so represents one of the possible permutations. Given the specified input, it is easy to see that P_n will be the set of all permutations of the first n natural numbers. As we shall see, production of the set P_n can be a useful subprocedure for other computations.

3.3 Algorithms for a selection of NP-complete problems.

We now describe a number of algorithms for graph-theoretic NP-complete problems (see [12], for example). Problems in the complexity class NP seem to have a natural expression and ease of solution within the model. We describe linear-time solutions although, of course, there is frequently an implication of an exponential number of processors available to execute any of the basic operations in unit time.

The 3-vertex-colourability problem.

Our first problem concerns proper vertex colouring of a graph. In a proper colouring, colours are assigned to the vertices in such a way that no two adjacent vertices are similarly coloured. The problem of whether 3 colours are sufficient to achieve such a colouring for an arbitrary graph is NP-complete [12].

– **Problem: Three colouring**
Given a graph $G = (V, E)$, find a 3-vertex-colouring if one exists, otherwise return the value *empty*.

– **Solution**

- *Input*: The input set U consists of all strings of the form $p_1 c_1 p_2 c_2 \ldots p_n c_n$ where $n = |V|$ is the number of vertices in the graph. Here, for all i, p_i uniquely encodes "position i" and each c_i is any one of the "colours" 1, 2 or 3. Each such string represents one possible assignment of colours to the vertices of the graph in which, for each i, colour c_i is assigned to vertex i.

- *Algorithm*

  for $j = 1$ **to** n **do**
     begin
       copy$(U, \{U_1, U_2, U_3\})$
       for $i=1$, 2 and 3, and all k such that $(j, k) \in E$
           in parallel do remove$(U_i, \{p_j \neg i, p_k i\})$
       union$(\{U_1, U_2, U_3\}, U)$
     end
  select(U)

- *Complexity*: $O(n)$ parallel time.

After the jth iteration of the **for** loop, the computation ensures that in the remaining strings vertex j (although it may be coloured 1, 2 or 3 depending on which of the sets U_i it survived in) has no adjacent vertices that are similarly coloured. Thus, when the algorithm terminates, U only encodes legal colourings if any exist. Indeed, every legal colouring will be represented in U.

The Hamiltonian Path problem.

A Hamiltonian path between any two vertices u, v of a graph is a path that passes through every vertex in $V - \{u, v\}$ precisely once [12].

– **Problem: Hamiltonian path**
Given a graph $G = (V, E)$ with n vertices, determine whether G contains a Hamiltonian path.

– **Solution**

- *Input*: The input set U is the set P_n of all permutations of the integers from 1 to n as output from **Problem: Permutations**. An integer i at position p_k in such a permutation is interpreted as follows: the string represents a candidate solution to the problem in which vertex i is visited at step k.

- *Algorithm*

  for $2 \leq i \leq n - 1$ and j, k such that $(j, k) \notin E$
     in parallel do remove $(U, \{j p_i k\})$
  select(U)

- *Complexity*: Constant parallel time given P_n.

In surviving strings there is an edge of the graph for each consecutive pair of vertices in the string. Since the string is also a permutation of the vertex set it must also be a Hamiltonian path. Of course, U will contain every legal solution to the problem.

The Subgraph isomorphism problem.

Given two graphs G_1 and G_2 the following algorithm determines whether G_2 is a subgraph of G_1.

- **Problem: Subgraph isomorphism**
  Is $G_2 = (V_2, E_2)$ a subgraph of $G_1 = (V_1, E_1)$? By $\{v_1, v_2, \ldots, v_s\}$ we denote the vertex set of G_1, similarly the vertex set of G_2 is $\{u_1, u_2, \ldots, u_t\}$ where, without loss of generality, we take $t \leq s$.
- **Solution**
  - *Input*: The input set U is the set P_s of permutations output from the Permutations algorithm. For $1 \leq j \leq t$ an element $p_1 i_1 p_2 i_2 \ldots p_s i_s$ of P_s is interpreted as associating vertex $p_j \in \{u_1, u_2, \ldots, u_t\}$ with vertex $i_j \in \{v_1, v_2, \ldots, v_s\}$. The algorithm is designed to remove any element which maps vertices in V_1 to vertices in V_2 in a way which does not reflect the requirement that if $(p_s, p_t) \in E_1$ then $(i_s, i_t) \in E_2$.
  - *Algorithm*
    ```
    for j=1 to t do
      begin
      copy(U, {U − 1, U₂, ..., Uₜ})
      for all l, j < l ≤ t such that (pⱼ, pₗ) ∈ E₂ and (ij, iₗ) ∉ E₁
              in parallel do remove(Uⱼ, {pₗiₗ})
      union({U − 1, U₂, ..., Uₜ}, U)
      end
    select(U)
    ```
  - *Complexity*: $O(|V_s|)$ parallel time.

for any remaining strings, the first t pairs $p_l i_l$ represent a one-to-one association of the vertices of G_1 with the vertices of G_2 indicating the subgraph of G_1 which is isomorphic to G_2. If select(U) returns the value *empty* then G_2 is not a subgraph of G_1.

The Maximum clique and maximum independent set problems.

A clique K_i is the complete graph on i vertices [12]. The problem of finding a maximum independent set is closely related to the maximum clique problem.

- **Problem: Maximum clique**
  Given a graph $G = (V, E)$ determine the largest i such that K_i is a subgraph of G. Here K_i is the complete graph on i vertices.

- **Solution**
    - In parallel run the subgraph isomorphism algorithm for pairs of graphs (G, K_i) for $2 \leq i \leq n$. The largest value of i for which a non- *empty* result is obtained solves the problem.
    - *Complexity*: $O(|V|)$ parallel time.

      A maximum independent set is a subset of vertices of a graph such that no two members of the set are adjacent [12].
- **Problem: Maximum independent set** Given a graph $G = (V, E)$ determine the largest i such that there is a set of i vertices in which no pair are adjacent.
- **Solution**

    Run the maximum clique algorithm on the *complement* of G.
    - *Complexity*: $O(|V|)$ parallel time.

The above examples fully illustrate the way in which the NP-complete problems have a natural mode of expression within the model.

4 Implementation of the model in DNA chemistry

Notice that the algorithms of the previous section work perfectly well if the basic data structure, *set*, is replaced by *multiset*. The **permutation** algorithm now outputs a multiset in which each permutation appears as many times as it was represented in the input set. However, since the *select* operation returns a single element, the output of the other algorithms is exactly as before.

In the proposed implementation outlined below, the algorithms are realised by multisets of single-stranded DNA. In practice, there would be a similar number of copies of each element in any such multiset.

4.1 The structure and manipulation of DNA

DNA (deoxyribonucleic acid) [8] encodes the genetic information of cellular organisms. It consists of *polymer chains*, commonly referred to as DNA *strands*. Each strand may be viewed as a chain of *nucleotides*, or *bases*. An n-letter sequence of consecutive bases is known as an n-mer or an *oligonucleotide* of length n. The four DNA nucleotides are adenine, guanine, cytosine and thymine, commonly abbreviated to A, G, C and T respectively.

Each strand has, according to chemical convention, a $5'$ and a $3'$ end, thus any single strand has a natural orientation. The classical double helix of DNA is formed when two separate strands bond. Bonding occurs by the pairwise attraction of bases; A bonds with T and G bonds with C. The pairs (A, T) and (G, C) are therefore known as *complementary* base pairs. In what follows we adopt the following convention: if x denotes an oligonucleotide, then $\overline{x}$ denotes the complement of x. The bonding process, known as *annealing*, is fundamental to our implementation. A strand will only anneal to its complement if they have opposite polarities. Therefore, one strand of the double helix extends from $5'$ to $3'$, and the other from $3'$ to $5'$, as, for example, in Figure 3.

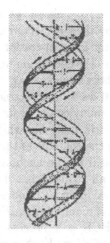

Fig. 2. DNA double helix

5' A T A G A G T T 3'
| | | | | | | |
3' T A T C T C A A 5'

Fig. 3. Structure of double-stranded DNA

4.2 Operations on DNA

All models of DNA computation apply a specific sequence of biological operations to a set of strands. These operations are all commonly used by molecular biologists. Note that some operations are specific to certain models of DNA computation.

- **Synthesis.** Oligonucleotides may be synthesised to order by a machine the size of a microwave oven. The synthesiser is supplied with the four nucleotide bases in solution, which are combined according to a sequence entered by the user. The instrument makes millions of copies of the required oligonucletide and places them in solution in a small vial.
- **Melting.** Double-stranded DNA may be dissolved into single strands by heating the solution to a specific temperature. Heating breaks the hydrogen bonds between complementary strands.
- **Annealing** is the reverse of melting, whereby a solution of single strands is cooled, allowing complementary strands to bind together.
- **Ligation.** In double-stranded DNA, if one of the single strands contains a discontinuity (ie, one nucleotide is not bonded to its neighbour) then this

may be repaired by DNA *ligase* [7]. This allows us to create a unified strand from several bound together by their respective complements.

- **Separation by hybridization** is an operation central to early models of DNA computation, and involves the extraction from a test tube of any *single* strands containing a specific short sequence (e.g., extract all strands containing the sequence $TAGACT$). If we want to extract single strands containing the sequence x we first create many copies of its complement, $\bar{x}$. We attach to these oligonucleotides a biotin molecule which bind in turn to a fixed avitin bead matrix. If we pour the contents of the test tube over this matrix, strands containing x will anneal to the anchored complementary strands. Washing the matrix removes all strands that did not anneal, leaving only strands containing x. These may then be removed from the matrix.

- **PCR.** Another useful method of manipulating DNA is the Polymerase Chain Reaction, or PCR [19, 20]. PCR is a process that quickly amplifies the amount of DNA in a given solution. Each cycle of the reaction doubles the quantity of each strand, giving an exponential growth in the number of strands.

- **Gel electrophoresis** is an important technique for sorting DNA strands by size [7]. This is achieved by applying an electrical field to a gel in which the DNA is placed. Since DNA has a negative charge strands are "pulled" through the gel. Small DNA strands have less hydrodynamic drag than longer ones, so they move quicker through the gel. The DNA is then made visible by flourescent or radioactive labelling. If the sample contains many strands of the same length they form a narrow band in the gel. If there is a continuous distribution of lengths a wide band, or streak, is visible.

- **Polymerization.** The DNA *polymerases* perform several functions, including the repair and duplication of DNA. Given a short *primer* oligonucleotide, p in the presence of nucleoside triphosphates, the polymerase extends p if and only if p is bound to a longer *template* oligonucleotide, t. For example, in Figure 4(a)), p is the oligonucleotide TCA which is bound to t, $ATAGAGTT$. In the presence of the polymerase, p is extended by a complementary strand of bases to the 3' end of t (Figure 4(b)).

- **Restriction enzymes** [24, page 33] recognize a specific sequence of DNA, known as a *restriction site*. Any *double-stranded* DNA that contains the restriction site within its sequence is cut by the enzyme at that point[4] For example, the double-stranded DNA in Figure 5(a) is cut by restriction enzyme $Sau3AI$, which recognizes the restriction site $GATC$. The resulting DNA is depicted in Figure 5(b).

4.3 Implementation of the model

Here we first describe how an initial multiset within the model may be constructed in DNA, and then how the set operations may be implemented.

---

[4] In reality, only certain enzymes cut specifically at the restriction site, but we take this factor into account when selecting an enzyme.

```
5' A T A G A G T T 3'
          | | |            (a)
3'        T C A    5'
```

```
5' A T A G A G T T 3'
   | | | | | | | |         (b)
3' T A T C T C A    5'
```

Fig. 4. (a) Primer anneals to longer template (b) Polymerase extends primer in the 5' to 3' direction

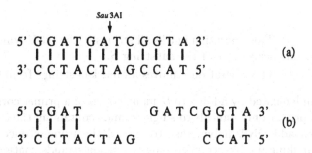

```
              Sau3AI
                ↓
5' G G A T G A T C G G T A 3'
   | | | | | | | | | | | |           (a)
3' C C T A C T A G C C A T 5'
```

```
5' G G A T            G A T C G G T A 3'
   | | | |            | | | |            (b)
3' C C T A C T A G            C C A T 5'
```

Fig. 5. (a) Double-stranded DNA being cut by *Sau*3AI (b) The result

An essential difficulty in the model is that initial multisets generally have a cardinality which is exponential in the problem size. It would be too costly in time, therefore, to generate these individually. What we do in practice is to construct an initial solution, or *tube*, containing a polynomial number of distinct strands. The design of these strands ensures that the exponentially large initial multisets of our model will be generated automatically. The following paragraph describes this process in detail.

Consider an initial set of all elements of the form $p_1 k1, p_2 k_2, \ldots, p_n k_n$. This may be constructed as follows. We generate an oligo uniquely encoding each possible subsequence $p_i k_i$ where $1 \leq i \leq n$ and $1 \leq k_i \leq k$. Embedded within the sequence representing p_i is our chosen restriction site. There are thus nk distinct oligos of this form. The task now is how to combine these to form the desired initial multiset. This is achieved as follows. For each pair $(p_i k_i, p_{i+1} k_{i+1})$ we construct an oligo which is the concatenation of the complement of the second half of the oligo representing $p_i k_i$ and the complement of the first half of the oligo representing $p_{i+1} k_{i+1}$. We also construct oligos that are the complement of the first half of the oligo representing $p_1 k_1$ and the last half of the oligo representing $p_n k_n$. There is therefore a total of $2nk+1$ oligos in solution.

The effect of adding adding these new oligos is that the tube will now contain double-strands of DNA, one strand in each will be an element of the desired initial set. The new oligos have, through annealing, acted as "splints" to join the first oligos in the desired sequences. What we really require in solution are only the single strands encoding elements of the initial set. This is achieved as follows. We ligate the double strands and then heat the tube to break the hydrogen bonds between the encoding strands and the splint strands. The splint strands are created with magnetic beads attached to them in order to facilitate their removal at this stage.

It remains in our implementation to describe how the set operations are realized. This is achieved as follows:

- **Remove** *remove* is implemented as a composite operation, comprised of the following:

  - $mark(U, S)$. This operation marks all strings in the set U which contains at least one occurrence of the substring S.
  - $destroy(U)$. This operation removes all marked strings from U.

  $mark$ is implemented by adding to U many copies of a primer corresponding to $\overline{S}$. This primer only anneals to single strands containing the subsequence S. We then add DNA polymerase to extend the primers once they have annealed, making double-stranded only the single strands containing S.

  We may then *destroy* strands containing S by adding the restriction enzyme *Sau*3AI. Double-stranded DNA (i.e. strands marked as containing S) is cut at the restriction sites embedded within, single strands remaining intact. We may then remove all intact strands by separating on length using *gel electrophoresis* [7]. However, this is not strictly necessary, and leaving the fragmented strands in solution will not affect the operation of the algorithm.

- **Union**

  We may obtain the *union* of two or more tubes by simply mixing them together, forming a single tube.

- **Copy**

  We obtain i "copies" of the set U by splitting U into i tubes of equal volume. We assume that, since the initial tube contains multiple copies of each candidate strand, each tube will also contain many copies.

- **Select** We can easily detect remaining DNA using PCR and then sequence strands to reveal the encoded solution to the given problem. One problem with this method is that there are often multiple correct solutions left in the soup which must be sequenced using nested PCR. A possible solution is to utilise *cloning* (see [2]).

Although the initial tube contain multiple copies of each strand, after many *remove* operations the volume of material may be depleted below an acceptable empirical level. This difficulty can be avoided by periodic amplification by PCR.

5   Advantages of the model

As we have shown, algorithms within our model perform successive "filtering" operations, keeping *good* strands (i.e., strands encoding a legal solution to the given problem) and destroying *bad* strands (i.e., those that do not). So long as the operations work correctly, the final set of strands will only consist of good solutions. However, as we have already stated, errors can take place. If either good strands are accidentally destroyed or bad strands are left to survive through to the final set then the algorithm will fail.

The main advantage of our model is that it doesn't repeatedly use the notoriously error-prone separation by DNA hybridization method to extract strands containing a certain subsequence. Restriction enzymes are *guaranteed* [6, page 9] [5] to cut any double-stranded DNA containing the appropriate restriction site, whereas hybridization separation is never 100% efficient. Instead of extracting *most* strands containing a certain subsequence we simply destroy *all* of them with absolute certainty, without harming those strands that *do not* contain the subsequence. Even if, in reality, restriction enzymes have a small non-zero error rate associated with them, we believe that it is far lower than that of hybridization separation.

Another advantage of our model is that it minimizes physical manipulation of tubes during a computation. Biological operations such as pipetting, filtering and extraction all lose a certain amount of material along the way. As the number of operations increases, the material loss rises and the probability of successful computation decays. Our implementation uses relatively benign physical manipulation, and avoids certain "lossy" operations.

6   Conclusions

Recent work in DNA computation has cast serious doubt on the reliability of proposed implementations of extant models of DNA computation. The use of error-prone operations such as DNA hybridization separation can lead to inconclusive final results. In this paper we described a feasible model of DNA computation whose implementation avoids the problems of those previously proposed. An abstract model of computation was introduced, and we showed that it is sufficiently strong to solve any of the problems in the class NC (which includes the NP-complete problems). We then described algorithms within the model for a selection of graph-theoretic NP-complete problems. It is an easy matter to extend the model to include some form of *splice* operation, which would ensure its full Turing-completeness. We show that the problems in the complexity class NP have a natural expression within our model, especially through the use of *Parallel Filtering* which was defined in the text.

At this time, the empirical basis for being optimistic about the future of DNA computation is very slim. We believe that what is most needed now is

---

[5] "New England Biolabs provides a color-coded 10X NEBuffer with each restriction endonuclease to ensure optimal (100%) activity."

more experimental work. As many authors have demonstrated, it is not difficult to design algorithms within a variety of models which, on a naive basis, appear to be realisable. The fact is that is a difficult matter to design experiments that are likely to be succesful in the laboratory. We hope that we have made a contribution in this direction. At the present time we have taken our model into the laboratory and hope to report in due course on the results of our experiments. These are essentially implementations of our algorithms described in this paper.

An important area of enquiry that is outstanding is the quest for algorithms which proceed through polynomial-sized volumes of DNA. In principle, this ought to be possible for problems which belong to the complexity class P. Success in this area would give real hope for solving problems with very large problem size through the exploitation of the potentially massive parallelism offered by the biochemistry of DNA.

References

1. Leonard Adleman. Molecular computation of solutions to combinatorial problems. *Science*, 266:1021–1024, 1994.
2. Martyn Amos, Alan Gibbons, and David Hodgson. Error-resistant implementation of DNA computations. Research Report CS-RR-298, Department of Computer Science, University of Warwick, Coventry, UK, January 1996.
3. Eric Bach, Anne Condon, Elton Glaser, and Celena Tanguay. DNA models and algorithms for NP-complete problems. Submitted, March 1996.
4. Eric B. Baum and Dan Boneh. Running dynamic programming algorithms on a DNA computer. Unpublished manuscript, February 1996.
5. C. H. Bennett. The thermodynamics of computation – a review. *International Journal of Theoretical Physics*, 21:905–940, 1982.
6. New England Biolabs. Catalog, 1996.
7. T.A. Brown. *Genetics: A molecular approach*. Chapman and Hall, 1993.
8. James D. Watson et. al. *Recombinant DNA*. Scientific American Books, 1992.
9. Richard P. Feynman. There's plenty of room at the bottom. In D. Gilbert, editor, *Miniaturization*, pages 282–296. Reinhold, 1961.
10. Rudolf Freund, Lila Kari, and Gheorghe Păun. DNA computation based on splicing: The existence of universal computers. Submitted.
11. Michael R. Garey and David S. Johnson. *Computers and Intractibility: A Guide to the Theory of NP-Completeness*. W. H. Freeman and Company, New York, 1979.
12. A. M. Gibbons. *Algorithmic Graph Theory*. Cambridge University Press, 1985.
13. David K. Gifford. On the path to computation with DNA. *Science*, 266:993–994, 1994.
14. Juris Hartmanis. On the weight of computations. *Bulletin of the European Association For Theoretical Computer Science*, 55:136–138, 1995.
15. Tom Head. Formal language theory and DNA: an analysis of the generative capacity of specific recombinant behaviors. *Bulletin of Mathematical Biology*, 49:737–759, 1987.
16. Peter Kaplan, Guillermo Cecchi, and Albert Libchaber. Molecular computation: Adleman's experiment repeated. Technical report, NEC Research Institute, 1995.

17. Richard M. Karp, Claire Kenyon, and Orli Waarts. Error-resilient DNA computation. In *7th ACM-SIAM Symposium on Discrete Algorithms*, pages 458–467. SIAM, 1996.
18. Richard J. Lipton. DNA solution of hard computational problems. *Science*, 268:542–545, 1995.
19. Kary B. Mullis. The unusual origin of the polymerase chain reaction. *Scientific American*, 262:36–43, 1990.
20. Kary B. Mullis, François Ferré, and Richard A. Gibbs, editors. *The polymerase chain reaction*. Birkhauser, 1994.
21. Robert Pool. A boom in plans for DNA computing. *Science*, 268:498–499, 1995.
22. John H. Reif. Parallel molecular computation: Models and simulations. In *Proceedings of the Seventh Annual ACM Symposium on Parallel Algorithms and Architectures (SPAA)*, Santa Barbara, June 1995.
23. Paul W. K. Rothemund. A DNA and restriction enzyme implementation of Turing Machines. Unpublished manuscript, 1995.
24. J. Williams, A. Ceccarelli, and N. Spurr. *Genetic Engineering*. βios Scientific Publishers, 1993.

Theory and Practice of Action Semantics

Peter D. Mosses*

BRICS,** Dept. of Computer Science, University of Aarhus,
Ny Munkegade bldg. 540, DK-8000 Aarhus C, Denmark

Abstract. Action Semantics is a framework for the formal description
of programming languages. Its main advantage over other frameworks
is pragmatic: action-semantic descriptions (ASDs) scale up smoothly to
realistic programming languages. This is due to the inherent extensibility and modifiability of ASDs, ensuring that extensions and changes to
the described language require only proportionate changes in its description. (In denotational or operational semantics, adding an unforeseen
construct to a language may require a reformulation of the entire description.)

After sketching the background for the development of action semantics,
we summarize the main ideas of the framework, and provide a simple
illustrative example of an ASD. We identify which features of ASDs
are crucial for good pragmatics. Then we explain the foundations of
action semantics, and survey recent advances in its theory and practical
applications. Finally, we assess the prospects for further development
and use of action semantics.

The action semantics framework was initially developed at the University of Aarhus by the present author, in collaboration with David Watt
(University of Glasgow). Groups and individuals scattered around five
continents have since contributed to its theory and practice.

1 Background

Readers of this paper are presumably familiar with the main ideas of denotational, operational, and axiomatic semantics. Action semantics was developed in
response to problems with application of these frameworks—especially denotational semantics—to practical programming languages.

1.1 Some Problems with Denotational Semantics

First, a bit about my own background: I was fortunate to be studying at the
University of Oxford when Scott and Strachey started their collaboration on
the development of denotational semantics in 1969. I became an enthusiastic

---

* E-mail address: pdmosses@brics.dk, WWW URL: http://www.brics.dk/~pdm
** Centre for Basic Research in Computer Science, Danish National Research
Foundation.

follower of the approach, and the first paper I ever wrote provided a fairly complete denotational description of Algol60 [28]. My thesis work was on the use of denotational-semantic descriptions in compiler generation [29, 30], and I carried on to develop a prototype semantics implementation system called SIS [32].

SIS, which is now obsolete, took denotational descriptions as input. It transformed a denotational description into a λ-expression which, when applied to the abstract syntax of a program, reduced to a λ-expression that represented the semantics of the program in the form of an input-output function. This expression could be regarded as the 'object code' of the program for the λ-reduction machine that SIS provided. By applying this code to some input, and reducing again, one could get the output of the program according to the semantics.

The intended use of SIS was two-fold: 'debugging' semantic descriptions, by empirical testing of whether they gave the *intended* input-output behaviour for programs; and automatic generation of correct (prototype) implementations of programming languages from their semantic descriptions.

In the mid 1970's, denotational semantics was generally regarded as the most promising framework for semantic description. Adequate techniques had been developed for representing the semantics of all common (and many uncommon) constructs of programming languages. It was expected that before long, every major programming language would have a complete denotational description, which could then be given as input to SIS to provide (inefficient but) correct implementations. However, this expectation was not fulfilled—far from it.

It turned out that, despite the elegant and powerful theory of denotational semantics, there are severe pragmatic problems with applying it to languages of the scale of Pascal, C, etc. These problems can be observed already in the descriptions of small illustrative languages given in pedagogical texts on denotational semantics [40, 61, 64]: when changes or extensions to the described language require changes to the definitions of the semantic domains, the original semantic equations may need to be completely reformulated.

This is admittedly only a minor annoyance when dealing with small examples, but it becomes a serious hindrance when developing descriptions of larger languages: making extensions and changes to large denotational descriptions is simply too tedious and error-prone. It also prevents reuse of parts of a denotational description when describing a related language. Better modifiability and reusability are essential, especially if formal semantics is to be usable during the process of language design.

Another purely pragmatic problem is the difficulty of recovering fundamental concepts, such as order of execution or scopes for bindings, from their denotational description. The writer of the description may have a clear conceptual understanding of a programming language, but it gets obscured by the representation of the concepts directly in terms of higher-order functions on domains. In particular, implementors of programming languages are unlikely to refer to semantic descriptions unless the latter provide clear specification of the intended *operational* properties of language constructs.

The main causes of these pragmatic problems appear to be as follows:

1. The definitions of semantic domains are globally visible throughout a denotational description. (Denotational semantics was developed before the use of information-hiding modules and abstract data types became accepted practice for coping with problems of scale in software engineering.) Changes in domain definitions are often required when extending the described language with further constructs—e.g., a change from the direct style to the continuation style when adding jumps, or to power domains when adding nondeterminism. If we could anticipate all such changes, we could *start* with the more complex domains, but that would be unreasonable, as well as notationally burdensome.

2. The way λ-notation is used for specifying semantic entities depends strongly on the details of domain definitions. The classic example of this problem is provided by the semantics of statement sequences: with domains defined for the direct style of semantics, the denotations of the statements have to be composed normally; with the continuation style, the composition has essentially to be reversed. Also, the notation used for constructing elements of sum domains $D_1 + \cdots + D_n$ from the summands, and for case selection on such elements, is sensitive to the positions of the D_i in the sum: inserting a new summand in the middle can involve major changes throughout, just to preserve well-formedness of the notation.

3. All programming concepts have to be reduced to pure functions. This corresponds to translating arbitrary programs into a (lazy) functional programming language, and the amount of encoding required can be large. E.g., assigning values to variables is represented by composing functions that map stores to stores; the fact that the store remains 'single-threaded' is thoroughly obscured by the encoding.

What is needed to remedy the problems listed above?

To alleviate problem 1, one might think it would be sufficient to introduce explicit modular structure into denotational descriptions. This, however, doesn't help if all the module bodies still depend on the details of the domain definitions. It is essential for the modules to be specified in the style of abstract data types, providing notation for the required operations on elements of domains independently of their internal structure. Once this has been done, the explicit modular structure serves mainly as a *reminder* of the independence that has been obtained.

Regarding problem 2, one seems forced to introduce notation for *combining* denotations of program phrases without exploiting knowledge of their domain structure. Then only the definition of this notation, not its use, needs changing when the domains of denotations are changed. E.g., one might introduce notation for sequencing statement executions (as with monads [25, 27]), the use of this notation being independent of whether direct or continuation style domains are used. The pragmatic problems with positional notation for sums of domains could be addressed by using labelled sums (although there are still problems concerning sensitivity to nesting levels of summands).

Finally, to remedy problem 3, it appears necessary to introduce notation not just for sequencing but for all the *fundamental computational concepts* found in programming languages: scopes of bindings, storage of values, communication between concurrent processes, nondeterminism, etc. The use of such notation in semantic equations allows the conceptual analysis of constructs to be expressed directly, and its 'coding' in λ-notation is hidden in the definition of the notation.

1.2 Abstract Semantic Algebras

The above considerations of the pragmatic problems with denotational semantics led to the gradual development of the action semantics framework. At first [31] the idea was to keep as close as possible to denotational semantics, merely avoiding dependence on the structure of domains by introducing a few *combinators* for sequencing and data flow and defining them as auxiliary notation.

It soon became clear, however, that the combinators formed an interesting algebraic structure. The sequencing combinator was of course reminiscent of the composition of a category. Some of other combinators could be given familiar categorical interpretations, e.g., as source and target tupling; but it was not clear how best to accommodate further compositions, such as sequencing combined with dataflow, i.e., strict functional composition. So a pure categorical formulation was rejected in favour of a more general notion of *semantic algebra*— analogous to a data type, but with operations being combinators and primitives corresponding somehow to fundamental concepts of programming languages.

A series of papers on semantic algebras [33, 34, 35, 36] presented various sets of combinators, together with algebraic laws that they were supposed to obey, giving so-called *abstract* semantic algebras. The elements of abstract semantic algebras were always intended to have a clear operational interpretation; they were referred to as *actions* starting from around 1985 [11, 12, 49, 62, 70]. The combinators and primitives of the *action notation* have been rather stable since then—although the symbols used to denote them didn't stabilize until 1991 [42, 47, 63, 71].

1.3 Structural Operational Semantics

How should the intended interpretation of action notation be defined? Throughout the development of action semantics, much emphasis has been placed on the algebraic laws that are to be satisfied by the action combinators and primitives. It is however problematic to take such a set of laws as a definition: they might be inconsistent[3] or incomplete. Even if a consistent and complete set of laws could be found, it might well be difficult to see that they ensured the *intended* operational interpretation of action notation. Note that algebraic laws are also used in action semantics for specifying the data processed by actions, but there the problems of consistency and completeness are much less severe.

Instead, the current definition of action notation [42] is given using Structural Operational Semantics [60]. The required laws are then supposed to hold for a

---

[3] Inconsistent in the sense of having only trivial models.

derived testing equivalence (although in practice most of them can be verified using bisimulation).

Note that it would be difficult to provide a satisfactory denotational semantics for the full action notation: not only is it doubtful that a denotational model satisfying all the laws could be constructed using standard domains (one would need something close to a fully abstract model) but also action notation involves unbounded nondeterministic choice, which poses problems for ordinary continuity. To cite Abramsky (arguing for an intensional semantics based on games): "... once languages with features beyond the purely functional are considered, the appropriateness of modelling programs by functions is increasingly open to question. Neither concurrency nor 'advanced' imperative features have been captured denotationally in a fully convincing fashion." [1] Some of the action combinators and primitives are however quite easy to define as auxiliary notation in denotational semantics [25, 41].

It should be apparent from the above sketch of the development of action semantics that, formally, this framework is just a mixture of techniques from denotational and operational semantics, together with algebraic laws. *The only real originality of action semantics lies in the design of the action notation.*

In the next sections, we shall consider the concepts of action semantics more closely, give a simple illustration, and explain how the design of action notation ensures the desired pragmatic benefits.

2 Action Semantics

As indicated in Section 1, the starting point for the development of action semantics was denotational semantics. Action semantics has retained two of the main features of denotational semantics: the use of context-free grammars to define abstract-syntax trees; and the use of semantic equations to give inductive definitions of compositional semantic functions mapping such trees to semantic entities. (Action semantics may also be viewed as initial-algebra semantics [13].)

The essential deviation from conventional denotational semantics concerns the universe of semantic entities, and the notation used to specify individual entities.

Semantic entities are used to represent the implementation-independent behaviour of programs, as well as the contributions that parts of programs make to overall behaviour. There are actually three kinds of semantic entity used in action semantics: *actions*, *data*, and *yielders*. The main kind is actions; data and yielders are subsidiary. The notation used for specifying actions and the subsidiary semantic entities is called, unsurprisingly, *action notation*.

Actions are essentially dynamic, computational entities. The *performance* of an action directly represents information processing behaviour and reflects the gradual, step-wise nature of computation: each step of an action performance may access and/or change the *current information*. Yielders occurring in actions may access, but not change, the current information. The *evaluation* of a yielder always results in a data entity (including a special entity used to represent undefinedness). For example, a yielder might always evaluate to the datum currently stored in a particular cell, which could change during the performance of an action, and become undefined when the cell is freed.

2.1 Actions

A performance of an action, which may be part of an enclosing action, either: *completes*, corresponding to normal termination; or *escapes*, corresponding to exceptional termination; or *fails*, corresponding to abandoning an alternative; or *diverges*. Actions can be used to represent the semantics of programs: action performances correspond to possible program behaviours. Furthermore, actions can represent the (perhaps indirect) contribution that *parts* of programs, such as statements and expressions, make to the semantics of entire programs.

An action may be nondeterministic, having different possible performances for the same initial information. Nondeterminism represents implementation-dependence, where the behaviour of a program (or the contribution of a part of it) may vary between different implementations—or even between different instants of time on the same implementation.

The information processed by action performance may be classified as follows:

- *transient*: tuples of data, corresponding to intermediate results;
- *scoped*: bindings of tokens to data, corresponding to symbol tables;
- *stable*: data stored in cells, corresponding to the values assigned to variables;
- *permanent*: data communicated between distributed actions.

Transient information is made available to an action for immediate use. Scoped information, in contrast, may generally be referred to throughout an entire action, although it may also be hidden locally in a sub-action. Stable information can be changed, but not hidden, in the action, and it persists until explicitly destroyed. Permanent information cannot even be changed, merely augmented.

When an action is performed, transient information is given only on completion or escape, and scoped information is produced only on completion. In contrast, changes to stable information and extensions to permanent information are made *during* action performance.

The different kinds of information give rise to so-called *facets* of actions, focusing on the processing of at most one kind of information at a time:

- the *basic* facet, processing independently of information (control flows);
- the *functional* facet, processing transient information (actions are *given* and *give* data);
- the *declarative* facet, processing scoped information (actions *receive* and *produce* bindings);
- the *imperative* facet, processing stable information (actions *reserve* and *unreserve* cells of storage, and *change* the data stored in cells); and
- the *communicative* facet, processing permanent information (actions *send* messages, *receive* messages in buffers, and offer *contracts* to *agents*).

These facets of actions are independent. For instance, changing the data stored in a cell—or even unreserving the cell—does not affect bindings involving that cell.

The standard notation for specifying actions consists of *primitive actions* and action *combinators*. Each primitive action is single-faceted, affecting information in only one facet—although any yielders that it contains may refer to several kinds of information.

An action combinator determines control and information flow for each facet of the combined actions, allowing the expression of multi-faceted actions, such as an action that both (imperatively) reserves a cell of storage and then (functionally) gives the identity of the reserved cell. For instance, one combinator determines left-to-right sequencing together with left-to-right transient data flow, letting received bindings flow to its sub-actions; another combinator differs from that only regarding data flow: it concatenates any transients that the sub-actions give when completing, not passing transients between the actions at all. Some selections of control and information flow are disallowed, e.g., interleaving together with transient data flow between the interleaved sub-actions. In particular, the combination of imperative and communicative facets always follows the flow of control.

Note that actions with only a functional facet correspond quite closely to pure partial mathematical functions, the difference being that performance of a functional action may escape or fail, as well as completing or diverging.

2.2 Data

The information processed by actions consists of items of data, organized in structures that give access to the individual items. Data can include various familiar mathematical entities, such as truth values, numbers, characters, strings, lists, sets, and maps. It can also include entities with purely computational usage, such as tokens, cells, and agents—all used for accessing data from the current information—and some compound entities with data components, such as messages and contracts. Actions themselves are not data, but they can be incorporated in so-called *abstractions*, which are data, and subsequently *enacted* back into actions. (Abstraction and enaction are a special case of so-called *reification* and *reflection*.) New sorts of data can be introduced *ad hoc*, for representing special pieces of information.

2.3 Yielders

Yielders are entities that can be evaluated to yield data during action performance. The data yielded may depend on the current information, i.e., the given transients, the received bindings, and the current state of the storage and message buffer. Evaluation cannot affect the current information. Compound yielders can be formed by the application of data operations to yielders.

2.4 Action Notation

The standard symbols used in action notation are ordinary English *words*. In fact action notation mimics natural language: terms standing for actions form imperative verb phrases involving conjunctions and adverbs, e.g., check it and then escape, whereas terms standing for data and yielders form noun phrases, e.g., the items of the given list. Definite and indefinite articles can be exploited for readability, e.g., choose a cell then reserve the given cell (formally, 'a' and 'the' denote the identity function).

These simple principles for choice of symbols provide a surprisingly grammatical fragment of English, allowing specifications of actions to be made fluently readable—without sacrificing formality at all. To specify grouping unambiguously, parentheses may be used.[4]

Compared to other formalisms, such as λ-notation, action notation may appear to lack conciseness: each symbol generally consists of several letters, rather than a single sign. But the comparison should also take into account that each action combinator usually corresponds to a complex pattern of applications and abstractions in λ-notation. For instance, (under the simplifying assumption of determinism) the action term 'A1 then A2' might correspond to something like $\lambda\varepsilon_1.\lambda\rho.\lambda\kappa.A_1\varepsilon_1\rho(\lambda\varepsilon_2.A_2\varepsilon_2\rho\kappa)$. In any case, the increased length of each symbol seems to be far outweighed by its increased perspicuity.

For some applications, however, such as formal reasoning about program equivalence on the basis of their action semantics, optimal conciseness may be highly desirable, and it would be appropriate to use abbreviations for our verbose symbols. The choice of abbreviations is left to the discretion of the user. Such changes of symbols do not affect the *essence* of action notation, which lies in the standard primitives and combinators, rather than in the verboseness of the standard symbols.

The informal appearance and suggestive words of action notation should encourage programmers to read it, at first, rather casually, in the same way that they might read reference manuals. Having thus gained a broad impression of the intended actions, they may go on to read the specification more carefully, paying attention to the details. A more cryptic notation might discourage programmers from reading it altogether.

The intended interpretation of the standard notation for actions has been specified operationally, once and for all [42]. All that one has to do before using action notation is to specify the information that is to be processed by actions—it may vary significantly according to the programming language being described. This specification may involve *extending* data notation with further sorts of data, and *specializing* standard sorts, using sort equations. Furthermore, it may be convenient to introduce formal *abbreviations* for commonly-occurring, conceptually significant patterns of notation. Extensions, specializations, and abbreviations are all specified *algebraically*. The specification in Section 3 illustrates the use of sort equations to specialize some standard sorts of data, and to

---

[4] To avoid a plethora of parentheses in larger examples, typographic devices such as indentation and vertical lines may be used instead.

specify two nonstandard sorts of data for use in the semantic equations, namely
value and number.

3 Illustrative Example

The example of an action-semantic description provided in this section is a very
simple one. It serves merely as an illustration of the use of the main combinators
of action notation. Some more interesting and realistic examples of ASDs are
referenced in Section 7.

The example is divided into three modules, specifying abstract syntax, semantic functions, and semantic entities. The modules are written in the ASCII-
format accepted by the ASD Tools [68], which were used to check their well-
formedness.

```
module: Abstract Syntax.  grammar:

(*)  Stmt  = [[Id ":=" Expr]]
           | [["if" Expr "then" Stmts "else" Stmts]]
           | [["while" Expr "do" Stmts]].

(*)  Stmts = <Stmt <";" Stmt>*>.

(*)  Expr  = Num | Id | [[Expr Op Expr]].

(*)  Op    = "+" | "/=".

(*)  Num   = [[digit+]].

(*)  Id    = [[letter (letter|digit)*]].

endgrammar. closed. endmodule: Abstract Syntax.
```

Table 1. The SIMPL Illustrative Language

The grammar shown in Table 1 specifies several sorts of abstract-syntax trees,
using a variant of BNF grammar allowing regular expressions. The details are not
so important, but note that the double brackets [[...]] indicate node construction (in denotational semantics, they are used only to delimit syntactic phrases
in semantic equations). The angle brackets <...> group components (thus Stmts
is the sort of sequences of the form <S1 ";"..."; " Sn>).

The semantic equations in Table 2 define semantic functions mapping
abstract-syntax trees to semantic entities. Note the explicit specification of the
dependence of this module on the other two (indicated by needs:).

module: Semantic Functions. needs: Abstract Syntax, Semantic Entities.

introduces: execute_, evaluate_, the result of_.

variables: I:Id; N:Num; E,E1,E2:Expr; O:Op; S:Stmt; S1,S2:Stmts.

(*)   execute_ :: Stmts -> action[completing|diverging|storing].

[1:]   execute [[I ":=" E]] =
         evaluate E then store the given number in the cell bound to I.

[2:]   execute [["if" E "then" S1 "else" S2]] =
         evaluate E then
         ((check the given truth-value and then execute S1) or
           (check not the given truth-value and then execute S2)).

[3:]   execute [["while" E "do" S1]] =
         unfolding
         (evaluate E then
           ((check the given truth-value and then
               execute S1 and then unfold) or
             (check not the given truth-value))).

[4:]   execute <S ";" S2> = execute S and then execute S2.

(*)   evaluate_ :: Expr -> action[giving a value].

[5:]   evaluate N = give decimal N.

[6:]   evaluate I = give the number bound to I or
                    give the number stored in the cell bound to I.

[7:]   evaluate [[E1 O E2]] =
         (evaluate E1 and evaluate E2) then give the result of O.

(*)   the result of_ :: Op -> yielder[of a value]
                               [using given (value,value)].

[8:]   the result of "+" = the number yielded by
           the sum of (the given number#1, the given number#2).

[9:]   the result of "/=" = not (the given value#1 is the given value#2).

endmodule: Semantic Functions.

Table 2. SIMPL Action Semantics

The symbols introduced in ASDs may be prefix, postfix, infix, or more generally, 'mixfix'. There is a uniform precedence rule to allow omission of grouping parentheses: infixes have the weakest precedence (and associate to the left), then come prefixes, and finally postfixes. The place-holder '_' shows where the arguments go when operations are applied. Semantic functions, of course, take only one argument, and are conventionally denoted by prefix symbols.

The symbol action denotes the sort of all actions; a term action[O] denotes a subsort, including only those actions whose possible outcomes are contained in O. Similarly, a term of the form yielder[of D][using I] denotes the subsort of yielder that includes those yielders whose evaluation always returns a data item of sort D, referring at most to the current information indicated by I.

In equation 1, the functional action combination A1 then A2 represents ordinary functional composition of A1 and A2: the transients given by A1 on completion are given only to A2. Regarding control flow, A1 then A2 specifies normal left-to-right sequencing.

The yielder given D yields all the data given to its evaluation, provided that this is of the data sort D. For instance the given number (where 'the' is optional) yields a single individual of sort number, if such is given. (Otherwise it yields the special entity nothing, which represents undefinedness, and similarly in other cases below.) The yielder the D bound to T refers to the current binding of sort D for the token T.

The primitive action store Y1 in Y2 requires Y1 to yield a storable value, and Y2 to yield a cell.

In equation 2, the action check Y requires Y to yield a truth value; it completes when the value is true, otherwise it fails. The action A1 or A2 represents implementation-dependent choice between alternative actions. If the alternative currently being performed fails, it is abandoned and, if possible, some other alternative is performed instead, i.e., *back-tracking*. Here, A1 and A2 are guarded by complementary checks, so the choice is deterministic.

The basic action combination A1 and then A2 combines the actions A1 and A2 into a compound action that represents their normal, left-to-right sequencing, performing A2 only when A1 completes.

Equation 3 shows how iteration is specified in action semantics. The action combination unfolding A performs A, but whenever it reaches the dummy action unfold, it performs A instead.

Note that the semantics of a statement sequence is well-defined by equation 4, because the restriction on the sort of S ensures that the argument on the left-hand side of the equation can only match a statement sequence in one way.

In equation 5, the primitive action give Y completes, giving the data yielded by evaluating the yielder Y. The operation decimal is a standard data operation on strings.[5]

The two alternatives in equation 6 correspond to the cases that I is a constant or a variable identifier. The disjointness of the sorts number and cell ensures that

---

[5] An abstract-syntax tree whose direct components are all single characters is identified with the string of those characters.

the choice of an unfailing alternative is deterministic. When no binding for the identifier I to a number or to a cell is received, the action fails.[6]

The action A1 and A2 used in equation 7 represents implementation-dependent order of performance of the indivisible sub-actions of A1, A2. When these sub-actions cannot 'interfere' with each other, as here, it indicates that their order of performance is simply irrelevant. Left-to-right order of evaluation can be specified by using the combinator A1 and then A2 instead of A1 and A2 above. In both cases, the values given by the sub-actions get tupled.

In equation 8, the yielder the number yielded by Y is used to insist that Y yields a value of sort number (the standard data operation sum might return an integer not in the number subsort).

The yielder given Y#n used in equations 8 and 9 yields the n'th individual component of a given tuple, provided that this component is of sort Y.

```
module: Semantic Entities.

includes: Action Notation.

introduces: value, number.

(*)  token    =  string.
(*)  bindable =  cell | number.
(*)  storable =  number.
(*)  value    =  number | truth-value.
(*)  number   =< integer.

endmodule: Semantic Entities.
```

Table 3. Specializing Action Notation for SIMPL Semantics

The illustrative example of an ASD is completed by the module in Table 3, which specifies some sorts (token, bindable, storable) that are left open in the standard specification of Action Notation. (The use of includes: rather than needs: specifies that the imported notation is also exported.) The sorts value and number are introduced just for use in this illustrative ASD, and have no predetermined interpretation in Action Notation. Just as in Table 1, a vertical bar expresses union of sorts. The sort inclusion number =< integer leaves open whether number is bounded.

So much for the example, which should have given a rough impression of the 'look and feel' of action notation and action-semantic descriptions.

---

[6] As there are no declarations at all in SIMPL, the semantics of a SIMPL statement depends on received bindings for pre-declared identifiers.

4 Pragmatics

Let us now assess some of the pragmatic aspects of action-semantic descriptions. In particular, how well may we expect ASDs to scale up from illustrative examples (like the one given in the preceeding section) to practical programming languages?

In marked contrast to the situation with denotational-semantic descriptions, *making extensions and changes to an ASD generally affects only those parts of the description dealing directly with the constructs involved.*

E.g., adding expressions that allow function or process activation would *not* require any changes at all to the semantic equations given in Section 3. The enrichment of the actions representing expression evaluation by, e.g., the potential for side-effects or communication, does not invalidate the use of the combinator A1 and A2 in equation 7, since it has a well-defined interpretation for actions with arbitrary information-processing capabilities (it interleaves the atomic steps).

This very desirable pragmatic aspect of action semantics depends on two crucial features of action notation:

- Each combinator is defined *universally* on actions. Contrast this with function composition in λ-notation, which requires exact matching of types between the composed functions.
- There is no mention of the presence or absence of any particular kind of information processing, except where creation or inspection of this information is required. For instance, stored information is referred to *only* in equations 1 and 6, whereas in a denotational semantics for the same language, every semantic equation would have to cater for the fact that all denotations are functions of the store.

It also depends on the fact that we may extend sorts of data (e.g., bindable) with new values or subsorts without disturbing the notation for creating or matching values. This is a feature of the algebraic specification framework used for the foundations of action semantics (summarized in the next section): it provides a genuine sort union operation, which behaves like set union, and avoids the pragmatic problems of the notation for sums of domains used in denotational semantics.

Since the above features ensure that ASDs have an *inherent* modularity, the use of explicit modules is almost redundant. In fact it is usual in ASDs to let all the semantic entities be visible throughout all the semantic equations, and this does not cause any problems with modifiability, etc.

Action semantics provides a high degree of extensibility, modifiability, and reusability, which is especially important when using semantic descriptions during language design, and highly significant when developing larger ASDs in general. An action-semantic description is also strongly suggestive of an operational understanding of the described language, as required by implementors. Moreover, it has been found to be very well suited for generating compilers [5, 56, 58] and interpreters [69].

Thus the pragmatic aspects of action semantics seem to be satisfactory.

5 Foundations

The foundations of action semantics are based on the framework of *unified algebras* [37, 38, 39]: each part of an ASD is interpreted as a unified algebraic specification.

The signature of a unified algebra is simply a ranked alphabet. The universe of a unified algebra is a (distributive) lattice with a bottom value, together with a distinguished subset of *individuals*. The operations of a unified algebra are required to be monotone (total) functions on the lattice; they are not required to be strict or additive, nor to preserve the property of individuality.

All the values of a unified algebra may be thought of as *sorts*, with the individuals corresponding to singleton sorts. The partial order of the lattice represents sort inclusion; join is sort union and meet is sort intersection. The bottom value (denoted by nothing) is a vacuous sort, often used to represent the lack of a result from applying an operation to unintended arguments. A special case of a unified algebra is a *power algebra*, whose universe is a power set, with the singletons as individuals.

The axioms of unified algebraic specifications are Horn clauses involving equations T1=T2, inclusions T1=<T2, and individual inclusions T1:T2. An equation holds when the terms have identical values, an inclusion holds when the values of the terms are in the partial order of the lattice, and an individual inclusion T1:T2 holds when the value of T1 is not only included in that of T2, but also in the distinguished subset of individuals.

Unified algebraic specifications always have initial models, because they are essentially just unsorted Horn clause logic (with equality) specifications, and the lattice structure and monotonicity of operations can all be captured by Horn clauses.

As illustrated in Section 3, an ASD consists of a grammar, some semantic equations, and a specification of the required universe of semantic entities. Thanks to the sort equations and inclusions allowed as axioms by unified algebras, all these parts have a straightforward interpretation, as follows.

Regarding grammars, each nonterminal symbol is interpreted as a sort constant, and the alternatives on the right-hand sides of productions as sort terms, combined with sort union; productions themselves are simply regarded as equational axioms [45], in contrast to the more elaborate interpretation of grammars as signatures usually taken in the literature [13].

The formal interpretation of a set of semantic equations is that the semantic functions are ordinary equationally-specified operations (taking a single syntactic argument and returning a semantic entity). The only use made here of the special features of unified algebras is in specifying subsorts of actions and yielders in the functionalities of the semantic functions. An alternative—but more complicated—interpretation would be to take account of the intended compositionality and inductiveness of the definitions, by regarding the semantic functions as the components of the unique homomorphism from the initial algebra of abstract syntax to a target algebra derived from the semantic equations.

The specification of the semantic entities consists of the specialization of action notation to particular sorts of data, together with the algebraic specification of abstract data types. The former involves sort equations and inclusions; the latter could be given in any decent algebraic specification framework. The operational semantics of the general action notation is fixed, and cannot be changed; it has been specified [42] in the style of structural operational semantics [60] (using an unorthodox presentation where the transition relation is a function from individual configurations to sorts, exploiting here unified algebras again). Notions of bisimulation and testing equivalence on actions are defined, completing the formalization of actions. The laws that action notation obeys are thus consequences of the definitions, rather than axioms. The next section surveys further work concerning the theory of actions.

6 Theory

After the experience with the development of denotational semantics, where much effort was spent on theoretical aspects, and the problems of applying the framework to practical languages were only realized after some time, I decided to proceed differently with action semantics: the first priority was to check that the pragmatic aspects of ASDs were satisfactory. Thus it is not surprising that a decent theory for action semantics has been slow to emerge. (Of course foundations of action semantics, such as those sketched in Section 5, were provided right from the start, otherwise it could not have been regarded as a formal framework at all.)

The theory of action semantics still has not been developed to the extent of, say, domain theory for denotational semantics. There may be many reasons for this: action notation may appear too large or unwieldy for theoretical analysis; or perhaps the notation conventionally used in ASDs is too verbose and informal-looking.

Nevertheless, significant work on several aspects of the theory of action semantics has already been done. The descriptions of it below are mostly adapted from the abstracts of the cited papers.

Note that the aim of this theoretical development is not only to investigate the properties of action notation *per se*, but also to allow reasoning about programs and programming languages by means of their action-semantic descriptions. For instance, algebraic laws established for action notation may be used to reason about program equivalence. (The situation is similar with regard to practical applications: flow analysis and code generation for action notation allow efficient compilation of programs according to their action semantics, as reported in Section 7.) The *possibility* of lifting analyses from action notation to programming languages depends on the compositionality of action semantics; its *usefulness* depends crucially on the simplicity of the actions that represent program constructs—and of course on the existence of action-semantic descriptions for programming languages of practical interest.

6.1 Type Inference

In papers in TCS and at the ESOP conference in 1990, Even and Schmidt [11, 12] formulated a model for action semantics based on Reynolds's *category-sorted algebra*. In the model, actions are natural transformations, and the composition operators are compositions in a 'category of actions'. They use the model to prove semantic soundness and completeness of a unification-based, decidable type-inference algorithm for action semantics expressions.

In a paper at ESOP'92, Doh and Schmidt [8] described a method that automatically extracts a type checking semantics, encoded as a set of type inference rules, from a category sorted algebra-based action-semantics definition of a programming language. The type inference rules are guaranteed to enforce strong typing, since they are based on an underlying meta-semantics for action semantics, which uses typing functions and natural transformations to give meaning. They use the type checking semantics to extract a dynamic semantics definition from the original action-semantics definition.

A distinguishing characteristic of action semantics is its facet system, which defines the variety on information flows in a language definition. The facet system can be analysed to validate the well-formedness of a language definition, and to calculate the operational semantics. At the Workshop on Action Semantics in 1994 [44] Doh and Schmidt [10] presented a single framework for doing all of this. The framework exploits the internal subsorting structure of the facets so that sort checking, static analysis, and operational semantics are related, sound instances of the same underlying analysis. The framework also suggests that action semantics' extensibility can be understood as a kind of 'weakening rule' in a 'logic' of actions. In the cited paper, the framework is used to perform type inference on specific programs, to justify meaning-preserving code transformations, and to 'stage' an ASD of a programming language into a static semantics stage and a dynamic semantics stage.

6.2 Provably-Correct Compiler Generation

As reported in his PhD thesis [58] and papers [57, 59] at ESOP and ICCL in 1992, Palsberg has designed, implemented, and proved the correctness of a compiler generator that accepts action-semantic descriptions of imperative programming languages. He has used it to generate compilers for both a toy language and a non-trivial subset of Ada. The generated compilers emit absolute code for an abstract RISC machine language that is assembled into code for the SPARC and the HP Precision Architecture. The generated code is an order of magnitude better than that produced by compilers generated by classical semantics-based compiler generators. His machine language needs no runtime type-checking and is thus more realistic than those considered in previous compiler proofs. He uses solely algebraic specifications; proofs are given in the initial model. The use of action semantics makes the processable language specification easy to read and pleasant to work with. His compiler generator may be seen as the first step towards user-friendly and automatic generation of realistic and provably correct compilers.

6.3 Action Analysis and Compiler Generation

The Actress system [5] accepts the action-semantic description of a source language, and from it generates a compiler. The generated compiler translates its source program to an action, performs sort inference on this action, (optionally) simplifies it by transformations, and finally translates it to object code. The sort inference provides valuable information for the subsequent transformation and code generation phases; Brown and Watt [4] reported their studies of the problem of sort inference for actions at the Workshop on Action Semantics. Transformations of the intermediate action greatly improve the efficiency of the object code; Moura's PhD thesis from 1993 [51] is concerned with these transformations, as reported at CC'94 [52].

Also at CC'94, Ørbæk [56] presented several analyses of actions. These allow his compiler generator (called OASIS) to generate efficient, optimizing compilers for procedural and functional languages with higher order recursive functions.

6.4 Action Equivalence

Lassen's PhD thesis work is devoted to developing a tractable theory for action equivalence. In a recent technical report [24] he has developed the foundations for a richer action theory, by bringing together concepts and techniques from process theory and from work on operational reasoning about functional programs. Semantic pre-orders and equivalences in the action semantics setting are studied and useful operational techniques for establishing contextual equivalences are presented. These techniques are applied to establish equational and inequational action laws and an induction rule for the basic facet of action notation.

Even more recently [23] Lassen has extended this theory to the functional and declarative facets, covering a substantial fragment of action notation involving transient and scoped information flow and higher-order, (unbounded) nondeterministic, and interleaving computation. Based on a reduction semantics for actions, operational reasoning techniques have been developed and used to establish an inequational theory for action notation. The potential of this theory is illustrated by proofs of various functional program equivalences via an action-semantic description of a functional language. He is currently extending the theory to the imperative facet in order to reason about practical, imperative programming languages.

7 Practice

The development of action semantics has so far involved rather modest resources, and there is still some way to go before a completely satisfactory set of examples and tools becomes available. The proceedings of the first International Workshop on Action Semantics [44] give a good impression of the level of activity in the area. This section lists some of the major applications so far.

7.1 ASDs of Programming Languages

Apart from the published ASDs mentioned below, various Master's theses at the University of Aarhus have covered large parts of various programming languages, e.g., ML, Amber, Joyce, Modula-3, and occam. Unfortunately, most of them were written in older versions of action notation, and are now out of print.

Pascal: The Pascal Action Semantics by Watt and the present author, available by FTP since 1993 [50], provides an almost complete formal specification of the dynamic semantics of Standard Pascal (Level 0). It is intended primarily as a 'showcase' example of the use of action semantics, rather than as a contribution to the understanding of Pascal. The current version is for readers who are already familiar with the action semantics framework. More work is needed before it is ready for a more general readership, including completion of an automated checking of the internal consistency and completeness of the specification, which is being done using the ASD Tools.

Standard ML: At the MFPS conference in 1988, Watt [70] reported on an ASD of the Standard ML 'bare' language in an early version of action notation. This has now been converted to use the current version of action notation, but not yet published. When ready—and extended to the whole of Standard ML—it will make an interesting basis for comparison between action semantics and structural operational semantics, as the latter approach was used for defining Standard ML [26].

The ANDF-FS: ANDF is an Architecture- and Language-Neutral Distribution Format developed by the Open Software Foundation (OSF) and other collaborators around the world. It is based on the TDF technology provided by the Defence Research Agency of the United Kingdom Ministry of Defence. It is a medium-level intermediate language, used as the target language of compilers for ordinary high-level languages. The production of the ANDF-FS, a formal specification of ANDF, was one of the tasks of the ESPRIT project OMI/GLUE [15].

In a technical report from 1993 [66], Toft assessed the feasibility of using various frameworks for the ANDF-FS: a hybrid of denotational and algebraic semantics; structural operational semantics, expressed as an algebraic specification in RSL; and action semantics, where the definition of action notation may either be expressed as higher-order functions or in terms of an operational semantics. His main conclusion was that "action semantics with an underlying structural operational semantics is the most qualified candidate for the ANDF-FS".

Unfortunately, at the time there was no mature tool support for action semantics. A compromise was found: the RAISE Specification Language RSL [65] is a general-purpose specification language with good tool support; by giving a structural operational semantics of the required subset of action notation in RSL, the RSL tools could be used for action semantics.

The ANDF-FS from 1994 [55], presented also at the Workshop on Action Semantics [16], was the first example of 'industrial' interest in action semantics.

Somewhat surprisingly, action semantics was used not only for the dynamic semantics of ANDF but also for its *static* semantics, exploiting action combinators to express static evaluation, type-checking, scopes of bindings, etc.

Particularly encouraging is the fact that the ANDF-FS wasn't put on the shelf to gather dust after completion. It was used by a group at the OSF Research Institute in Grenoble, for three purposes: to discuss refinement of the informal ANDF specifications, as an aid to develop a validation suite for ANDF, and as an aid to develop an interpreter. Despite little previous exposure to formal semantics, this group confirms that they found the ANDF-FS quite accessible—thanks at least partly to the verbosity of action notation.

Another significant aspect is that the project was conducted by persons who had not been involved in the development of action semantics. Even though the project took place in Denmark, my contact with it was limited to answering one question about my book.

7.2 Concurrent Languages

An ASD for a non-trivial sublanguage of Ada (including tasks) was given by the present author in his text book on action semantics [42]. He also reported the use of action semantics to describe concurrent languages in a REX workshop paper in 1992 [43]. His paper together with Krishnan at RTFT'92 [21] on specifying asynchronous transfer of control led to a further paper by Krishnan [19].

Musicante reported on an ASD for the Sun RPC protocol in 1992 [53]. With the present author he investigated a minor potential extension of the communicative facet of action notation to provide shared storage [54]. Together they presented an ASD of a small fragment of Standard ML at the FME'94 [48], and showed that adding concurrency primitives to the described language requires only extensions, not changes, to the description of the sequential constructs.

7.3 Tools

The ASD Tools have been developed since 1993 by van Deursen and the present author, and demonstrated at various conferences, e.g., AMAST'96 [68]. They provide a support environment for using action semantics, including facilities for parsing, syntax-directed (and textual) editing, checking, and interpretation of action-semantic descriptions. Such facilities significantly enhance accuracy and productivity when writing and maintaining large specifications, and are also particularly useful for those learning how to write ASDs. The notation supported by the ASD Tools, illustrated in Section 3, is a direct ASCII representation of the standard notation used for ASDs in the literature.

The ASD Tools are implemented using the ASF+SDF system [18], which is itself based on the Centaur system (developed by INRIA, among others). A licence for Centaur is required; this is available free of charge to academic institutions, and the entire ASD Tools implementation can be obtained by FTP. The ASD Tools were taken as a case study in the use of ASF+SDF by van Deursen in his PhD thesis [67].

The Actress system developed by Watt's group at Glasgow provides proto-type tools, reported at CC'92 [5] and CC'94 [52], for interpreting action notation and for compiling it into C. The interpreter deals with most of the standard action notation except for the communicative facet.

In his PhD thesis [6] in 1992, and later in a joint paper with Schmidt [9], Doh presented a methodology for compiler synthesis based on action semantics. Each symbol of action notation is assigned specific 'analysis functions', such as a typing function and a binding-time function. When a language is given an action semantics, the typing and binding-time functions for the individual actions compose into typing and binding-time analyses for the language; these are implemented as the type checker and static semantics processor, respectively, in the synthesized compiler. Other analyses can be similarly formalized and implemented.

Partial evaluation has been used extensively in connection with compiler generation. In 1993, Bondorf and Palsberg [3] used the Similix system to obtain an action compiler by partial evaluation of an action interpreter. In a paper at PEPM'95, Doh proposed using partial evaluation for action transformation [7].

The compiler generator OASIS, presented by Ørbæk at CC'94 [56], is capable of generating efficient, optimizing compilers for procedural and functional languages with higher-order recursive functions. The automatically generated compilers produce code that is comparable with code produced by handwritten compilers. This work is perhaps the most convincing evidence so far of the practical applicability of action-semantics-based compiler generation.

8 Prospects

We have reviewed the current state of theory and practice of action semantics. What prospects are there for the future development and use of this framework?

Regarding the theory of action semantics: despite recent progress with proof techniques for action equivalence, much remains to be done in that direction. In particular, it is urgent to develop a useful proof calculus for the communicative facet of actions, which is based on *asynchronous* processes and message-passing; the work of Agha, Mason, Smith, and Talcott [2] appears to be applicable here.

The 'official' notation for actions and data has remained the same for the past five years. Some potential improvements have recently been suggested [22], and this may be a good time to make a revised version, taking into account the experience gained through using the present version in large ASDs. One of the major issues here is whether it might be better to take 'yielders' merely as a subsort of actions, their evaluation then being a special case of action performance. More superficially, different vocabularies of symbols for action combinators and primitives might be developed—perhaps even using graphics for visualization [20].

It is unclear whether users of action semantics should be allowed to change the operational semantics of action notation, or add new action combinators and primitives, e.g. as required for letting agents share storage [54]. Such changes would require re-proving all the laws of action notation. In any case, the current structural operational semantics of action notation is not so easy to modify; alternative forms of operational semantics, such as evolving-algebra semantics [14], might be preferable in that respect.

The underlying algebraic specification language used in action semantics is currently the somewhat unorthodox and lesser-known framework of 'unified algebras', summarized in Section 5. It might be advisable to replace it by a framework with a more direct set-theoretic basis [17] or by a sublanguage of the common algebraic specification language that is currently being developed by the Common Framework Initiative [46]. In either case, algebraic data-type definitions should be included, as this would allow a significant reduction in the size of the modules that specify auxiliary semantic entities.

To further encourage the practical use of action semantics, an integrated tool set supporting the writing, editing, and testing of ASDs should be provided. In particular, navigating and searching in larger ASDs needs to be made easier than with the existing tools, e.g., by using hyper-text links and indexing.

Then there is the matter of completing the existing partial ASDs of a number of practical programming languages, and of developing new ones, such as for the Java language. The larger ASDs that have already been produced confirm that action semantics has (much) better applicability than denotational semantics, so it should be feasible to build up an on-line library of checked ASDs in the near future. Large parts of these ASDs could then be reused in the descriptions of further languages. Moreover, they would provide appropriate input for compiler generators based on action semantics, avoiding a recurrence of the problems experienced with exploiting SIS for denotational semantics some 20 years ago, and perhaps stimulating further research and development of semantics-based compiler generation.

Finally, it is important for the future of action semantics that it gets taught in semantics courses at the undergraduate level. For this purpose, it may be best to provide a cut-down version of action notation, including only those symbols needed for describing familiar languages such as Pascal and Standard ML; the operational semantics of such a subset would probably be significantly simpler than that defining the full notation.

Up-to-date information about action semantics may be found via the Action Semantics Home Page, URL: http://www.brics.dk/Projects/AS.

Acknowledgements

I would like to take this opportunity to thank all those who have contributed so far to the development of action semantics, and to colleagues at the University of Aarhus and elsewhere for their encouragement to continue with this work. Thanks especially to Olivier Danvy, Christian Fabre, Søren B. Lassen, Peter Ørbæk, and David A. Watt, who all suggested significant improvements to earlier versions of this paper.

I would also like to thank the organizers of MFCS'96 for inviting me to present this paper, and for letting me have some extra pages.

My own part in the development of action semantics has been supported by: the Department of Computer Science, University of Aarhus; the Danish National Research Foundation Centre BRICS; the Danish Science Research Council project DART (5.21.08.03); and ESPRIT Basic Research Working Group COMPASS (3264 and 6112).

References

1. S. Abramsky. Semantics of interaction. In *Trees in Algebra and Programming - CAAP'96, Proc. 21st Int. Coll., Linköping*, volume 1059 of *Lecture Notes in Computer Science*, page 1. Springer-Verlag, 1996.

2. G. Agha, I. A. Mason, S. F. Smith, and C. L. Talcott. A foundation for actor computation. To appear in *Journal of Functional Programming*, 1994.

3. A. Bondorf and J. Palsberg. Compiling actions by partial evaluation. In *FPCA'93, Proc. Sixth ACM Conf. on Functional Programming Languages and Computer Architecture, Copenhagen*, pages 308–317, 1993.

4. D. Brown and D. A. Watt. Sort inference in the Actress compiler generator. In [44], pages 81–98, 1994.

5. D. F. Brown, H. Moura, and D. A. Watt. Actress: an action semantics directed compiler generator. In *CC'92, Proc. 4th Int. Conf. on Compiler Construction, Paderborn*, volume 641 of *Lecture Notes in Computer Science*, pages 95–109. Springer-Verlag, 1992.

6. K.-G. Doh. *Action Semantics-Directed Prototyping*. PhD thesis, Kansas State University, 1992.

7. K.-G. Doh. Action transformation by partial evaluation. In *PEPM'95, Proc. ACM/SIGPLAN Symposium on Partial Evaluation and Semantics-based Program Transformation, La Jolla, California*, pages 230–240, 1995.

8. K.-G. Doh and D. A. Schmidt. Extraction of strong typing laws from action semantics definitions. In *ESOP'92, Proc. European Symposium on Programming, Rennes*, volume 582 of *Lecture Notes in Computer Science*, pages 151–166. Springer-Verlag, 1992.

9. K.-G. Doh and D. A. Schmidt. Action semantics-directed prototyping. *Comput. Lang.*, 19(4):213–233, 1993.

10. K.-G. Doh and D. A. Schmidt. The facets of action semantics: Some principles and applications (extended abstract). In [44], pages 1–15, 1994.

11. S. Even and D. A. Schmidt. Category sorted algebra-based action semantics. *Theoretical Comput. Sci.*, 77:73–96, 1990.

12. S. Even and D. A. Schmidt. Type inference for action semantics. In *ESOP'90, Proc. European Symposium on Programming, Copenhagen*, volume 432 of *Lecture Notes in Computer Science*, pages 118–133. Springer-Verlag, 1990.

13. J. A. Goguen, J. W. Thatcher, E. G. Wagner, and J. B. Wright. Initial algebra semantics and continuous algebras. *J. ACM*, 24:68–95, 1977.

14. Y. Gurevich. Evolving algebras 1993: Lipari guide. In E. Börger, editor, *Specification and Validation Methods*. Oxford University Press, 1995.

15. B. S. Hansen and J. Bundgaard. The role of the ANDF formal specification. Technical Report 202104/RPT/5, issue 2, DDC International A/S, Lundtoftevej 1C, DK–2800 Lyngby, Denmark, 1992.

16. B. S. Hansen and J. U. Toft. The formal specification of ANDF, an application of action semantics. In [44], pages 34–42, 1994.

17. C. Hintermeier, H. Kirchner, and P. D. Mosses. Combining algebraic and set-theoretic specification. In *Recent Trends in Data Type Specification, Proc. 11th Workshop on Specification of Abstract Data Types, Oslo, 1995, Selected Papers.* Lecture Notes in Computer Science, 1996. To appear.

18. P. Klint. A meta-environment for generating programming environments. *ACM Transactions on Software Engineering Methodology*, 2(2):176–201, 1993.

19. P. Krishnan. Specification of systems with interrupts. *J. Systems Software*, 21:291–304, 1993.

20. P. Krishnan, B. McKenzie, and S. Hunt. Guile: Graphical user interface for linguistic experiments. In *Proceedings of the 17th Annual Computer Science Conference, Australian Communications*, pages 309–320, 1994.

21. P. Krishnan and P. D. Mosses. Specifying asynchronous transfer of control. In *RTFT'92, Proc. Symp. on Formal Techniques in Real-Time and Fault-Tolerant Systems, Delft*, volume 571 of *Lecture Notes in Computer Science*. Springer-Verlag, 1992.

22. S. B. Lassen. Design and semantics of action notation. In [44], pages 34–42, 1994.

23. S. B. Lassen. Action semantics reasoning about functional programs. BRICS, Dept. of Computer Science, Univ. of Aarhus, Dec. 1995.

24. S. B. Lassen. Basic action theory. Technical Report RS-95-25, BRICS, Dept. of Computer Science, Univ. of Aarhus, 1995.

25. S. Liang and P. Hudak. Modular denotational semantics for compiler construction. In *Programming Languages and Systems – ESOP'96, Proc. 6th European Symposium on Programming, Linköping*, volume 1058 of *Lecture Notes in Computer Science*, pages 219–234. Springer-Verlag, 1996.

26. R. Milner, M. Tofte, and R. Harper. *The Definition of Standard ML*. The MIT Press, 1990.

27. E. Moggi. Computational lambda-calculus and monads. In *LICS'89, Proc. 4th Ann. Symp. on Logic in Computer Science*, pages 14–23. IEEE, 1989.

28. P. D. Mosses. The mathematical semantics of Algol60. Tech. Mono. PRG–12, Programming Research Group, Univ. of Oxford, 1974.

29. P. D. Mosses. *Mathematical Semantics and Compiler Generation*. D.Phil. dissertation, University of Oxford, 1975.

30. P. D. Mosses. Compiler generation using denotational semantics. In *MFCS'76, Proc. Symp. on Math. Foundations of Computer Science, Gdańsk*, volume 45 of *Lecture Notes in Computer Science*. Springer-Verlag, 1976.

31. P. D. Mosses. Making denotational semantics less concrete. In *Proc. Int. Workshop on Semantics of Programming Languages, Bad Honnef*, pages 102–109. Abteilung Informatik, Universität Dortmund, 1977. Bericht nr. 41.

32. P. D. Mosses. SIS, Semantics Implementation System: Reference manual and user guide. Tech. Mono. MD–30, Dept. of Computer Science, Univ. of Aarhus, 1979. Out of print.

33. P. D. Mosses. A constructive approach to compiler correctness. In *ICALP'80, Proc. Int. Coll. on Automata, Languages, and Programming, Noordwijkerhout*, volume 85 of *Lecture Notes in Computer Science*, pages 449–469. Springer-Verlag, 1980.

34. P. D. Mosses. A semantic algebra for binding constructs. In *Proc. Int. Coll. on Formalization of Programming Concepts, Peñiscola*, volume 107 of *Lecture Notes in Computer Science*, pages 408–418. Springer-Verlag, 1981.

35. P. D. Mosses. Abstract semantic algebras! In *Formal Description of Programming Concepts II, Proc. IFIP TC2 Working Conference, Garmisch-Partenkirchen, 1982*, pages 45–71. North-Holland, 1983.

36. P. D. Mosses. A basic abstract semantic algebra. In *Proc. Int. Symp. on Semantics of Data Types, Sophia-Antipolis*, volume 173 of *Lecture Notes in Computer Science*, pages 87–107. Springer-Verlag, 1984.

37. P. D. Mosses. Unified algebras and action semantics. In *STACS'89, Proc. Symp. on Theoretical Aspects of Computer Science, Paderborn*, volume 349 of *Lecture Notes in Computer Science*. Springer-Verlag, 1989.

38. P. D. Mosses. Unified algebras and institutions. In *LICS'89, Proc. 4th Ann. Symp. on Logic in Computer Science*, pages 304–312. IEEE, 1989.

39. P. D. Mosses. Unified algebras and modules. In *POPL'89, Proc. 16th Ann. ACM Symp. on Principles of Programming Languages*, pages 329–343. ACM, 1989.

40. P. D. Mosses. Denotational semantics. In *Handbook of Theoretical Computer Science*, volume B, chapter 11. Elsevier Science Publishers, Amsterdam; and MIT Press, 1990.

41. P. D. Mosses. A practical introduction to denotational semantics. In *Formal Description of Programming Concepts*, IFIP State-of-the-Art Report, pages 1–49. Springer-Verlag, 1991.

42. P. D. Mosses. *Action Semantics*. Number 26 in Cambridge Tracts in Theoretical Computer Science. Cambridge University Press, 1992.

43. P. D. Mosses. On the action semantics of concurrent programming languages. In *Semantics: Foundations and Applications, Proc. REX Workshop, Beekbergen, 1992*, volume 666 of *Lecture Notes in Computer Science*, pages 398–424. Springer-Verlag, 1993.

44. P. D. Mosses, editor. *Proc. 1st Intl. Workshop on Action Semantics, Edinburgh, 1994*, number NS-94-1 in BRICS Notes Series. BRICS, Dept. of Computer Science, Univ. of Aarhus, 1994.

45. P. D. Mosses. Unified algebras and abstract syntax. In *Recent Trends in Data Type Specification, Proc. 9th Workshop on Specification of Abstract Data Types, Caldes de Malavella, 1992, Selected Papers*, volume 785 of *Lecture Notes in Computer Science*, pages 280–294. Springer-Verlag, 1994.

46. P. D. Mosses, editor. *CoFI: Initiative for a Common Framework for Algebraic Specification*, URL: http://www.brics.dk/Projects/CoFI, 1996.

47. P. D. Mosses. A tutorial on action semantics. 50pp. Tutorial notes for FME'94 (Formal Methods Europe, Barcelona, 1994) and FME'96 (Formal Methods Europe, Oxford, 1996), Mar. 1996.

48. P. D. Mosses and M. A. Musicante. An action semantics for ML concurrency primitives. In *FME'94, Proc. Formal Methods Europe: Symposium on Industrial Benefit of Formal Methods, Barcelona*, volume 873 of *Lecture Notes in Computer Science*, pages 461–479. Springer-Verlag, 1994.

49. P. D. Mosses and D. A. Watt. The use of action semantics. In *Formal Description of Programming Concepts III, Proc. IFIP TC2 Working Conference, Gl. Avernæs, 1986*, pages 135–166. North-Holland, 1987.

50. P. D. Mosses and D. A. Watt. Pascal action semantics, version 0.6. URL: ftp://ftp.brics.dk/pub/BRICS/Projects/AS/Papers/MossesWatt93DRAFT.ps.Z, Mar. 1993.

51. H. Moura. *Action Notation Transformations*. PhD thesis, Dept. of Computing Science, Univ. of Glasgow, 1993.

52. H. Moura and D. A. Watt. Action transformations in the Actress compiler genera-
    tor. In *CC'94, Proc. 5th Intl. Conf. on Compiler Construction, Edinburgh*, volume
    786 of *Lecture Notes in Computer Science*, pages 16–30. Springer-Verlag, 1994.
53. M. A. Musicante. The Sun RPC language semantics. In *Proceedings of PANEL'92,
    XVIII Latin-American Conference on Informatics*. Universidad de Las Palmas de
    Gran Canaria, 1992.
54. M. A. Musicante and P. D. Mosses. Communicative action notation with shared
    storage. Tech. Mono. PB–452, Dept. of Computer Science, Univ. of Aarhus, 1993.
55. J. P. Nielsen and J. U. Toft. Formal specification of ANDF, existing subset. Tech-
    nical Report 202104/RPT/19, issue 2, DDC International A/S, Lundtoftevej 1C,
    DK–2800 Lyngby, Denmark, 1994.
56. P. Ørbæk. OASIS: An optimizing action-based compiler generator. In *CC'94,
    Proc. 5th Intl. Conf. on Compiler Construction, Edinburgh*, volume 786 of *Lecture
    Notes in Computer Science*, pages 1–15. Springer-Verlag, 1994.
57. J. Palsberg. An automatically generated and provably correct compiler for a sub-
    set of Ada. In *ICCL'92, Proc. Fourth IEEE Int. Conf. on Computer Languages,
    Oakland*, pages 117–126. IEEE, 1992.
58. J. Palsberg. *Provably Correct Compiler Generation*. PhD thesis, Dept. of Com-
    puter Science, Univ. of Aarhus, 1992. xii+224 pages.
59. J. Palsberg. A provably correct compiler generator. In *ESOP'92, Proc. European
    Symposium on Programming, Rennes*, volume 582 of *Lecture Notes in Computer
    Science*, pages 418–434. Springer-Verlag, 1992.
60. G. D. Plotkin. A structural approach to operational semantics. Lecture Notes
    DAIMI FN–19, Dept. of Computer Science, Univ. of Aarhus, 1981.
61. D. A. Schmidt. *Denotational Semantics: A Methodology for Language Develop-
    ment*. Allyn & Bacon, 1986.
62. D. A. Schmidt. *The Structure of Typed Programming Languages*. The MIT Press,
    1994.
63. K. Slonneger and B. L. Kurtz. *Formal Syntax and Semantics of Programming
    Languages: A Laboratory Based Approach*. Addison-Wesley, 1995.
64. J. E. Stoy. *Denotational Semantics: The Scott-Strachey Approach to Programming
    Language Theory*. The MIT Press, 1977.
65. The RAISE Language Group. *The RAISE Specification Language*. BCS Practi-
    tioner Series. Prentice-Hall, 1992.
66. J. U. Toft. Feasibility of using RSL as the specification language for the ANDF for-
    mal specification. Technical Report 202104/RPT/12, issue 2, DDC International
    A/S, Lundtoftevej 1C, DK–2800 Lyngby, Denmark, 1993.
67. A. van Deursen. *Executable Language Definitions: Case Studies and Origin Track-
    ing Techniques*. PhD thesis, Univ. of Amsterdam, 1994.
68. A. van Deursen and P. D. Mosses. ASD: The action semantic description tools.
    In *AMAST'96, Proc. 5th Intl. Conf. on Algebraic Methodology and Software Tech-
    nology, Munich*, Lecture Notes in Computer Science. Springer-Verlag, 1996. To
    appear.
69. D. A. Watt. Executable semantic descriptions. *Software – Practice and Experi-
    ence*, 16:13–43, 1986.
70. D. A. Watt. An action semantics of Standard ML. In *Proc. Third Workshop on
    Math. Foundations of Programming Language Semantics, New Orleans*, volume
    298 of *Lecture Notes in Computer Science*, pages 572–598. Springer-Verlag, 1988.
71. D. A. Watt. *Programming Language Syntax and Semantics*. Prentice-Hall, 1991.

Linear Time Temporal Logics over Mazurkiewicz Traces

Madhavan Mukund and P.S. Thiagarajan

School of Mathematics, SPIC Science Foundation,
92 G N Chetty Rd, Madras 600 017, India
E-mail: {madhavan,pst}@ssf.ernet.in

Abstract. Temporal logics are a well-established tool for specifying and reasoning about the computations performed by distributed systems. Although temporal logics are interpreted over sequences, it is often the case that such sequences can be gathered together into equivalence classes where all members of an equivalence class represent the same partially ordered stretch of behaviour of the system. This appears to have important implications for improving the practical efficiency of automated verification methods based on temporal logics. With this as motivation, we study logics that are directly interpreted over partial orders. We survey a number of linear time temporal logics whose underlying frames are Mazurkiewicz traces. We describe automata theoretic methods for solving the satisfiability and model checking problems for these logics. It turns out that we still do not know what the "canonical" linear time temporal logic over Mazurkiewicz traces looks like. We identify here the criteria that should be met by this elusive logic.

Introduction

Propositional Linear time Temporal Logic (LTL) proposed by Pnueli [Pnu] has become a well established tool for specifying and reasoning about complex distributed behaviours [MP]. A central feature of LTL is that its formulas are interpreted over infinite sequences. In applications of LTL, the infinite sequences consist of the runs of a distributed system with each run being an infinite sequence of states assumed by the system or an infinite sequence of actions executed by the system during the course of a computation. Interesting distributed systems consist of a number of autonomous sequential agents that coordinate their behaviour with the help of some communication mechanism. In such systems, substantial portions of a computation will consist of causally independent tasks performed by different agents at separate locations. Consequently a single partially ordered stretch of behaviour of the system will be modelled by many different runs that differ from each other only in the order in which they record causally independent occurrences of actions. This kind of run-based view is often referred to as an interleaved semantics of distributed systems.

The interleaved view of the behaviour of distributed systems has proved to be very successful and popular. However it has been known for some time that the practical effectiveness of LTL and related formalisms can be often enhanced by modelling and analyzing the concerned behaviours in terms of partial orders rather than sequences.

In typical applications, an LTL formula constitutes the specification of the system behaviour and the verification problem consists of checking whether every run of the system is a model of the formula and therefore whether the system meets the specification. The property expressed by the specification is very often of the kind where either *all* the interleaved runs corresponding to a single partially ordered computation have the property or *none* of the interleavings have the property. A typical example of such a property is freedom from deadlock, as pointed out by Valmari [Val]. As a result, it suffices to verify the desired property for just one representative run of each partially ordered computation. The resulting saving in running time and memory usage can be substantial in practice [GW]. This is the background and motivation underlying the so called partial order based verification methods which are a subject of active research [GW, KP, Val].

There is an alternative way to exploit non-sequential behaviours and the attendant partial order based verification methods. It consists of developing temporal logics and related techniques that can be *directly* applied to specify and reason about the properties of partial order based runs of a distributed system. In this paper we survey linear time temporal logics that have arisen from this approach.

In going from sequences to partial orders it is easy to go overboard because so many possibilities are available. Fortunately, in the context of distributed behaviours, Mazurkiewicz has formulated a tractable and yet very fruitful way of passing from sequences to partial orders [Maz]. The resulting restricted partial orders are known as Mazurkiewicz traces, often called—as we shall do here— just traces. The theory of traces is well developed [Die, DR] and is strongly related to the theory of other well known formalisms such as Petri nets and event structures. Further, the classical theory of ω-regular (word) languages in terms of its logical, algebraic and automata-theoretic aspects has been successfully extended to ω-regular trace languages [EM, GP]. Finally, the structures that underlie the partial order based verification methods being developed recently can be almost always be viewed as traces.

Hence there is a good deal of motivation for formulating linear time temporal logics that are to be directly interpreted over traces. Many such logics are now available. In the present survey, we will mainly concentrate on the ones that fulfill two criteria:

(i) The logic should be expressible within the first order theory of traces.
(ii) The satisfiability problem for the logic should admit a treatment in terms of asynchronous Büchi automata.

This seemingly arbitrary choice of criteria can be justified as follows. LTL is *the* linear time temporal logic over sequences in that it is equivalent in expressive power to the first order theory of sequences [Zuc]. We consider the task of identifying the counterpart of LTL for traces to be an important one both from a theoretical and practical standpoint (see the last portion of Section 4). At present we do not know what this counterpart of LTL looks like. However, it seems a good starting point to concentrate on those linear time temporal logics that are at least no more expressive than the first order theory of traces.

As for the second criterion, an appealing feature of LTL is that its satisfiability and model checking problems can be transparently solved using Büchi automata [VW]. This has led to a clean separation of the logical and combinatorial aspects of these problems, thus contributing to the development of automated verification methods and related optimization techniques. The evidence available at present suggests that asynchronous Büchi automata are an appropriate machine model for dealing with ω-regular trace languages. Hence it seems worthwhile to lift the interplay between LTL and Büchi automata to the level of traces.

In the next section we review the basic aspects of traces. In Section 2 we describe asynchronous Büchi automata and present our version of these automata called, for want of a better name, A2-automata. In Section 3, the heart of the paper, we present the logic TrPTL (Trace based Propositional Temporal logic of Linear time) and two of its sublogics TrPTL^{con} and TrPTL$^{\otimes}$. The logic TrPTL is directly interpreted over traces. We show that the satisfiability and model checking problems for TrPTL can be solved using A2-automata. We then show that the syntactic restrictions imposed to obtain TrPTL^{con} and TrPTL$^{\otimes}$ lead to corresponding simplifications in the world of automata. After presenting these results we survey a number of other temporal logics that use traces as their underlying frames. In Section 4 we show that TrPTL is expressible within the first order theory of traces. The final section contains concluding remarks.

Most of the results will be presented without proofs. The proofs are either available in the literature or can be easily manufactured using the results available in the literature.

1 Traces

The starting point for trace theory is a trace alphabet (Σ, I), where Σ, the alphabet, is a finite set and $I \subseteq \Sigma \times \Sigma$ is an irreflexive and symmetric independence relation. In most applications, Σ consists of the actions performed by a distributed system while I captures a strong static notion of causal independence between actions. The idea is that contiguous independent actions occur with no causal order between them. Thus, every sequence of actions from Σ corresponds to an interleaved observation of a partially-ordered stretch of system behaviour. This leads to a natural equivalence relation over execution sequences: two sequences are equated iff they correspond to different interleavings of the same partially-ordered stretch of behaviour.

To formulate this equivalence relation precisely, we need some terminology. For the rest of the section we fix a trace alphabet (Σ, I) and let a, b range over Σ. $D = (\Sigma \times \Sigma) - I$ is called the dependency relation. Note that D is reflexive and symmetric. A set $p \subseteq \Sigma$ is called a D-clique iff $p \times p \subseteq D$. We set $\Sigma^{\infty} = \Sigma^* \cup \Sigma^{\omega}$

where Σ^* is the set of finite words over Σ and Σ^ω is the set of infinite words over Σ. We let σ, σ' with or without subscripts range over Σ^∞ and τ, τ' with or without subscripts range over Σ^*. The equivalence relation $\sim_I \subseteq \Sigma^\infty \times \Sigma^\infty$ induced by I is given by:

$$\sigma \sim_I \sigma' \text{ iff } \sigma \lceil p = \sigma' \lceil p \text{ for every } D\text{-clique } p.$$

Here and elsewhere, if A is a finite set, $\rho \in A^\infty$ and $B \subseteq A$ then $\rho \lceil B$ is the sequence obtained by erasing from ρ all occurrences of letters in $A - B$.

Clearly $\sim_I$ is an equivalence relation. Notice that if $\sigma = \tau a b \sigma_1$ and $\sigma' = \tau b a \sigma_1$ with $(a, b) \in I$ then $\sigma \sim_I \sigma'$. Thus σ and σ' are identified if they differ only in the order of appearance of a pair of adjacent independent actions. In fact, for finite words, an alternative way to characterize $\sim_I$ is to say that $\sigma \sim_I \sigma'$ iff σ' can be obtained from σ by a finite sequence of permutations of adjacent independent actions. Unfortunately, the definition of $\sim_I$ in terms of permutations is too naïve to be transported to infinite words, which is why we work with the less intuitive definition presented here.

The equivalence classes generated by $\sim_I$ are called *(Mazurkiewicz) traces*. The theory of traces is well developed and documented—see [Die, DR] for basic material as well as a substantial number of references to related work.

Traces have many equivalent representations. We shall view traces as special kinds of labelled partial orders. Since sequences can be viewed as labelled *total* orders, this representation emphasizes that traces are an elegant and non-trivial generalization of sequences.

Recall that a Σ-labelled poset is a structure $F = (E, \leq, \lambda)$ where $\leq$ is a partial order on the set E and $\lambda : E \to \Sigma$ is a labelling function. The *covering relation* $\lessdot \subseteq E \times E$ is given by: $e \lessdot e'$ iff $e < e'$ (i.e., $e \leq e'$ and $e \neq e'$) and for every $e'' \in E$, $e \leq e'' \leq e'$ implies $e = e''$ or $e'' = e'$.

For $X \subseteq E$ we define $\downarrow X$ to be the set $\{y \mid y \leq x \text{ for some } x \in X\}$. If X is a singleton $\{x\}$, we write $\downarrow x$ instead of $\downarrow\{x\}$.

We can now formulate traces in terms of labelled partial orders. A *trace* over (Σ, I) is a Σ-labelled poset $F = (E, \leq, \lambda)$ which satisfies the following conditions.

- E is a countable set.
- For each $e \in E$, $\downarrow e$ is a finite set.
- For all $e, e' \in E$, if $e \lessdot e'$ then $(\lambda(e), \lambda(e')) \in D$.
- For all $e, e' \in E$, if $(\lambda(e), \lambda(e')) \in D$ then $e \leq e'$ or $e' \leq e$.

Let $TR(\Sigma, I)$ denote the set of Σ-labelled posets that satisfy the definition above. We now sketch briefly the proof that $\Sigma^\infty / \sim_I$ and $TR(\Sigma, I)$ represent the same class of objects. We construct representation maps $\text{str} : \Sigma^\infty \to TR(\Sigma, I)$ and $\text{trs} : TR(\Sigma, I) \to \Sigma^\infty / \sim_I$ and state some results which show that these maps are "inverses" of each other. We shall not prove these results. The details can be easily obtained using the constructions developed in [WN] for relating traces and event structures.

Henceforth, we will not distinguish between isomorphic elements in $TR(\Sigma, I)$. In other words, whenever we write $F = F'$ for traces $F = (E, \leq, \lambda)$ and $F' =$

$(E', \leq', \lambda')$, we mean that there is a label-preserving isomorphism between F and F'.

For $\sigma \in \Sigma^\infty$, $[\sigma]$ stands for the $\sim_I$-equivalence class containing σ. We use $\preceq$ to describe the usual prefix ordering over sequences. Let $\text{prf}(\sigma)$ denote the set of finite prefixes of σ.

We now define $\text{str} : \Sigma^\infty \to TR(\Sigma, I)$. Let $\sigma \in \Sigma^\infty$. Then $\text{str}(\sigma) = (E, \leq, \lambda)$ where:

- $E = \{\tau a \mid \tau a \in \text{prf}(\sigma)\}$. Recall that $\tau \in \Sigma^*$ and $a \in \Sigma$. Thus $E = \text{prf}(\sigma) - \{\varepsilon\}$, where ε is the null string.
- $\leq \subseteq E \times E$ is the least partial order which satisfies:
  For all $\tau a, \tau' b \in E$, if $\tau a \preceq \tau' b$ and $(a, b) \in D$ then $\tau a \leq \tau' b$.
- For $\tau a \in E$, $\lambda(\tau a) = a$.

The map str induces a natural map str' from $\Sigma^\infty / \sim_I$ to $TR(\Sigma, I)$ defined by $\text{str}'([\sigma]) = \text{str}(\sigma)$. One can show that if $\sigma, \sigma' \in \Sigma^\infty$, then $\sigma \sim_I \sigma'$ iff $\text{str}(\sigma) = \text{str}(\sigma')$. This observation guarantees that str' is well defined. In fact, henceforth we shall write str to denote both str and str'.

To go in the other direction let $F = (E, \leq, \lambda)$ be a trace over (Σ, I). Then $\rho \in E^\infty$ is called a linearization of F iff every $e \in E$ appears exactly once in ρ and, moreover, whenever $e, e' \in E$ and $e < e'$, e appears before e' in ρ.

As usual, we can extend the labelling function $\lambda : E \to \Sigma$ to words over E in a canonical way. If $\rho = e_0 e_1 \ldots$ is a word in E^∞ then $\lambda(\rho)$ denotes the corresponding word $\lambda(e_0)\lambda(e_1)\ldots$ in Σ^∞. We can now define the map $\text{trs} : TR(\Sigma, I) \to \Sigma^\infty / \sim_I$ as follows:

$$\text{trs}(F) = \{\lambda(\rho) \mid \rho \text{ is a linearization of } F\}.$$

Proposition 1.1

(i) *For every $\sigma \in \Sigma^\infty$, $\text{trs}(\text{str}(\sigma)) = [\sigma]$.*
(ii) *For every $F \in \Sigma^\infty$, $\text{str}(\text{trs}(F)) = F$.*

This result justifies our claim that $\Sigma^\infty / \sim_I$ and $TR(\Sigma, I)$ are indeed two equivalent ways of talking about the same class of objects.

In the poset representation of traces, finite configurations play the same role that finite prefixes do in sequences. Let $F = (E, \leq, \lambda)$ be a trace over (Σ, I). Then $c \subseteq E$ is a *configuration* iff c is finite and $\downarrow c = c$. We let $\mathcal{C}_F$ denote the set of configurations of F. Notice that $\emptyset$, the empty set, is a configuration. It is the least configuration under set inclusion. More importantly, $\downarrow e$ is a configuration for every e. These "pointed" configurations associated with the events are also called *prime configurations*. They constitute the building blocks for the Scott domains induced by traces [NPW]. We shall see that they also play a fundamental role in defining linear time temporal logics over traces.

We now turn our attention to distributed alphabets. Distributed alphabets can be viewed as "implementations" of trace alphabets. They form the basis for defining machine models with a built-in notion of independence which recognize trace languages.

Let $\mathcal{P}$ be a finite set of sequential agents called *processes*. A distributed alphabet is a family $\{\Sigma_p\}_{p \in \mathcal{P}}$ where Σ_p is a finite non-empty alphabet for each $p \in \mathcal{P}$. The idea is that whenever an action from Σ_p occurs, the agent p must participate in it. Hence the agents can constrain each other's behaviour, both directly *and* indirectly.

Trace alphabets and distributed alphabets are closely related to each other. Let $\widetilde{\Sigma} = \{\Sigma_p\}_{p \in \mathcal{P}}$ be a distributed alphabet. Then $\Sigma_{\mathcal{P}}$, the global alphabet associated with $\widetilde{\Sigma}$, is the collection $\bigcup_{p \in \mathcal{P}} \Sigma_p$. The distribution of $\Sigma_{\mathcal{P}}$ over $\mathcal{P}$ can be described using a *location function* $\mathrm{loc}_{\widetilde{\Sigma}} : \Sigma_{\mathcal{P}} \to 2^{\mathcal{P}}$ defined as follows:

$$\mathrm{loc}_{\widetilde{\Sigma}}(a) = \{p \mid a \in \Sigma_p\}.$$

This in turn induces the relation $I_{\widetilde{\Sigma}} \subseteq \Sigma_{\mathcal{P}} \times \Sigma_{\mathcal{P}}$ given by:

$$(a, b) \in I_{\widetilde{\Sigma}} \text{ iff } \mathrm{loc}_{\widetilde{\Sigma}}(a) \cap \mathrm{loc}_{\widetilde{\Sigma}}(b) = \emptyset.$$

Clearly $I_{\widetilde{\Sigma}}$ is irreflexive and symmetric and hence $(\Sigma_{\mathcal{P}}, I_{\widetilde{\Sigma}})$ is a trace alphabet. Thus every distributed alphabet canonically induces a trace alphabet. Two actions are independent according to $\widetilde{\Sigma}$ if they are executed by disjoint sets of processes. Henceforth, we write loc for $\mathrm{loc}_{\widetilde{\Sigma}}$ whenever $\widetilde{\Sigma}$ is clear from the context.

Going in the other direction there are, in general, many different ways to implement a trace alphabet as a distributed alphabet. A standard approach is to create a separate agent for each maximal D-clique generated by (Σ, I). Recall that a D-clique of (Σ, I) is a non-empty subset $p \subseteq \Sigma$ such that $p \times p \subseteq D$. Let $\mathcal{P}$ be the set of maximal D-cliques of (Σ, I). This set of processes induces the distributed alphabet $\widetilde{\Sigma} = \{\Sigma_p\}_{p \in \mathcal{P}}$ where $\Sigma_p = p$ for every process p. The alphabet $\widetilde{\Sigma}$ implements (Σ, I) in the sense that the canonical trace alphabet induced by it is exactly (Σ, I). In other words, $\Sigma_{\mathcal{P}} = \Sigma$ and $I_{\widetilde{\Sigma}} = I$.

For example, consider the trace alphabet (Σ, I) where $\Sigma = \{a, b, d\}$ and $I = \{(a, b), (b, a)\}$. The canonical D-clique implementation of (Σ, I) yields the distributed alphabet $\widetilde{\Sigma} = \{\{a, d\}, \{d, b\}\}$.

As mentioned earlier, distributed alphabets play a crucial role in the automata-theoretic aspects of trace theory. The fundamental result of Zielonka [Zie] says that every regular trace language over (Σ, I) can be recognized by an asynchronous automaton over a distributed alphabet $\widetilde{\Sigma}$ which implements (Σ, I). This result has been extended to ω-regular trace languages in terms of asynchronous Büchi automata by Gastin and Petit [GP].

Distributed alphabets arise naturally in a variety of models of distributed systems. In particular they are associated with the restricted but very useful model of a distributed system consisting of a network of sequential agents that coordinate their behaviour by performing common actions together. The linear time temporal logics that we consider in this paper will be based on distributed alphabets.

We conclude this section with a technical remark. Most of the theory of traces presented in this paper, including the automata-theoretic and logical aspects, constitutes a natural and conservative extension of the existing theory in the sequential setting. The sequential theory can almost always be recovered by setting $I = \emptyset$ when dealing with trace alphabets. Correspondingly, when dealing with distributed alphabets, the sequential case corresponds to having just one agent—i.e., $|\mathcal{P}| = 1$.

2 Automata over Infinite Traces

From now on we shall focus on infinite traces. With a little additional work most of the material we shall present on automata and logics can be extended to handle finite traces as well. Through the rest of this section we fix a distributed alphabet $\tilde{\Sigma} = \{\Sigma_p\}_{p \in \mathcal{P}}$ with the induced trace alphabet (Σ, I), where $\Sigma = \bigcup_{p \in \mathcal{P}} \Sigma_p$ and $I = \{(a, b) \mid \mathrm{loc}(a) \cap \mathrm{loc}(b) = \emptyset\}$.

The terminology and notational conventions developed in the previous section are assumed here as well. We will be dealing with many $\mathcal{P}$-indexed families. For convenience we shall often write $\{X_p\}$ to denote the $\mathcal{P}$-indexed family $\{X_p\}_{p \in \mathcal{P}}$. A similar convention will be followed in dealing with Σ-indexed families: $\{Y_a\}$ will denote the family $\{Y_a\}_{a \in \Sigma}$.

Asynchronous Büchi automata, due to Gastin and Petit [GP], are the basic class of automata operating over infinite traces. They constitute a common generalization of the asynchronous automata of Zielonka [Zie] operating over finite traces and a mild variant of the classical Büchi automata operating over infinite sequences. We shall consider here a number of variants of asynchronous Büchi automata, each with a slightly different acceptance condition.

We begin with a brief and slightly non-standard presentation of Büchi automata. A word ω-automaton over Σ is a pair $\mathcal{B} = (TS, T)$ where

- $TS = (S, \{\rightarrow_a\}, S_{in})$ is a finite state transition system over Σ. In other words, S is a finite set of states, $\rightarrow_a \subseteq S \times S$ is an a-labelled transition relation for each $a \in \Sigma$ and $S_{in} \subseteq S$ is a set of initial states.
- T is an acceptance table accompanied by an acceptance condition.

Before considering a number of possibilities for T, let us define the notion of a run. The Σ-indexed family of transition relations $\{\rightarrow_a\}$ induces a global transition relation $\rightarrow_\mathcal{B} \subseteq S \times \Sigma \times S$ given by $s \xrightarrow{a}_\mathcal{B} s'$ iff $(s, s') \in \rightarrow_a$. Where $\mathcal{B}$ is clear from the context $\rightarrow_\mathcal{B}$ will be written as $\rightarrow$.

Let $\sigma \in \Sigma^\omega$ (i.e., $\sigma : \omega \rightarrow \Sigma$ where, as usual, $\omega = \{0, 1, 2, \ldots\}$ is the set of natural numbers). A run of TS over σ is a map $\rho : \omega \rightarrow S$ such that $\rho(0) \in S_{in}$ and $\rho(i) \xrightarrow{\sigma(i)} \rho(i+1)$ for every $i \geq 0$.

The set of states encountered infinitely along the run ρ is denoted $\inf(\rho)$: $\inf(\rho) = \{s \mid \text{for infinitely many } i, \rho(i) = s\}$.

Let us now consider just two of the various possibilities for T.

(B0) $T = F \subseteq S$.

A run ρ over σ is accepting with respect to B0 iff $\inf(\rho) \cap F \neq \emptyset$. We shall say that $\mathcal{A}$ is a B0-automaton if it uses B0 as its acceptance criterion. Of course, we shall also refer to these by their standard name; Büchi automata.

$L(\mathcal{A})$, the language accepted (recognized) by $\mathcal{A}$, is the set of infinite words σ such that there is an accepting run of $\mathcal{A}$ on σ. A language $L \subseteq \Sigma^\omega$ is said to be ω-regular iff there exists a Büchi automaton $\mathcal{A}$ over Σ such that $L(\mathcal{A}) = L$. As is well known, ω-regular languages have equivalent algebraic and logical presentations, as detailed in the excellent survey [Tho].

A second possibility for T is:

(B1) $T \subseteq 2^S$.

A run ρ over σ is accepting with respect to B1 iff there exists $F \in T$ such that $\inf(\rho) \supseteq F$. It is easy to show that $L \subseteq \Sigma^\omega$ is ω-regular iff there exists a B1-automaton $\mathcal{A}$ (i.e., an automaton $\mathcal{A}$ that uses B1 as its acceptance criterion) such that $L = L(\mathcal{A})$. Thus at the level of sequences there is no difference in expressive power between Büchi automata and B1-automata. As we shall see, at the level of traces, B0 is weaker than B1.

For defining automata on infinite traces we need to develop some notation. Let $F = (E, \leq, \lambda) \in TR(\Sigma, I)$. Then F is an infinite trace iff E is an infinite set. Let $TR^\omega(\Sigma, I)$ denote the subclass of infinite traces over (Σ, I). Often, we shall write TR^ω instead of $TR^\omega(\Sigma, I)$.

Let $F \in TR^\omega$ with $F = (E, \leq, \lambda)$ and let $p \in \mathcal{P}$. Then $e \in E$ is a p-event iff $\lambda(e) \in \Sigma_p$. Similarly, e is an a-event iff $\lambda(e) = a$. We let E_p denote the set of p-events and E_a denote the set of a-events.

There are two natural transition relations that one can associate with F. The event based transition relation $\Rightarrow_F \subseteq \mathcal{C}_F \times E \times \mathcal{C}_F$ is defined as $c \stackrel{e}{\Rightarrow}_F c'$ iff $e \notin c$ and $c \cup \{e\} = c'$. The action-based transition relation $\rightarrow_F \subseteq \mathcal{C}_F \times \Sigma \times \mathcal{C}_F$ is defined as $c \stackrel{a}{\rightarrow}_F c'$ iff there exists $e \in E$ such that $\lambda(e) = a$ and $c \stackrel{e}{\Rightarrow}_F c'$.

To define automata on infinite traces, we have to first define a distributed version of transition systems. The distributed transition systems we work with here are essentially the *asynchronous automata* of Zielonka [Zie]. We begin with some notation involving local and global states.

Let $\mathcal{P}$ be a set of processes. We equip each process $p \in \mathcal{P}$ with a finite non-empty set of local p-states, denoted S_p. We set $S = \bigcup_{p \in \mathcal{P}} S_p$ and call S the set of *local states*.

We let P, Q range over non-empty subsets of $\mathcal{P}$ and let p, q range over $\mathcal{P}$. A Q-state is a map $s : Q \rightarrow S$ such that $s(q) \in S_q$ for every $q \in Q$. We let S_Q denote the set of Q-states. We call $S_\mathcal{P}$ the set of *global states*.

If $Q' \subseteq Q$ and $s \in S_Q$ then $s_{Q'}$ is s restricted to Q'. In other words $s_{Q'}$ is the Q'-state s' which satisfies $s'(q') = s(q')$ for every q' in Q'. We use a to abbreviate $\text{loc}(a)$ when talking about states (recall that $\text{loc}(a) = \{p \mid a \in \Sigma_p\}$). Thus an a-state is just a $\text{loc}(a)$-state and S_a denotes the set of all $\text{loc}(a)$-states. If $\text{loc}(a) \subseteq Q$ and s is a Q-state we shall write s_a to mean $s_{\text{loc}(a)}$.

A *distributed transition system* TS over $\tilde{\Sigma}$ is a structure $(\{S_p\}, \{\to_a\}, S_{in})$ where

- S_p is a finite non-empty set of p-states for each process p.
- For $a \in \Sigma$, $\to_a \subseteq S_a \times S_a$ is a transition relation between a-states.
- $S_{in} \subseteq S_{\mathcal{P}}$ is a set of initial global states.

The idea is that an a-move by TS involves only the local states of the agents which participate in the execution a. This is reflected in the global transition relation $\to_{TS} \subseteq S_{\mathcal{P}} \times \Sigma \times S_{\mathcal{P}}$ which is defined as:

$$s \xrightarrow{a}_{TS} s' \text{ iff } (s_a, s'_a) \in \to_a \text{ and } s_{\mathcal{P}-\mathrm{loc}(a)} = s'_{\mathcal{P}-\mathrm{loc}(a)}.$$

From the definition of $\to_{TS}$, it is clear that actions which are executed by disjoint sets of agents are processed independently by TS.

A trace ω-automaton over $\tilde{\Sigma} = \{\Sigma_p\}$ is a pair $\mathcal{A} = (TS, \mathcal{T})$ where $TS = (\{S_p\}, \{\to_a\}, S_{in})$ is a distributed transition system over $\tilde{\Sigma}$ and $\mathcal{T}$ is an acceptance table (which we will elaborate on later).

A *trace run* of TS over $F \in TR^\omega$ is a map $\rho : C_F \to S_{\mathcal{P}}$ such that $\rho(\emptyset) \in S_{in}$ and for every $(c, a, c') \in \to_F$, $\rho(c) \xrightarrow{a}_{TS} \rho(c')$.

To define acceptance we must now compute $\inf_p(\rho)$, the set of p-states that are encountered infinitely often along ρ. The obvious definition, namely $\inf_p(\rho) = \{s_p \mid \rho(c)(p) = s_p$ for infinitely many $c \in C_F\}$, will not work. The complication arises because some processes may make only finitely many moves, even though the overall trace consists of an infinite number of events.

For instance, consider the distributed alphabet $\tilde{\Sigma}_0 = \{\{a\}, \{b\}\}$. In the corresponding distributed transition system, there are two processes p and q which execute a's and b's completely independently. Consider the trace $F = (E, \leq, \lambda)$ where $|E_p| = 1$ and E_q is infinite—i.e., all the infinite words in $\mathrm{trs}(F)$ contain one a and infinitely many b's. Let s_p be the state of p after executing a. Then, there will be infinitely many configurations whose p-state is s_p, even though p only moves a finite number of times.

Continuing with the same example, consider another infinite trace $F' = (E', \leq', \lambda')$ over the same alphabet where both E_p and E_q are infinite. Once again, let s_p be the local state of p after reading one a. Further, let us suppose that after reading the second a, p never returns to the state s_p. It will still be the case that there are infinitely many configurations whose p-state is s_p: consider the configurations $c_0, c_1, c_2, \ldots$ where c_j is the finite configuration after one a and j b's have occurred.

So, we have to define $\inf_p(\rho)$ carefully in order to be able to distinguish whether or not process p is making progress. The appropriate formulation is as follows:

Case 1 E_p *is finite:* $\inf_p(\rho) = \{s_p\}$, where $\rho(\downarrow E_p) = s$ and $s_p = s(p)$.
Case 2 E_p *is an infinite set:*
$\quad \inf_p(\rho) = \{s_p \mid$ for infinitely many $e \in E_p, s_e(p) = s_p$, where $\rho(\downarrow e) = s_e\}$.

We can now begin to consider various acceptance tables.

(A0) $T = \{F_p\}$ with $F_p \subseteq S_p$ for each p.

A run ρ over F is accepting with respect to A0 iff $\inf_p(\rho) \cap F_p \neq \emptyset$ for every p. The trace language accepted by the A0-automaton $\mathcal{A}$ (i.e., where T is of the form A0) is the set $L_{Tr}(\mathcal{A}) = \{F \mid \exists$ an accepting run of TS over $F\}$. A0-automata are the obvious common generalization of asynchronous automata and Büchi automata. It turns out that A0-automata are not expressive enough: the acceptance criterion cannot distinguish whether or not an agent executes infinitely many actions.

To bring this out and to motivate the acceptance condition we are after, we will put down a crude definition of ω-regular trace languages.

A trace language over $\widetilde{\Sigma}$ is just a subset of TR^ω. To define ω-regular trace languages, we exploit the result from the previous section linking $\Sigma^\infty / \sim_I$ and $TR(\Sigma, I)$ which permits us to associate a language of infinite words with each trace language. We can then transport the definition of ω-regularity from subsets of Σ^ω to infinite traces.

Let $L \subseteq \Sigma^\omega$. Then L is I-consistent iff for every $\sigma \in \Sigma^\omega$, if $\sigma \in L$ then $[\sigma] \subseteq L$. Thus if L is I-consistent either all members of the $\sim_I$-equivalence class $[\sigma]$ are in L or none of them are in L.

Let $L' \subseteq TR^\omega$. We say that L' is an ω-regular trace language iff there exists an I-consistent ω-regular language $L \subseteq \Sigma^\omega$ such that $L' = \{\text{str}(\sigma) \mid \sigma \in L\}$. Stated differently, $L' \subseteq TR^\omega$ is a ω-regular trace language iff $L = \bigcup\{\text{trs}(F) \mid F \in L'\}$ is a ω-regular subset of Σ^ω. As in the word case, algebraic and logical presentations of ω-regular trace languages have been worked out [EM, GP]. These presentations have a flavour which is pleasingly similar to the classical algebraic and logical characterisations of ω-regular subsets of Σ^ω.

Returning to the distributed alphabet $\widetilde{\Sigma}_0 = (\{a\}, \{b\})$, let (Σ_0, I_0) denote the corresponding trace alphabet. Consider $L \subseteq TR^\omega(\Sigma_0, I_0)$ consisting of the single trace $F = (E, \leq, \lambda)$ such that E_a and E_b are both infinite sets. It is easy to check that L is a ω-regular trace language but, as argued in [GP], no A0-automaton over $\widetilde{\Sigma}$ can recognize L.

It is worth noting that having multiple entries in the acceptance table does not help. In other words, one might consider the following acceptance criterion.

(A0') $T = \{T_0, T_1, \ldots, T_n\}$ with $T_i = \{F_p^i\}_{p \in \mathcal{P}}$ and $F_p^i \subseteq S_p$ for each $i \in \{1, 2, \ldots, n\}$ and each $p \in \mathcal{P}$. A run ρ of TS over $F \in TR^\omega$ is accepting with respect to A0' iff there exists i such that $\inf_p(\rho) \cap F_p^i \neq \emptyset$ for each p.

The reason why A0' does not help is that the class of languages accepted by A0-automata is closed under union, thanks to the presence of multiple global initial states. We can construct an A0-automaton $\mathcal{A}_i = (TS, T_i)$ for each entry T_i from the table of an A0'-automaton $\mathcal{A} = (TS, T)$. If $T = \{T_0, T_1, \ldots, T_n\}$, it is clear that $L(\mathcal{A}) = \bigcup_{i \in \{1, 2, \ldots, n\}} L(\mathcal{A}_i)$. Thus, every A0'-automaton can be simulated by an A0-automaton.

Gastin and Petit showed that the following acceptance condition provides a suitable generalization of classical Büchi automata to the setting of infinite traces.

(A1) $T = \{T_1, T_2, \ldots, T_n\}$ with $T_i = \{F_p^i\}_{p \in \mathcal{P}}$ and $F_p^i \subseteq S_p$ for each $i \in \{1, 2, \ldots, n\}$ and each $p \in \mathcal{P}$. A run ρ of TS over $F \in TR^\omega$ is accepting with respect to A1 iff there exists i such that $\inf_p(\rho) \supseteq F_p^i$ for each p.

The condition A1 is an extension of the sequential condition B1 in a distributed setting. Notice that A1 "couples" together final sets of the components in each entry $T_i \in T$.

Theorem 2.1 ([GP]) $L \subseteq TR^\omega$ *is a ω-regular trace language iff there exists an A1-automaton $\mathcal{A}$ such that $L_{Tr}(\mathcal{A}) = L$.*

Subsequently, Niebert has shown that the A1 condition can be modified to avoid coupling final sets across processes [Nie]. In effect, it is possible to have a local B1 table for each process and define a run ρ to be accepting if for each process p, $\inf_p(\rho)$ satisfies p's B1 table. Going one step further, we arrive at the acceptance criterion A2, which is the one we will use in connection with the logics to be studied in the next section.

(A2) $T = \{(F_p^\omega, F_p)\}_{p \in \mathcal{P}}$ with $F_p^\omega, F_p \subseteq S_p$ for each p.

A run ρ over $F = (E, \leq, \lambda)$ is accepting with respect to A2 iff for each process p the following conditions are met.

Case 1 E_p *is finite:* Then $\inf_p(\rho) \cap F_p \neq \emptyset$.

Case 2 E_p *is an infinite set:* Then $\inf_p(\rho) \cap F_p^\omega \neq \emptyset$.

Thus, on an input F, the decision as to whether a process p uses F_p or F_p^ω to determine acceptance depends on whether or not p executes infinitely many actions in F.

Theorem 2.2

(i) *The class of languages accepted by A2-automata is closed under union.*

(ii) *The class of languages accepted by A1-automata is identical to the class of languages accepted by A2-automata.*

Proof Sketch.

(i) Suppose $\mathcal{A}_1$ and $\mathcal{A}_2$ are two A2-automata. Then we construct an A2-automaton $\mathcal{A}$ which is the *disjoint* union of $\mathcal{A}_1$ and $\mathcal{A}_2$. The global initial states of $\mathcal{A}$ will determine for each run whether $\mathcal{A}_1$ or $\mathcal{A}_2$ (but not both!) is going to be explored. It is easy to check that $L(\mathcal{A}) = L(\mathcal{A}_1) \cup L(\mathcal{A}_2)$.

(ii) Let $\mathcal{A} = (TS, T)$ be an A1-automaton. From part (i), it suffices to consider the case where T has just one entry. So assume that $T = \{T_1\}$ and $T_1 = \{F_p\}$. Let $TS = (\{S_p\}, \{\rightarrow_a\}, S_{in})$. Define the A2-automaton $\mathcal{A}' = (TS', T')$ as follows. $TS' = (\{S_p'\}, \{\Rightarrow_a\}, S_{in}')$ where:

- $S_p' = S_p \times 2^{F_p} \times \{\text{on}, \text{off}\}$ for each p.
- Let s_a', t_a' be a-states in TS' such that $s_a'(p) = (s_p, X_p, u_p)$ and $t_a'(p) = (t_p, Y_p, v_p)$ for each $p \in \mathcal{P}$. Then $(s_a', t_a') \in \Rightarrow_a$ iff there exists $(s_a, t_a) \in \rightarrow_a$ such that the following conditions are satisfied for each $p \in \text{loc}(a)$.
  (1) $u_p = \text{on}$, $s_p = s_a(p)$ and $t_p = t_a(p)$.

(2) If $X_p = \emptyset$ then $Y_p = F_p$. Otherwise, $Y_p = X_p - \{t_p\}$.

- $S'_{in} = \{s' \in S'_p \mid \exists s \in S_{in}. \forall p \in \mathcal{P}. \exists u_p \in \{on, off\}. \ s'(p) = (s(p), \emptyset, u_p)\}$
- $T' = \{(G_p^\omega, G_p)\}$ where for each p,

$$G_p^\omega = S_p \times \{\emptyset\} \times \{on\}$$
$$G_p = F_p \times 2^{F_p} \times \{off\}$$

It is easy to check that $L_{Tr}(\mathcal{A}) = L_{Tr}(\mathcal{A}')$.

Conversely, let $\mathcal{A} = (TS, T)$ be an A2-automaton with $TS = (\{S_p\}, \{\rightarrow_a\}, S_{in})$ and $T = \{(F_p^\omega, F_p)\}$. We say that $\mathcal{A}$ is in standard form if it satisfies:

- $F_p^\omega \cap F_p = \emptyset$ for each p.
- If $(s_a, t_a) \in \rightarrow_a$ and $p \in loc(a)$, then $s_a(p) \notin F_p$.

Thus, if $\mathcal{A}$ is in standard form, the p-states in F_p are "dead" and are disjoint from F_p^ω. It is a simple exercise to verify that every A2-automaton $\mathcal{A}$ can be converted to an A2-automaton $\mathcal{A}'$ in standard form such that $L_{Tr}(\mathcal{A}) = L_{Tr}(\mathcal{A}')$.

So, let $\mathcal{A} = (TS, T)$ be an A2-automaton in standard form with $TS = (\{S_p\}, \{\rightarrow_a\}, S_{in})$ and $T = \{(F_p^\omega, F_p)\}$. Let G be the set of functions of the form $g : \mathcal{P} \rightarrow S$ such that $g(p) \in F_p^\omega \cup F_p$ for each p. Define the A1-automaton $\mathcal{A}' = (TS', T')$ where $TS' = TS$ and $T' = \{T'_g\}_{g \in G}$, such that for each $g \in G$, $T'_g = \{\{g(p)\}\}_{p \in \mathcal{P}}$. It is easy to verify that $L_{Tr}(\mathcal{A}) = L_{Tr}(\mathcal{A}')$. $\square$

We now argue that the emptiness problem for A2-automata is decidable. This will be required to settle the satisfiability problem for the logics considered in the next section. Let $\mathcal{A} = (TS, T)$ be an A2-automaton be with $TS = (\{S_p\}, \{\rightarrow_a\}, S_{in})$ and $T = \{(F_p^\omega, F_p)\}$. Though it is not strictly necessary, it will be illuminating to first associate a language of infinite words with $\mathcal{A}$.

Let $\sigma \in \Sigma^\omega$. Then a (word) run of TS over σ is a map $\rho : \omega \rightarrow S_\mathcal{P}$ such that $\rho(0) \in S_{in}$ and $\rho(i) \xrightarrow{\sigma(i)}_{TS} \rho(i+1)$ for each $i \geq 0$. The run ρ over σ is accepting iff the following conditions are satisfied for each p.

(i) If $i \in \omega$ such that $\sigma(j) \notin \Sigma_p$ for every $j \geq i$ then $s_i(p) \in F_p$, where $s_i = \rho(i)$.
(ii) If $\sigma(j) \in \Sigma_p$ for infinitely many j then for infinitely many i it is the case that $s_i(p) \in F_p^\omega$, where $s_i = \rho(i)$.

We define $L_{seq}(\mathcal{A})$, the language of infinite words accepted by $\mathcal{A}$ to be the set of all words σ such that there exists an accepting run of $\mathcal{A}$ over σ. The distributed nature of TS together with the basic properties of the maps str and trs defined earlier lead to the next result.

Theorem 2.3 *For any A2-automaton $\mathcal{A}$, $L_{Tr}(\mathcal{A}) = \{str(\sigma) \mid \sigma \in L_{seq}(\mathcal{A})\}$. Consequently $L_{Tr}(\mathcal{A}) \neq \emptyset$ iff $L_{seq}(\mathcal{A}) \neq \emptyset$.*

Similar statements hold, of course, for A0-automata and A1-automata.

All the A2-automata that we construct in the next section will be in standard form. So assume that $\mathcal{A} = (TS, \mathcal{T})$ is an A2-automaton in standard form with $TS = (\{S_p\}, \{\rightarrow_a\}, S_{in})$ and $\mathcal{T} = \{(F_p^\omega, F_p)\}$. Construct the directed graph $G_\mathcal{A} = (S_\mathcal{P}, E_\mathcal{A})$ where $S_\mathcal{P}$ is the set of global states of TS and $(s, s') \in E_\mathcal{A}$ if there exists $a \in \Sigma$ such that $s \xrightarrow{a}_{TS} s'$. We also label each edge in $G_\mathcal{A}$ with a set of processes. Let $\pi : E_\mathcal{A} \rightarrow 2^\mathcal{P}$ be given by $\pi((s, s')) = \bigcup \{\mathrm{loc}(a) \mid s \xrightarrow{a}_{TS} s'\}$.

We call $X \subseteq S_\mathcal{P}$ a *good component* iff X is a maximal strongly connected component in $G_\mathcal{A}$ which meets one the following conditions for each p.

(i) There exists $s \in X$ such that $s(p) \in F_p$. (Because $\mathcal{A}$ is in standard form this implies that $s'(p) = s''(p) \in F_p$ for every $s', s'' \in X$).
(ii) There exists $s \in X$ such that $s(p) \in F_p^\omega$ and for some $s' \in X$, $(s', s) \in E_\mathcal{A}$ and $p \in \pi((s', s))$.

From Theorem 2.3 we know that $L_{Tr}(\mathcal{A})$ is non-empty iff $L_{seq}(\mathcal{A})$ is. It is not difficult to prove that $L_{seq}(\mathcal{A})$ is non-empty iff $G_\mathcal{A}$ has a good component. It is known that the maximal strongly connected components of a digraph can be computed in time which is linear in the size of the digraph [AHU]. Clearly, the size of $G_\mathcal{A}$ is bounded by the number of global states of $\mathcal{A}$. As a consequence it is easy to derive the next result.

Theorem 2.4 *Let $\mathcal{A}$ be an A2-automaton in standard form. Then $L_{Tr}(\mathcal{A}) \neq \emptyset$ iff $G_\mathcal{A}$ has a good component. For $p \in \mathcal{P}$, let $n_p = |S_p|$ denote the number of p-states. Let $n = \max\{n_p\}_{p \in \mathcal{P}}$ and $m = |\mathcal{P}|$. Then checking that $G_\mathcal{A}$ has a good component can be done in time $O(m^{2n})$.*

We conclude this section with a few remarks on deterministic automata over infinite traces. As with automata on infinite words, non-deterministic A2-automata on infinite traces are strictly more expressive than deterministic A2-automata. In the absence of determinacy, complementation is difficult. When applying these automata to settle questions in logic, complementation is often required to handle negation in formulas. (Fortunately, the automata-theoretic treatment of linear time temporal logic on traces which we will describe here does *not* require complementation.)

To obtain determinacy without loss of expressive power one must use a more sophisticated acceptance criterion corresponding to the Muller, Rabin or Streett acceptance conditions for infinite words. Here, we will look only at the Muller acceptance condition.

(M) $\mathcal{T} = \{\mathcal{T}_1, \dots, \mathcal{T}_n\}$ with $\mathcal{T}_i = \{F_p^i\}$ and $F_p^i \subseteq S_p$ for each i and each p. A run ρ over $F \in TR^\omega$ is accepting with respect to M iff there exists $\mathcal{T}_i \in \mathcal{T}$ such that $\inf_p(\rho) = F_p^i$ for each p.

Diekert and Muscholl [DM] showed that *deterministic* M-automata are as expressive as non-deterministic A1-automata. Their proof however does not lead to a determinization construction for A1-automata.

There are two independent solutions available in the literature for the difficult problem of complementing A1-automata. Muscholl first showed how to directly construct a non-deterministic A1-automaton which is the complement of the given automaton [Mus]—this approach does not yield a determinization construction for A1-automata. In [Mus] the complementation is carried out for asynchronous *cellular* Büchi automata, in which there is one agent for each letter. To transport this complementation result to A1-automata, one has to resort to a simulation which carries non-trivial overheads in the size of the alphabet. The second solution due to Klarlund, Mukund and Sohoni [KMS] is a direct determinization construction for A1-automata which then easily leads to the complementation result. In both cases, the blow-up in the local state space of each process is exponential in the global state space of the original automaton, which is essentially optimal. Surprisingly in both [Mus] and [KMS], the A1 acceptance condition must be first transformed into an equivalent one which describes in considerable detail the communication patterns established by the infinite trace that is being examined for acceptance.

3 Linear Time Temporal Logics over Traces

A variety of linear time temporal logics to be interpreted over traces have been proposed in the literature. As mentioned in the Introduction, our focus here will be on those logics which meet the following criteria:

(i) The logic should be expressible within the first order theory of traces.
(ii) The logic should admit a treatment in terms of asynchronous Büchi automata of one kind or the other.

We begin with the logic TrPTL (Trace based Propositional Temporal logic of Linear time). This is the earliest and—to date—the most expressive linear time logic of the chosen kind. For a detailed treatment of this logic the reader is referred to [Thi1]. After presenting TrPTL we will consider two subsystems denoted TrPTL^{con} (connected TrPTL) and TrPTL$^{\otimes}$ (product TrPTL). These subsystems are obtained by placing suitable syntactic restrictions on the formulas. The interesting point is that these restrictions result in proportionate simplification of the automata theoretic constructions associated with the logics. Towards the end of the section we will take a quick look at other temporal logics that have been proposed with traces as the underlying frames.

Henceforth, it will be notationally convenient to deal with distributed alphabets in which the names of the processes are positive integers. Through this section and the next, we fix a distributed alphabet $\widetilde{\Sigma} = \{\Sigma_i\}_{i \in \mathcal{P}}$ with $\mathcal{P} = \{1, 2, \ldots, K\}$ and $K \geq 1$. We let i, j and k range over $\mathcal{P}$. As before, let P, Q range over non-empty subsets of $\mathcal{P}$. The trace alphabet induced by $\widetilde{\Sigma}$ is denoted (Σ, I). We assume the terminology and notations developed in the previous sections. In particular when dealing with a $\mathcal{P}$-indexed family $\{X_i\}_{i \in \mathcal{P}}$, we will often write just $\{X_i\}$.

The logic TrPTL is parameterized by the class of distributed alphabets. Having fixed $\widetilde{\Sigma}$ we shall often almost always write TrPTL to mean $\text{TrPTL}(\widetilde{\Sigma})$, the logic associated with $\widetilde{\Sigma}$. Fix a set of atomic propositions AP with p, q ranging over AP. Then $\Phi_{\text{TrPTL}(\widetilde{\Sigma})}$, the set of formulas of $\text{TrPTL}(\widetilde{\Sigma})$, is defined inductively via:

- For $p \in AP$ and $i \in \mathcal{P}$, $p(i)$ is a formula (which is to be read "p at i").
- If α and β are formulas, so are $\neg\alpha$ and $\alpha \vee \beta$.
- If α is a formula and $a \in \Sigma_i$ then $\langle a \rangle_i \alpha$ is a formula.
- If α and β are formulas so is $\alpha\, \mathcal{U}_i \beta$.

From now on, we denote $\Phi_{\text{TrPTL}(\widetilde{\Sigma})}$ as just Φ. In the semantics of the logic which will be based on infinite traces, the i-view of a configuration will play a crucial role. Let $F \in TR^{\omega}$ with $F = (E, \leq, \lambda)$. Recall that $E_i = \{e \mid e \in E$ and $\lambda(e) \in \Sigma_i\}$. Let $c \in \mathcal{C}_F$ and $i \in \mathcal{P}$. Then $\downarrow^i(c)$ is the i-view of c and it is defined as:

$$\downarrow^i(c) = \downarrow(c \cap E_i).$$

We note that $\downarrow^i(c)$ is also a configuration. It is the "best" configuration that the agent i is aware of at c. We say that $\downarrow^i(c)$ is an i-local configuration. Let $\mathcal{C}_F^i = \{\downarrow^i(c) \mid c \in \mathcal{C}_F\}$ be the set of i-local configurations. For $Q \subseteq \mathcal{P}$ and $c \in \mathcal{C}_F$, we let $\downarrow^Q(c)$ denote the set $\bigcup\{\downarrow^i(c) \mid i \in Q\}$. Once again, $\downarrow^Q(c)$ is a configuration. It represents the collective knowledge of the processes in Q about the configuration c.

The following basic properties of traces follow directly from the definitions.

Proposition 3.1 *Let $F = (E, \leq, \lambda)$ be an infinite trace. The following statements hold.*

(i) *Let $\leq_i = \leq \cap (E_i \times E_i)$. Then $(E_i, \leq_i)$ is a linear order isomorphic to ω if E_i is infinite and isomorphic to a finite initial segment of ω if E_i is finite.*

(ii) *$(\mathcal{C}_F^i, \subseteq)$ is a linear order. In fact $(\mathcal{C}_F^i - \{\emptyset\}, \subseteq)$ is isomorphic to $(E_i, \leq_i)$.*

(iii) *Suppose $\downarrow^i(c) \neq \emptyset$ where $c \in \mathcal{C}_F$. Then there exists $e \in E_i$ such that $\downarrow^i(c) = \downarrow e$. In fact e is the $\leq_i$-maximum event in $(c \cap E_i)$.*

(iv) *Suppose $Q \subseteq Q' \subseteq \mathcal{P}$ and $c \in \mathcal{C}_F$. Then $\downarrow^Q(c) = \downarrow^Q(\downarrow^{Q'}(c))$. In particular, for a single process i, $\downarrow^i(c) = \downarrow^i(\downarrow^i(c))$.*

We can now present the semantics of TrPTL. A model is a pair $M = (F, \{V_i\}_{i \in \mathcal{P}})$ where $F = (E, \leq, \lambda) \in TR^{\omega}$ and $V_i : \mathcal{C}_F^i \to 2^{AP}$ is a valuation function which assigns a set of atomic propositions to i-local configurations for each process i. Let $c \in \mathcal{C}_F$ and $\alpha \in \Phi$. Then $M, c \models \alpha$ denotes that α is satisfied at c in M and it is defined inductively as follows:

- $M, c \models p(i)$ for $p \in AP$ iff $p \in V_i(\downarrow^i(c))$.
- $M, c \models \neg\alpha$ iff $M, c \not\models \alpha$.
- $M, c \models \alpha \vee \beta$ iff $M, c \models \alpha$ or $M, c \models \beta$

- $M, c \models \langle a \rangle_i \alpha$ iff there exists $e \in E_i - c$ such that $\lambda(e) = a$ and $M, \downarrow e \models \alpha$. Moreover, for every $e' \in E_i$, $e' < e$ iff $e' \in c$.
- $M, c \models \alpha \, U_i \beta$ iff there exists $c' \in C_F$ such that $c \subseteq c'$ and $M, \downarrow^i(c') \models \beta$. Moreover, for every $c'' \in C_F$, if $\downarrow^i(c) \subseteq \downarrow^i(c'') \subset \downarrow^i(c')$ then $M, \downarrow^i(c'') \models \alpha$.

Thus TrPTL is an action based agent-wise generalization of LTL. Indeed both in terms of its syntax and semantics, LTL corresponds to the case where there is only one agent and where this agent can execute only one action at any time. With $\mathcal{P} = \{1\}$ and $\Sigma_1 = \{a_0\}$ one then writes p instead of $p(1)$, $O\alpha$ instead of $\langle a_0 \rangle \alpha$ and $\alpha \, U \beta$ instead of $\alpha \, U_i \beta$. The semantics of TrPTL when specialized down to this case yields the usual LTL semantics. In the next section we will say more about the relationship between TrPTL and LTL.

Returning to TrPTL, the assertion $p(i)$ says that the i-view of c satisfies the atomic proposition p. Observe that we could well have $p(i)$ satisfied at c but not $p(j)$ (with $i \neq j$). It is interesting to note that all atomic assertions (that we know of) concerning distributed behaviours are local in nature. Indeed, it is well-known that global atomic propositions will at once lead to an undecidable logic in the current setting [LPRT, Pen].

Suppose $M = (F, \{V_i\})$ is a model and $c \xrightarrow{a}_F c'$ with $j \notin \mathrm{loc}(a)$. Then $M, c \models p(j)$ iff $M, c' \models p(j)$. In this sense the valuation functions are local. There are, of course, a number of equivalent ways of formulating this idea which we will not get into here.

The assertion $\langle a \rangle_i \alpha$ says that the agent i will next participate in an a-event. Moreover, at the resulting i-view, the assertion α will hold. The assertion $\alpha \, U_i \beta$ says that there is a future i-view (including the present i-view) at which β will hold and for all the intermediate i-views (if any) starting from the current i-view, the assertion α will hold.

Before considering examples of TrPTL specifications, we will introduce some notation. We let α, β with or without subscripts range over Φ. Abusing notation, we will use loc to denote the map which associates a set of locations with each formula.

- $\mathrm{loc}(p(i)) = \mathrm{loc}(\langle a \rangle_i \alpha) = \mathrm{loc}(\alpha \, U_i \beta) = \{i\}$.
- $\mathrm{loc}(\neg \alpha) = \mathrm{loc}(\alpha)$.
- $\mathrm{loc}(\alpha \vee \beta) = \mathrm{loc}(\alpha) \cup \mathrm{loc}(\beta)$.

In what follows, $\Phi^i = \{\alpha \mid \mathrm{loc}(\alpha) = \{i\}\}$ is the set of i-type formulas. A basic observation concerning the semantics of TrPTL can be phrased as follows:

Proposition 3.2 Let $M = (F, \{V_i\})$ be a model, $c \in C_F$ and α a formula such that $\mathrm{loc}(\alpha) \subseteq Q$. Then $M, c \models \alpha$ iff $M, \downarrow^Q(c) \models \alpha$.

A corollary to this result is that in case $\alpha \in \Phi^i$ then $M, c \models \alpha$ iff $M, \downarrow^i(c) \models \alpha$. As a result, the formulas in Φ^i can be used in exactly the same manner as one would use LTL (in the setting of sequences) to express properties of the agent i. Boolean combinations of such local assertion can be used to capture various interaction patterns between the agents implied by the logical connectives as well as the coordination enforced by the distributed alphabet $\widetilde{\Sigma}$.

For writing specifications, apart from the usual derived connectives of propositional calculus such as $\wedge$, $\Rightarrow$ and $\equiv$, the following operators are also available

- $\top \triangleq p_1(1) \vee \neg p_1(1)$ denotes the constant "True", where $AP = \{p_1, p_2, \ldots\}$. We use $\perp = \neg\top$ to denote "False"..
- $\Diamond_i \alpha \triangleq \top \, \mathcal{U}_i \alpha$ is a local version of the $\Diamond$ modality of LTL.
- $\square_i \alpha \triangleq \neg \Diamond_i \neg \alpha$ is a local version of the $\square$ modality of LTL..
- Let $X \subseteq \Sigma_i$ and $\overline{X} = \Sigma_i - X$. Then $\alpha \, \mathcal{U}_i^X \beta \triangleq (\alpha \wedge \bigwedge_{a \in \overline{X}} [a]_i \perp) \, \mathcal{U}_i \beta$. In other words $\alpha \, \mathcal{U}_i^X \beta$ is fulfilled using (at most) actions taken from X. We set $\Diamond_i^X \alpha \triangleq \top \, \mathcal{U}_i^X \alpha$ and $\square_i^X \alpha \triangleq \neg \Diamond_i^X \neg \alpha$.
- $\alpha(i) \triangleq \alpha \, \mathcal{U}_i \alpha$ (or equivalently $\perp \mathcal{U}_i \alpha$). $\alpha(i)$ is to be read as "α at i". If $M = (F, \{V_i\})$ is a model and $c \in C_F$ then $M, c \models \alpha(i)$ iff $M, \downarrow^i(c) \models \alpha$. It could of course be the case that $\text{loc}(\alpha) \neq \{i\}$.

A simple but important observation is that every formula is a boolean combination of formulas taken from $\bigcup_{i \in \mathcal{P}} \Phi^i$. In TrPTL we can say that a specific global configuration is reachable from the initial configuration. Let $\{\alpha_i\}_{i \in \mathcal{P}}$ be a family with $\alpha_i \in \Phi^i$ for each i. Then we can define a derived connective $\Diamond(\alpha_1, \alpha_2, \ldots, \alpha_K)$ which has the following semantics at the empty configuration. Let $M = (F, \{V_i\})$ be a model. Then $M, \emptyset \models \Diamond(\alpha_1, \alpha_2, \ldots, \alpha_k)$ iff there exists $c \in C_F$ such that $M, c \models \alpha_1 \wedge \alpha_2 \wedge \cdots \wedge \alpha_K$.

To define this derived connective set $\Sigma_1' = \Sigma_1$ and, for $1 < i \leq K$, set $\Sigma_i' = \Sigma_i - \cup \{\Sigma_j \mid 1 \leq j < i\}$. Then $\Diamond(\alpha_1, \alpha_2, \ldots, \alpha_K)$ is the formula:

$$\Diamond_1^{\Sigma_1'}(\alpha_1 \wedge \Diamond_2^{\Sigma_2'}(\alpha_2 \wedge \Diamond_3^{\Sigma_3'}(\alpha_3 \wedge \cdots \Diamond_K^{\Sigma_K'} \alpha_K)) \cdots).$$

The idea is that the sequence of actions leading up to the required configuration can be reordered so that one first performs all the actions in Σ_1, then all the actions in $\Sigma_2 - \Sigma_1$ etc. Hence, if now is an atomic proposition, the formula $\Diamond(\text{now}(1), \text{now}(2), \ldots, \text{now}(K))$ is satisfied at the empty configuration iff there is a reachable configuration at which all the agents assert now.

Dually, safety properties that hold at the initial configuration can also be expressed. For example, let $\text{crt}(i)$ be the atomic assertion declaring that the agent i is currently in its critical section. Then it is possible to write a formula φ_{ME} which asserts that at all reachable configurations at most one agent is in its critical section, thereby guaranteeing that the system satisfies the mutual exclusion property. We omit the details of how to specify φ_{ME}.

On the other hand, it seems difficult to express nested global and safety properties in TrPTL. This is mainly due to the local nature of the modalities which results in information about the past sneaking into the semantics even though there are no explicit past operators in the logic. In particular, TrPTL admits formulas that are satisfiable but not root-satisfiable.

A formula α is said to be root-satisfiable iff there exists a model M such that $M, \emptyset \models \alpha$. On the other hand, α is said to be satisfiable iff there exists a model

$M = (F, \{V_i\})$ and $c \in C_F$ such that $M, c \models \alpha$. It turns out that these two notions are not equivalent. Consider the distributed alphabet $\tilde{\Sigma}_0 = \{\Sigma_1, \Sigma_2\}$ with $\Sigma_1 = \{a, d\}$ and $\Sigma_2 = \{b, d\}$. Then it is not difficult to verify that the formula $p(2)(1) \wedge \square_2 \neg p(2)$ is satisfiable but not root-satisfiable. (Recall that $p(2)(1)$ abbreviates $\bot \, \mathcal{U}_1 p(2)$). One can however transform every formula α into a formula α' such that α is satisfiable iff α' is root satisfiable.

This follows from the observation that every α can be expressed as a boolean combination of formulas taken from the set $\bigcup_{i \in \mathcal{P}} \Phi^i$. Hence the given formula α can be assumed to be of the form $\alpha = \bigvee_{j=1}^m (\alpha_{j1} \wedge \alpha_{j2} \wedge \cdots \wedge \alpha_{jK})$ where $\alpha_{ji} \in \Phi^i$ for each $j \in \{1, 2, \ldots, m\}$ and each $i \in \mathcal{P}$. Now convert α to the formula α' where $\alpha' = \bigvee_{j=1}^m \Diamond(\alpha_{j1}, \alpha_{j2}, \cdots, \alpha_{jK})$. (Recall the derived modality $\Diamond(\alpha_1, \alpha_2, \ldots, \alpha_K)$ introduced earlier.) From the semantics of $\Diamond(\alpha_1, \alpha_2, \ldots, \alpha_K)$ it follows that α is satisfiable iff α' is root-satisfiable.

Hence, in principle, it suffices to consider only root-satisfiability in developing a decision procedure for TrPTL. There is of course a blow-up involved in converting satisfiable formulas to root-satisfiable formulas. If one wants to avoid this blow-up then the decision procedure for checking root-satisfiability can be suitably modified to yield a direct decision procedure for checking satisfiability as done in [Thi1]. In any case, it is root satisfiability which is of importance from the standpoint of model checking. Hence here we shall only develop a procedure for deciding if a given formula of TrPTL is root-satisfiable.

As a first step we augment the syntax of our logic by one more construct.

- If α is a formula, so is $O_i \alpha$. In the model $M = (F, \{V_i\})$, at the configuration $c \in C_F$, $M, c \models O_i \alpha$ iff $M, c \models \langle a \rangle_i \alpha$ for some $a \in \Sigma_i$. We also define $\text{loc}(O_i \alpha) = \{i\}$.

Thus $O_i \alpha \equiv \bigvee_{a \in \Sigma_i} \langle a \rangle_i \alpha$ is a valid formula and O_i is expressible in the former syntax. It will be however more efficient to admit O_i as a first class modality.

Fix a formula α_0. Our aim is to effectively associate an A2-automaton $\mathcal{A}_{\alpha_0}$ with α_0 such that α_0 is root-satisfiable iff $L_{Tr}(\mathcal{A}_{\alpha_0}) \neq \emptyset$. Since the emptiness problem for A2-automata is decidable (Theorem 2.4), this will yield the desired decision procedure. Let $CL'(\alpha_0)$ be the least set of formulas containing α_0 which satisfies:

- $\neg \beta \in CL'(\alpha_0)$ implies $\beta \in CL'(\alpha_0)$.
- $\alpha \vee \beta \in CL'(\alpha_0)$ implies $\alpha, \beta \in CL'(\alpha_0)$.
- $\langle a \rangle_i \alpha \in CL'(\alpha_0)$ implies $\alpha \in CL'(\alpha_0)$.
- $O_i \alpha \in CL'(\alpha_0)$ implies $\alpha \in CL'(\alpha_0)$.
- $\alpha \, \mathcal{U}_i \beta \in CL'(\alpha_0)$ implies $\alpha, \beta \in CL'(\alpha_0)$. In addition, $O_i(\alpha \, \mathcal{U}_i \beta) \in CL'(\alpha_0)$.

We then define $CL(\alpha_0)$ to be the set $CL'(\alpha_0) \cup \{\neg\beta \mid \beta \in CL'(\alpha_0)\}$.

Thus $CL(\alpha_0)$, sometimes called the Fisher-Ladner closure of α_0, is closed under negation with the convention that $\neg\neg\beta$ is identified with β. From now we shall write CL instead of $CL(\alpha_0)$.

$A \subseteq CL$ is called an i-type atom iff it satisfies:

- $\forall \alpha \in CL.\ \alpha \in A$ iff $\neg\alpha \notin A$.
- $\forall \alpha \vee \beta \in CL.\ \alpha \vee \beta \in A$ iff $\alpha \in A$ or $\beta \in A$.
- $\forall \alpha\, \mathcal{U}_i \beta \in CL.\ \alpha\, \mathcal{U}_i \beta \in A$ iff $\beta \in A$ or $(\alpha \in A$ and $O_i(\alpha\, \mathcal{U}_i \beta) \in A)$.
- If $\langle a \rangle_i \alpha,\ \langle b \rangle_i \beta \in A_i$ then $a = b$.

AT_i denotes the set of i-type atoms. We now need to define the notion of a formula in CL being a member of a collection of atoms. Let $\alpha \in CL$ and $\{A_i\}_{i \in Q}$ be a family of atoms with $\mathrm{loc}(\alpha) \subseteq Q$ and $A_i \in AT_i$ for each $i \in Q$. Then the predicate $\alpha \in \{A_i\}_{i \in Q}$ is defined inductively as:

- If $\mathrm{loc}(\alpha) = \{j\}$ then $\alpha \in \{A_i\}_{i \in Q}$ iff $\alpha \in A_j$.
- If $\alpha = \neg\beta$ then $\alpha \in \{A_i\}_{i \in Q}$ iff $\beta \notin \{A_i\}_{i \in Q}$.
- If $\alpha = \alpha_1 \vee \alpha_2$ then $\alpha_1 \vee \alpha_2 \in \{A_i\}_{i \in Q}$ iff $\alpha_1 \in \{A_i\}_{i \in Q}$ or $\alpha_2 \in \{A_i\}_{i \in Q}$.

The construction of the A2-automaton $\mathcal{A}_{\alpha_0}$ is guided by the construction due to Vardi and Wolper for LTL [VW]. However in the much richer setting of traces it turns out that one must make crucial use of the latest information that the agents have about each other when defining the transitions of $\mathcal{A}_{\alpha_0}$. It has been shown by Mukund and Sohoni [MS] that this information can be kept track of by a deterministic A2-automaton whose size depends only on $\widetilde{\Sigma}$. (Actually the automaton described in [MS] operates over finite traces but it is a trivial task to convert it into A2-automaton having the desired properties). To bring out the relevant properties of this automaton, let $F \in TR^\omega$ with $F = (E, \leq, \lambda)$. For each subset Q of processes, the function $\mathrm{latest}_{F,Q} : \mathcal{C}_F \times \mathcal{P} \to Q$ is given by $\mathrm{latest}_{F,Q}(c, j) = \ell$ iff ℓ is the least member of Q (under the usual ordering over the integers) with the property $\downarrow^j(\downarrow^q(c)) \subseteq \downarrow^j(\downarrow^\ell(c))$ for every $q \in Q$. In other words, among the agents in Q, ℓ has the best information about j at c, with ties being broken by the usual ordering over integers.

Theorem 3.3 ([MS]) *There exists an effectively constructible deterministic A2-automaton $\mathcal{A}_\Gamma = (TS, \mathcal{T})$ with $TS = (\{\Gamma_i\}, \{\Rightarrow_a\}, \Gamma_{in})$ such that:*

(i) $L_{Tr}(\mathcal{A}_\Gamma) = TR^\omega$.

(ii) *For each $Q = \{i_1, i_2, \ldots, i_n\}$, there exists an effectively computable function $\mathrm{gossip}_Q : \Gamma_{i_1} \times \Gamma_{i_2} \times \cdots \times \Gamma_{i_n} \times \mathcal{P} \to Q$ such that for every $F \in TR^\omega$, every $c \in \mathcal{C}_F$ and every $j \in \mathcal{P}$, $\mathrm{latest}_{F,Q}(c, j) = \mathrm{gossip}_Q(\gamma(i_1), \ldots, \gamma(i_n), j)$ where $\rho_F(c) = \gamma$ and ρ_F is the unique (accepting) run of $\mathcal{A}_\Gamma$ over F.*

Henceforth, we refer to $\mathcal{A}_\Gamma$ as the *gossip automaton*. Each process in the gossip automaton has $2^{O(K^2 \log K)}$ local states, where $K = |\mathcal{P}|$. Moreover the function g_Q can be computed in time which is polynomial in the size of K.

Each i-state of the automaton $\mathcal{A}_{\alpha_0}$ will consist of an i-type atom together with an appropriate i-state of the gossip automaton. Two additional component will be used to check for liveness requirements. One component will take values from the set $N_i = \{0, 1, 2, \ldots, |U_i|\}$ where $U_i = \{\alpha\, \mathcal{U}_i \beta \mid \alpha\, \mathcal{U}_i \beta \in CL\}$. This component will be used to ensure that all "until" requirements are met. The other component will take values from the set {on,off}. This will be used to detect when an agent has quit.

The automaton $\mathcal{A}_{\alpha_0}$ can now be defined.

Definition 3.4 $\mathcal{A}_{\alpha_0} = (TS, \mathcal{T})$, where $TS = (\{S_i\}, \{\rightarrow_a\}, S_{in})$ and $\mathcal{T} = \{(F_i^\omega, F_i)\}$ are defined as follows:

(i) For each i, $S_i = AT_i \times \Gamma_i \times N_i \times \{on,off\}$. Recall that Γ_i is the set of i-states of the gossip automaton and $N_i = \{0, 1, 2, \ldots, |U_i|\}$ with $U_i = \{\alpha\, U_i \beta \mid \alpha\, U_i \beta \in CL\}$.

(ii) Let $s_a, s_a' \in S_a$ with $s_a(i) = (A_i, \gamma_i, n_i, v_i)$ and $s_a'(i) = (A_i', \gamma_i', n_i', v_i')$ for each $i \in loc(a)$. Then $(s_a, s_a') \in \rightarrow_a$ iff the following conditions are met.

    (1) $(\gamma_a, \gamma_a') \in \Rightarrow_a$ (recall that $\{\Rightarrow_a\}$ is the family of transition relations of the gossip automaton) where $\gamma_a, \gamma_a' \in \Gamma_a$ such that $\gamma_a(i) = \gamma_i$ and $\gamma_a'(i) = \gamma_i'$ for each $i \in loc(a)$.

    (2) $\forall i, j \in loc(a)$, $A_i' = A_j'$.

    (3) $\forall i \in loc(a)\ \forall \langle a \rangle_i \alpha \in CL$. $\langle a \rangle_i \alpha \in A_i$ iff $\alpha \in A_i'$.

    (4) $\forall i \in loc(a)\ \forall O_i \alpha \in CL$. $O_i \alpha \in A$ iff $\alpha \in A_i'$.

    (5) $\forall i \in loc(a) \forall \langle b \rangle_i \beta \in CL$. If $\langle b \rangle_i \beta \in A_i$ then $b = a$.

    (6) Suppose $j \notin loc(a)$ and $\beta \in CL$ with $loc(\beta) = \{j\}$. Further suppose that $loc(a) = \{i_1, i_2, \ldots, i_n\}$. Then $\beta \in A_i'$ iff $\beta \in A_\ell$ where $\ell = gossip_{loc(a)}(\gamma_{i_1}, \gamma_{i_2}, \ldots, \gamma_{i_n}, j)$.

    (7) Let $i \in loc(a)$, $U_i = \{\alpha_1\, U_i \beta_1, \alpha_2\, U_i \beta_2, \ldots, \alpha_{n_i}\, U_i \beta_{n_i}\}$. Then u_i' and u_i are related to each other via:

$$u_i' = \begin{cases} (u_i + 1) \bmod (n_i + 1), & \text{if } u_i = 0 \text{ or } \beta_{u_i} \in A_i \text{ or } \alpha_{u_i}\, U_i \beta_{u_i} \notin A_i \\ u_i, & \text{otherwise} \end{cases}$$

    (8) For each $i \in loc(a)$, $v_i = on$. Moreover, if $v_i' = off$ then $\langle a \rangle_i \alpha \notin A_i'$ for every $i \in loc(a)$ and every $\langle a \rangle_i \alpha \in CL$.

(iii) Let $s \in S_P$ with $s(i) = (A_i, \gamma_i, u_i, v_i)$ for every i. Then $s \in S_{in}$ iff $\alpha_0 \in \{A_i\}_{i \in P}$ and $\gamma \in \Gamma_{in}$ where $\gamma \in \Gamma_P$ satisfies $\gamma(i) = \gamma_i$ for every i. Furthermore, $u_i = 0$ for every i. Finally, for every i, $v_i = off$ implies that $\langle a \rangle_i \alpha \notin A_i$ for every $\langle a \rangle_i \alpha \in CL$.

(iv) For each i, $F_i^\omega \subseteq S_i$ is given by $F_i^\omega = \{(A_i, \gamma_i, u_i, v_i) \mid u_i = 0 \text{ and } v_i = on\}$ and $F_i \subseteq S_i$ is given by $F_i = \{(A_i, \gamma_i, u_i, v_i) \mid v_i = off\}$.

The automaton $\mathcal{A}_{\alpha_0}$ extends the automata theoretic construction for LTL described in [VW] to the setting of TrPTL. The main new feature is the use of the gossip automaton in step (ii)(6) when dealing with formulas located at agents not taking part in the current action. A detailed explanation of $\mathcal{A}_{\alpha_0}$ can be found in [Thi1].

This construction differs from the original construction for TrPTL presented in [Thi1] in a number of ways. Each S_i in [Thi1] was defined to be $AT_1 \times AT_2 \times \cdots \times AT_K \times \tilde{U}_i \times \{act_i^+, act_i^-, stop_i\}$ with $\tilde{U}_i$ as the set of subsets of U_i. The acceptance condition used was $A1$. Using $A2$, we need just two elements {on, off} to record when an agent has quit. Using the counter N_i instead of $\tilde{U}_i$ leads to a more compact description of $\mathcal{A}_{\alpha_0}$. The significant improvement, namely, replacing $AT_1 \times AT_2 \times \cdots \times AT_K$ by just AT_i is due to Narayan Kumar [Nar]. The arguments described in [Thi1] go through in the present setting with minor modifications. These arguments lead to the next set of results.

Theorem 3.5

(i) α_0 is root-satisfiable iff $L_{Tr}(\mathcal{A}_{\alpha_0}) \neq \emptyset$.

(ii) The number of local states of $\mathcal{A}_{\alpha_0}$ is bounded by $2^{O(\max(n, m^2 \log m))}$ where $n = |\alpha_0|$ and m is the number of agents mentioned in α_0. Clearly, $m \leq n$. It follows that the root-satisfiability problem (and in fact the satisfiability problem) for TrPTL is solvable in time $2^{O(\max(n, m^2 \log m) \cdot m)}$.

The number of local states of each process in $\mathcal{A}_{\alpha_0}$ is determined by two quantities: the length of α_0 and the size of the gossip automaton $\mathcal{A}_\Gamma$. As far as the size of $\mathcal{A}_\Gamma$ is concerned, it is easy to verify that we need to consider only those agents in $\mathcal{P}$ that are mentioned in $loc(\alpha_0)$, rather than all agents in the system.

The model checking problem for TrPTL can be phrased as follows. A finite state distributed program over $\tilde{\Sigma}$ is a pair $Pr = (\mathcal{A}_{Pr}, V_{Pr})$ where $\mathcal{A}_{Pr} = ((\{S_i^{Pr}\}, \{\Rightarrow_a^{Pr}\}, S_{in}^{Pr}), \{(S_i, S_i)\})$ is an A2-automaton modelling the state space of Pr and $V_{Pr} : S \rightarrow 2^{AP}$ is an interpretation of the atomic propositions over the local states of the program. (In this context, one assumes AP to be a finite set.)

Let ρ be a run of $\mathcal{A}_{Pr}$ over $F = (E, \leq, \lambda)$. Then ρ induces the model M_ρ via V_{Pr} as follows: $M_\rho = (F, \{V_i^\rho\})$ where for each i and each $c \in C_F$, $V_i^\rho(\downarrow^i(c)) = V_{Pr}(s_i) \cap P$, where $s = \rho(c)$ and $s_i = s(i)$. Viewing a formula α_0 as a specification, we say that Pr meets the specification α_0—denoted $Pr \models \alpha_0$—if for every $F \in Tr^\omega$ and for every run ρ of $\mathcal{A}_{Pr}$ over F, it is the case that $M_\rho, \emptyset \models \alpha_0$.

The model checking problem is to determine whether $Pr \models \alpha_0$. This problem can be solved by "intersecting" the program automaton $\mathcal{A}_{Pr}$ with the formula automaton $\mathcal{A}_{\neg\alpha_0}$ to yield an automaton $\mathcal{A}$ such that $L_{Tr}(\mathcal{A}) = L_{Tr}(\mathcal{A}_{Pr}) \cap L_{Tr}(\mathcal{A}_{\neg\alpha_0})$. It turns out that $L_{Tr}(\mathcal{A}) = \emptyset$ iff $Pr \models \alpha_0$. It is easy to construct $\mathcal{A}$. The only point to care of is that the i-local states of $\mathcal{A}$ should consist of only those pairs (s_i, s_i') (where s_i is an i-local state of $\mathcal{A}_{Pr}$ and $s_i' = (A_i', \gamma_i', n_i', v_i')$ is an i-local state of $\mathcal{A}_{\neg\alpha_0}$) such that $V_{Pr}(s_i) \cap AP = A_i \cap AP$. The details can be found in [Thi1].

It turns out that this model checking problem has time complexity $O(|\mathcal{A}_{Pr}| \cdot 2^{O(\max(n, m^2 \log m) \cdot m)})$ where $|\mathcal{A}_{Pr}|$ is the size of the global state space of the A2-automaton modelling the behaviour of the given program Pr and, as before, $n = |\alpha_0|$ and m is the number of agents mentioned in α_0, where α_0 is the specification formula.

We now turn to two interesting sublogics of TrPTL. The first is the sublogic TrPTL^{con}, which consists of the so called connected formulas of TrPTL. We define $\Phi^{\text{con}}_{\text{TrPTL}}$ (from now on written as Φ^{con}) to be the least subset of Φ satisfying the following conditions:

(i) $p(i) \in \Phi^{\text{con}}$ for every $p \in P$ and every $i \in \mathcal{P}$.
(ii) If $\alpha, \beta \in \Phi^{\text{con}}$, so are $\neg\alpha$ and $\alpha \vee \beta$.
(iii) If $\alpha \in \Phi^{\text{con}}$ and $a \in \Sigma_i$ such that $\text{loc}(\alpha) \subseteq \text{loc}(a)$ then $\langle a \rangle_i \alpha \in \Phi^{\text{con}}$.
(iv) If $\alpha, \beta \in \Phi^{\text{con}}$ with $\text{loc}(\alpha) = \text{loc}(\beta) = \{i\}$ then $\alpha \, \mathcal{U}_i \beta \in \Phi^{\text{con}}$. Actually one need only demand that $\text{loc}(\alpha), \text{loc}(\beta) \subseteq \bigcap\{\text{loc}(a) \mid a \in \Sigma_i\}$ but this leads to notational complications that we wish to avoid here.
(v) If $\alpha \in \Phi^{\text{con}}$ and $\text{loc}(\alpha) = \{i\}$ then $O_i\alpha \in \Phi^{\text{con}}$. (Once again one needs to just demand that $\alpha \subseteq \bigcap\{\text{loc}(a) \mid a \in \Sigma_i\}$.)

Connected formulas were first identified by Niebert and used by Huhn [Huh]. They have also been independently identified by Ramanujam [Ram]. Thanks to the syntactic restrictions imposed on the next state and until formulas, past information is not allowed to creep in. Indeed one can prove the following:

Proposition 3.6 *Let $\alpha \in \Phi^{\text{con}}$. Then α is satisfiable iff α is root-satisfiable.*

Yet another pleasing feature of TrPTL^{con} is that the gossip automaton can be eliminated in the construction of the automaton $\mathcal{A}_{\alpha_0}$ whenever $\alpha_0 \in \Phi^{\text{con}}$. In fact one can do a bit more.

Let $\alpha_0 \in \Phi^{\text{con}}$ and let $CL_i = CL \cap \Phi^i$ for each i (recall that CL is an abbreviation for $CL(\alpha_0)$). We redefine an i-type atom to be a subset A of CL_i such that:

- $\forall \beta \in CL_i \; \beta \in A$ iff $\neg\beta \notin A$.
- $\forall \alpha \vee \beta \in CL_i$. $\alpha \vee \beta \in A$ iff $\alpha \in A$ or $\beta \in A$.
- $\forall \alpha \, \mathcal{U}_i \beta \in CL_i \; \alpha \, \mathcal{U}_i \beta \in A$ iff $\beta \in A$ or $\alpha \in A$ and $O_i(\alpha \, \mathcal{U}_i \beta) \in A$.

As before (but with the new definition in operation!), AT_i is the set of i-type atoms.

Let $\alpha \in CL$ with $\text{loc}(\alpha) \subseteq Q$. The notion of α belonging to a family of atoms $\{A_i\}_{i \in Q}$, with $A_i \in AT_i$ for each $i \in Q$, is defined inductively in the obvious way—if $\text{loc}(\alpha) = \{i\}$ then $\alpha \in \{A_i\}_{i \in Q}$ iff $\alpha \in A_i$ etc. etc. The construction of $\mathcal{A}_{\alpha_0}$ is as specified in Definition 3.4 with the following modifications:

(i) $S_i = AT_i \times N_i \times \{\text{on,off}\}$ for each $i \in \mathcal{P}$. Thus the gossip automaton is eliminated and AT_i is the set of i-type atoms of the new kind.
(ii) (1) This condition is obviously dropped.
    (2) Interestingly enough, this condition is also dropped.
    (3) This condition is modified to $\forall \langle a \rangle_i \alpha \in CL_i . \langle a \rangle_i \alpha \in A_i$ iff $\alpha \in \{A'_j\}_{j \in \text{loc}(a)}$.

In addition, condition (ii)(6) is dropped, while conditions (ii)(4), (ii)(5), (ii)(7) and (ii)(8) remain unchanged. Parts (iii) and (iv) are modified to eliminate all references to the gossip automaton. After these alterations, it is not difficult to prove the following result.

Theorem 3.7 *Let $\alpha_0 \in \Phi^{\text{con}}$ and $\mathcal{A}_{\alpha_0}$ be constructed as detailed above.*

(i) *α_0 is satisfiable iff $L_{Tr}(\mathcal{A}_{\alpha_0}) \neq \emptyset$.*
(ii) *The satisfiability problem for TrPTL^{con} is solvable in time $2^{O(|\alpha_0|)}$.*

Once again, a suitably modified statement can be made about the associated model checking problem. At present we do not know whether or not TrPTL is strictly more expressive than TrPTL^{con}. We shall formulate this question more rigorously in the next section.

Yet another sublogic of TrPTL is called product TrPTL and is denoted as TrPTL$^{\otimes}$. Let $\Phi^{\otimes}$, the set of formulas of TrPTL$^{\otimes}$, be the least subset of Φ which satisfies:

(i) $p(i) \in \Phi^{\otimes}$ for every $p \in P$ and every $i \in \mathcal{P}$.
(ii) If $\alpha, \beta \in \Phi^{\otimes}$ then so are $\neg\alpha$ and $\alpha \vee \beta$.
(iii) If $\alpha \in \Phi^{\otimes}$ with $\text{loc}(\alpha) = \{i\}$ and $a \in \Sigma_i$ then $\langle a \rangle_i \alpha \in \Phi^{\otimes}$.
(iv) If $\alpha, \beta \in \Phi^{\otimes}$ with $\text{loc}(\alpha) = \text{loc}(\beta) = \{i\}$ then $\alpha \, \mathcal{U}_i \beta \in \Phi^{\otimes}$.

Clearly $\Phi^{\otimes} \subseteq \Phi^{\text{con}} \subseteq \Phi$. In case $\alpha_0 \in \Phi^{\otimes}$, the automaton $\mathcal{A}_{\alpha_0}$ can be simplified even further (than the case when $\alpha_0 \in \Phi^{\text{con}}$). $\mathcal{A}_{\alpha_0}$ essentially consists of a synchronized product of Büchi automata. A detailed treatment of TrPTL$^{\otimes}$ is provided in [Thi2]. The interest in this subsystem lies in the fact that the accompanying program model is particularly simple and commonplace. Namely, it consists of a fixed set of finite state transition systems that coordinate their behaviour by performing common actions together. Here we shall just sketch the construction for $\mathcal{A}_{\alpha_0}$.

A product Büchi automaton over $\widetilde{\Sigma}$ is a structure $\mathcal{A} = (\{TS_i\}_{i \in \mathcal{P}}, S_{in}, T)$ where $TS_i = (S_i, \rightarrow_i)$ for each i with $\rightarrow_i \subseteq S_i \times \Sigma_i \times S_i$ as the local transition relation of the agent i. Everything else is as in the definition of an A2-automaton. Thus the key difference is that each agent comes with its own local transition relation. From these agent transition relations, one can derive the action indexed transition relations $\{\rightarrow_a\}$ as follows: $(s_a, s'_a) \in \rightarrow_a$ iff $s_a(i) \xrightarrow{a} s'_a(i)$ for every $i \in \text{loc}(a)$. Thus product Büchi automata are a (strict) subclass of the class of A2-automata.

Given $\alpha_0 \in \Phi^{\otimes}$, the construction of $\mathcal{A}_{\alpha_0}$ proceeds as in the case where $\alpha_0 \in \Phi^{\text{con}}$. The only difference is, we must define the transition relations $\{\rightarrow_i\}_{i \in \mathcal{P}}$ instead of the transition relations $\{\rightarrow_a\}_{a \in \Sigma}$. This can be done as follows:

Let $s_i, s'_i \in S_i$ with $s_i = (A_i, u_i, v_i)$ and $s'_i = (A'_i, u'_i, v'_i)$. Let $a \in \Sigma_i$. Then $s_i \xrightarrow{a_i} s'_i$ iff the following conditions are satisfied:

(i) $\forall \langle a \rangle_i \alpha \in CL$. $\langle a \rangle_i \alpha \in A_i$ iff $\alpha \in A'_i$.
(ii) $\forall O_i \alpha \in CL$. $O_i \alpha \in A$ iff $\alpha \in A'_i$.
(iii) If $\langle b \rangle_i \alpha \in A_i$ then $b = a$.
(iv) u_i and u'_i are related to each other just as in part (ii)(7) of Definition 3.4.
(v) v_i and v'_i satisfy part (ii)(8) of Definition 3.4.

As shown in [Thi2] one can establish the following result for TrPTL$^{\otimes}$.

Theorem 3.8 *Let* $\alpha_0 \in \Phi^\otimes$ *and* $\mathcal{A}_{\alpha_0}$ *be constructed as above.*

(i) α_0 *is satisfiable iff* $L_{Tr}(\mathcal{A}_{\alpha_0}) \neq \emptyset$.

(ii) *The satisfiability problem for* $TrPTL^\otimes$ *can be solved in time* $2^{O(|\alpha_0|)}$.

Once again, one can make suitably modified statements about the accompanying model checking problem. As mentioned earlier, the program model in this setting consists of a fixed set (one for each i) finite state transition systems.

We conclude this section with a quick look at some related logics. Katz and Peled introduced the logic ISTL [KP] which can be easily viewed as a temporal logic over traces. However, it has branching time modalities which permit quantification over the so called observations of a trace. ISTL uses global atomic propositions rather than local atomic propositions. Penczek has also studied a number of temporal logics (including a version of ISTL) with branching time modalities and global atomic propositions [Pen]. His logics are interpreted directly over the space of configurations of a trace resulting in a variety of axiomatizations and undecidability results. We feel that local atomic propositions (as used in TrPTL) are crucial for obtaining tractable partial order based temporal logics. Niebert has considered a μ-calculus version of TrPTL [Nie] and has obtained a decidability result using a variant of asynchronous Büchi automata. Since this logic uses "local" fixed points, it is not clear at present what is the expressive power of this logic. The four linear time temporal logics studied by Ramanujam in a closely related setting [Ram] can be easily captured as four sublogics of TrPTL through purely syntactic restrictions. Two of the resulting sublogics are TrPTL$^\otimes$ and TrPTL^{con}. It is not clear at present whether the other two logics admit a simpler treatment in terms of asynchronous Büchi automata (than the one for TrPTL).

The temporal logic of causality (TLC) proposed by Alur, Peled and Penczek is basically a temporal logic over traces [APP]. The concurrent structures used in [APP] as frames for TLC can be easily represented as traces over an appropriately chosen trace alphabet. The interesting feature of TLC is that its branching time modalities are interpreted over causal paths. In a trace $(E, \leq, \lambda)$, the sequence $e_0 e_1 \cdots \in E^\infty$ is a causal path if $e_0 < e_1 < e_2 \cdots$. This logic is almost certainly not expressible within the first order theory of traces although it admits an elementary time (in fact essentially exponential time) decision procedure.

Finally, Ebinger has also proposed a linear time temporal logic to be interpreted over traces [Ebi]. An interesting property of this logic is that when its frames are restricted to be *finite* traces then it is exactly equivalent to the first order theory of *finite* traces. Unfortunately the decidability of this logic is settled using a translation into the first order theory of infinite traces. Hence the decision procedure has non-elementary time complexity.

4 Expressiveness Issues

Our main aim here is to show that TrPTL is expressible within the first order theory of traces. In order to simplify the presentation, we shall eliminate atomic propositions and instead use the single constant $\top$ standing for "True" (and $\perp = \neg\top$ standing for "False"). The resulting logic will also be called TrPTL accompanied by the notations and terminology developed in the previous section. The function loc which assigns a set of processes to a formula works exactly as before except that we start with $\text{loc}(\top) = \emptyset$. As will be seen later, this will entail minor changes in the definition of the syntax of TrPTL^{con} and TrPTL$^{\otimes}$. For now, we repeat that the syntax of Φ, the set of formulas of TrPTL is now given by:

$$\Phi ::= \top \mid \neg\alpha \mid \alpha \vee \beta \mid \langle a \rangle_i \alpha \mid \alpha \, \mathcal{U}_i \beta.$$

As before, for $\langle a \rangle_i \alpha$ to be a formula we require $a \in \Sigma_i$. Local atomic propositions can be coded up into the actions and hence their elimination does not result in loss of expressive power.

A model is just an infinite trace $F \in TR^{\omega}$. We set $F, c \models \top$ for every $c \in \mathcal{C}_F$. The rest of the semantics is as before. L_α, the ω-trace language defined by the formula α is given by, $L_\alpha = \{F \mid F \in TR^{\omega} \text{ and } F, \emptyset \models \alpha\}$. We say that $L \subseteq TR^{\omega}$ is TrPTL-definable iff there exists $\alpha \in \Phi$ such that $L = L_\alpha$.

First we shall compare the expressive powers of TrPTL, TrPTL^{con} and TrPTL$^{\otimes}$. In order to do so, we must define the syntax of the two sublogics in the present setting. For TrPTL^{con} the only changes that are required are:

- $\top \in \Phi^{\text{con}}$.
- If $\alpha, \beta \in \Phi^{\text{con}}$ such that $\text{loc}(\alpha), \text{loc}(\beta) \subseteq \{i\}$ then $\alpha \, \mathcal{U}_i \beta \in \Phi^{\text{con}}$.

For TrPTL$^{\otimes}$, the only changes that are required are

- $\top \in \Phi^{\otimes}$.
- If $\alpha \in \Phi^{\otimes}$ such that $\text{loc}(\alpha) \subseteq \{i\}$ and if $a \in \Sigma_i$ then $\langle a \rangle_i \alpha \in \Phi^{\otimes}$.
- If $\alpha, \beta \in \Phi^{\otimes}$ with $\text{loc}(\alpha), \text{loc}(\beta) \subseteq \{i\}$ then $\alpha \, \mathcal{U}_i \beta \in \Phi^{\otimes}$.

The notion of $L \subseteq TR^{\omega}$ being TrPTL^{con}-definable or TrPTL$^{\otimes}$-definable is formulated in the obvious way. Since $\Phi^{\otimes} \subseteq \Phi^{\text{con}} \subseteq \Phi$ it is clear that TrPTL is at least as expressive as $TrPTL^{\text{con}}$ which in turn is at least as expressive as $TrPTL^{\otimes}$. As mentioned earlier we do not know at present if TrPTL is strictly more expressive than $TrPTL^{\text{con}}$, though we conjecture that this the case.

We do know however that TrPTL^{con} is strictly more expressive than TrPTL$^{\otimes}$. To illustrate this it will be convenient to extend the notion of definability to subsets of Σ^{ω}. We say that $L \subseteq \Sigma^{\omega}$ is TrPTL-definable iff L is I-consistent and $\{\text{str}(\sigma) \mid \sigma \in L\}$ is TrPTL-definable. This notion is defined for TrPTL^{con} and TrPTL$^{\otimes}$ in the obvious way. Hence in order to show that TrPTL^{con} is more expressive than TrPTL$^{\otimes}$ it suffices to exhibit some $L \subseteq \Sigma^{\omega}$ which is I-consistent and is TrPTL^{con}-definable but not TrPTL$^{\otimes}$-definable.

Let $\tilde{\Gamma} = \{\Gamma_1, \Gamma_2\}$ with $\Gamma_1 = \{a, a', d\}$ and $\Gamma_2 = \{b, b', d\}$. Let $\Gamma = \{a, a', b, b', d\}$. Consider $L \subseteq \Gamma^{\omega}$ given by:

$$L = (d(ab + ba + a'b' + b'a'))^{\omega}.$$

It turns out that L is *not* TrPTL$^{\otimes}$-definable. Clearly L is I-consistent. As shown in [Thi2], for L to be TrPTL$^{\otimes}$-definable, it must be a so-called (synchronized) product language. As a result, it would have to possess the following property:

(PR) Suppose $\sigma \in \Gamma^\omega$. Then $\sigma \in L$ iff there exist $\sigma_1, \sigma_2 \in L$ such that $\sigma \lceil \Gamma_1 = \sigma_1 \lceil \Gamma_1$ and $\sigma \lceil \Gamma_2 = \sigma_2 \lceil \Gamma_2$.

Now let $\sigma = (dab')^\omega$, $\sigma_1 = (dab)^\omega$ and $\sigma_2 = (da'b')^\omega$. Clearly $\sigma \lceil \Gamma_1 = \sigma_1 \lceil \Gamma_1$ and $\sigma \lceil \Gamma_2 = \sigma_2 \lceil \Gamma_2$. Since $\sigma_1, \sigma_2 \in L$, this implies that $\sigma \in L$ which it is not. Hence L cannot be a product language and therefore is not $\mathrm{TrPTL}^\otimes$-definable. On the other hand, it is a simple exercise to come up with a formula $\alpha \in \Phi^{con}$ such that $\{\mathrm{str}(\sigma) \mid \sigma \in L\} = L_\alpha$.

We now turn to $FO(\widetilde{\Sigma})$, the first order theory of infinite traces over $\widetilde{\Sigma}$. One starts with a countable set of individual variables $X = \{x_0, x_1, \ldots\}$ with x, y, z with or without subscripts ranging over X. For each $a \in \Sigma$ there is a unary predicate symbol R_a. There is also a binary predicate symbol $\leq$.

$R_a(x)$ and $x \leq y$ are atomic formulas. If φ and φ' are formulas, so are $\neg\varphi$, $\varphi \vee \varphi'$ and $(\exists x)\varphi$. The structures for this first order theory are elements of TR^ω. Let $F \in TR^\omega$ with $F = (E, \leq, \lambda)$ and let $\mathcal{I} : X \to E$ be an interpretation. Then $F \models_{\mathcal{I}}^{FO} R_a(x)$ iff $\lambda(\mathcal{I}(x)) = a$ and $F \models_{\mathcal{I}}^{FO} x \leq y$ iff $\mathcal{I}(x) \leq \mathcal{I}(y)$. The remaining semantic definitions go along the expected lines. Each sentence φ (i.e., a formula with no free occurrences of variables) defines the ω-trace language $L_\varphi = \{F \mid F \models^{FO} \varphi\}$.

We say that $L \subseteq TR^\omega$ is FO-definable iff there exists a sentence φ in $FO(\widetilde{\Sigma})$ such that $L = L_\varphi$. As before we will say that $L \subseteq \Sigma^\omega$ is FO-definable iff L is I-consistent and $\{\mathrm{str}(\sigma) \mid \sigma \in L\}$ is FO-definable.

Using the fact that LTL has the same expressive power as the first order theory of sequences, one can show that $L \subseteq \Sigma^\omega$ is FO-definable iff it is I-consistent and LTL-definable [EM]. It will be worthwhile to pin down the notion of LTL-definability. In the current setting, remembering that (Σ, I) is the trace alphabet induced by $\widetilde{\Sigma}$, we define the syntax of the logic $\mathrm{LTL}(\Sigma)$ as follows:

$$\mathrm{LTL}(\Sigma) ::= \top \mid \neg\alpha \mid \alpha \vee \beta \mid \langle a \rangle \alpha \mid \alpha \, \mathcal{U} \beta.$$

A model is a infinite word σ. For $\sigma \in \Sigma^\omega$ and $n \in \omega$, the notion of $\alpha \in \mathrm{LTL}(\Sigma)$ being satisfied at stage n is denoted by $\sigma, n \models \alpha$. This satisfaction relation is defined in the usual manner. The only point of interest might be that $\sigma, n \models \langle a \rangle \alpha$ iff $\sigma(n+1) = a$ and $\sigma, n+1 \models \alpha$. We say that $L \subseteq \Sigma^\omega$ is LTL-definable iff there exists $\alpha \in \mathrm{LTL}(\Sigma)$ such $L = L_\alpha$ where $L_\alpha = \{\sigma \in \Sigma^\omega \mid \sigma, 0 \models \alpha\}$.

The result in [EM] relating FO-definable subsets of TR^ω and LTL-definable subsets of Σ^ω can now be phrased as follows.

Proposition 4.1 *Let $L \subseteq \Sigma^\omega$. Then, the following statements are equivalent.*

(i) *L is I-consistent and LTL-definable.*
(ii) *$\{\mathrm{str}(\sigma) \mid \sigma \in L\}$ is an FO-definable subset of TR^ω.*

We now wish to concentrate on showing that TrPTL is expressible within the first order theory of infinite traces.

To show this, we will freely use the standard derived connectives of Propositional Calculus, together with universal quantification and abbreviations such as $x = y$ for $(x \leq y) \wedge (y \leq x)$, $x \leq y \leq z$ for $(x \leq y) \wedge (y \leq z)$ etc.

An event e is an i-event iff $\lambda(e) \in \Sigma_i$. With this in mind, we let $x \in E_i$ stand for the formula $\bigvee_{a \in \Sigma_i} R_a(x)$. The key to the result we are after is the observation that configurations of a trace can be described using predicates of *bounded* dimension. In what follows we let Q, Q', Q'' range over the non-empty subsets of $\mathcal{P}$. For $Q = \{i_1, i_2, \ldots, i_n\}$, the formula $\mathrm{config}(\{x_i\}_{i \in Q})$ is defined as:

$$\mathrm{config}(\{x_i\}_{i \in Q}) = (\varphi_1 \wedge \varphi_2 \wedge \varphi_3), \text{ where}$$
$$\varphi_1 = \bigwedge_{i \in Q} x_i \in E_i,$$
$$\varphi_2 = \bigwedge_{i,j} \bigwedge_a (R_a(x_i) \wedge R_a(x_j)) \Rightarrow x_i = x_j,$$
$$\varphi_3 = \bigwedge_{i,j} (\forall y)\, (y \in E_j \wedge y \leq x_i) \Rightarrow y \leq x_j.$$

We can now write down a formula describing prime configurations—recall that a prime configuration is one of the form $\downarrow e$, where $e \in E$. Let $\mathrm{loc}(a) \subseteq Q$. Then the formula $\mathrm{prime}_a(\{x_i\}_{i \in Q})$ is defined as

$$\mathrm{config}(\{x_i\}_{i \in Q}) \wedge \bigwedge_{i \in \mathrm{loc}(a)} \bigwedge_{j \in Q - \mathrm{loc}(a)} R_a(x_i) \wedge (x_j \leq x_i).$$

A careful examination of this formula along with the basic properties of traces at once leads to the next result.

Proposition 4.2 *Let $F = (E, \leq, \lambda) \in TR^\omega$ and let $\mathcal{I} : X \to E$ be an interpretation. Then $F \models^{FO}_{\mathcal{I}} \mathrm{prime}_a(\{x_i\}_{i \in Q})$ iff there exists an a-event e such that for each $j \in Q$, $\mathcal{I}(x_j)$ is the $\leq_j$-maximum event in $\downarrow e \cap E_j$ and for each $j \notin Q$, $\downarrow e \cap E_j = \emptyset$.*

For each $\alpha \in \Phi$ we now define the sentence $\mathrm{SAT}(\emptyset, \alpha)$ and the set of formulas $\{\mathrm{SAT}(\{x_i\}_{i \in Q}, \alpha) \mid \{x_i\}_{i \in Q} \subseteq X \text{ and } \emptyset \neq Q \subseteq \mathcal{P}\}$ through simultaneous induction as follows:

- $\mathrm{SAT}(\emptyset, \top) = \mathrm{SAT}(\{x_i\}_{i \in Q}, \top) = (\exists x)\, x = x.$

- $\mathrm{SAT}(\emptyset, \neg \alpha) = \neg \mathrm{SAT}(\emptyset, \alpha).$
  $\mathrm{SAT}(\{x_i\}_{i \in Q}, \neg \alpha) = \neg \mathrm{SAT}(\{x_i\}_{i \in Q}, \alpha).$

- $\mathrm{SAT}(\emptyset, \alpha \vee \beta) = \mathrm{SAT}(\emptyset, \alpha) \vee \mathrm{SAT}(\emptyset, \beta).$
  $\mathrm{SAT}(\{x_i\}_{i \in Q}, \alpha \vee \beta) = \mathrm{SAT}(\{x_i\}_{i \in Q}, \alpha) \vee \mathrm{SAT}(\{x_i\}_{i \in Q}, \beta).$

- $\mathrm{SAT}(\emptyset, \langle a \rangle_j \alpha) = \bigvee_{Q \supseteq \mathrm{loc}(a} (\exists x_{i_1}, \exists x_{i_2}, \ldots, \exists x_{i_n})\, \varphi_1 \wedge \varphi_2 \wedge \varphi_3$
  where $Q = \{i_1, i_2, \ldots, i_n\}$ and
  $$\varphi_1 = \mathrm{prime}_a(\{x_i\}_{i \in Q}),$$
  $$\varphi_2 = \mathrm{SAT}(\{x_i\}_{i \in Q}, \alpha),$$
  $$\varphi_3 = (\forall y)\, (y \in E_j \wedge y \leq x_j) \Rightarrow y = x_j.$$

$\text{SAT}(\{x_i\}_{i \in Q}, \langle a \rangle; \alpha)$ is defined according to two cases.

Case 1 $j \notin Q$: $\text{SAT}(\{x_i\}_{i \in Q}, \langle a \rangle; \alpha) = \text{SAT}(\emptyset, \langle a \rangle; \alpha)$.

Case 2 $j \in Q$
$$\text{SAT}(\{x_i\}_{i \in Q}, \langle a \rangle; \alpha) = \bigvee_{Q' \supseteq \text{loc}(a)} (\exists y_{k_1}, \exists y_{k_2}, \ldots \exists y_{k_n}) \, \varphi_1 \wedge \varphi_2 \wedge \varphi_3$$
where $Q' = \{k_1, k_2, \ldots, k_n\}$ and $\{y_k\}_{k \in Q'}$ is disjoint from $\{x_i\}_{i \in Q}$ and

$$\varphi_1 = \text{prime}_a(\{y_k\}_{k \in Q'}),$$
$$\varphi_2 = \text{SAT}(\{y_k\}_{k \in Q'}, \alpha),$$
$$\varphi_3 = \forall y \, (y \in E_j \Rightarrow (y < y_j \Leftrightarrow y \leq x_j)).$$

- $\text{SAT}(\emptyset, \alpha \, U_j \, \beta) = \text{SAT}(\emptyset, \beta) \vee (\text{SAT}(\emptyset, \alpha) \wedge \text{SAT}(\emptyset, \bigvee_{a \in \Sigma_j} \langle a \rangle; \alpha \, U_j \, \beta))$.

$\text{SAT}(\{x_i\}_{i \in Q}, \alpha \, U_j \, \beta)$ is defined according to two cases.

Case 1 $j \notin Q$: $\text{SAT}(\{x_i\}_{i \in Q}, \alpha \, U_j \, \beta) = \text{SAT}(\emptyset, \alpha \, U_j \, \beta)$.

Case 2 $j \in Q$:
$$\text{SAT}(\{x_i\}_{i \in Q}, \alpha \, U_j \, \beta) = \bigvee_{a \in \Sigma_j} \bigvee_{Q' \supseteq \text{loc}(a)} (\exists y_{k_1}, \exists y_{k_2}, \ldots \exists y_{k_n}) \, \varphi_1 \wedge \varphi_2 \wedge \varphi_3 \wedge$$
φ_4
where $Q' = \{k_1, k_2, \ldots, k_n\}$ and $\{y_k\}_{k \in Q'}$ is disjoint from $\{x_i\}_{i \in Q}$ and

$$\varphi_1 = \text{prime}_a(\{y_k\}_{k \in Q'}),$$
$$\varphi_2 = x_j \leq y_j,$$
$$\varphi_3 = \text{SAT}(\{y_k\}_{k \in Q'}, \beta),$$
$$\varphi_4 = \forall z (z \in E_j \wedge x_j \leq z < y_j) \Rightarrow \varphi_4'.$$

where $\varphi_4' = \bigvee_{a \in \Sigma_j} \bigvee_{Q'' \supseteq \text{loc}(a)} (\exists z_{\ell_1}, \exists z_{\ell_2}, \ldots, \exists z_{\ell_m}) \, \varphi_{41}' \wedge \varphi_{42}' \wedge \varphi_{43}'$
with $Q'' = \{\ell_1, \ell_2, \ldots, \ell_m\}$ and $\{z_\ell\}_{\ell \in Q''}$ disjoint from both $\{x_i\}_{i \in Q}$ and
$\{y_k\}_{k \in Q'}$ and

$$\varphi_{41}' = \text{prime}_a(\{z_\ell\}_{\ell \in Q''}),$$
$$\varphi_{42}' = (z = z_j),$$
$$\varphi_{43}' = \text{SAT}(\{z_\ell\}_{\ell \in Q''}, \alpha).$$

Let f be the map which sends each formula in Φ to a sentence in $FO(\widetilde{\Sigma})$ via $f(\alpha) = \text{SAT}(\emptyset, \alpha)$. Using the previous proposition and the semantics of TrPTL, it is not difficult to prove the following:

Theorem 4.3
(i) *For every* $F \in TR^\omega$, $F, \emptyset \models \alpha$ *iff* $F \models^{FO} f(\alpha)$.
(ii) *If* $L \subseteq TR^\omega$ *is TrPTL definable then it is also* $FO(\widetilde{\Sigma})$-*definable*.

As mentioned earlier we do not know at present if TrPTL is expressively complete—i.e., whether every $L \subseteq TR^\omega$ which is $FO(\widetilde{\Sigma})$-definable is also TrPTL-definable. Clearly from Proposition 4.1 it follows that the expressive completeness of TrPTL can be characterized as follows:

Corollary 4.4 *The following statements are equivalent:*

(i) TrPTL *is expressively complete.*
(ii) *For every $L \subseteq \Sigma^\omega$, if L is I-consistent and L is LTL-definable then L is TrPTL definable.*

We believe that TrPTL is *not* expressively complete. This leads to the following question: What is *the* linear time temporal logic of infinite traces? Such a logic should possess the following properties:

(TR1) It should be expressively complete.
(TR2) It should admit a decision procedure (preferably in terms of asynchronous Büchi automata) whose time complexity is $2^{p(n,m)}$ where n is the size of the input formula, $m = |\Sigma|$ and p is a (low degree) polynomial in n and m.
(TR3) It should be possible to *transparently* express global liveness and safety properties in the logic.

It is worth noting that TrPTL and most of the decidable temporal logics over traces mentioned earlier such as [Nie] and [APP] cannot express all global invariant properties. The somewhat awkward semantics of the logic in [Ebi] also makes it event-based and hence not suitable for expressing invariant properties. However we believe that it should be possible to define a logic with a variant of the until operator defined in [Ebi] which will be able to capture global liveness and safety properties in a straightforward manner.

Any linear time temporal logic over traces which fulfills the properties (PR1)–(PR3) will be a very useful specification tool. In particular it will exactly capture properties that are expressible by I-consistent formulas in LTL—($\alpha \in \mathrm{LTL}(\Sigma)$ is I-consistent iff L_α is I-consistent). This is important because it is such properties which can be verified efficiently using partial order based verification methods [GW, Val].

5 Conclusion

In this paper we have considered linear time temporal logics over traces. Our emphasis has been on TrPTL and its two sublogics $\mathrm{TrPTL}^{\mathrm{con}}$, $\mathrm{TrPTL}^\otimes$. The choice of these logics has been mainly motivated by the fact that they are expressible within the first order theory of traces and the fact that they can be studied using asynchronous Büchi automata.

Our formulation of asynchronous Büchi automata in terms of the acceptance condition A2 appears to be particularly suited for logical studies. The present constructions are much more compact and transparent than the ones in [Thi1] which used A1 as the acceptance condition. We feel that, in the future, alternating versions of our automata will play an important role in the study of temporal logics over traces.

As we have mentioned a number of times, an important open problem is to pin down a linear time temporal logic for traces (assuming it exists!) which will fulfill the properties set out in the previous section. A solution to this problem will at once open up the possibility of investigating branching time temporal logics where path quantification is over traces.

References

[AHU] A.V. AHO, J.E. HOPCROFT AND J.D. ULLMAN: *The Design and Analysis of Algorithms*, Addison-Wesley, Reading (1974).

[APP] R. ALUR, D. PELED AND W. PENCZEK: Model-Checking of Causality Properties, *Proc. 10th IEEE LICS* (1994).

[Die] V. DIEKERT: *Combinatorics on traces, LNCS* **454** (1990).

[DM] V. DIEKERT AND A. MUSCHOLL: Deterministic Asynchronous Automata for Infinite Traces, *Acta Inf.*, **31** (1993) 379-397.

[DR] V. DIEKERT, G. ROZENBERG (Eds.): *The Book of Traces*, World Scientific, Singapore (1995).

[Ebi] W. EBINGER: *Charakterisierung von Sprachklassenunendlicher Spuren durch Logiken*, Ph.D. Thesis, Institut für Informatik, Universität Stuttgart, Stuttgart, Germany (1994).

[EM] W. EBINGER AND A. MUSCHOLL: Logical Definability on Infinite Traces, *Proc. ICALP '93, LNCS* **700** (1993) 335-346.

[GP] P. GASTIN AND A. PETIT: Asynchronous Cellular Automata for Infinite Traces, *Proc. ICALP '92, LNCS* **623** (1992) 583-594.

[GW] P. GODEFROID AND P. WOLPER: A Partial Approach to Model Checking, *Inform. and Comput.*, **110** (1994) 305-326.

[Huh] M. HUHN: On Semantic and Logical Refinement of Actions, Technical Report, Institut für Informatik, Universität Hildesheim, Germany (1996).

[KP] S. KATZ AND D. PELED: Interleaving Set Temporal Logic, *Theor. Comput. Sci.*, **75** (3) (1992) 21-43.

[KMS] N. KLARLUND, M. MUKUND AND M. SOHONI: Determinizing Büchi asynchronous automata, *Proc. FST&TCS 1995, LNCS* **1026** (1995) 456-470.

[LPRT] K. LODAYA, R. PARIKH, R. RAMANUJAM AND P.S. THIAGARAJAN: A logical study of distributed transition systems, *Inform. and Comput.*, **119** (1995) 91-118.

[MP] Z. MANNA AND A. PNUELI: *The Temporal Logic of Reactive and Concurrent Systems (Specification)*, Springer-Verlag, Berlin (1991).

[Maz] A. MAZURKIEWICZ: Concurrent Program Schemes and their Interpretations, *Report DAIMI-PB-78*, Computer Science Department, Aarhus University, Denmark (1978).

[MS] M. MUKUND AND M. SOHONI: Keeping track of the latest gossip: Bounded time-stamps suffice, *Proc. FST&TCS '93, LNCS* **761** (1993) 388-399.

[Mus] A. MUSCHOLL: On the complementation of Büchi asynchronous cellular automata, *Proc. ICALP '94, LNCS* **820** (1994) 142-153.

[Nar] K. NARAYAN KUMAR: An Improved Decision Procedure for TrPTL, Unpublished Manuscript, Tata Institute of Fundamental Research, Bombay, India (1994).

[Nie] P. NIEBERT: A ν-Calculus with Local Views for Systems of Sequential Agents, *Proc. MFCS'95, LNCS* **969** (1995) 563-573.

[NPW] M. NIELSEN, G.D. PLOTKIN AND G. WINSKEL: Petri Nets, Event Structures and Domains I, *Theor. Comput. Sci.*, **13** (1980) 86-108.

[Pen]   W. PENCZEK: Temporal Logics for Trace Systems: On Automated Verification, *Int. J. Found. of Comput. Sci.*, 4(1) (1993) 31-68.

[Pnu]   A. PNUELI: The Temporal Logic of Programs, *Proc. 18th IEEE FOCS* (1977) 46-57.

[Ram]   R. RAMANUJAM: Locally Linear Time Temporal Logic, To appear in *Proc. 11th IEEE LICS* (1996).

[Tho]   W. THOMAS: Automata on infinite objects, in J. van Leeuwen (ed.), *Handbook of Theoretical Computer Science, Volume B*, North-Holland, Amsterdam (1990) 133-191.

[Thi1]   P.S. THIAGARAJAN: A Trace Based Extension of Linear Time Temporal Logic, *Proc. 9th IEEE LICS* (1994) 438-447. Full version available as: TrPTL: A Trace Based Extension of Linear Time Temporal Logic, *Report TCS-93-6*, School of Mathematics, SPIC Science Foundation, Madras, India (1993).

[Thi2]   P.S. THIAGARAJAN: A Trace Consistent Subset of PTL, *Proc. CONCUR'95*, *LNCS* **962** (1995) 438-452. Full version available as: PTL over Product State Spaces, *Report TCS-95-4*, School of Mathematics, SPIC Science Foundation, Madras, India (1995).

[Val]   A. VALMARI: Stubborn Sets for Reduced State Space Generation, *LNCS* **483** (1990) 491-515.

[VW]   M. VARDI, P. WOLPER: An automata theoretic approach to automatic program verification, *Proc. 1st IEEE LICS* (1986) 332-345.

[WN]   G. WINSKEL AND M. NIELSEN: Models for Concurrency, In: S. Abramsky and D. Gabbay (Eds.), *Handbook of Logic in Computer Science*, Vol 3, Oxford University Press, Oxford (1994).

[Zie]   .W. ZIELONKA: Notes on finite asynchronous automata, *R.A.I.R.O.—Inf. Théor. et Appl.*, **21** (1987) 99-135.

[Zuc]   L. ZUCK: *Past Temporal Logic*, Ph.D. Thesis, Weizmann Institute, Rehovot, Israel (1986).

Partial Order Reduction:
Model-Checking Using Representatives

Doron Peled
Bell Laboratories
600 Mountain Avenue
Murray Hill, NJ 07974
doron@research.bell-labs.com

Abstract

Partial order reductions is a family of techniques for diminishing the state-space explosion problem for model-checking concurrent programs. It is based on the observation that execution sequences of a concurrent program can be grouped together into equivalence classes that are indistinguishable by the property to be checked. Applying the reduction constructs a reduced state-space that generates at least one representative for each equivalence class. This paper surveys some algorithms for partial order model-checking. The presentation focuses on the *verification using representatives* approach. The reduction approach is extended to branching specifications.

1 Introduction

Being able to check the correctness of a concurrent system is both important and challenging. It is hard to predict all the possible interactions of concurrent components; serious problems may be caused by unlikely schedulings of concurrent components. Even the most thorough types of testing may not be enough.

Total order semantics, also called *interleaving semantics*, are traditionally considered easier to work with, as they lend themselves to simple representations, e.g., using finite state machines. Partial order semantics is more recent in modeling concurrent programs. It is argued by its supporters that it can reflect the executions of concurrent systems more accurately, and hence is sometimes called *true concurrency*. Until recently, though, using partial order semantics in verification was scarce, as it was considered to be more difficult. In recent years, new research showed that partial order semantics can be as simple to use as total order semantics, and in many cases partial order verification methods perform significantly better than total order based methods.

Partial order reduction techniques are used to alleviate the state-space explosion in automatically verifying concurrent programs [31, 11, 14, 13, 32, 27, 28, 17, 9, 13]. The list of potential applications of partial order techniques in industrial applications is substantial. The success of partial order techniques in the domain of software verification may be compared to the successes of BDDs (binary decision diagrams) in the domain of hardware verification. In particular, partial order reduction techniques were integrated in tools such as SPIN [17] and VFSM-valid [13]. Using the partial order reduction techniques, it has become possible to analyze problems of larger size, which did not lend themselves to automatic verification before. The simplicity of the principles behind these methods suggest that they can be integrated into any state-based automatic verification tool.

In this paper we survey a family of partial order reduction methods. We show how equivalence relations can be used to group together sequences that are undistinguishable with respect to the specification. This allows to construct a reduced state-space for the checked system. A reduced state-space for a concurrent system contains only representative sequences from each equivalence classes rather than all the sequences in the class. An algorithm for deciding whether a specification cannot distinguish between equivalent sequences for such an equivalence relation is described. We also show how this approach can be extended to deal with branching-time specification.

2 Modeling Concurrent Systems

2.1 State Spaces of Concurrent Systems

A *finite state system* $\mathcal{F}$ is a triple $\langle S, T, \iota \rangle$, where

- S is a finite set of *states*,
- $T \subseteq S \times S$ is a finite set of deterministic *transitions*, i.e., satisfying that if (s, t) and (s, t') are both in T, then $t = t'$, and
- $\iota \in S$ is an *initial state*.

The domain of each transition $\tau \in T$ is denoted by $en_\tau \subseteq S$. These are the states where τ is *executable* or *enabled*. The set of transitions enabled at a state S is denoted by $enabled(s)$. When $(s, t) \in \tau$ we also write $t = \tau(s)$. Executing the transitions $\alpha_0 \alpha_1 \ldots \alpha_i$ hence obtains the state $\alpha_i(\alpha_{i-1}(\ldots \alpha_1(\alpha_0(\iota))\ldots))$.

An *interpreted* system is a triple $\mathcal{I} = \langle \mathcal{F}, P, M \rangle$, where

- $\mathcal{F} = \langle S, T, \iota \rangle$ is a finite state system,
- P is a finite set of *propositions*, and
- $M : S \mapsto 2^P$ is the *state labeling* function.

In the sequel we will use the term *system* for interpreted finite state systems.

The *state-space* $SP(\mathcal{I})$ of a system $\mathcal{I} = \langle \mathcal{F}, P, M \rangle$ where $\mathcal{F} = \langle S, T, \iota \rangle$, is a labeled graph $\langle V, E \rangle$ such that

- $V \subseteq S$ is the minimal set of *reachable states* satisfying:

  1. $\iota \in V$,
  2. If $s \in V$ and $(s, t) \in \tau \in T$, then $t \in V$.

- $E = \{(s, \tau, t) | (s, t) \in \tau \in T\}$

Thus, the state-space of $\mathcal{I}$ contains the states reachable from the initial state ι by repeatedly executing the transitions T of $\mathcal{I}$. The *label* of $e = (s, \tau, t)$ is τ.

The *transitions sequences* generated by $\mathcal{I}$ correspond to edge labels along the maximal paths of $SP(\mathcal{I})$ that start from the initial state ι. Hence, a transitions sequence is a finite or infinite sequence of transitions $\tau_0 \tau_1 \tau_2 \ldots$ such that there exists a sequence of states $s_0 s_1 s_2 \ldots$ satisfying

- $s_0 = \iota$ [The first state is the initial state.]

- for each $i \geq 0$, $(s_i, s_{i+1}) \in \tau_i$. [Each adjacent pairs of states correspond to the execution of a transition.]

- The sequence is maximal, namely it is either infinite, or ends with a state s such that $enabled(s) = \phi$.

The states sequence that correspond to a transitions sequence v is denoted by $states(v)$. For simplicity, it is possible to assume that all transitions sequences are infinite. This can be achieved by adding a special transition τ' such that $en_{\tau'} = S \setminus \cup_{\tau \in T} en_\tau$, and $\tau' = \{(s, s) | s \in en_\tau\}$.

Notice that the state-space of a system $\mathcal{I}$ can be considered as a more explicit representation of $\mathcal{I}$; $\mathcal{I}$ contains in S all the *potential* states of $\mathcal{I}$, while $SP(\mathcal{I})$ contains in V only the *actual* states that can be reached. The partial order reduction algorithms are aimed at exploiting a graph smaller than $SP(\mathcal{I})$ that represents enough information about the property that we want to check. Notice that not every model-checking algorithm is based upon using an explicit representation of the state-space, e.g., using symbolic model-checking algorithms [3].

For each transitions sequence v of $SP(\mathcal{I})$ there is a sequence $prop(v)$ of propositions obtained in the following way: if $states(v) = s_0 s_1 s_2 \ldots$, then $prop(v)$ is the sequence $M(s_0) M(s_1) M(s_2) \ldots$. Thus, there are three languages defined for an interpreted system $\mathcal{I}$:

- The language $\mathcal{L}(\mathcal{I}) \subseteq T^\omega$ of transitions sequences.

- The language $\mathcal{L}_{states}(\mathcal{I}) \subseteq S^\omega$ of states sequences.

- The language $\mathcal{L}_{prop}(\mathcal{I}) \subseteq 2^{P^\omega}$ of propositional sequences.

A *specification* for a system $\mathcal{I}$ can be given as a language over one of the three domains T, S and 2^P. Most specifications use transitions or states sequences. In the rest of this section we will usually treat the latter case; the others can be dealt with similarly. In model-checking, the specification is often given using a regular automaton over infinite words, e.g., as a Büchi automaton, or using a logic, such as linear temporal logic (LTL) [30]. A system $\mathcal{I}$ *satisfies* the specification φ, corresponding to the language L_φ, where both are using the same set of propositions P, iff $\mathcal{L}(\mathcal{I}) \subseteq L_\varphi$. Graph-theoretical algorithms [23] can then be applied to state space graphs to check that $\mathcal{I}$ satisfies φ.

2.2 Traces and Trace Equivalence

Using interleaving semantics has a lot of advantages for modeling concurrent systems. In particular, its simplicity and use of strings allows exploiting automata and language theory. On the other hand, interleaving semantics is often criticized for distinguishing between entities that are basically the same. Namely, it can distinguish between executions which differ from each other only by the order of some concurrently executed transition. This order is largely artificial. Trace semantics groups transitions sequences into equivalence classes, allowing a higher abstraction of the specified system. One can exploits this for model-checking properties that do not distinguish between different sequences that are trace-equivalent.

A *concurrent alphabet* is a pair (T, D), where T is a finite set (representing transitions in our context), and $D \subseteq T \times T$ is a symmetric and reflexive relation called the *dependency relation*.

We define trace equivalence in several steps:

1. Define the relation $\overset{1}{\equiv} \subseteq T^* \times T^*$ such that $v \overset{1}{\equiv} v'$ iff $v = v'$ or $v = uabw$, $v' = ubaw$ for some $u, w \in T^*, (a, b) \in D$.

2. Define the *trace equivalence* relation for finite sequences as the reflexive and transitive closure of $\overset{1}{\equiv}$. Thus, $v \equiv w$ iff one can obtain v from w by repeatedly commuting the order of adjacent independent letters.

3. Define trace preorder relation $\preceq$ among infinite strings as follows: $v \preceq v'$ iff for each finite prefix u of v, there exists a finite prefix u' of v' and a finite string w such that $uw \equiv u'$.

4. Define trace equivalence among infinite strings such that $v \equiv v'$ iff $v \preceq v'$ and $v' \preceq v$.

Thus, for the concurrent alphabet $(\{a, b\}, \{(a, a), (b, b)\})$ we have $aabb \overset{1}{\equiv} abab$, $aabb \equiv bbaa$, $aaab^\omega \preceq (ab)^\omega$, and $(ab)^\omega \equiv (aab)^\omega$.

Traces are then the equivalence classes of the relation $\equiv$ over finite or infinite strings.

To achieve that when $v \equiv w$, v is a transitions sequence of $\mathcal{I}$ iff w is a transitions sequence of $\mathcal{I}$, we enforce the following two conditions for independent transitions $(a, b) \notin \mathcal{D}$:

D1 if $s \in en_\alpha$, then $s \in en_b$ iff $a(s) \in en_b$. [executing a does not affect the enabledness of b].

D2 If $s \in en_a \cap en_b$ then $a(b(s)) = b(a(s))$. [When both a and b are enabled, executing them in either order results in the same state].

2.3 Stuttering Equivalence

Denote $\Sigma^\infty = \Sigma^* \cup \Sigma^\omega$. The stuttering removal operator $\natural : \Sigma^\infty \mapsto \Sigma^\infty$ applied to a string v replaces every maximal finite subsequence of identical elements by a single copy of this element. For example, $\natural(aabaaacc) = abac$, $\natural(aabaac^\omega) = abac^\omega$.

Two sequences v, w will be considered *stuttering-equivalent* iff $\natural v = \natural w$. We denote this by $v \rightleftharpoons w$. Lamport argued [22] that a specification should not distinguish between two propositional sequences that are stuttering equivalent.

2.4 Fairness Issues

The *total order semantics* or *interleaving semantics* of a program identifies transitions (or states) sequences as *executions* of a program. Sometimes, the transitions sequences that are considered to be executions are constrained using a *fairness assumption*. Such a constraint can be given as a language **R**. If a fairness assumption **R** is imposed, only sequences that are fair are considered to be execution of a system. Hence, the fair transitions sequences $\mathcal{L}^{\mathbf{R}}(\mathcal{I})$ of a system $\mathcal{I}$ are $\mathcal{L}(\mathcal{I}) \cap R$.

The following fairness assumption is in particular natural when using partial order semantics:

F-fairness. If a transition a is enabled from a state reached by the execution the transitions that appear in some finite prefix of a fair transitions sequence then some transition that is dependent on α must appear later in this sequence.

This fairness assumption was shown in [21, 26] to be equivalent to restricting the set of sequences to those that are maximal with respect to the order relation $\preceq$.

3 Verification Using Representatives

We are interested in generating a reduced state-space for a system $\mathcal{I}$. Although we want the reduced state-space to be as small as possible, it must still contain enough information to preserve the checked property. The aim is that the model-checking algorithm would be applicable to the reduced state-space instead of the full one. Besides preserving the truth of the checked specification, the reduced state-space needs also to be able to suply a counter-example in the case that the specification does not hold for the checked system.

3.1 Ample Sub-state-spaces

A *sub-state-space* S for a system $\mathcal{I} = \langle \mathcal{F}, P, M \rangle$ is a labeled subgraph $\langle S', E' \rangle$ of $SP(\mathcal{I}) = \langle S, E \rangle$ such that

- $\iota \in S'$ [S' includes the initial state],

- $S' \subseteq S$, and

- $E' \subseteq E \cap (S' \times T \times S')$.

Similar to state-spaces, a sub-state-space S generates a set of transitions sequences $\mathcal{L}(S)$, a set of states sequences $\mathcal{L}_{states}(S)$ and a set of propositional sequences $\mathcal{L}_{prop}(S)$. In fact, we have:

$$\mathcal{L}(S) \subseteq \mathcal{L}(\mathcal{I}), \; \mathcal{L}_{states}(S) \subseteq \mathcal{L}_{states}(\mathcal{I}), \; \mathcal{L}_{prop}(S) \subseteq \mathcal{L}_{prop}(\mathcal{I})$$

Definition 1 *A language $\mathcal{L}$ is said to be* closed *under an equivalence relation $\sim$, if for every equivalence class C of $\sim$, either $C \subseteq \mathcal{L}$ or $C \cap \mathcal{L} = \phi$. We also say that $\sim$* saturates $\mathcal{L}$.

Definition 2 *A sub-state-space of a system $\mathcal{I}$ is said to be* ample *with respect to the equivalence relation $\sim$ if it generates at least one (states, transitions or propositional) sequence for every equivalence class C of $\sim$ such that $C \cap \mathcal{L}(\mathcal{I}) \neq \phi$.*

The following simple observation suggests the use of equivalences in conjunction with sub-state-spaces:

> Let L_φ be the language of a specification φ that is closed under an equivalence relation $\sim$. Let S be an ample sub-state-space for a system $\mathcal{I}$ with respect to $\sim$. Then, $\mathcal{L}(S) \subseteq L_\varphi$ iff $\mathcal{L}(\mathcal{I}) \subseteq L_\varphi$.

To exploit the above Lemma, we need an equivalence relation $\sim$ where the following exist:

1. An effective way to decide whether a given specification φ is closed under $\sim$.

2. An effective way to construct an ample sub-state-space for $\mathcal{I}$ with respect to $\sim$.

3.2 Checking Equivalence Closedness

Section 3.1 motivated the need for checking whether a specification φ is closed under a given equivalence relation $\sim$. In [29], an algorithm is given for deciding the closure of a specification for a given class of equivalence relations, represented as either a non-deterministic automaton (over infinite words) or as linear temporal logic formula. This class includes in particular trace and stuttering equivalence. It is characterized by having a symmetric and reflexive relation $\overset{1}{\sim}$ on finite strings such that

- $\sim^{fin}$ is the transitive closure of $\overset{1}{\sim}$ (hence $\sim^{fin}$ is an equivalence relation).

- $\overset{1}{\sim} \subseteq \Sigma^* \times \Sigma^*$ is a regular language (i.e., recognizable by a finite automaton) over the alphabet $\Sigma \times \Sigma$. Thus, $\overset{1}{\sim}$ is defined between strings of equal lengths.

- $\sim^{fin}$ is a left cancellative relation, i.e., if $vw \sim^{fin} vw'$, then $w \sim^{fin} w'$.

- $\sim$ is defined as the *limit extension* of $\sim^{fin}$, namely $v \sim v'$ iff

  - for each finite prefix u of v, there exists a finite prefix u' of v' and a finite string w such that $uw \sim^{fin} u'$, and
  - for each finite prefix u' of v', there exists a finite prefix u of v and a finite string w' such that $u'w' \sim^{fin} u$.

The definition of trace equivalence $\equiv$ in Section 2.2 already uses the relation $\overset{1}{\equiv}$, which satisfies the above conditions.

For stuttering equivalence, there is a small technical complication in obtaining a relation $\overset{1}{\rightleftharpoons}$, as it needs to be defined between pairs of strings of equal length. We achieve this by extending the alphabet into $\Sigma \cup \{\$\}$, where $\$$ serves only to force the strings to have the same length. Then, we can relate u with itself, and $uav\$$ with $uaav$, where $u, v \in \Sigma^*$ and $a \in \Sigma$.

Checking that an ω-regular language L, represented by a Büchi automaton A_L, is closed under an equivalence relation $\sim$ that satisfies the above conditions can be done using the following algorithm, introduced in [29]. The algorithm checks the emptiness of the intersection of the following three languages over the alphabet $\Sigma \times \Sigma$ ($(\Sigma \cup \{\$\}) \times (\Sigma \cup \{\$\})$ for stuttering equivalence, respectively). Hence, each infinite word $\mathbf{w} = (w_1, w_2)$ over this alphabet has a left component w_1 and a right component w_2. The three languages are:

1. The language where the left component w_1 of the input is in L (after removing the $\$$ symbols, respectively).

2. The language where the right component w_2 of the input is not in L (after removing the $\$$ symbols, respectively).

3. An automaton that checks that the input can be decomposed into infinitely many factors that are all elements of $\overset{1}{\sim}$.

The naive way to implement the algorithm by constructing the automata for the three languages and then intersecting them can take space exponentially bigger than A_L. However, the algorithm can be implemented in PSPACE [29]. The idea is that there is no need to fully construct the automaton for the complement of the language L; instead, one can use a binary search through the state-space of such a complement automaton [33].

When the specification L is given as a temporal formula φ_L, it is not necessary to translate first the formula into a Büchi automaton. Such a translation requires again in the worse case space exponential in the size of the formula. It is again possible to conduct a binary search through the state-space of the corresponding automata, for φ_L and for $\neg\varphi_L$. This requires space only polynomial in the size of the checked formula. For the stuttering and trace equivalences, checking closeness is in PSPACE-complete, by a reduction from universality of ω-regular automata [29].

3.3 Syntax and Semantics of CTL*

Let P be a finite set of propositions. The set of CTL* state and path formulas are defined inductively:

S1. every member of P is a state formula,

S2. if φ and ψ are state formulas, then so are $\neg\varphi$ and $\varphi \wedge \psi$,

S3. if φ is a path formula, then Aφ is a state formula,

P1. any state formula φ is also a path formula,

P2. if φ, ψ are path formulas, then so are $\varphi \wedge \psi$ and $\neg\varphi$,

P3. if φ, ψ are path formulas, then so is $\varphi U \psi$.

The modal operator A has the intuitive meaning: "for all paths". U denotes the standard strong "until". CTL* consists of the set of all state formulae. The following abbreviations will be used: $E\varphi = \neg A\neg\varphi$, $F\varphi = trueU\varphi$, $G\varphi = \neg F\neg\varphi$.

The logic LTL is obtained by restricting the set of formulas to the form Aφ, where φ does not contain A and E. We write φ instead of Aφ, when confusion is unlikely. We purposely avoided using the nexttime operator X, which can express that a change is made from one specific state to another.

A *model* for CTL* is a quadruple $\mathcal{M} = \langle V, E, \iota, M \rangle$, where V are states, E are edges, $\iota \in S$ is a distinguished initial state, and M is an interpretation function, mapping S into subsets of a set of propositions P. The edge relation is assumed to be total; i.e. $\forall s \in V \exists t \in V \exists \tau \in T : (s, \tau, t) \in E$. The labels on the edges in the definition of the graph are only used for the benefit of the description of the suggested algorithm, but are ignored by the interpretation of the temporal logics.

Denote by $\pi = (s_0, s_1, \ldots)$ an infinite path of S, starting at $s_0 \in S$. Denote the first state of π by $first(\pi)$. The suffix of π, starting from state s_i will be denoted π_i. The satisfaction of a formula φ in a state s of S is written $\mathcal{M}, s \models \varphi$, or just $s \models \varphi$. It is defined inductively as follows:

S1. $s \models q$ iff $q \in M(s)$, for $q \in P$,

S2. $s \models \neg\varphi$ iff not $s \models \varphi$, $s \models \varphi \wedge \psi$ iff $s \models \varphi$ and $s \models \psi$,

S3. $s \models A\varphi$ iff $\pi \models \varphi$ for every path π starting at s,

P1. $\pi \models \varphi$ iff $first(\pi) \models \varphi$ for any state formula φ,

P2. $\pi \models \neg\varphi$ iff not $\pi \models \varphi$, $\pi \models \varphi \wedge \psi$ iff $\pi \models \varphi$ and $\pi \models \psi$,

P3. $\pi \models \varphi U \psi$ iff there is an $i \geq 0$ such that $\pi_i \models \psi$ and $\pi_j \models \varphi$ for all $0 \leq j < i$.

When using a fairness assumption to limit the execution sequences, we replace "path" by "fair path" in the above definition. (As usual, we require that a fairness assumption satisfies that an infinite sequence is fair iff each suffix of it is fair). We write $\mathcal{M} \models \varphi$ iff $\mathcal{M}, \iota \models \varphi$. Notice that for an LTL specification Aφ, $\mathcal{M} \models$ Aφ iff every (fair) sequence of $\mathcal{M}$ satisfies φ.

4 Partial Order Reduction

Partial order reduction methods is a generic name for a family of model-checking methods that avoid constructing the full state-space of the checked program. The family of methods are historically related to partial order because of the connection between traces and partial order semantics [24]. The basic ideas of the reduction is to generate at least one transitions sequence for each such trace. However, as will be seen later, this is not always the case, i.e., there are cases where there is a single sequence that represents a collection of traces.

4.1 The Ample-Sets Reduction Method

Partial order reduction is based upon modifying the depth first search (DFS) construction of a state-space, depicted in Figure 1. The DFS creates a node for a global state (starting with the initial state ι), pushes this node into its stack, then recursively creates nodes for all the successors of this node, and pops the node from the stack after all their successors were created. When a new node is generated, the value is hashed using a hashing table (using the procedure **create_node** at lines 1 and 11). Checking if a node is new is facilitated by checking if it already exists in the hashing table (using the function **new** at line 10). A node that is already discovered during the search is said to be 'open' if its on the stack (lines 2 and 12) and 'closed' once it is removed from the stack (line 16). Although the information about whether a node is open or closed is not used here, it will be used in the sequel for detecting cycles. Recall that a cycle is detected exactly when an edge is created (at line 14) pointing to a node that is open (hence not new).

The partial order reduction algorithm (and other reduction algorithms, e.g., symmetry reduction [6]) modifies the DFS by expanding only a subset of the enabled transitions from each state:

5   working_set(s):=ample(s);

where $ample(s) \subseteq enabled(s)$. This obviously generates a sub-state-space. The problem is how to select these ample sets of successors such that the sub-state-space will be ample with respect to a given effective equivalence relation.

The *ample sets* method provides a set of constraints for selecting the successors of a state. The set of constraints depends on the effective equivalence relation used. This

```
1    create_node(ι);
2    push ι; /* ι is open */
3    expand_node(ι);

4    proc expand_node(s);
5        working_set(s):=enabled(s);
6        while working_set(s)≠ φ do
7          let τ ∈working_set(s);
8          working_set(s):=working_set(s)\{τ};
9          t:=τ(s);
10         if new(t) then
11           create_node(t);
12           push t; /* t is open */
13           expand_node(t) fi;
14         create_edge(s, τ, t);
15       end while;
16       pop s; /* s is closed */
17   end expand_node.
```

Figure 1: Using DFS to construct the state-space graph of a program

in turn can depend on the specification to be checked and whether a fairness constraint is assumed.

In order to present such a set of constraints, define a *visible* transition [32] to be a transition $\tau \in T$ that can change the propositional interpretation of a state:

Definition 3 *Given a system* $\langle \mathcal{I}, P, M \rangle$, *a transition* $\tau \in T$ *is visible if there are two states* s, $t \in S$ *such that* $M(s) \neq M(t)$ *and* $t \in \tau(s)$.

We will consider the following constraints:

C0 [Non-emptiness condition] $ample(s)$ is empty iff $enabled(s)$ is empty.

C1 [Faithful decomposition [19, 31, 27, 13]] For every path of $SP(\mathcal{I})$, starting from the state s and labeled with the transitions $\tau_0 \tau_1 \tau_2 \ldots$, a transition that is dependent on some transition in $ample(s)$ cannot appear before a transition from $ample(s)$.

C2 [Cycle closing condition [27]] If $ample(s)$ is a *proper* subset of $enabled(s)$, then for no transition $\alpha \in ample(s)$ it holds that $\alpha(s)$ is on the search stack (i.e., is open).

C3 [Non-visibility condition [32]] If $ample(s)$ is a *proper* subset of the transitions enabled from s, then none of the transitions in it is visible.

We have the following results concerning sub-states-space constructed using ample sets:

Theorem 1 ([27]) *The sub-state-space constructed using conditions* **C0–C2** *is ample with respect to trace equivalence under* **F**-*fairness.*

Hence, if the specification is given as a language that is closed under trace equivalence, and F-fairness is assumed, one can use a sub-state-space that is constructed while conditions C0–C2 are satisfied at each constructed state. Several temporal logics were devised for expressing properties that are closed under trace equivalence, e.g., the logics TrPTL [34] and TLC [1]. Alternatively, one can use the decision procedure of [29] and presented in Section 3.2 to check whether a given LTL or Büchi automaton specification is closed under trace equivalence.

If the specification is not closed under trace equivalence, one can keep adding new dependencies, until it becomes closed. Of course, adding dependencies can ultimately completely prohibit the reduction, e.g., when all transitions are made interdependent.

There is a subtle point to notice about adding dependencies: the definition of F-fairness is sensitive to the dependency relation used. By adding more dependencies, more sequences would become F-fair. Hence, at worst, representatives for sequences that were not originally fair are generated. Since the model-checking algorithm applied to the reduced state-space will ignore unfair sequences, correctness is preserved.

To understand why Theorem 1 holds, observe the following Lemmas, assuming the sub-state-space are constructed under conditions C0–C2:

Lemma 1 ([28]) *Let s be a state in the sub-state-space $S = \langle S', E' \rangle$ of $SP(\mathcal{I}) = \langle S, E \rangle$. Let v be a sequence of transitions labeling a path of $SP(\mathcal{I})$, staring at s. Then there exists a transition $\tau \in ample(s)$ such that $v \equiv \tau w$, for some $w \in T^{\omega}$.*

Proof. According to C1, only transitions that are independent of those in $ample(s)$ can appear in v before some transition of $ample(s)$. According to **F**, it is not possible that v does not contain transitions of $ample(s)$. Therefore, some transition $\tau \in ample(s)$ appears after transitions independent of it, hence can be commuted to the beginning. ∎

We aim at simulating each fair path of $\mathcal{I}$ by a fair path of the reduced sub-state-space S. The basic simulation step is based on the following:

Lemma 2 ([28]) *Let s and v be as in Lemma 1. Let α be the first transition of v. Then, the reduced sub-state-space S contains a finite path labeled with $\beta_1 \beta_2 \ldots \beta_n \alpha$, such that each β_i is independent of α, and $\alpha \beta_1 \beta_2 \ldots \beta_n w \equiv v$ for some $w \in T^{\omega}$.*

Proof. The proof is by induction on the order in which nodes are removed from the stack (at line 16 in Figure 1), i.e., are closed. There are two cases. In the first case, $\alpha \in ample(s)$, hence the corresponding path has length of one. In the second case,

$\alpha \notin ample(s)$. Hence, according to Lemma 1, there is a transition $\beta_0 \in ample(s)$ that is independent of α and appears in v after a sequence of transitions that are independent of β_0. We can look now at the state $s' = \beta_0(s)$. Since $\alpha \notin ample(s)$, we know from Condition C2 that the transition β_0 could not close a cycle. Hence, s' is created after s and thus according to the DFS order, will be removed from the stack before s. Therefore, we can assume the induction hypothesis from s', i.e., there exists a sequence $\beta_1\beta_2 \ldots \beta_n \alpha$ from s' such that each β_i is independent of α. The required sequence is then $\beta_0\beta_1\beta_2 \ldots \beta_n \alpha$. ∎

Lemma 2 can be used to show that for each sequence v of $\mathcal{I}$ there exists a sequence w such that $w \equiv v$ in S, proving Theorem 1. The proof in [28] constructs the path w: each transition α_i, taken in its turn from $v = \alpha_0\alpha_1\alpha_2 \ldots$, either (a) appears in w after some 'deficit' sequence of independent transitions $\beta_1\beta_2 \ldots \beta_n$, according to Lemma 2, or (b) has already appeared as part of the so far accumulated deficit.

Unfortunately, when the fairness condition **F** (or any stronger fairness condition) is not assumed, Lemma 1 does not hold. Hence, also Lemma 2 and Theorem 1 do not hold. To see this, assume there is a transition α which is enabled at a state s, and independently, a loop starts at s, consisting of the transitions β and γ, which are independent of α. Thus, $enabled(s) = \{\alpha, \beta\}$. Then, without assuming **F**-fairness, the transitions sequence $v = (\beta\gamma)^\omega$, starting at state s is allowed. Choosing $ample(s) = \{\alpha\}$ satisfies the conditions C0-C2, hence no sequence equivalent to v starts from s in the constructed sub-state-space.

To recover the situation, observe that although the sequence $w = \alpha(\beta\gamma)^\omega$ is not trace-equivalent to v, α appears before a sequence of independent transitions. If α is invisible, then no stuttering-closed specification can distinguish between v and w. We have the following:

Theorem 2 ([28]) *The sub-state-space constructed using conditions* **C0–C3** *is ample with respect to stuttering equivalence.*

4.2 Implementation Issues

Finding ample sets that satisfy condition C1 is based on analyzing the current global state. Knowledge about the type of transitions that are currently enabled or can become enabled can allow accepting a given subset of the enabled transitions as an ample set. Observing condition C1, the full knowledge about the future enabledness of transitions (e.g., by construct the state-space $SP(\mathcal{I})$) allows deciding which subsets of the enabled transitions are ample sets. However, obtaining this knowledge is as hard as the original model-checking problem (PSPACE-complete in the size of the checked system).

It is easier to enforce C1 by analyzing the enabled transitions from the current state s, allowing only part of subsets that satisfy C1 as ample sets. The analysis is based on information about the type of transitions currently enabled. It allows efficient algorithms for finding appropriate subsets of the enabled transitions.

We will discuss two types of systems, with matching algorithms. In both cases, we assume that each system consists of a set of *processes*, with each process containing a (not necessarily disjoint) set of transitions. Each process has a set of local variables that can be changed only by transitions that belong to the process. Transitions whose effect is only to change the process variables are called *local transitions*. The local state of each process includes the values of its local variables. Each (global) system state is a combination of the local states of all the processes.

Synchronous Communication

Synchronous communication systems incorporate CSP or ADA-like communication. Communication is done cooperatively at the same time by the sender and the receiver. Sending and receiving can thus be considered a single transition, shared by two processes. Hence, the communication transition belongs to both the sending and the receiving process. We say that a communication transition τ between a pair of processes P_i and P_j is *locally enabled* by a process P_i at state s if it can be executed at the current state s, or any state s' such that the local states of P_i in s and s' are the same. This means that P_i is willing to do his part in the communication transition τ. We assume that such a system includes only local and synchronous communication transitions.

The dependency relation for synchronous communication systems relates transitions that belong to the same process. Hence, two transitions are interdependent iff they belong to the same process. Notice that a communication transition belongs to and hence is dependent on transitions of two processes. Choosing a subset of the enabled transitions that satisfy condition C3 can be done as follows:

> Choose all the transitions enabled in the current state s that belong to a subset P of the processes, such that there is no communication transition between a process P_i in P and a process outside P that is locally enabled by P_i.

The above rule prevents the case where, by executing transitions outside the selected ample set, a communication that is dependent on transitions in the set will become (globally) enabled and will execute before any transition in the ample set, contradicting C1. Such a set of transitions can be found by choosing initially the currently enabled transitions that belong to a single process. If the above rule does not hold, repeat adding transitions of additional processes, until the rule holds.

Asynchronous Communication

In this communication model, we have separate sends and receives. In addition to the local variables of each process, pairs of processes that can communicate with each other share fifo queues, through which the communication is handled. The sender does not have to wait for the receiver, unless the message queue it uses is full. Similarly, the receiver does not have to wait for the sender unless there is no message in its input queue.

Send and receive transitions are matching if they share the same communication queue. We will assume that for each queue there is only a single (different) process that can send, and a single process that can receive.

It is evident that matching sends and receives do not satisfy the conditions on the dependency relation from Section 2.2. However, one can weaken condition **D1**, allowing transitions α and β to be independent when executing one cannot disable the other (but can enabled the other, as oppose to condition **D1**). In this case, it is no longer true that when $v \equiv w$ and v is a transitions sequence of a system $\mathcal{I}$, then w is also a a transitions sequence of $\mathcal{I}$. However, a close look at Definition 2 reveals that ample sub-state-spaces do not need to be saturated by a given equivalence relation in order to exploit the equivalence for reduction. This is indeed the case here.

Choosing a subset of the enabled transitions at s that satisfy condition **C3** can be done as follows:

> Choose all the transitions enabled in the current state that belong to a subset $\mathcal{P}$ of the processes, such that
>
> - there is no send transition of a process $\mathcal{P}_i$ in $\mathcal{P}$ that could send a message to a process outside $\mathcal{P}$ if its queue was not full in s.
> - there is no receive transition of a process $\mathcal{P}_i$ in $\mathcal{P}$ that could receive a message from a process outside $\mathcal{P}$ if its queue was not empty in s.

Separate Process Analysis

As explained above, additional knowledge about the future enabledness of transitions allows certifying more subsets as ample sets. As an example, in synchronous communication, we can weaken the requirement that the subset of processes $\mathcal{P}$ does not contain a locally enabled communication transition α, communicating with a processes that is outside $\mathcal{P}$; the existence of such a transition α does not prohibits the enabled transitions of $\mathcal{P}$ from being an ample set if the process $\mathcal{P}_j$ can not participate in such a communication in every state that is reachable from the current one. A similar weakening is possible for the asynchronous communication case.

The future disabledness of a transition from a given state is as hard to check as the model-checking problem itself. Thus, we may be satisfied with a solution that would not identify *every* transition that can no longer become enabled from the current state, but would identify at least a subset of such transitions. This can be done using a *separate process reachability*. In the above example for synchronous communication, we will check whether process $\mathcal{P}_j$ could have reached the matching communication from its current *local state*. This search looks at the process $\mathcal{P}_j$ in isolation. It assumes all transitions that are joint with other processes to be locally enabled by the other processes. Furthermore, we may even choose to ignore data values, reverting to static analysis.

Such a search can be done in a preparatory stage, identifying from each local state 'offending' transitions (which can include synchronous communication transitions, asynchronous communication transitions or use of global variables) that are *not* reachable.

This information can be used then to improve the reduction by identifying more subsets as ample sets.

On-the-fly Reduction

In previous sections, the model-checking algorithm was explained as a two-phase process, where at the first phase, the state-space (or reduced state-space) is constructed, and in the second phase, a graph-theoretic algorithm is applied to it. In practice, many model-checking tools work in a slightly different, more efficient, way. They combine the construction of the graph with checking that it satisfies the specification. Then, it is sometimes possible to identify 'on-the-fly' that the system violates the specification, before completing the construction. We will describe how partial order reduction can be applied while doing on-the-fly model-checking.

Obtaining an on-the-fly model-checking algorithm can be done by using a Büchi automaton A that corresponds to the complement of the specification φ. Namely, A would recognize the sequences that are not allowed by the specification. A translation from LTL formulas to Büchi automata can be found e.g., in [35, 10].

A Büchi automaton is a fivetuple $\langle Q, i, \Sigma, \delta, F \rangle$, where Q is a finite state of *automaton states*, $i \in Q$ is the *initial automaton state*, Σ is a finite set of *input values*, which is in our case 2^P, $\delta \subseteq Q \times \Sigma \times Q$ is a non-deterministic *transition function*, and $F \subseteq Q$ is the set of *accepting states*. A *run* of the automaton A over an infinite sequence $\sigma \in \Sigma^\omega$, where $\sigma = r_0 r_1 r_2 \ldots$ is an infinite sequence of automaton states $q_0 q_1 \ldots$ such that for each $i \geq 0$, $(q_i, r_i, q_{i+1}) \in \delta$. A run is *accepting* iff at least one automaton state from F appears on it infinitely many times.

Verfying that a system $\mathcal{I}$ satisfies a specification φ is thus done by checking whether there are execution sequences of $\mathcal{I}$ that are accepted by runs of A. If there are such sequences, they correspond to counter-examples (since A accept the sequences disallowed by the specification). Otherwise, $\mathcal{I}$ satisfies φ.

To carry out the above task, we can generate the *product automaton* $\mathcal{I} \times A$: the states of the product are pairs from $S \times Q$. We will refer to such pairs simply as *states*. The transitions are pairs from $T \times \delta$. The accepting states are fixed by the automaton state component, i.e., are pairs $\langle s, q \rangle$ such that $q \in F$. The initial state is the pair $\langle \iota, i \rangle$. To make the sequences of $\mathcal{I} \times A$ correspond to runs of A over sequences of $\mathcal{I}$, we make the following correspondence: $\langle s, q \rangle \xrightarrow{(\alpha, \beta)} \langle s', q' \rangle$ is a transition of $\mathcal{I} \times A$ iff (1) $s' = \alpha(s)$, (2) $(q, \beta, q') \in \delta$, and (3) $M(s) = \beta$. The last requirement means that the A transition β agrees with the labeling of the outgoing system state s.

We can now construct $\mathcal{I} \times A$ on-the-fly: from the current pair $\langle s, q \rangle \in S \times Q$, generate all possible transitions $\langle \alpha, \beta \rangle$ that satisfy (1), (2) and (3) above. Better yet, we can employ the partial order reduction and restrict the first component such that $\alpha \in ample(s)$.

The only condition that appears to be problematic is the cycle closing condition C2: the cycles in the product are not necessarily the same as the ones in the reduced state-space for $\mathcal{I}$. However, in [28] it is shown that it is correct to use the cycles of $\mathcal{I} \times A$.

Using Tarjan's DFS algorithm, we can find the maximal strongly connected components of $\mathcal{I} \times \mathcal{A}$. A strongly connected component that is reachable from the initial state and contains an accepting state means that the property φ does not hold for $\mathcal{I}$, and can be used to construct a counter-example.

An even more efficient model-checking procedure is obtained by observing that an accepting run exists iff there is a cycle through a reachable accepting state. The procedure [16, 7] applies an interleaved double DFS procedure: when the first DFS retracts to an accepting state, the second DFS starts searching for a cycle through this state. If the second DFS fails to find a cycle, the first DFS resumes from the point it has stopped. We can use the following bits for every state of the product that is put in the hash table:

- The state was found during the first DFS.
- The state was found during the second DFS.
- The state is in the first DFS stack.
- The state is in the second DFS stack.

Notice that these bits allow information about the two different (virtual) copies of the same state in the two searches. Notice further that there is no need to explicitly store the edges.

Applying the partial order reduction to the improved search requires a subtle change in the algorithm: it is important to guarantee that the second DFS uses the states that were already found in the first DFS. Repeating exactly the same reduction from every state is thus important to achieve this goal. However, notice that when the second search reaches a state that is on the stack of the first DFS, it may continue to search new states that were not encountered yet during the first DFS. But notice that once a state x that is on the stack of the first DFS is reached in the second DFS, the search can terminate: it is guaranteed that there is a path from x to the accepting state from which the second DFS has begun, hence completing a cycle through it. Hence, the algorithm in [16, 7] can be changed as follows [18]:

Upon reaching during the second DFS a state that is on the stack of the first DFS, terminate the search. Use the concatenation of the states in the first and second DFS as a counter-example.

This early termination of the algorithm can be applied to the full search as well and can result in shorter counter-examples.

It is important to repeat the same reduction in both searches. Additional state information can be used to check during the second DFS in which states condition C2 was used in the first DFS. This allows the second DFS to generate the same sets of successors as the first one, as it is not guaranteed that the second DFS would close cycles at the same states as the first one.

The SPIN Implementation

The model-checking tool SPIN [16] contains an implementation of the ample sets method. SPIN allows a variety of communication mechanisms, including synchronous and asynchronous message communication. It also allows global transitions, which change values of variables that belong to all the processes. Hence, the rules to achieve ample sets that satisfy condition C1 are more complicated. SPIN includes the on-the-fly partial order reduction [17], with the double DFS described above [18].

5 Extensions

5.1 Reduction for Branching Properties

We will deal now with reductions that preserve the branching properties of a system, as expressed e.g., using the temporal logic CTL^* [5]. When expressing the branching properties of a system, it is not sufficient to generate a sub-state-space with at least one representative from each equivalence class of an effective equivalence relation. The reason is that branching properties refer to a single tree (or DAG) which represent the *combination* of the execution sequences. Generating a sub-state-space that is ample with respect to a given equivalence relation does not help here. Instead, we use simulation relations.

Let $\mathcal{M}_1 = \langle V_1, E_1, \iota_1, M_1 \rangle$ and $\mathcal{M}_2 = \langle V_2, E_2, \iota_1, M_2 \rangle$ be a pair of structures where the range of M_1 is the same as the range of M_2. Simulations relations $\cong \subseteq V_1 \times V_2$ are used to show that such structures are indistinguishable with respect to some specification language.

Definition 4 ([2]) *A relation* $\cong \subseteq V_1 \times V_2$ *is a* stuttering equivalence *between the states of two structures* $S = \langle V, E, \iota, M \rangle$ *and* $S' = \langle V', E', \iota', M' \rangle$ *if the following conditions hold:*

1. $\iota \cong \iota'$,

2. if $s \cong s'$, *then* $M(s) = M'(s')$ *and for every path* π *of* S *that starts at* s, *there is a path* π' *in* S' *that starts at* s', *a partition* $B_1, B_2 \ldots$ *of* π, *and a partition* $B'_1, B'_2 \ldots$ *of* π' *such that for each* $j \geq 0$, B_j *and* B'_j *are nonempty and finite, and every state in* B_j *is related by* $\cong$ *to every state in* B'_j, *and*

3. the same condition as (2) where S, π *and* s *are interchanged with* S', π' *and* s'.

In [2], it was proved that one cannot distinguish between two branching structures using any CTL^* formula iff these two structures are related by the simulation relation of *bisimilarity*[25]. Another variant of this theorem, also proved in [2], is even more useful for establishing a reduction; it connects the distinguishing power of the above temporal logics, when restricted from using the nexttime operator X, with a stuttering version of bisimulation:

Theorem 3 ([2]) *Let φ be a CTL^* formula with the set of atomic propositions P. Let S_1 and S_2 be two models, where the range of the labeling function M_1 and M_2 is the subsets of atomic propositions P. Let the relation $\cong$ be a stuttering equivalence between the states of S_1 and S_2. Then for every pair of stuttering bisimilar states $s_1 \cong s_2$ it holds that $S_1, s_1 \models \varphi$ iff $S_2, s_2 \models \varphi$.*

It is thus useful to find a set of conditions for ample sets that will guarantee that the generated sub-structure is stuttering bisimilar to the full state-space. The following condition C4, when added to the previous C0–C3, is sufficient, as shown in [9]:

C4 [Singleton condition [9]] $E(q)$ contains either all transitions enabled in state q, or exactly one transition.

Theorem 4 ([9]) *The structure $S = \langle V, E, \iota, M \rangle$ obtained by applying a reduced DFS to an interpreted system $\mathcal{I}$, with ample sets satisfying the conditions C0-C4 is stuttering bisimilar to the full state-space of $\mathcal{I}$.*

5.2 Verifying the Partial Order Reduction Algorithm

Due to the fact that the partial order reduction algorithm has a non-trivial correctness proof, it was mechanically verified by Chou and Peled [4] using the HOL proof system [15]. HOL, which stands for Higher Order Logic, allows the human assisted verification of complicated theorems. The verification effort, consisting of one of the largest such efforts to date, consists of over 7000 lines of HOL code, which translates to over 250,000 simple inferences. It was done over a period of 10 weeks.

The verified version of the algorithm is the F-fair case, consisting of conditions C0–C2. Hence, the verification consisted of proving Theorem 1. The "purist" approach adopted in the proof assumed no axioms: theorems for trace theory, finite and infinite sequences, state-spaces and other auxiliary theorems where defined and proved using type-theory primitives. As a side effect of the verification, the partial order reduction was shown also to be applicable also to search strategies different than DFS, such as Breadth First Search (BFS).

Acknowledgement I would like to thank the people that I had the pleasure of working with on various aspects of partial order verification: Rajeev Alur, Ching-Tsun Chou, Rob Gerth, Patrice Godefroid, Gerard Holzmann, Shmuel Katz, Ruurd Kuiper, Wojciech Penczek, Amir Pnueli, Mark Staskauskas, Thomas Wilke, Pierre Wolper and Mihalis Yannakakis. I would like to thank Bob Kurshan for many illuminating discussions.

References

[1] R. Alur, D. Peled, W. Penczek, Model-Checking of Causality Properties, *10th Symposium on Logic in Computer Science*, IEEE, 1995, San Diego, California, USA, 90–100.

[2] M.C. Browne, E.M. Clarke, O. Grümberg, Characterizing Finite Kripke Structures in Propositional Temporal Logic, *Theoretical Computer Science* 59 (1988), Elsevier, 115–131.

[3] J.R. Burch, E.M. Clarke, K.L. McMillan, D.L. Dill, J. Hwang, Symbolic model checking: 10^{20} states and beyond, *5th Annual IEEE* Symposium on Logic in Computer Science, 1990, 428–439.

[4] C.T. Chou, D. Peled, Verifying a Model-Checking Algorithm, *Tools and Algorithms for the Construction and Analysis of Systems*, LNCS 1055, Springer-Verlag, 1996, Passau, Germany, to appear March 1996, 241–257.

[5] E.M. Clarke, E.A. Emerson, and A.P. Sistla, Automatic verification of finite-state concurrent systems using temporal-logic specifications, *ACM Transactions on Programming Languages and Systems*, 8(1986), 244–263.

[6] E.M. Clarke, A.P. Sistla, Symmetry and model checking, *5th International Conference on Computer-Aided Verification*, 1993.

[7] C. Courcoubetis, M. Vardi, P. Wolper, M, Yannakakis, Memory-efficient algorithms for the verification of temporal properties, Formal methods in system design 1 (1992) 275–288.

[8] V. Diekert, P. Gastin, A. Petit, Rational and Recognizable Trace Languages, *Information and Computation*, 116 (1995), 134–153.

[9] R. Gerth, R. Kuiper, W. Penczek, D. Peled, A Partial Order Approach to Branching Time Logic Model Checking, ISTCS'95, *3rd Israel Symposium on Theory on Computing and Systems*, IEEE press, 1995, Tel Aviv, Israel, 130-139.

[10] R. Gerth, D. Peled, M.Y. Vardi, P. Wolper, Simple On-the-fly Automatic Verification of Linear Temporal Logic, *PSTV95, Protocol Specification Testing and Verification*, 3-18, Chapman & Hall, 1995, Warsaw, Poland.

[11] P. Godefroid. Using partial orders to improve automatic verification methods. In Proc. *2nd Workshop on Computer Aided Verification*, LNCS 531, Springer-Verlag, New Brunswick, NJ, 1990, 176–185.

[12] P. Godefroid, D. Pirottin, Refining dependencies improves partial order verification methods, *5th Conference on Computer Aided Verification*, LNCS 697, Elounda, Greece, 1993, 438–449.

[13] P. Godefroid, D. Peled, M. Staskauskas, Using Partial Order Methods in the Formal Validation of Industrial Concurrent Programs, 1996, ISSTA'96, *International Symposium on Software Testing and Analysis*, ACM Press, San Diego, California, USA, 261-269.

[14] P. Godefroid, P. Wolper, A Partial Approach to Model Checking, *6th Annual IEEE Symposium on Logic in Computer Science*, 1991, Amsterdam, 406–415.

[15] M.J.C. Gordon, T.F. Melham, *Introduction to HOL: A Theorem-Proving Environment for Higher-Order Logic*, Cambridge University Press, 1993.

[16] G. J. Holzmann, *Design and Validation of Computer Protocols*, Prentice Hall Software Series, 1992.

[17] G.J. Holzmann, D. Peled, An Improvement in Formal Verification, *7th International Conference on Formal Description Techniques*, Berne, Switzerland, 1994, 177–194.

[18] G.J. Holzmann, D. Peled, M. Yannakakis, On Nested Depth First Search, in preparation.

[19] S. Katz, D. Peled, Verification of Distributed Programs using Representative Interleaving Sequences, *Distributed Computing* 6 (1992), 107–120. A preliminary version appeared in Temporal Logic in Specification, UK, 1987, LNCS 398, 21–43.

[20] S. Katz, D. Peled, Defining conditional independence using collapses, Theoretical Computer Science 101 (1992), 337-359, a preliminary version appeared in *BCS–FACS Workshop on Semantics for Concurrency*, Leicester, England, July 1990, Springer, 262-280.

[21] M. Z. Kwiatkowska, Event Fairness and Non-Interleaving Concurrency, *Formal Aspects of Computing* 1 (1989), 213-228.

[22] L. Lamport, What good is temporal logic, *Information Processing 83*, Elsevier Science Publishers, 1983, 657-668.

[23] O. Lichtenstein, A. Pnueli, Checking that finite-state concurrent programs satisfy their linear specification, 11 th Annual ACM Symposium on Principles of Programming Languages, 1984, 97-107.

[24] A. Mazurkiewicz, Trace Theory, *Advances in Petri Nets 1986*, Bad Honnef, Germany, LNCS 255, Springer, 1987, 279-324.

[25] R. Milner, *A Calculus of Communicating System*, LNCS, Springer-Verlag, 92.

[26] D. Peled, A. Pnueli, Proving Partial Order Properties, *Theoretical Computer Science*, 126(1994), 143-182.

[27] D. Peled, All from one, one for all, on model-checking using representatives, 5 th Conference on Computer Aided Verification, Greece, 1993, LNCS, Springer, 409-423.

[28] D. Peled. Combining partial order reductions with on-the-fly model-checking. *Formal Methods in System Design* 8 (1996), 39-64.

[29] D. Peled, Th. Wilke, P. Wolper, An Algorithmic Approach for Checking Closure Properties of ω-Regular Languages, submitted.

[30] A. Pnueli, The temporal logic of programs, *18th FOCS, IEEE Symposium on Foundation of Computer Science*, 1977, 46-57.

[31] A. Valmari, Stubborn sets for reduced state space generation, 10 th International Conference on Application and Theory of Petri Nets, Vol. 2, Bonn, Germany, 1989, 1-22.

[32] A. Valmari, A stubborn attack on state explosion. *Formal Methods in System Design*, 1 (1992), 297-322.

[33] A.P. Sistla, M.Y. Vardi, P. Wolper, The Complementation Problem for Büchi Automata with Applications to Temporal Logic, *Theoretical Computer Science*, 49 (1987), 217—237.

[34] P.S. Thiagarajan, A Trace Based Extension of Linear Time Temporal Logic. *Proc. 10th IEEE Conference on Logic In Computer Science*, 1994, 438-447.

[35] M.Y. Vardi, P. Wolper, An automata-theoretic approach to automatic program verification, *1st Annual IEEE Symposium on Logic in Computer Science*, 1986, Cambridge, England, 322-331.

Nonmonotonic Rule Systems: Forward Chaining, Constraints, and Complexity

Jeffrey B. Remmel

Department of Mathematics
University of California at San Diego
La Jolla, CA 92093
and
Sagent Corporation
11201 SE 8th Street, Bldg J., Suite 140
Bellevue, WA 98004

Abstract. Nonmontonic rule systems are simple algebraic structures which capture all the essential features of many nonmonotonic reasoning systems including Reiter's default logic, Moore's autoepistemic logic, nonmonotonic modal logics of McDermott and Doyle, the stable semantics of general logic programs, and the answer set semantics for logic programs with classical negation. Because there are easy translations between all these nonmonotonic rule systems and nonmonotonic rules systems, theorems proved about nonmonotonic rule systems apply to all of the formalisms above. We will survey some of the recent work of Marek, Nerode, and Remmel on various complexity issues in nonmonotonic reasoning and how to construct a generalization of Reiter's normal default logic in nonmontonic rule systems. We will also discuss an extension of nonmontonic rule systems due to Marek, Nerode, and Remmel in which we incorporate arbitrary constraints into nonmonotonic reasoning formalisms and some recent work of Pollett and Remmel on succinctness hierarchies of default theoreies and general logic programs with Boolean quantified constraints.

Mind the Gap!
Abstract Versus Concrete Models of Specifications

Donald Sannella*   Andrzej Tarlecki[†]

Abstract

In the theory of algebraic specifications, many-sorted algebras are used to model programs: the representation of data is arbitrary and operations are modelled as ordinary functions. The theory that underlies the formal development of programs from specifications takes advantage of the many useful properties that these models enjoy.

The models that underlie the semantics of programming languages are different. For example, the semantics of Standard ML uses rather concrete models, where data values are represented as closed constructor terms and functions are represented as "closures". The properties of these models are quite different from those of many-sorted algebras.

This discrepancy brings into question the applicability of the theory of specification and formal program development in the context of a concrete programming language, as has been attempted in the Extended ML framework for the formal development of Standard ML programs. This paper is a preliminary study of the difference between abstract and concrete models of specifications, inspired by the kind of concrete models used in the semantics of Standard ML, in an attempt to determine the consequences of the discrepancy.

1   Introduction

The starting point for work on algebraic specification is the use of *many-sorted algebras* as models of programs. Thus the representation of data values is arbitrary and operations on data are modelled as ordinary set-theoretic functions. This is a natural choice since the primary aim of this work is to provide foundations for the development of programs that are *correct* with respect to a given specification of requirements. Correctness is a property of the input/output behaviour of a program, and in studying correctness it is convenient to abstract away from all other aspects of program behaviour. Models of programs more complicated than many-sorted algebras have

*Laboratory for Foundations of Computer Science, Edinburgh University, Edinburgh, Scotland; e-mail dts@dcs.ed.ac.uk. This research was supported by the EC-funded COMPASS Basic Research working group and MeDiCiS Scientific Cooperation Network, by EPSRC grants GR/J07303 and GR/H73103 and by an EPSRC Advanced Fellowship.

†Institute of Informatics, Warsaw University, and Institute of Computer Science, Polish Academy of Sciences, Warsaw, Poland; e-mail tarlecki@mimuw.edu.pl. This research was supported by the EC-funded COMPASS Basic Research working group and MeDiCiS Scientific Cooperation Network, and by EPSRC grant GR/J07303.

been used to deal with advanced features of programming languages (higher-order functions, infinite behaviour, etc.) but the idea is still to abstract away from details of code and algorithms insofar as this is possible.

There is a well-developed theory of algebraic specification which provides syntax for specifications, a formalization of concepts involved in going from specifications to programs by stepwise refinement, methods for proving correctness of refinement steps, etc. Some recent references are [Wir90], [BKLOS91], [LEW96], [ST95] and [ST9?]. This theory takes full advantage of the properties of many-sorted algebras and uses various standard constructions (quotient, free extension, etc.) for building new algebras by adding to and/or combining others. It is mature enough to be applied to the specification and formal development of programs written using "pure" fragments of programming languages.

Of course, programs that have been formally developed are entirely useless unless they are expressed in some implemented programming language. Since the models that underlie the semantics of programming languages are often rather different from many-sorted algebras, there is a potential problem in ensuring that arguments about correctness in the theory of formal development are valid for the semantics of the programming language.

A concrete instance of this problem arises in our work on specification and formal development of Standard ML (SML) programs [Pau91] using the Extended ML (EML) framework [ST89], [San91], [KST95]. SML has an operational semantics [MTH90] which uses rather concrete models of programs, with data represented as closed constructor terms and functions represented as *closures* (a closure is a λ-term together with an environment binding its free variables). Although it is not difficult to map these concrete models to many-sorted algebras, there is *a priori* no guarantee that this map would preserve and/or reflect important properties of algebras, There are concrete models that do not correspond to any SML program, as well as many algebras that do not correspond to any concrete model, so one question concerns closure of these classes under various algebraic constructions. It is by no means obvious that the concrete models enjoy the particular properties of algebras that are used to explain why formal development in EML yields correct programs.

The situation in SML/EML is an example of a possibly worrying discrepancy between the abstract models used to reason about correctness and the more or less concrete models used to give semantics to programming languages. It seems necessary to check that the process of abstraction has not led us to make false conclusions about the underlying reality. This paper attempts to study this issue, looking to SML and EML for inspiration. This is an awkward thing to study because of the complexity of real programming languages (and SML is no exception) so that some degree of abstraction is unavoidable in the study itself. We therefore proceed by analogy, studying what happens when computability restrictions inspired by those arising in SML are placed on algebras. Even when we refer to SML, we will be considering a "pure" fragment which excludes complications like real numbers, input/output, and polymorphic types. We argue that the analogy is strong enough so that the answers we obtain are valid for SML.

This issue is by no means unique to the specification and formal development of programs using algebraic specifications. The same question arises for any theory of specification and formal development, and more generally whenever an abstract view is taken of a real system, whether it is a set of differential equations describing the forces acting on a bridge or a mathematical model of program behaviour.

Much of the paper is devoted to the development of a satisfactory notion of "algebra coded by a program" and an investigation of its properties. We begin with *modest algebras*: partial algebras that come equipped with a computable implementation using natural number encodings of data (Sect. 2). A special case is the *A0-extensions*, modest algebras in which the implementation of certain types is fixed (Sect. 3), which appear to capture SML-codable algebras closely enough for the purposes of this paper. The meaning of formulae in modest algebras is defined in Sect. 4, and then in Sect. 5 we explore which constructions on partial algebras carry over to modest algebras. In Sect. 7 we consider the question of which partial algebras can be given the structure of a modest algebra. We proceed to define a simple specification language (Sect. 8) which can be interpreted in various institutions of partial and modest algebras (cf. Sect. 6) and then finally examine the gap between the interpretation of specifications using partial algebras and using modest algebras or modest *A0*-extensions (Sect. 9). Our conclusion is that the discrepancy has no alarming consequences, but the analysis reveals some interesting choices which influence the completeness and flexibility of the formal program development process (Sect. 10).

2 Modest algebras

This section is devoted to the search for a formalization of the concept of "algebra coded by a program". Since this is sensitive to (at least) what we mean by "program", we cannot give a definitive answer. The notion of *modest A0-extension*, presented in the next section, is an approximation that seems to be adequate for our purposes.

The reader is assumed to be familiar with usual notions of many-sorted signature, signature morphism, many-sorted algebra, homomorphism, reduct of a Σ'-algebra A' along a signature morphism $\sigma : \Sigma \to \Sigma'$ to yield the Σ-algebra $A'|_\sigma$ (written $A'|_\Sigma$ when σ is an inclusion) and similarly for Σ'-homomorphisms, etc.; see e.g. [Wir90]. We use ordinary many-sorted signatures with first-order operation symbols, but restrict to signatures having a countable number of operation symbols. Call the category of such signatures **AlgSig**, and let $\Sigma = \langle S, \Omega \rangle$ be a signature in $|\mathbf{AlgSig}|$. We use partial algebras having countable carrier sets, and so-called *strong* homomorphisms between such algebras, which preserve *and reflect* definedness of operations. The category of such Σ-algebras and Σ-homomorphisms is called **PAlg**(Σ). We also use *strong* congruences on partial algebras, which are closed under application of operations and preserve and reflect their definedness. Throughout, we use the arrow $\to$ for total functions (and in types of operation symbols) and the arrow $\rightharpoonup$ for partial functions.

A modest algebra is a partial algebra equipped with a computable "implementation": its data values are encoded as natural numbers, and its operations are mirrored by partial recursive functions over these encodings.

Definition 2.1 *A* modest Σ-algebra $\mathcal{A}$ *is a partial Σ-algebra A together with:*

- *a recursive set $\bar{s} \subseteq \mathbb{N}$ of codes for each $s \in S$;*

- *a total surjective decoding function $[\cdot]_s : \bar{s} \to |A|_s$ for each $s \in S$; and*

- a partial recursive tracking function $\bar{f} : \bar{s}_1 \times \cdots \times \bar{s}_n \rightharpoonup \bar{s}$ for each $f : s_1 \times \cdots \times s_n \rightharpoonup s$ in Ω

such that for any such f and $m_1 \in \bar{s}_1, \ldots, m_n \in \bar{s}_n$, $\bar{f}(m_1, \ldots, m_n)$ is defined iff $f_A([m_1]_{s_1}, \ldots, [m_n]_{s_n})$ is defined and then $[\bar{f}(m_1, \ldots, m_n)]_s = f_A([m_1]_{s_1}, \ldots, [m_n]_{s_n})$. We drop subscripts from decoding functions when they are determined by the context. We write $|A|$ for A and say that $\mathcal{A}$ is over A, and we write $\bar{s}_A$, $[\cdot]_A$ and $\bar{f}_A$ when $\mathcal{A}$ is not obvious.

This definition is intended to reflect the way that programs in a language like Standard ML "code" algebras. Think of each sort s as an SML type, with the codes in $\bar{s}$ being Gödel-style encodings of constructor terms of this type. If s is a function type, the codes in $\bar{s}$ are encodings of λ-terms. The requirement that $\bar{s}$ be recursive stems from the fact that typechecking in SML is decidable: given an SML term, we can decide whether or not it is a constructor term (or λ-term) of a given type over a given set of constructors. The tracking functions need to be partial recursive to make them SML-implementable. For any built-in type or user-defined concrete data type s whose definition does not involve abstract types or function types (in SML parlance, an "eqtype"), $[\cdot]$ is a bijection and so equality in $|A|_s$ corresponds to identity of constructor terms of type s. SML then provides a function $= : s \times s \rightarrow \text{bool}$ that decides the equality. But for a function type $s \rightarrow t$, if the interpretation taken in $|A|_{s \rightarrow t}$ is the usual set-theoretic function space (restricted to denotable functions) then $[\cdot]_{s \rightarrow t} : \overline{s \rightarrow t} \rightarrow |A|_{s \rightarrow t}$ is in general no longer a bijection since it maps extensionally equal λ-terms to the same set-theoretic function. The *kernel* of $[\cdot]_{s \rightarrow t}$, $\equiv_{[\cdot]_{s \rightarrow t}} =_{\text{def}} \{\langle m, m' \rangle \in \overline{s \rightarrow t} \times \overline{s \rightarrow t} \mid [m] = [m']\}$, is typically not even semi-decidable.

The term "modest algebra" comes from the term "modest set" [Ros90] where however the requirement that the sets of codes be recursive is absent. The function space used there would not satisfy such an assumption even if its domain and range did, since it is not decidable whether or not a given partial recursive function on $\mathbb{N}$ is a function between two given recursive sets. As explained above, we think of (codes of) statically well-typed λ-terms as codes for values of the function space.

The restriction to countable algebras is forced by the definition of modest algebra: obviously, uncountable carrier sets cannot be encoded using natural numbers (although see [SHT95] for a theory of computable approximations of uncountable algebras). Some other simple properties of modest algebras follow directly from the definition. We start with an alternative formulation of the definition itself.

Proposition 2.2 *The families* $\bar{S} = \langle \text{recursive } \bar{s} \subseteq \mathbb{N} \rangle_{s \in S}$, $[\cdot] = \langle [\cdot]_s : \bar{s} \rightarrow |A|_s \rangle_{s \in S}$ *and* $\bar{\Omega} = \langle \text{partial recursive } \bar{f} : \bar{s}_1 \times \cdots \times \bar{s}_n \rightharpoonup \bar{s} \rangle_{f : s_1 \times \cdots \times s_n \rightarrow s \in \Omega}$ *form a modest Σ-algebra $\mathcal{A}$ over a partial Σ-algebra A iff $[\cdot] : \bar{A} \rightarrow A$ is a surjective Σ-homomorphism, where $\bar{A}$ is the partial Σ-algebra with carriers $\bar{S}$ and operations $\bar{\Omega}$.* □

This means that we can view any modest algebra $\mathcal{A}$ as a surjective homomorphism from $\bar{A}$ to $|A|$, and vice versa. It also says that one may view $\bar{A}$ as a concrete implementation of $|A|$, with $[\cdot]$ as the so-called "abstraction function" [Hoa72].

Corollary 2.3 *Let $\bar{A}$ together with $[\cdot]$ form a modest Σ-algebra over A. If $j : A \rightarrow B$ is a surjective Σ-homomorphism then $\bar{A}$ together with $j \circ [\cdot] : \bar{A} \rightarrow B$ forms a modest Σ-algebra over B.* □

Proposition 2.4 *Let A be a partial Σ-algebra and $\bar{A}$ be as above. There is a surjective Σ-homomorphism $[\cdot] : \bar{A} \to A$ (i.e. a modest Σ-algebra over A with codes $\bar{A}$ and decoding function $[\cdot]$) iff there is a Σ-congruence $\approx$ on $\bar{A}$ such that $A \cong \bar{A}/\approx$.* ☐

Proposition 2.5 *Let $\bar{A}$ and $[\cdot] : \bar{A} \to A$ be as in Prop. 2.2, but with $\bar{S} = \langle r.e. \ \bar{s} \subseteq \mathbb{N}\rangle_{s \in S}$. Then there is a modest Σ-algebra over A.*

PROOF: *Since each $\bar{s}$ is r.e. there is a recursive bijection $e_s : \bar{s}' \to \bar{s}$ where $\bar{s}' \subseteq \mathbb{N}$ is recursive. Define the decoding functions $[\cdot]' = \langle[\cdot]'_s : \bar{s}' \to |A|_s\rangle_{s \in S}$ by $[\cdot]'_s = [\cdot]_s \circ e_s$, and the tracking functions $\bar{\Omega}' = \langle \bar{f}'\rangle_{f : s_1 \times \cdots \times s_n \to s \in \Omega}$ by $\bar{f}'(m'_1, \ldots, m'_n) = e_s^{-1}(\bar{f}(e_{s_1}(m'_1), \ldots, e_{s_n}(m'_n)))$ for $m'_1 \in \bar{s}'_1, \ldots, m'_n \in \bar{s}'_n$. It is easy to see that $\bar{S}' = \langle \bar{s}'\rangle_{s \in S}$, $\bar{\Omega}'$ and $[\cdot]'$ form a modest Σ-algebra over A.* ☐

Definition 2.6 *A modest Σ-homomorphism $h : A \to B$ is a Σ-homomorphism between the underlying partial algebras $|h| : |A| \to |B|$ together with a family of partial recursive tracking functions $\langle \bar{h}_s : \mathbb{N} \rightharpoonup \mathbb{N}\rangle_{s \in S}$ such that for each $s \in S$ and $m \in \bar{s}_A$, $\bar{h}_s(m)$ is defined, $\bar{h}_s(m) \in \bar{s}_B$, and $[\bar{h}_s(m)]_B = |h|_s([m]_A)$. We say that h is over $|h|$. Modest Σ-algebras and modest Σ-homomorphisms (with the obvious composition) form a category called $\mathbf{MAlg}(\Sigma)$, and $|\cdot| : \mathbf{MAlg}(\Sigma) \to \mathbf{PAlg}(\Sigma)$ is a functor.*

Proposition 2.7 *A modest Σ-isomorphism $h : A \to B$ (i.e. an isomorphism in $\mathbf{MAlg}(\Sigma)$) is an isomorphism $|h| : |A| \to |B|$ in $\mathbf{PAlg}(\Sigma)$ together with a (total) recursive S-sorted bijection on $\mathbb{N}$ that tracks $|h|$.* ☐

The usual definitions of reduct of a partial Σ'-algebra and Σ'-homomorphism along a signature morphism $\sigma : \Sigma \to \Sigma'$ extend to modest Σ'-algebras and their homomorphisms, e.g. using the formulation of modest algebras given by Prop. 2.2: the σ-reduct of $[\cdot] : \bar{A}' \to A'$ is $[\cdot]|_\sigma : \bar{A}'|_\sigma \to A'|_\sigma$. We write $A'|_\sigma$, or $A'|_\Sigma$ when σ is an inclusion. This, together with the category $\mathbf{MAlg}(\Sigma)$ for each signature Σ, gives a contravariant functor $\mathbf{MAlg} : \mathbf{AlgSig}^{op} \to \mathbf{Cat}$.

There is a well-developed theory of computable algebra, beginning with the work of Mal'cev [Mal61] (cf. [Rab60]) and including many papers on the expressive power of algebraic specification methods by Bergstra and Tucker, see e.g. [BT87]. For a recent overview see [SHT95]. Although the *total* modest algebras are exactly the so-called "effectively numbered algebras" of [SHT95], there has been little attention paid to the counterpart for partial algebras. Furthermore, it is common to impose computability restrictions on the decoding functions (e.g. the "computably numbered algebras" of [SHT95] are total modest algebras such that the kernel $\equiv_{[\cdot]_s}$ is decidable for all sorts $s \in S$); these apply to *all* sorts while we have seen that in the SML context such restrictions are appropriate for some sorts (for example, those corresponding to eqtypes) but not for others (for example, those corresponding to function types). Another difference is that while [SHT95] concentrates on *modestizable* algebras, where the encoding structure remains implicit (see Sect. 7 below), we deal with *modest* algebras, which contain the encoding structure explicitly.

3 Modest $A0$-extensions

Despite the comments above concerning the relationship between modest algebras and SML, modest algebras do not capture only SML-codable algebras.

Counterexample 3.1 *Let A be a two-sorted algebra with carriers $|A|_s = |A|_t = \mathbf{N}$ such that $f_A : |A|_s \to |A|_t$ is a total non-computable bijection and f is the only operation symbol. A modest algebra over A is given by taking $\bar{s} = \bar{t} = \mathbf{N}$, $[\cdot]_s = f^{-1}$ and $[\cdot]_t$ and the tracking function $\bar{f}$ to be the identity.* □

The question of which partial algebras can and cannot be given the structure of a modest algebra will be treated in Sect. 7.

A further reason for the mismatch is that SML insists on a particular interpretation of certain types, the so-called "pervasive" types like **string** and **bool** whose implementation is fixed by the system. Suppose that $\Sigma0 = \langle S0, \Omega0 \rangle$ is a signature containing the sorts and operation symbols whose interpretation we want to fix as the one given by a particular modest $\Sigma0$-algebra $\mathcal{A}0$. We assume that $\equiv_{[\cdot]_s}$ is decidable for all $s \in S0$; then there is no loss of generality in assuming that $[\cdot]_s$ is a bijection for all $s \in S0$ so we make this assumption. We assume that $\Sigma0$ contains at least the sorts *bool* (Booleans) and *nat* (natural numbers) with the usual operations $(true, false :\to bool, 0 :\to nat, + : nat \times nat \to nat$, etc.) and that the interpretations of these in $|\mathcal{A}0|$ are as usual. It follows that all values in $|\mathcal{A}0|$ are $\Sigma0$-reachable. We restrict attention to signatures Σ that extend $\Sigma0$. Let $\mathbf{AlgSig}_{\Sigma0}$ be the category of signatures extending $\Sigma0$, with signature morphisms that are the identity on $\Sigma0$.

Definition 3.2 *Let Σ be a signature extending $\Sigma0$. A modest Σ-algebra $\mathcal{A}$ extends $\mathcal{A}0$ ($\mathcal{A}$ is a modest $\mathcal{A}0$-extension) if there is a modest $\Sigma0$-isomorphism $h : \mathcal{A}|_{\Sigma0} \to \mathcal{A}0$ such that $\bar{h}$ is the $S0$-sorted identity function on $\mathbf{N}$.*

Note that this fixes the *concrete implementation* of $\Sigma0$ while fixing its interpretation on the *abstract* level only up to isomorphism: if $\mathcal{A}$ is an $\mathcal{A}0$-extension then $\bar{\mathcal{A}}|_{\Sigma0} = \overline{\mathcal{A}0}$ while $|\mathcal{A}|_{\Sigma0}| \cong |\mathcal{A}0|$.

If $\mathcal{A}0$ is the built-in implementation of the pervasive types of SML, then for the purposes of this paper we will identify SML-codable algebras with modest $\mathcal{A}0$-extensions. The correspondence is not exact — there are modest $\mathcal{A}0$-extensions that are not SML-codable, since we have placed no computability restrictions on the kernel $\equiv_{[\cdot]_s}$ for $s \notin S0$ — but because of the complexity of the semantics of SML [MTH90], capturing exactly the class of SML-codable algebras (and proving that we have done so) would be very difficult. However, the match is close enough that the ideas and results in the sequel should apply to SML. On one hand, the class of modest $\mathcal{A}0$-extensions is small enough to expose some problems when compared with the class of partial algebras (see Sects. 7 and 9 below); on the other hand it is large enough to cover the SML-codable algebras. The reader is encouraged to check that the class of SML programs is closed under the main constructions on modest $\mathcal{A}0$-extensions we consider: reduct, amalgamation and definitional extension, cf. Sect. 5.

Definition 3.3 *Let Σ be a signature extending $\Sigma0$. $\mathbf{MAlg}_{\mathcal{A}0}(\Sigma)$ is the subcategory of $\mathbf{MAlg}(\Sigma)$ with modest $\mathcal{A}0$-extensions as objects, and as morphisms modest Σ-homomorphisms h such that $h|_{\Sigma0}$ is tracked by the $S0$-sorted identity function. This extends to a contravariant functor $\mathbf{MAlg}_{\mathcal{A}0} : \mathbf{AlgSig}_{\Sigma0}^{op} \to \mathbf{Cat}$.*

At the abstract level of partial algebras we will similarly concentrate on extensions (up to isomorphism) of the abstract interpretation of built-in sorts and operations given by $A0 =_{\text{def}} |\mathcal{A}0|$.

Definition 3.4 *Let* $\Sigma \in |\mathbf{AlgSig}_{\Sigma 0}|$ *be a signature extending* $\Sigma 0$. *A partial* Σ-*algebra extends* $A0$ *if* $A|_{\Sigma 0} \cong A0$. $\mathbf{PAlg}_{A0}(\Sigma)$ *is the subcategory of* $\mathbf{PAlg}(\Sigma)$ *with partial* $A0$-*extensions as objects, and as morphisms all* Σ-*homomorphisms* h *such that* $h|_{\Sigma 0}$ *is a* $\Sigma 0$-*isomorphism. This extends to a contravariant functor* $\mathbf{PAlg}_{A0} : \mathbf{AlgSig}_{\Sigma 0}^{op} \to \mathbf{Cat}$.

Proposition 3.5 *For any signature* $\Sigma \in |\mathbf{AlgSig}_{\Sigma 0}|$ *extending* $\Sigma 0$, $|\cdot| : \mathbf{MAlg}(\Sigma) \to \mathbf{PAlg}(\Sigma)$ *from Def. 2.6 restricts to a functor* $|\cdot| : \mathbf{MAlg}_{A0}(\Sigma) \to \mathbf{PAlg}_{A0}(\Sigma)$. $\qquad\square$

4 Terms and formulae in modest algebras

Let X be an S-sorted set of variables. The (total) algebra $T_\Sigma(X)$ of Σ-terms with variables in X and the value t_A^v of a term t in a partial Σ-algebra A under a (total) valuation $v : X \to |A|$ are as usual. With evaluation of Σ-terms in a modest Σ-algebra $\mathcal{A}$, we have a choice between evaluation in the underlying partial algebra or using the tracking functions. The latter corresponds to ordinary evaluation in the partial algebra $\bar{A}$ given by Prop. 2.2.

Definition 4.1 *A valuation of variables* X *in* $\mathcal{A}$ *is an* S-*sorted (total) function* $v = \langle v_s : X_s \to \bar{s} \rangle_{s \in S}$. *The (extensional) value of* t *under* v *in* $\mathcal{A}$, *written* $t_{\mathcal{A}}^v$, *is* $t_{|A|}^{[\cdot] \circ v}$. *If* t *is ground (does not contain any variables) then we can write* $t_{\mathcal{A}}$.

We use formulae of first-order logic where the atomic formulae are definedness formulae and (strong) equations. Satisfaction of Σ-formulae in a modest Σ-algebra $\mathcal{A}$ may be defined either on the level of the underlying partial algebra or on the level of the tracking functions; the latter corresponds to satisfaction in $\bar{A}$.

Definition 4.2 *Let* φ *be a formula with free variables in* X *and let* v *be a valuation of* X *in* $\mathcal{A}$. *We define (extensional) satisfaction of* φ *by* $\mathcal{A}$ *under* v, *written* $\mathcal{A} \models_v \varphi$, *by induction on the structure of* φ. *Here are the cases for atomic formulae:*

$\mathcal{A} \models_v D(t)$ *iff* $t_{\mathcal{A}}^v$ *is defined*

$\mathcal{A} \models_v t = u$ *iff* $t_{\mathcal{A}}^v, u_{\mathcal{A}}^v$ *are both undefined, or are both defined and equal*

Thus $\mathcal{A} \models_v \varphi$ *iff* $|A| \models_{[\cdot] \circ v} \varphi$ *in the usual sense. As usual,* $\mathcal{A} \models \varphi$ *iff* $\mathcal{A} \models_v \varphi$ *for all valuations* v.

5 Constructions on modest algebras

In this section we will discuss some constructions on modest algebras (reduct, definitional extension, and amalgamation) that are used in the formal development process to be presented in Sect. 10. We will also have a look at some other standard constructions (quotient and subalgebra). Although these do not appear explicitly in our formalization of the development process, they are taken for granted in work on algebraic specification. Some of these standard constructions carry over easily to modest algebras; others carry over only under certain additional conditions.

We have already seen that the reduct of a partial Σ'-algebra and Σ'-homomorphism along a signature morphism $\sigma : \Sigma \to \Sigma'$ extend to modest Σ'-algebras and modest Σ'-homomorphisms. Likewise for modest $A0$-extensions and their homomorphisms, provided that σ is the identity on $\Sigma 0$.

Definition 5.1 *Let $\approx$ be a congruence on $|\mathcal{A}|$, where $\mathcal{A}$ is given by the surjective homomorphism $[\cdot] : \bar{A} \to |\mathcal{A}|$. The quotient of $\mathcal{A}$ by $\approx$, written $\mathcal{A}/\approx$, is the modest Σ-algebra given by the surjective homomorphism $[\cdot]_\approx \circ [\cdot] : \bar{A} \to |\mathcal{A}|/\approx$.*

Proposition 5.2 *If $\mathcal{A}$ extends $\mathcal{A}0$ and $\approx_s$ is the identity for all $s \in S0$ then $\mathcal{A}/\approx$ extends $\mathcal{A}0$.* $\quad\square$

Definition 5.3 *$\mathcal{B}$ is a modest subalgebra of a modest Σ-algebra $\mathcal{A}$ if: $|\mathcal{B}|$ is a subalgebra of $|\mathcal{A}|$; $\bar{B}$ is a subalgebra of $\bar{A}$; and $([\cdot]_\mathcal{B})_s = ([\cdot]_\mathcal{A})_s \restriction \bar{s}_\mathcal{B}$ for all $s \in S$. A modest Σ-algebra $\mathcal{A}$ is reachable if for any $s \in S$ and $m \in \bar{s}$, there is a ground Σ-term t such that $t_\mathcal{A} = m$.*

Proposition 5.4 *If $\mathcal{A}$ is reachable then $\mathcal{A}$ has no proper modest subalgebra.* $\quad\square$

For partial Σ-algebras, the converse of Prop. 5.4 holds as well, but this does not extend to modest Σ-algebras as the following counterexample shows.

Counterexample 5.5 *Let Σ have sorts s and s' and operation symbols $0 : \to s$, $succ : s \to s$, $f : s \to s'$ and $g : s' \times s \to s'$. For any two r.e. sets $X, Y \subseteq \mathbf{N}$, let $A_{X,Y}$ be the Σ-algebra such that $|A|_s = |A|_{s'} = \mathbf{N}$ with 0 and $succ$ interpreted as usual and*

$$f_{A_{X,Y}}(n) = \begin{cases} n & \text{if } n \in X \\ \text{undefined} & \text{otherwise} \end{cases} \qquad g_{A_{X,Y}}(n,m) = \begin{cases} m & \text{if } n \in Y \\ \text{undefined} & \text{otherwise} \end{cases}$$

Let $\mathcal{A}$ be the modest Σ-algebra over A with $\bar{s} = \bar{s}' = \mathbf{N}$, $[\cdot]_s$ and $[\cdot]_{s'}$ the identity, and the evident tracking functions for 0, $succ$, f and g ($f_{A_{X,Y}}$ and $g_{A_{X,Y}}$ are partial recursive since X and Y are r.e.). $A_{X,Y}$ is not reachable if X is a proper subset of $\mathbf{N}$.

   Now choose X and Y so that they are disjoint and not recursively separable, that is, for any recursive set $Q \subseteq \mathbf{N}$, if $X \subseteq Q$ then $Q \cap Y \neq \emptyset$ (such sets X and Y exist, see e.g. [Rog67], Chap. 7, Th. XII). Let $\mathcal{B}$ be a modest subalgebra of $A_{X,Y}$. Then $X \subseteq |\mathcal{B}|_{s'}$, and so there exists $y \in Y$ such that $y \in |\mathcal{B}|_{s'}$ (since $|\mathcal{B}|_{s'} = \bar{s}'_\mathcal{B}$ is a recursive subset of $\mathbf{N}$ and X and Y are not recursively separable). But then, for each $n \in \mathbf{N}$, $g_{|\mathcal{B}|}(y, n) = n$, and so $n \in |\mathcal{B}|_{s'}$. Thus $|\mathcal{B}|_{s'} = \mathbf{N} = |A_{X,Y}|_{s'}$, and so $\mathcal{B} = A_{X,Y}$. Therefore $A_{X,Y}$ has no proper modest subalgebra. $\quad\square$

Any partial algebra has a unique reachable subalgebra. Counterexample 5.5 shows that this is not the case for modest algebras. For any modest Σ-algebra $\mathcal{A}$, the sets $\langle\{t_\mathcal{A} \mid t \in |T_\Sigma|_s\} \subseteq \bar{s}\rangle_{s \in S}$ are r.e., and so using Prop. 2.5 it is possible to construct a reachable modest algebra from $\mathcal{A}$ and this family of r.e. sets. But the construction involves a re-coding so the result will not be a modest subalgebra of $\mathcal{A}$ in general.

The amalgamated union construction, used to combine algebras over different signatures having a common reduct, generalizes to modest algebras.

Proposition 5.6 *Consider a pushout in the category* **AlgSig**:

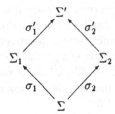

Then, for any modest Σ_1-algebra A_1 and modest Σ_2-algebra A_2 such that $A_1|_{\sigma_1} = A_2|_{\sigma_2}$, there exists a unique modest Σ'-algebra A' such that $A'|_{\sigma_1'} = A_1$ and $A'|_{\sigma_2'} = A_2$. If A_1, A_2 and $A_1|_{\sigma_1}$ are all modest $A0$-extensions then so is A'. □

In Sect. 4 we have introduced formulae, which can be used as axioms to *specify* required properties of modest algebras. The quotient construction and amalgamation as discussed above are examples of how modest algebras can be modified and combined. To build modest algebras "from scratch" we need some elementary ways to *define* new modest algebras and/or their components. As a simple example, let us now consider *definitions* of operations.

Given a signature Σ, a Σ-*definition* has the form

$$\textsf{fun } f(x_1{:}s_1, \ldots, x_n{:}s_n) = t{:}s$$

where f does not occur in Σ, t is a term of sort s over the signature $\Sigma[f] =_{\text{def}} \Sigma \cup \{f : s_1 \times \cdots \times s_n \to s\}$ (so t may refer to f) with variables $\{x_1 : s_1, \ldots, x_n : s_n\}$. Mutual recursion is not provided but adding it should not introduce any new problems.

Such a definition determines an extension of an ordinary partial algebra, which can be used to explain the effect of definitions on modest algebras.

Definition 5.7 *Given a partial Σ-algebra A, a Σ-definition*

$$\textsf{fun } f(x_1{:}s_1, \ldots, x_n{:}s_n) = t{:}s$$

determines a sequence of partial $\Sigma[f]$-algebras $A_0, A_1, A_2, \ldots$ extending A. For each $j \geq 0$, $A_j|_\Sigma = A$, and for every $a_1 \in |A|_{s_1}, \ldots, a_n \in |A|_{s_n}$, $f_{A_0}(a_1, \ldots, a_n)$ is undefined and for $j \geq 0$, $f_{A_{j+1}}(a_1, \ldots, a_n) = t_{A_j}^{\{x_1 \mapsto a_1, \ldots, x_n \mapsto a_n\}}$. The extension of A by $\textsf{fun } f(x_1{:}s_1, \ldots, x_n{:}s_n) = t{:}s$, *written (somewhat imprecisely) $A[f \leftarrow t]$, is the partial $\Sigma[f]$-algebra such that $A[f \leftarrow t]|_\Sigma = A$ and for every $a_1 \in |A|_{s_1}, \ldots, a_n \in |A|_{s_n}$ and $a \in |A|_s$, $f_{A[f \leftarrow t]}(a_1, \ldots, a_n) = a$ iff $f_{A_j}(a_1, \ldots, a_n) = a$ for some $j \geq 0$.*

(This is well-defined, since $f_{A_j}, j \geq 0$, form a chain w.r.t. the evident ordering between partial functions.)

Proposition 5.8 *Given a Σ-homomorphism $h : A \to B$ between partial Σ-algebras and a Σ-definition* $\textsf{fun } f(x_1{:}s_1, \ldots, x_n{:}s_n) = t{:}s$. $h : A[f \leftarrow t] \to B[f \leftarrow t]$ *is a $\Sigma[f]$-homomorphism.* □

Notice that for any modest Σ-algebra A, $f_{\tilde{A}[f \leftarrow t]}$ is a partial recursive function. Then, since by Prop. 2.2 any modest Σ-algebra A may be identified with the surjective homomorphism $[\cdot] : \tilde{A} \to |A|$, Prop. 5.8 justifies the following definition:

Definition 5.9 *Given any modest Σ-algebra A and Σ-definition*

$$\textsf{fun } f(x_1{:}s_1, \ldots, x_n{:}s_n) = t{:}s$$

the extension of A by $\textsf{fun } f(x_1{:}s_1, \ldots, x_n{:}s_n) = t{:}s$, *written (somewhat imprecisely) $A[f \leftarrow t]$, is the modest $\Sigma[f]$-algebra given by the surjective $\Sigma[f]$-homomorphism $[\cdot] : \tilde{A}[f \leftarrow t] \to |A|[f \leftarrow t]$.*

Proposition 5.10 *If A is a modest $A0$-extension then so is $A[f \leftarrow t]$.* □

6 Institutions

In the previous sections we have introduced concepts which can be put together to form a number of logical systems we might want to use in the process of software specification and development. To systematize this a bit, let us first recall the notion of an *institution*, which formalizes the concept of a logical system based on a model-theoretic view of logic. This will also allow us to take advantage of "institution-independent" concepts and results in the literature.

Definition 6.1 ([GB92]) *An* institution I *consists of: a category* $\mathbf{Sign}_I$ *of signatures; functors* $\mathbf{Sen}_I\colon \mathbf{Sign}_I \to \mathbf{Set}$ *and* $\mathbf{Mod}_I\colon \mathbf{Sign}_I^{op} \to \mathbf{Cat}$, *giving for each signature* $\Sigma \in |\mathbf{Sign}_I|$ *a set* $\mathbf{Sen}_I(\Sigma)$ *of* Σ-*sentences and a category* $\mathbf{Mod}_I(\Sigma)$ *of* Σ-*models respectively; and a family* $\langle \models_{I,\Sigma} \subseteq |\mathbf{Mod}_I(\Sigma)| \times \mathbf{Sen}_I(\Sigma)\rangle_{\Sigma\in|\mathbf{Sign}_I|}$ *of satisfaction relations such that for any signature morphism* $\sigma\colon \Sigma \to \Sigma'$, Σ-*sentence* $\varphi \in \mathbf{Sen}_I(\Sigma)$ *and* Σ'-*model* $M' \in |\mathbf{Mod}_I(\Sigma')|$, $M' \models_{I,\Sigma'} \mathbf{Sen}_I(\sigma)(\varphi)$ *iff* $\mathbf{Mod}_I(\sigma)(M') \models_{I,\Sigma} \varphi$.

We will omit the subscript I when the institution is obvious, and then for any signature morphism $\sigma : \Sigma \to \Sigma'$ and Σ'-model $M' \in |\mathbf{Mod}(\Sigma')|$, we will write $M'|_\sigma$ for $\mathbf{Mod}(\sigma)(M')$. Then, for any Σ-model $M \in |\mathbf{Mod}(\Sigma)|$, Σ-sentence $\varphi \subseteq \mathbf{Sen}(\Sigma)$ and set of Σ-sentences $\Phi \subseteq \mathbf{Sen}(\Sigma)$, we will write $M \models \Phi$ and $\Phi \models \varphi$ with the usual meaning. The latter notation introduces the crucial concept of *semantic entailment*.

In the previous sections we have in effect defined four institutions:

I_P **(partial algebras):** The category of signatures is $\mathbf{AlgSig}$; sentences are closed first-order formulae with equality and definedness formulae; the model functor is $\mathbf{PAlg} : \mathbf{AlgSig}^{op} \to \mathbf{Cat}$; and the satisfaction relation is the usual satisfaction of first-order formulae in partial algebras.

I_{A0} **(partial $A0$-extensions):** Like I_P, but the category of signatures is $\mathbf{AlgSig}_{\Sigma 0}$ and the model functor is $\mathbf{PAlg}_{A0} : \mathbf{AlgSig}_{\Sigma 0}^{op} \to \mathbf{Cat}$.

I_M **(modest algebras):** Like I_P, but the model functor is $\mathbf{MAlg} : \mathbf{AlgSig}^{op} \to \mathbf{Cat}$ and the satisfaction relation is the extensional satisfaction of first-order formulae in modest algebras.

I_{A0} **(modest $A0$-extensions):** Like I_M, but the category of signatures is $\mathbf{AlgSig}_{\Sigma 0}$ and the model functor is $\mathbf{MAlg}_{A0} : \mathbf{AlgSig}_{\Sigma 0}^{op} \to \mathbf{Cat}$.

An important property of institutions is whether they admit amalgamation of models, used in the process of modular composition of programs, cf. [EM85], [ST88b].

Definition 6.2 *An institution* I *admits* amalgamation[1] *if its category of signatures* $\mathbf{Sign}$ *has pushouts and its model functor* $\mathbf{Mod} : \mathbf{Sign}^{op} \to \mathbf{Cat}$ *maps pushouts in* $\mathbf{Sign}$ *to pullbacks in* $\mathbf{Cat}$.

The most important consequence of this property is that amalgamation of models (and of model morphisms), as spelled out in Prop. 5.6 for modest algebras, is then unambiguously defined.

Proposition 6.3 I_P, I_{A0}, I_M *and* I_{A0} *admit amalgamation.* □

---

[1]Such institutions were called *semiexact* in [DGS93].

7 Modestizable algebras

In Sect. 2 we suggested that modest algebras (or, to be more precise, modest $\mathcal{A}0$-extensions) can be viewed as a representation of the structures that arise in a real programming language like SML. Another approach would be to use partial algebras, but restrict attention only to those that are codable in the real programming language.

Definition 7.1 *A partial Σ algebra $A \in |\mathbf{PAlg}(\Sigma)|$ is modestizable if there exists a modest Σ-algebra $\mathcal{A} \in |\mathbf{MAlg}(\Sigma)|$ so that $|\mathcal{A}| = A$. $\mathbf{PAlg}_{\exists M}(\Sigma)$ denotes the category of modestizable partial Σ-algebras, and there is an obvious functor $\mathbf{PAlg}_{\exists M}$: $\mathbf{AlgSig}^{op} \to \mathbf{Cat}$.*

The first question to investigate is how many partial algebras are modestizable. We begin with two easy positive statements:

Proposition 7.2 *Any finite partial algebra is modestizable.* □

Proposition 7.3 *Any total algebra $A \in \mathbf{PAlg}(\Sigma)$ is modestizable.*

PROOF: *Suppose $A \in |\mathbf{PAlg}(\Sigma)|$ is total. Fix an enumeration $N_s = \{a_1, a_2, \dots\}$ of $|A|_s$ for each $s \in S$; this is possible since $|A|_s$ is required to be countable. Let $N = \langle N_s \rangle_{s \in S}$. Define $\mathcal{A} \in |\mathbf{MAlg}(\Sigma)|$: $|\mathcal{A}| = A$; for any $s \in S$, let $\bar{s}$ be the set of Gödel-style encodings $\ulcorner t \urcorner$ of terms $t \in |T_\Sigma(N)|_s$ and $[\ulcorner t \urcorner]_s = t_A^{id}$; and for any $f : s_1 \times \cdots \times s_n \to s$ in Ω, $\bar{f}(\ulcorner t_1 \urcorner, \dots, \ulcorner t_n \urcorner) = \ulcorner f(t_1, \dots, t_n) \urcorner$.* □

Of course, the carrier of any modestizable algebra is countable — this is why we have restricted attention to countable algebras only. But even under this assumption, not all partial algebras are modestizable:[2]

Counterexample 7.4 *Let Σ_{nonmod} have sort s and operation symbols $0 : \to s$ and $succ : s \to s$. Consider the partial Σ_{nonmod}-algebra A with $|A|_s = \mathbb{N}$, 0 and $succ$ interpreted as usual, and where $f_A : \mathbb{N} \rightharpoonup \mathbb{N}$ is a function with non-r.e. domain, e.g.*

$$f_A(n) = \begin{cases} 0 & \text{if the nth TM doesn't halt} \\ \text{undefined} & \text{otherwise.} \end{cases}$$

Now suppose that A is modestizable. Then $n \in dom(f_A)$ iff $f_A(n)$ is defined iff $f_A(succ_A^n(0_A))$ is defined iff $\bar{f}(\overline{succ}^n(\bar{0}))$ is defined iff $n \in dom(\lambda x. \bar{f}(\overline{succ}^x(\bar{0})))$. But $\bar{f}$ and $\overline{succ}$ are partial recursive, so $dom(\lambda x. \bar{f}(\overline{succ}^x(\bar{0})))$ is r.e., and this contradicts the assumption that $dom(f_A)$ is non-r.e. □

The essence of this counterexample is that in any modest algebra, the set of ground terms with defined values is always a recursively enumerable subset of all terms, and so partial algebras for which this property fails are not modestizable.

The situation is even more delicate when modest $\mathcal{A}0$-extensions are considered. One might expect that any modestizable partial $\mathcal{A}0$-extension $A \in \mathbf{PAlg}_{\mathcal{A}0}(\Sigma)$ is modestizable so as to extend $\mathcal{A}0$. Unfortunately, the following counterexample shows that this is not the case, even if $A|_{\Sigma 0} = \mathcal{A}0$.

---

[2] This example is due to Martin Hofmann and Thomas Streicher.

Counterexample 7.5 *Consider an algebra A_{halt} with $A_{halt}|_{\Sigma_0} = A0$ which contains a total function halt : nat $\rightarrow$ bool that solves the halting problem. A_{halt} is modestizable by Prop. 7.3 (assuming that A0 is total), but there is no modest algebra extending A0 over A_{halt}.* $\square$

On an abstract level, what is happening when we try to work with modestizable algebras is that we start with an institution **I** (in this case, the institution **I$_P$**) and then, for each signature $\Sigma \in |\textbf{Sign}|$, identify a class of "admissible" Σ-models (in this case, the modestizable algebras) which then may be identified with a full subcategory of $\textbf{Mod}(\Sigma)$. In general, this need not yield an institution: for some signature morphisms $\sigma : \Sigma \rightarrow \Sigma'$ and admissible Σ'-models $M' \in |\textbf{Mod}(\Sigma')|$, the reduct $M'|_\sigma \in |\textbf{Mod}(\Sigma)|$ may not be admissible. Fortunately, in our case this is not a problem since reducts of modest algebras are well-defined and "commute" with reducts of partial algebras. Consequently, we have an institution:

I$_{\exists M}$ (modestizable partial algebras): Like **I$_P$**. but the model functor is **PAlg$_{\exists M}$** : **AlgSig^{op}** $\rightarrow$ **Cat**.

However, the institution induced by choosing a class of admissible models of an institution may lose some of the properties enjoyed by the original institution:[3]

Proposition 7.6 **I$_{\exists M}$** *does not admit amalgamation.*

PROOF: *The partial Σ_{nonmod}-algebra A of Counterexample 7.4 is not modestizable, but can easily be presented as the amalgamated union of two modestizable partial algebras: its reduct to the subsignature of Σ_{nonmod} with f removed and its reduct to the subsignature of Σ_{nonmod} with 0 and succ removed.* $\square$

Similarly, in general we might lose the existence of reachable subalgebras, free extensions, quotients, etc.

When we restrict the class of models of an institution, its logical properties may also change. In general, the logical entailments of the original institution remain valid in the institution of "admissible models".

Proposition 7.7 *For any signature Σ. set of Σ-sentences Φ and Σ-sentence φ, if $\Phi \models_{I_P} \varphi$ then $\Phi \models_{I_{\exists M}} \varphi$.* $\square$

However, the opposite implication does not hold. since over a smaller class of models more entailments might become true:

Counterexample 7.8 *Consider any enumeration $\langle \mathcal{R}_n \rangle_{n \in \mathbb{N}}$ of all the recursively enumerable subsets of $\mathbb{N}$. Let Σ be a signature with one sort s, constant $0 :\ \rightarrow s$ and two unary operations succ, $f : s \rightarrow s$. Put:*

$$\Phi = \{D(f(succ^n(0))) \mid n \in \mathbb{N} \wedge n \notin \mathcal{R}_n\} \cup$$
$$\{\neg D(f(succ^n(0))) \mid n \in \mathbb{N} \wedge n \in \mathcal{R}_n\} \cup$$
$$\{D(0), \forall x{:}s. D(succ(x))\}.$$

*Then, Φ is a set of sentences that is consistent in **I$_P$**, and so e.g. $\Phi \not\models_{I_P}$ false.*

---

[3]This point arose in a discussion with José Fiadeiro.

But Φ is inconsistent in $\mathbf{I_{3M}}$, since no partial algebra satisfying Φ is modestizable, by an argument similar to that in Counterexample 7.4: given any modest Σ-algebra $\mathcal{A}$, $X = dom(\lambda n.\bar{f}(\overline{succ}^n(\bar{0})))$ is an r.e. subset of $\mathbf{N}$. Moreover, $n \in X$ iff $\mathcal{A} \models D(f(succ^n(0)))$, and if $\mathcal{A} \models \Phi$ then this holds iff $n \notin \mathcal{R}_n$. So, for $\mathcal{A} \models \Phi$, $n \in X$ iff $n \notin \mathcal{R}_n$ for all $n \in \mathbf{N}$. Therefore X cannot be r.e., and this contradiction proves that Φ has no modest model. Thus $\Phi \models_{\mathbf{I_{3M}}} \varphi$ for any Σ-sentence φ. □

8 Specifications

In [ST88a] we presented a powerful specification framework based on ASL [SW83] which can be used for writing specifications in an arbitrary institution. For the purposes of this paper, let us concentrate on a fragment of this formalism:

Definition 8.1 *The syntax of specifications in an institution* **I** *is given by the following grammar:*

$$
\begin{array}{llll}
SP & ::= & \langle \Sigma, \Phi \rangle & \textit{basic specification} \\
& | & SP \cup SP' & \textit{combination of specifications} \\
& | & \textbf{translate } SP \textbf{ by } \sigma & \textit{renaming and/or adding symbols} \\
& | & \textbf{derive from } SP \textbf{ by } \sigma & \textit{hiding and/or renaming symbols}
\end{array}
$$

where Σ ranges over signatures, Φ ranges over sets of sentences, and σ ranges over signature morphisms, all from **I**.

As usual, $SP \cup SP'$ combines the requirements that are expressed separately in SP and SP', and then in typical institutions like the ones defined above, **translate** SP **by** σ renames sorts and operation symbols in SP and/or adds new sorts and operation symbols without constraining them at all, and **derive from** SP **by** σ hides and/or renames sorts and operation symbols in SP.

Definition 8.2 *The* signature $Sig(SP)$ *of a specification* SP *is determined as follows:*

$$
\begin{array}{ll}
Sig(\langle \Sigma, \Phi \rangle) = \Sigma & \textit{if } \Phi \subseteq Sen(\Sigma) \\
Sig(SP \cup SP') = Sig(SP) & \textit{if } Sig(SP) = Sig(SP') \\
Sig(\textbf{translate } SP \textbf{ by } \sigma) = \Sigma' & \textit{if } \sigma : Sig(SP) \rightarrow \Sigma' \\
Sig(\textbf{derive from } SP' \textbf{ by } \sigma) = \Sigma & \textit{if } \sigma : \Sigma \rightarrow Sig(SP')
\end{array}
$$

A specification is well-formed *if it has a signature; otherwise it is* ill-formed. *We will implicitly require all specifications below to be well-formed.*

The semantics of specifications is defined by associating a class of $Sig(SP)$-models to every well-formed specification SP, as follows.

Definition 8.3 *The* model class *of a specification* SP *in an institution* **I** *is the class of $Sig(SP)$-models $Mod(SP) \subseteq |\mathbf{Mod}(Sig(SP))|$ determined as follows:*

$$
\begin{array}{l}
Mod(\langle \Sigma, \Phi \rangle) = \{M \in |\mathbf{Mod}(\Sigma)| \mid M \models_{\Sigma} \Phi\} \\
Mod(SP \cup SP') = Mod(SP) \cap Mod(SP') \\
Mod(\textbf{translate } SP \textbf{ by } \sigma) = \{M' \mid M'|_{\sigma} \in Mod(SP)\} \\
Mod(\textbf{derive from } SP' \textbf{ by } \sigma) = \{M'|_{\sigma} \mid M' \in Mod(SP')\}
\end{array}
$$

This semantics is compositional: the class of models of a specification is determined from the class of models of its immediate constituents.

The above definitions of specifications in an arbitrary institution can now be instantiated to the framework of each of the institutions of interest here, as defined in Sect. 6. These institutions share a common "syntax" and so they share the class of specifications. The model classes of specifications differ though. For each specification SP in $\mathbf{I_P}$, we will write $Mod_{\mathbf{P}}(SP)$ for the class of models of SP in $\mathbf{I_P}$ and $Mod_{\mathbf{M}}(SP)$ for its class of models in $\mathbf{I_M}$. Moreover, if SP is a specification in $\mathbf{I_{A0}}$ then we will write $Mod_{\mathbf{A0}}(SP)$ for its class of models in $\mathbf{I_{A0}}$ and $Mod_{\mathcal{A0}}(SP)$ for its class of models in $\mathbf{I_{\mathcal{A0}}}$.

The model classes $Mod_{\mathbf{A0}}(SP)$ and $Mod_{\mathbf{P}}(SP)$, and similarly $Mod_{\mathcal{A0}}(SP)$ and $Mod_{\mathbf{M}}(SP)$, are related in the obvious way:

Proposition 8.4 *For any specification SP in $\mathbf{I_{A0}}$,*

1. $Mod_{\mathbf{A0}}(SP) = Mod_{\mathbf{P}}(SP) \cap |\mathbf{PAlg}_{A0}(Sig(SP))|$, *and*

2. $Mod_{\mathcal{A0}}(SP) = Mod_{\mathbf{M}}(SP) \cap |\mathbf{MAlg}_{A0}(Sig(SP))|$. □

The distinction between $Mod_{\mathbf{P}}(SP)$ and $Mod_{\mathbf{M}}(SP)$ is more interesting, and will be studied in detail in Sect. 9. Here, let us only recall that for any signature $\Sigma \in |\mathbf{AlgSig}|$, we have introduced a functor $|\cdot|_\Sigma : \mathbf{MAlg}(\Sigma) \to \mathbf{PAlg}(\Sigma)$ (in Def. 2.6 written without the subscript Σ). This extends to classes of models in the obvious way, which allows us to compare $Mod_{\mathbf{P}}(SP)$ and $Mod_{\mathbf{M}}(SP)$.

Proposition 8.5 *For any specification SP in $\mathbf{I_P}$, $|Mod_{\mathbf{M}}(SP)| \subseteq Mod_{\mathbf{P}}(SP)$.*

PROOF: *By easy induction on the structure of SP. The key to the proof is that the family of functors $|\cdot|_\Sigma : \mathbf{MAlg}(\Sigma) \to \mathbf{PAlg}(\Sigma)$, $\Sigma \in |\mathbf{AlgSig}|$, is "smooth" w.r.t. change of signature, so that we have a natural transformation $|\cdot| : \mathbf{MAlg} \to \mathbf{PAlg}$. Moreover, for any signature Σ, Σ-sentence φ and modest Σ-algebra A, $A \models_\Sigma \varphi$ (in the institution $\mathbf{I_M}$) iff $|A|_\Sigma \models_\Sigma \varphi$ (in the institution $\mathbf{I_P}$).* □

In fact, the remarks in the proof show that we have an "institution representation" $\rho : \mathbf{I_P} \to \mathbf{I_M}$ [Tar96], or "plain map of institutions" [Mes89] (which here is trivial on signatures and sentences, but non-trivial on models) and a similar fact may be proved for structured specifications translated by an arbitrary institution representation.

Corollary 8.6 *For any specification SP in $\mathbf{I_{A0}}$, $|Mod_{\mathcal{A0}}(SP)| \subseteq Mod_{\mathbf{A0}}(SP)$.* □

9 Mind the gap!

The institution $\mathbf{I_P}$ of partial algebras provides a basic abstract framework for program specification and development, and has for this reason been extensively studied. But if we consider the fact that actual programs are written in real programming languages, and accordingly take issues of computability into account, then we are in fact working in the more restricted framework of the institution of modest (or at least modestizable) algebras. This raises the question of whether there is an appropriate correspondence between the world of programs and the abstract world of algebras studied in the literature. If there is a mismatch, then in the worst case there is the danger of a program being certified as "verified" even though it contains errors. We will try to answer these questions in this section by comparing the meanings of specifications in the institutions of partial and modest algebras.

First, for any specification SP we have $|Mod_{\mathbf{M}}(SP)| \subseteq Mod_{\mathbf{P}}(SP)$ (Prop. 8.5) and $|Mod_{\mathbf{A0}}(SP)| \subseteq Mod_{\mathbf{A0}}(SP)$ (Cor. 8.6), and so any program represented as a modest algebra that correctly realizes a specification SP in the framework of the institution $\mathbf{I_M}$ (or $\mathbf{I_{A0}}$), correctly realizes SP in the more abstract framework of the institution $\mathbf{I_P}$ (or $\mathbf{I_{A0}}$) as well. This shows that it is *sound* to develop programs in the framework of modest algebras, and so that there is no *dangerous gap* between the two frameworks.

Definition 9.1 *A specification SP witnesses a gap between $\mathbf{I_P}$ and $\mathbf{I_M}$ if $Mod_{\mathbf{P}}(SP) \neq |Mod_{\mathbf{M}}(SP)|$. The gap is worrying if for some modestizable algebra A, $A \in Mod_{\mathbf{P}}(SP)$ but $A \notin |Mod_{\mathbf{M}}(SP)|$. SP witnesses a consistency gap if $Mod_{\mathbf{P}}(SP) \neq \emptyset$ while $Mod_{\mathbf{M}}(SP) = \emptyset$. Similar definitions apply to gaps between $\mathbf{I_{A0}}$ and $\mathbf{I_{A0}}$.*

Clearly, gaps between $\mathbf{I_P}$ and $\mathbf{I_M}$ exist and are witnessed by the trivial specification $\langle \Sigma, \emptyset \rangle$ for any signature $\Sigma \in |\mathbf{AlgSig}|$ that is sufficiently rich to ensure that the functor $| \cdot |_{\Sigma} : \mathbf{MAlg}(\Sigma) \to \mathbf{PAlg}(\Sigma)$ is not surjective (cf. Counterexample 7.4). However, such a gap is in itself not a cause for alarm, since it might be that in the process of constructing modest algebras it would remain invisible. This is in contrast with worrying gaps: if for some specification SP there is a modest algebra $A \notin Mod_{\mathbf{M}}(SP)$ such that $|A| \in Mod_{\mathbf{P}}(SP)$ then the gap exhibited by SP should worry us, as there is a danger that by interpreting the specification SP in the institution of modest algebras we exclude some perfectly acceptable realizations of SP.

Unfortunately, we have consistency gaps between $\mathbf{I_P}$ and $\mathbf{I_M}$ (Counterexample 7.8; translating this to a signature including $\Sigma0$ we can also exhibit a consistency gap between $\mathbf{I_{A0}}$ and $\mathbf{I_{A0}}$). The specification formalism described in Sect. 8 is rich enough to exploit consistency gaps in a worrying way:

Proposition 9.2 *If there is a consistency gap between $\mathbf{I_P}$ and $\mathbf{I_M}$ (resp. between $\mathbf{I_{A0}}$ and $\mathbf{I_{A0}}$) then there is also a worrying consistency gap between $\mathbf{I_P}$ and $\mathbf{I_M}$ (resp. between $\mathbf{I_{A0}}$ and $\mathbf{I_{A0}}$).*

PROOF: *Let SP witness a consistency gap between $\mathbf{I_P}$ and $\mathbf{I_M}$, and let $\iota : \Sigma_{\emptyset} \hookrightarrow Sig(SP)$ be the inclusion of the empty signature $\Sigma_{\emptyset}$ into $Sig(SP)$. Then the specification* derive from SP by ι *witnesses a worrying consistency gap between $\mathbf{I_P}$ and $\mathbf{I_M}$:* $Mod_{\mathbf{M}}($derive from SP by $\iota) = \emptyset$ *while* $Mod_{\mathbf{P}}($derive from SP by $\iota) = |\mathbf{PAlg}(\Sigma_{\emptyset})|$, *and the latter contains the trivial empty algebra, which is modestizable.*

The construction of a worrying consistency gap out of a consistency gap between $\mathbf{I_{A0}}$ and $\mathbf{I_{A0}}$ is similar: just let $\iota : \Sigma0 \hookrightarrow Sig(SP)$ be the inclusion of $\Sigma0$ into the signature of the specification which witnesses the consistency gap. □

Corollary 9.3 *There are worrying consistency gaps between $\mathbf{I_P}$ and $\mathbf{I_M}$, as well as between $\mathbf{I_{A0}}$ and $\mathbf{I_{A0}}$.* □

The reader might feel that the use of the empty algebra over the empty signature in the proof of Prop. 9.2 is a little dubious, but the same pattern applies in the case of any other signature. More significantly, Counterexample 7.8 is not very convincing:

we doubt that a specification like this would ever be written in practice (especially since it is essentially infinite). Unfortunately, we do not know at present if there is a *finite* first-order specification witnessing a worrying gap between $\mathbf{I_P}$ and $\mathbf{I_M}$.

However, we can exhibit quite natural specifications that witness a worrying consistency gap between $\mathbf{I_{A0}}$ and $\mathbf{I_{A0}}$. For instance, looking at Counterexample 7.5, since the halting function is arithmetical and so is first-order axiomatizable over the standard model of the natural numbers, there is a sentence φ_{halt} (over the signature Σ_{halt} which extends $\Sigma 0$ by the operation symbol $halt : nat \rightarrow bool$) such that A_{halt} is (up to isomorphism) the only partial $A0$-extension satisfying φ_{halt}. Thus, repeating the argument from Counterexample 7.5, $Mod_{A0}(\langle \Sigma_{halt}, \{\varphi_{halt}\}\rangle) \neq \emptyset$ while $Mod_{A0}(\langle \Sigma_{halt}, \{\varphi_{halt}\}\rangle) = \emptyset$ and then by Prop. 9.2 a worrying consistency gap can be obtained. Similarly but perhaps more convincingly:

Example 9.4 *Let Σ_{equiv} extend $\Sigma 0$ by $equiv : nat \times nat \rightarrow bool$, and let φ_{equiv} be a sentence such that for any $A \in Mod_{A0}(\langle \Sigma_{equiv}, \{\varphi_{equiv}\}\rangle)$, $equiv_A(n, m) = true$ iff the nth and mth Turing machines are equivalent (such a sentence exists, since the equivalence of TMs is arithmetical).*

This can be used as the basis for a specification of functions that perform transformations on Turing machines. For example, let Σ_{opt} extend $\Sigma 0$ by $opt : nat \rightarrow nat$, and let $\iota_1 : \Sigma_{equiv} \hookrightarrow (\Sigma_{equiv} \cup \Sigma_{opt})$ and $\iota_2 : \Sigma_{opt} \hookrightarrow (\Sigma_{equiv} \cup \Sigma_{opt})$ be signature inclusions. Then a specification SP_{opt} defined as:

> **derive from**
> > **translate** $\langle \Sigma_{equiv}, \{\varphi_{equiv}\}\rangle$ **by** ι_1 $\quad\cup$
> > $\langle \Sigma_{equiv} \cup \Sigma_{opt}, \{\forall n{:}nat.\,equiv(opt(n), n) = true\}\rangle$
> **by** ι_2

specifies an optimizing function opt transforming TMs. (Axioms could be added to require that the output of opt is at least as efficient as its input.) This specification witnesses a worrying consistency gap between $\mathbf{I_{A0}}$ and $\mathbf{I_{A0}}$: $Mod_{A0}(SP_{opt}) = \emptyset$ since $Mod_{A0}(\langle \Sigma_{equiv}, \{\varphi_{equiv}\}\rangle) = \emptyset$, while $Mod_{A0}(SP_{opt})$ is not empty and contains many modestizable algebras. □

In contrast to some of our other examples, the pattern in Example 9.4 actually arises in practice. The notation used in SP_{opt} hides the idea; in a higher-level specification language it might look like this:

> **local** **val** $equiv : nat \times nat \rightarrow bool$
> > **axiom** φ_{equiv}
> **in** **val** $opt : nat \rightarrow nat$
> > **axiom** $\forall n{:}nat.\,equiv(opt(n), n) = true$
> **end**

The example is expressed in terms of Gödel encodings of Turing machines where its practical utility may not be immediately obvious, but exactly the same example could be phrased in terms of program fragments in a real programming language and the above specification could then appear as part of the specification of an optimizing compiler or program transformation system.

Examples of realistic specifications where a similar worrying consistency gap arises lead us to the conclusion that we do in fact want to interpret specifications at the abstract level of partial algebras, i.e. in the institution $\mathbf{I_P}$. However, the construction of programs satisfying requirements specifications happens at the more concrete level of programs, modelled here by modest algebras. In effect, the development process (see Sect. 10) will use yet another semantics of specifications:

Definition 9.5 *For any specification SP in $\mathbf{I_P}$, define*

$$Mod_{\mathbf{P \to M}}(SP) = \{\mathcal{A} \in |\mathbf{MAlg}(Sig(SP))| \mid |\mathcal{A}| \in Mod_{\mathbf{P}}(SP)\}.$$

Similarly, for any specification SP in $\mathbf{I_{A0}}$, define

$$Mod_{\mathbf{A0 \to A0}}(SP) = \{\mathcal{A} \in |\mathbf{MAlg}_{A0}(Sig(SP))| \mid |\mathcal{A}| \in Mod_{\mathbf{A0}}(SP)\}.$$

Neither of these two semantics is compositional, as is shown by the examples we have just been discussing. However, the validation of specifications (proving their logical consequences) and the verification of correctness of refinement steps may be carried out at the abstract level, where the semantics of specifications is compositional:

Proposition 9.6 *For any specification SP in the institution $\mathbf{I_P}$ and $Sig(SP)$-sentence φ, $Mod_{\mathbf{P \to M}}(SP) \models \varphi$ whenever $Mod_{\mathbf{P}}(SP) \models \varphi$.*

For any two specifications SP and SP' in $\mathbf{I_P}$, $Mod_{\mathbf{P \to M}}(SP') \subseteq Mod_{\mathbf{P \to M}}(SP)$ whenever $Mod_{\mathbf{P}}(SP') \subseteq Mod_{\mathbf{P}}(SP)$. $\square$

Similar facts hold for specifications in $\mathbf{I_{A0}}$ and their model classes given by $Mod_{\mathbf{A0 \to A0}}$ and $Mod_{\mathbf{A0}}$.

10 Formal software development

In [ST88b] (cf. [ST95]) we have proposed to view the process of software development as the production of a sequence of *constructor implementation* steps, starting from a specification of requirements and gradually adding design and implementation decisions expressed as constructions on algebras, and leading to a stage where nothing is left to be implemented. Then the sequence of constructions used in the development determines a program that correctly implements the original specification.

In this section we will recast these ideas in the framework of this paper, paying special attention to the effects of considering both abstract and concrete models of programs (partial and modest algebras, respectively). We will explicitly work with modest $\mathcal{A}0$-extensions only; but of course everything can be repeated for arbitrary modest algebras if needed.

In the framework discussed here, constructions involved in implementation steps work on modest $\mathcal{A}0$-extensions and so can be modelled as functions on modest $\mathcal{A}0$-extensions. But in view of the gaps exhibited in Sect. 9, correctness of constructor implementation steps should involve the more permissive semantics of specifications from the institution of partial algebras to ensure that some possible constructions are not excluded:

Definition 10.1 *Given specifications SP and SP' in $\mathbf{I_{A0}}$ and a construction $\kappa :$ $|\mathbf{MAlg}_{A0}(Sig(SP'))| \to |\mathbf{MAlg}_{A0}(Sig(SP))|$ on modest $\mathcal{A}0$-extensions, we say that SP' implements SP via κ, written $SP \underset{\kappa}{\leadsto} SP'$, if for all modest $\mathcal{A}0$-extensions $\mathcal{A} \in$ $|\mathbf{MAlg}_{A0}(Sig(SP'))|$, $|\kappa(\mathcal{A})| \in Mod_{\mathbf{A0}}(SP)$ whenever $|\mathcal{A}| \in Mod_{\mathbf{A0}}(SP')$, that is: $\kappa(Mod_{\mathbf{A0 \to A0}}(SP')) \subseteq Mod_{\mathbf{A0 \to A0}}(SP).$*

Developments viewed as sequences of such steps ensure that correctness of the final program may be inferred from the correctness of all the individual steps:

Theorem 10.2 *Given a sequence $SP_0 \underset{\kappa_1}{\rightsquigarrow} SP_1 \underset{\kappa_2}{\rightsquigarrow} \cdots \underset{\kappa_n}{\rightsquigarrow} SP_n = \langle \Sigma 0, \emptyset \rangle$, we have* $|\kappa_1(\kappa_2(\ldots \kappa_n(A0)\ldots))| \in Mod_{A0}(SP_0)$. $\qquad \Box$

We have defined various constructions on modest $A0$-extensions, including definitional extension, reduct along a signature morphism (available in any institution) and amalgamation. The latter typically arises when the task of implementing a specification is decomposed into a number of subtasks, each to implement some simpler specification, and the results must then be combined to build an implementation of the original specification. Of course, this is the essence of a modular approach to software development, and consequently institutions which do not admit amalgamation can only be of limited use as frameworks for such a development methodology. In fact, this is an important technical argument against the use of the institution $\mathbf{I_{3M}}$ of modestizable algebras. However, in the framework we adopted above — the institution $\mathbf{I_{A0}}$ of modest $A0$-extensions with the semantics of specifications based on their interpretation in the institution $\mathbf{I_{A0}}$ of partial $A0$-extensions — all these constructions are well defined and everything works fine!

A further refinement of the development methodology, enhancing its practical flexibility in an essential way, involves taking a *behavioural* interpretation of specifications [ST88b]. This is based on the notion of behavioural equivalence between partial algebras, which can easily be generalized to modest algebras.

Definition 10.3 *Let OBS be a set of* observable sorts *in a signature Σ.*

Partial Σ-algebras A and B are behaviourally equivalent *(with respect to OBS), written $A \equiv_{OBS} B$, if there is an OBS-sorted set X of variables and valuations v_A in A and v_B in B that are surjective on sorts in OBS such that: for any term $t \in T_\Sigma(X)$, $A \models_{v_A} D(t)$ iff $B \models_{v_B} D(t)$; and for any terms $t, u \in T_\Sigma(X)$ of the same sort in OBS, $A \models_{v_A} t = u$ iff $B \models_{v_B} t = u$.*

Modest Σ-algebras A and B are (extensionally) behaviourally equivalent *(with respect to OBS), written $A \equiv_{OBS} B$, if there is an OBS-sorted set X of variables and valuations v_A in A and v_B in B that are surjective on sorts in OBS such that: for any term $t \in T_\Sigma(X)$, $A \models_{v_A} D(t)$ iff $B \models_{v_B} D(t)$; and for any terms $t, u \in T_\Sigma(X)$ of the same sort in OBS, $A \models_{v_A} t = u$ iff $B \models_{v_B} t = u$.*

We will omit the set of observable sorts if it is unimportant or obvious. For instance, when we work with $A0$-extensions it is natural to choose the sorts $S0$ as observable.

Proposition 10.4 *For any modest Σ-algebras A and B, $A \equiv B$ iff $|A| \equiv |B|$.* $\qquad \Box$

The behavioural interpretation of a specification is simply the closure of its usual class of models under behavioural equivalence:

Definition 10.5 *The* behavioural interpretation *of any specification SP in $\mathbf{I_{A0}}$ is $Beh_{A0 \to A0}(SP) = \{A \in |\mathbf{MAlg}_{A0}(Sig(SP))| \mid A \equiv B$ for some $B \in Mod_{A0 \to A0}(SP)\}$.*

Prop. 10.4 might suggest that it is not important whether the behavioural closure of specifications is taken at partial algebra or at modest algebra level. But this is not the case: given a modest algebra A, a partial algebra that is behaviourally equivalent to $|A|$ need not be modestizable. A consequence of this is that behavioural correctness of implementations must be based on behavioural equivalence of modest algebras.

Definition 10.6 *Given two specifications SP and SP' in* $\mathbf{I_{A0}}$ *and a construction* $\kappa : |\mathbf{MAlg}_{A0}(Sig(SP'))| \rightarrow |\mathbf{MAlg}_{A0}(Sig(SP))|$ *on modest A0-extensions, we say that SP' behaviourally implements SP via* κ, *written* $SP \underset{\kappa}{\overset{\equiv}{\rightsquigarrow}} SP'$, *if for all modest A0-extensions* $\mathcal{A} \in |\mathbf{MAlg}_{A0}(Sig(SP'))|$, $\kappa(\mathcal{A}) \in Beh_{A0\to A0}(SP)$ *whenever* $|\mathcal{A}| \in Mod_{A0}(SP')$, *that is* $\kappa(Mod_{A0\to A0}(SP')) \subseteq Beh_{A0\to A0}(SP)$.

Developments viewed as sequences of such steps should again ensure that correctness of the final program may be inferred from the correctness of all the individual steps. But this holds only if the constructions available are stable [Sch87], [ST89]:

Definition 10.7 *A construction* $\kappa : |\mathbf{MAlg}(\Sigma')| \rightarrow |\mathbf{MAlg}(\Sigma)|$ *on modest algebras is stable if* $\kappa(\mathcal{A}) \equiv \kappa(\mathcal{B})$ *for all modest* Σ'-*algebras* $\mathcal{A}, \mathcal{B}$ *such that* $\mathcal{A} \equiv \mathcal{B}$.

Theorem 10.8 *If* $SP_0 \underset{\kappa_1}{\overset{\equiv}{\rightsquigarrow}} SP_1 \underset{\kappa_2}{\overset{\equiv}{\rightsquigarrow}} \cdots \underset{\kappa_n}{\overset{\equiv}{\rightsquigarrow}} SP_n = \langle \Sigma0, \emptyset \rangle$ *and all the constructions used are stable, then* $\kappa_1(\kappa_2(\ldots \kappa_n(A0)\ldots)) \in Beh_{A0\to A0}(SP_0)$. □

In the above we have been rather sloppy in omitting the subscript *OBS* everywhere. The subtleties of stability and local and global correctness of constructions, as introduced in [Sch87] and discussed in [ST89], apply here without change.

11 Future work

We have studied the apparent conflict between the use of abstract models in theories of specification and formal program development (here, partial algebras) and the use of more or less concrete models of programs in semantics of real programming languages (here, approximated by modest algebras and modest A0-extensions). We conclude that all is well, but note the need for certain adjustments to the theory of specification and formal development to take account of the difference between the world of programs and the world of abstract models.

The picture is not yet complete. For instance, we have shown that the constructions needed for formal program development using constructor implementations (definitional extension, reduct, amalgamation) extend from partial algebras to modest algebras. But we have not yet explored the consequences of the fact that some other familiar constructions on partial algebras (e.g. existence of reachable subalgebras) do not carry over. Further constructions should be studied, such as closure under arrow types. The difficulty here is that (as explained in Sect. 2) we would like the underlying values of arrow type to be extensional functions, restricted to those that can be expressed using well-typed λ-terms.

The framework of specification and formal development on which the presentation in Sects. 8 and 10 is based was generalized to specification and development of (higher-order) parameterized algebras in [SST92]. A similar generalization should go through here, but this would have to be based on the $Mod_{P\to M}$ (or $Mod_{A0\to A0}$) semantics of specifications. However, the computability restrictions in modest algebras do not seem to extend easily to computability restrictions on parameterised modest algebras. For the case of first-order parameterised algebras, this corresponds to the (unposed but interesting) question of the computability of constructions like those in Sect. 5.

In the case of EML, the picture we have been painting is complicated by the fact that the semantics of EML [KST94] uses models that are based on, but more expressive than, those used in the semantics of SML. The main difference between the two is that the closures used to represent functions in EML may contain "logical" constructs such as universal and existential quantifiers. Thus we need to consider three levels: that of SML programs and their models; that of EML specifications and their models; and that of partial algebras. It seems that EML models are captured by relaxing the requirement in modest algebras that the tracking functions be partial recursive, and require them to be merely arithmetical. Call these *immodest algebras*; then EML models are approximated by immodest $\mathcal{A}0$-extensions. Since closures in EML models may be formed using a recursion operator as well as quantifiers, it is not clear if this is expressive enough. But it is possible to show, using Gödel's Fixpoint Theorem, that if the satisfaction of closed EML formulae can be defined in EML, then EML is inconsistent.[4] Thus in any case there is a gap between the level of EML models and partial algebras. This means that we have two separate gaps. It seems likely that our analysis applies to both, but there may be interesting complications in dealing with both at once.

Finally, this paper treats just one kind of abstract model, namely partial algebras. The same study could be repeated for other kinds of abstract models. Although we expect that the results would be much the same, there may be interesting variations in the details in some cases.

Acknowledgements: Thanks to Stefan Kahrs, Martin Hofmann and Thomas Streicher for useful discussions, and to Martin Wirsing and Jan Bergstra for providing pointers to the literature. Thanks especially to Stefan for service as an oracle concerning the semantics of SML and EML and recursive function theory.

References

[BT87]   J. Bergstra and J. Tucker. Algebraic specifications of computable and semicomputable data types. *Theoretical Computer Science* 50:137–181 (1987).

[BKLOS91] M. Bidoit, H.-J. Kreowski, P. Lescanne, F. Orejas and D. Sannella (eds.) *Algebraic System Specification and Development: A Survey and Annotated Bibliography.* Springer LNCS 501 (1991).

[DGS93]   R. Diaconescu, J.A. Goguen and P. Stefaneas. Logical support for modularisation. In: *Logical frameworks* (G. Huet and G. Plotkin, eds.), 83–130. Cambridge Univ. Press (1993).

[EM85]   H. Ehrig and B. Mahr. *Fundamentals of Algebraic Specification 1: Equations and Initial Semantics.* Springer (1985).

[GB92]   J. Goguen and R. Burstall. Institutions: abstract model theory for specification and programming. *Journal of the ACM* 39:95–146 (1992).

[Hoa72]   C.A.R. Hoare. Proofs of correctness of data representations. *Acta Informatica* 1:271–281 (1972).

[KST94]   S. Kahrs, D. Sannella and A. Tarlecki. The definition of Extended ML. Report ECS-LFCS-94-300, Univ. of Edinburgh (1994).

[KST95]   S. Kahrs, D. Sannella and A. Tarlecki. The definition of Extended ML: a gentle introduction. Report ECS-LFCS-95-322, Univ. of Edinburgh (1995). *Theoretical Computer Science*, to appear (1996).

---

[4]Thanks to Martin Hofmann for this observation, and to Stefan Kahrs for attempting (and, luckily, failing!) to specify this. His conclusion is that in EML such a predicate cannot be total (predicates in EML are functions into bool, and functions need not be total).

[LEW96] J. Loeckx, H.-D. Ehrich and M. Wolf. *Specifications of Abstract Data Types.* Wiley (1996).

[Mal61] A.I. Mal'cev. Constructive algebras I. *Russian Mathematical Surveys* 16:77–129 (1961). Also in: *The Metamathematics of Algebraic Systems. Collected papers 1936–1967* (B. Wells ed.), 148–212. North-Holland (1971).

[Mes89] J. Meseguer. General logic. *Logic Colloquium '87* (H.-D. Ebbinghaus et al., eds.), 279–329. North-Holland (1989).

[MTH90] R. Milner, M. Tofte and R. Harper. *The Definition of Standard ML.* MIT Press (1990).

[Pau91] L. Paulson. *ML for the Working Programmer.* Cambridge Univ. Press (1991).

[Rab60] M. Rabin. Computable algebra, general theory and theory of computable fields. *Trans. of the AMS* 95:341–360 (1960).

[Rog67] H. Rogers, Jr. *Theory of Recursive Functions and Effective Computability.* McGraw-Hill (1967).

[Ros90] G. Rosolini. About modest sets. *Intl. Journal of Foundations of Computer Science* 1:341–353 (1990).

[San91] D. Sannella. Formal program development in Extended ML for the working programmer. *Proc. 3rd BCS/FACS Workshop on Refinement*, Hursley Park. Springer Workshops in Computing, 99–130 (1991).

[SST92] D. Sannella, S. Sokolowski and A. Tarlecki. Toward formal development of programs from algebraic specifications: parameterisation revisited. *Acta Informatica* 29:689–736 (1992).

[ST88a] D. Sannella and A. Tarlecki. Specifications in an arbitrary institution. *Information and Computation* 76:165–210 (1988).

[ST88b] D. Sannella and A. Tarlecki. Toward formal development of programs from algebraic specifications: implementations revisited. *Acta Informatica* 25:233–281 (1988).

[ST89] D. Sannella and A. Tarlecki. Toward formal development of ML programs: foundations and methodology. *Proc. 3rd Joint Conf. on Theory and Practice of Software Development*, Barcelona. Springer LNCS 352, 375–389 (1989).

[ST95] D. Sannella and A. Tarlecki. Model-theoretic foundations for formal program development: basic concepts and motivation. ICS PAS Report 791, Institute of Computer Science PAS, Warsaw (1995).

[ST9?] D. Sannella and A. Tarlecki. *Foundations of Algebraic Specifications and Formal Program Development.* Cambridge Univ. Press. to appear (199?).

[SW83] D. Sannella and M. Wirsing. A kernel language for algebraic specification and implementation. *Proc. 1983 Intl. Conf. on Foundations of Computation Theory*, Borgholm. Springer LNCS 158, 413–427 (1983).

[Sch87] O. Schoett. Data Abstraction and the Correctness of Modular Programming. Ph.D. thesis; Report CST-42-87, Univ. of Edinburgh (1987).

[SHT95] V. Stoltenberg-Hansen and J. Tucker. Effective algebras. In: *Handbook of Logic in Computer Science, Vol. 4* (S. Abramsky, D. Gabbay and T. Maibaum, eds.), 357–526. Oxford Univ. Press (1995).

[Tar96] A. Tarlecki. Moving between logical systems. *Recent Trends in Data Type Specifications. 11th Workshop on Specification of Abstract Data Types* (M. Haveraaen et al., eds.). Springer LNCS, to appear (1996).

[Wir90] M. Wirsing. Algebraic specification. In: *Handbook of Theoretical Computer Science, Vol. B* (J. van Leeuwen, ed.), 675–788. North-Holland (1990).

A Sequent Calculus for Subtyping Polymorphic Types

Jerzy Tiuryn*
Institute of Informatics
Warsaw University

Abstract

We present two complete systems for subtyping polymorphic types. One system is in the style of natural deduction, while another is a Gentzen style sequent calculus system. We prove several metamathematical properties for these systems including cut elimination, subject reduction, coherence, and decidability of type reconstruction. Following the approach by J.Mitchell, the sequents are given a simple semantics using logical relations over applicative structures. The systems are complete with respect to this semantics. The logic which emerges from this paper can be seen as a successor to the original Hilbert style system proposed by J. Mitchell in 1988, and to the "half way" sequent calculus of G. Longo, K. Milsted and S. Soloviev proposed in 1995.

1 Introduction

The notion of subtyping is one of the most important concepts introduced recently into the theory of functional languages (see [CMMS94, CW85, CG92]). One of the basic motivations comes from the area of object-oriented programming via the concept of *inheritance*. Another, not necessarily disjoint, way of viewing subtyping is the possibility of treating objects of a given type σ as legal objects of any type which contains σ as a subtype. This is known as the *subsumption rule*.

The relation of subtyping for polymorphic types has been axiomatized by J. Mitchell in 1988 [Mit88]. This axiomatization was proved complete with respect

*This work is partly supported by NSF Grants CCR-9417382, CCR-9304144, and by Polish KBN Grant 2 P301 031 06. Authors address: Institute of Informatics, Warsaw University, Banacha 2, 02-097 Warsaw, POLAND. E-mail: tiuryn@mimuw.edu.pl

to the semantics based on set-theoretic containment of logical relations over applicative structures.[1] This relation of subtyping has been proved recently undecidable by P. Urzyczyn and the author of this paper (see [TU96]).[2]

An important progress towards understanding subtyping was done in 1995 by G. Longo, K. Milsted and S. Soloviev (see [LMS95]) by looking at the subtyping expression $\sigma \sqsubseteq \tau$ as a sequent $\sigma \vdash \tau$ and designing a certain sequent calculus equivalent to the original Mitchell's system. The main contribution of [LMS95] was to bring to the surface the logical contents of the relation of subtyping. However, the sequent calculus of [LMS95] is neither a Gentzen style sequent calculus, nor a natural deduction system. We discuss this point in Section 4 in some detail.

The aim of the present paper is to propose two systems which are strongly related with subtyping. One system, $\vdash_{ND}$, is in the natural deduction style, while another, $\vdash_S$ is a Gentzen style sequent calculus. We show that the system of [LMS95] is a proper sybsystem of $\vdash_{ND}$ in the sense that all rules of the former system are admissible in the latter. We also show several metamathematical properties of $\vdash_{ND}$ and $\vdash_S$, including their equivalence.

The paper is organized as follows. Section 2 contains basic definitions including the polymorphic lambda calculus, the Mitchell's system and the Longo-Milsted-Soloviev system, $\vdash_{LMS}$. In Section 3 we recall the semantics using logical relations over applicative structures. In Section 4 we introduce $\vdash_{ND}$, discuss cut rules and show a containment of $\vdash_{LMS}$ in $\vdash_{ND}$. We also give semantics of sequents with finitely many premises by using the logical relations and show completeness of $\vdash_{ND}$ with respect to this semantics. In Section 5 we discuss subject reduction and strong normalization for terms typable in $\vdash_{ND}$. Section 6 is devoted to the coherence property. We use an adaptation of the equational theory of [LMS95] and of [CMMS94]. In Section 7 we introduce the Gentzen style sequent calculus $\vdash_S$ and prove cut elimination for $\vdash_S$. Finally, in Section 8 we discuss the issue of type inhabitation (provability) and type reconstruction. As we mentioned earlier provability is undecidable. However, we show in this section that type reconstruction problem for $\vdash_{ND}$ is decidable. This is a corollary of the chracterization of pure lambda terms which are erasures of the terms of $\vdash_{ND}$. In Conclusion (Section 9) we briefly discuss the relationship of $\vdash_S$ with *unified calculus*, of J.-Y. Girard [Gir93]. Most of the proofs are omitted from the proceedings version of the paper.[3]

---

[1] These relations are called in [Mit88] *simple inference models*. This semantics is based on the ideas of recursive realizability due to S.C. Kleene, see [TvD88].

[2] There is available another proof of this result due to J. Wells [Wel95].

[3] The full version of this paper is available at URL
(http://zls.mimuw.edu.pl/~tiuryn/papers.html) and
(ftp://ftp.mimuw.edu.pl/pub/users/tiuryn/sequent.{dvi,ps}.gz).

2 Formal Systems

The notation for *polymorphic types* is given in the following grammar.

$$\sigma \ ::= \ X \mid (\sigma \rightarrow \sigma) \mid (\forall X.\sigma)$$

where X ranges over *type variables*. We will use X, Y, Z (possibly with indices) as metavariables which denote type variables and σ, ρ, τ are metavariables which denote types. Paretheses are usually omitted if this doesn't lead to a confusion. We will denote by $\mathcal{T}$ the set of all types.

Typed lambda terms are defined by the following grammar.

$$M \ ::= \ x \mid (MM) \mid (\lambda x : \sigma.M) \mid (M\sigma) \mid (\Lambda X.M)$$

where x ranges over *term variables* and σ ranges over polymorphic types. We use x, y, z, u, v and M, N, P, Q as metavariables ranging over term variables and over terms, respectively.

Pure lambda terms are obtained from the typed terms by erasing all the type information. Hence pure lambda terms are generated by the following grammar.

$$M \ ::= \ x \mid (MM) \mid (\lambda x.M)$$

For each typed lambda term M we have a pure lambda term, $erase(M)$, it is obtained from M by erasing all the type information.

For terms M and N and a term variable x, $N[M/x]$ denotes the result of substituting M for all free occurences of x in N (bound variables in N may have to be α-renamed in order to avoid capture of free variables in M). Similarly, $\tau[\sigma/X]$ denotes the result of substituting a type σ for all free occurences of X in τ.

We have two kinds of β reductions and two kinds of η reductions.

$$(\beta_1) \ (\lambda x : \sigma. \ M)N \ \rightarrow_\beta \ M[N/x] \qquad (\beta_2) \ (\Lambda X. \ M)\rho \ \rightarrow_\beta \ M[\rho/X]$$

and

$$(\eta_1) \ (\lambda x : \sigma. \ Mx) \ \rightarrow_\eta \ M \qquad (\eta_2) \ (\Lambda X. \ MX) \ \rightarrow_\eta \ M$$

In (η_1) we assume that x doesn't occur free in M. In (η_2) we assume that X doesn't occur free in M. We write $M \rightarrow_\beta^* N$ to indicate that N can be obtained from M by performing a certain number of β-reductions. We use a similar notation for $M \rightarrow_\eta^* N$. We can also mix reductions, $M \rightarrow_{\beta,\eta}^* N$ means that N can be obtained from M by a finite sequence of $\beta\eta$ reductions. Similar notions apply to pure lambda terms.[4]

---

[4] Then we have just one notion of β-reduction and one notion of η-reduction.

2.1 The system F

The system **F** of *polymorphic λ calculus* is due to J.-Y. Girard (see [Gir71, Gir72]) and independently to J. Reynolds (see [Rey74]). It is a system for deriving *typing judgements* which are triples of the form $A \vdash M : \sigma$, where A is a finite partial function from object variables to types, (A is called an *environment*), M is a typed term and σ is a type. The environment A can be thought as containing typing assumptions about variables which may occur free in M. We represent A as a finite set of ordered pairs of the form $(x : \tau)$, where x is an object variable and τ is a type. $Dom(A)$ is the domain of A, and $FV(A)$ is the set of all type variables which occur free in types of A. By $A(x : \sigma)$ we mean a new environment which assigns to x the type σ, and is defined the same as A for all other object variables.

Axiom: $\qquad A \vdash (x : \sigma)$, provided $(x : \sigma) \in A$.

Rules:

$$(\rightarrow\text{-intro}) \qquad \frac{A(x : \sigma) \vdash M : \tau}{A \vdash (\lambda x : \sigma.M) : \sigma \rightarrow \tau}$$

$$(\rightarrow\text{-elim}) \qquad \frac{A \vdash M : \sigma \rightarrow \tau \qquad A \vdash N : \sigma}{A \vdash (MN) : \tau}$$

$$(\forall\text{-intro}) \qquad \frac{A \vdash M : \sigma}{A \vdash (\Lambda X.M) : \forall X.\sigma} \qquad (X \notin FV(A))$$

$$(\forall\text{-elim}) \qquad \frac{A \vdash M : \forall X.\sigma}{A \vdash M\tau : \sigma[\tau/X]}$$

We write $A \vdash_{\mathbf{F}} M : \sigma$ to indicate that the typing judgement $A \vdash M : \sigma$ is derivable in the above system.

2.2 Mitchell's containment system

The system which we present here derives epressions of the form $\sigma \sqsubseteq \tau$, which are pronounced "type σ is contained in (or is a subtype of) a type τ". This system is due to J. Mitchell ([Mit88]).

Axioms:

(refl) $\sigma \sqsubseteq \sigma$

(inst) $\forall X.\sigma \sqsubseteq \sigma[\rho/X]$

(dummy) $\sigma \sqsubseteq \forall X.\sigma, \quad (X \notin FV(\sigma))$

(distr) $(\forall X.\sigma \rightarrow \tau) \sqsubseteq (\forall X.\sigma) \rightarrow (\forall X.\tau)$

Rules:

$$(\rightarrow) \quad \frac{\sigma' \sqsubseteq \sigma \qquad \tau \sqsubseteq \tau'}{\sigma \rightarrow \tau \sqsubseteq \sigma' \rightarrow \tau'} \qquad\qquad (\forall) \quad \frac{\sigma \sqsubseteq \tau}{\forall X.\sigma \sqsubseteq \forall X.\tau}$$

$$(\text{trans}) \quad \frac{\sigma \sqsubseteq \rho \qquad \rho \sqsubseteq \tau}{\sigma \sqsubseteq \tau}$$

We write $\vdash \sigma \sqsubseteq \tau$ to indicate that the formula $\sigma \sqsubseteq \tau$ is derivable in the Mitchell's system.

An important result due to J. Mitchell (see [Mit88]) relating the system **F** and the Mitchell's system is the following equivalence.

Theorem 1 *For all polymorphic types σ and τ the following conditions are equivalent:*
(i) $\vdash \sigma \sqsubseteq \tau$.
(ii) *There is a typed term M such that $(x : \sigma) \vdash_{\mathbf{F}} M : \tau$ and $erase(M) \rightarrow^*_\eta x$.*

2.3   The Longo-Milsted-Soloviev system

This system (see [LMS95]) derives sequents of the form $(x : \sigma) \vdash M : \tau$, where σ and τ are polymorphic types, x is an object variable and M is a typed term.

Axiom: $(x : \sigma) \vdash x : \sigma$.

Rules:

$$(\rightarrow) \quad \frac{(x : \tau_1) \vdash M : \sigma_1 \qquad (y : \sigma_2) \vdash N : \tau_2}{(y : \sigma_1 \rightarrow \sigma_2) \vdash (\lambda x : \tau_1.\ N[(yM)/y]) : \tau_1 \rightarrow \tau_2}$$

$$(\forall_L) \quad \frac{(x : \sigma[\rho/X]) \vdash M : \tau}{(x : \forall X.\sigma) \vdash M[(x\rho)/x] : \tau}$$

For $0 \le k \le n$

$$(\forall_{k,n}) \quad \frac{(x : \sigma) \vdash \lambda x_1 : \tau_1 \ldots \lambda x_k : \tau_k. \ M : \tau_1 \to \ldots \tau_n \to \tau}{(x : \sigma) \vdash \lambda x_1 : \tau_1 \ldots \lambda x_n : \tau_n.\Lambda X. \ M x_{k+1} \ldots x_n : \tau_1 \to \ldots \tau_n \to \forall X.\tau}$$

In the above rule we assume that X doesn't occur freely in $\sigma, \tau_1, \ldots \tau_n$, and either $k = n$, or $k < n$ and M doesn't start with λ.[5]

We write $(x : \sigma) \vdash_{LMS} M : \tau$ to denote that the sequent $(x : \sigma) \vdash M : \tau$ is derivable in the above system.

The main result connecting Mitchell's system and $\vdash_{LMS}$ was obtained by Longo, Milsted and Soloviev in [LMS95].

Theorem 2 *The following conditions are equivalent:*
(i) $\vdash \sigma \sqsubseteq \tau$.
(ii) *There is a typed term M such that $(x : \sigma) \vdash_{LMS} M : \tau$.*

The above result implies that (trans) rule of the Mitchell's system is admisssible in $\vdash_{LMS}$. This can be seen as a sort of cut elimination theorem for $\vdash_{LMS}$.

The technical restriction[6] in rule $(\forall_{k,n})$ implies that terms typable in $\vdash_{LMS}$ are in β normal form. The consequence of dropping this restriction is that more terms would be typable but the stronger system would lack the subject reduction property, i.e, typable terms need not be closed under β reduction.

3 Semantics via Logical Relations

For the purposes of this section it will be enough to work with very general structures. Let $\mathcal{D} = \langle D, \cdot \rangle$ be a groupoid, i.e. a nonempty set D with a binary operation $\cdot$. Such structures are called *applicative structures*. The intuition behind $\mathcal{D}$ is that every element of D can be viewed both as a code of a certain function or as an argument to a function. Then $d \cdot e$ can be viewed as a result of applying the function whose code is d to the value e.

---

[5] In the original formulation of the system in [LMS95] there are two rules $(\forall_{k,n})$, the first, as defined here, and the second defined for the case $0 \le n < k$. The reader may check that the second rule is a special case of the first rule.

[6] The restriction which says that the term starts with strictly fewer λ's than the arity of the type on the right of $\vdash$.

The operation · can be extended to vectors componentwise. For $(d_1, \ldots, d_n) \in D^n$ and $(e_1, \ldots, e_n) \in D^n$, we define $(d_1, \ldots, d_n) \cdot (e_1, \ldots, e_n) = (d_1 \cdot e_1, \ldots, d_n \cdot e_n)$. Given an applicative structure $\mathcal{D}$ and $n \geq 1$, we can define a binary operation $\Rightarrow$ on n-ary relations over D,

$$R_1 \Rightarrow R_2 = \{ \vec{d} \in D^n \mid \text{ for all } \vec{e} \in R_1, \vec{d} \cdot \vec{e} \in R_2 \}$$

We can further extend · to n-ary relations elementwise: $R_1 \cdot R_2 = \{ \vec{d} \cdot \vec{e} \mid \vec{d} \in R_1, \vec{e} \in R_2 \}$. Let us observe that the following property holds for n-ary relations R, R_1, R_2 over $\mathcal{D}$. The proof is left for the reader.

Lemma 3 *If R is a relation over an applicative structure $\mathcal{D} = \langle D, \cdot \rangle$, then $R \subseteq (R_1 \Rightarrow R_2)$ iff $(R \cdot R_1) \subseteq R_2$.*

Hence $R_1 \Rightarrow R_2$ is the largest relation R which satisfies $(R \cdot R_1) \subseteq R_2$.

A n-ary *logical relation* over $\mathcal{D}$ is a family $\mathcal{R} = \{ R^\sigma \}_{\sigma \in \mathcal{T}}$ of n-ary relations indexed by types, subject to the following two conditions.

$$(\rightarrow) \qquad R^{\sigma \rightarrow \tau} = R^\sigma \Rightarrow R^\tau$$

$$(\forall) \qquad R^{\forall X. \sigma} = \bigcap_{\tau \in \mathcal{T}} R^{\sigma[\tau/X]}$$

We say that type σ is a subtype of a type τ under a logical relation $\mathcal{R}$, written $\mathcal{R} \models \sigma \sqsubseteq \tau$, if $R^\sigma \subseteq R^\tau$. We say that σ is a *subtype* of τ, written $\models \sigma \sqsubseteq \tau$, if σ is a subtype of τ under every logical relation.

The following result, due to J. Mitchell ([Mit88]), establishes soundness and completeness of Mitchell's system with respect to logical relations.

Theorem 4 *For arbitrary types σ and τ, the following two conditions are equivalent:*
(i) $\vdash \sigma \sqsubseteq \tau$.
(ii) $\models \sigma \sqsubseteq \tau$.
Moreover there is a unary logical relation $\mathcal{R}_*$ over a certain applicative structure[7] *$\mathcal{D}$, such that for all types σ, τ, we have $\vdash \sigma \sqsubseteq \tau$ iff $\mathcal{R}_* \models \sigma \sqsubseteq \tau$.*

There is a well known approach to the semantics of system F which uses *partial equivalence relations* (PER), see [BFSS89, BL90, CL90, BM92]. A partial equivalence relation over D is a relation which is symmetric and transistive. The

---

[7]This structure is even a λ model.

importatnt property of PER's is that they are closed under $\Rightarrow$. There is a one-to-one correspondence between subsets of D (i.e. unary relations) and PER's with at most one equivalence class — given $R \subseteq D$ we define a PER $\hat{R}$ such that $(d, e) \in \hat{R}$ iff $d, e \in R$. This correspondence preserves $\Rightarrow$ and arbitrary intersections. Hence, it follows from the second part of Theorem 4 that the Mitchell's system is also complete with respect to PER semantics.

4 Natural Deduction System

The Michell's containment system presented in Section 2.2 is in Hilbert style, i.e., its main emphasis is on the specific axioms. On the other hand, $\vdash_{LMS}$ presented in Section 2.3 can be seen as "half way" sequent calculus in the style of Gentzen — it has just one basic axiom and several rules which introduce the type constructors: $\rightarrow$ and $\forall$. Rule $(\forall_L)$ is a typical Gentzen style rule of introducing $\forall$ on the left of $\vdash$. However, the rule $(\rightarrow)$ introduces $\rightarrow$ on both sides of $\vdash$ simultaneously, while the rule $(\forall_{k,n})$ introduces $\forall$ on the right of $\vdash$ in the special context of a type, i.e., the $\forall$ is introduced to the right of n arrows. The reason for having $(\forall_{k,n})$ in $\vdash_{LMS}$ for each n is that the sequents have only one premise.[8] These additional restrictions are not Gentzen style. A typical Gentzen style system has rules, one for each type constructor, introducing it on the left of $\vdash$, and one for introducing it on the right of $\vdash$. The reader should compare $\vdash_{LMS}$ with the Gentzen style sequent calculus introduced in Section 7.

There is another style of presenting a logical system. It is called *natural deduction* style. The typical feature of this style is that for each type constructor there is one rule which introduces it,[9] and one which eliminates it. An example of a natural deduction system is the system **F** presented in Section 2.1.

In this section we will introduce a natural deduction system $\vdash_{ND}$ which can be seen as an extension of $\vdash_{LMS}$ in the sense that all proofs in $\vdash_{LMS}$ can be carried over in $\vdash_{ND}$. This extension is proper in several respects. First of all it is dealing with a broader class of sequents than $\vdash_{LMS}$ does. But even restricted to the $\vdash_{LMS}$ sequents it has more proofs than $\vdash_{LMS}$ does. The system is remarkably similar to the system **F**, though a few differences, notably non-commutativity and linearity, make the metamathematical theory quite simple. Nevertheless, as we will see, $\vdash_{ND}$ (and also $\vdash_{LMS}$) is undecidable.

In this approach we assume that environments are sequences rather than sets. Hence an *environment* E is a finite non-empty sequence of expressions of the form $(x : \sigma)$ such that no term variable occurs twice in E. By $Dom(E)$ we denote the set of term variables occuring in E, and by $FV(E)$ we denote the set

---

[8] Usually dealing with sequents having more than one premise makes $(\forall_{k,n})$ admissible, as a consequence of the usual rules which introduce $\rightarrow$ and $\forall$ on the right.

[9] On the right of $\vdash$.

of all type variables which ocur free in types of E. $E_1 * E_2$ denotes concatenation of environments.

We stress two important points in which this definition of an environment differs from the usual definition, e.g. in system **F**:

- E is non-commutative, e.g. $(x : \sigma) * (y : \tau) \neq (y : \tau) * (x : \sigma)$.

- E is non-empty.

The first assumption is crucial. The second can be eliminated at the expense of complication of the system.

The following rules define our natural deduction system for coercions.

Axiom: $\qquad (x : \sigma) \vdash (x : \sigma)$

Rules:

$(\rightarrow\text{-intro}) \qquad \dfrac{E * (x : \sigma) \vdash M : \tau}{E \vdash (\lambda x : \sigma.M) : \sigma \rightarrow \tau}$

$(\rightarrow\text{-elim}) \qquad \dfrac{E \vdash M : \sigma \rightarrow \tau \quad (x : \rho) \vdash N : \sigma}{E * (x : \rho) \vdash (MN) : \tau} \qquad (x \notin Dom(E))$

$(\forall\text{-intro}) \qquad \dfrac{E \vdash M : \sigma}{E \vdash (\Lambda X.M) : \forall X.\sigma} \qquad (X \notin FV(E))$

$(\forall\text{-elim}) \qquad \dfrac{E \vdash M : \forall X.\sigma}{E \vdash M\tau : \sigma[\tau/X]}$

In the above rules we assume that $E \neq \emptyset$. We will write $E \vdash_{ND} M : \sigma$ to denote that the sequent $E \vdash M : \sigma$ is derivable in the above system.

4.1 Cut rules

Depending on whether the cut is performed on the leftmost formula of the environment, or somwhere inside, we have two cut rules. When cut is performed on the leftmost formula we have the following rule

$$(\text{cut}^l) \quad \frac{(y : \rho) * R \vdash M : \sigma \quad (x : \sigma) * E \vdash N : \tau}{(y : \rho) * R * E \vdash N[M/x] : \tau}$$

When the cut formula is not necessarily the leftmost formula of the environment the rule becomes more restrictive.

$$(\text{cut}^r) \quad \frac{(y : \rho) \vdash M : \sigma \quad L * (x : \sigma) * E \vdash N : \tau}{L * (y : \rho) * E \vdash N[M/x] : \tau}$$

Let us observe that when $R = \emptyset = L$, then both rules coincide. These two rules can be conveniently represented by the following single rule.

$$(\text{cut}) \quad \frac{(y : \rho) * R \vdash M : \sigma \quad L * (x : \sigma) * E \vdash N : \tau}{L * (y : \rho) * R * E \vdash N[M/x] : \tau} \quad (L = \emptyset \text{ or } R = \emptyset)$$

By taking $L = \emptyset$ in (cut) we obtain (cut^r), while $R = \emptyset$ gives us (cut^l). We have the following result.

Proposition 5 *Rule* (cut) *is admissible in* $\vdash_{ND}$.

The reader may have started wondering why do we assume that the environment E in the rules of $\vdash_{ND}$ is non-empty. If we allow $E = \emptyset$, then all the rules make perfect sense and the system can be shown to be equivalent to $\vdash_{ND}$ in the sense that it doesn't prove more sequents of the form $E \vdash M : \sigma$, with $E \neq \emptyset$. The problem with this extension is that (cut^l) is not admissible in it. To see this let us observe that assuming rule (cut^l) and empty environments, the following derivation is legal. Let $E = (z_1 : X \to X) * (z_2 : X)$.

$$\frac{E \vdash z_1 z_2 : X \quad \dfrac{\dfrac{\vdash (\lambda x : X.\ x) : X \to X \quad (y : X) \vdash y : X}{(y : X) \vdash (\lambda x : X.\ x) y : X} (\to\text{-elim})}{}}{E \vdash (\lambda x : X.\ x)(z_1 z_2) : X} (\text{cut}^l)$$

On the other hand it is easy to show that the sequent $(z_1 : X \to X) * (z_2 : X) \vdash (\lambda x : X.\ x)(z_1 z_2) : X$ is not derivable in $\vdash_{ND}$ which allows empty environments.

In order to have the admissibility of (cut) restored one needs an addtional rule which extends ($\rightarrow$-elim).

$$(\rightarrow \text{-elim-empty}) \qquad \frac{\vdash M : \sigma \rightarrow \tau \qquad E \vdash N : \sigma}{E \vdash (MN) : \tau}$$

Environment E in the above rule is arbitrary. In fact, one can prove that (cut^l) and ($\rightarrow$ -elim-empty) are equivalent in the extended system.

4.2 From $\vdash_{LMS}$ to $\vdash_{ND}$

It turns out that if $(x : \sigma) \vdash_{LMS} M : \tau$, then also $(x : \sigma) \vdash_{ND} M : \tau$, i.e. every derivation in $\vdash_{LMS}$ can be viewed as being in $\vdash_{ND}$.

Theorem 6 *If* $(x : \sigma) \vdash_{LMS} M : \tau$, *then* $(x : \sigma) \vdash_{ND} M : \tau$.

Before passing farther let us observe that the converse to Theorem 6 doesn't hold. It follows from the observation that the way $\vdash_{LMS}$ is setup, it types only terms which contain no β_1-redexes (i.e. object-redexes). The reader may easily check that the following term is typable in $\vdash_{ND}$. (Let $\bot$ denote type $\forall X.X$.)

$$(x : \bot \rightarrow \bot) \vdash_{ND} \lambda y : \bot \, \Lambda X. \, ((\lambda z : \bot. \, xzX)y) : \bot \rightarrow \bot$$

4.3 Soundness and completeness

We first present semantics for sequents. Let $\mathcal{R}$ be a logical relation over an applicative structure. $\mathcal{R}$ *satisfies* a sequent $\sigma * \tau_1 * \ldots * \tau_n \vdash \tau$, written $\sigma * \tau_1 * \ldots * \tau_n \models_{\mathcal{R}} \tau$, if $R^\sigma \cdot R^{\tau_1} \cdot \ldots R^{\tau_n} \subseteq R^\tau$.[10] We say that the sequent $\sigma * \tau_1 * \ldots * \tau_n \vdash \tau$ is *valid*, written $\sigma * \tau_1 * \ldots * \tau_n \models \tau$, if for every applicative structure $\mathcal{D}$ and for every logical relation $\mathcal{R}$ over $\mathcal{D}$, $\sigma * \tau_1 * \ldots * \tau_n \models_{\mathcal{R}} \tau$ holds.

The next result is a straight generalization of Mitchell's Theorems 1 and 4.

Theorem 7 *For arbitrary types* $\sigma, \tau_1, \ldots, \tau_n, \tau$ *the following are equivalent:*
(i) *There is a typed term M such that* $(x : \sigma) * (y_1 : \tau_1) * \ldots * (y_n : \tau_n) \vdash_{ND} M : \tau$.
(ii) *There is a typed term M such that* $\{(x : \sigma), (y_1 : \tau_1), \ldots, (y_n : \tau_n)\} \vdash_{\mathbf{F}} M : \tau$ *and* $\text{erase}(M) \rightarrow_\eta^* xy_1 \ldots y_n$.
(iii) $\vdash \sigma \sqsubseteq \tau_1 \rightarrow \ldots \rightarrow \tau_n \rightarrow \tau$.
(iv) $\sigma * \tau_1 * \ldots * \tau_n \models \tau$.

---

[10]Partentheses associate to the left, i.e., $R^\sigma \cdot R^{\tau_1} \cdot \ldots R^{\tau_n}$ stands for $(\ldots (R^\sigma \cdot R^{\tau_1}) \cdot \ldots R^{\tau_n})$.

Proof: (i)⇒(ii) is proved by an easy induction on M. We use the observation that every derivation in $\vdash_{ND}$ can be easily transformed into a derivation in $\vdash_{\mathbf{F}}$. The property of the system F used in this transformation is that the following weakening rule is admissible in $\vdash_{\mathbf{F}}$: if $A \vdash_{\mathbf{F}} M : \tau$, and B is an environment extending A, then $B \vdash_{\mathbf{F}} M : \tau$. We leave the details for the reader.

(ii)⇒(iii). Assuming (ii) we have $\{(x : \sigma)\} \vdash_{\mathbf{F}} (\lambda y_1 : \tau_1 \ldots \lambda y_n : \tau_n.M) : \tau_1 \to \ldots \to \tau_n \tau$ and $erase(\lambda y_1 : \tau_1 \ldots \lambda y_n : \tau_n.M) \to_\eta^* x$. Hence, by Theorem 1 we get (iii).

(iii)⇒(iv) follows from Theorem 4 and Lemma 3.

(iv)⇒(i). Assuming (iv), by Lemma 3 we have $\models \sigma \sqsubseteq \tau_1 \to \ldots \to \tau_n \to \tau$. Hence by Theorem 4, Theorem 2 and Theorem 6 we have that there is a typed term M such that

$$(x : \sigma) \vdash_{ND} M : \tau_1 \to \ldots \to \tau_n \to \tau$$

Applying n times ($\to$-elim) rule we obtain

$$(x : \sigma) * (y_1 : \tau_1) * \ldots * (y_n : \tau_n) \vdash_{ND} (M y_1 \ldots y_n) : \tau$$

Hence $M y_1 \ldots y_n$ is the sought term. ∎

5 Subject Reduction and Strong Normalization

Subject reduction is a property of the system which says that each type is closed under β reduction of terms typable in that type. We have the following result.

Theorem 8 (Subject Reduction)
Let $E \vdash_{ND} M : \tau$ and let $M \to_\beta N$. Then $E \vdash_{ND} N : \tau$.

It turns out that terms typable in $\vdash_{ND}$ are also closed under η-reductions.

Proposition 9 *Let $E \vdash_{ND} M : \tau$ and let $M \to_\eta N$. Then $E \vdash_{ND} N : \tau$.*

We conclude this section with a result which states *strong normalization* of terms typable in $\vdash_{ND}$. Since typability in $\vdash_{ND}$ can be easily seen as a part of typability in $\mathbf{F}$, strong normalization of terms typable in $\vdash_{ND}$ follows from strong normalization for system $\mathbf{F}$ (see [Gir72]). However, the result for $\vdash_{ND}$ has an obvious direct proof due to the special form of terms typable in $\vdash_{ND}$.

Theorem 10 (Strong Normalization)
If $E \vdash_{ND} M : \tau$, then every sequence of β-reductions starting from M is finite, i.e., leads to a normal form.

Proof. It is enough to observe that all terms typable in $\vdash_{ND}$ are linear, i.e. every lambda abstraction in such a term binds just one occurence of a variable. Hence every β-reduction decreases the size of the term. ∎

6 Coherence

Coherence of $\vdash_{ND}$ is the property which says that any two terms which represent coercions between the same types are provably equal. For this we need an equational theory for term typable in $\vdash_{ND}$.

We start with a formal system for deriving expressions of the form $E \vdash M = N : \tau$, which we read as 'terms M and N are provably equal in type τ'.

Axioms for equality:
We assume that for every axiom $E \vdash M = N : \tau$ stated below we assume that both $E \vdash_{ND} M : \tau$ and $E \vdash_{ND} N : \tau$ hold.

(eq-refl) $\qquad E \vdash M = M : \tau$

(eq-β) $\qquad E \vdash (\lambda x : \sigma.M)N = M[N/x] : \tau$

(eq-$\beta2$) $\qquad E \vdash (\Lambda X.M)\sigma = M[\sigma/X] : \tau$

(eq-η) $\qquad E \vdash \lambda x : \sigma.Mx = M : \sigma \to \tau, \quad x \notin FV(M)$

(eq-$\eta2$) $\qquad E \vdash \Lambda X.MX = M : \forall X.\sigma, \quad X \notin FV(M)$

Rules for equality:

(eq-appl) $\qquad \dfrac{E \vdash M = N : \sigma \to \tau \qquad (x : \xi) \vdash P = Q : \sigma}{E * (x : \xi) \vdash MP = NQ : \tau} \qquad x \notin dom(E)$

(eq appl2) $\qquad \dfrac{E \vdash M = N : \forall X.\sigma}{E * E' \vdash P[M\tau_1/y] = Q[N\tau_2/y] : \xi}$

where in (eq appl2) we assume that $(y : \sigma[\tau_1/X]) * E' \vdash_{ND} P : \xi$ and $(y : \sigma[\tau_2/X]) * E' \vdash_{ND} Q : \xi$.

(eq abs) $\qquad \dfrac{E * (x : \sigma) \vdash M = N : \tau}{E \vdash (\lambda x : \sigma.M) = (\lambda x : \sigma.N) : \sigma \to \tau}$

$$(\text{eq abs2}) \quad \frac{E \vdash M = N : \tau}{E \vdash (\Lambda X.M) = (\Lambda X.N) : \forall X.\tau} \quad X \notin FV(E)$$

$$(\text{eq symm}) \quad \frac{E \vdash M = N : \tau}{E \vdash N = M : \tau}$$

$$(\text{eq trans}) \quad \frac{E \vdash M = P : \tau \qquad E \vdash P = N : \tau}{E \vdash N = M : \tau}$$

We will use notation $E \vdash_{ND} M = N : \sigma$ to denote that the terms M, N are provably equal (in the environment E and in the type σ) in the above system, i.e. the expression $E \vdash M = N : \sigma$ is derivable in the above system.

The above axioms and rules are pretty obvious and need no explanation, except perhaps the rule (eq appl2). This rule is a direct translation into our framework of explicit coercions of the rule (Eq appl2) from [CMMS94] which is a congruence rule for polymorphic type application. This is quite a powerful rule having many intriguing consequences. The reader may also consult [LMS95] for a discussion of this rule and its "coercion version" for $\vdash_{LMS}$.

The next result says that for a given environment Γ and a type σ there is at most one term, modulo provable equality, typable into σ under Γ.

Theorem 11 (Coherence)
If $E \vdash_{ND} M_1 : \tau$ and $E \vdash_{ND} M_2 : \tau$, then $E \vdash_{ND} M_1 = M_2 : \tau$.

7 Gentzen Style Sequent Calculus and Cut Elimination

In this section we will present a Gentzen-style sequent calculus for the logic of coercions.

Axiom: $\quad (x : \sigma) \vdash (x : \sigma)$

Rules:

$$(\rightarrow\text{-L}) \quad \frac{(y : \rho) \vdash M : \sigma_1 \qquad (x : \sigma_2) * E \vdash N : \tau}{(x : \sigma_1 \rightarrow \sigma_2) * (y : \rho) * E \vdash N[xM/x] : \tau} \quad y \neq x \text{ and } y \notin Dom(E)$$

$$(\rightarrow\text{-R}) \quad \frac{E * (x : \sigma) \vdash M : \tau}{E \vdash (\lambda x : \sigma.M) : \sigma \rightarrow \tau} \quad (E \neq \emptyset)$$

$$(\forall\text{-L}) \quad \frac{(x : \sigma[\rho/X]) * E \vdash M : \tau}{(x : \forall X.\sigma) * E \vdash M[(x\rho)/x] : \tau}$$

$$(\forall\text{-R}) \quad \frac{E \vdash M : \sigma}{E \vdash (\Lambda X.M) : \forall X.\sigma} \quad (X \notin FV(E))$$

We write $E \vdash_S M : \tau$ to denote that the sequent $E \vdash M : \tau$ is derivable in the above system.

The immediate goal is to prove that the natural deduction system and the above introduced Gentzen-style sequent calculus are equivalent. We will achieve this in two steps. First we show that $\vdash_{ND}$ is equivalent to $\vdash_S$ extended with (cut) and then we show that (cut) is admissible in $\vdash_S$. Let $\vdash_{S+}$ denote the system $\vdash_S$ augmented with (cut).

Proposition 12 *The systems* $\vdash_{ND}$ *and* $\vdash_{S+}$ *are of equal power, i.e., for every sequent* $E \vdash M : \tau$ *we have* $E \vdash_{ND} M : \tau$ *iff* $E \vdash_{S+} M : \tau$.

Theorem 13 (Cut Elimination)
If $E \vdash_{S+} M : \tau$, *then* $E \vdash_S M' : \tau$, *where* M' *is a* β-*normal form of* M.

Proof.   Let $E \vdash_{S+} M : \tau$. By Proposition 12, subject reduction (Theorem 8) and strong normalization (Theorem 10) we can assume without loss of generality that M is in β-normal form. It follows that in the derivation of $E \vdash M : \tau$ in $\vdash_{S+}$ all terms are in β-normal form — an easy proof by induction on the structure of this derivation is left for the reader.[11]

Next we introduce a cut measure on derivations Δ of $E \vdash M : \tau$. For each occurence of cut rule in Δ, cut measure of this occurence is the number of nodes in Δ which are above that occurence. *Cut measure* of a derivation Δ is the sum of cut measures of all occurences of cut rules in Δ. We prove that if $E \vdash_{S+} M : \tau$ and M is in β-normal form, then $E \vdash_S M : \tau$. The proof is by induction on the cut measure of a derivation of $E \vdash M : \tau$ in $\vdash_{S+}$.

---

[11] An important property of the type system used in this proof is that if $E \vdash M : \tau$ is derivable in $\vdash_{S+}$, then every object variable which occurs in E also occurs freely in M.

Take a derivation Δ of $E \vdash M : \tau$ in $\vdash_{S^+}$ with positive cut measure. Consider an occurence of cut rule in Δ such that no cut rule occurs above it. Let $\hat{\Delta}$ be the derivation which consists of all nodes of Δ which are above this occurence of the cut rule, including that occurence. Thus $\hat{\Delta}$ looks as follows.

$$
\frac{\begin{array}{cc} \vdots \, \Delta_1 & \vdots \, \Delta_2 \\ E_1 \vdash N_1 : \tau_1 \qquad & E_2 \vdash N_2 : \tau_2 \end{array}}{E \vdash N : \tau_2} \quad \text{(cut)}
$$

If $E_1 \vdash N_1 : \tau_1$ is an axiom, then $E_2 \vdash N_2 : \tau_2$ and $E \vdash N : \tau_2$ are identical, up to variable renaming, and we replace $\hat{\Delta}$ by the following derivation

$$
\begin{array}{c} \vdots \, \Delta_2' \\ E \vdash N : \tau_2 \end{array}
$$

where derivation Δ_2' is obtained from Δ_2 by variable renaming so as to yield $E \vdash N : \tau_2$. The resulting derivation has smaller cut measure and we can apply the induction hypothesis.

If the last rule in Δ_1 is $(\rightarrow\text{-L})$ then Δ_1 looks as follows

$$
\frac{\begin{array}{cc} \vdots \, \Delta_1' & \vdots \, \Delta_1'' \\ E_1' \vdash N_1' : \tau_1' \qquad & E_1'' \vdash N_1'' : \tau_1 \end{array}}{E_1 \vdash N_1 : \tau_1} \quad (\rightarrow\text{-L})
$$

Thus we can replace $\hat{\Delta}$ by the following derivation.

$$
\frac{\begin{array}{cc} \vdots \, \Delta_1' & \dfrac{\dfrac{\vdots \, \Delta_1''}{E_1'' \vdash N_1'' : \tau_1} \quad \dfrac{\vdots \, \Delta_2}{E_2 \vdash N_2 : \tau_2}}{E_1' \vdash N_1''' : \tau_2} \, \text{(cut)} \\ E_1' \vdash N_1' : \tau_1' \qquad & \end{array}}{E \vdash N : \tau_2} \quad (\rightarrow\text{-L})
$$

where $E_1''' \vdash N_1''' : \tau_2$ is a suitably chosen sequent. For this transformation to work we have to notice that if y is a variable which doesn't occur freely in a term P, then for arbitrary terms Q_1, Q_2 and for an arbitrary variable x we have

$$
P[Q_1/x][Q_2/y] = P[Q_1[Q_2/y]/x]
$$

We leave the easy details of this transformation for the reader. The derivation which results from this transformation has smaller cut measure and we can apply the induction hypothesis.

If the last rule in Δ_1 is ($\forall$-L) then we proceed in a similar way, pushing the cut rule one level up through the left branch.

If the last rule in Δ_1 is ($\to$-R) or ($\forall$-R) then, since N is in β-normal form, it follows that either $E_2 \vdash N_2 : \tau_2$ is an axiom, or the last rule in Δ_2 is either ($\to$-R) or ($\forall$-R) .[12] We proceed similarly as in the previous part of the proof, eliminating this cut, if $E_2 \vdash N_2 : \tau_2$ is an axiom, or pushing it up through the right branch. In any case this results in decreasing the cut measure of the derivation. We leave the easy details for the reader. ∎

8 Type Inhabitation and Type Reconstruction

Despite simplicity of the metatheory of the sequent calculus presented in this paper, the following problem is undecidable: given a sequent $\sigma_1 * \ldots * \sigma_n \vdash \tau$, is there a term M such that $(y_1 : \sigma_1) * \ldots * (y_n : \sigma_n) \vdash_{ND} M : \tau$ holds? This problem is known as the *inhabitation problem*: assuming that we have objects of types $\sigma_1, \ldots, \sigma_n$, can we construct an object of type τ. This result has been proved by P. Urzyczyn and the author of this paper (see [TU96]). It is proved there that for types σ and τ the relation $\vdash \sigma \sqsubseteq \tau$ is undecidable. The reader may also consult [Wel95] for another proof of this result. Thus none of the systems: Mitchell's containment relation, $\vdash_{LMS}$, $\vdash_{ND}$, $\vdash_S$, and $\vdash_{S+}$ is decidable. The reader may compare this result with undecidability of type inhabitation for the system **F** which was proved by D.M. Gabbay ([Gab74]) and independently by M.H. Löb ([Löb76]).

It is perhaps interesting to contrast the undecidability result of [TU96] with decidability of the following problem: given two types σ and τ, does $\vdash \sigma \sqsubseteq \tau$ and $\vdash \tau \sqsubseteq \sigma$ hold? This problem turns out to be decidable (see [Tiu95]). This is a consequence of a simple equational axiomatization of the above relation. This axiomatization is given in [Tiu95].

Type inhabitation problem is concerned with decidability of the set of theorems of a given logic. Hence it has a clear logical motivation. There is another problem which has a computational motivation and is in a ceratin sense dual to the above mentioned problem. It is known as *type reconstruction problem*. An instance of this problem is a pure lambda term M and we are interested whether it can be typed in $\vdash_{ND}$, i.e. whether there is a typed term Q, a noncommutative environment E and a type τ such that $erase(Q) = M$, and $E \vdash_{ND} M : \tau$. We will prove that this problem is decidable. The reader may find it interesting to compare it with undecidability of type reconstruction for system **F** (see [Wel94]).

We are going to give a simple syntactic characterization of pure lambda terms

---

[12] It can be neither ($\to$-L) , nor ($\forall$-L) , since this would result in a β-redex in M.

which are typable in $\vdash_{ND}$. Call a term M *strongly linear* if for every subterm N of M, $N \to^*_{\eta\beta} x_1 \ldots x_n$, where $x_1, \ldots, x_n$ are pairwise different variables, and $n \geq 1$.

Proposition 14 *If $\lambda x.M$ is a strongly linear term, then M has exactly one free occurence of x. Hence all β-reductions performed on strongly linear terms shrink their size. In particular, it follows that the set of strongly linear terms is decidable and that strongly linear terms are strongly normalizing.*

Proof. This is a routine induction on M of the following more general satement: if M strongly linear, then every variable occuring free in M has exactly one occurence and for every subterm of M of the form $\lambda y.N$, N has exactly one free occurence of y. We leave the easy details for the reader.

Decidability of the property "being strongly linear" follows from the previous part — given a term M, if it is strongly linear it has to shrink its size during every step of $\beta\eta$-reductions, hence the normal form (if it exists), or the information that there is none, is computable in polynomial time. ∎

The next result gives a complete characterization of terms which are typable in $\vdash_{ND}$. It also yields decidability of type reconstruction for $\vdash_{ND}$.

Theorem 15 (Type Reconstruction)
The following conditions are equivalent:
(i) *M is strongly linear;*
(ii) *There is a typed term Q, such that $erase(Q) = M$ and for some E and τ, $E \vdash_{ND} Q : \tau$.*
Hence the type reconstruction problem for $\vdash_{ND}$ is decidable.

9 Conclusion

We have introduced a sequent calculus for deriving sequents of the form $\sigma_1 * \ldots * \sigma_n \vdash \tau$, with $n \geq 1$. The calculus comes in two flavors: as a natural deduction system and as a Gentzen style sequent calculus. The rules of the systems are annotated with terms in a smooth and natural way. Hence we have an instance of the Curry-Howard isomorphism which links logical and computational aspects of the logic. The logic which emerges from these systems has many properties including: coherence, subject reduction, strong normalization and decidable problem of type reconstruction. This logic is undecidable.

It is interesting to compare our logic with the *unified calculus*, **LU** of J.-Y. Girard [Gir93]. This system deals with sequents of the form $\Gamma; \Gamma' \vdash \Delta'; \Delta$, i.e., sequents have four zones: two *linear zones* Γ and Δ and two *classical zones* Γ', Δ'. Our logic can be embedded in a certain sense[13] into **LU** by representing the sequent $\sigma_1 * \ldots * \sigma_n \vdash \tau$ by $\sigma_1, \ldots, \sigma_n; \vdash ; \tau$ (i.e. classical zones are empty). Thus our logic can be qualified as non-commutative linear intuitionistic logic.

Acknowledgment:

Part of the results presented in this paper was obtained by the author during his visits in 1994 and 1995 at Ecole Normale Superieure in Paris. He would like to thank French Ministry of Research, the French Embassy and Polish State Research Agency (KBN) for generous grants to support these visits. He would also like to thank Pierre-Louis Curien, Jean-Yves Girard, Giuseppe Longo and Sergei Soloviev for stimulating discussions on the topics presented here. Thanks are also for Paweł Urzyczyn for his comments on the previous version of this paper.

References

[BFSS89] E.S. Bainbridge, P.J. Freyd, A. Scedrov, and P.J. Scott. Functorial polymorphism. *Theoretical Computer Science*, 70:35–64, 1989. Corrigendum *ibid.*, 71:431, 1990. Preliminary report in *Logical Foundations of Functional Programming*, ed. G. Huet, Addison-Wesley (1990) 315–327.

[BL90] K. Bruce and G. Longo. A modest model of records, inheritance and bounded quantification. *Information and Computation*, 87(1/2):196–240, 1990.

[BM92] K. Bruce and J.C. Mitchell. PER models of subtyping, recursive types and higher-order polymorphism. In *Proc. 19th ACM Symp. on Principles of Programming Languages*, pages 316–327, January 1992.

[CG92] P.-L. Curien and G. Ghelli. Coherence of subsumption, minimum typing and type-checking in $F_\le$. *Math. Struct. in Comp. Sci.*, 2:55–91, 1992.

[CL90] L. Cardelli and G. Longo. A semantic basis for Quest. Technical Report 55, DEC Systems Research Center, 1990. To appear in *J. Functional Programming*.

---

[13] There is problem here since **LU** is a commutative logic, i.e. it has exchange rules, while our logic is non-commutative. The reader has to be careful with this embedding since some rules of LU, notably introduction of linear implication on the left, have to be rewritten in order to accomodate to the non-commutative situation.

[CMMS94] L. Cardelli, S. Martini, J. Mitchell, and A. Scedrov. An extension of System F with subtyping. *Information and Computation*, 109:4–56, 1994. Preliminary version appeared *Proc. Theor. Aspects of Computer Software*, Springer LNCS 526, September 1991, pages 750–770.

[CW85] L. Cardelli and P. Wegner. On understanding types, data abstraction, and polymorphism. *Computing Surveys*, 17(4):471–522, 1985.

[Gab74] D.M. Gabbay. On 2nd order intuitionistic propositional calculus with full comprehension. *Arch. Math. Logik*, 16:177–186, 1974.

[Gir71] J.-Y. Girard. Une extension de l'interpretation de Gödel à l'analyse, et son application à l'élimination des coupures dans l'analyse et la théorie des types. In J.E. Fenstad, editor, *2nd Scandinavian Logic Symposium*, pages 63–92. North-Holland, Amsterdam, 1971.

[Gir72] J.-Y. Girard. Interpretation fonctionelle et elimination des coupures de l'arithmetique d'ordre superieur. These D'Etat, Universite Paris VII, 1972.

[Gir93] J.-Y. Girard. On the unity of logic. *Annals of Pure and Applied Logic*, 59:201–217, 1993.

[LMS95] G. Longo, K. Milsted, and S. Soloviev. A logic of subtyping. In *Proc. IEEE Symp. on Logic in Computer Science*, pages 292–299, 1995.

[Löb76] M.H. Löb. Embedding first order predicate logic in fragments of intuitionistic logic. *Journal of Symbolic Logic*, 41:705–718, 1976.

[Mit88] J.C. Mitchell. Polymorphic type inference and containment. *Information and Computation*, 76(2/3):211–249, 1988. Reprinted in *Logical Foundations of Functional Programming*, ed. G. Huet, Addison-Wesley (1990) 153–194.

[Rey74] J.C. Reynolds. Towards a theory of type structure. In *Paris Colloq. on Programming*, pages 408–425, Berlin, 1974. Springer-Verlag LNCS 19.

[Tiu95] J. Tiuryn. Equational axiomatization of bicoercibility for polymorphic types. In Ed. P.S. Thiagarajan, editor, *Proc. 15th Conference Foundations of Software Technology and Theoretical Computer Science*, volume 1026 of *Lecture Notes in Computer Science*, pages 166–179. Springer Verlag, 1995.

[TU96] J. Tiuryn and P. Urzyczyn. The subtyping problem for second-order types is undecidable. In *Proc. LICS 96, to appear*, 1996.

[TvD88] A.S. Troelstra and D. van Dalen. *Constructivism in Mathematics. An Introduction. Volume 1*. North-Holland, 1988. ISBN-0-444-705066.

[Wel94]   J. Wells. Typability and type checking in the second-order λ-calculus are equivalent and undecidable. In *Proc. 9th Ann. IEEE Symp. Logic in Comput. Sci.*, pages 176–185, 1994.

[Wel95]   J. Wells. The undecidability of Mitchell's subtyping relation. Technical report, Computer Sci. Dept., Boston University, December 1995.

Kolmogorov Complexity:
Recent Research in Moscow

Vladimir A. Uspensky

Department of Mathematical Logic and Theory of Algorithms,
Faculty of Mechanics and Mathematics,
Lomonosov Moscow State University,
V-234 Moscow, GSP-3, 119899 Russia
email: uspensky@lpcs.math.msu.ru *

Introduction

The Kolmogorov complexity theory emerged in sixties; main definitions were independently found by Ray Solomonoff, A.N. Kolmogorov and G. Chaitin. The motivations for the definition were quite different. For Solomonoff the main goal was inductive inference theory. Kolmogorov's work was closely connected with foundations of probability theory and information theory. Chaitin's studied the length of programs for computing binary sequences.

For the historical account as well as for the exact definitions of the main notions of Kolmogorov complexity theory we refer the reader to the book of Ming Li and Paul Vitanyi [6]. Let us say only that most basic questions connected with Kolmogorov complexity were solved in seventies; the main achievement, probably, was the definition of randomness for individual random sequence and the complexity characterization of random sequences (A.N. Kolmogorov, P. Martin-Löf, L.A. Levin, C.P. Schnorr).

At the same time it became clear that the philosophical notion of randomness should be studied in the framework of computational complexity theory. The notions of a pseudorandom number generator plays here a central role; since eighties a lot of philosophically interesting and technically deep results were obtained.

Nevertheless, there are some natural question about Kolmogorov complexity which are still unsolved. We believe that they deserve attention (for mathematical, if not philosophical reasons); the goal of this talk is to state some of the questions and the results obtained in Moscow during the last years.

1 Semilattice of Complexity Degrees

This semilattice may be considered as a finite version of the semilattice of Turing degrees thoroughly studied by recursion theorists. The semilattice of Turing

* The research described in this publication was made possible in part by Grant # MQ3300 from the International Science Foundation and Russian Government and by INTAS grant #93–983.

degrees is formed by sets of natural numbers; the ordering is defined as follows:

$$(A \leq_T B) \quad \Leftrightarrow \quad A \text{ is Turing-reducible to } B$$

The set A is Turing-reducible to B if there exists an oracle Turing machine which computes the characteristic function of A using B as an oracle. One can say in this case that A has no more information than B because A can be effectively reconstructed when B is given.

We want to construct the finite version of that semilattice. Roughly speaking, we want to introduce a partial ordering on binary strings: Now $A \leq B$ means that A contains no new information compared to B (can be easily obtained from B).

To give an exact definition, we need the notion of (conditional) Kolmogorov complexity. The *conditional complexity of A when B is known*, denoted by $K(A|B)$, is defined as the length of the shortest program which produces A applied to the input string B. Of course, this definition depends on the choice of the programming language, but different reasonable languages lead to complexity measures which differ only by a bounded additive term. We assume that some reasonable programming language is fixed.

The conditional complexity $K(A|A)$ (here A is an empty string) is denoted by $K(A)$ and is called (unconditional) Kolmogorov complexity of a binary string A.

Speaking on Kolmogorov complexity here, we were speaking actually on one particular version of that, namely on the so-called *simple* (conditional as well as unconditional) *Kolmogorov complexity*. If one wants to distinguish that version from the others, one should denote it by KS. But as it will be explained in few lines below, for our goals there is no need to distinguish between the various versions of Kolmogorov complexity, so we can use the notation K which is good for any of them.

Consider the set S whose elements are infinite sequences $\langle s_1, s_2, \ldots \rangle$ of binary strings such that length of s_i is bounded by a polynomial in i. (This polynomial may be different for different elements of S.) Now we define a quasi-ordering on S:

$$\langle a_1, a_2, \ldots \rangle \leq \langle b_1, b_2, \ldots \rangle$$

if $K(a_i|b_i) = O(\log i)$.

Here two things should be explained.

First, why we use sequences of strings instead of strings? It seems unavoidable because of the asymptotic nature of the definition of Kolmogorov complexity. Second, why we require that length of strings is bounded by a polynomial and complexity is bounded by $O(\log i)$? The reason is that there are different versions of Kolmogorov complexity (prefix complexity KP, monotone complexity KM, etc., see [11], [12] for a detailed account); all the differences $KP - K$, $KM - K$ etc. are bounded by $O(\log n)$ if the length of the string does not exceed n (or a polynomial in n). Therefore, our definition would not change if we replace K by other version of Kolmogorov complexity.

Now we consider the equivalence classes ($\sigma, \tau \in S$ are equivalent if $\sigma \leq \tau$ and $\tau \leq \sigma$); the equivalence classes form an upper semilattice (like Turing degrees). This semilattice has a minimal element $0 = (0, 0, \ldots)$ but no maximal elements. Very few things are known about it; the only non-trivial result is the following example obtained by A.E. Romashchenko and N.K. Vereshchagin:

Theorem 1. *There are two elements in that semilattice which have no greatest lower bound.*

Informally speaking, this example can be constructed as follows. Consider $2n$ independent random binary strings $A_1, \ldots, A_n, B_1, \ldots, B_n$ of length n and another binary string $c_1 \ldots c_n$ which is independent of A_i and B_i. (This means that the catenation $A_1 \ldots A_n B_1 \ldots B_n c_1 \ldots c_n$ is a random string of length $2n^2 + n$.) Now consider the string $T = C_1 \ldots C_n$ which is a catenation of strings $C_i = $ **if** $c_i = 0$ **then** A_i **else** B_i **fi**. It turns out that the strings T and $A_1 \ldots A_n B_1 \ldots B_n$ (to be exact, we should speak about sequences composed of those strings for $n = 1, 2, 3 \ldots$) have no greatest lower bound. (For any i the string C_i is a lower bound but there is no lower bound which is bigger than all of C_i's)

2 Inequalities for Complexities

To define the complexity of a pair of strings $\langle A, B \rangle$ we fix some computable encoding of pairs by binary strings; the complexity $K(A, B)$ of a pair is defined as the complexity of its encoding. Of course, different encodings lead to different complexity measures; nevertheless, the difference (for computable encodings) is bounded. We assume that some encoding is fixed.

It is easy to prove that

$$K(A, B) \leq K(A) + K(B) + O(\log n) \tag{1}$$

if A and B are binary strings of length not exceeding n. Here the logarithmic term is due to the necessity to separate somehow the programs for A and B. If we use the so-called prefix complexity where programs are self-delimiting, that term disappears. However, we do not want now to take into account subtle differences between the different versions of complexity; so all our considerations will be modulo $O(\log n)$ term.

It is easy to see that $K(A, B)$ may be significantly smaller than $K(A) + K(B)$: indeed, this is the case if $A = B$. More generally, it happens when A and B share a lot of information, as the following equality shows:

$$K(A, B) = K(A) + K(B|A) + O(\log n) \tag{2}$$

(again, A and B are strings of length not exceeding n). This rather natural fact ("to generate A and B we should generate A and then generate B using known A") was established in sixties by Kolmogorov and Levin.

This remark makes natural the following definition of the amount of *mutual information* in A and B:

$$I(A : B) = K(A) + K(B) - K(A, B) \qquad (3)$$

If we ignore logarithmic terms, this expression can be rewritten as $K(A) - K(A|B)$ (or $K(B) - K(B|A)$); this difference measures how much the description of A may be simplified if B is known.

The inequality (1) can now be rewritten as $I(A : B) \geq 0$ (since now we systematically omit all logarithmic terms for brevity). It has a "conditional version"

$$I(A : B|C) = K(A|C) + K(B|C) - K(A, B|C) \geq 0 \qquad (4)$$

which is also valid up to a logarithmic term. Using the Kolmogorov – Levin theorem mentioned above, we may rewrite (4) using only "unconditional" complexities;

$$K(A, B, C) + K(C) \leq K(A, C) + K(B, C) \qquad (5)$$

Here the complexity $K(A, B, C)$ of a triple $\langle A, B, C \rangle$ is defined in a natural way (similar to the definition of the complexity of a pair.) (Indeed, $K(A|C) + K(B|C) - K(A, B|C) = K(A, C) - K(C) + K(B, C) - K(C) - K(A, B, C) + K(C) = K(A, C) + K(B, C) - K(A, B, C) - K(C)$ up to logarithmic terms.)

The main question discussed in this section is whether other inequalities for Kolmogorov complexities can be found. By "other" we mean inequalities which are not consequences of the inequality (5) and the trivial inequalities

$$0 \leq K(A) \leq K(A, B) \qquad (6)$$

Here we allow A, B, C be tuples of strings, so, e.g., the inequality

$$K(X, Y, Z, U) + K(Z, U) \leq K(X, Z, U) + K(Y, Z, U) \qquad (7)$$

is considered as a special case of (5) for $A = X$, $B = Y$, $C = \langle Z, U \rangle$.

Let us state this question in a more precise way. Assume that some natural k is fixed. Consider k variables $A_1, \ldots, A_k$ whose values are binary strings. Then we have $2^k - 1$ non-empty subsets of the set $\{A_1, \ldots, A_k\}$. For each subset u consider the complexity $K(A^u)$ of the corresponding tuple A^u composed of A_i such that $i \in u$. (For example, if $u = \{2, 4\}$, $K(A^u) = K(A_2, A_4)$.)

Now consider a linear inequality of the form

$$\sum_u \lambda_u K(A^u) \geq 0 \qquad (8)$$

Here the summation is over all $2^k - 1$ non-empty subsets. Consider all vectors $\{\lambda_u\}$ for which this inequality is valid (up to a logarithmic term, as usual). These vectors form a convex cone in $(2^k - 1)$-dimensional space.

Problem 2. Find the description of that cone for any natural k

The inequalities of type (5) and (6) belong to the cone in question. More precisely, consider the inequalities

$$K(A^u) \geq 0 \qquad (9)$$

$$K(A^u) - K(A^v) \geq 0 \qquad (10)$$

$$K(A^u) + K(A^v) - K(A^{u \cup v}) - K(A^{u \cap v}) \geq 0 \qquad (11)$$

for any u in (9), for any u and v such that $u \supset v$ in (10) and for any u and v (11). All the inequalities (9), (10) and (11) belong to the cone considered.

Conjecture 3. They generate the cone: any valid linear inequality for Kolmogorov complexities of tuples is a positive linear combination of inequalities (9), (10), (11).

The following observations were made by D. Hammer, A. Shen, N.K. Vereshchagin:

Theorem 4. *The conjecture 3 is true for $k = 1, 2, 3$*

However, for $k = 4$ we do not know whether it is true. One can only prove that

Theorem 5. *The conjecture 3 is true for $k = 4$ if for any n there exist binary strings A, B, C, D such that*

$$K(A) = K(B) = K(C) = K(D) = 2n,$$
$$K(AB) = K(AC) = K(AD) = K(BC) = K(BD) = 3n,$$
$$K(CD) = 4n,$$
$$K(BCD) = K(ACD) = K(ABD) = K(ABC) = 4n,$$
$$K(ABCD) = 4n \qquad (12)$$

Recall that all the equalities are up to a $O(\log n)$-term.

This vector appears as the extreme point of the cone which is dual to the cone generated by the inequalities (9), (10), (11). (See also [9].) We do not know if such A, B, C, D can be constructed.

Here is one of the reasons why this construction is difficult. The easiest method to construct binary strings with given complexity is to consider a random point in a finite-dimensional vector space over a finite field and the values of some (fixed) linear operators at that point. However, this method is limited by the Ingleton inequality for the dimensions of vector spaces which can be rewritten as

$$I(A : B) \leq I(A : B|C) + I(A : B|D) + I(C : D) \qquad (13)$$

which is valid for strings constructed in that way. However, as A.E. Romashchenko showed, this inequality is not true in the general case. He found that

Theorem 6. *There exist random variables a, b, c, d with finite range such that $I(a : b|c) = 0, I(a : b|d) = 0, I(c : d) = 0$ but $I(a : b) > 0$.*

Here I denotes mutual information in Shannon sense. Shannon entropy for a random variable α taking values $a_1, \ldots, a_n$ with probabilities $p_1, \ldots, p_n$ is defined as

$$H(\alpha) = -p_1 \log p_1 - p_2 \log p_2 - \ldots - p_n \log p_n \tag{14}$$

The conditional entropy is defined as

$$H(\alpha|\beta) = H(\langle \alpha, \beta \rangle) - H(\beta) \tag{15}$$

After that the mutual information in two random variables α and β is defined as

$$I(\alpha : \beta) = H(\alpha) + H(\beta) - H(\langle \alpha, \beta \rangle) \tag{16}$$

(unconditional) and

$$I(\alpha : \beta|\gamma) = H(\alpha|\gamma) + H(\beta|\gamma) - H(\langle \alpha, \beta \rangle|\gamma) \tag{17}$$

The conditional mutual information $I(\alpha : \beta|\gamma)$ is non-negative; it is equal to 0 if and only if α and β are independent for any fixed value of γ.

Romashchenko's example shows that for random variables the inequality of type (13) is not valid. Therefore, it is not valid for Kolmogorov complexity too. To find the counterexample for Kolmogorov complexity, we take N independent copies of random variables a, b, c, d and consider the sequences of outcomes $A = a_1 \ldots a_N$, $B = b_1 \ldots b_N$ etc. With probability close to 1 the strings A, B, C, D will provide counterexample to (13).

The same argument shows that any inequality which is valid for Kolmogorov complexities must correspond to a valid inequality for Shannon entropies. We do not know whether the converse is true or there are inequalities valid only for Shannon entropies.

One of the reasons why it is interesting to find all the inequalities which are true for Kolmogorov complexities is that they are closely connected with some geometric facts. For example, the inequality

$$2K(A, B, C) \leq K(A, B) + K(A, C) + K(B, C) \tag{18}$$

(as usual, the logarithmic term is omitted) corresponds to the following geometric fact: if V is volume of some set $X \in \mathbf{R}^3$ and S_1, S_2, S_3 are areas of three projections (along OX, OY and OZ), then

$$V^2 \leq S_1 \cdot S_2 \cdot S_3 \tag{19}$$

This correspondence allows to prove the geometric fact using the inequality for complexities and vice versa (see [3])

For the inequality

$$K(A, B, C) + K(C) \leq K(A, C) + K(B, C) \tag{20}$$

the corresponding geometric fact (pointed out by A. Shen) is not so simple. Here it is. If a set X has projections along OX and OY of area S_1 and S_2 and V and l are two positive numbers such that $V \cdot l = S_1 \cdot S_2$ then the set X can be covered by two sets; one has volume V and other has projection onto OZ (along OXY) of length l.

However, the correspondence here is not so straightforward in this case. The inequality is a corollary of that geometric fact (more precisely, of some algorithmic version of it) but we do not know if this argument can be reversed.

3 Common and Mutual Information

Assume that we want to construct (for some n) two strings A and B such that $K(A) = 2n$, $K(B) = 2n$ and $I(A : B) = n$. (Therefore, $K(A, B)$ should be around $3n$. Recall that we ignore terms of order $O(\log n)$.) The easiest way to do that is to consider three independent random strings X, Y, Z of length n (this means that the string XYZ is a string of length $3n$ whose complexity is close to its length) and then let A be XY and B be YZ.

In this case the mutual information in A and B is represented by the string Y which is a part of both A and B. In general, if for two binary strings A and B there exists a string C such that $K(C|A)$ and $K(C|B)$ are close to 0, then $K(C)$ cannot be much bigger than $I(A : B)$. Indeed, the inequality

$$K(C) \leq I(A : B) + K(C|A) + K(C|B) \tag{21}$$

(which is valid up to $O(\log n)$ term; n is the maximal length of A, B, C) is a consequence of the inequalities (9)–(11) of the previous section.

If $K(C|A)$ and $K(C|B)$ are close to 0, the string C may be considered as containing some part of mutual information in A and B. If at the same time $K(C)$ is close to $I(A : B)$ one can say that C contains *all* the mutual information in A and B.

P. Gács and J. Körner discovered [2] that in some cases the mutual information cannot be extracted (a string C with the mentioned properties does not exist). In our terminology, their results imply the following

Proposition 7. *There exist two sequences $A = (a_1, a_2, \ldots)$ and $B = (b_1, b_2, \ldots)$ of binary strings such that a_n and b_n have length $2n$, $K(a_n) = 2n + O(\log n)$, $K(b_n) = 2n + O(\log n)$, $I(a_n : b_n) = n$ but the greatest lower bound of A and B in the semilattice S is 0 (the minimal element of S).*

Different attempts to understand this rather strange result were performed. It was interesting to find a simple alternative proof of that proposition. It was done by An.A. Muchnik [8] who suggested a simple (though highly non-constructive) proof.

More constructive example (also suggested by An.A. Muchnik) is provided by the following theorem.

Theorem 8. *Let F be a finite field containing 2^n elements. Consider a projective plane over F. Take a random line l and a random point $p \in l$. Then $K(l) = 2n$, $K(p) = 2n$, $I(p : l) = n$ but p and l have no common information:*

$$K(C) \leq O(K(C|p) + K(C|l) + \log n)$$

To be exact, we should speak not about p and l but about their coordinates (considered as binary strings); "random" means that the complexity of p and l is close to the maximal possible ($2n$ in our case).

Another question about strings A and B with $K(A) = 2n$, $K(B) = 2n$, $I(A : B) = n$ can be stated as follows: for which positive u, v, w one can find a string C such that $K(C) \le un$, $K(A|C) \le vn$, $K(B|C) \le wn$? For different pairs A, B the set of all $\langle u, v, w \rangle$ with that property is different. This set is maximal (in a reasonable sense) for the strings with extractable common information (the first example in this section). Using the An.A. Muchnik approach [8], one can find the pair for which this set is minimal (A. Shen, N.K. Vereshchagin).

The original example of string where mutual information cannot be extracted [2] uses the following construction. Consider two random variables α and β with range 0 and 1. Make N independent trials for the pair α, β and get two random binary strings $A = a_1 \ldots a_n$ and $B = b_1 \ldots b_n$. If the random variables α and β are dependent (i.e., $H(\alpha : \beta) > 0$) then the strings A and B with high probability will have about $I(\alpha : \beta) \cdot N$ bit of mutual information.

Gács and Körner proved that the mutual information cannot be extracted, but their proof is rather complicated. It turns out that for some cases an easy proof of that result can be obtained using the example of theorem 6 and the following observation (made by A.E. Romashchenko, A. Shen):

Theorem 9. *For any strings A, B, C, D, X of length not exceeding n*

$$K(X) \le I(A : B|C) + I(A : B|D) + I(C : D) + 2K(X|A) + 2K(X|B) + O(\log n)$$

This inequality implies that if for a given A and B one can find C and D such that $I(A : B|C) = 0$, $I(A : B|D) = 0$ and $I(C : D) = 0$, then A and B have no extractable common information (though mutual information $I(A : B)$ may be significant).

4 Complexity of Problems

In his old paper [5] A.N. Kolmogorov suggested the interpretation of propositional connectives as operations on problems. Assume that propositional variables A, B are problems. Then the formula $A \wedge B$ may be considered as a problem "solve both A and B"; the formula $A \vee B$ may be interpreted as the problem "choose one of the problems A and B and solve it"; the formula $A \Rightarrow B$ may be interpreted as the problem "solve B assuming that A is solved. This rather vague description can be made precise in different ways. One of them is based on the notion of Kleene's realizability [4]. Another one is based on Yu.T. Medvedev's paper [7]. Both approaches can be used to define the complexity of a problem. The second one was developed by A. Shen [10] (using notions from domains theory); his goal was to provide the uniform approach to existing versions of Kolmogorov complexity. However, for our purposes the first (realizability) approach is enough (and definitions are simpler).

Here is it. By a *problem* we mean any set of binary strings. (Intuitively, the problem is to find any of the strings from the set.) Now we define propositional operations on problems as follows:

- $A \wedge B$ is the set of all $\langle a, b \rangle$ where $a \in A$, $b \in B$. (We fix some computable one-to-one correspondence between pairs of strings and strings and denote it by $\langle \cdot, \cdot \rangle$;
- $A \vee B$ is the set $\{0a | a \in A\} \cup \{1b | b \in B\}$;
- $A \Rightarrow B$ is the set of all programs p such that $p(a)$ is defined and $p(a) \in B$ for any string $a \in A$

The complexity of a problem can be defined as the minimal complexity of its elements. Now we can easily define conditional complexity in our terms: complexity:

Proposition 10. *For any binary strings A and B*

$$K(B|A) = K(\{A\} \Rightarrow \{B\})$$

However, for more complicated propositional formulae we do not know whether the complexity of the resulting problem can be expressed in terms of string complexity.

Problem 11. Find whether the complexity of problem

$$(\{A\} \vee \{B\}) \Rightarrow \{C\}$$

can be expressed (up to a logarithmic term) in terms of complexities $K(A)$, $K(B)$, $K(C)$, $K(A, B)$, $K(A, C)$, $K(B, C)$, $K(A, B, C)$.

Here A, B, C are binary strings. There are some interesting examples where problems complexity can be expressed in terms of string complexities.

Theorem 12.

$$K(\{A\} \leftrightarrow \{B\}) = \max(K(A|B), K(B|A)); \qquad (22)$$
$$K((\{A\} \Rightarrow \{B\}) \Rightarrow \{B\}) = \min(K(A), K(B)); \qquad (23)$$
$$K(((\{A\} \Rightarrow \{B\}) \Rightarrow \{A\}) \leftrightarrow \{A\}) = 0; \qquad (24)$$

(All equalities are up to a logarithmic term.) The first equality is a reformulation of a theorem of C.H. Bennett, P. Gács, M. Li, P.M.B. Vitányi and W. Zurek [1]; others were found by N.K. Vereshchagin.

5 Classification of Random Strings

Let n be a natural number. Consider the set of all strings of length n. Most of the strings in this set have complexity close to n and hence may be considered as "random" strings of length n. (So "random" means "incompressible".) The question arises whether we can consider all random strings as indistinguishable or some of them have some special properties which distinguish them. This is a rather vague question. Of course, some random strings have first bit 0 while other have 1, but this is not the distinction we look for. We want the property in question to be expressed in "invariant terms".

The exact statement of this question (suggested by An.A. Muchnik and S. Positselsky) requires a rather technical definition which we omit here. It turns out that a property distinguishing some random string from others does exist. The property in question is, roughly speaking, that $0'$-complexity of a string is small (while ordinary complexity is large). Here $0'$-complexity (denoted by $K^{0'}$) is relativized Kolmogorov complexity (all functions are computable with respect to $0'$, the degree of m-complete recursively enumerable set).

It turns out that that property can be expressed in terms of ordinary complexity using the following observation made by K.Yu. Gorbunov, An.A. Muchnik:

Theorem 13.

$$\limsup_{|Y| \to \infty} K(X|Y) \leq K^{0'}(X) \leq 2 \limsup_{|Y| \to \infty} K(X|Y)$$

Here $|Y|$ denotes the length of Y; as usual, logarithmic terms are omitted.

References

1. C.H. Bennett, P. Gács, M. Li, P.M.B. Vitányi, W. Zurek. Thermodynamics of computation and information distance. In *Proc. 25th ACM Symp. Theory of Computing*, p. 21–30, 1993.
2. P. Gács, J. Körner. Common information is far less than mutual information. *Problems of Control and Information Theory*, 1973, v. 2, p. 149–162.
3. D. Hammer, A. Shen. A Strange Application of Kolmogorov complexity. *Mathematical Systems Theory*, in press. (Also CWI Tech. Rep. CS-R9328.)
4. S.C. Kleene. *Introduction to Metamathematics*. D. van Nostrand company, 1952.
5. A. Kolmogoroff. Zur Deutung der intuitionistischen Logik. *Mathematische Zeitschrift*, 1932, Bd. 35, H. 1, S. 58–65.
6. M. Li, P. Vitányi. *An Introduction to Kolmogorov Complexity and Its Applications*. Springer, 1993.
7. Yu.T. Medvedev. Finite problems (in Russian) *Doklady Akademii nauk SSSR*, 1962, v. 142, no. 5, p. 1015–1018.
8. An.A. Muchnik. On the extraction of common information of two words (in Russian) *Proceedings of the First World Bernoulli congress = Trudy Pervogo Vsemirnogo Kongressa Obshchestva Matem. Statistiki i Teorii Verojatnostei imeni Bernoulli*. Moscow, 1986, p. 453.
9. L.S. Shapley. Cores of convex games. *Intern J. Game Theory*, 1971, v. 1, p. 1–26.

10. A.Kh. Shen. Algorithmic variants of the notion of entropy. *Soviet Math. Doklady*, 1984, v. 29, p. 569–573.
11. V.A. Uspensky. Complexity and entropy: an introduction to the theory of Kolmogorov complexity. In O.Watanabe, editor, *Kolmogorov Complexity and Computational Complexity*, p. 85–102. Springer-Verlag, 1992.
12. V.A. Uspensky and A.Kh. Shen. Relations between varieties of Kolmogorov complexities. *Mathematical Systems Theory*, in press. (Also CWI Tech. Rep. CS-R9329.)

A Modal Logic for Data Analysis

Philippe BALBIANI

Laboratoire d'informatique de Paris-Nord
Institut Galilée
Université Paris-Nord
Avenue Jean-Baptiste Clément
F-93430 Villetaneuse

Abstract

Modal logic is a natural framework for the representation and the mechanization of reasoning with incomplete information about objects in terms of attributes. Its use in the theory of the systems of information rests on the concept of a Kripke frame with relative accessibility relations. This paper presents the proof of the completeness of a logic based on these frames.

1   Introduction

The system of information is a good tool for the representation of knowledge on a collection of objects and constitutes a nice formalization of some aspects of reasoning with incomplete information about objects in terms of attributes [9]. It is usually made up of a collection of pieces of information which have the form "object, a list of properties of the object" [10]. More formally, let PAR be a nonempty set of *attributes* or *parameters*. A *system of information* is a structure of the form $S = (OB, \{VAL_a\}_{a \in PAR}, f)$ where OB is a nonempty set of *objects*, $\{VAL_a\}_{a \in PAR}$ is a family of sets of values of attributes and $f \colon OB \times PAR \to \bigcup \{VAL_a \colon a \in PAR\}$ is an *assignment* of properties to objects such that, for every object $x \in OB$ and for every attribute $a \in PAR$, $f(x, a) \in VAL_a$.

The use of modal logic in the theory of the systems of information has been accepted by several authors: Demri [2], Demri and Orlowska [3], Fariñas del Cerro and Orlowska [4], Konikowska [7] [8], Orlowska [9] [10], Vakarelov [11]. The frames of modal logic are structures of the form $(W, \mathcal{R})$ where W is a nonempty set of possible worlds and $\mathcal{R}$ is a family of binary relations on W. Since the system of information is not a structure of this form, it cannot constitute the semantic basis of a modal logic for reasoning about knowledge. A *frame of information* is a structure of the form $\mathcal{F} = (W, \{R(P)\}_{P \subseteq PAR})$ where W is a nonempty set of *possible worlds* and $\{R(P)\}_{P \subseteq PAR}$ is a family of relations of equivalence on W such that $R(P \cup Q) = R(P) \cap R(Q)$, for every subset P, Q of PAR.

Let $S = (OB, \{VAL_a\}_{a \in PAR}, f)$ be a system of information. Let $R(P)$ be the relation of equivalence on OB defined in the following way: $x\ R(P)\ y$ iff, for every attribute $a \in P$, $f(x, a) = f(y, a)$, for every subset P of PAR. Direct calculations would lead to the conclusion that $R(P \cup Q) = R(P) \cap R(Q)$, for every subset P, Q of PAR. Therefore, $\phi(S) = (OB, \{R(P)\}_{P \subseteq PAR})$ is a frame of information. Now, let $\mathcal{F} = (W, \{R(P)\}_{P \subseteq PAR})$ be a frame of information. For every attribute $a \in PAR$, let $VAL_a = \{R(\{a\})(x): x \in W\}$. For every possible world $x \in W$ and for every attribute $a \in PAR$, let $f(x, a) = R(\{a\})(x)$. Direct calculations would lead to the conclusion that, for every possible world $x \in W$ and for every attribute $a \in PAR$, $f(x, a) \in VAL_a$. Consequently, $\sigma(\mathcal{F}) = (W, \{VAL_a\}_{a \in PAR}, f)$ is a system of information. Finally, let $\mathcal{F}$ be a frame of information and $\mathcal{F}' = \phi(\sigma(\mathcal{F}))$. Direct calculations would lead to the conclusion that $\mathcal{F}$ and $\mathcal{F}'$ are isomorphic. Therefore, the class of all frames of information constitutes a semantic alternative to the class of all systems of information on which a modal logic of knowledge representation can be defined.

A frame of information is but a special case of the more general concept of a Kripke frame with relative accessibility relations [10]. Let $\bowtie \in \{\cup, \cap\}$. A *Kripke frame with $\bowtie$-relative accessibility relations* is a structure of the form $\mathcal{F} = (W, R)$ where W is a nonempty set of *possible worlds* and R is a mapping on the set of the subsets of PAR to the set of the binary relations on W such that $R(P \bowtie Q) = R(P) \cap R(Q)$, for every subset P, Q of PAR. Direct calculations would lead to the conclusion that frames of information are Kripke frames with $\cup$-relative accessibility relations of equivalence. According to Orlowska [10], "axiomatization of logics based on Kripke models with relative accessibility relations is an open problem". This is not any longer true since Balbiani [1] gives a complete axiomatization of the set of the formulas valid in every Kripke models with $\bowtie$-relative accessibility relations. A Kripke frames with $\cup$-relative ($\cap$-relative) accessibility relations is *cosy* when $R(P \cap Q) = R(P) \cup R(Q)$ $(R(P \cup Q) = R(P) \cup R(Q))$, for every subset P, Q of PAR. Balbiani [1] gives a complete axiomatization of the set of the formulas valid in every cosy Kripke models with $\bowtie$-relative accessibility relations.

A Kripke frames with $\cup$-relative ($\cap$-relative) accessibility relations of equivalence is *sweet* when $R(P \cap Q) = R(P) \cup^* R(Q)$ $(R(P \cup Q) = R(P) \cup^* R(Q))$, for every subset P, Q of PAR. Sweet Kripke frames with $\cap$-relative accessibility relations of equivalence have been introduced by Fariñas del Cerro and Orlowska [4] in the context of DAL, the logic for the analysis of data. According to Orlowska [10], "to find a Hilbert-style complete axiomatization of logic DAL is an open problem". This is not any longer true since this paper gives a complete axiomatization of the set of the formulas valid in every sweet Kripke models with $\bowtie$-relative accessibility relations. The paper is organized in the following way. Section 2 is devoted to the language of the modal logic for data analysis while section 3 introduces its semantics. Section 4 presents the axiomatization $DAL^{\cup}$ of the set of the formulas valid in every sweet Kripke models with $\cup$-relative accessibility relations of equivalence while sections 5 and 6 gives the proof of the completeness of $DAL^{\cup}$ for the class of all sweet Kripke models with $\cup$-relative accessibility relations of equivalence. Section 7 extends this result to the class of all sweet Kripke models with $\cap$-relative accessibility relations of equivalence.

2 Language

The linguistic basis of the modal logic for data analysis is the propositional calculus. Let VAR be the set of its *atomic formulas*. Let PAR be a nonempty set of *parameters*. Let $\mathcal{P}(PAR)^{\circ}$ be the set of the cofinite subsets of PAR. For every parameter $a \in$

PAR, let $\underline{a} = PAR \setminus \{a\}$. Let $\underline{P} = PAR \setminus P$, for every cofinite subset P of PAR. Direct calculations would lead to the conclusion that:

Lemma 1 *If* $|\underline{P}| \geq 2$ *then* $P = \bigcap_{a \in \underline{P}} \underline{a}$, *for every cofinite subset P of PAR.*

The modal operator $[P]$ is added to the standard propositional formalism, for every cofinite subset P of PAR. A set Φ of formulas is *closed* when:

- Φ is closed for the subformula,

- for every parameter $a \in PAR$ and for every formula $[\underline{a}]A \in \Phi$, $[PAR][\underline{a}]A \in \Phi$,

- if $|\underline{P}| \geq 2$ then, for every parameter $a \in \underline{P}$ and for every formula $[P]A \in \Phi$, $[\underline{a}][P]A \in \Phi$, for every cofinite subset P of PAR.

3 Semantics

This section introduces the semantics of the modal logic for data analysis. The technics of the filtration and of the copying will be used in sections 5 and 6 for the proof of the completeness of $DAL^{\cup}$.

3.1 Frame

A *frame* is a structure of the form $\mathcal{F} = (W, R)$ where W is a nonempty set of *possible worlds* and R is a mapping on the set of the cofinite subsets of PAR to the set of the relations of equivalence on W such that:

- $R(P \cup Q) \subseteq R(P) \cap R(Q)$,

for every cofinite subset P, Q of PAR. Direct calculations would lead to the conclusion that:

Lemma 2 *Let* $\mathcal{F} = (W, R)$ *be a frame. For every cofinite subset P of PAR, if* $|\underline{P}| \geq 2$ *then* $R(P) \supseteq \bigcup^{*}_{a \in \underline{P}} R(\underline{a})$.

$\mathcal{F}$ is *normal* when:

- if $|\underline{P}| \geq 2$ then $R(P) = \bigcup^{*}_{a \in \underline{P}} R(\underline{a})$,

for every cofinite subset P of PAR. Direct calculations would lead to the conclusion that:

Lemma 3 *Let* $\mathcal{F} = (W, R)$ *be a normal frame.* $R(P \cap Q) = R(P) \cup^{*} R(Q)$, *for every cofinite subset P, Q of PAR.*

$\mathcal{F}$ is *quasi-standard* when if $P \cup Q = PAR$ then:

- $R(P \cup Q) = R(P) \cap R(Q)$,

for every cofinite subset P, Q of PAR. $\mathcal{F}$ is *standard* when:

- $R(P \cup Q) = R(P) \cap R(Q)$,

for every cofinite subset P, Q of PAR. Direct calculations would lead to the conclusion that if PAR is finite then the normal standard frames defined in this section are exactly the sweet Kripke frames with $\cup$-relative accessibility relations of equivalence defined in section 1.

3.2 Model

Let $\mathcal{F} = (W, R)$ be a frame. A mapping m on the set of the atomic formulas to the set of the subsets of W is called *assignment* on $\mathcal{F}$. A *(normal, quasi-standard, standard) model* is a structure of the form $\mathcal{M} = (W, R, m)$ where $\mathcal{F} = (W, R)$ is a (normal, quasi-standard, standard) frame and m is an assignment on $\mathcal{F}$.

3.3 Satisfiability

The *satisfiability relation* in a model $\mathcal{M} = (W, R, m)$ between a formula A and a possible world x is defined in the following way:

- $x \models_\mathcal{M} A$ iff $x \in m(A)$, A atomic formula,
- $x \models_\mathcal{M} \neg A$ iff $x \not\models_\mathcal{M} A$,
- $x \models_\mathcal{M} A \wedge B$ iff $x \models_\mathcal{M} A$ and $x \models_\mathcal{M} B$,
- $x \models_\mathcal{M} [P]A$ iff, for every possible world $y \in W$, $x\ R(P)\ y$ only if $y \models_\mathcal{M} A$, for every cofinite subset P of PAR.

A formula is *valid* in a model when it is satisfied in every possible world of this model. A schema is *valid* in a frame when every instance of the schema is valid in every model of the frame.

3.4 Filtration

Let $\mathcal{M} = (W, R, m)$ be a model. Let Φ be a closed set of formulas and $\equiv_\Phi$ be the binary relation on W defined in the following way:

- $x \equiv_\Phi y$ iff, for every formula $A \in \Phi$, $x \models_\mathcal{M} A$ iff $y \models_\mathcal{M} A$.

$\equiv_\Phi$ is an equivalence relation on W. A *filtration* of $\mathcal{M}$ through Φ is a model $\mathcal{M}° = (W°, R°, m°)$ such that:

- $W°$ is a subset of W such that, for every equivalence class modulo $\equiv_\Phi$, exactly one element of the class belongs to $W°$,
- for every possible world $x, y, z, t \in W$, if $x \in W°$, $x \equiv_\Phi y$, $y\ R(P)\ z$, $z \equiv_\Phi t$ and $t \in W°$ then $x\ R°(P)\ t$, for every cofinite subset P of PAR,
- for every possible world $x, y \in W°$, if $x\ R°(P)\ y$ then, for every formula $[P]A \in \Phi$, if $x \models_\mathcal{M} [P]A$ then $y \models_\mathcal{M} A$, for every cofinite subset P of PAR,
- $m°(A) = m(A) \cap W°$, A atomic formula.

Let us notice that if Φ is finite then $W°$ is finite. Filtration preserves the satisfiability of a formula of Φ according to the following lemma:

Lemma 4 *Let* $\mathcal{M} = (W, R, m)$ *be a model,* Φ *be a closed set of formulas and* $\mathcal{M}° = (W°, R°, m°)$ *be a filtration of* $\mathcal{M}$ *through* Φ. *For every formula* $A \in \Phi$ *and for every possible world* $x \in W°$, $x \models_\mathcal{M} A$ *iff* $x \models_{\mathcal{M}°} A$.

Its proof can be done by induction on the complexity of A. The technics of the filtration are found in the Lemmon notes. They are mainly used for proving that a modal logic has the finite model property [6].

3.5 Copying

Let $\mathcal{F} = (W, R)$ be a frame. Let $\mathcal{F}' = (W', R')$ be a frame and I be a set of mappings on W to W'. I is a *copying* from $\mathcal{F}$ into $\mathcal{F}'$ when:

- for every mapping $f, g \in I$ and for every possible world $x, y \in W$, $f(x) = g(y)$ only if $x = y$,

- for every possible world $x' \in W'$, there exists a mapping $f \in I$ and a possible world $x \in W$ such that $f(x) = x'$,

- for every mapping $f \in I$ and for every possible world $x, y \in W$, $x\ R(P)\ y$ only if there exists a mapping $g \in I$ such that $f(x)\ R'(P)\ g(y)$, for every cofinite subset P of PAR,

- for every mapping $f, g \in I$ and for every possible world $x, y \in W$, $f(x)\ R'(P)\ g(y)$ only if $x\ R(P)\ y$, for every cofinite subset P of PAR.

Copying preserves the satisfiability of a formula according to the following lemma:

Lemma 5 *Let $\mathcal{F} = (W, R)$ be a frame. Let $\mathcal{F}' = (W', R')$ be a frame and I be a copying from $\mathcal{F}$ into $\mathcal{F}'$. Let m be an assignment on $\mathcal{F}$ and m' be the assignment on $\mathcal{F}'$ defined in the following way:*

- $m'(A) = \bigcup_{f \in I} f(m(A))$, A atomic formula.

Let $\mathcal{M} = (W, R, m)$, $\mathcal{M}' = (W', R', m')$. Then, for every formula A, for every mapping $f \in I$ and for every possible world $x \in W$, $x \models_{\mathcal{M}} A$ iff $f(x) \models_{\mathcal{M}'} A$.

Its proof can be done by induction on the complexity of A. The technics of the copying have been introduced by Vakarelov [11]. They are mainly used for proving that a modal logic which is characterized by a certain class of frames is also characterized by another class. See also Vakarelov [12] as well as Vakarelov [13].

4 $DAL^{\cup}$

This section presents the axiomatization $DAL^{\cup}$ of the set of the formulas valid in every normal quasi-standard model. Together with the classical tautologies, all the instances of the following schemata are axioms of $DAL^{\cup}$:

- $[P](A \rightarrow B) \rightarrow ([P]A \rightarrow [P]B)$, for every cofinite subset P of PAR,

- $[P]A \rightarrow A$, for every cofinite subset P of PAR,

- $A \rightarrow [P] < P > A$, for every cofinite subset P of PAR,

- $[P]A \rightarrow [P][P]A$, for every cofinite subset P of PAR,

- $[P]A \vee [Q]A \rightarrow [P \cup Q]A$, for every cofinite subset P, Q of PAR,

- if $|\underline{P}| \geq 2$ then $[P](A \rightarrow \bigwedge_{a \in \underline{P}} [\underline{a}]A) \rightarrow (A \rightarrow [P]A)$, for every cofinite subset P of PAR.

Together with the *modus ponens*, the following schema is a rule of $DAL^{\cup}$:

- if $\vdash_{DAL^{\cup}} A$ then $\vdash_{DAL^{\cup}} [P]A$,

for every cofinite subset P of PAR. Direct calculations would lead to the conclusion that:

Theorem 1 *The theorems of $DAL^{\cup}$ are valid in every normal model.*

Let $\mathcal{M}$ be the canonical model of $DAL^{\cup}$. Direct calculations would lead to the conclusion that $\mathcal{M}$ is a model. Therefore:

Theorem 2 *$DAL^{\cup}$ is complete for the class of all models, that is to say: the formulas valid in every model are theorems of $DAL^{\cup}$.*

5 Normal completeness

This section presents the completeness proof of $DAL^{\cup}$ for the class of all normal models. Let $\mathcal{M} = (W, R, m)$ be a model. Let Φ be a closed set of formulas. We use filtration through Φ to transform $\mathcal{M}$ into a normal model. Let W° be a subset of W such that, for every equivalence class modulo $\equiv_{\Phi}$, exactly one element of the class belongs to W°. Let $R^{\circ}(PAR)$ be the binary relation on W° defined by $x \, R^{\circ}(PAR) \, y$ iff:

- for every formula $[PAR]A \in \Phi$, $x \models_{\mathcal{M}} [PAR]A$ iff $y \models_{\mathcal{M}} [PAR]A$.

Direct calculations would lead to the conclusion that:

Lemma 6 $R^{\circ}(PAR)$ *is a relation of equivalence on W°.*

For every parameter $a \in PAR$, let $R^{\circ}(\underline{a})$ be the binary relation on W° defined by $x \, R^{\circ}(\underline{a}) \, y$ iff:

- for every formula $[\underline{a}]A \in \Phi$, $x \models_{\mathcal{M}} [\underline{a}]A$ iff $y \models_{\mathcal{M}} [\underline{a}]A$.

Direct calculations would lead to the conclusion that:

Lemma 7 *For every parameter $a \in PAR$, $R^{\circ}(\underline{a})$ is a relation of equivalence on W°.*

Moreover:

Lemma 8 *For every parameter $a \in PAR$, $R^{\circ}(PAR) \subseteq R^{\circ}(\underline{a})$.*

If $|\underline{P}| \geq 2$ then let $R^{\circ}(P) = \bigcup_{a \in \underline{P}}^{*} R^{\circ}(\underline{a})$, for every cofinite subset P of PAR. Direct calculations would lead to the conclusion that:

Lemma 9 *If $|\underline{P}| \geq 2$ then $R^{\circ}(P)$ is a relation of equivalence on W°, for every cofinite subset P of PAR.*

Moreover:

Lemma 10 *If $|\underline{P}| \geq 2$ then $R^{\circ}(PAR) \subseteq R^{\circ}(P)$, for every cofinite subset P of PAR.*

Lemma 11 *If $|\underline{P}| \geq 2$ then, for every parameter $a \in \underline{P}$, $R^{\circ}(\underline{a}) \subseteq R^{\circ}(P)$, for every cofinite subset P of PAR.*

Let it be proved that $R^{\circ}(P \cup Q) \subseteq R^{\circ}(P) \cap R^{\circ}(Q)$, for every cofinite subset P, Q of PAR.

First case: $|P \cup Q| \geq 2$. If $x \, R^{\circ}(P \cup Q) \, y$ then there exists $c_1, \ldots, c_k \in \underline{P \cup Q}$ such that $x \, R^{\circ}(\underline{c_1}) \circ \cdots \circ R^{\circ}(\underline{c_k}) \, y$. Therefore, there exists $a_1, \ldots, a_i \in \underline{P}$ such that $x \, R^{\circ}(\underline{a_1}) \circ \cdots \circ R^{\circ}(\underline{a_i}) \, y$ and there exists $b_1, \ldots, b_j \in \underline{Q}$ such that $x \, R^{\circ}(\underline{b_1}) \circ \cdots \circ R^{\circ}(\underline{b_j}) \, y$. Consequently, $x \, R^{\circ}(P) \, y$ and $x \, R^{\circ}(Q) \, y$.

Second case: There exists a parameter $a \in PAR$ such that $P \cup Q = \underline{a}$. Direct consequence of lemma 11.

Third case: $P \cup Q = PAR$. Direct consequence of lemmas 8 and 10.

Therefore:

Lemma 12 $R^\circ(P \cup Q) \subseteq R^\circ(P) \cap R^\circ(Q)$, for every cofinite subset P, Q of PAR.

Direct calculations would lead to the conclusion that:

Lemma 13 For every possible world $x, y, z, t \in W$, if $x \in W^\circ$, $x \equiv_\Phi y$, $y \ R(PAR)$ z, $z \equiv_\Phi t$ and $t \in W^\circ$ then $x \ R^\circ(PAR) \ t$.

Lemma 14 For every possible world $x, y \in W^\circ$, if $x \ R^\circ(PAR) \ y$ then, for every formula $[PAR]A \in \Phi$, if $x \models_\mathcal{M} [PAR]A$ then $y \models_\mathcal{M} A$.

Moreover:

Lemma 15 For every parameter $a \in PAR$ and for every possible world $x, y, z, t \in W$, if $x \in W^\circ$, $x \equiv_\Phi y$, $y \ R(\underline{a}) \ z$, $z \equiv_\Phi t$ and $t \in W^\circ$ then $x \ R^\circ(\underline{a}) \ t$.

Lemma 16 For every parameter $a \in PAR$ and for every possible world $x, y \in W^\circ$, if $x \ R^\circ(\underline{a}) \ y$ then, for every formula $[\underline{a}]A \in \Phi$, if $x \models_\mathcal{M} [\underline{a}]A$ then $y \models_\mathcal{M} A$.

The proof of lemmas 17 and 18 is exactly the same as the proof of the *ancestral lemma* described by Goldblatt [5] in the context of linear temporal logic or in the context of propositional dynamic logic.

Lemma 17 If $|\underline{P}| \geq 2$ then, for every possible world $x, y, z, t \in W$, if $x \in W^\circ$, $x \equiv_\Phi y$, $y \ R(P) \ z$, $z \equiv_\Phi t$ and $t \in W^\circ$ then $x \ R^\circ(P) \ t$, for every cofinite subset P of PAR.

Lemma 18 If $|\underline{P}| \geq 2$ then, for every possible world $x, y \in W^\circ$, if $x \ R^\circ(P) \ y$ then, for every formula $[P]A \in \Phi$, if $x \models_\mathcal{M} [P]A$ then $y \models_\mathcal{M} A$, for every cofinite subset P of PAR.

Therefore:

Lemma 19 For every model $\mathcal{M}$ and for every (finite) closed set Φ of formulas, there exists a (finite) normal filtration of $\mathcal{M}$ through Φ.

Let A be a consistent formula of $DAL^\cup$. According to theorem 2, A is satisfied in a possible world of some model. Let Φ be a finite closed set of formulas containing A. According to lemmas 4 and 19, A is satisfied in a possible world of some finite normal model. Consequently:

Theorem 3 $DAL^\cup$ is complete for the class of all finite normal models, that is to say: the formulas valid in every finite normal model are theorems of $DAL^\cup$.

6 Normal quasi-standard completeness

This section presents the completeness proof of $DAL^\cup$ for the class of all normal quasi-standard models. Let $\mathcal{F} = (W, R)$ be a normal frame. We use copying to transform $\mathcal{F}$ into a normal quasi-standard frame. Let $\pi(P)$ be the mapping on $W \times W$ to $\mathcal{P}(W)$ defined in the following way:

- $\pi(P)(x, y) = R(P)(x) + R(P)(y)$,

for every cofinite subset P of PAR. Direct calculations would lead to the conclusion that:

Lemma 20 For every possible world $x, y \in W$, $\pi(P)(x, y) = \emptyset$ iff $x \ R(P) \ y$, for every cofinite subset P of PAR.

Let I be the set of the mappings f on $\mathcal{P}(PAR)^\circ \times PAR$ to $\mathcal{P}(W)$ such that the set $\{\alpha : \alpha \in P$ and $f(P,\alpha) \neq \emptyset\}$ is finite, for every cofinite subset P of PAR. Let us notice that $\mathcal{B} = \mathcal{P}(W)$ is a boolean ring where:

- $0_{\mathcal{B}} = \emptyset$,
- $1_{\mathcal{B}} = W$,
- $A +_{\mathcal{B}} B = (A \setminus B) \cup (B \setminus A)$,
- $A \times_{\mathcal{B}} B = A \cap B$.

Let $W' = W \times I$. Let us remark that if PAR is finite then W is finite only if W' is finite. Let $R'(PAR)$ be the binary relation on W' defined by $(x,f) \; R'(PAR) \; (y,g)$ iff:

- $x \; R(PAR) \; y$,
- for every parameter $\alpha \in PAR$ and for every cofinite subset P of PAR, if $\alpha \in P$ then $f(P,\alpha) + g(P,\alpha) = \emptyset$.

Direct calculations would lead to the conclusion that:

Lemma 21 $R'(PAR)$ *is a relation of equivalence on* W'.

For every parameter $a \in PAR$, let $R'(\underline{a})$ be the binary relation on W' defined by $(x,f) \; R'(\underline{a}) \; (y,g)$ iff:

- $x \; R(\underline{a}) \; y$,
- for every parameter $\alpha \in \underline{a}$ and for every cofinite subset P of PAR, if $\alpha \in P$ then $f(P,\alpha) + g(P,\alpha) = \emptyset$,
- for every cofinite subset P of PAR, if $P \cap \underline{a} \neq \emptyset$ then $\Sigma_{\alpha \in P} f(P,\alpha) + g(P,\alpha) = \pi(P)(x,y)$.

Direct calculations would lead to the conclusion that:

Lemma 22 *For every parameter* $a \in PAR$, $R'(\underline{a})$ *is a relation of equivalence on* W'.

Moreover:

Lemma 23 *For every parameter* $a \in PAR$, $R'(PAR) \subseteq R'(\underline{a})$.

If $|\underline{P}| \geq 2$ then let $R'(P) = \bigcup^{\bullet}_{a \in \underline{P}} R'(\underline{a})$, for every cofinite subset P of PAR. Direct calculations would lead to the conclusion that:

Lemma 24 *If* $|\underline{P}| \geq 2$ *then* $R'(P)$ *is a relation of equivalence on* W', *for every cofinite subset* P *of* PAR.

Moreover:

Lemma 25 *If* $|\underline{P}| \geq 2$ *then* $R'(PAR) \subseteq R'(P)$, *for every cofinite subset* P *of* PAR.

Lemma 26 *If* $|\underline{P}| \geq 2$ *then, for every parameter* $a \in \underline{P}$, $R'(\underline{a}) \subseteq R'(P)$, *for every cofinite subset* P *of* PAR.

Let it be proved that $R'(P \cup Q) \subseteq R'(P) \cap R'(Q)$, for every cofinite subset P, Q of PAR.

First case: $|\underline{P \cup Q}| \geq 2$. If $(x,f) \; R'(P \cup Q) \; (y,g)$ then there exists $c_1, \ldots, c_k \in \underline{P \cup Q}$ such that $(x,f) \; R'(\underline{c_1}) \circ \cdots \circ R'(\underline{c_k}) \; (y,g)$. Therefore, there exists $a_1, \ldots, a_i \in \underline{P}$ such that $(x,f) \; R'(\underline{a_1}) \circ \cdots \circ R'(\underline{a_i}) \; (y,g)$ and there exists $b_1, \ldots, b_j \in \underline{Q}$ such that $(x,f) \; R'(\underline{b_1}) \circ \cdots \circ R'(\underline{b_j}) \; (y,g)$. Consequently, $(x,f) \; R'(P) \; (y,g)$ and $(x,f) \; R'(Q) \; (y,g)$.

Second case: There exists a parameter $a \in PAR$ such that $P \cup Q = \underline{a}$. Direct consequence of lemma 26.

Third case: $P \cup Q = PAR$. Direct consequence of lemmas 23 and 25.

Therefore:

Lemma 27 $R'(P \cup Q) \subseteq R'(P) \cap R'(Q)$, *for every cofinite subset* P, Q *of* PAR.

Let it be proved that, for every mapping $f \in I$ and for every possible world $x, y \in W$, $x \, R(PAR) \, y$ only if there exists a mapping $g \in I$ such that $(x, f) \, R'(PAR) \, (y, g)$. If $x \, R(PAR) \, y$ then let g be the mapping on $\mathcal{P}(PAR)^\circ \times PAR$ to $\mathcal{P}(W)$ defined in the following way:

- for every parameter $\alpha \in PAR$ and for every cofinite subset P of PAR, if $\alpha \notin P$ then $g(P, \alpha) = \emptyset$,

- for every parameter $\alpha \in PAR$ and for every cofinite subset P of PAR, if $\alpha \in P$ then $g(P, \alpha) = f(P, \alpha)$.

Direct calculations would lead to the conclusion that $g \in I$ and $(x, f) \, R'(PAR) \, (y, g)$. Consequently:

Lemma 28 *For every mapping* $f \in I$ *and for every possible world* $x, y \in W$, x $R(PAR) \, y$ *only if there exists a mapping* $g \in I$ *such that* $(x, f) \, R'(PAR) \, (y, g)$.

Let it be proved that, for every parameter $a \in PAR$, for every mapping $f \in I$ and for every possible world $x, y \in W$, $x \, R(\underline{a}) \, y$ only if there exists a mapping $g \in I$ such that $(x, f) \, R'(\underline{a}) \, (y, g)$. If $x \, R(\underline{a}) \, y$ then let g be the mapping on $\mathcal{P}(PAR)^\circ \times PAR$ to $\mathcal{P}(W)$ defined in the following way:

- for every parameter $\alpha \in PAR$ and for every cofinite subset P of PAR, if $\alpha \notin P$ then $g(P, \alpha) = \emptyset$,

- for every parameter $\alpha \in \underline{a}$ and for every cofinite subset P of PAR, if $\alpha \in P$ then $g(P, \alpha) = f(P, \alpha)$,

- for every cofinite subset P of PAR, if $a \in P$ then $g(P, a) = f(P, a) + \pi(P)(x, y)$.

Direct calculations would lead to the conclusion that $g \in I$ and $(x, f) \, R'(\underline{a}) \, (y, g)$. Consequently:

Lemma 29 *For every parameter* $a \in PAR$, *for every mapping* $f \in I$ *and for every possible world* $x, y \in W$, $x \, R(\underline{a}) \, y$ *only if there exists a mapping* $g \in I$ *such that* (x, f) $R'(\underline{a}) \, (y, g)$.

Let it be proved that if $|\underline{P}| \geq 2$ then, for every mapping $f \in I$ and for every possible world $x, y \in W$, $x \, R(P) \, y$ only if there exists a mapping $g \in I$ such that $(x, f) \, R'(P)$ (y, g), for every cofinite subset P of PAR. If $x \, R(P) \, y$ then there exists $a_1, \dots, a_i \in \underline{P}$ such that $x \, R(\underline{a_1}) \circ \cdots \circ R(\underline{a_i}) \, y$. Therefore, according to lemma 29, there exists a mapping $g \in I$ such that $(x, f) \, R'(\underline{a_1}) \circ \cdots \circ R'(\underline{a_i}) \, (y, g)$. Consequently, there exists a mapping $g \in I$ such that $(x, f) \, R'(P) \, (y, g)$. Therefore:

Lemma 30 *If* $|\underline{P}| \geq 2$ *then, for every mapping* $f \in I$ *and for every possible world* $x, y \in W$, $x \, R(P) \, y$ *only if there exists a mapping* $g \in I$ *such that* $(x, f) \, R'(P) \, (y, g)$, *for every cofinite subset* P *of* PAR.

Every mapping $f \in I$ can equally be considered as the mapping on W to W' defined in the following way:

- $f(x) = (x, f)$.

Consequently:

Lemma 31 *I is a copying from $\mathcal{F}$ into $\mathcal{F}'$.*

Let it be proved that, for every parameter $a, b \in PAR$, if $\underline{a} \cup \underline{b} = PAR$ then $R'(\underline{a} \cup \underline{b}) = R'(\underline{a}) \cap R'(\underline{b})$. Suppose that $(x, f)\ R'(\underline{a})\ (y, g)$ and $(x, f)\ R'(\underline{b})\ (y, g)$. If $\underline{a} \cup \underline{b} = PAR$ then, for every parameter $\alpha \in PAR$, either $\alpha \in \underline{a}$ or $\alpha \in \underline{b}$. Consequently, for every cofinite subset O of PAR, if $\alpha \in O$ then $f(O, \alpha) + g(O, \alpha) = \emptyset$. Moreover, $\Sigma_{\alpha \in PAR} f(PAR, \alpha) + g(PAR, \alpha) = \pi(PAR)(x, y)$. Therefore, $\pi(PAR)(x, y) = \emptyset$ and $x\ R(PAR)\ y$. Consequently, $(x, f)\ R'(PAR)\ (y, g)$. Therefore:

Lemma 32 *For every parameter $a, b \in PAR$, if $\underline{a} \cup \underline{b} = PAR$ then $R'(\underline{a} \cup \underline{b}) = R'(\underline{a}) \cap R'(\underline{b})$.*

Let it be proved that, for every parameter $a \in PAR$, if $|Q| \geq 2$ and $\underline{a} \cup Q = PAR$ then $R'(\underline{a} \cup Q) = R'(\underline{a}) \cap R'(Q)$, for every cofinite subset Q of PAR. Suppose that $(x, f)\ R'(\underline{a})\ (y, g)$ and $(x, f)\ R'(Q)\ (y, g)$. Consequently, there exists $b_1, \ldots, b_j \in Q$ such that $(x, f)\ R'(\underline{b}_1) \circ \cdots \circ R'(\underline{b}_j)\ (y, g)$. If $\underline{a} \cup Q = PAR$ then, for every parameter $\alpha \in PAR$, either $\alpha \in \underline{a}$ or $\alpha \in Q$ — in which case, for every $l \in (j)$, $\alpha \in \underline{b}_l$. Therefore, for every cofinite subset O of PAR, if $\alpha \in O$ then $f(O, \alpha) + g(O, \alpha) = \emptyset$. Moreover, $\Sigma_{\alpha \in PAR} f(PAR, \alpha) + g(PAR, \alpha) = \pi(PAR)(x, y)$. Therefore, $\pi(PAR)(x, y) = \emptyset$ and $x\ R(PAR)\ y$. Consequently, $(x, f)\ R'(PAR)\ (y, g)$. Therefore:

Lemma 33 *For every parameter $a \in PAR$, if $|Q| \geq 2$ and $\underline{a} \cup Q = PAR$ then $R'(\underline{a} \cup Q) = R'(\underline{a}) \cap R'(Q)$, for every cofinite subset Q of PAR.*

Let it be proved that, if $|P| \geq 2$, $|Q| \geq 2$ and $P \cup Q = PAR$ then $R'(P \cup Q) = R'(P) \cap R'(Q)$, for every cofinite subset P, Q of PAR. Suppose that $(x, f)\ R'(P)\ (y, g)$ and $(x, f)\ R'(Q)\ (y, g)$. Consequently, there exists $a_1, \ldots, a_i \in P$ such that $(x, f)\ R'(\underline{a}_1) \circ \cdots \circ R'(\underline{a}_i)\ (y, g)$ and there exists $b_1, \ldots, b_j \in Q$ such that $(x, f)\ R'(\underline{b}_1) \circ \cdots \circ R'(\underline{b}_j)\ (y, g)$. If $P \cup Q = PAR$ then, for every parameter $\alpha \in PAR$, either $\alpha \in P$ — in which case, for every $k \in (i)$, $\alpha \in \underline{a}_k$ — or $\alpha \in Q$ — in which case, for every $l \in (j)$, $\alpha \in \underline{b}_l$. Therefore, for every cofinite subset O of PAR, if $\alpha \in O$ then $f(O, \alpha) + g(O, \alpha) = \emptyset$. Moreover, $\Sigma_{\alpha \in PAR} f(PAR, \alpha) + g(PAR, \alpha) = \pi(PAR)(x, y)$. Therefore, $\pi(PAR)(x, y) = \emptyset$ and $x\ R(PAR)\ y$. Consequently, $(x, f)\ R'(PAR)\ (y, g)$. Therefore:

Lemma 34 *If $|P| \geq 2$, $|Q| \geq 2$ and $P \cup Q = PAR$ then $R'(P \cup Q) = R'(P) \cap R'(Q)$, for every cofinite subset P, Q of PAR.*

Consequently:

Lemma 35 *For every (finite) normal frame $\mathcal{F}$, (if PAR is finite then) there exists a (finite) normal quasi-standard frame $\mathcal{F}'$ and a copying I from $\mathcal{F}$ into $\mathcal{F}'$.*

Let A be a consistent formula of $DAL^{\cup}$. According to theorem 3, A is satisfied in a possible world of some finite normal model. According to lemmas 5 and 35, (if PAR is finite then) A is satisfied in a possible world of some (finite) normal quasi-standard model. Therefore:

Theorem 4 *(If PAR is finite then) $DAL^{\cup}$ is complete for the class of all (finite) normal quasi-standard models, that is to say: the formulas valid in every (finite) normal quasi-standard model are theorems of $DAL^{\cup}$.*

7 $DAL^{\cap}$

This section presents a rough sketch of the proof of the completeness of the axiomatization $DAL^{\cap}$ of the set of the formulas valid in every sweet Kripke models with $\cap$-relative accessibility relations of equivalence. Only the differences with the language, the semantics and the proof of the completeness of $DAL^{\cup}$ are described.

7.1   Language

Let PAR be a nonempty set of *parameters*. Let $\mathcal{P}(PAR)^{\bullet}$ be the set of the finite subsets of PAR. For every parameter $a \in PAR$, let $\overline{a} = \{a\}$. The modal operator $[P]$ is added to the standard propositional formalism, for every finite subset P of PAR. A set Φ of formulas is *closed* when:

- Φ is closed for the subformula,

- for every parameter $a \in PAR$ and for every formula $[\overline{a}]A \in \Phi$, $[\emptyset][\overline{a}]A \in \Phi$,

- if $|P| \geq 2$ then, for every parameter $a \in P$ and for every formula $[P]A \in \Phi$, $[\overline{a}][P]A \in \Phi$, for every finite subset P of PAR.

7.2   Semantics

A *frame* is a structure of the form $\mathcal{F} = (W, R)$ where W is a nonempty set of *possible worlds* and R is a mapping on the set of the finite subsets of PAR to the set of the relations of equivalence on W such that:

- $R(P \cap Q) \subseteq R(P) \cap R(Q)$,

for every finite subset P, Q of PAR. Direct calculations would lead to the conclusion that:

Lemma 36 *Let $\mathcal{F} = (W, R)$ be a frame. If $|P| \geq 2$ then $R(P) \supseteq \bigcup_{a \in P}^{\bullet} R(\overline{a})$, for every finite subset P of PAR.*

$\mathcal{F}$ is *normal* when:

- if $|P| \geq 2$ then $R(P) = \bigcup_{a \in P}^{\bullet} R(\overline{a})$,

for every finite subset P of PAR. Direct calculations would lead to the conclusion that:

Lemma 37 *Let $\mathcal{F} = (W, R)$ be a normal frame. $R(P \cup Q) = R(P) \cup^{\bullet} R(Q)$, for every finite subset P, Q of PAR.*

$\mathcal{F}$ is *quasi-standard* when if $P \cap Q = \emptyset$ then:

- $R(P \cap Q) = R(P) \cap R(Q)$,

for every finite subset P, Q of PAR. $\mathcal{F}$ is *standard* when:

- $R(P \cap Q) = R(P) \cap R(Q)$,

for every finite subset P, Q of PAR. Direct calculations would lead to the conclusion that if PAR is finite then the normal standard frames defined in this section are exactly the sweet Kripke frames with $\cap$-relative accessibility relations of equivalence defined in section 1.

7.3 $DAL^\cap$

Together with the classical tautologies, all the instances of the following schemata are axioms of $DAL^\cap$:

- $[P](A \to B) \to ([P]A \to [P]B)$, for every finite subset P of PAR,
- $[P]A \to A$, for every finite subset P of PAR,
- $A \to [P] < P > A$, for every finite subset P of PAR,
- $[P]A \to [P][P]A$, for every finite subset P of PAR,
- $[P]A \vee [Q]A \to [P \cap Q]A$, for every finite subset P, Q of PAR,
- if $|P| \geq 2$ then $[P](A \to \bigwedge_{a \in P}[\overline{a}]A) \to (A \to [P]A)$, for every finite subset P of PAR.

Together with the *modus ponens*, the following schema is a rule of $DAL^\cap$:

- if $\vdash_{DAL^\cap} A$ then $\vdash_{DAL^\cap} [P]A$,

for every finite subset P of PAR.

7.4 Normal completeness

Let $R^\circ(\emptyset)$ be the binary relation on W° defined by $x\ R^\circ(\emptyset)\ y$ iff:

- for every formula $[\emptyset]A \in \Phi$, $x \models_{\mathcal{M}} [\emptyset]A$ iff $y \models_{\mathcal{M}} [\emptyset]A$.

For every parameter $a \in PAR$, let $R^\circ(\overline{a})$ be the binary relation on W° defined by $x\ R^\circ(\{a\})\ y$ iff:

- for every formula $[\overline{a}]A \in \Phi$, $x \models_{\mathcal{M}} [\overline{a}]A$ iff $y \models_{\mathcal{M}} [\overline{a}]A$.

If $|P| \geq 2$ then let $R^\circ(P) = \bigcup^*_{a \in P} R^\circ(\overline{a})$, for every finite subset P of PAR.

7.5 Normal quasi-standard completeness

Let I be the set of the mappings f on $\mathcal{P}(PAR)^\bullet \times PAR$ to $\mathcal{P}(W)$ such that the set $\{\alpha \colon \alpha \in \underline{P}$ and $f(P, \alpha) \neq \emptyset\}$ is finite, for every finite subset P of PAR. Let $R'(\emptyset)$ be the binary relation on W' defined by $(x, f)\ R'(\emptyset)\ (y, g)$ iff:

- $x\ R(\emptyset)\ y$,
- for every parameter $\alpha \in PAR$ and for every finite subset P of PAR, if $\alpha \in \underline{P}$ then $f(P, \alpha) + g(P, \alpha) = \emptyset$.

For every parameter $a \in PAR$, let $R'(\overline{a})$ be the binary relation on W' defined by $(x, f)\ R'(\overline{a})\ (y, g)$ iff:

- $x\ R(\overline{a})\ y$,
- for every parameter $\alpha \in \underline{a}$ and for every finite subset P of PAR, if $\alpha \in \underline{P}$ then $f(P, \alpha) + g(P, \alpha) = \emptyset$,
- for every finite subset P of PAR, if $\underline{P} \cap \underline{a} \neq \emptyset$ then $\Sigma_{\alpha \in \underline{P}} f(P, \alpha) + g(P, \alpha) = \pi(P)(x, y)$.

If $|P| \geq 2$ then let $R'(P) = \bigcup^*_{a \in P} R'(\overline{a})$, for every finite subset P of PAR.

8 Conclusion

Our proof of the completeness of DAL^n for the class of all normal quasi-standard frames solves an open problem brought up by Fariñas del Cerro and Orlowska [4] [10] in the context of the logic for the analysis of data. The problem of the "Hilbert-style complete axiomatization of logic DAL" is considered by Demri and Orlowska [3]. Our result is more general than theirs since we do not impose any condition of bounded chain or local agreement to the Kripke frames with relative accessibility relations. However, our result is only partial since it does not say anything about the axiomatizability of the set of the formulas valid in every normal *standard* model. As well, the decidability of this set of formulas is still unknown, despite a result of Demri [2] who considers the decidability of the set of the formulas valid in every normal standard model satisfying the restrictive condition of the local agreement.

References

[1] P. Balbiani. *Modal logics with relative accessibility relations*. E. Orlowska (editor), Reasoning with Incomplete Information. To appear.

[2] S. Demri. *The validity problem for the logic DALLA is decidable*. Bulletin of the Polish Academy of Sciences, to appear.

[3] S. Demri and E. Orlowska. *Logical analysis of indiscernability*. E. Orlowska (editor), Reasoning with Incomplete Information. To appear.

[4] L. Fariñas del Cerro and E. Orlowska. *DAL - a logic for data analysis*. Theoretical Computer Science, Volume 36, 251-264, 1985.

[5] R. Goldblatt. *Logics of Time and Computation*. Center for the Study of Language and Computation, Lecture Notes Number 7, 1987.

[6] G. Hughes and M. Cresswell. *A Companion to Modal Logic*. Methuen, 1984.

[7] B. Konikowska. *A formal language for reasoning about indiscernability*. Bulletin of the Polish Academy of Sciences, Volume 35, 239-249, 1987.

[8] B. Konikowska. *A logic for reasoning about relative similarity*. E. Orlowska (editor), Reasoning with Incomplete Information. To appear.

[9] E. Orlowska. *Modal logics in the theory of information systems*. Zeitschr. f. math. Logik und Grundlagen d. Math., Volume 30, 213-222, 1984.

[10] E. Orlowska. *Kripke semantics for knowledge representation logics*. Studia Logica, Volume 49, 255-272, 1990.

[11] D. Vakarelov. *Modal logics for knowledge representation systems*. Theoretical Computer Science, Volume 90, 433-456, 1991.

[12] D. Vakarelov. *A modal theory of arrows. Arrow logics I*. D. Pearce and G. Wagner (editors), Logics in AI, European Workshop JELIA '92, Berlin, Germany, September 1992, Proceedings. Lecture Notes in Artificial Intelligence 633, 1-24, Springer-Verlag, 1992.

[13] D. Vakarelov. *Many-dimensional arrow structures. Arrow logics II*. Journal of Applied Non-Classical Logics, to appear.

From Specifications to Programs:
A Fork-Algebraic Approach to Bridge the Gap

*Gabriel A. Baum, †Marcelo F. Frias, †Armando M. Haeberer,
*Pablo E. Martínez López

*Universidad Nacional de La Plata,
LIFIA, Departamento de Informática,
C.C.11, Correo Central, 1900, La Plata,
Provincia de Buenos Aires, República Argentina.
e–mail: {gbaum, fidel}@info.unlp.edu.ar.
†Laboratório de Métodos Formais,
Departamento de Informática,
Pontifícia Universidade Católica do Rio de Janeiro,
Rua Marquês de São Vicente 225, 22453–900,
Rio de Janeiro, RJ, Brazil.
e–mail: {mfrias, armando}@inf.puc-rio.br.

Abstract. The development of programs from first–order specifications
has as its main difficulty that of dealing with universal quantifiers. This
work is focused in that point, i.e., in the construction of programs whose
specifications involve universal quantifiers. This task is performed within
a relational calculus based on *fork algebras*. The fact that first–order theories can be translated into equational theories in abstract fork algebras
suggests that such work can be accomplished in a satisfactory way. Furthermore, the fact that these abstract algebras are representable guarantees that all properties valid in the standard models are captured by
the axiomatization given for them, allowing the reasoning formalism to
be shifted back and forth between any model and the abstract algebra.
In order to cope with universal quantifiers, a new algebraic operation
— *relational implication* — is introduced. This operation is shown to
have deep significance in the relational statement of first–order expressions involving universal quantifiers. Several algebraic properties of the
relational implication are stated showing its usefulness in program calculation. Finally, a non–trivial example of derivation is given to asses the
merits of the relational implication as an specification tool, and also in
calculation steps, where its algebraic properties are clearly appropriate
as transformation rules.

1 Introduction

The last few years have witnessed a renewed interest of the computing science
community in relational programming calculi. This interest is mainly due to the
advantage of relational calculi for dealing with non–determinism. At the same
time, the postulated advantage of relational calculi for describing, in early steps
of specification construction, *what–to–do* instead of *how–to–do*, gives relevance

to such calculi. Even while the first advantage cannot be denied (for example when comparing relational calculi against functional calculi), the second claimed advantage must be thoroughly discussed, since in relational frameworks, everything that can be said can be said easily, but not everything can be said. Thus relational calculi may fall for short when dealing with the problem of software specification.

This paper is divided in three main parts. The first part introduces fork algebras and some well known results about their expressiveness. The second one deals with the fork algebras representation problem, and its relevance in the process of program construction from specifications. Finally, in the third part — containing the main results regarding formal program construction —, the relational implication is introduced and some algebraic rules are provided. Also, a case study is presented to show the suitability of relational implication in dealing with problems whose natural specifications involve universal quantifiers. While relational implication was already introduced in [7], where a simple case study is also given, it is here where for the first time a thorough study about it is carried on. We will present properties of the relational implication which result essential when deriving algorithms from specifications containing universal quantifiers.

2 Fork Algebras: Models, Axiomatization and Expressiveness

Fork Algebras arose in computer science as an equational calculus for formal program specification and development within a relational framework [1, 3, 7, 8]; they showed however to have inherence also in the fields of algebra and logic, as shown in [5, 6, 11, 12, 13]. In this section we will present the models that motivated the usage of fork algebras as a calculus for program specification and development. Also given is an abstract axiomatization, which constitutes a finite base for those models as proved in Theo. 6. Finally, some known results on the expressiveness of fork algebras are mentioned, and a discussion on the implications of these results in the field of program specification is carried on.

Proper fork algebras (PFAs for short) are algebras of relations [2, 15] extended with a new operator — called *fork* — devised to deal with complex objects with a tree–like structure. These "trees" may have branches of infinite height. Even though this may seem a drawback, it can be useful for modeling infinite processes or infinite computations as it is done in [3]. Furthermore, this tree–like structure allows handling of as many variables as necessary when representing first–order formulas as fork algebra terms.

In order to define PFA's[1], we will first define the class of powerset ⋆PFAs by

---

[1] It is important to remark that PFA's are quasi–concrete structures since, as was pointed out by Andréka and Németi in a private communication, concrete structures must be fully characterized by their underlying domain, something that does not happen with proper fork algebras because of the (hidden) operation ⋆.

Definition 1. A powerset $\star$PFA is a two sorted structure with domains $\mathcal{P}(V)$ and U

$$< \mathcal{P}(V), U, \cup, \cap, ', \emptyset, V, |, Id, \sim, \underline{\nabla}, \star >$$

such that

1. V is an equivalence relation with domain U,
2. $|$, Id and $\sim$ stand respectively for composition between binary relations, the diagonal relation on U and the converse of binary relations, thus making the reduct $< \mathcal{P}(V), \cup, \cap, ', \emptyset, V, |, Id, \sim >$ an algebra of binary relations,
3. $\star : U \times U \to U$ is an injective function when its domain is restricted to V,
4. whenever xVy and xVz, also $xV \star (y,z)$,
5. $R\underline{\nabla}S = \{< x, \star(y,z) >: xRy \text{ and } xSz\}$.

Definition 2. The class of PFA's is defined as $\mathbf{S}\,\mathbf{Rd}\,[\star \text{PFA}]$, where $\mathbf{Rd}$ takes reducts to the similarity type $< \cup, \cap, ', \emptyset, V, |, Id, \sim, \underline{\nabla} >$ and $\mathbf{S}$ takes subalgebras.

It is important to notice that, in the characterization of proper fork algebras, we use variables ranging over two different classes of entities. Some variables range over relations (like R and S in Def. 1) and others range over individuals (like x, y, z also in Def. 1), thus leading to a two sorted theory. In the same way that dummy variables ranging over individuals are avoided in functional languages, we are going to present an abstract characterization of proper fork algebras as a first-order class whose variables range only over relations. This relationship between proper fork algebras and the class of abstract fork algebras — defined below —, is made precise in Sect. 3.

Definition 3. An abstract fork algebra (AFA for short) is an algebraic structure

$$< R, +, \bullet, {}^-, 0, \infty, ;, 1, \smile, \nabla >$$

satisfying the following set of axioms

1. Axioms stating that $< R, +, \bullet, {}^-, 0, \infty, ;, 1, \smile >$ is a relation algebra[2] where $< R, +, \bullet, {}^-, 0, \infty >$ is the Boolean reduct, $< R, ;, 1 >$ is the monoid reduct, and $\smile$ stands for relational converse,
2. $r\nabla s = (r; (1\nabla\infty)) \bullet (s; (\infty\nabla 1))$,
3. $(r\nabla s) ; (t\nabla q)^\smile = (r;t^\smile) \bullet (s;q^\smile)$,
4. $(1\nabla\infty)^\smile \nabla (\infty\nabla 1)^\smile \preceq 1$, where $\preceq$ is the partial ordering induced by the Boolean structure.

Next we will introduce some non–fundamental fork algebra operation that will be used in the specification and derivation process.

Definition 4. Let the operations π, ρ and $\otimes$ be defined by

1. $\pi = (1\nabla\infty)^\smile$ (first projection)

---

[2] Equational axiomatic systems for relation algebras are given in [2, 15].

2. $\rho = (\infty \nabla 1)^{\smile}$ (second projection)
3. $R \otimes S = (\pi;R)\nabla(\rho;S)$ (cross product)
4. $2 = 1\nabla 1$ (equality filter)

Operations π and ρ stand for projecting relations (see Fig.1) regarding the underlying pair formation operation $\star$. Since 1 stands for the diagonal relation, the relation $2^{\smile}$ acts like a filter, filtering those pairs $< x \star y >$ such that $x \neq y$.

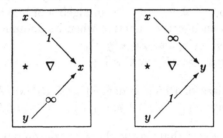

Fig. 1. The projections π and ρ.

2.1 The Expressiveness of Abstract Fork Algebras

In order to describe shortly the relationship existing between first order logic with equality and fork algebras, we can say that first order theories can be *interpreted* as equational theories in fork algebras. More formally, let L be a first order language. Let us denote by $< A, L' >$ the extension of the similarity type of abstract fork algebras with a sequence of constant symbols whose names are sequentially assigned from the symbols in L. Then the following theorem holds.

Theorem 5. *There exists a recursively defined mapping T, translating formulas in L into equations in $< A, L' >$, satisfying*

$$\Gamma \vdash \alpha \qquad \Longleftrightarrow \qquad \{T(\gamma) = \infty : \gamma \in \Gamma\} \vdash_{\nabla} T(\alpha) = \infty.$$

The symbol $\vdash_{\nabla}$ in Th.5 stands for provability in fork algebras, i.e., proofs are made in equational logic and the extralogical axioms defining the fork algebra operators are assumed to hold.

The result shown in Th.5 was already known for other algebraic systems closely related with fork algebras, as quasi–projective relation algebras and pairing relation algebras. The work on the interpretability of first–order theories into quasi–projective relation algebras was extensively developed by Tarski and Givant in [15], while the version for pairing relation algebras was developed by Maddux in [10].

Theorem 5 has an strong application in program specification within the framework of abstract fork algebras. If we use as our primitive specification language some first–order theories — assumption more than reasonable since first–order languages are simple and expressive formal languages — Th.5 guarantees that by applying the mapping T to a first–order specification of a given problem, we obtain a faithful abstract relational specification of it.

Regarding the issue presented in Sect. 1 on the suitability of relation algebras for expressing *what–to–do* instead of *how–to–do*, it must be noted that the answer is not so straight. Although relations have some natural operations as converse or complement which are powerful in the specification construction process (both allow to specify non–functional problems), the expressiveness of the involved relational language cannot be ignored. It is, for example, a well known fact that the language of relation algebras — as presented in [15] — has a very limited expressive power [9], and thus is not adequate for complex program specification. On the other hand, the language of abstract fork algebras has been shown to be expressive enough as to cope with the specification and derivation process, as shown in [7].

3 The Representation Problem for Fork Algebras

Fork algebras' expressiveness theorem establish that the specifications and the properties of the application domain, which may be expressed in first–order logic, can also be expressed in the *equational* theory of abstract fork algebras. However, this expressibility is insufficient for one to formulate, within the theory, many of the fundamental aspects of the program construction process. The process of program construction by calculations within relational calculi requires more than the possibility to express the specification of requirements. It also requires the calculus to provide heuristics that guide the syntactic manipulations in the development process. If every AFA would be representable as a PFA, then we could look at elements from AFAs as binary relations, thus inheriting all the heuristic power of binary relations into the abstract calculus. Fortunately, as will be shown in the next theorem, AFAs are representable.

Theorem 6. *Every abstract fork algebra is isomorphic to some proper fork algebra.*

A first proof of this theorem for complete and atomistic AFAs is given in [4]. Later on Gyuris [6] and Frias et al. [5] proved a representation theorem for the whole class. Strictly speaking, since PFAs are not a concrete class because of the hidden operation $\star$, this can be considered as a weak–representation theorem[3]. Anyway, the theorem provides all the machinery we need for the process of program construction. A strong representation theorem for fork algebras was proved by Németi in [12], although in non–well founded set theories.

The representation theorem may seem of interest only for theoreticians, but it has a great impact in program specification and development within the framework of fork algebras. An immediate corollary of the theorem is that

---

[3] This was pointed out by Andréka and Németi in a private communication.

$Cn(\mathsf{PFA}) \subseteq Cn(\mathsf{AFA})^4$, and thus any first–order property valid in the standard models is reflected in the abstract ones. This simple fact, along with the suitability of proper fork algebras for reasoning about programs, makes the pair $<$abstract, proper$>$ fork algebras a framework with particular heuristic power not shared by some other calculi, either relational or functional.

4 Calculating Algorithms from First–Order Specifications Involving Universal Quantifiers

In the course of this section, we will introduce the *relational implication*, a slight modification of the right residual, which is shown to be very adequate for representing first–order formulas involving universal quantifiers. Furthermore, relational implication has a nice representation in the standard model which, together with the discussion carried on in the last paragraph of Sect. 3, leads to an abstract operation with deep significance for the process of program specification and development.

Finally, we will present a problem whose first–order specification involves universal quantification. We will obtain from it a relational specification in terms of the relational implication. The main steps of a smooth derivation of a recursive specification will be given, showing the adequacy of the relational implication in the task of software specification and construction.

After some experiences with problems whose first–order specification involves universal quantification, the formula

$$\phi(x,\ y) := (\forall z)(x \ R \ z \ \Rightarrow \ y \ S \ z)$$

(or some minor variants of it) recurrently appears. It is used, for example, for specifying minimization and sorting problems [7] — two classical case studies — and in this paper it will be used in the specification of a non trivial problem about binary trees.

It is not difficult to see that the formula above is equivalent to the formula

$$\Psi(x,y) := x \ \overline{R \ ; \ \overline{S^{\smile}}} \ y,$$

thus, in our abstract framework we can define the relational implication of relations R and S by

$$R{\rightarrow}S := \overline{R \ ; \ \overline{S^{\smile}}}.$$

In the standard models, the relational implication can be understood as x is related with y via $R{\rightarrow}S$ if and only if the image of x by R is contained in the image of y by S — see Fig. 2.

Since the relation algebraic characterization of "$\rightarrow$" is hard to manipulate — complement doesn't behave nicely over composition —, we will present some properties of "$\rightarrow$" that allow us to avoid using its definition in derivations.

---

[4] Cn denotes the set of first–order valid formulas in the class.

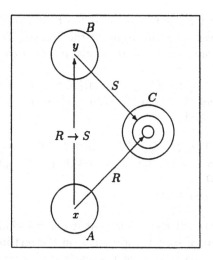

Fig. 2. The *relational implication.*

Besides of some easy properties, such as $((P+Q)\to R) = (P\to R) \bullet (Q\to R)$ and $(P\to(Q\bullet R)) = (P\to Q) + (P\to R)$, — which follow directly from the definition — we will present some more elaborated properties, leading to recursive relational expressions for computing the relational implication.

Proposition 7. *For any relations P, Q and R, we have*
$$Dom\,(P+Q);\,((P+Q)\to R) =$$
$$Dom\,(P);\,\Big((P \to R) \bullet \Big(Dom\,(Q);\,(Q \to R) + \Big(\overline{Dom\,(Q)\bullet 1}\Big);\infty\Big)\Big)$$
$$+$$
$$Dom\,(Q);\,\Big((Q \to R) \bullet \Big(Dom\,(P);\,(P \to R) + \Big(\overline{Dom\,(P)\bullet 1}\Big);\infty\Big)\Big).$$

Proposition 8. *If A is a functional relation, then $Dom\,(A);\,(A\to P) = A;P^{\smile}$.*

Proposition 9. *If B is a functional relation, then $Dom\,(B);\,((B;P)\to Q) = B;\,(P\to Q)$.*

Proposition 10. *Let A and B be functional relations, and let $P = A+B;P$ and $T = P\to Q$. Moreover, let also $Dom\,(P) = Dom\,(A)+Dom\,(B)$ and $Dom\,(A)\bullet Dom\,(B) = 0$. Then*
$$Dom\,(P);T = A;Q^{\smile}+B;T.$$

Proposition 11. *Let A, B and C be functional relations, and let $P = A+B;P+C;P$ and $T = P\to Q$. Moreover, let us suppose that $Dom\,(P) = Dom\,(A)+Dom\,(B)+Dom\,(C)$ and $Dom\,(A)\bullet Dom\,(B) = 0$, $Dom\,(A)\bullet Dom\,(C) = 0$, and $Dom\,(B)\bullet Dom\,(C) = 0$. Then*
$$Dom\,(P);T = A;Q^{\smile}+B;T+C;T.$$

¿From Props. 10 and 11 we can see that the recursiveness of the relation P allows to obtain a recursive specification for the relation T.

Some other nice properties of the relational implication are[5]

Proposition 12. *Let* $T = P{\to}Q$. *If* $P = \inf\{X : X = A{+}B{;}X\}$, *A and B are functional relations, and* $Dom(A)\bullet Dom(B) = 0$, *then*
$$T = A^{\smile};Q + B;T.$$

Proposition 13. *Let* $T = P{\to}Q$. *If* $P = \inf\{X : X = A{+}B{;}X{+}C{;}X\}$, *A, B and C are functional relations, and* $Dom(A)\bullet Dom(B) = Dom(A)\bullet Dom(C) = Dom(B)\bullet Dom(C) = 0$, *then*
$$T = A^{\smile};Q + B;T + C;T.$$

In this section we present a problem that is easily specified using first–order logic, and whose relational specification relies on the relational implication. We then proceed with the main steps of a smooth derivation that leads us to a recursive expression for our problem.

The problem we present as example is stated as

"Given a Binary Tree T without repeated nodes, and two elements x and y belonging to T, find the *Minimum Common Ancestor* of x and y, i.e., that node in T which is the closest ancestor to both x and y".

Let us take a relation HA (abbreviating **has_ancestor**), which is meant to give — given a tree T and an element x in T — the ancestors of x in T. A formal specification of a relation MCA capturing the problem in the language of first–order logic is given by

$$[T, x, y]\text{MCA } a \iff [T, x]\text{HA } a \wedge [T, y]\text{HA } a \wedge$$
$$(\forall z)(([T, x]\text{HA } z \wedge [T, y]\text{HA } z) \Rightarrow [T, a]\text{HA } z).$$

A first–order specification of HA is given by the following formula:

$$[T, x] \text{ HA } a \text{ iff } (\exists T')(T' \sqsubseteq T \wedge T' \text{ root } a \wedge x \text{ in } T'),$$

where the relations HA, in, root and $\sqsubseteq$ are the basic relations on the underlying data type Binary–trees. The relation $\sqsubseteq$ is meant to give all the trees that contain a given tree T' as subtree; we will also use the converse of $\sqsubseteq$, noted by $\sqsupseteq$. The relation in relates a given tree with its elements and the relation **root** gives the root of the tree.

The specification in first–order logic is translated into an equation in AFA almost directly using the relational implication. If we define the relation CA (abbreviating **common_anc**) by

$$\text{CA} = \left(\begin{pmatrix} 1 \\ \otimes \\ \pi \end{pmatrix};\text{HA}\right) \bullet \left(\begin{pmatrix} 1 \\ \otimes \\ \rho \end{pmatrix};\text{HA}\right), \tag{1}$$

---

[5] Props. 12 and 13 were proved by G. Schmidt and M. Frias at the Workshop of the Relational Methods in Computer Science Group, held in Rio de Janeiro, August 1994

then the following equation provides a relational specification for our problem:
$\mathsf{MCA} = \mathsf{MCA}^\diamond;\rho$, where

$$\mathsf{MCA}^\diamond = \begin{pmatrix} \pi \\ \nabla \\ \mathsf{CA} \end{pmatrix} \bullet (\mathsf{CA}\rightarrow\mathsf{HA}). \qquad (2)$$

We also have an equation specifying the relation HA, which is obtained by pattern matching between its first-order specification and the definition of the relational operators in the elementary theory of binary relations [14]:

$$\mathsf{HA} = \begin{pmatrix} \underline{\underline{\beth}} \\ \otimes \\ 1 \end{pmatrix}; \left(\begin{pmatrix} \pi;\mathsf{root} \\ \nabla \\ \begin{pmatrix} \mathsf{in} \\ \otimes \\ 1 \end{pmatrix} ; \smile \\ 2 \end{pmatrix} \right);\pi. \qquad (3)$$

¿From this relational specification, unfolding a recursive definition for $\underline{\underline{\beth}}$, making simple manipulations in fork algebras, and folding HA, we obtain the following recursive equation[6].

$$\mathsf{HA} = \begin{pmatrix} 1_{T_1} \\ \otimes \\ 1 \end{pmatrix}; \begin{pmatrix} \mathsf{root} \\ \otimes \\ 1 \end{pmatrix}; \underset{2}{\smile} + \begin{pmatrix} 1_{T>1} \\ \otimes \\ 1 \end{pmatrix}; \left(\begin{pmatrix} \pi;\mathsf{root} \\ \nabla \\ \begin{pmatrix} \mathsf{in} \\ \otimes \\ 1 \end{pmatrix} ; \smile \\ 2 \end{pmatrix} \right);\pi + \begin{pmatrix} 1_{T>1} \\ \otimes \\ 1 \end{pmatrix}; \begin{pmatrix} \mathsf{right+left} \\ \otimes \\ 1 \end{pmatrix};\mathsf{HA}$$

$$(4)$$

¿From the recursive specification for HA, we can proceed with the derivation of CA. By unfolding Eq. 4 in Eq. 1, making some fork algebra manipulation, and then folding CA, we finally obtain:

$$\mathsf{CA} = \begin{pmatrix} 1_{T_1};\mathsf{root} \\ \otimes \\ 2 \end{pmatrix}; \underset{2}{\smile} + \qquad (5)$$

$$\begin{pmatrix} 1_{T>1} \\ \otimes \\ 1 \end{pmatrix}; \underbrace{\left(Dom\left(\begin{pmatrix} \mathsf{in} \\ \otimes \\ \pi \end{pmatrix} ; \underset{2}{\smile} \right) \bullet Dom\left(\begin{pmatrix} \mathsf{in} \\ \otimes \\ \rho \end{pmatrix} ; \underset{2}{\smile} \right) \right)}_{\phi} ;\pi;\mathsf{root} +$$

$$\begin{pmatrix} 1_{T>1} \\ \otimes \\ 1 \end{pmatrix}; \left(\begin{pmatrix} \mathsf{right} \\ \otimes \\ 1 \end{pmatrix} + \begin{pmatrix} \mathsf{left} \\ \otimes \\ 1 \end{pmatrix} \right);\mathsf{CA}$$

---

[6] 1_{T_1} stands for the relation $\{< x, x >: \mathrm{heigth}(x) = 1\}$, and $1_{T>1}$ stands for the relation $\{< x, x >: \mathrm{heigth}(x) > 1\}$. Given a binary tree, right and left give respectivelly its right and left subtree.

Calling A the first term, B the second term, $C = (C_1+C_2)$ the non-recursive part of the third term and P to CA, Eq. 5 has the shape:

$$P = A + B + (C_1+C_2);P \qquad (6)$$

Now, unfolding Eq. 6 on Eq. 2, and distributing fork over join, we have

$$MCA^\circ = \left(\begin{pmatrix} \pi \\ \nabla \\ A \end{pmatrix} + \begin{pmatrix} \pi \\ \nabla \\ B \end{pmatrix} + \begin{pmatrix} \pi \\ \nabla \\ (C_1+C_2);P \end{pmatrix} \right) \bullet ((A + B + (C_1+C_2);P) \to HA). \quad (7)$$

Applying Props.7, 8, 9, and making elementary manipulations in fork algebras, the term on the right hand side of Eq.7 equals

$$\begin{pmatrix} \pi \\ \nabla \\ A \end{pmatrix} \bullet (A;HA^\smile) \qquad (I)$$

$$+ \begin{pmatrix} \pi \\ \nabla \\ B \end{pmatrix} \bullet (B;HA^\smile) \bullet (Dom\,(C_1;P);C_1;(P \to HA)) \qquad (II)$$

$$+ \begin{pmatrix} \pi \\ \nabla \\ B \end{pmatrix} \bullet (B;HA^\smile) \bullet (Dom\,(C_2;P);C_2;(P \to HA)) \qquad (III)$$

$$+ \begin{pmatrix} \pi \\ \nabla \\ B \end{pmatrix} \bullet (B;HA^\smile) \bullet \left(Dom\,(B);((\overline{Dom\,((C_1+C_2);P)}\bullet 1)\ ;\infty)\right) \qquad (IV)$$

$$+ \begin{pmatrix} \pi \\ \nabla \\ (C_1+C_2);P \end{pmatrix} \bullet (B;HA^\smile) \bullet (Dom\,(C_1;P);C_1;(P \to HA)) \qquad (V)$$

$$+ \begin{pmatrix} \pi \\ \nabla \\ (C_1+C_2);P \end{pmatrix} \bullet (B;HA^\smile) \bullet (Dom\,(C_2;P);C_2;(P \to HA)) \qquad (VI)$$

$$+ \begin{pmatrix} \pi \\ \nabla \\ (C_1+C_2);P \end{pmatrix} \bullet (B;HA^\smile) \bullet \left(Dom\,(B);((\overline{Dom\,((C_1+C_2);P)}\bullet 1)\ ;\infty)\right). \qquad (VII)$$

Since[7] $\pi\nabla A \preceq A;HA^\smile$, $\pi\nabla B \preceq B;HA^\smile$, $\pi\nabla(C_1+C_2);P \leq B;HA^\smile$, and II, III and VII equal 0, we obtain:

$$MCA = \begin{pmatrix} \pi \\ \nabla \\ A \end{pmatrix} ;\rho + \left(\begin{pmatrix} \pi \\ \nabla \\ B \end{pmatrix} \bullet \left(Dom\,(B);((\overline{Dom\,((C_1+C_2);P)}\bullet 1)\ ;\infty)\right) \right) ;\rho \quad (a)$$

---

[7] These properties were proved algebraically. The reader may convince himself of their validity by thinking about the standard models. This is totally valid because of the representation theorem.

$$+ \left(\left(\begin{matrix} \pi \\ \nabla \\ (C_1 + C_2);P \end{matrix} \right) \bullet (Dom(C_1;P);C_1;(P \rightarrow HA)) \right) ; \rho \qquad (b)$$

$$+ \left(\left(\begin{matrix} \pi \\ \nabla \\ (C_1 + C_2);P \end{matrix} \right) \bullet (Dom(C_2;P);C_2;(P \rightarrow HA)) \right) ; \rho. \qquad (c)$$

¿From (a) we obtain the base case of the recursive specification for MCA, while (b) and (c) lead to the recursive parts of the algorithm. We finally obtain (replacing A, B and C by their definitions),

$$MCA = \left(\begin{matrix} 1_{T_1};root \\ \otimes \\ 2 \end{matrix} \right) ; \breve{2}$$

$$+ Dom \left(\dfrac{Dom\left(\left(\begin{matrix} right \\ \otimes \\ 1 \end{matrix} \right) ;CA \right) \bullet 1}{Dom\left(\left(\begin{matrix} left \\ \otimes \\ 1 \end{matrix} \right) ;CA \right) \bullet 1} \right) ; \left(Dom \left(\left(\begin{matrix} in \\ \otimes \\ \pi \end{matrix} \right) ; \breve{2} \right) \bullet Dom \left(\left(\begin{matrix} in \\ \otimes \\ \rho \end{matrix} \right) ; \breve{2} \right) \right) ;\pi;root$$

$$+ Dom \left(\left(\begin{matrix} 1_{T>1} \\ \otimes \\ 1 \end{matrix} \right) ; \left(\begin{matrix} right \\ \otimes \\ 1 \end{matrix} \right) ;CA ; \left(\begin{matrix} right \\ \otimes \\ 1 \end{matrix} \right) ;MCA \right)$$

$$+ Dom \left(\left(\begin{matrix} 1_{T>1} \\ \otimes \\ 1 \end{matrix} \right) ; \left(\begin{matrix} left \\ \otimes \\ 1 \end{matrix} \right) ;CA ; \left(\begin{matrix} left \\ \otimes \\ 1 \end{matrix} \right) ;MCA \right)$$

$$(8)$$

It is important to notice that the terms involving domains are tests for *if–then–else*–like statements. Hence, Eq. 8 leads to the following function:

MCA(t, x, y)

$$\begin{aligned} &= root(t) & , &if \ \ height(t) = 1 \ \wedge \ x = y = root(t) \\ &= root(t) & , &if \ \{ \text{“ x and y are not in the same subtree”} \} \\ &= MCA(right(t), x, y) \ , &if \ \{ \text{“ x and y are in the right subtree”} \} \\ &= MCA(left(t), x, y) \ \ , &if \ \{ \text{“ x and y are in the left subtree”} \} \end{aligned}$$

5 Conclusions

We have discussed three relevant concepts about fork algebras, namely, the applications of the expressiveness and representability theorems to program development, and also the heuristic power of the relational implication in "breaking" recursively, universally quantified expressions.

With respect to the derivation itself, it is important to note that it was developed by strict calculation from the axioms both of fork algebra and the particular abstract data type Binary–trees. In other words, the only means we have used to introduce semantics along the calculation were Binary-trees formally expressed properties. At no point we have used non formal knowledge about the discourse domain.

References

1. Berghammer, R., Haeberer, A.M., Schmidt, G., and Veloso, P.A.S,. *Comparing Two Different Approaches to Products in Abstract Relation Algebras*, in Proceedings of the Third International Conference on Algebraic Methodology and Software Technology, AMAST '93, Springer Verlag, 1993, 167–176.
2. Chin, L.H. and Tarski, A., *Distributive and Modular Laws in the Arithmetic of Relation Algebras*, in University of California Publications in Mathematics. University of California, 1951, 341–384.
3. Frias, M.F., Aguayo N.G. and Novak B., *Development of Graph Algorithms with Fork Algebras*, in Proceedings of the XIX Latinamerican Conference on Informatics, 1993, 529–554.
4. Frias, M.F., Baum, G.A., Haeberer, A.M. and Veloso, P.A.S., *Fork Algebras are Representable*, in Bulletin of the Section of Logic, University of Lódź, (24)2, 1995, pp.64–75.
5. Frias, M.F., Haeberer, A.M. and Veloso, P.A.S., *A Finite Axiomatization for Fork Algebras*, to appear in Journal of the IGPL, 1996.
6. Gyuris, V., *A Short Proof for Representability of Fork Algebras*, Journal of the IGPL, vol 3, N.5, 1995, pp.791–796.
7. Haeberer, A.M., Baum, G.A. and Schmidt G., *On the Smooth Calculation of Relational Recursive Expressions out of First-Order Non-Constructive Specifications Involving Quantifiers*, in Proceedings of the Intl. Conference on Formal Methods in Programming and Their Applications, LNCS 735, Springer Verlag, 1993, 281–298.
8. Haeberer, A.M. and Veloso, P.A.S., *Partial Relations for Program Derivation: Adequacy, Inevitability and Expressiveness*, in Constructing Programs from Specifications – Proceedings of the IFIP TC2 Working Conference on Constructing Programs from Specifications. North Holland., 1991, 319–371.
9. Löwenheim, L., *Über Möglichkeiten im Relativkalkül*, Math. Ann. vol. 76, 1915, 447–470.
10. Maddux, R., *Finitary Algebraic Logic*, Zeitschr. f. math. Logik und Grundlagen d. Math., vol. 35, 1989, 321–332.
11. Mikulás, S., Sain, I., Simon, A., *Complexity of Equational Theory of Relational Algebras with Projection Elements*. Bulletin of the Section of Logic, Vol.21, N.3, 103–111, University of Lódź, October.1992.
12. Németi I., *Strong Representability of Fork Algebras, a Set Theoretic Foundation*, to appear in Journal of the IGPL, 1996.
13. Sain, I. and Németi, I., *Fork Algebras in Usual as well as in Non-well-founded Set Theories*, preprint of the Mathematical Institute of the Hungarian Academy of Sciences, 1994.
14. Tarski, A., *On the Calculus of Relations*, Journal of Symbolic Logic, vol. 6, 1941, 73–89.
15. Tarski, A. and Givant, S.,*A Formalization of Set Theory without Variables*, A.M.S. Coll. Pub., vol. 41, 1987.

Logic of Predicates with Explicit Substitutions

Marek A. Bednarczyk

Institute of Computer Science
P.A.S., Gdańsk

We present a non-commutative linear logic — the logic of predicates with
equality and *explicit substitutions*. Thus, the position of linear logic with
respect to the usual logic is given a new explanation.

1 The world according to Girard

A recent introduction to linear logic, cf. [13], starts with the following explanation
of the position of usual logic with respect to the linear.

> Linear logic is not an alternative logic ; it should rather be seen as an
> extension of usual logic.

This paper aims at supporting the same idea. Our justification of the claim is,
however, quite different from the one offered by Girard. The latter, cf. [9], translates every sequent of the usual propositional logic (classical, or intuitionistic)
into a sequent of commutative linear logic. Then one shows that a sequent can be
proved classically, resp., intuitionistically, iff its translation can be proved linearly.

By contrast, our embedding only works on the level of predicate logic. We show
that every theory of classical logic of predicates with equality lives as a theory
within a non-commutative intuitionistic substructural logic: *the logic of predicates
with equality and explicit substitution*. Also, our explanation does not require to
call upon so called *exponentials* — the modalities introduced by Girard just to
facilitate his embedding.

Our construction is also different from other proposals to move substitutions
from the level of metatheory to the theory of logic, cf. [16]. They add substitutions
as modal constructions. Here, substitutions are considered new atomic formulæ.

1.1 Linear logic

In classical or intuitionistic logics there are several equivalent ways of saying
what *conjunction* or *disjunction* is. It has been argued by Avron, cf. [2], that
logical connectives do not exist outside the context provided by the underlying
logic, i.e., the underlying *consequence relation*. In other words, only after defining
a consequence relation it is possible to talk about connectives associated with
it. The connectives, then, are classified according to properties they satisfy with
respect to this consequence relation. It should come as no surprise that the same
operation on formulæ may play different connectives for different consequence
relations. A simple example of this phenomenon is based on the duality of the
notion of the multi-conclusion consequence relation. Thus, an operation which
is conjunction with respect to one consequence relation is at the same time a
disjunction with respect to the dual consequence relation.

Presentation of logics in terms of *invertible* rules offers many technical advantages, see e.g., [8, 9]. Avron, cf. [2], has made of it a dogma saying that the (proof-theoretic) meaning of a logical connective should always be given in terms of an invertible rule. The dogma limits the number of potential definitions of conjunction, say, to the following two clauses only.

$$\Gamma, A \otimes B \vdash \Delta \quad \text{iff} \quad \Gamma, A, B \vdash \Delta \tag{1}$$

$$\Gamma \vdash A \,\&\, B, \Delta \quad \text{iff} \quad \Gamma \vdash A, \Delta \ \text{and} \ \Gamma \vdash B, \Delta \tag{2}$$

Clause (1) explains conjunction as a way of putting together two consecutive assumptions in a sequent. The other way is to say that to infer a conjunction means to infer both of the conjuncts separately, see (2).

Equivalent, Gentzen-style formulation of conditions (1) and (2) can be found in [2]. Thus, (1) is equivalent to the assumption that the consequence relation is closed under the rules (3). Similarly, (2) is equivalent to the assumption that the consequence relation is closed under the rules (4).

$$\frac{\Gamma, A, B \vdash \Delta}{\Gamma, A \otimes B \vdash \Delta} \qquad \frac{\Gamma \vdash A, \Delta \quad \Gamma' \vdash B, \Delta'}{\Gamma, \Gamma' \vdash A \otimes B, \Delta, \Delta'} \tag{3}$$

$$\frac{\Gamma \vdash A, \Delta \quad \Gamma \vdash B, \Delta}{\Gamma \vdash A \,\&\, B, \Delta} \qquad \frac{\Gamma, A \vdash \Delta}{\Gamma, A \,\&\, B \vdash \Delta} \qquad \frac{\Gamma, B \vdash \Delta}{\Gamma, A \,\&\, B \vdash \Delta} \tag{4}$$

In the presence of the structural rules, i.e., exchange, weakening and contraction, conditions (1) and (2) are equivalent. As a result, the two conjunctions are *logically equivalent*, i.e., the following holds.

$$A \otimes B \dashv\vdash A \,\&\, B \tag{5}$$

Hence, any occurrence of $A \otimes B$ in a formula can always be replaced by $A \,\&\, B$, and vice versa, without changing the meaning of that formula.

The presence of the structural rules is essential. As soon as one of them is dropped the equivalence (5) breaks down. Linear logic considered by Girard in [10, 13] admits exchange as the only structural rule. Consequently, conjunction splits in two: multiplicative $\otimes$ and additive $\&$. Same story goes for disjunction which splits into multiplicative $\#$ and additive $\oplus$.

1.2 Exponentials

The elimination of contraction and weakening seriously limits the expressive power of linear logic. Without some extra provisos it would be impossible to represent in it either classical or intuitionistic logic. Thus, to increase the expressiveness, Girard introduced two exponentials: ! (*Of course!*) and ? (*Why not?*), together with the following rules.

$$\frac{\Gamma, A \vdash \Delta}{\Gamma, !A \vdash \Delta} \ (dereliction) \qquad \frac{\Gamma \vdash \Delta}{\Gamma, !A \vdash \Delta} \ (weakening)$$

$$\frac{!\Gamma \vdash A, ?\Delta}{!\Gamma \vdash !A, ?\Delta} \ (ofcourse) \qquad \frac{\Gamma, !A, !A \vdash \Delta}{\Gamma, !A \vdash \Delta} \ (contraction)$$

Above, only the inference rules for ! are given. Those for ? are dual.

Notation $!\Gamma$ and $?\Delta$ in (*of course*) rule is used to express the side condition that all the assumptions and conclusions have the form $!C$, and $?D$, respectively. The following logical equivalences hold in linear logic.

$$!A \otimes !B \dashv\vdash !(A \,\&\, B) \quad and \quad ?A \,\#\, ?B \dashv\vdash ?(A \oplus B) \tag{6}$$

1.3 Representing classical logics into linear logic

The addition of exponentials helps — classical and intuitionistic logics can now be represented in linear logic. The exposition given below is based on [9]. Let

$$\phi_1, \ldots, \phi_k \vdash \psi_1, \ldots, \psi_\ell \tag{7}$$

be a classical sequent. Its translation into linear logic treats in a different way formulæ on the left and on the right. Thus, one defines *two* translations on the level of formulæ: $n(\phi)$ and $p(\psi)$.

$$
\begin{aligned}
n(a) &= a & p(a) &= a & \text{for atomic } a \\
n(\neg\phi) &= p(\phi)^\perp & p(\neg\psi) &= n(\psi)^\perp \\
n(\phi \wedge \psi) &= n(\phi) \,\&\, n(\psi) & p(\phi \wedge \psi) &= ?p(\phi) \,\&\, ?p(\psi) \\
n(\phi \vee \psi) &= !n(\phi) \oplus !n(\psi) & p(\phi \vee \psi) &= p(\phi) \oplus p(\psi) \\
n(\phi \supset \psi) &= !(p(\phi)^\perp) \oplus !n(\psi) & p(\phi \supset \psi) &= n(\phi)^\perp \oplus p(\psi)
\end{aligned}
$$

Then, the linear counterpart of (7) is defined as follows.

$$!n(\phi_1), \ldots, !n(\phi_k) \vdash ?p(\psi_1), \ldots, ?p(\psi_\ell) \tag{8}$$

In this way classical logic is embedded in linear logic as the following result shows.

Proposition 1 Lemma 3.2 of [9]. *Provability of a sequent in classical logic is equivalent to the provability of its translation in linear logic.*

1.4 Objections against Girard's translation

Girard's explanation of the relationship between the usual and linear logics raises several objections.

What is the linear counterpart of the classical conjunction and disjunction? The translation is schizophrenic when it comes to answer the above question. Indeed, two different answers are given.

– The translation of formulæ uses the additives, plus exponentials when needed.
– On the level of sequents the multiplicatives are used, plus exponentials again.

Thus, two classically equivalent sequents: $\phi, \phi' \vdash \psi, \psi'$ and $\phi \wedge \phi' \vdash \psi \vee \psi'$ are translated into $!n(\phi), !n(\phi') \vdash ?p(\psi), ?p(\psi')$ and $!(n(\phi) \,\&\, n(\phi')) \vdash ?(p(\psi) \oplus p(\psi'))$, respectively. The validity of Proposition 1 crucially depends on (6).

The rôle of the exponentials is not quite clear. The introduction of exponentials adds another level of complexity to logical system that has already been made quite complicated after splitting finitary conjunctions and disjunctions into multiplicative and additive versions. Are they *really* needed?

Moreover, one cannot say that $!A$ and $?A$ capture the *classical content* of a linear formula A. This could be demonstrated even in intuitionistic case where only one exponential is present. Namely, given $!A$ and $!B$ it need not be the case that $!A$ & $!B$ is logically equivalent to a formula of the form $!C$ again. Thus, formulæ of the form $!A$ do not constitute the classical sublogic of linear logic.

The inadequacy of the explanation of the rôle played by the exponentials in linear logic is also felt by others. For example Galier, cf. [9], suggested a variation obtained by adding the following rules.

$$\frac{!A, !B, \Gamma \vdash \Delta}{!A \ \& \ !B, \Gamma \vdash \Delta} \qquad \frac{\Gamma \vdash ?A, ?B, \Delta}{\Gamma \vdash ?A \oplus ?B, \Delta}$$

The rule on the left gives $!A$ & $!B \dashv\vdash \ !(A$ & $B)$. This solves one of the problems mentioned above.

Exponentials are non-universal. That major flow of the construction has already been stressed by Girard. The point is that the choice of rules governing the exponentials does not define them up to logical equivalence. That is, given two pairs of exponentials $!_1, ?_1$ and $!_2, ?_2$ it is impossible to prove, e.g., $!_1 A \dashv\vdash !_2 A$.

Our goal is to demonstrate that another explanation of the connection between the classical and the linear logics is also possible. We offer a translation that does not resort to exponentials at all.

2 Logic of Predicates with Equality

Logicians, see for example [9, 13], often like to simplify the presentation of logics by rejecting negation as a primitive connective. Instead, it is assumed that the atomic formulas are split evenly into *positive* and *negative*. Then negation is defined as (meta)operation on formulæ with the help of de Morgan rules. Similarly, one also considers implication as a defined operation if multiplicative conjunction and disjunction are at hand.

We resort to this trick not for convenience sake, but out of real necessity. The ban on negation and derived operations, like classical implication, is crucial for our development to go through.

2.1 A sequent-style presentation

The syntax of *logic of predicates* $\text{LP}^=$ — a fragment of the classical logic of predicates with equality considered here — is given by the following grammar.

$$\phi ::= a \quad \text{(atoms)} \quad | \ \textbf{tt} \quad \text{(truth)} \quad | \ \phi \wedge \phi \quad \text{(conjunction)}$$
$$| \ \textbf{ff} \quad \text{(falsity)} \quad | \ \phi \vee \phi \quad \text{(disjunction)}$$

Thus, we consider quantifier free fragment of classical logic. The structure of atoms as predicates and the way in which they allow to discuss negation is explained later in greater detail.

Let us turn to a sequent-style presentation of the logic $\text{LP}^=$. Let φ, ϕ, ψ, etc., be metavariables ranging over the formulæ, and let Φ, Ψ, etc., range over sequences of formulæ of $\text{LP}^=$.

A sequent system $\mathcal{K}$ for the logic is given in Table 1 in the appendix. In the sequel $\Phi \vdash_{\mathcal{K}} \Psi$ means that the sequent $\Phi \vdash \Psi$ can be derived in $\mathcal{K}$.

In $\mathcal{K}$ conjunctions (nullary and binary) and disjunctions are presented as additive conjunction and disjunction, respectively. It is well-known that in the presence of the structural rules the same effect is obtained by taking the rules for multiplicatives. The preference given to the additive rules facilitates the process of translation of $\text{LP}^=$ to linear logic. There, the classical conjunction and disjunction are mapped to the additive conjunction and disjunction, respectively.

2.2 Predicates and admissible negation

Let e be a vector of k expressions, $e = e_1, \ldots, e_k$. Each predicate symbol R of arity k generates *two* kinds of atomic formulæ:

- $R^+(e)$, an *affirmative* atom, states that the predicate R *does* hold on e.
- $R^-(e)$, a *refutative* atom, states that the predicate R *does not* hold on e.

Notation $R^*(e)$ is used to denote either $R^+(e)$ or $R^-(e)$.

Now, negation can be seen as a definable operation.

$$\neg R^+(e) = R^-(e) \qquad\qquad \neg R^-(e) = R^+(e)$$
$$\neg(\phi \wedge \psi) = (\neg\phi) \vee (\neg\psi) \qquad\qquad \neg\mathbf{tt} = \mathbf{ff}$$
$$\neg(\phi \vee \psi) = (\neg\phi) \wedge (\neg\psi) \qquad\qquad \neg\mathbf{ff} = \mathbf{tt}$$

It follows easily from the definition that $\neg(\neg\varphi) \equiv \varphi$, where $\varphi \equiv \psi$ means the syntactic equality of formulæ. Thus, negation is an involution. However, the structural and logical rules in Table 1 do not capture the idea that $\neg\varphi$ is indeed a negation of φ. One way of enforcing this is to accept the following axioms.

$$(\neg\vdash) \; \frac{}{a, \neg a \vdash} \qquad (\vdash\neg) \; \frac{}{\vdash a, \neg a}$$

The first says that assuming a and $\neg a$ leads to contradiction; the second asserts that either a or $\neg a$ always holds. It follows by induction that $\varphi, \neg\varphi \vdash$ and $\vdash \varphi, \neg\varphi$ hold for all φ.

The implication connective is also meta-definable: $\varphi \Rightarrow \phi \mathrel{\hat{=}} \neg\varphi \vee \phi$.

2.3 Equality

Equality predicate plays a very special role in mathematics and in computer science. From the perspective of the intended embedding it presence is, simply, indispensable.

Our axiomatization is equivalent to the well-established tradition, cf. [15, 8].

$$(=) \; \frac{}{\vdash e = e} \qquad (=s) \; \frac{}{e = e', \varphi[e/x] \vdash \varphi[e'/x]}$$

The first axiom schema asserts transitivity of equality.

The second axiom schema relates substitution to equality. It captures the idea that equals may be substituted for equals. Formally, $(=s)$ says that under assumption that terms e and e' are equal one can replace some occurrences of e in a formula $\varphi[e/x]$ by e'. Here, $\varphi[e/x]$ denotes the formula obtained as a result of syntactic substitution of e for all (free) occurrences of x in φ. Thus, just like negation, substitution is presented as a part of the metatheory of $LP_\sigma^=$.

2.4 Theories

The logic of predicates with equality, i.e., its consequence relation, is generated by the set $\mathcal{K}$ of rules together with axiom schemas $(\vdash\neg)$, $(\neg\vdash)$, $(=)$ and $(=s)$.

Any consequence relation which is obtained in such a way is called a *theory*. In case of classical logic an equivalent definition is obtained by saying that a theory is determined by fixing a proof system like $\mathcal{K}$ and a *set* of formulæ, viz. the 'axioms' of the theory. The reason is that, in the presence of implication, validity of any sequent is equivalent to validity of a sequent of the form $\vdash \psi$. However, as we shall see in section 3, not all logics allow such simplification. That is why a more general notion of an axiom is accepted in this paper.

Axioms are used also in many other situations. Typically, they are a convenient way of formalizing properties of the domain of objects under consideration.

3 Logic of Predicates with Explicit Substitutions

Substitution is normally considered, just like in section 2, as a part of the metatheory of a logic. This applies not only to logics, but to λ-calculæ and type theories as well. It has been recently realized that more efficient implementations of functional languages can be achieved if one better controls the process of performing a substitution. This calls for frameworks with substitution as a *primitive operation*.

Indeed, a variety of λ-calculæ with explicit substitutions have already been considered, cf [1, 14]. All of them are 2-sorted — the old syntactic class of λ-*terms* is retained while a new class of *substitutions* is added.

The logic of predicates has already two sorts: the sort of *terms* and, built over terms, the sort of *formulæ*. So far nobody has considered adding explicit substitutions to it via a new syntactic sort. All attempts known to the author use the idea that substitutions behave as modal operators, see e.g., [16]. We have good reasons to consider substitutions as a new kind of atomic formulæ, cf. [5].

Predicates are eternal. They represent *facts* the truth of which does not depend on the context, or *state*, in which their truth is evaluated. Therefore, we call them *Platonic* here. Substitutions provide a formalization of the idea of an *action* or *change of state* in logic. Hence, we call them *dynamic* atoms.

Since a substitution σ is an atom, a new logical connective is needed to express (pending) application of σ to, say, a predicate a. Let us use the tensor symbol $\otimes$ to denote the postulated connective. Thus, the above situation could be written as $\sigma \otimes a$. Consequently, assuming $\otimes$ is associative, $\sigma \otimes (\sigma' \otimes a) \dashv\vdash (\sigma \otimes \sigma') \otimes a$. Hence, $\sigma \otimes \sigma'$ would corresponds to *composition* of substitutions.

Now, the question is this: Does such a *logical* connective exist? It turns out that we can view $\otimes$ as a *non-commutative* multiplicative conjunction. The identity substitutions $\left[\frac{x}{x}\right]$, for all x, should be neutral elements of $\otimes$. It is therefore natural to identify them all with I — the *multiplicative truth* constant. Since I captures the idea that "nothing changes" it is a good specification for a "do nothing" program, cf. [5].

Altogether, we are led to discover that the logic of predicates with explicit substitution, called $\text{LP}_\sigma^=$, is a fragment of *intuitionistic non-commutative linear logic*. Formulæ of $\text{LP}_\sigma^=$ are given by the following grammar.

$$A ::= \sigma \mid a \mid A \otimes A \mid I \mid A \& A \mid \top \mid A \oplus A \mid \bot$$

Unlike Girard we use $\bot$ instead of $\mathbf{0}$ to denote the neutral element of $\oplus$. This is to stress that it is the least element w.r.t. the derivability relation, and a dual to $\top$.

3.1 A sequent system for $\text{LP}_\sigma^=$

A sequent-style presentation of the logic is given in Table 2 in the appendix.

With one exception, the rules in Table 2 are the natural generalisations of the rules given by Girard for the *commutative* intuitionistic linear logic, cf. [10, 11, 12], to the non-commutative case, cf [7]. The exceptional axiom is $(\bot)$. Its expected generalisation is $\Gamma, \bot, \Delta \vdash A$, as in [7]. However, the stronger axiom is not valid in our intended interpretation in *quasi quantales* as described in section 4 and in. [4]. Embedding the usual logic into $LP_\sigma^=$ gives a good reason for *not* assuming that having falsehood as one of the assumptions always logically implies anything.

3.2 Platonic Formulæ

The idea underlying our embedding is that the logic of predicates lives as a *Platonic sublanguage* of $LP_\sigma^=$. More precisely, a formula is called Platonic if it contains neither explicit substitutions, nor $\otimes$ nor I. To put the same statement in positive form, a Platonic formula is built from Platonic atoms and from additive conjunctions and disjunctions.

Clearly, the Platonic formulæ are in 1–1 correspondence with formulæ of the predicate logic considered in section 2. Thus, with a slight abuse of notation, we let φ range over Platonic forumlæ of $LP_\sigma^=$.

3.3 Platonic axioms

Bear in mind that in a classical sequent $\Phi \vdash \Psi$ sequences Φ and Ψ stand for the additive conjunctions and disjunction of formulæ in Φ and Ψ, respectively. Thus, axioms $(\neg\vdash)$ and $(\vdash\neg)$ responsible for negation correspond to

$$(\&\neg) \; \frac{}{R^\star(e) \, \& \, \neg R^\star(e) \vdash \bot} \quad and \quad (\oplus\neg) \; \frac{}{\top \vdash R^\star(e) \oplus \neg R^\star(e)}$$

The linear counterpart of the reflexivity of equality axiom schema is the following.

$$(=) \; \frac{}{\top \vdash e = e}$$

The other facet of equality, that equals can be substituted for equals, reflects the dynamic nature of substitution. This is treated in subsection 3.5.

In first order logic the conjunction distributes over disjunction, i.e.,

$$(A \oplus B) \, \& \, C \dashv\vdash (A \, \& \, C) \oplus (B \, \& \, C) \quad and \quad A \oplus (B \, \& \, C) \dashv\vdash (A \oplus B) \, \& \, (A \oplus C).$$

In linear logic not all of the above follow from the rules of Table 2. Consequently, we add those missing as axioms.

$$(\oplus\&) \; \frac{}{(A \oplus B) \, \& \, C \vdash (A \, \& \, C) \oplus (B \, \& \, C)} \qquad (\&\oplus) \; \frac{}{A \oplus (B \, \& \, C) \dashv (A \oplus B) \, \& \, (A \oplus C)}$$

3.4 Dynamic atoms

Substitutions, ranged over by σ, have the form $\left[\frac{e}{x}\right]$.

3.5 Dynamic axioms

The additive connectives and predicate atoms are the Platonic ingredients of $LP_\sigma^=$. The dynamic ingredients are: the substitutions, multiplicative conjunction, and multiplicative truth. The dynamic features of the logic are captured by the following set of axioms. We add them all to the theory we build.

Substitution versus equality The principal property that equals may be substituted for equals can now be expressed as the following axiom.

$$(=\sigma) \ \overline{e_1 = e_2 \ \& \ \left[\tfrac{e_1}{x}\right] \vdash \left[\tfrac{e_2}{x}\right]}$$

Quick comparison with its predecessor in the logic of predicates reveals that the formula φ which played a dummy there is now, simply, removed.

Pending substitutions Idea: $\sigma \otimes B$ represents σ *pending* upon B.

Substitution pending on a Platonic atom is explained via meta-substitution.

$$(\sigma R^*) \ \overline{\left[\tfrac{e}{x}\right] \otimes R^*(e_1, \ldots, e_k) \ \dashv\vdash \ R^*(e_1[e/x], \ldots, e_k[e/x])}$$

One substitution pending upon another corresponds to their composition. Thus, composing two substitutions which concern the same variable results in aggregating the effects of both in a single substitution. Just as in (σR^*) pending explicit substitution in the world of formulæ is reduced/explained by metalevel substitution in the world of terms. All identity substitutions are equal I.

$$(xx) \ \overline{\left[\tfrac{e_1}{x}\right] \otimes \left[\tfrac{e_2}{x}\right] \ \dashv\vdash \ \left[\tfrac{e_2[e_1/x]}{x}\right]} \qquad (I) \ \overline{\left[\tfrac{x}{x}\right] \dashv\vdash I}$$

Since simultaneous substitutions are not allowed, all we can do to explain the effect of composition of two substitutions for different variables is to say how they commute. Again, the equality predicate is needed to state the axioms.

$$(xz) \ \overline{c_1 = e_2[e_1/x] \ \& \ e_1 = c_2[c_1/z] \ \& \ \left[\tfrac{e_1}{x}\right] \otimes \left[\tfrac{e_2}{z}\right] \vdash \left[\tfrac{c_1}{z}\right] \otimes \left[\tfrac{c_2}{x}\right]}$$

Axiom (xz) has the side-condition that x and z are different variables.

The meaning of $\sigma \otimes B$ in other cases is guided by distributivity axioms in subsection 3.5.

Platonic formulæ are facts The meaning of $A \otimes B$ where A is a Platonic formulæ has not been described yet. Predicates, i.e., the Platonic atoms, were described as facts, i.e., formulæ the truth of which does not depend on the state. The same property holds for $\bot$, and is enforced on the additive truth. The following axiom schemas capture the idea.

$$(R^*\otimes) \ \overline{R^*(e_1, \ldots, e_k) \vdash R^*(e_1, \ldots, e_k) \otimes \bot} \qquad (\top\otimes) \ \overline{\top \vdash \top \otimes \bot}$$

Notice that axiom $(\top\otimes)$ does not co-exist with the more general axiom $\Gamma, \bot, \Delta \vdash A$ for the additive false. Together they give inconsistency: $\top \vdash \bot$.

As a consequence of the above axioms one obtains the following result.

Proposition 2. *Platonic ϕ's are a constant predicate transformers:* $\phi \dashv\vdash \phi \otimes B$.

The above proposition provides a technical justification of the idea that the Platonic formulæ do not depend on the state.

Distributivity axioms In the fragment of linear logic described in Table 2 the only distributivity law which is guaranteed is that of $\otimes$ over binary additive disjunction $\oplus$.

In general, as exemplified by $(R^*\otimes)$ and $(\top\otimes)$, $\otimes$ does not distribute over nullary additive disjunction $\perp$ on the right. But it does for dynamic atoms.

$$(\sigma\perp) \; \frac{}{\sigma \otimes \perp \vdash \perp}$$

Let us recall that the right-sided distributivity $\perp \otimes A \vdash \perp$ always holds by $(\perp)$.

Distributivity of $\otimes$ over additive conjunctions is not guaranteed in general. But we want it at least for the binary &.

$$(\otimes\&) \; \frac{}{A \otimes (B \& C) \dashv (A \otimes B) \& (A \otimes C)} \qquad (\&\otimes) \; \frac{}{(A \& B) \otimes C \dashv (A \otimes C) \& (B \otimes C)}$$

For nullary conjunction we impose left-sided distributivity only for dynamic atoms.

$$(\sigma\top) \; \frac{}{\sigma \otimes \top \dashv \top}$$

Its general form is not valid, cf. prop. 2. Its right-sided form always holds by $(\top\otimes)$.

Elimination of substitutions The consequences converse to those above follow from the logical rules, so the above give a method for elimination of substitutions and other non-Platonic ingredients. Consider the following inductive definition which shows how it can be done when a formula has a form $A \otimes \varphi$.

Lemma 3. *For any Platonic φ and any A the formula $\lceil A \otimes \varphi \rceil$ is a well defined Platonic formula such that $A \otimes \varphi \dashv\vdash \lceil A \otimes \varphi \rceil$.*

4 Translation of Logic of Predicates to $LP_\sigma^=$

The general idea underlying our translation should be clear by now:

Logic of predicates lives as the Platonic fragment of a $LP_\sigma^=$.

Formally, given a classical formula φ its translation into $LP_\sigma^=$, denoted $\lfloor \varphi \rfloor$, is defined by induction on the structure of φ as follows.

$$\lfloor R^*(e_1,\ldots,e_k) \rfloor = R^*(e_1,\ldots,e_k) \qquad \lfloor \mathbf{tt} \rfloor = \top \qquad \lfloor \varphi \wedge \psi \rfloor = \lfloor \varphi \rfloor \& \lfloor \psi \rfloor$$
$$\lfloor \mathbf{ff} \rfloor = \perp \qquad \lfloor \varphi \vee \psi \rfloor = \lfloor \varphi \rfloor \oplus \lfloor \psi \rfloor$$

Now, one can show that the function that eliminates tensor defined in subsection 3.5 captures the performing of substitution.

Lemma 4. $\lceil \sigma \otimes \lfloor \varphi \rfloor \rceil = \lfloor \varphi\sigma \rfloor$.

Hence, $\sigma \otimes \lfloor \varphi \rfloor \dashv\vdash \lfloor \varphi\sigma \rfloor$ in $LP_\sigma^=$, by Lemma 3.

In LP the following holds

$$\Phi \vdash \Psi \quad \text{iff} \quad \bigwedge \Phi \vdash \bigvee \Psi \tag{9}$$

where $\bigwedge \Phi$ and $\bigvee \Psi$ denote the conjunction and disjunction of all formulæ in Φ and Ψ, respectively.

Accordingly, let $\lfloor \Phi \rfloor$ denote the sequence obtained from Φ by applying the translation elementwise. Keeping in mind the connection between the classical connectives and the linear connectives, and the equivalence (9), one defines the translation on the level of sequents as follows.

$$\lfloor \Phi \vdash \Psi \rfloor \overset{\text{def}}{=} \&\lfloor \Phi \rfloor \vdash \bigoplus \lfloor \Psi \rfloor.$$

4.1 Soundness of the translation

The reader should notice that the axioms responsible for negation adopted in $LP_\sigma^=$ are translations of the negation axioms in predicate logic. Same applies to reflexivity of equality axiom.

This generalises to arbitrary first order theories since every axiom of predicate logic corresponds to a Platonic axiom of $LP_\sigma^=$ via translation.

Somewhat more subtle situation is in the case of the axiom $(=s)$. Its translation is $e=e' \& \lfloor \varphi[e/x] \rfloor \vdash \lfloor \varphi[e'/x] \rfloor$. From prop. 2 it follows that $e=e' \vdash e=e' \otimes \varphi$. From lemma 4 it follows that $\lfloor \varphi[e/x] \rfloor \vdash \left[\frac{e}{x}\right] \otimes \lfloor \varphi \rfloor$ and $\left[\frac{e'}{x}\right] \otimes \lfloor \varphi \rfloor \vdash \lfloor \varphi[e'/x] \rfloor$. Thus, the task substantially simplifies. Now, it can be accomplished by the following proof.

$$
\cfrac{
\cfrac{}{
\text{by } (\&\otimes) - \text{distributivity}
}
\qquad
\cfrac{
\cfrac{e=e' \& \left[\frac{e}{x}\right] \vdash \left[\frac{e'}{x}\right] \quad \lfloor \varphi \rfloor \vdash \lfloor \varphi \rfloor}{e=e' \& \left[\frac{e}{x}\right], \lfloor \varphi \rfloor \vdash \left[\frac{e'}{x}\right] \otimes \lfloor \varphi \rfloor}
}{
(e=e' \& \left[\frac{e}{x}\right]) \otimes \lfloor \varphi \rfloor \vdash \left[\frac{e'}{x}\right] \otimes \lfloor \varphi \rfloor
}
}{}
$$

by $(=\sigma)$

$$\text{by } (\&\otimes) - \text{distributivity}$$
$$(e=e' \otimes \lfloor \varphi \rfloor) \& (\left[\frac{e}{x}\right] \otimes \lfloor \varphi \rfloor) \vdash (e=e' \& \left[\frac{e}{x}\right]) \otimes \lfloor \varphi \rfloor$$

$$(e=e' \otimes \lfloor \varphi \rfloor) \& (\left[\frac{e}{x}\right] \otimes \lfloor \varphi \rfloor) \vdash \left[\frac{e'}{x}\right] \otimes \lfloor \varphi \rfloor$$

Before we formulate the main result of this section let us introduce a bit of notation. Given a set $\mathcal{A}$ of $LP^=$ axioms let $\lfloor \mathcal{A} \rfloor = \{ \&\lfloor \Phi \rfloor \vdash_\lambda \bigoplus \lfloor \Psi \rfloor \mid \Phi \vdash \Psi \in \mathcal{A} \}$ denote its translation to $LP_\sigma^=$. With the above notation we can formulate the following result.

Theorem 5. *Let $\vdash_\kappa$ be a LP theory obtained by extending $\mathcal{K}$ with negation and equality axioms, and an arbitrary set of axioms A. Suppose that $\vdash_\lambda$ is a $LP_\sigma^=$ theory obtained by extending $\mathcal{L}$ and such that the corresponding negation and equality axioms are satisfied together with $(\oplus\&)$, $(\&\oplus)$ and $(\&\otimes)$ distributivity laws. Finally, suppose that $\vdash_\lambda$ admits all axioms in $\lfloor \mathcal{A} \rfloor$. Then*

$$\Phi \vdash_\kappa \Psi \quad \text{implies} \quad \&\lfloor \Phi \rfloor \vdash_\lambda \bigoplus \lfloor \Psi \rfloor.$$

4.2 Conservativity of the translation

The result converse to Theorem 5 also holds. The proof presented in this section is based on semantical considerations.

For the remaining part of this section let $\vdash_\kappa$ be a LP theory obtained by extending $\mathcal{K}$ with negation and equality axioms, and an arbitrary set of axioms $\mathcal{A}$. Let $\vdash_\lambda$ be obtained by extending $\mathcal{L}$ with all axioms listed in section 3, i.e., negation, equality, substitutivity, distributivity and, finally, all axioms from the set $\lfloor \mathcal{A} \rfloor$.

Denotational semantics of LP The notion of a *model* M for LP theory is given as follows.

First, a set D is chosen, and for each function symbol f of arity k a function $f_M : D^k \to D$ called f's *interpretation*. These data induce expression interpretation function $\mathcal{E}_M[\_] : \mathsf{Exp} \to (\mathsf{Var} \to D) \to D$, where Var and Exp are the syntactic classes of variables and expressions, respectively.

Call $\mathsf{Val} \doteq \mathsf{Var} \to D$ the set of *valuations* of variables in D. For $v : \mathsf{Val}$, $x : \mathsf{Var}$ and $d : D$ define a new valuation, denoted $v[x \mapsto d]$, by $v[x \mapsto d]y \doteq d$ whenever x and y are the same variables, and $v[x \mapsto d]y \doteq vy$ otherwise.

It is assumed here that, given a particular vocabulary for building expressions, the reader is capable of filling in the details required to give the definition of $\mathcal{E}_M[\_]$. The denotation of the formulæ of the logic of predicates in M is given by function

$$\mathcal{F}_M[\_] : \mathsf{Pred} \to \mathcal{P}ow(\mathsf{Val})$$

where Pred is the set of LP$^=$ formulæ. Thus, each formula is identified with a set of valuations, viz., the valuations that *satisfy* the formula. This denotation function is determined once an interpretation $R_M \subseteq D^k$ for each k-ary predicate symbol is given. It is assumed, that the interpretation $=_M \subseteq D \times D$ is the diagonal binary relation on D. Let $\mathbf{e} \doteq e_1, \dots, e_k$.

$$
\begin{aligned}
\mathcal{F}_M[R^+(\mathbf{e})] &= \{v : \mathsf{Val} \mid (\mathcal{E}_M[e_1]v, \dots, \mathcal{E}_M[e_k]v) \in R_M\} & \mathcal{F}_M[\mathbf{ff}]) &= \emptyset \\
\mathcal{F}_M[R^-(\mathbf{e})] &= \{v : \mathsf{Val} \mid (\mathcal{E}_M[e_1]v, \dots, \mathcal{E}_M[e_k]v) \notin R_M\} & \mathcal{F}_M[\mathbf{tt}]) &= \mathsf{Val} \\
\mathcal{F}_M[\varphi \vee \psi] &= \mathcal{F}_M[\varphi] \cup \mathcal{F}_M[\psi] & \mathcal{F}_M[\varphi \wedge \psi] &= \mathcal{F}_M[\varphi] \cap \mathcal{F}_M[\psi]
\end{aligned}
$$

We may consider the substitution as an operation on denotations of formulæ. Formally, given a set S of valuations and a substitution $[\frac{e}{x}]$ we define

$$\left[\frac{e}{x}\right] S \doteq \{v : \mathsf{Val} \mid v[x \mapsto \mathcal{E}[e]v] \in S\}.$$

With the above notation one obtains the Substitution Lemma.

Lemma 6 Substitution. $\mathcal{F}_M[\varphi[e/x]] = [\frac{e}{x}] \mathcal{F}_M[\varphi]$.

Every model M determines a consequence relation $\models_M$ given by

$$\Phi \models_M \Psi \quad \text{iff} \quad \bigcap \mathcal{F}_M[\Phi] \subseteq \bigcup \mathcal{F}_M[\Psi]$$

Call M a *model of* $\vdash_\kappa$ if $\vdash_\kappa \subseteq \models_M$, i.e., whenever $\Phi \vdash_\kappa \Psi$ implies $\Phi \models_M \Psi$.

Let $\mathcal{M}_\kappa$ be the set of all models of $\vdash_\kappa$. Then Gödel completeness result follows.

Theorem 7 Completeness. $\Phi \vdash_\kappa \Psi$ *iff* $\Phi \models_M \Psi$, *for all* $M \in \mathcal{M}_\kappa$.

Denotational semantics of LP$^=_\sigma$ The notion of a *model* for a LP$^=_\sigma$ theory is obtained by a trivial adaptation of the idea of LP-model. That is, each LP$^=_\sigma$-model is characterised by exactly the same set of ingredients as before. We take the same denotations for expressions. All the difference are focused on the area of denotations of formulæ. Whereas before these were *set* of valuations, now they are functions between valuations.

Formally, the denotation of the formula of the logic of predicates with explicit substitutions in model M is given by function

$$S_M[\_] : \mathsf{Pred}\sigma \to (\mathcal{P}ow(\mathsf{Val}) \to \mathcal{P}ow(\mathsf{Val}))$$

where $\mathsf{Pred}\sigma$ is the set of $\mathrm{LP}_\sigma^=$ formulæ.

$$S_M[R^\star(\mathbf{e})]S = \mathcal{F}_M[R^\star(\mathbf{e})] \qquad\qquad S_M[[\tfrac{e}{x}]]S = [\tfrac{e}{x}]\,S$$
$$S_M[A \otimes B]S = S_M[A](S_M[B]S) \qquad\qquad S_M[I]S = S$$
$$S_M[A \oplus B]S = S_M[A]S \cup S_M[B]S \qquad\qquad S_M[\bot])S = \emptyset$$
$$S_M[A \,\&\, B]S = S_M[A]S \cap S_M[B]S \qquad\qquad S_M[\top])S = \mathsf{Val}$$

The semantic function extends to sequences in accord to the intuition that the meaning of a sequence stands for iterated tensor.

Let us write $\Gamma \models_M^\sigma A$ iff $S_M[\Gamma]S \subseteq S_M[A]S$, for all $S \subseteq \mathsf{Val}$.

Proposition 8. *1. For any LP formula φ its dynamic meaning (via translation) is a constant function given by: $S_M[\lfloor\varphi\rfloor]S = \mathcal{F}_M[\varphi]$.*
2. $\lfloor\varphi\rfloor \models_M^\sigma \lfloor\psi\rfloor$ implies $\varphi \models_M \psi$.

It is time to come back to the issue of negation as a 'second class citizen' in the logic of predicates. Taking it to the first class league' would result in many problems. Simply, there can be no negation in our intended dynamic denotational semantics.

At the moment we have not defined how to interpret non-intuitionistic sequents $\Gamma \vdash \Delta$, where Δ is not a singleton. But, assuming that we have done it, somehow, the problem remains. Negation $\neg A$ of a linear formula A should satisfy at least $\neg A, A \vdash \bot$. But then $\neg I, I \vdash \bot$, and hance $\neg I \dashv\vdash \bot$. Since $\neg\top \dashv\vdash \bot$ should also hold the negation could not be involution.

The result First, notice that under assumptions about $\vdash_\kappa$ and $\vdash_\lambda$, every model of the former is also a model of the latter.

Proposition 9. *Let M be a model of $\vdash_\kappa$. Then it is also a model of $\vdash_\lambda$.*

And now we are ready to formulate the main result.

Theorem 10. $\vdash_\lambda$ *is a conservative extension of $\vdash_\kappa$*

Proof. Let $\&\lfloor\Phi\rfloor \vdash_\lambda \oplus\lfloor\Psi\rfloor$. Without loss of generality we may assume that Φ and Ψ consist of a single formulæ φ and ψ, respectively.

Let M be a model of $\vdash_\kappa$. By prop. 9 it is also a model of $\vdash_\lambda$. So, $\lfloor\varphi\rfloor \models_M^\sigma \lfloor\psi\rfloor$. Hence, by prop. 8 one obtains $\varphi \models_M \psi$. Finally, $\varphi \vdash_\kappa \psi$, by prop. 7. $\qquad\square$

Conclusions

We have shown that it is possible to give an explanation of the connections between the usual logic and linear logics based on the idea that the former is a platonic fragment of a specific linear theory $\mathrm{LP}_\sigma^=$ — the logic of predicates with explicit substitutions.

Admittedly, the target linear logic is rather weak: it is non-commutative, has only the additives and multiplicative conjunctions, and even lacks one of the implications — see [4] for details.

The interest in studying $\text{LP}_{\sigma}^{\equiv}$ stems from the fact that its formulæ are natural specifications of imperative programs in a mechanised variant of Hoare logic. Thus, I is a natural and best specification for **skip**. Similarly, $\left[\frac{e}{x}\right]$ captures the meaning of the assignment $x := e$. Finally, sequential composition of programs is mimiced by tensor of their specifications. See [5] for details.

Acknowledgements

I would like to acknowledge stimulating discussions with colleagues at ICS PAS, and in particular I would like to thank Beata Konikowska and Andrzej Tarlecki. The research presented here was partially sponsored by The State Committee for Scientific Research, Grant No 2P301 007 04.

References

1. Abadi, M., Cardelli, L., Curien, P.-L. and J-J. Lévy. Explicit substitutions. ACM Conf. on Principles of Programming Languages, San Francisco, 1990.

2. Avron, A. Simple consequence relations. *Information & Computation*, Vol 92, No 1, pp. 105–139, May 1991.

3. Basin, D., Matthews, S., L.Vigano. A modular presentation of modal logics in a logical framework. Proc. 1st Isabelle User Workshop, pp.137–148, Cambridge, 1995.

4. Bednarczyk, M. A. Quasi quantales. GDM Research Report, ICS PAS, 1994.

5. Bednarczyk, M. A. and T. Borzyszkowski Towards program development with Isabelle. Proc. 1st Isabelle User Workshop, pp.101–121, Cambridge, 1995.

6. Bednarczyk, M. A. Logic of predicates versus linear logic. ICS PAS Reports, N° 795, December 1995.

7. Brown, C. and D. Gurr. Relations and non-commutative linear logic. To appear in *J. of Pure and Applied Algebra*

8. Gallier, J. *Logic for Computer Science. Foundations of Automatic Theorem Proving.* Harper & Row, New York, 1986.

9. Gallier, J. Constructive logics. Part II: linear logic and proof nets. Research Report, CIS, University of Pennsylvania, 1992.

10. Girard, J.-Y. Linear logic. *Theoretical Computer Science*, 50 (1987), pp 1–102.

11. Girard, J.-Y. and Y. Lafont. Linear logic and lazy computation. In: *Proc. TAP-SOFT'87 (Pisa), vol. 2*, LNCS 250, Springer Verlag, 1987, pp 52–66.

12. Girard, J.-Y., Lafont, Y. and P. Taylor. *Proofs and Types.* volume 7 of *Cambridge Tracts in Theoretical Computer Science*, Cambridge University Press, 1989.

13. Girard, J.-Y. Linear logic : its syntax and semantics. In *Advances of Linear Logic*, 1995.

14. Lescane, P. From $\lambda\sigma$ to $\lambda\upsilon$ a journey through calculi of explicit substitutions. ACM Conf. on Principles of Programming Languages, San Portland, 1994.

15. Mendelson, E. *Introduction to Mathematical Logic.* D. Van Nostrand Company, 1964.

16. Poigné, A. Basic Category Theory. In *Handbook of Logic in Computer Science. Vol I..* Clarendon Press, Oxford, 1992.

Appendix

(Id)	$$\overline{\varphi \vdash \varphi}$$	(Cut)	$$\frac{\Phi \vdash \Psi, \varphi \quad \varphi, \Phi' \vdash \Psi'}{\Phi, \Phi' \vdash \Psi, \Psi'}$$
(Contr $\vdash$)	$$\frac{\varphi, \varphi, \Phi \vdash \Psi}{\varphi, \Phi \vdash \Psi}$$	($\vdash$ Contr)	$$\frac{\Phi \vdash \Psi, \psi, \psi}{\Phi \vdash \Psi, \psi}$$
(Weak $\vdash$)	$$\frac{\Phi \vdash \Psi}{\varphi, \Phi \vdash \Psi}$$	($\vdash$ Weak)	$$\frac{\Phi \vdash \Psi}{\Phi \vdash \Psi, \psi}$$
(Exch $\vdash$)	$$\frac{\Phi, \varphi, \psi, \Phi' \vdash \Psi}{\Phi, \psi, \varphi, \Phi' \vdash \Psi}$$	($\vdash$ Exch)	$$\frac{\Phi \vdash \Psi, \varphi, \psi, \Psi'}{\Phi \vdash \Psi, \psi, \varphi, \Psi'}$$
(ff $\vdash$)	$$\overline{\mathbf{ff}, \Phi \vdash \Psi}$$	($\vdash$ tt)	$$\overline{\Phi \vdash \Psi, \mathbf{tt}}$$
($\vee \vdash$)	$$\frac{\varphi, \Phi \vdash \Psi \quad \phi, \Phi \vdash \Psi}{\varphi \vee \phi, \Phi \vdash \Psi}$$	($\vdash \wedge$)	$$\frac{\Phi \vdash \Psi, \varphi \quad \Phi \vdash \Psi, \phi}{\Phi \vdash \Psi, \varphi \wedge \phi}$$
($\wedge \vdash l$)	$$\frac{\varphi, \Phi \vdash \Psi}{\varphi \wedge \phi, \Phi \vdash \Psi}$$	($\vdash \vee l$)	$$\frac{\Phi \vdash \Psi, \varphi}{\Phi \vdash \Psi, \varphi \vee \phi}$$
($\wedge \vdash r$)	$$\frac{\phi, \Phi \vdash \Psi}{\varphi \wedge \phi, \Phi \vdash \Psi}$$	($\vdash \vee r$)	$$\frac{\Phi \vdash \Psi, \phi}{\Phi \vdash \Psi, \varphi \vee \phi}$$

Table 1. A sequent system $\mathcal{K}$ for PL$^=$.

(Refl)	$$\overline{A \vdash A}$$	(Cut)	$$\frac{\Gamma \vdash A \quad \Delta, A, \Delta' \vdash B}{\Delta, \Gamma, \Delta' \vdash B}$$
(LI)	$$\frac{\Gamma, \Delta \vdash A}{\Gamma, I, \Delta \vdash A}$$	(RI)	$$\overline{\vdash I}$$
(L$\otimes$)	$$\frac{\Gamma, A, B, \Delta \vdash C}{\Gamma, A \otimes B, \Delta \vdash C}$$	(R$\otimes$)	$$\frac{\Gamma \vdash A \quad \Delta \vdash B}{\Gamma, \Delta \vdash A \otimes B}$$
(L&-L)	$$\frac{\Gamma, A, \Delta \vdash C}{\Gamma, A \mathbin{\&} B, \Delta \vdash C}$$	($\top$)	$$\overline{\Gamma \vdash \top}$$
(L&-R)	$$\frac{\Gamma, B, \Delta \vdash C}{\Gamma, A \mathbin{\&} B, \Delta \vdash C}$$	(R&)	$$\frac{\Gamma \vdash A \quad \Gamma \vdash B}{\Gamma \vdash A \mathbin{\&} B}$$
($\bot$)	$$\overline{\bot, \Gamma \vdash A}$$	(L$\oplus$)	$$\frac{\Gamma, A, \Delta \vdash C \quad \Gamma, B, \Delta \vdash C}{\Gamma, A \oplus B, \Delta \vdash C}$$
(R$\oplus$-L)	$$\frac{\Gamma \vdash A}{\Gamma \vdash A \oplus B}$$	(R$\oplus$-R)	$$\frac{\Gamma \vdash B}{\Gamma \vdash A \oplus B}$$

Table 2. A sequent system $\mathcal{L}$ for NILL.

On the Query Complexity of Sets

Richard Beigel[*1], William Gasarch[**2], Martin Kummer[3], Georgia Martin[2], Timothy McNicholl[4], Frank Stephan[***5]

[1] Department of Computer Science, Yale University, P.O. Box 208285, New Haven, CT 06520-8285, U.S.A. (email: beigel@cs.yale.edu)
[2] Department of Computer Science, University of Maryland, College Park, MD 20742, U.S.A. (email: {gasarch; gam}@cs.umd.edu)
[3] Institut für Logik, Komplexität und Deduktionssysteme, Universität Karlsruhe, 76128 Karlsruhe, Germany (email: kummer@ira.uka.de)
[4] Department of Mathematics, Ottawa University, Ottawa, KS 66067, U.S.A. (email: mcnichol@iso.ott.edu)
[5] Mathematisches Institut, Ruprecht-Karls-Universität, Im Neuenheimer Feld 294, 69120 Heidelberg, Germany (email: fstephan@math.uni-heidelberg.de)

Abstract. There has been much research over the last eleven years that considers the *number of queries* needed to compute a function as a measure of its complexity. We are interested in the complexity of certain *sets* in this context. We study the sets
$\text{ODD}_n^A = \{(x_1, \ldots, x_n) : |A \cap \{x_1, \ldots, x_n\}| \text{ is odd}\}$ and
$\text{WMOD}(m)_n^A = \{(x_1, \ldots, x_n) : |A \cap \{x_1, \ldots, x_n\}| \not\equiv 0 \pmod{m}\}$.
If $A = K$ or A is semirecursive, we obtain tight bounds on the query complexity of ODD_n^A and $\text{WMOD}(m)_n^A$. We obtain lower bounds for A r.e. The lower bounds for A r.e. are derived *from* the lower bounds for A semirecursive. We obtain that every tt-degree has a set A such that ODD_n^A *requires* n parallel queries to A, and a set B such that ODD_n^B can be decided with one query to B. Hence for bounded-query complexity, how information is packaged is more important than Turing degree.
We investigate when extra queries add power. We show that, for several nonrecursive sets A, the more queries you can ask, the more sets you can decide; however, there are sets for which more queries do not help at all.

1 Introduction

One concern of theoretical computer science is the classification of problems in terms of how hard they are. The natural measure of difficulty of the function f is the amount of *time* needed to compute $f(x)$. Other resources, such as *space*, have also been considered.

In recursion theory, by contrast, a function is considered to be 'easy' if there exists *some* algorithm that computes it. We wish to classify functions that are

---

* Supported in part by NSF Grant CCR-8952528 and NSF Grant CCR-9415410.
** Supported in part by NSF Grant CCR-9301339.
*** Supported by the Deutsche Forschungsgemeinschaft (DFG) Grant Am 60/9-1.

'hard,' i.e., not computable, in a quantitative way. We cannot use time or space, since the functions are not even computable. We cannot use Turing degree, since this notion is not quantitative. Hence we use a notion of complexity that is quantitative like time or space, yet captures difficulty within recursion theory like Turing degree. It is for these reasons that the theory of bounded queries in recursion theory was developed [1, 2, 6].

Given that a function f is recursive in a set X, our main concern will be *the number of queries to X required to compute f*. This will be our measure of complexity. Several natural *functions* have been classified this way (see [2, 3, 7, 11, 12]). In this paper we investigate the complexity of *sets* (i.e., the complexity of their characteristic functions). We informally summarize some of our results; a more formal and complete summary is contained at the end of Section 2.

We are concerned with the following sets.

1. $\mathrm{ODD}_n^A = \{(x_1, \ldots, x_n) : |A \cap \{x_1, \ldots, x_n\}| \text{ is odd}\}$, for $n \geq 1$.
2. $\mathrm{WMOD}(m)_n^A = \{(x_1, \ldots, x_n) : |A \cap \{x_1, \ldots, x_n\}| \not\equiv 0 \pmod{m}\}$, for $m \geq 2, n \geq 1$.

The set ODD_n^A is similar to the PARITY problem, which has been well studied [5, 9, 16]. In addition, Yao has proved [8, 13] (in a different framework) that if a set A is unpredictable, then ODD_n^A is also unpredictable. Hence the set ODD_n^A is of interest. $\mathrm{WMOD}(m)_n^A$ is a generalization of ODD_n^A. The acronym WMODm stands for 'Weak-MODm.' MODm is the numeric function that, given x, returns $y \in \{0, \ldots, m-1\}$ such that $x \equiv y \pmod{m}$. Weak-MODm yields less information: just the answer to the question "Is MODm$(x) = 0$?".

Clearly, ODD_n^A can be decided with n parallel queries to A. We show that if A is semirecursive or r.e., this is optimal (unless A is recursive). Since there is a semirecursive set in every tt-degree, this shows that every nonrecursive tt-degree has a set A such that ODD_n^A requires n parallel queries to A. By contrast, we also show that every tt-degree has a set B such that ODD_n^B can be decided with just *one* query to B. This contrast bolsters a recurring theme in bounded-query complexity: How information is packaged (e.g., as a semirecursive set) will affect the complexity, but the information itself (e.g., the tt-degree) will not.

We also obtain upper and lower bounds on the complexity of ODD_n^A using sequential queries, and for $\mathrm{WMOD}(m)_n^A$ using both parallel and sequential queries. These results, and others, are summarized at the end of Section 3. We have other results on bounded-query complexity that are also summarized there.

2 Notation, Definitions, and Useful Lemmas

We use standard notation from recursion theory [15, 17]. $M_0, M_1, \ldots$ is an effective list of Turing machines. $M_0^{()}, M_1^{()}, \ldots$ is an effective list of oracle Turing machines. W_e is the domain of M_e. Hence $W_0, W_1, \ldots$ is an effective list of all r.e. sets. $D_e = \{i : \text{the } i^{\text{th}} \text{ bit of } e \text{ is } 1\}$. Hence $D_0, D_1, \ldots$ is a list of all finite sets. We often use W_e and D_e to denote sets of strings. Let $e, x, s \in \mathsf{N}$. $M_e(x) \downarrow$ means that $M_e(x)$ converges. $M_e(x) \downarrow= b$ means that $M_e(x)$ converges and the output is b. $M_e(x) \uparrow$ means that $M_e(x)$ diverges. $M_{e,s}(x) \downarrow$ means that $M_e(x)$ converges within s steps. $M_{e,s}(x) \uparrow$ means that $M_e(x)$ does not converge within s steps. Let A be a set. A' is $\{e : M_e^A(e) \downarrow\}$ (i.e., A' is the halting problem relative to A). Define $A^{(0)} = A$ and $A^{(i+1)} = (A^{(i)})'$. Let A^ω be $\bigcup_{i=1}^\omega \{\langle i, x\rangle : x \in A^{(i)}\}$. We will be using $\emptyset^\omega$. Note that if A is in the arithmetic hierarchy, then $A \leq_m \emptyset^\omega$.

We define several classes of functions. At the end of this section we will state a known lemma that relates these classes.

Definition 1. Let $n \in \mathsf{N}$, and let A be a set. FQ(n, A) is the collection of all total functions f such that f is recursive in A via an algorithm that makes at most n sequential (adaptive) queries to A. FQC(n, A) is the collection of all functions f such that f is recursive in A via an algorithm $M^{()}$ such that (1) for all x, $M^A(x)$ makes at most n sequential queries to A, and (2) for all x, Y, the $M^Y(x)$ computation converges. (Note that if $f \in$ FQC(n, A) via M^A, then even if false answers are supplied to the queries, the $M^{()}(x)$ computation will converge, though perhaps not to $f(x)$.) FQ$_{||}(n, A)$ is the collection of all total functions f such that f is recursive in A via an algorithm that makes at most n parallel (nonadaptive) queries to A.

Definition 2. Let $n \geq 1$, and let f be a total function. Then f is *n-enumerable* (*strongly n-enumerable*), denoted $f \in$ EN(n) ($f \in$ SEN(n)), if there exists a recursive function g such that, for all x, $|W_{g(x)}| \leq n$ and $f(x) \in W_{g(x)}$ ($|D_{g(x)}| \leq n$ and $f(x) \in D_{g(x)}$). If $f \in$ EN(n) then, given x, we can—over time—list out $\leq n$ possibilities for $f(x)$, one of which is correct. We never know that we have obtained all the possibilities (unless we eventually get n possibilities). If $f \in$ SEN(n) then, given x, we can list out $\leq n$ possibilities for $f(x)$, one of which is correct; moreover, we *do* have all the possibilities—and we *know* we have them all. (This concept first appeared in a recursion-theoretic framework in [1]. Some very general theorems about EN(n) were proved in [12]. The term 'enumerable' is from [4], where it was defined in a polynomial-time-bounded framework.)

Definition 3. Let B be a set. $B \in$ Q(n, A) if $\chi_B \in$ FQ(n, A). $B \in$ Q$_{||}(n, A)$ if $\chi_B \in$ FQ$_{||}(n, A)$. $B \in$ QC(n, A) if $\chi_B \in$ FQC(n, A).

Definition 4. [2] Let $n \geq 1$, let $x_1, \ldots, x_n \in \mathsf{N}$, and let $A \subseteq \mathsf{N}$. Then C_n^A is the string-valued function defined by $C_n^A(x_1, \ldots, x_n) = A(x_1) \cdots A(x_n)$.

Definition 5. Let A be a set. A is *supportive* if $(\forall n)[$Q$(n, A) \subset$ Q$(n + 1, A)]$. A is *parallel supportive* if $(\forall n)[$Q$_{||}(n, A) \subset$ Q$_{||}(n + 1, A)]$.

We will prove several results about semirecursive sets. Semirecursive sets are interesting because (1) there is a semirecursive set in every tt-degree, (2) there is an r.e. semirecursive set in every r.e. T-degree, and (3) we will derive results about r.e. sets *from* results about semirecursive sets.

Definition 6. [10] Let X be a set. If $\sqsubseteq$ is a linear ordering, then we write $\sqsubseteq \leq_T X$ to indicate that the problem of determining how two elements compare is recursive in X. A set A is *semirecursive in* X if there exists a linear ordering $\sqsubseteq \leq_T X$ such that A is closed downward under $\sqsubseteq$. This is equivalent to the existence of a function $f \leq_T X$ such that (1) $f(x, y) \in \{x, y\}$, and (2) $A \cap \{x, y\} \neq \emptyset \Rightarrow f(x, y) \in A$. A set A is *semirecursive* if it is semirecursive in $\emptyset$.

Definition 7. If A is a set, then A^{tt} is the set of all quantifier-free statements using $\wedge, \vee, \neg, \in$ that are true of A. (Example: $((12 \in A) \wedge (14 \notin A))$.)

The following lemmas from [2] will be useful.

Lemma 8. *If* $n \in \mathbb{N}$, *then* $(\exists X)[f \in \mathrm{FQ}(n, X)]$ *iff* $f \in \mathrm{EN}(2^n)$.

Lemma 9. *If* $(\exists n)[C_n^A \in \mathrm{EN}(n)]$, *then* A *is recursive.*

Lemma 10. *If* $(\exists n)[C_n^A \in \mathrm{FQ}_{||}(n-1, A)]$, *then* A *is recursive.*

Lemma 11. *[2] If* $(\exists k)[C_k^A \in \mathrm{SEN}(2^k - 1)]$, *then* $(\exists c)(\forall n)[C_n^A \in \mathrm{SEN}(cn^k)]$.

3 Summary of Results

We summarize our results. Let $A = K$ or any nonrecursive semirecursive set. Let $n \geq 1$, let $m \geq 2$, and assume that m divides n. Then the following hold.

1. $\mathrm{ODD}_n^A \in \mathrm{Q}_{||}(n, A) - \mathrm{Q}_{||}(n-1, A)$.
2. $\mathrm{ODD}_{2^n}^A \in \mathrm{Q}(n+1, A) - \mathrm{Q}(n, A)$.
3. $\mathrm{WMOD}(m)_n^A \in \mathrm{Q}_{||}(\frac{2n}{m}, A) - \mathrm{Q}_{||}(\frac{2n}{m} - 1, A)$.
4. $\mathrm{WMOD}(m)_n^A \in \mathrm{Q}(\lceil \log(\frac{2n}{m} + 1) \rceil, A) - \mathrm{Q}(\lceil \log(\frac{2n}{m} + 1) \rceil - 1, A)$.
5. A is both supportive and parallel supportive.
6. All the lower bounds implied by the above items hold for any nonrecursive r.e. set. The upper bounds *do not* hold for every r.e. set [2].

There are also sets that are not supportive and/or not parallel supportive.

1. There exist sets that are neither supportive nor parallel supportive.
2. The sets K^{tt} and $\emptyset^\omega$ are supportive but not parallel supportive.

We also obtain results about when $\mathrm{Q}(n, A) = \mathrm{QC}(n, A)$.

1. $(\forall n)[\mathrm{Q}(n, K) = \mathrm{QC}(n, K)]$.
2. If A is a nonrecursive semirecursive r.e. set, then $\mathrm{Q}(1, A) - \bigcup_{i=0}^{\infty} \mathrm{QC}(i, A) \neq \emptyset$.

4 Supportive Sets and the Complexity of ODD_n^A

The following lemma will help in proving that sets are supportive.

Lemma 12. *Let* A *be a set. If* $(\forall n)(\exists m \geq 1)[\mathrm{ODD}_m^A \notin \mathrm{Q}(n, A)]$, *then* A *is supportive and parallel supportive.*

Proof sketch. We show that A is parallel supportive. The proof that A is supportive is similar. Fix n. Note that $A \in Q_{||}(1, A) - Q_{||}(0, A)$, since A is not recursive, so assume that $n > 0$. Let $m \geq 1$ be the least number such that $\mathrm{ODD}_m^A \notin Q_{||}(n, A)$. Note that $m > 1$ and $\mathrm{ODD}_{m-1}^A \in Q_{||}(n, A)$. One can use this to show that $\mathrm{ODD}_m^A \in Q_{||}(n+1, A)$. Hence $\mathrm{ODD}_m^A \in Q_{||}(n+1, A) - Q_{||}(n, A)$, so A is parallel supportive.

Our next two theorems pin down the complexity of ODD_n^A in terms of both parallel and sequential queries to A, for A semirecursive.

Theorem 13. *Let A be a nonrecursive semirecursive set, and let $n \geq 1$. Then $\mathrm{ODD}_n^A \in Q_{||}(n, A) - Q_{||}(n - 1, A)$.*

Proof sketch. Clearly, $\mathrm{ODD}_n^A \in Q_{||}(n, A)$.

Assume that A is semirecursive via recursive linear ordering $\sqsubseteq$. Assume that $\mathrm{ODD}_n^A \in Q_{||}(n - 1, A)$ via M^A. The following algorithm shows that $C_{2n+1}^A \in FQ_{||}(2n, A)$. Hence by Lemma 10, A is recursive. This contradicts the hypothesis. Let the input be $(x_1, \ldots, x_{2n+1})$, where $x_1 \sqsubseteq \cdots \sqsubseteq x_{2n+1}$. Note that $C_{2n+1}^A(x_1, \ldots, x_{2n+1}) \in 1^*0^*$. Use M^A to obtain $z_1, \ldots, z_{n-1}$ such that we can determine $\mathrm{ODD}_n^A(x_2, x_4, \ldots, x_{2n})$ from $C_{n-1}^A(z_1, \ldots, z_{n-1})$. Ask $C_{2n}^A(x_1, x_3, \ldots, x_{2n+1}, z_1, z_2, \ldots, z_{n-1})$. Now compute $\mathrm{ODD}_n^A(x_2, x_4, \ldots, x_{2n})$ and $C_{n+1}^A(x_1, x_3, \ldots, x_{2n+1})$. Since $C_{2n+1}^A(x_1, \ldots, x_{2n+1}) \in 1^*0^*$, we can easily compute $C_{2n+1}^A(x_1, \ldots, x_{2n+1})$.

Theorem 14. *Let A be a nonrecursive semirecursive set, and let $n \in \mathbb{N}$. Then $\mathrm{ODD}_{2^n}^A \in Q(n+1, A) - Q(n, A)$.*
Hence by Lemma 12, A is supportive and parallel supportive.

Proof sketch. Let A be semirecursive via recursive linear ordering $\sqsubseteq$. A binary-search algorithm shows that $C_{2^n}^A \in Q(n+1, A)$, hence $\mathrm{ODD}_{2^n}^A \in Q(n+1, A)$.

Assume that $\mathrm{ODD}_{2^n}^A \in Q(n, A)$ via M^A. We use $M^{()}$ in the following algorithm to show that $C_{2^n}^A \in EN(2^n)$. By Lemma 9, A is recursive, contrary to hypothesis.

Input $(x_1, \ldots, x_{2^n})$, where $x_1 \sqsubseteq \cdots \sqsubseteq x_{2^n}$. The algorithm will search (forever, if need be) for evidence that a possibility for $C_{2^n}^A(x_1, \ldots, x_{2^n})$ deserves to be output. It outputs the possibility $1^i 0^{2^n - i}$ if it finds $\sigma_i \in \{0,1\}^n$ and $y_1, \ldots, y_n \in \mathbb{N}$ such that the following occur.

1. If $M^{()}(x_1, \ldots, x_{2^n})$ is run and the jth query is answered with $\sigma_i(j)$ (for $j = 1, \ldots, n$), then the computation converges to $b \in \{0, 1\}$, where $b = 1$ iff i is odd.
2. The queries made are (in order) $y_1, \ldots, y_n$.
3. The assignment $C_{2^n+n}^A(x_1, \ldots, x_{2^n}, y_1, \ldots, y_n) = 1^i 0^{2^n - i} \sigma_i$ is consistent with $\sqsubseteq$. (E.g., if $x_2 \sqsubseteq y_7 \sqsubseteq x_3$ and x_2, x_3 are both assigned to 1, then y_7 must be assigned to 1.)

Clearly, the value of $C_{2^n}^A(x_1, \ldots, x_{2^n})$ will be enumerated. We need to show that $\leq 2^n$ possibilities are enumerated. Let Y be the set of all queries that could be asked on any converging query path of the computation. Note that $|Y| \leq 2^n - 1$, hence there are at most 2^n possible assignments with domain Y that are consistent with $\sqsubseteq$.

Assume, by way of contradiction, that all $2^n + 1$ elements of the form $1^i 0^{2^n - i}$ are enumerated as possibilities for $C_{2^n}^A(x_1, \ldots, x_{2^n})$. We show that there are $2^n + 1$ consistent assignments with domain Y, a contradiction.

Let $i \leq 2^n$. Since $1^i 0^{2^n - i}$ is enumerated, we know that σ_i is an assignment with domain a *subset* of Y that is compatible with $C_{2^n}^A(x_1, \ldots, x_{2^n}) = 1^i 0^{2^n - i}$. Let ρ_i be the minimal assignment with domain Y that is consistent with $\sqsubseteq$ and extends σ_i in such a way that ρ_i is compatible with $C_{2^n}^A(x_1, \ldots, x_{2^n}) = 1^i 0^{2^n - i}$ (minimal in the number of elements of Y that are assigned to 1). Then the number of elements of Y that are assigned to 1 by ρ_i is a nondecreasing function of i.

If $i < 2^n$, then $\rho_i \neq \rho_{i+1}$, since ρ_i and ρ_{i+1} lead to different answers about the value of $\text{ODD}_{2^n}^A(x_1, \ldots, x_{2^n})$. Therefore ρ_i is a strictly increasing function of i. Hence $\rho_0, \ldots, \rho_{2^n}$ are distinct.

We now show that if A is a nonrecursive r.e. set, then $\text{ODD}_n^A \notin Q_{||}(n - 1, A)$ and $\text{ODD}_{2^n}^A \notin Q(n, A)$. Our plan is to make r.e. sets 'look like' semirecursive sets and then apply a relativized version of Theorems 13 and 14.

Definition 15. A set X is *nice* if every partial recursive 0,1-valued function g has a total extension $h \leq_T X$.

Definition 16. A set X is *recursively bounded* (abbreviated r.b.) if, for every (total) function $f \leq_T X$, there exists a recursive function g such that $(\forall x)[f(x) < g(x)]$. A Turing degree is *r.b.* if it has an r.b. set. Note that all the sets in an r.b. degree are r.b. (In the literature, r.b. degrees are called 'hyperimmune free.' This term comes from an equivalent definition that we will not be using. See [14] or [15, 17].)

Lemma 17. *Let A, X be sets, and assume that A is r.e. and X is nice. Then A is semirecursive in X. (Hence r.e. sets 'look' semirecursive.)*

Proof. We will show that A is semirecursive in X by exhibiting a function $f \leq_T X$ such that $f(x, y) \in \{x, y\}$ and $A \cap \{x, y\} \neq \emptyset \Rightarrow f(x, y) \in A$.

Let g be the partial recursive 0,1-valued function defined by

$$g(x, y) = \begin{cases} 1, & \text{if } (\exists s)[x \in A_s \wedge y \notin A_s]; \\ 0, & \text{if } (\exists s)[y \in A_s \wedge x \notin A_s]; \\ \uparrow, & \text{otherwise.} \end{cases}$$

Since X is nice, there is a 0,1-valued total function $h \leq_T X$ such that h extends g. A is semirecursive in X via the function f defined by

$$f(x, y) = \begin{cases} x, & \text{if } h(x, y) = 1; \\ y, & \text{if } h(x, y) = 0. \end{cases}$$

Lemma 18. *Let A, X be sets, and assume that A is r.e., X is r.b., and $A \leq_T X$. Then A is recursive.*

Proof. Let f be the function defined by

$$f(x) = \begin{cases} \mu s[x \in A_s], & \text{if } x \in A; \\ 0, & \text{otherwise.} \end{cases}$$

Clearly, $f \leq_T A \leq_T X$. Since X is r.b., there is a recursive function g such that $(\forall x)[f(x) < g(x)]$. Note that, for all x, $x \in A$ iff $x \in A_{g(x)}$. Hence A is recursive.

The following result is known.

Lemma 19. *There exist sets which are both r.b. and nice.*

Theorem 20. *Let A be an r.e. set, and let $n \geq 1$. If $\mathrm{ODD}_n^A \in Q_{||}(n-1, A)$ or $\mathrm{ODD}_{2n}^A \in Q(n, A)$, then A is recursive.*

Proof. Let X be a nice r.b. set (such exist, by Lemma 19). A is semirecursive in X, by Lemma 17. If $\mathrm{ODD}_n^A \in Q_{||}(n-1, A)$ or $\mathrm{ODD}_{2n}^A \in Q(n, A)$ then, by a relativized version of Theorems 13 and 14, $A \leq_T X$. By Lemma 18, A is recursive.

Note 21. We have shown, by other techniques, that if A and B are r.e. and either $\mathrm{ODD}_n^A \in Q_{||}(n-1, B)$ or $\mathrm{ODD}_{2n}^A \in Q(n, B)$, then A is recursive.

5 The Complexity of $\mathrm{WMOD}(m)_n^A$

We prove a very general theorem which applies to semirecursive sets and K.

Definition 22. Let A be a set. GE^A is the set of $(x_1, \ldots, x_n, j)$ such that $n \geq 1$, $j \in \mathbb{N}$, and at least j of $x_1, \ldots, x_n$ are in A. (GE is a mnemonic for 'Greater than or Equal to.')

Theorem 23. *Let A be a set, and let m, n be such that $2 \leq m \leq n$ and m divides n.*

1. *If $\mathrm{GE}^A \in Q(1, A)$, then $\mathrm{WMOD}(m)_n^A \in Q_{||}(\frac{2n}{m}, A) \cap Q(\lceil \log(\frac{2n}{m} + 1) \rceil, A)$.*
2. *If $\mathrm{GE}^A \leq_m A$ and $\mathrm{ODD}_{2n/m}^A \notin Q_{||}(\frac{2n}{m} - 1, A) \cup Q(\lceil \log(\frac{2n}{m} + 1) \rceil - 1, A)$, then $\mathrm{WMOD}(m)_n^A \notin Q_{||}(\frac{2n}{m} - 1, A) \cup Q(\lceil \log(\frac{2n}{m} + 1) \rceil - 1, A)$.*

Proof sketch. We prove the statements about $Q_{||}(-, A)$. The ones about $Q(-, A)$ are similar.

(1) Assume that $\mathrm{GE}^A \in Q(1, A)$. The following algorithm shows that $\mathrm{WMOD}(m)_n^A \in Q_{||}(\frac{2n}{m}, A)$. We use the premise $\mathrm{GE}^A \in Q(1, A)$ implicitly.

On input $(x_1, \ldots, x_n)$, make the queries

$$"(x_1, \ldots, x_n, 1) \in GE^A?",$$

$$"(x_1, \ldots, x_n, m) \in GE^A?", \qquad "(x_1, \ldots, x_n, m+1) \in GE^A?",$$
$$"(x_1, \ldots, x_n, 2m) \in GE^A?", \qquad "(x_1, \ldots, x_n, 2m+1) \in GE^A?",$$

$$\vdots \qquad\qquad\qquad\qquad \vdots$$

$$"(x_1, \ldots, x_n, (\tfrac{n}{m} - 1)m) \in GE^A?", \ "(x_1, \ldots, x_n, (\tfrac{n}{m} - 1)m + 1) \in GE^A?",$$
$$"(x_1, \ldots, x_n, \tfrac{n}{m}m) \in GE^A?".$$

From this information, one can determine $WMOD(m)_n^A(x_1, \ldots, x_n)$.

(2) Assume that $GE^A \leq_m A$ and $ODD_{2n/m}^A \notin Q_{||}(\tfrac{2n}{m} - 1, A)$. The following algorithm shows that $ODD_{2n/m}^A \in Q(1, WMOD(m)_n^A)$.
This implies that $WMOD(m)_n^A \notin Q_{||}(\tfrac{2n}{m} - 1, A)$, since otherwise $ODD_{2n/m}^A \in Q_{||}(\tfrac{2n}{m} - 1, A)$, which contradicts our assumption.

Let the input be $(x_1, \ldots, x_{2n/m})$. Using the fact that $GE^A \leq_m A$, obtain numbers $z_1, \ldots, z_{2n/m}$ such that, for all i with $1 \leq i \leq 2n/m$,

$$(x_1, \ldots, x_{2n/m}, i) \in GE^A \text{ iff } z_i \in A.$$

Ask whether

$$(z_1, z_2, \ldots, z_{2n/m}, z_2, \ldots, z_2, z_4, \ldots, z_4, \ldots, z_{2n/m}, \ldots, z_{2n/m}) \in WMOD(m)_n^A.$$

(The expression $z_i, \ldots, z_i$ means that the element z_i appears as a component $m - 2$ times in succession.) If the answer is YES, then output 1; otherwise, output 0.

Clearly, this algorithm makes 1 query to $WMOD(m)_n^A$. One can show that it decides $ODD_{2n/m}^A$.

If $A = K$ or if A is a nonrecursive semirecursive set, then $GE^A \leq_m A$. Combining this fact with Theorems 13, 14, and 20, we obtain the following corollary.

Corollary 24. *Let m, n be such that $2 \leq m \leq n$ and m divides n.*

1. *If a set A is nonrecursive and semirecursive, then*
   $WMOD(m)_n^A \in Q_{||}(\tfrac{2n}{m}, A) - Q_{||}(\tfrac{2n}{m} - 1, A)$, *and*
   $WMOD(m)_n^A \in Q(\lceil \log(\tfrac{2n}{m} + 1) \rceil, A) - Q(\lceil \log(\tfrac{2n}{m} + 1) \rceil - 1, A)$.
2. $WMOD(m)_n^K \in Q_{||}(\tfrac{2n}{m}, K) - Q_{||}(\tfrac{2n}{m} - 1, K)$, *and*
   $WMOD(m)_n^K \in Q(\lceil \log(\tfrac{2n}{m} + 1) \rceil, K) - Q(\lceil \log(\tfrac{2n}{m} + 1) \rceil - 1, K)$.
3. *Every nonzero r.e. degree has an r.e. set A with*
   $WMOD(m)_n^A \in Q_{||}(\tfrac{2n}{m}, A) - Q_{||}(\tfrac{2n}{m} - 1, A)$, *and*
   $WMOD(m)_n^A \in Q(\lceil \log(\tfrac{2n}{m} + 1) \rceil, A) - Q(\lceil \log(\tfrac{2n}{m} + 1) \rceil - 1, A)$.

Theorem 25. *Every tt-degree has a set B such that $(\forall n \geq 1)[ODD_n^B \in Q(1, B)]$.*

Proof. Let A be some set in the tt-degree. Let $B = A^{tt}$, and let $n \geq 1$. Clearly, $ODD_n^B \in Q(1, B)$, by making a query involving an exponentially large Boolean combination constructed from the n statements in the input.

6 Supportive and Non-supportive Sets

We show that there exists a set that is neither supportive nor parallel supportive.

Theorem 26. *Let A be an r.b. set. Then (1) $(\forall n)[Q(n, A) = QC(n, A)]$, and (2) A^{tt} is neither supportive nor parallel supportive.*

Proof sketch. Let $n \in \mathbb{N}$. Assume that $B \in Q(n, A)$ via M^A. We show that $B \in QC(n, A)$. Let f be the function $f(x) = \mu s[M^A(x)$ halts in s steps]. Clearly, $f \leq_T A$. By the definition of an r.b. set, there exists a recursive function g such that $(\forall x)[f(x) < g(x)]$. We use g to define an oracle Turing machine $N^{()}$ such that $B \in QC(n, A)$ via N^A.

$N^{()}$ is defined as follows: On input x and with X as oracle, simulate $M^X(x)$ for $g(x)$ steps. If the computation halts within $g(x)$ steps, then output the result; else output 0. Clearly, $B \in QC(n, A)$ via N^A.

Note that A^{tt} is r.b., hence $Q(n, A^{tt}) = QC(n, A^{tt})$. It is straightforward to show that $QC(n, A^{tt}) \subseteq QC(1, A^{tt})$ (hence that $QC(n, A^{tt}) = QC(1, A^{tt})$), by making a query involving an exponentially large Boolean combination of statements used as queries in the $QC(n, A^{tt})$ algorithm. Thus $Q(n, A^{tt}) = QC(1, A^{tt})$.

Theorem 27. *K^{tt} and $\emptyset^\omega$ are supportive but not parallel supportive.*

Proof sketch. We prove this for K^{tt}; the proof for $\emptyset^\omega$ is similar.

K^{tt} not parallel supportive: Let $n \in \mathbb{N}$, and let $A \in Q_{\parallel}(n, K^{tt})$. We can assume that $n \geq 1$. On input x, we can find $x_1, \ldots, x_n$ such that knowing $C_n^{K^{tt}}(x_1, \ldots, x_n)$ would yield $A(x)$. We can easily formulate a query y to K^{tt}, using $x_1, \ldots, x_n$, such that $x \in A$ iff $y \in K^{tt}$.

K^{tt} supportive: For all n and all sets X, let

$$H(n, X) = \{e : M_e^X(e) \downarrow \text{ and asks} \leq n \text{ queries}\}.$$

It is easy to show that, for all n, X, $H(n, X) \notin Q(n, X)$. One can also show that $H(n, K^{tt}) \in Q(n + 1, K^{tt})$.

7 The Power of Not Converging

The class $Q(n, A)$ seems to be more powerful than $QC(n, A)$, since the machine used to decide a set $B \in Q(n, A)$ need not converge on every query path. In this section we examine which sets have $(\forall n)[Q(n, A) = QC(n, A)]$. Most of our theorems involve sets that are extreme: For some A, $(\forall n)[Q(n, A) = QC(n, A)]$; for other A, $Q(1, A) - \bigcup_{n=0}^{\infty} QC(n, A) \neq \emptyset$. Our definitions reflect this behavior.

Definition 28. A set A is *oracle convergent* (abbreviated o.c.) if $(\forall n)[Q(n, A) = QC(n, A)]$. A (T-, tt-) degree is *o.c.* if it has an o.c. set. A (T-, tt-) degree is *completely o.c.* if every set in it is o.c. A is *oracle non-convergent* (abbreviated o.n.c.) if $Q(1, A) - \bigcup_{n=0}^{\infty} QC(n, A) \neq \emptyset$. A (T-, tt-) degree is *o.n.c.* if it has an o.n.c. set. A (T-, tt-) degree is *completely o.n.c.* if every set in it is o.n.c. (By Theorem 26, any r.b. degree is completely o.c.)

Definition 29. Let A be an r.e. set, and let $e, x, m \in \mathbb{N}$. $M_e^A(x)$ *changes its mind at least m times* if there exist $s_0, \ldots, s_m$ with $s_0 < \cdots < s_m$ (if $m > 0$) such that, for all $i \leq m$, $M_{e,s_i}^{A_{s_i}}(x) \downarrow \in \{0, 1\}$ and (if $i < m$) $M_{e,s_i}^{A_{s_i}}(x) \neq M_{e,s_{i+1}}^{A_{s_{i+1}}}(x)$.

Theorem 30. K *is o.c.*

Proof sketch. Let $X \in Q(n, K)$, and assume that $X \in Q(n, K)$ via M_e^K. We show that $X \in \mathrm{QC}(n, K)$. Let

$$B = \{(x, m) : M_e^K(x) \text{ changes its mind at least } m \text{ times}\}.$$

B is r.e., hence $B \leq_m K$. Given x, we can perform a binary search, with n queries to K, to determine how many times $M_e^K(x)$ changes its mind. Then determine the least s such that $M_{e,s}^K(x) \downarrow \in \{0, 1\}$, and let $b = M_{e,s}^K(x)$. Output b if the number of mindchanges is even; otherwise, output $1 - b$. Note that even if the answers supplied were incorrect, the algorithm converges.

We will show that every nonrecursive semirecursive r.e. set is o.n.c.

Lemma 31. *Let A, B be sets, and assume that $(\exists k \geq 1)[C_k^A \in \mathrm{SEN}(2^k - 1)]$ and $(\forall k \geq 1)[C_k^B \notin \mathrm{SEN}(2^k - 1)]$. Then $B \notin \bigcup_{n=0}^{\infty} \mathrm{QC}(n, A)$. (The intuition is that A is 'easy' to decide and B is 'hard,' so it is reasonable that B cannot be reduced to A in certain ways.)*

Proof. We show that $B \notin \bigcup_{n=1}^{\infty} \mathrm{QC}(1, C_n^A)$. It is easy to show that $\bigcup_{n=1}^{\infty} \mathrm{QC}(1, C_n^A) = \bigcup_{n=0}^{\infty} \mathrm{QC}(n, A)$; hence we obtain our result.

Assume, by way of contradiction, that $(\exists n_0 \geq 1)[B \in \mathrm{QC}(1, C_{n_0}^A)]$. By making queries in parallel, we have $(\forall n \geq 1)[C_n^B \in \mathrm{FQC}(1, C_{n_0 \cdot n}^A)]$. Since $(\exists k \geq 1)[C_k^A \in \mathrm{SEN}(2^k - 1)]$ and n_0 is a constant, it follows (by Lemma 11) that there exists c such that $(\forall n)[C_{n_0 \cdot n}^A \in \mathrm{SEN}(cn^k)]$. Hence for large enough n, $C_n^B \in \mathrm{FQC}(1, C_{n_0 \cdot n}^A) \subseteq \mathrm{SEN}(cn^k) \subseteq \mathrm{SEN}(2^n - 1)$. This contradicts the hypothesis on B.

Definition 32. A function f is *dominated* if there is a recursive function g such that $(\forall x)[f(x) < g(x)]$.

Lemma 33. *Let A be a set such that (1) $(\exists k \geq 1)[C_k^A \in \mathrm{SEN}(2^k - 1)]$ and (2) there exists a function $f \in \mathrm{FQ}(1, A)$ that is not dominated. Then A is o.n.c.*

Proof sketch. We construct a set B so that $(\forall k \geq 1)[C_k^B \notin \mathrm{SEN}(2^k - 1)]$ and $B \in Q(1, A)$, and then apply Lemma 31 to obtain that $B \notin \bigcup_{n=0}^{\infty} \mathrm{QC}(n, A)$.

Let the sets $X_{(e,i,k)}$ (for $e, i, k \in \mathbb{N}$) be a recursive partition of $\mathbb{N}$ so that $|X_{(e,i,k)}| = k$. Let $\mathbf{x}_{(e,i,k)}$ be the k-tuple of numbers formed by taking the elements of $X_{(e,i,k)}$ in increasing order. We use the $\mathbf{x}_{(e,i,k)}$ to make sure that C_k^B is not strongly $(2^k - 1)$-enumerable via M_e. In particular, we construct B so that, for all e, k, we satisfy requirement

$$R_{(e,k)} : M_e \text{ total} \Rightarrow (\exists i)[|D_{M_e(\mathbf{x}_{(e,i,k)})}| \geq 2^k \vee C_k^B(\mathbf{x}_{(e,i,k)}) \notin D_{M_e(\mathbf{x}_{(e,i,k)})}].$$

We construct $B \in Q(1, A)$ via the following algorithm for it. We will use f, a function in $FQ(1, A)$ that is not dominated, in the algorithm.

On input x, find e, i, k such that $x \in X_{\langle e,i,k \rangle}$. Compute $t = f(i)$. (This will require at most one query to A.) If $M_{e,t}(\mathbf{x}_{\langle e,i,k \rangle}) \uparrow$, then output 0 and halt; else compute $z = M_{e,t}(\mathbf{x}_{\langle e,i,k \rangle})$. (Note that if C_k^B is strongly $(2^k - 1)$-enumerable via M_e and $M_{e,t}(\mathbf{x}_{\langle e,i,k \rangle}) \downarrow$, then $C_k^B(\mathbf{x}_{\langle e,i,k \rangle}) \in D_z$ and $|D_z| \leq 2^k - 1$. We ensure that this does not happen.) If $|D_z| \geq 2^k$, then output 0 and halt. Otherwise, let $b_1 \cdots b_k$ be the lexicographically least element of $\{0,1\}^k - D_z$, and let j be such that x is the jth component of $\mathbf{x}_{\langle e,i,k \rangle}$. Output b_j and halt.

Let $e, k \in \mathbb{N}$. We show that $R_{\langle e,k \rangle}$ is satisfied. Assume, by way of contradiction, that $R_{\langle e,k \rangle}$ is not satisfied. We obtain a recursive function g that dominates f, contrary to our assumption about f. Since $R_{\langle e,k \rangle}$ is not satisfied, we know that M_e is total and, for all i, the number of steps required for the computation of $M_e(\mathbf{x}_{\langle e,i,k \rangle})$ to halt is strictly greater than $f(i)$ (otherwise, $R_{\langle e,k \rangle}$ would have been satisfied when the elements of $X_{\langle e,i,k \rangle}$ were input to the algorithm). Hence the recursive function $g(i) = \mu t[M_{e,t}(\mathbf{x}_{\langle e,i,k \rangle}) \downarrow]$ dominates f.

Corollary 34. *Let A be a nonrecursive r.e. set which is semirecursive. Then A is o.n.c.*

Proof sketch. We show that A satisfies the two premises of Lemma 33.

$(\forall k)[C_k^A \in \text{SEN}(k+1)]$: Let A be semirecursive via recursive linear ordering $\sqsubseteq$. On input $(x_1, \ldots, x_k)$, we enumerate exactly $k+1$ candidates for the value of $C_k^A(x_1, \ldots, x_k)$. We can assume that $x_1 \sqsubseteq \cdots \sqsubseteq x_k$. Enumerate the candidates $\{1^i 0^{k-i} : 0 \leq i \leq k\}$.

We exhibit a function $f \in FQ(1, A)$ that is not dominated. Define f as follows. If $x \in A$, then $f(x) = \mu s[x \in A_s]$, else $f(x) = 0$. Clearly, $f \in FQ(1, A)$. If f is dominated by a recursive function g, then $x \in A$ iff $x \in A_{g(x)}$; hence A is recursive, contrary to hypothesis.

We state three theorems without proof.

Theorem 35. *There exists an r.e. degree none of whose r.e. sets is o.c.*

Theorem 36. *There exist r.e. sets A such that $\emptyset <_T A <_T K$ and A is o.c.*

Theorem 37. *Let A be a set. The tt-degree of A is completely o.c. iff $(\forall f)[f \leq_{\text{wtt}} A \Rightarrow f$ is dominated].*
The tt-degree of A is o.n.c. iff $(\exists f)[f \leq_{\text{wtt}} A$ and f is not dominated]. Hence every tt-degree is either o.n.c. or completely o.c. ($f \leq_{\text{wtt}} B$ if there exists an oracle Turing machine $M^{()}$ and a recursive function h such that (1) M^B computes f and (2) for all x, $h(x)$ is an upper bound on the queries made in the $M^B(x)$ computation.)

8 Acknowledgments

We would like to thank Richard Chang for proofreading.

References

1. R. Beigel. *Query-Limited Reducibilities*. PhD thesis, Stanford University, 1987. Also available as Report No. STAN-CS-88-1221.
2. R. Beigel, W. Gasarch, J. T. Gill, and J. C. Owings. Terse, superterse, and verbose sets. *Information and Computation*, 103:68–85, 1993.
3. R. Beigel, W. Gasarch, and E. Kinber. Frequency computation and bounded queries. *Theoretical Computer Science*, 1996 (to appear).
4. J. Cai and L. A. Hemachandra. Enumerative counting is hard. *Information and Computation*, 82(1):34–44, July 1989.
5. M. Furst, J. B. Saxe, and M. Sipser. Parity, circuits, and the polynomial-time hierarchy. *Mathematical Systems Theory*, 17(1):13–27, April 1984.
6. W. I. Gasarch. A hierarchy of functions with applications to recursive graph theory. Technical Report 1651, University of Maryland, Dept. of Computer Science, 1985.
7. W. Gasarch. Bounded queries in recursion theory: A survey. In *Proc. of the 6th Ann. Conf. on Structure in Comp. Theory*, IEEE Comp. Soc. Press, June 1991.
8. O. Goldreich, N. Nisan, and A. Wigderson. On Yao's xor-lemma. Technical Report TR95-050, Electronic Colloquium on Computational Complexity, 1995.
9. J. Hastad. *Computational Limitations of Small-Depth Circuits*. MIT Press, 1987.
10. C. G. Jockusch. Semirecursive sets and positive reducibility. *Transactions of the AMS*, 131:420–436, May 1968.
11. M. Kummer. A proof of Beigel's cardinality conjecture. *Journal of Symbolic Logic*, 57(2):677–681, June 1992.
12. M. Kummer and F. Stephan. Effective search problems. *Math. Logic Quarterly*, 40, 1994.
13. L. Levin. One way fns and pseudorandom generators. *Combinatorica*, 7, 1987.
14. W. Miller and D. A. Martin. The degree of hyperimmune sets. *Zeitsch. f. math. Logik und Grundlagen d. Math.*, 14:159–166, 1968.
15. P. Odifreddi. *Classical Recursion Theory (Volume I)*. North-Holland, 1989.
16. R. Smolensky. Algebraic methods in the theory of lower bounds for Boolean circuit complexity. In *Proc. of the 19th ACM Sym. on Theory of Comp.*, 1987.
17. R. I. Soare. *Recursively Enumerable Sets and Degrees*. Perspectives in Mathematical Logic. Springer-Verlag, Berlin, 1987.

A Lambda Calculus of Incomplete Objects

Viviana Bono*   Michele Bugliesi   Luigi Liquori***

*Dipartimento di Informatica, Università di Torino
C.so Svizzera 185, I-10149 Torino, Italy
e-mail: {bono,liquori}@di.unito.it

**Dipartimento di Matematica, Università di Padova
Via Belzoni 7, I-35131 Padova, Italy
e-mail: michele@math.unipd.it

Abstract. This paper extends the Lambda Calculus of Objects as proposed in [5] with a new support for *incomplete* objects. Incomplete objects behave operationally as "standard" objects; their typing, instead, is different, as they may be typed even though they contain references to methods that are yet to be added. As a byproduct, incomplete objects may be typed independently of the order of their methods and, consequently, the operational semantics of the untyped calculus may be soundly defined relying on a permutation rule that treats objects as sets of methods. The new type system is a conservative extension of the system of [5] that retains the *mytype* specialization property for inherited methods peculiar to [5], as well as the ability to statically detect run-time errors such as *message not understood*.

1 Introduction

Object-oriented languages have been classified as either *class-based* or *delegation-based* according to the underlying object-oriented model. In class-based languages, such as *Smalltalk* [7] and *C++* [4], the implementation of an object is specified by a template, the class of the object, and every object is created by instantiating its class. In contrast, delegation-based languages, such as *Self* [9], are centered around the idea that objects are created dynamically by modifying existing objects used as *prototypes*. An object created from a given prototype may add new methods or redefine methods supplied by the prototype, and any message sent to the object is handled directly by that object, if it contains a corresponding method, or it is "passed back", i.e. *delegated*, to the prototype.

Delegation-based languages have gained renewed popularity in the last years. In [1], an object calculus is presented that supports (destructive) method override and inheritance by *object subsumption*. In [5], a functional model of a delegation-based calculus is presented that extends previous foundational work from [8]. The *Lambda Calculus of Objects* of [5] is an untyped lambda calculus enriched with object forms and three primitive operations on objects: *method addition*, to define new methods, *method override*, to redefine methods, and *method call*, to send a message to (i.e., invoke a method on) an object. The resulting calculus allows a natural encoding of the object-oriented notion of *self* directly by lambda abstraction, and it provides a powerful and simple inheritance mechanism based on a dynamic method-lookup semantics. Furthermore, the static type system supports an elegant form of *mytype* method specialization whereby the type of inherited methods may be specialized to the type of the inheriting objects. Subsequent work [3, 6] on this calculus has shown that the type system is amenable to extensions with different forms of subtyping.

In the present paper we extend the original calculus of [5] along a direction orthogonal to subtyping. One weakness of [5] is that the typing rules impose a rather rigid discipline on the way that objects may be created from a prototype. In particular, the addition of an m method to an object can be typed correctly only if all the methods that are referenced to (via message sends or method overrides to *self*) in the body of m are already available from that object. Besides making it difficult to write mutually recursive methods, this constraint leads to a somewhat involved formulation of the operational semantics where a notion of *objects in standard form* is needed to extract the appropriate method upon evaluation of a message.

Our extension is based on a new encoding of the types of objects that allows us to treat objects as *sets of methods* (as opposed to ordered sequences) and, consequently, to rely on a simpler operational semantics. The new encoding also gives additional flexibility to the type system, by allowing a method invocation to be typed correctly even though the receiver of the message is an *incomplete* object, i.e. an object that contains references to methods that are yet to be added. This flexibility appears to be desirable for prototyping languages, such as delegation-based languages, where prototypes may reasonably be defined, and operated with as well, while part of their implementation (i.e. their methods) are yet to be defined. The new features are achieved at the expense of little additional complexity in the typing rules, and they provide a conservative extension of the original system: the new type system retains the property of method specialization of [5], as well as the ability to statically detect run-time errors such as *message not understood*.

The rest of the paper is organized as follows. In Section 2 we briefly overview the untyped calculus of [5] and present the new operational semantics. In Section 3 we present the new typing rules for objects. In Section 4 we first prove Subject Reduction and then use it to show Type Soundness. We conclude in Section 5 with some final remarks.

2 The Untyped Calculus

The syntax of the untyped calculus is as in [5]. An expression can be any of the following:

$$e ::= x \mid c \mid \lambda x.e \mid e_1 e_2 \mid \langle \rangle \mid e \Leftarrow m \mid \langle e_1 \longleftrightarrow m{=}e_2 \rangle \mid \langle e_1 \leftarrow m{=}e_2 \rangle,$$

where x is a variable, c a constant and m a method name. The object-related forms are as in [5], namely:

$\langle\rangle$ is the empty object,

$e \Leftarrow m$ sends message m to object e,

$\langle e_1 \longleftrightarrow\!\!+\ m{=}e_2 \rangle$ extends object e_1 with a new method m having body e_2,

$\langle e_1 \leftarrow m{=}e_2 \rangle$ replaces e_1's method body for m with e_2.

The expression $\langle e_1 \longleftrightarrow\!\!+\ m{=}e_2 \rangle$ is defined only when e_1 denotes an object that does not have an m method, whereas $\langle e_1 \leftarrow m{=}e_2 \rangle$ is defined only when e_1 denotes an object that *does* contain an m method. Both these conditions are enforced statically by the type system.

The other main operation on object is method invocation. Methods are invoked by means of message sends according to the following semantics: when the object $\langle e_1 \longleftrightarrow\!\!+\ m{=}e_2 \rangle$ is sent the message m, the result is obtained by "self-applying" e_2 to $\langle e_1 \longleftrightarrow\!\!+\ m{=}e_2 \rangle$. Therefore, in defining the operational semantics of the calculus, we must give, besides the rules of β-reduction and method invocation, also a mechanism for extracting the appropriate method of an object. As it turns out, the following three rules suffice ($\leftarrow\!\circ$ denotes either $\longleftrightarrow\!\!+$ or $\leftarrow$):

$$(\beta) \qquad (\lambda x.e_1)\,e_2 \qquad \xrightarrow{ev} \quad [e_2/x]\,e_1$$

$$(\Leftarrow) \qquad \langle e_1 \leftarrow\!\circ\ m{=}e_2 \rangle \Leftarrow m \quad \xrightarrow{ev} \quad e_2\,\langle e_1 \leftarrow\!\circ\ m{=}e_2 \rangle$$

$$(perm) \qquad \langle\langle e \leftarrow\!\circ\ m{=}e_1 \rangle \leftarrow\!\circ\ n{=}e_2 \rangle \quad \xrightarrow{ev} \quad \langle\langle e \leftarrow\!\circ\ n{=}e_2 \rangle \leftarrow\!\circ\ m{=}e_1 \rangle.$$

Note that we allow permutations only for methods with different names, whereas different definitions for the same name maintain their respective position through subsequent overrides. Accordingly, a message send for any m method always selects the definition provided by the last override for that method.

The operational semantics of the calculus is defined as the least reflexive, transitive and contextual closure $\xrightarrow{ev}$ generated by the reduction rules above. As a remark, we note that the $(perm)$ rule above is justified in our calculus since the following equational rule for objects is sound with respect to the type system:

$$\langle\langle e \leftarrow\!\circ\ m{=}e_1 \rangle \leftarrow\!\circ\ n{=}e_2 \rangle = \langle\langle e \leftarrow\!\circ\ n{=}e_2 \rangle \leftarrow\!\circ\ m{=}e_1 \rangle.$$

This equality did not hold for the system of [5] because in that case the typing rules allow objects to be typed only when methods are added in the appropriate order. As a consequence, in order for method extraction to be performed correctly, a series of *bookkeeping* rules are needed, that make the definition of the operational semantics rather involved.

3 Static Type System

The type of an incomplete object is defined by a type expression of the form:

$$\text{class}\,t.\langle\!\langle m_1{:}\alpha_1, \ldots, m_k{:}\alpha_k \rangle\!\rangle \bullet \langle\!\langle p_1{:}\gamma_1, \ldots, p_l{:}\gamma_l \rangle\!\rangle,$$

where the m_i's and p_i's are method names, and the α_i's and the γ_i's are *labeled-type* expressions (whose role is discussed below). Given the above type, we refer to the two components $\langle\!\langle m_1{:}\alpha_1, \ldots, m_k{:}\alpha_k \rangle\!\rangle$ and $\langle\!\langle p_1{:}\gamma_1, \ldots, p_l{:}\gamma_l \rangle\!\rangle$ as, respectively, the *interface-* and *completion-rows* of the type. The order of methods within each row is irrelevant and we rely on the following equational rule for rows throughout the paper:

$$\langle\langle R \mid n{:}\tau_1\rangle \mid m{:}\tau_2\rangle = \langle\langle R \mid m{:}\tau_2\rangle \mid n{:}\tau_1\rangle.$$

The binder **class** scopes over the two rows of the type, and the bound variable t may occur free within the scope of the binder, with every free occurrence referring to the class-type itself; thus, as in [5], class-types are a form of recursively-defined types.

The intuitive reading of these types is as follows. The interface-row describes all the methods (and their types) that are contained in the objects of the current type, and that may be invoked by means of corresponding messages. The completion-row, instead, lists the methods (and their types) that are referenced to by the methods of the object (whose types are listed in the interface-row) even though they are not yet available from the object. Accordingly, given an object containing the m_i methods of the interface-row, the completion-row lists the methods (and associated types) that are needed to "complete" the object. As we anticipated, this encoding of types leads us to formulate typing rules that allow object expressions to be formed by sequences of method additions and overrides where the order of such operations does not matter. Using labeled-types within the rows of our class-types, also enables us to type a method invocation even though the receiver of the message is incomplete (i.e., its class-type has a non-empty completion-row).

Labeled-types were first introduced in [3], to model a form of *width* subtyping for the original calculus of [5]. Here they bear (almost) the same meaning: if τ_Δ is the type of, say, an m method within an object, then Δ provides an (approximate) representation of the remaining methods of that object upon which m depends: in other words, Δ includes the names of the methods referenced to by m in a send or an override for *self*, together with the methods referenced to by these methods and so on.

The use of transitive (or indirect) references within labels is enforced by the typing rules and, as we shall se in Section 3.3, it is crucial to ensure the correctness of method invocation on incomplete objects. Having direct as well as indirect references within labels, the typing rule for a method invocation may be stated as follows:

$$\frac{\Gamma \vdash e : \mathbf{class}\, t.\langle R \mid \overline{n{:}\alpha}, m{:}\tau_{\{\overline{n}\}}\rangle \bullet C}{\Gamma \vdash e \Leftarrow m : [(\mathbf{class}\, t.\langle R \mid \overline{n{:}\alpha}, m{:}\tau_{\{\overline{n}\}}\rangle \bullet C)/t]\tau} \ (send)$$

In order to type a method invocation for an m method in an object e, we require (*i*) that e contain (in its interface-row) the method name m, and (*ii*) that all of the methods contained in the label associated with m be contained in the interface-row of the object's type.

3.1 Types, Rows, and Kinds

The type expressions include type-constants, type variables, function-types and class-types. The symbols α, β, ... range over labeled-types. We also use the notation $\overline{m{:}\alpha}$, as shorthand for $m_1{:}\alpha_1, \ldots, m_k{:}\alpha_k$, for some k. The sets of types, rows and kinds are defined by the following productions:

$$
\begin{array}{lll}
\text{Types} & \tau ::= \iota \mid t \mid \tau{\to}\tau \mid \textbf{class}\, t.R{\bullet}R & \\
\text{Rows} & R ::= r \mid \langle\rangle \mid \langle R \mid m{:}\tau_\Delta\rangle \mid R\tau \mid \lambda t.R & (m \notin \Delta) \\
\text{Labels} & \Delta ::= \{m_1, \ldots, m_k\} & (k \geq 0) \\
\text{Kinds} & \kappa ::= T \mid [m_1, \ldots, \mathfrak{m}_k] \mid T^n{\to}[m_1, \ldots, m_k] & (n \geq 1,\ k \geq 0).
\end{array}
$$

Although the interface- and completion-rows of a class-type are structurally equivalent, we will often find it convenient to distinguish their role by choosing different notations, namely: R and r to denote respectively interface-rows and interface-row variables, whereas C and c to stand for arbitrary completion-rows and completion-row variables.

Our definition of class-type generalizes the original definition of [5] in that types of the form $\textbf{class}\, t.R$ from [5] are represented here simply as $\textbf{class}\, t.R{\bullet}\langle\rangle$. As we shall see, a corresponding generalization applies to the typing of method bodies: a method body will be built as a polymorphic function whose type is defined in terms of a (universally quantified) row variable, as in [5], and of a (universally quantified) completion variable.

This generalization requires a few changes in the type system, in order to characterize the interdependence between the interface- and completion-rows of a class-type: the intention is to give a more precise definition of the type $\textbf{class}\, t.R{\bullet}$ C by requiring that R and C be *disjoint*, i.e. that methods occurring in R do not occur in C and vice-versa. To formalize this idea, we redefine the meaning of the kinds $[m_1, \ldots, m_k]$ and $T^n{\to}[m_1, \ldots, m_k]$ as follows. The elements of the kind $[m_1, \ldots, m_k]$ are pairs of disjoint rows neither of which contains any of the method names $m_1, \ldots, m_k$. A corresponding interpretation applies to the kinds $T^n{\to}[m_1, \ldots, m_k]$, for $n \geq 1$, that are used to infer polymorphic types for method bodies.

The structure of valid context (see Appendix A) is defined as follows:

$$
\Gamma ::= \varepsilon \mid \Gamma, x : \tau \mid \Gamma, t : T \mid \Gamma, r{\bullet}c : \kappa,
$$

where x, t, and r, c are, respectively, term, type and row variables. Correspondingly, the judgement are: $\Gamma \vdash *$, $\Gamma \vdash R{\bullet}C : \kappa$, $\Gamma \vdash \tau : T$ and $\Gamma \vdash e : \tau$. The judgement $\Gamma \vdash *$ can be read as "Γ is a well-formed context" and the meaning of the other judgements is the usual one.

3.2 Typing Rules for Objects

For the most part, the type system is routine. The object-related rules are discussed below. The first rule defines the type of the empty object: having no methods, the empty object needs no further method to be complete. Hence:

$$\frac{\Gamma \vdash *}{\Gamma \vdash \langle \rangle : \mathbf{class}\, t.\langle \rangle \bullet \langle \rangle} \quad (empty\ object)$$

The typing rule to invoke messages has the format described in the previous subsection. The rule for method addition is defined as follows:

$$\frac{\begin{array}{ll} \Gamma \vdash e_1 : \mathbf{class}\, t.R \bullet \langle C \lfloor n{:}\tau_\Delta \rfloor \rangle & \{\overline{m{:}\alpha}\} \in R \bullet C \\ \Gamma, t : T \vdash R \bullet C : [n, \overline{p}] & \Delta = \{\overline{m}, \overline{p}\} \\ \Gamma, r \bullet c : T \to [\overline{m}, n, \overline{p}] \vdash & \\ \quad e_2 : [(\mathbf{class}\, t.\langle rt \mid \overline{m{:}\alpha}, n{:}\tau_\Delta, \overline{p{:}\gamma} \rangle \bullet ct)/t](t \to \tau) \quad r, c \notin (\tau, \overline{\gamma}) & \end{array}}{\Gamma \vdash \langle e_1 \longleftarrow n{=}e_2 \rangle : \mathbf{class}\, t.\langle R \mid n{:}\tau_\Delta \rangle \bullet \langle C \mid \overline{p{:}\gamma} \rangle} \quad (obj\ ext)$$

where $\{\overline{m{:}\alpha}\} \in R \bullet C$ indicates that the $\overline{m{:}\alpha}$ methods are contained in $R \bullet C$, and $R \bullet \langle C \lfloor n{:}\tau_\Delta \rfloor \rangle$ indicates that the completion-row of the type of e_1 may or may not contain n. Whether or not n must be included in the completion of the type of e_1 depends on whether or not the methods in e_1 contain n in the labels associated with their types. As for the n method being added, the set of its dependences may, in general, include the methods that are already contained in the object as well as methods that are yet to be added: the former are a subset of the $\overline{m}$ methods occurring in R, whereas the latter are the $\overline{p}$ methods together with the subset of $\overline{m}$ occurring in C. Note that all of the dependences of n are assumed to be part of the interface-row in the type of e_2: this, together with the condition $\Delta = \{\overline{m}, \overline{p}\}$, either checks (if n belongs to the type of e_1) – or otherwise it enforces – the constraint that Δ contains all methods that are referenced to (either directly, or indirectly) by e_2. To see this, consider the case when $e_2 \stackrel{def}{=} \lambda self.(self \Leftarrow p)$ for a given method p. Then, an inspection of the (send) rule shows that, in order for the invocation $self \Leftarrow p$ to be typeable, the interface-row of the type of $self$ must include not only p, but also all of the, say, $\overline{q}$ methods in the label of the type of p. But then Δ, the label of n, must include p, a direct reference, as well as the $\overline{q}$ methods that n references indirectly via p.

Note, finally, that as in [5], the type of n has the form $t \to \tau$ (with a class type substituted for t) to conform with the self-application semantics of method invocation. Here, however, this type is polymorphic both in r and in c, and so that e_2 will have the indicated type for every R and C provided that R and C have the correct kind. Hence, invocations of n on future objects derived from $\langle e_1 \longleftarrow n{=}e_2 \rangle$ will be well-typed just in case these objects are complete with respect to n, i.e. they contain all of the methods upon which n depends.

The rule for method override is similar but simpler (see Appendix A): as for (obj-ext), the side-conditions on the labeled type of the method being overridden enforce the correct propagation of transitive references within labels. This may be observed as in the example above, taking $e_2 \stackrel{def}{=} \lambda self.\langle self \longleftarrow p{=}\lambda s.(s \Leftarrow q) \rangle$ where q is, say, a constant method (whose type has an empty label).

3.3 Examples of Type Derivations

Example 1. This example shows that methods can be added in any order, regardless of the interdependences between them. Consider the following expression from [5]:

$$\text{pt} \overset{def}{=} \langle\langle\langle\rangle \longleftarrow \text{plus1} = \lambda s.(s \Leftarrow x) + 1\rangle\longleftarrow x = \lambda \text{self}.3\rangle.$$

The above object cannot be typed with the system of [5] because the subexpression $\langle\langle\rangle\longleftarrow \text{plus1} = \lambda s.(s \Leftarrow x) + 1\rangle$ is not well-typed. Using the typing rule introduced in the previous section, instead, we may proceed as follows. Let $\Gamma_1 = r \bullet c : T\rightarrow[x, \text{plus1}]$, $s : \text{class}\, t.\langle rt \mid \text{plus1}:int_{\{x\}}, x:int\rangle\bullet ct$. It is now easy to see that the following judgements are all derivable:

$$\varepsilon \vdash \langle\rangle : \text{class}\, t.\langle\rangle\bullet\langle\rangle$$
$$\Gamma_1 \vdash (s \Leftarrow x) + 1 : int,$$
$$\Gamma_1 - s \vdash \lambda s.(s \Leftarrow x) + 1 : \text{class}\, t.\langle rt \mid \text{plus1}:int_{\{x\}}, x:int\rangle\bullet ct\rightarrow int,$$
$$\varepsilon \vdash \langle\langle\rangle\longleftarrow \text{plus1} = \lambda s.(s \Leftarrow x) + 1\rangle : \text{class}\, t.\langle \text{plus1}:int_{\{x\}}\rangle\bullet\langle x:int\rangle.$$

where the occurrence of int in the completion-row stands for $int_{\{\}}$. Now, letting $\Gamma_2 = r \bullet c : T\rightarrow[x, \text{plus1}]$, $\text{self} : \text{class}\, t.\langle rt \mid \text{plus1}:int_{\{x\}}, x:int\rangle\bullet ct$, with a sequence of steps similar to the previous one, we may conclude with:

$$\varepsilon \vdash \text{pt} : \text{class}\, t.\langle \text{plus1}:int_{\{x\}}, x:int\rangle\bullet\langle\rangle.$$

Clearly, the same typing discipline also allows us to add mutually recursive methods with no need to resort to the use of dummy methods as in [5].

Example 2. The last example motivates the requirement that the label associated with a method type contains both direct and indirect dependences for that method. Consider extending the pt object of the previous example as shown below:

$$\text{newpt} \overset{def}{=} \langle p\longleftarrow \text{plus} = \lambda s.s \Leftarrow \text{plus1}\rangle.$$

It can be verified that the following judgement may be derived:

$$\varepsilon \vdash \text{newpt} : \text{class}\, t.\langle \text{plus}:int_{\{\text{plus1},x\}}, \text{plus1}:int_{\{x\}}\rangle\bullet\langle x:int\rangle.$$

The point to notice is that the typing rules force the label of plus to include the x method, although plus depends on x indirectly via plus1. The reason why indirect references in the labels are need should now be clear: having only plus1 in the label of plus, we would be able to type the invocation newpt $\Leftarrow$ plus which, instead, causes a *message not understood* error because newpt does not contain x.

4 Soundness of the Type System

The soundness of the type system is proved following the same schema as in [5]. We first show that types are preserved by the reduction process. Then we introduce an evaluation strategy that allows us to formalize the notion of error, and we show that that errors are detected statically by the type system.

To prove subject reduction, we need the following results that help isolate some interesting properties of the type system (proofs are omitted for the lack of space). Lemma 1 is used to specialize class-types to contain additional methods.

Lemma 1. *If $\Gamma, r \bullet c : T^n \to [\overline{m}], \Gamma' \vdash e : \tau$ and $\Gamma \vdash R \bullet C : T^n \to [\overline{m}]$ are both derivable, then so is $\Gamma, [R/r, C/c]\Gamma' \vdash e : [R/r, C/c]\tau$.*

The next two lemmas are used for building well–formed row expressions that can be substituted for row variables in typing derivations.

Lemma 2. *If $\Gamma \vdash \mathtt{class}\, t. \langle R \mid \overline{m{:}\alpha} \rangle \bullet \langle C \mid \overline{p{:}\gamma} \rangle : T$ is derivable, then so are $\Gamma, t : T \vdash \alpha_i : T$ for each α_i in $\overline{\alpha}$, and $\Gamma, t : T \vdash \gamma_i : T$ for each γ_i in $\overline{\gamma}$, and $\Gamma, t : T \vdash R \bullet C : [\overline{m}, \overline{p}]$.*

Lemma 3. *If $\Gamma \vdash e : \tau$ is derivable, then so is $\Gamma \vdash \tau : T$.*

Besides these results, in the following we will also assume that our type derivations be in normal form (the reader is referred to [3] for the definition of such normal form).

Theorem 4 (Subject Reduction). *If $\Gamma \vdash e : \tau$ is derivable, and $e \overset{ev}{\to} e'$, then $\Gamma \vdash e' : \tau$ is also derivable.*

Proof. The proof is by induction on the number of steps in $e \overset{ev}{\to} e'$. For the basic step (i.e. one $\overset{ev}{\to}$ step), we proceed by cases on the definition of $\overset{ev}{\to}$ and use induction on the context of the redex. The proof for (β) is standard, whereas for $(perm)$ the proof is by induction on the structure of the first judgement: it distinguishes the four cases that arise when each occurrence of $\leftarrow\!\circ$ is either $\leftarrow\!\!+$ or $\leftarrow$ and, in each such case, it further distinguishes four sub-cases according to the possible mutual dependences between n and m.

The remaining case is $(\Leftarrow)$: what we need to show is that if $\Gamma \vdash \langle e \leftarrow\!\circ\; n{=}e_n \rangle \Leftarrow n : \tau$ is derivable, then so is $\Gamma \vdash e_n \langle e \leftarrow\!\circ\; n{=}e_n \rangle : \tau$. Again, we need to distinguish the possible instances of the $\leftarrow\!\circ$ operator: below we consider the case when $\leftarrow\!\circ$ is $\leftarrow\!\!+$, the case when $\leftarrow\!\circ$ is $\leftarrow$ being similar. The proof is by induction on the derivation of $\Gamma \vdash \langle e \leftarrow\!\circ\; n{=}e_n \rangle \Leftarrow n : \tau$. For rule $(weak)$ it follows directly from the induction hypothesis, so we consider only rule $(send)$.

If the last applied rule is $(send)$, then the derivation has the following form:

$$\frac{\dfrac{\Xi}{\Gamma \vdash \langle e \leftarrow\!\!+\; n{=}e_n \rangle : \mathtt{class}\, t. \langle R \mid \overline{m{:}\alpha}, n{:}\tau_{\{\overline{m}\}} \rangle \bullet C}}{\Gamma \vdash \langle e \leftarrow\!\!+\; n{=}e_n \rangle \Leftarrow n : [(\mathtt{class}\, t. \langle R \mid \overline{m{:}\alpha}, n{:}\tau_{\{\overline{m}\}} \rangle \bullet C)/t]\tau} \;(send)$$

The interesting case is when Ξ ends up with (*obj ext*), the only other possibility being (*weak*). Note, further, that the (*obj ext*) may only have the form:

$$\frac{\begin{array}{l} \Gamma \vdash e : \mathbf{class}\, t.\langle R \mid \overline{m{:}\alpha}\rangle \bullet \langle C \lfloor n{:}\tau_{\{\overline{m}\}} \rfloor\rangle \\ \Gamma, t : T \vdash \langle R \mid \overline{m{:}\alpha}\rangle \bullet C : [n] \\ \Gamma, r \bullet c : T{\rightarrow}[\overline{m}, n] \vdash e_n : [(\mathbf{class}\, t.\langle rt \mid \overline{m{:}\alpha}, n{:}\tau_{\{\overline{m}\}}\rangle \bullet ct)/t](t{\rightarrow}\tau) \end{array}}{\Gamma \vdash \langle e {\longleftrightarrow} n {=} e_n\rangle : \mathbf{class}\, t.\langle R \mid \overline{m{:}\alpha}, n{:}\tau_{\{\overline{m}\}}\rangle \bullet C} \quad (obj\ ext)$$

Now we show that a derivation exists for $\Gamma \vdash e_n\langle e {\longleftrightarrow} n {=} e_n\rangle : [(\mathbf{class}\, t.\langle R \mid \overline{m{:}\alpha}, n{:}\tau_{\{\overline{m}\}}\rangle \bullet C)/t]\tau$ regardless of whether n is in the type of e_1 or not. First note that the judgment: $\Gamma, t : T \vdash R \bullet C : [\overline{m}, n]$ is derivable. If n is in the type of e, from $\Gamma \vdash e : \mathbf{class}\, t.\langle R \mid \overline{m{:}\alpha}\rangle \bullet \langle C \mid n{:}\tau_{\{\overline{m}\}}\rangle$, by Lemma 3, we have that $\Gamma \vdash \mathbf{class}\, t.\langle R \mid \overline{m{:}\alpha}\rangle \bullet \langle C \mid n{:}\tau_{\{\overline{m}\}}\rangle : T$ is derivable and the claim follows by Lemma 2. Otherwise, the claim derives directly from $\Gamma, t : T \vdash \langle R \mid \overline{m{:}\alpha}\rangle \bullet C : [n]$. By an application of (*rabs*), we then derive $\Gamma \vdash \lambda t.R \bullet \lambda t.C : T{\rightarrow}[\overline{m}, n]$. From this, by applying Lemma 1 to the typing of e_n, we next derive the judgement: $\Gamma \vdash e_n : [(\mathbf{class}\, t.\langle R \mid \overline{m{:}\alpha}, n{:}\tau_{\{\overline{m}\}}\rangle \bullet C)/t](t{\rightarrow}\tau)$. From this, an application of (*eapp*) completes the derivation of the desired judgement. $\square$

4.1 Type Soundness

We conclude formalizing the notion of *message not understood* errors. Intuitively, an error occurs when a message n is sent to an object that does not contain n. The structural operational semantics that formalizes this situation is defined as in [5] in terms of two functions, *eval* and get_n: the *eval* function is a variation of the standard *lazy* evaluator for the λ-calculus that, when fed with an expression of the form $e \Leftarrow n$, calls the get_n function to extract a definition for the n method from e. To perform method extraction, get_n inspects e (possibly evaluating it) until it either finds a definition for the n method, or it determines that e does not contain any definition for n: in the first case get_n returns the body of n as a result, in the second it returns *error*. The complete set of evaluation rules is presented in Appendix B.

Due to the lack of space, we only give the statement of the soundness theorem (and of the main lemmas) and omit proofs. Proofs are carried out exactly as in [5] by induction on the definition of the *eval* and *get* functions, using Subject Reduction. As in that case, since the result of evaluation may be undefined besides being successful or error, it is easier to prove the contrapositive of soundness, i.e. that if $eval(e) = error$, then e cannot be typed.

Lemma 5. *If* $ev(e) = e'$ *then* $e \xrightarrow{ev} e'$, *where ev is either eval or get.*

Let the notation $\Gamma \not\vdash A$ indicate that the judgement $\Gamma \vdash A$ is not derivable. Then the following holds:

Lemma 6. *i) If* $get_n(e) = error$, *then* $\varepsilon \not\vdash e : \mathbf{class}\, t.\langle R \mid n{:}\alpha\rangle \bullet C$, *for any R, C rows, and α labeled-type.*
ii) If $eval(e) = error$, *then* $\varepsilon \not\vdash e : \tau$ *for any type τ.*

Theorem 7 (Soundness). *If* $\varepsilon \vdash e : \tau$ *is derivable, then* $eval(e) \neq error$.

5 Conclusions

We have presented an extension of the *Lambda Calculus of Objects* [5] with a new type system that supports a novel and more flexible typing discipline, while preserving all of the interesting properties of the original system. The new calculus enjoys a more elegant operational semantics and the additional expressive power that derives from the possibility of computing with objects whose implementation is only partially specified. The new features are accounted for by extending the class-types from [5] with two technical tools: *completion-rows* and *labeled-types*. Completion-rows convey information about methods yet to be added, thus allowing sequences of method additions to be typed regardless their order. Labeled-types, in turn, encode information on the structure of a method body that allows a method invocation to be type correctly even though the receiver of the message is an incomplete object. The new features induce only little additional complexity for the typing rules and, furthermore, as we show in [2], the use of labeled-types makes the type system amenable to a smooth integration with the notion of *width* subtyping of [3].

Acknowledgements We wish to thank Mariangiola Dezani-Ciancaglini for her insightful suggestions in countless discussions. Thanks are also due to Laurent Dami for pointing out a flaw in the preliminary version of this paper, and to the anonymous referees, whose remarks helped improve the technical presentation of the final version substantially.

References

1. M. Abadi and L. Cardelli. A Theory of Primitive Objects. In *Proceedings of Theoretical Aspect of Computer Software*, volume 789 of *LNCS*, pages 296–320. Springer-Verlag, 1994.
2. V. Bono, M. Bugliesi, and L. Liquori. A Calculus of Incomplete objects with Subtyping. In preparation.
3. V. Bono and L. Liquori. A Subtyping for the Fisher-Honsell-Mitchell Lambda Calculus of Objects. In *Proceedings of International Conference of Computer Science Logic*, volume 933 of *LNCS*, 1995.
4. E. Ellis and B. Stroustrop. *The Annotated C^{++} Reference Manual*. ACM Press, 1990.
5. K. Fisher, F. Honsell, and J. C. Mitchell. A Lambda Calculus of Objects and Method Specialization. *Nordic Journal of Computing*, 1(1):3–37, 1994.
6. K. Fisher and J. C. Mitchell. A Delegation-based Object Calculus with Subtyping. In *Proceedings of FCT-95*, Lecture Notes in Computer Science. Springer-Verlag, 1995. To appear.
7. A. Goldberg and D. Robson. *Smalltalk-80, The Language and its Implementation*. Addison Wesley, 1983.
8. J. C. Michell. Toward a Typed Foundation for Method Specialization and Inheritance. In *Proc. 17th ACM Symp. on Principles of Programming Languages*, pages 109–124. ACM, 1990.

9. D. Ungar and R. B. Smith. Self: the power of simplicity. In *Proceedings of ACM Symp. on Object-Oriented Programming Systems, Languages, and Applications*, pages 227–241. ACM Press, 1987.

A Typing Rules

General Rules ($a : b$ is either $e : \tau$, or $t : T$, or $r{\bullet}c : \kappa$)

$$\frac{}{\varepsilon \vdash *}\ (ax) \qquad\qquad \frac{\Gamma \vdash * \quad a:b \in \Gamma}{\Gamma \vdash a:b}\ (proj)$$

$$\frac{\Gamma \vdash * \quad a \notin Dom(\Gamma)}{\Gamma, a{:}b \vdash *}\ (var) \qquad\qquad \frac{\Gamma \vdash a:b \quad \Gamma, \Gamma' \vdash *}{\Gamma, \Gamma' \vdash a:b}\ (weak)$$

Rules for Types

$$\frac{\Gamma \vdash \tau_1 : T \quad \Gamma \vdash \tau_2 : T}{\Gamma \vdash \tau_1 \rightarrow \tau_2 : T}\ (t{-}app) \qquad\qquad \frac{\Gamma, t{:}T \vdash R{\bullet}C : [\overline{m}]}{\Gamma \vdash \text{class}\, t.R{\bullet}C : T}\ (class)$$

Types and Row Equality

$$\frac{\Gamma \vdash \tau : T \quad \tau \rightarrow_\beta \tau'}{\Gamma \vdash \tau' : T}\ (t{-}\beta) \qquad\qquad \frac{\Gamma \vdash R{\bullet}C : \kappa \quad R{\bullet}C \rightarrow_\beta R'{\bullet}C'}{\Gamma \vdash R'{\bullet}C' : \kappa}\ (r{-}\beta)$$

$$\frac{\Gamma \vdash e : \tau \quad \Gamma \vdash \tau' : T \quad \tau \leftrightarrow_\beta \tau'}{\Gamma \vdash e : \tau'}\ (t{-}eq)$$

Rules for Rows

$$\frac{\Gamma \vdash *}{\Gamma \vdash ()\bullet() : [\overline{m}]}\ (er) \qquad\qquad \frac{\Gamma \vdash C{\bullet}R : \kappa}{\Gamma \vdash R{\bullet}C : \kappa}\ (rexc)$$

$$\frac{\Gamma \vdash R{\bullet}C : T^n \rightarrow [\overline{m}] \quad \{\overline{n}\} \subseteq \{\overline{m}\}}{\Gamma \vdash R{\bullet}C : T^n \rightarrow [\overline{n}]}\ (rlab) \qquad\qquad \frac{\Gamma \vdash \tau : T \quad \Gamma \vdash R{\bullet}C : [\overline{m}, n] \ n \notin \Delta}{\Gamma \vdash (R \mid n{:}\tau_\Delta){\bullet}C : [\overline{m}]}\ (rext)$$

$$\frac{\Gamma, t{:}T \vdash R{\bullet}C : T^n \rightarrow [\overline{m}]}{\Gamma \vdash \lambda t.R{\bullet}\lambda t.C : T^{n+1} \rightarrow [\overline{m}]}\ (rabs) \qquad\qquad \frac{\Gamma \vdash \tau : T \quad \Gamma \vdash R{\bullet}C : T^{n+1} \rightarrow [\overline{m}]}{\Gamma \vdash R\tau{\bullet}C\tau : T^n \rightarrow [\overline{m}]}\ (rapp)$$

Rules for Expressions

$$\frac{\Gamma, x{:}\tau_1 \vdash e : \tau_2}{\Gamma \vdash \lambda x.e : \tau_1 \rightarrow \tau_2}\ (eabs) \qquad\qquad \frac{\Gamma \vdash e_1 : \tau_1 \rightarrow \tau_2 \quad \Gamma \vdash e_2 : \tau_2}{\Gamma \vdash e_1 e_2 : \tau_2}\ (eapp)$$

$$\frac{\Gamma \vdash *}{\Gamma \vdash \langle\rangle : \text{class}\, t.()\bullet()}\ (\langle\rangle) \qquad\qquad \frac{\Gamma \vdash e : \text{class}\, t.(R \mid \overline{n{:}\alpha}, m{:}\tau_{\{\overline{n}\}}){\bullet}C}{\Gamma \vdash e \Leftarrow m : [(\text{class}\, t.(R \mid \overline{n{:}\alpha}, m{:}\tau_{\{\overline{n}\}}){\bullet}C)/t]\tau}\ (send)$$

$$\frac{
\begin{array}{l}
\Gamma \vdash e_1 : \mathbf{class}\, t.R \bullet (C\lfloor n{:}\tau_\Delta\rfloor) \qquad\qquad \{\overline{m{:}\alpha}\} \in R \bullet C \\
\Gamma, t:T \vdash R \bullet C : [n, \overline{p}] \qquad\qquad\qquad \Delta = \{\overline{m}, \overline{p}\} \\
\Gamma, \mathbf{rec}:T \to [\overline{m}, n, \overline{p}] \vdash \\
\quad e_2 : [(\mathbf{class}\, t.\langle rt \mid \overline{m{:}\alpha}, n{:}\tau_\Delta, \overline{p{:}\gamma}\rangle \bullet ct)/t](t \to \tau) \quad r, c \notin (\tau, \overline{\gamma})
\end{array}
}{
\Gamma \vdash \langle e_1 \hookleftarrow\!\!+\, n{=}e_2\rangle : \mathbf{class}\, t.\langle R \mid n{:}\tau_\Delta\rangle \bullet \langle C \mid \overline{p{:}\gamma}\rangle
} \ (obj\ ext)$$

$$\frac{
\begin{array}{l}
\Gamma \vdash e_1 : \mathbf{class}\, t.\langle R \mid n{:}\tau_\Delta\rangle \bullet C \qquad\qquad \{\overline{m{:}\alpha}\} \in R \bullet C \\
\Gamma, t:T \vdash R \bullet C : [n] \qquad\qquad\qquad\quad \Delta = \{\overline{m}\} \\
\Gamma, \mathbf{rec}:T \to [\overline{m}, n] \vdash \\
\quad e_2 : [(\mathbf{class}\, t.\langle rt \mid \overline{m{:}\alpha}, n{:}\tau_\Delta\rangle \bullet ct)/t](t \to \tau)
\end{array}
}{
\Gamma \vdash \langle e_1 \leftarrow n{=}e_2\rangle : \mathbf{class}\, t.\langle R \mid n{:}\tau_\Delta\rangle \bullet C
} \ (obj\ over)$$

B Evaluation Strategy

Inference rules for get (z is either an expression or err)

$$\frac{}{get_n(x) = err}\ (get_n\ var) \qquad\qquad \frac{}{get_n(\langle\rangle) = err}\ (get_n\ \langle\rangle)$$

$$\frac{}{get_n(\lambda x.e) = err}\ (get_n\ abs)$$

$$\frac{}{get_n(\langle e_1 \hookleftarrow\!\circ\ n{=}e_2\rangle) = e_2}\ (get_n\ succ) \qquad \frac{get_n(e_1) = z \quad (m \neq n)}{get_n(\langle e_1 \hookleftarrow\!\circ\ m{=}e_2\rangle) = z}\ (get_n\ next)$$

$$\frac{\begin{array}{l} get_n(e) = e_1 \\ get_n(e_1\, e) = z \end{array}}{get_n(e \Leftarrow n) = z}\ (get_n \Leftarrow) \qquad\qquad \frac{get_n(e) = err}{get_n(e \Leftarrow n) = err}\ (get_n \Leftarrow err)$$

$$\frac{\begin{array}{l} eval(e_1) = \lambda x.e \\ get_n([e_2/x]\, e) = z \end{array}}{get_n(e_1\, e_2) = z}\ (get_n\ app) \qquad\qquad \frac{eval(e_1) = err}{get_n(e_1\, e_2) = err}\ (get_n\ app\ err)$$

Inference rules for eval

$$\frac{}{ev(x) = x}\ (eval\ var) \qquad \frac{}{ev(\langle\rangle) = \langle\rangle}\ (eval\ \langle\rangle) \qquad \frac{}{eval(\lambda x.e) = \lambda x.e}\ (eval\ abs)$$

$$\frac{}{eval(\langle e_1 \hookleftarrow\!\circ\ m{=}e_2\rangle) = \langle e_1 \hookleftarrow\!\circ\ m{=}e_2\rangle}\ (eval\ obj)$$

$$\frac{\begin{array}{l} get_n(e) = e_1 \\ eval(e_1 e) = z \end{array}}{eval(e \Leftarrow n) = z}\ (eval \Leftarrow) \qquad\qquad \frac{get_n(e) = err}{eval(e \Leftarrow n) = err}\ (eval \Leftarrow err)$$

$$\frac{\begin{array}{l} eval(e_1) = \lambda x.e \\ eval([e_2/x]e) = z \end{array}}{eval(e_1 e_2) = z}\ (eval\ app) \qquad\qquad \frac{eval(e_1) = err}{eval(e_1 e_2) = err}\ (eval\ app\ err)$$

Bisimilarity Problems Requiring Exponential Time (Extended Abstract)

Michele Boreale* and Luca Trevisan

Università di Roma "La Sapienza". Dipartimento di Scienze dell'Informazione. Via Salaria 113, 00198 Roma. Email {michele,trevisan}@dsi.uniroma1.it

Abstract. We study the complexity of deciding bisimilarity between non-deterministic processes. In particular, we consider a calculus with recursive definitions of processes, value passing (i.e. input/output of data) and an equality test over data. We show that the bisimilarity problem is EXP-complete over this calculus and thus that exponential time is *provably necessary* in order to solve it. We then prove that, if we add a *parallel composition* operator to the calculus, and we impose that parallel composition is never used inside recursive definitions, then the bisimilarity problem is still EXP-complete, thus no harder than in the fragment without parallel composition.

1 Introduction

A major field of research in theoretical computer science is concerned with the formal description and analysis of concurrent systems. Well-developed theories exist for calculi such as Milner's CCS [12], which permits naturally describing systems where different agents can interact via synchronization, without passing of data values involved. Recently, there has been much interest around extensions of CCS with explicit primitives for handling data values [7, 8, 15]. In these formalisms, referred to as *value-passing calculi*, data, beside being sent or received, can also be used as parameters in recursive definitions of processes and tested by means of certain predicates. As an example, the recursively defined process:

$$M(x) \Leftarrow [x < v](\bar{r}x.M(x) + w(y).M(y)) + [x \geq v]E(x)$$

specifies an updatable memory cell, containing an initial value x. As long as x remains less than a certain value v, the memory can either output its content at channel r, $\bar{r}x$., or input a new value at channel w, $w(y)$. (the symbol $+$ represents non-deterministic choice). As soon as x equals or exceeds v, a recovery procedure E is called. A peculiar kind of value-passing calculus is the π-calculus [13, 15, 1],

* This research was done while the first author was at the Istituto per l'Elaborazione dell'Informazione of the CNR (Italian Research Council). Work partially supported by EEC, within HCM Project Express, and by CNR, within the project "Specifica ad Alto Livello e Verifica di Sistemi Digitali".

where the values being exchanged among processes are channels themselves: this makes it possible to dynamically reconfigure processes' communication topology.

In this setting, a central problem is that of *verification*, which consists in establishing whether two given descriptions (representing, e.g., a specification and an implementation) are "equivalent" or not, according to a chosen notion of *behavioural equivalence*. By now, the algebraic aspects of this problem have become well-understood also for value-passing calculi [7, 1, 2, 8, 13, 15], but much is left to do concerning computational complexity. A fundamental issue is that of classifying the relative computational power of the different mechanisms for handling data values, also in the presence of other primitives of process calculi, such as parallelism. This could in practice provide useful indications to build more effective verification tools.

In the present paper, we shall tackle some of these issues, for one of the most widely studied equivalence, Milner's bisimulation equivalence (also called "bisimilarity"), written $\sim$ [12].

Previous works about decidability and complexity of value-passing bisimilarity are [9] and [3]. In [9], Jonsson and Parrow consider a particular class of non-deterministic value-passing processes, the *data-independent* ones: here, data can be sent, received and used in parametric definitions, but no predicate or function over them is allowed. Jonsson and Parrow show the decidability of bisimilarity for these processes by reducing the problem to bisimilarity of certain non-value-passing programs, for which verification methods exist [14, 10]. In [3], the authors prove that bisimilarity is PSPACE-hard (in the syntactical size of the terms) for data-independent processes.

Having data values without being able to test them is, in practice, of little use. It is therefore natural to ask what happens to complexity when a simple form of predicate is added to data-independent processes. In [3], a simple equality predicate over data was considered. Equality is perhaps the most elementary form of predicate one would admit over data: not even negative tests, to check inequality of two values, are permitted. The computational power of equality was indirectly showed in [3] by proving that, relying on it, input-output primitives ($a(x)$. and $\bar{a}v$.) can be polynomially reduced to the remaining operators. Here, we prove that, when equality is added, value-passing bisimilarity is EXP-complete.

We also consider adding other primitives typical of process calculi, such as parallelism [12]. It is known that, for full CCS, bisimilarity is undecidable; however, it is decidable for certain meaningful restricted formats, such as *finite control* processes, where parallelism does not occur inside the scope of recursive definitions. For processes in such format, we prove that, in the presence of equality, the overall computational complexity does not increase, i.e. bisimilarity remains EXP-complete. It is worth to stress that EXP-complete problems are *provably intractable*, that is, any algorithm solving an EXP-complete problem must have an exponential worst-case running time, and this can be shown without relying on unproven complexity-theoretic conjectures (conversely, NP-complete and PSPACE-complete problems are just *supposed* to be intractable). Meaningful EXP-complete problems are quite rare in the literature. In particular, we are not aware of any other EXP-complete problem concerned with verification of process equivalences.

The used proof techniques are also worth to mention. We rely on the characterization of the class EXP in terms of *Alternating Turing Machines (ATM)*. (see [5]). We show how to define processes that somehow *simulate* an arbitrary linear-space ATM. The more difficult technical step consists in simulating *alternation*.

The rest of the paper is organized as follows. In Section 2 we define the syntax and the operational and bisimulation semantics of the basic calculus, without parallel composition. In Section 3 we recall some basics facts about the class EXP and its characterization in terms of ATM's. Section 4 contains the EXP-completeness proofs. In Section 5 we deal with the parallel composition operator. A few conclusive remarks are contained in Section 6.

2 The Language

Below, we present first the syntax and then operational and bisimulation semantics of the calculi. The notation we use is that of value-passing CCS [11, 12] and of π-calculus [13]. We assume the following disjoint sets:

- a countable set *Act* of *pure actions* or *communications ports*, ranged over by $a, a', \ldots$;
- a countable set *Var* of *variables*, ranged over by $x, y, \ldots$;
- a countable set *Val* of *values*, ranged over by $v, v', \ldots$;
- a countable set *Ide* of *identifiers* each having a non-negative *arity*. *Ide* is ranged over by *Id* and capital letters.

A *value expression* is either a variable or a value. Value expressions are ranged over by $e, e', \ldots$. We also consider the set $\overline{Act} = \{\overline{a} \mid a \in Act\}$ of *co-actions*, which represent output synchronizations. The set $Act \cup \overline{Act}$ will be ranged over by c.

Following the notation of [3], we let $\mathcal{L}_{v,r}$ be the set of *terms* (ranged over by $P, Q, \ldots$) given by the operators of *pure synchronization prefix, input prefix, output prefix, non-determinism, matching* and *identifier*, according to the following grammar:

$$ P ::= c.P \mid a(x).P \mid \overline{a}e.P \mid \sum_{i \in I} P_i \mid [e_1 = e_2]P \mid Id(e_1, \ldots, e_k) $$

where k is the arity of Id. We also let $\mathcal{L}_r$ be the restriction of $\mathcal{L}_{v,r}$ to terms without input and output prefixes. We always assume that the index set I in $\sum_{i \in I} P_i$ is finite and sometimes write $P_1 + \cdots + P_n$ for $\sum_{i \in \{1,\ldots,n\}} P_i$. When I is empty, we use the symbol 0: $0 \stackrel{\text{def}}{=} \sum_{i \in \emptyset} P_i$.

An occurrence of a variable x in a term P is said to be *bound* if it is within the scope of an input prefix $a(x)$; otherwise it is said a *free* occurrence. The set of variables which have a free occurrence in P is denoted by $fvar(P)$. The *size* of a term P, indicated by $|P|$, is the number of symbols appearing in it; e.g., if $P = a(x).\bar{a}x.a'.0 + Id(x)$ then $|P| = 9$.

We presuppose an arbitrarely fixed *finite* set Eq of *identifiers definitions*, each of the form

$$Id(x_1, \ldots, x_k) \Leftarrow P$$

where $k \geq 0$ is the arity of the identifier Id, the x_i's are pairwise distinct and $fvar(P) \subseteq \{x_1, \ldots, x_k\}$. In Eq, each identifier has a single definition. The requirement for the set Eq to be finite is motivated by the fact that we are only interested in syntactically finite processes.

A process term P is said to be *closed* if $fvar(P) = \emptyset$; in this case, P is said to be a *process*. Processes are the terms we are most interested in. As we shall see, bisimulation semantics will be defined only over the set of processes.

The operational behaviour of our processes is defined by means of a transition relation. Its elements are triples (P, μ, P') written as $P \xrightarrow{\mu} P'$. Here, μ can be of three different forms: c, $\bar{a}v$ or $a(v)$. A *pure action* c represents a synchronization through the port c, without passing of data involved. An *output action* $\bar{a}v$ means transmission of the datum v through the port a. An *input action* $a(v)$ represents receipt of the datum v through the port a. We let μ range over actions. The transition relation is defined by the inference rules in Table 2. Note that $\xrightarrow{\mu}$ leads processes into processes. On the top of the transition relation $\xrightarrow{\mu}$, we define *strong bisimulation equivalence* $\sim$, [12, 13, 15] as usual:

$$(Sync)\ c.P \xrightarrow{c} P$$

$$(Inp)\ a(x).P \xrightarrow{a(v)} P\{v/x\},\ v \in Val \qquad (Out)\ \bar{a}v.P \xrightarrow{\bar{a}v} P$$

$$(Match)\frac{P \xrightarrow{\mu} P'}{[v = v]P \xrightarrow{\mu} P'} \qquad (Sum)\frac{P_j \xrightarrow{\mu} P'}{\sum_{i \in I} P_i \xrightarrow{\mu} P'}\ j \in I$$

$$(Ide)\frac{P\{\tilde{v}/\tilde{x}\} \xrightarrow{\mu} P'}{Id(\tilde{v}) \xrightarrow{\mu} P'} \qquad \text{if } Id(\tilde{x}) \Leftarrow P \text{ is in } Eq$$

Table 1. Inference rules for the transition relation $\xrightarrow{\mu}$.

Definition 1 (Strong bisimulation equivalence). A binary symmetric relation $\mathcal{R}$ over processes is a *bisimulation* if, whenever $P\mathcal{R}Q$ and $P \xrightarrow{\mu} P'$, there exists Q' s.t. $Q \xrightarrow{\mu} Q'$ and $P'\mathcal{R}Q'$. We let $P \sim Q$, and say that P *is bisimilar to* Q, if and only if $P\mathcal{R}Q$, for some bisimulation $\mathcal{R}$.

From now on, we will omit the adjective "strong".

3 Alternating Turing Machines and the Class EXP

In this paper, we will measure the complexity of deciding bisimilarity between P and Q with a set of identifier definitions Eq, in function of the sum of the syntactical sizes of P, Q and of the terms occurring in Eq. We will deal with the complexity classes P, LIN-EXP and EXP and with the notions of *polynomial-time reducibility*, *hardness* and *completeness*.

It is known that P $\subset$ LIN-EXP $\subset$ EXP and that the these three classes are provably distinct. A problem is hard for a class $\mathcal{C}$ if every problem in $\mathcal{C}$ is polynomial-time reducible to it; a $\mathcal{C}$-hard problem is said to be $\mathcal{C}$-complete if it belongs to $\mathcal{C}$. It is easy to show that a problem is LIN-EXP-hard if and only if it is EXP-hard. See e.g. [4] for a more complete introduction to complexity classes. Here we recall the following result due to Hartmanis and Stearns that states the provable intractability of LIN-EXP-hard problem.

Theorem 2 ([6]). *For any* LIN-EXP-*hard problem* A, *a constant* c_A *exists such that no algorithm can solve* A *with a worst case running time smaller than* $2^{n^{c_A}}$.

In the following we shall outline the characterization of LIN-EXP as the class of languages decided by *alternating Turing machines* (in short, ATM) working with linear space. This characterization will be exploited in Section 4 in order to prove the EXP-hardness of bisimilarity in $\mathcal{L}_r$.

Definition 3 (Alternating Turing Machine). An alternating Turing machine (ATM in short) AT is a five-tuple $AT = (Q, q_0, g, \Sigma, \delta)$ where

- Q is the set of *states*;
- q_0 is the *initial state*;
- $g : Q \rightarrow \{\wedge, \vee, \text{accept}, \text{reject}\}$;
- Σ is the *tape alphabet*;
- $\delta \subseteq Q \times (\Sigma \cup \{\square\}) \times Q \times \Sigma \times \{L, R\}$ is the *next move relation*.

Here $\square$ is a distinguished *blank* symbol, not belonging to Σ that represents unused parts of the tape. The function g partitions Q into four sets: the set $Q_U = \{q \in Q | g(q) = \wedge\}$ of *universal states*, the set $Q_E = \{q \in Q | g(q) = \vee\}$ of *existential states*, the set $Q_A = \{q \in Q | g(q) = \text{accept}\}$ of *accepting states*, and the set $Q_R = \{q \in Q | g(q) = \text{reject}\}$ of *rejecting states*.

Definition 4 (Configuration). A *configuration* of an ATM AT is a string

$$\bar{c} = (q_1, s_1, \ldots, q_n, s_n) \in ((Q \cup \{\bot\}) \cdot \Sigma)^*$$

such that exactly one index $j \in \{1, \ldots, n\}$ exists such that $q_j \neq \bot$.

Intuitively, $\bar{c} = (\bot, s_1, \ldots, \bot, s_{j-1}, q, s_j, \bot, s_{j+1}, \ldots, \bot, s_n)$ represents the global state of machine AT when n cells of the tape have been used, the head is on the j-th cell, the content of the tape is $s_1, \ldots, s_n$, and the finite control is in state q. We will denote by $\mathcal{GC}_{AT}$ the set of configurations of machine AT. A configuration is said to be *halting* (respectively, *existential, universal*) if it contains a halting (respectively, existential, universal) state. The *initial* configuration of AT with input $x = (x_1, \ldots, x_k)$ is $(q_0, x_1, \perp, x_2, \ldots, \perp, x_k)$.

With a slight abuse of notation, we will denote by $\delta(\bar{c})$ the set of configurations $\bar{c}'$ such that $\bar{c}$ can evolve in one step into $\bar{c}'$ according to the relation δ. Whenever $\bar{c}' \in \delta(\bar{c})$ we will write $\bar{c} \vdash \bar{c}'$; let $\vdash^*$ be the transitive and reflexive closure of $\vdash$, let $\bar{c}_0(x)$ be the initial configuration of machine AT with input x, we will denote by $\mathcal{GC}_{AT(x)}$ the set $\{\bar{c} \in \mathcal{GC}_{AT} \mid \bar{c}_0(x) \vdash^* \bar{c}\}$ and call it the *computation tree* of AT with input x. In this paper we shall only consider *time-bounded* ATM's, that is, machines having a finite computation tree for any input.

Acceptance is defined in a quite involved way for general ATM's (see [5]). In the case of time bounded ATM's, however, a much simpler inductive definition can be given.

Definition 5 (Acceptance). Let AT be a time-bounded ATM, x be a string, $\bar{c} \in \mathcal{GC}_{AT(x)}$ be a configuration.

1. If $\bar{c}$ is a halting configuration, then we say that $\bar{c}$ is an *accepting* configuration if it contains an accepting state, otherwise we say that it is a *rejecting* configuration.
2. If $\bar{c}$ is a universal configuration, then we say that it is accepting if all the configurations in $\delta(\bar{c})$ are accepting, otherwise we say that it is rejecting.
3. If $\bar{c}$ is an existential configuration, then we say that it is accepting if at least one configuration in $\delta(\bar{c})$ is accepting, otherwise we say that it is rejecting.

We say that AT *accepts* input x if the initial configuration of AT with input x is accepting. A language L is *decided* by an alternating Turing machine AT if AT accepts x if and only if $x \in L$. The following theorem has been proved by Chandra, Kozen, and Stockmeyer.

Theorem 6 ([5]). *Every language $L \in$ LIN-EXP is decidable by an alternating Turing machine AT_L working with linear space and exponential time.*

With standard techniques from the theory of Turing machines, we may assume without loss of generality that AT_L is such that, for any input x of size n, only the cells of the tape containing x are accessed, all the computation paths of $AT(x)$ have the same length and if $\bar{c} \in \mathcal{GC}_{AT(x)}$ is a universal (respectively, existential) configuration, then all the configurations in $\delta(\bar{c})$ are existential (respectively, universal). In the following, we shall call such a machine a *canonical linear-space alternating machine*.

4   The EXP-completeness Result

In the following we shall prove the following result.

Lemma 7. *Let AT be a canonical linear space ATM. Then, for any string x of length n, we can compute, in time polynomial in n, two processes P and Q of $\mathcal{L}_r$ such that $P \sim Q$ if and only if x is accepted by AT.*

In order to prove the above lemma, we shall define three identifiers A, S, F[2]. As a first approximation, if $\bar{c}$ is a configuration of AT, then $A(\bar{c})$ is a process that *simulates* the computation of AT starting from configuration $\bar{c}$. In particular, if $\bar{c}_0$ is the starting configuration of AT with input x, then the labeled transition system of $A(\bar{c}_0)$ is "isomorphic" to $\mathcal{GC}_{AT(x)}$: there is a correspondence between configurations $\bar{c} \in \mathcal{GC}_{AT(x)}$ and processes $A(\bar{c})$, and between nondeterministic branching of the ATM and nondeterministic choice in the process. Furthermore, the processes corresponding to halting configurations $\bar{c} \in \mathcal{GC}_{AT(x)}$ can do a single action: a if $\bar{c}$ is accepting and b if $\bar{c}$ is rejecting. S (respectively, F) is defined to be identical to A, except that states corresponding to halting configurations always do a (respectively, b). Thus, intuitively, A would be bisimilar to S in case AT accepts, and bisimilar to F otherwise.

Indeed, the above straightforward construction fails to express alternation of quantifiers in terms of bisimulation, and has to be slightly modified. For example, assume that an existential configuration $\bar{c}$ of $AT(x)$ branches into two configurations, one accepting and one rejecting. Then $\bar{c}$ is accepting, and we would like the corresponding process $A(\bar{c})$ to be bisimilar to $S(\bar{c})$. But $A(\bar{c})$ branches into both a rejecting and an accepting state, while $S(\bar{c})$ branches into two accepting states: thus $A(\bar{c})$ and $S(\bar{c})$ could not be bisimilar. This inconvenience can be overcome if we assume that each state corresponding to an existential (respectively, universal) configuration always branches into at least one state corresponding to a rejecting (respectively, accepting) configuration.

The actual definition of identifiers A, S, and F is given in Tables 2–4. It is convenient to split into three lemmas the proof that those processes indeed exhibit the desired behaviour.

Since no confusion can arise, in the following we will use $\mathcal{GC}$ as a shorthand for $\mathcal{GC}_{AT(x)}$.

Lemma 8. *Let $\bar{c}_1, \bar{c}_2 \in \mathcal{GC}$ be any two (not necessarily distinct) configurations that halt within the same number of steps, then the following holds.*

1. *$S(\bar{c}_1) \not\sim F(\bar{c}_2)$.*
2. *$S(\bar{c}_1) \sim S(\bar{c}_2)$ and $F(\bar{c}_1) \sim F(\bar{c}_2)$.*

Lemma 9. *For any $\bar{c} \in \mathcal{GC}$, either $A(\bar{c}) \sim S(\bar{c})$ or $A(\bar{c}) \sim F(\bar{c})$.*

The above two lemmas can be proved by induction on the number of steps required to move from $\bar{c}_1$ (respectively, $\bar{c}$) to a halting configurations. Canonicity plays an essential role in the proofs.

Lemma 10. *For any $\bar{c} \in \mathcal{GC}$, $A(\bar{c}) \sim S(\bar{c})$ if and only if $\bar{c}$ is an accepting configuration.*

---

[2] A stands for ATM, S for *success* and F for *failure*.

$$A(x_1, y_1, \ldots, x_n, y_n) \Leftarrow$$

$$\sum_{\substack{i \in \{1, \ldots, n\}, \, q \in Q_E \\ (q, s, q', s', L) \in \delta}} [x_i = s][y_i = q]a.A(x_1, y_1, \ldots, x_{i-1}, q', s', \bot, x_{i+1}, y_{i+1}, \ldots, x_n, y_n)$$

$$+ [x_i = s][y_i = q]a.F(x_1, y_1, \ldots, x_{i-1}, q', s', \bot, x_{i+1}, y_{i+1}, \ldots, x_n, y_n)$$

$$+ \sum_{\substack{i \in \{1, \ldots, n\}, \, q \in Q_E \\ (q, s, q', s', R) \in \delta}} [x_i = s][y_i = q]a.A(x_1, y_1, \ldots, x_{i-1}, y_{i-1}, s', \bot, x_{i+1}, q', \ldots, x_n, y_n)$$

$$+ [x_i = s][y_i = q]a.F(x_1, y_1, \ldots, x_{i-1}, y_{i-1}, s', \bot, x_{i+1}, q', \ldots, x_n, y_n)$$

$$+ \sum_{\substack{i \in \{1, \ldots, n\}, \, q \in Q_U \\ (q, s, q', s', L) \in \delta}} [x_i = s][y_i = q]a.A(x_1, y_1, \ldots, x_{i-1}, q', s', \bot, x_{i+1}, y_{i+1}, \ldots, x_n, y_n)$$

$$+ [x_i = s][y_i = q]a.S(x_1, y_1, \ldots, x_{i-1}, q', s', \bot, x_{i+1}, y_{i+1}, \ldots, x_n, y_n)$$

$$+ \sum_{\substack{i \in \{1, \ldots, n\}, \, q \in Q_U \\ (q, s, q', s', R) \in \delta}} [x_i = s][y_i = q]a.A(x_1, y_1, \ldots, x_{i-1}, y_{i-1}, s', \bot, x_{i+1}, q', \ldots, x_n, y_n)$$

$$+ [x_i = s][y_i = q]a.S(x_1, y_1, \ldots, x_{i-1}, y_{i-1}, s', \bot, x_{i+1}, q', \ldots, x_n, y_n)$$

$$+ \sum_{q \in Q_A} [y_i = q]a$$

$$+ \sum_{q \in Q_R} [y_i = q]b$$

Table 2. Definition of A.

Proof. We proceed again by induction on the number of residual steps. If $\bar{c}$ is a halting configuration, then the proof is trivial.

Otherwise, let us assume that, for any $\bar{c}' \in \delta(\bar{c})$, $\bar{c}'$ is an accepting configuration if and only if $A(\bar{c}') \sim S(\bar{c}')$. We have to distinguish two cases: either $\bar{c}$ is existential or it is universal. We show only the first case, as the second one is very similar.

Since $\bar{c}$ is existential, then it is accepting if and only if $\delta(\bar{c})$ contains at least one accepting configuration. By induction hypothesis, the latter statement holds if and only if:

$$\text{there is } \bar{c}' \in \delta(\bar{c}) \text{ s.t. } A(\bar{c}') \sim S(\bar{c}'). \tag{1}$$

We now prove that (1) holds if and only if $A(\bar{c}) \sim S(\bar{c})$. Let us assume that (1) holds. There are two non-trivial cases.

$$S(x_1, y_1, \ldots, x_n, y_n) \Leftarrow$$

$$\sum_{\substack{i \in \{1,\ldots,n\}, \ q \in Q_E \\ \langle q,s,q',s',L \rangle \in \delta}} [x_i = s][y_i = q]a.S(x_1, y_1, \ldots, x_{i-1}, q', s', \bot, x_{i+1}, y_{i+1}, \ldots, x_n, y_n)$$

$$+ [x_i = s][y_i = q]a.F(x_1, y_1, \ldots, x_{i-1}, q', s', \bot, x_{i+1}, y_{i+1}, \ldots, x_n, y_n)$$

$$+ \sum_{\substack{i \in \{1,\ldots,n\}, \ q \in Q_E \\ \langle q,s,q',s',R \rangle \in \delta}} [x_i = s][y_i = q]a.S(x_1, y_1, \ldots, x_{i-1}, y_{i-1}, s', \bot, x_{i+1}, q', \ldots, x_n, y_n)$$

$$+ [x_i = s][y_i = q]a.F(x_1, y_1, \ldots, x_{i-1}, y_{i-1}, s', \bot, x_{i+1}, q', \ldots, x_n, y_n)$$

$$+ \sum_{\substack{i \in \{1,\ldots,n\}, \ q \in Q_U \\ \langle q,s,q',s',L \rangle \in \delta}} [x_i = s][y_i = q]a.S(x_1, y_1, \ldots, x_{i-1}, q', s', \bot, x_{i+1}, y_{i+1}, \ldots, x_n, y_n)$$

$$+ [x_i = s][y_i = q]a.S(x_1, y_1, \ldots, x_{i-1}, y_{i-1}, s', \bot, x_{i+1}, q', \ldots, x_n, y_n)$$

$$+ \sum_{q \in Q_A \cup Q_R} [y_i = q]a$$

Table 3. Definition of S.

First, consider the case when $A(\bar{c}) \xrightarrow{a} A(\bar{c}_1)$, for any $\bar{c}_1 \in \delta(\bar{c})$. From canonicity and Lemma 9, it is $A(\bar{c}_1) \sim S(\bar{c}_1)$ or $A(\bar{c}_1) \sim F(\bar{c}_1)$: in the first case the matching move for $S(\bar{c})$ is $A(\bar{c}) \xrightarrow{a} S(\bar{c}_1)$, in the second case it is $S(\bar{c}) \xrightarrow{a} F(\bar{c}_1)$. Consider the case when $A(\bar{c})$ has to match a transition $S(\bar{c}) \xrightarrow{a} S(\bar{c}_1)$ from $A(\bar{c})$. Then we have:

$$A(\bar{c}) \xrightarrow{a} A(\bar{c}') \sim S(\bar{c}') \sim S(\bar{c}_1)$$

where the first $\sim$ follows from (1) and the second one from canonicity and Lemma 8.

Conversely, assume that $A(\bar{c}) \sim S(\bar{c})$. Take any $c_1 \in \delta(\bar{c})$ and consider the transition $S(\bar{c}) \xrightarrow{a} S(\bar{c}_1)$. Then we must have for some $\bar{c}'$ $A(\bar{c}) \xrightarrow{a} K(\bar{c}') \sim S(\bar{c}_1)$, where $K = F$ or $K = A$. The case $K = F$ cannot arise, due to canonicity and Lemma 8. Thus it must be $K = A$. From canonicity and Lemma 8 it follows $A(\bar{c}') \sim S(\bar{c}_1) \sim S(\bar{c}')$, which validates (1). $\square$

Proof. (Of Lemma 7) Let $\bar{c}_0$ be the initial configuration of AT with input x. Since the definition of the identifiers can be clearly constructed in time polynomial in n, the lemma follows by setting $P \stackrel{\text{def}}{=} A(\bar{c}_0)$ and $Q \stackrel{\text{def}}{=} S(\bar{c}_0)$. $\square$

$$F(x_1, y_1, \ldots, x_n, y_n) \Leftarrow$$

$$\sum_{\substack{i \in \{1,\ldots,n\},\ q \in Q_E \\ (q,s,q',s',L) \in \delta}} [x_i = s][y_i = q]a.F(x_1, y_1, \ldots, x_{i-1}, q', s', \bot, x_{i+1}, y_{i+1}, \ldots, x_n, y_n)$$

$$+ \sum_{\substack{i \in \{1,\ldots,n\},\ q \in Q_E \\ (q,s,q',s',R) \in \delta}} [x_i = s][y_i = q]a.F(x_1, y_1, \ldots, x_{i-1}, y_{i-1}, s', \bot, x_{i+1}, q', \ldots, x_n, y_n)$$

$$+ \sum_{\substack{i \in \{1,\ldots,n\},\ q \in Q_U \\ (q,s,q',s',L) \in \delta}} [x_i = s][y_i = q]a.S(x_1, y_1, \ldots, x_{i-1}, q', s', \bot, x_{i+1}, y_{i+1}, \ldots, x_n, y_n)$$

$$+ [x_i = s][y_i = q]a.F(x_1, y_1, \ldots, x_{i-1}, q', s', \bot, x_{i+1}, y_{i+1}, \ldots, x_n, y_n)$$

$$+ \sum_{\substack{i \in \{1,\ldots,n\},\ q \in Q_U \\ (q,s,q',s',R) \in \delta}} [x_i = s][y_i = q]a.S(x_1, y_1, \ldots, x_{i-1}, y_{i-1}, s', \bot, x_{i+1}, q', \ldots, x_n, y_n)$$

$$+ [x_i = s][y_i = q]a.F(x_1, y_1, \ldots, x_{i-1}, y_{i-1}, s', \bot, x_{i+1}, q', \ldots, x_n, y_n)$$

$$+ \sum_{q \in Q_A \cup Q_R} [y_i = q]b$$

Table 4. Definition of F.

Lemma 7 and the results recalled in the previous section immediately imply the EXP-hardness of $\mathcal{L}_r$.

Theorem 11. *The bisimilarity problem in the language $\mathcal{L}_r$ is EXP-hard.*

We shall indeed see in the next section that the problem is in EXP. From the above theorem and from Theorem 2 the intractability of bisimilarity in $\mathcal{L}_r$ follows.

Corollary 12. *A constant $c > 0$ exists such that any algorithm that decides the bisimilarity problem in $\mathcal{L}_r$ has a worst-case running time no better than 2^{n^c}, where n is the size of the input.*

5 The Parallel Composition Operator

In this section we consider adding the *parallel composition* operator | (see e.g. [12]) to the language described in Section 2. We will show that, for a certain restricted syntactic format, the bisimilarity problem with parallel composition is decidable and EXP-complete. As a consequence, the bisimilarity problem is in EXP for all the fragments we have considered in the paper.

The syntax of the language $\mathcal{L}_{v,r}$ is extended with the clause

$$P ::= P \mid P.$$

All definitions and notions given for $\mathcal{L}_{v,r}$ (such as free variables, subterms etc.) are extended to the new language in the expected way. Following [12], the operational semantics of the new operator is given by the rules:

$$(Par)\frac{P_1 \xrightarrow{\mu} P_1'}{P_1 \mid P_2 \xrightarrow{\mu} P_1' \mid P_2} \qquad (Com)\frac{P_1 \xrightarrow{\mu} P_1', \, P_2 \xrightarrow{\mu'} P_2'}{P_1 \mid P_2 \xrightarrow{\tau} P_1' \mid P_2'}$$

with ($\mu = a$ and $\mu' = \bar{a}$) or ($\mu = a(v)$ and $\mu' = \bar{a}v$), *plus* the symmetric versions of the above rules, where the roles of P_1 and P_2 are exchanged. Here $\tau \notin Act$ is a new kind of action, called the *silent* action.

An important class of parallel processes is that *finite-control* processes, where parallel composition never occurs inside recursive definitions. The corresponding sub-language is indicated with $\mathcal{L}_{v,r,p}$, while the sublanguage of $\mathcal{L}_{v,r,p}$ without input/output primitives is indicated by $\mathcal{L}_{r,p}$. By confining ourselves to finite-control processes, we are able to extend a reduction from $\mathcal{L}_{v,r}$ to $\mathcal{L}_r$ given in [3] and prove:

Proposition 13. *The bisimilarity problem in $\mathcal{L}_{v,r,p}$ is equivalent to the bisimilarity problem in $\mathcal{L}_{r,p}$, up to polynomial-time reduction.*

Observe that every process $P \in \mathcal{L}_{r,p}$, defined with respect to a set of identifier definitions Eq, has a finite transition system. More precisely, first note that the size of every term reachable from P cannot exceed k, where

$$k \stackrel{\text{def}}{=} |P| * \max\{|R| : R \text{ appears in } Eq\} \cup \{1\}.$$

Therefore, there are at most n^k different states in the transition system, where n is the number of distinct values appearing in P and in Eq. Both n and k are easily seen to be at most polynomial functions of the size of the problem. It follows that the bisimilarity problem in $\mathcal{L}_{r,p}$ can be solved in exponential time using, for example, the algorithm by Page and Tarjan [14]. Putting together the latter fact, Theorem 11, the cited result of [3] and the above Proposition 13, we have the following result of equivalence between languages.

Theorem 14. *The bisimilarity problems for the languages $\mathcal{L}_r$, $\mathcal{L}_{v,r}$, $\mathcal{L}_{r,p}$ and $\mathcal{L}_{v,r,p}$ are all EXP-complete.*

It is worth to notice that, even in the absence of values, the presence of parallel composition implies an exponential blow-up of the number of states. This is implicitly present, for example, in the so-called "expansion law" [12]: $a \mid b \sim a.b + b.a$. The above theorem tells us that the computational complexity due to parallel composition itself is not greater than that caused by the handling of data-values.

6 Conclusions

We studied the complexity of deciding bisimulation equivalence over calculi with recursive definitions, value-passing, and/or limited parallel composition. All the considered calculi are EXP-complete, and thus provably intractable and computationally equivalent.

It would be interesting to consider a calculus $\mathcal{L}_{v,p}$ with value-passing and unrestricted parallel composition, but with no recursion. Classifying the computational complexity of this calculus would give some insight into the intrinsic hardness of parallel composition, in a setting where it would not be overwhelmed by the expressiveness of recursive definitions.

References

1. M. Boreale and R. De Nicola. A symbolic semantics for the π-calculus. Short version in *Proc. of CONCUR'94*, LNCS, Springer Verlag. Full version to appear on *Information and Computation*.
2. M. Boreale and R. De Nicola. Testing equivalence for mobile processes. *Information and Computation*, 2(120):279–303, 1995.
3. M. Boreale and L. Trevisan. On the complexity of bisimilarity for value-passing processes. In *Proceedings of the 15th Conference on Foundations of Software Technology and Theoretical Computer Science*. LNCS, Springer Verlag, 1995.
4. D.P. Bovet and P. Crescenzi. *Introduction to the Theory of Complexity*. Prentice Hall, 1993.
5. A.K. Chandra, D.C. Kozen, and L.J. Stockmeyer. Alternation. *Journal of the ACM*, 28(1):114–133, 1981.
6. J. Hartmanis and R.E. Stearns. On the computational complexity of algorithms. *Transactions of the AMS*, 117:285–306, 1965.
7. M. Hennessy and A. Ingolfsdottir. A theory of communicating processes with value passing. *Information and Computation*, 2(107):202–236, 1993.
8. M. Hennessy and H. Lin. Symbolic bisimulations. *Theoretical Computer Science*, 138:353–389, 1995.
9. B. Jonsson and J. Parrow. Deciding bisimulation equivalences for a class of non-finite state programs. *Information and Computation*, 107:272–302, 1993.
10. P.C. Kanellakis and S.A. Smolka. CCS expressions, finite sate processes, and three problems of equivalence. *Information and Computation*, 86:43–68, 1990.
11. R. Milner. *A Calculus of Communicating Systems*. LNCS, 92. Springer-Verlag, Berlin, 1980.
12. R. Milner. *Communication and Concurrency*. Prentice-Hall, 1989.
13. R. Milner, J. Parrow, and D. Walker. A calculus of mobile processes, part 1 and 2. *Information and Computation*, 100:1–78, 1992.
14. R. Paige and R.E. Tarjan. Three partition refinement algorithms. *SIAM Journal on Computing*, 16(6):973–989, 1987.
15. J. Parrow and D. Sangiorgi. Algebraic theories for name-passing calculi. *Information and Computation*, 120(2):174–197, 1995.

Linear Dynamic Kahn Networks Are Deterministic

Arie de Bruin   S. H. Nienhuys-Cheng

`arie@cs.few.eur.nl` `cheng@cs.few.eur.nl`

Departement of Computer Science, H4-19
Erasmus University Rotterdam
P.O. Box 1738, 3000DR, Rotterdam, the Netherlands

Abstract. The (first part of the) Kahn principle states that networks with deterministic nodes are deterministic on the I/O level: for each network, different executions provided with the same input streams deliver the same output stream. The Kahn principle has thus far not been proved for dynamic, non-deterministic networks.

We consider a simple language L containing the fork-statement. For this language we define a non-deterministic transition system which defines all interleavings consisting of basic steps, for all possible executions of a program. We prove that, although on the execution level there is much nondeterminism, this nondeterminism disappears because all executions deliver the same output stream (or a prefix of it), given the same input stream. This proves the Kahn principle for linear, non-deterministic dynamic networks.

1   Introduction

A dataflow network consists of a number of parallel processes which are interconnected by directed channels. Processes communicate with each other only through these channels, they do not share variables. The channels act as possibly infinite FIFO queues. The Kahn principle concerns networks with deterministic nodes (Kahn networks), i.e. each process selects the channel it uses next in a deterministic way. Which process gets the chance to execute a step at a certain moment is described by a transition system. Such a system can be deterministic or nondeterministic. In a nondeterministic system, the process that will execute next is not pre-determined. Thus different executions (computations) are possible.

In his seminal paper [K74], Kahn characterizes such processes abstractly as functions, transforming input histories into output histories, where a history models a stream of values which appears on a channel during a computation. He states a property which has since then become known as the Kahn principle: a network consisting of deterministic nodes computes a function from input histories to output histories. This function can be obtained as the smallest solution of a set of equations derived from the network.

Thus the Kahn principle contains two parts. Part 1 states that the nondeterminism caused by the asynchronicity of the computing processes does not lead to global nondeterminism in the history level I/O-behaviour of the network. This means that given an initial state of the variables in the processes and a set of input streams, one for each input channel of the network, the set of output streams which will be delivered on the output channels, is uniquely determined, independent of the computations. This I/O function can be used to define the operational semantics for the program which has induced these processes. Part 2 can be reformulated as follows. A meaning (denotational semantics) can be given to every program in the language by defining it as the smallest solution of a set of equations. The function defined in part 1 is in fact this solution. In other words, the operational and denotational semantics are the same.

Kahn's paper has been quite influential, and has been the basis of much subsequent research, e.g. evaluation strategies or implementations of dataflow networks have been defined, e.g. [KM77, AG78, F82].

The Kahn principle has been proved for certain types of networks [C72, A81, LS89], i.e., for static networks, but not for dynamic ones where it is possible that a single process may expand into a new network of processes during execution. The Kahn principle in this situation shares perhaps the fate of many conjectures in mathematics: it is intuitively correct and no counterexample has been found so far, yet no one has been able to prove it.

We have investigated the Kahn principle for a simple imperative language L in which it is possible to dynamically create linear arrays of processes, not unlike a Unix pipeline which can be built up using the Unix primitives 'pipe' and 'fork'. Because every process has only one input channel and one output channel, there is only one input history and one output history for a process.

In [B86] this language L has been introduced and a denotational semantics has been defined for it, along the lines of [K74]. In [BBB93], an operational semantics has been given for this language. However, this operational semantics is limited, because it captures only one possible execution, and therefore it is deterministic. It is based on the demand driven (call by need) approach and it formalizes the so called coroutine- model proposed in [KM77]. In the same paper, a proof is given for the equivalence of this operational semantics with a denotational one.

In this paper we prove the first part of the Kahn principle for the linear dynamic networks induced by programs from L. For details, see [BN94]. We introduce a transition system **NT** defining a fully nondeterministic interleaving semantics. We then represent each computation defined by the transition system by a graph of one computation. To prove the first part of the Kahn principle, we show that all these graphs can be combined into one graph of all computations. Then we can prove that the global output streams of different computations are either the maximal output stream or a prefix of it.

1.1 An Example of a Linear Dynamic Network

We first introduce the language L informally. For a given input stream and an initial state, the execution of a program in L produces an output stream. Initially the program is executed by one process, using precisely one input and one output channel. Execution of the statement $read(x)$ will fetch the next value from the input channel and assign it to the variable x. Execution of the statement $write(e)$ will evaluate the expression e and write the resulting value to the output channel.

A process can split up into two new, nearly identical subprocesses, the mother process and the daughter process. This effect is achieved by a statement of the form $fork(w)$. The original input channel becomes the mother's input channel and likewise the original output channel becomes the daughter's output channel. Between mother and daughter a new channel is created, which is the mother's output and the daughter's input channel. This channel will originally be empty. Both processes proceed by executing the statement following $fork(w)$. We distinguish between mother and daughter by giving the variable w in the mother process the value 0, in the daughter the value 1.

Here we give an example which shows that more than one execution of the same program is possible and that an unbounded number of linearly connected processes may be created during one execution. This example describes the *Sieve of Eratosthenes*, which selects the sequence of prime numbers from its input sequence $2 \cdot 3 \cdot 4 \cdot 5 \ldots$.

```
while true do read(v); write(v); fork(w);
      if w = 0{mother} then while true do read(x);
                                  if (x mod v) ≠ 0 then write(x)
                                                    else skip fi
              od
            else skip{daughter}
      fi
od
```

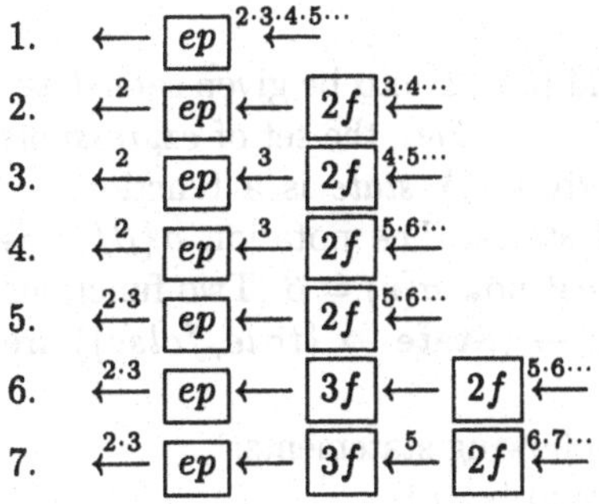

Fig. 1. One possible execution

At the beginning, the process expander (denoted by *ep* in Fig. 1 and described by the above program) reads 2 from the global input channel and writes it to the global output channel. It will then split into two processes by the 'fork'. The mother process $2f$ will be engaged in the second while-loop forever. Its task is to read the inputs from its input channel and to write the numbers which are not multiples of 2 to its output channel. For example, 3 is the first number read by $2f$ and it is passed to the daughter. The daughter is the new expander and it will split itself after the 3 is written. Thus a new expander and new filter $3f$ are created in a similar way. The new mother $3f$ will filter out the multiples of 3. In every stage there are many filters and one expander which can split itself into two new subprocesses. In fact, when a number n comes in from the rightmost channel, it has to pass through all filters mf where $m < n$ and m is prime. It can reach the expander and be written only if it is prime. A new filter nf will then be formed by the fork-statement. In Fig. 1, we show some stages of a particular execution of this program. Notice that after stage 3, either ep can read 3 or $2f$ can read 4. In this execution, stage 4 results from the second alternative. Afterwards ep reads 3 and writes 3 as shown in stage 5. Stage 6 and stage 7 also show some particular choices in the order of execution.

This example shows that a process can expand into an unlimited number of subprocesses. Since there is more than one subprocess at a certain stage, there are also different choices for the next stage. The first part of Kahn's principle states that different choices will not affect the (maximal) output stream. In our example, this means we will get the sequence of prime numbers or a prefix thereof in all cases. For this example, the first part of Kahn's principle is intuitively correct. How to prove this formally and in general?

2 A Nondeterministic Transition System

We will first give definitions of some of the concepts used in this article.

2.1 Preliminaries

- Let $(v \in)$ **Var**, $(\alpha, \beta, \gamma \in)$ **Val**, $(e \in)$ **Exp** and $(b \in)$ **Bexp** be given sets. They are usually called the set of *variables*, the set of *values*, the set of *expressions* and the set of *boolean expressions*, respectively. A *state* is a function $\sigma :$ **Var** $\to$ **Val**. Let **State** be the set of all states. The notation $\sigma\{\beta/v\}$ is used to denote a state equal to σ, except that now $\sigma(v) = \beta$. Two functions $V :$ **Exp** $\to$ (**State** $\to$ **Val**) and $B :$ **Bexp** $\to$ (**State** $\to \{true, false\}$) are assumed to be available.
- Let **L** be the language which contains the following statements:
  $s ::= v := e \mid skip \mid write(e) \mid read(v) \mid fork(v) \mid s; s \mid$
  if b **then** s **else** s **fi** $\mid$ **while** b **do** s **od**
- Let $(\eta, \zeta, \xi \in)$ **Val**$^\infty$ be the set of finite or infinite sequences of elements from **Val**. Such a sequence is called a *stream*. Let ϵ be the *empty stream* and let $rest(\alpha\eta) = \eta$. For two streams ξ and ζ, let $\xi \cdot \zeta(= \xi\zeta)$ be their concatenation.
- A process is a function $P :$ **State** $\to$ (**Val**$^\infty \to$ **Val**$^\infty$).
- Let E denote termination. A resumption is recursively defined by $r ::= E \mid s : r$, where s in **L**. Let **Res** be the set of all resumptions.

2.2 Configurations

We will use 'configurations' to describe snapshots of a network at work. Later we will define a nondeterministic transition system **NT** which models the basic steps a configuration can take in an execution. In the following figure, a box contains information of a process (a subprocess of the network) at a certain state. The state and the resumption which determine the process are written in the box.

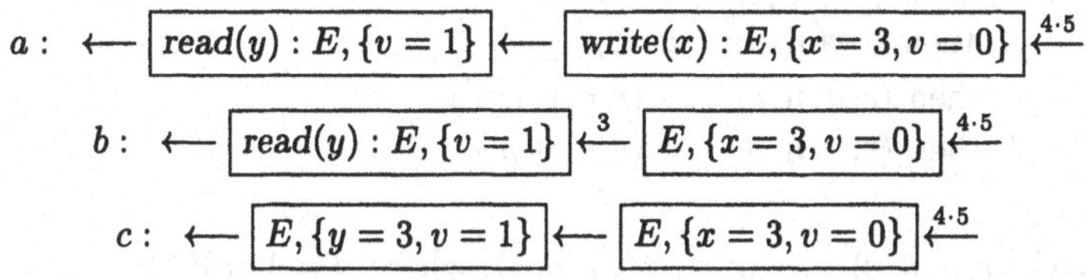

$$a: \quad \longleftarrow \boxed{read(y) : E, \{v = 1\}} \longleftarrow \boxed{write(x) : E, \{x = 3, v = 0\}} \overset{4 \cdot 5}{\longleftarrow}$$

$$b: \quad \longleftarrow \boxed{read(y) : E, \{v = 1\}} \overset{3}{\longleftarrow} \boxed{E, \{x = 3, v = 0\}} \overset{4 \cdot 5}{\longleftarrow}$$

$$c: \quad \longleftarrow \boxed{E, \{y = 3, v = 1\}} \longleftarrow \boxed{E, \{x = 3, v = 0\}} \overset{4 \cdot 5}{\longleftarrow}$$

Fig. 2. Snapshots of a network at work.

We begin with the situation shown in Fig. 2a. We allow 3 to be written without bothering whether it is needed by the next subprocess (Fig 2b). To describe such situations by configurations, we need a buffer stream containing what is written and not yet used in the input channel of a subprocess. For consistency we also introduce a buffer stream for the first process from the right. Initially this stream equals ϵ. We can view this situation as if there is still another subprocess (the 0-th subprocess) which contains hidden $read + write$ statements. All it does, is to fetch the next element from the global input stream and to write it into the buffer of the first subprocess. The configurations we use to describe the stages of the transitions in **NT** corresponding to a, b, c in Fig. 2 are:

$\underline{a} : \langle read(y) : E, \{v = 1\}, \epsilon, \langle write(x) : E, \{x = 3, v = 0\}, \epsilon, 4 \cdot 5 \rangle \rangle$

$\underline{b} : \langle read(y) : E, \{v = 1\}, 3, \langle E, \{x = 3, v = 0\}, \epsilon, 4 \cdot 5 \rangle \rangle$

$\underline{c} : \langle E, \{y = 3, v = 1\}, \epsilon, \langle E, \{x = 3, v = 0\}, \epsilon, 4.5 \rangle \rangle$

Since we will define a nondeterministic transition system, there is also a possible transition from $\underline{a}$ to the following configuration $\underline{b}'$, caused by 0-th subprocess:

$\underline{b}' : \langle read(y) : E, \{v = 1\}, \epsilon, \langle write(x) : E, \{x = 3, v = 0\}, 4, 5 \rangle \rangle$

Definition 1. A configuration ρ is recursively defined as $\rho ::= \zeta | \langle r, \sigma, \eta, \rho' \rangle$, where r is a resumption, $\zeta, \eta \in \mathbf{Val}^\infty, \sigma \in \mathbf{State}$ and ρ' is a configuration. Let **Config** be the set of all configurations. A configuration of nesting number $n > 0$ has the form

$$\rho = \langle r_n, \sigma_n, \eta_n, \langle r_{n-1}, \sigma_{n-1}, \eta_{n-1}, \langle \ldots \langle r_1, \sigma_1, \eta_1, \zeta \rangle \ldots \rangle$$

2.3 A Nondeterministic Transition System NT for L

A transition system is a relation $\rightarrow$ on **Config**, i.e. $\rightarrow \subseteq \mathbf{Config} \times \mathbf{Config}$. A configuration ρ is called a terminal if there is no ρ' such that $(\rho, \rho') \in \rightarrow$. This is denoted by $\rho \longrightarrow \otimes$. A transition can be accompanied by a label $\alpha \in \mathbf{Val}$, in which case we write $\rho \overset{\alpha}{\longrightarrow} \rho'$ (If not, we write $\rho \longrightarrow \rho'$). We say α is the output of the transition. We define this relation by induction on the nesting number of configurations. The following transition system is based on the principle that every subprocess is a candidate to take the next step (cf. case 2 below). Thus different executions are possible.

1. $\epsilon \longrightarrow \otimes$
   If $\rho \longrightarrow \otimes$, then $\langle E, \sigma, \eta, \rho \rangle \longrightarrow \otimes$
   If $\rho \longrightarrow \otimes$, then $\langle \mathrm{read}(v) : r, \sigma, \epsilon, \rho \rangle \longrightarrow \otimes$
2. If $\rho \overset{\alpha}{\longrightarrow} \rho'$, then $\langle r, \sigma, \eta, \rho \rangle \longrightarrow \langle r, \sigma, \eta\alpha, \rho' \rangle$
   If $\rho \longrightarrow \rho'$, then $\langle r, \sigma, \eta, \rho \rangle \longrightarrow \langle r, \sigma, \eta, \rho' \rangle$
3. $\beta\zeta' \overset{\beta}{\longrightarrow} \zeta'$
4. $\langle (v := e) : r, \sigma, \eta, \rho \rangle \longrightarrow \langle r, \sigma\{\beta/v\}, \eta, \rho \rangle$, where $\beta = V(e)(\sigma)$.
5. $\langle \mathrm{skip} : r, \sigma, \eta, \rho \rangle \longrightarrow \langle r, \sigma, \eta, \rho \rangle$
6. $\langle \mathrm{write}(e) : r, \sigma, \eta, \rho \rangle \overset{\beta}{\longrightarrow} \langle r, \sigma, \eta, \rho \rangle$ where $\beta = V(e)(\sigma)$
7. $\langle \mathrm{read}(v) : r, \sigma, \beta\eta, \rho \rangle \longrightarrow \langle r, \sigma\{\beta/v\}, \eta, \rho \rangle$
8. $\langle \mathrm{fork}(v) : r, \sigma, \eta, \rho \rangle \longrightarrow \langle r, \sigma\{1/v\}, \epsilon, \langle r, \sigma\{0/v\}, \eta, \rho \rangle \rangle$
9. $\langle (s_1;\ s_2) : r, \sigma, \eta, \rho \rangle \longrightarrow \langle s_1 : (s_2 : r), \sigma, \eta, \rho \rangle$
10. if $B(b)(\sigma)$, then $\langle \textbf{if } b \textbf{ then } s_1 \textbf{ else } s_2 \textbf{ fi} : r, \sigma, \eta, \rho \rangle \longrightarrow \langle s_1 : r, \sigma, \eta, \rho \rangle$
    if $\neg B(b)(\sigma)$, then $\langle \textbf{if } b \textbf{ then } s_1 \textbf{ else } s_2 \textbf{ fi} : r, \sigma, \eta, \rho \rangle \longrightarrow \langle s_2 : r, \sigma, \eta, \rho \rangle$
11. $\langle \textbf{while } b \textbf{ do } s \textbf{ od} : r, \sigma, \eta, \rho \rangle \longrightarrow$
    $\langle \textbf{if } b \textbf{ then } s;\ \textbf{while } b \textbf{ do } s \textbf{ od else } skip \textbf{ fi} : r, \sigma, \eta, \rho \rangle$

Example 1. The following is a possible transition sequence in **NT**.

$$\langle \mathrm{write}(1) : (\mathrm{write}(2) : E), \sigma, \epsilon, 3 \rangle \overset{1}{\longrightarrow} \langle \mathrm{write}(2) : E, \sigma, \epsilon, 3 \rangle \longrightarrow$$
$$\langle \mathrm{write}(2) : E, \sigma, 3, \epsilon \rangle \overset{2}{\longrightarrow} \langle E, \sigma, 3, \epsilon \rangle \longrightarrow \otimes$$

2.4 Enabledness and Computations

Given a configuration $\rho = \langle r_m, \sigma_m, \eta_m, \dots, \langle r_i, \sigma_i, \eta_i, \langle \dots \langle r_1, \sigma_1, \eta_1, \zeta \rangle \dots \rangle$, the *input stream* ζ is defined as the *0-th subconfiguration*. For $i = 1, \dots, m$, we define the *i-th subconfiguration* as $\langle r_i, \sigma_i, \eta_i, \langle \dots \langle r_1, \sigma_1, \eta_1, \zeta \rangle \dots \rangle$. Informally we say that the (r_i, σ_i) pair is the *i-th subprocess* of ρ. Here r_i, σ_i and η_i are called the *resumption, state* and *buffer of* the *i*-th subprocess, respectively. The input stream can also be viewed as the buffer stream of the 0-th subprocess. The transition $\langle \dots \eta_{i+1}, \langle r_i, \sigma_i, \eta_i, \langle \dots \rangle \dots \rangle \longrightarrow \langle \dots \eta'_{i+1}, \langle r'_i, \sigma'_i, \eta'_i, \langle \dots \rangle \dots \rangle$ is said to be *determined* or *caused by* the *i*-th subprocess. The *i*-th subprocess can cause a transition only if it is enabled. We define that the 0-th subprocess is *enabled* if the input stream is not empty. The *i*-th subprocess is *enabled* if one of the rules from 4 to 11 can be applied to $\langle r_i, \sigma_i, \eta_i, \epsilon \rangle$.

If a subprocess is not enabled, then it is called *disabled*. For an initial configuration ρ, a maximal transition sequence is called a *computation* of ρ:

$$c(\rho) : \rho = \rho_0 \longrightarrow \rho_1 \longrightarrow \dots \rho_{n-1} \longrightarrow \rho_n \longrightarrow \dots$$

Here 'maximal' means that either $c(\rho)$ is infinite or the final configuration is a terminal. The transition from ρ_{n-1} to ρ_n is called the n-th *step* of the computation. ρ_n is called the n-th *stage* of the computation. The *output stream* of $c(\rho)$ is the sequence of labels produced by $c(\rho)$. In Example 1 above, we have a computation with $1 \cdot 2$ as output. Due to the nondeterminism, different transition sequences starting from ρ are possible. Consider a sequence $\rho_n, n = 1, 2, \ldots$ of a computation $c(\rho)$. Suppose some subprocess of ρ_n is enabled. If $\rho_n, \rho_{n+1}, \ldots, \rho_{n+k}$ involve only transitions caused by the other subprocesses, then this subprocess stays enabled in ρ_{n+k}. We say a computation is *fair* if every enabled subprocess of ρ_n will make a transition eventually.

As we will prove later, all fair computations give the same maximal output. However, some unfair computations may deliver only a prefix of the maximal output. Consider for instance computations starting in

$$\rho = \langle write(1) : E, \sigma, \epsilon, \langle \textbf{while } true \textbf{ do } skip \textbf{ od} : E, \sigma, \epsilon \rangle \rangle$$

If a computation involves an execution of $write(1)$, then the output stream will be nonempty. If every transition is caused by the first subprocess, then there will be no output at all.

3 The Graph Of One Computation

Given a configuration ρ, there may be different transition sequences starting from it. We want to prove that different computations give essentially the same output stream. People may consider proving this by induction on the nesting number of ρ. However, this is difficult since we have dynamic networks, where the nesting number may increase during execution due to the fork-statements. To find the relation between the streams of labels produced by these different transition sequences, we first analyse how one such sequence is built up. To this end, we introduce in this section the notion of a graph of one computation . Later we will construct the graph of all computations, which can be proved to contain each individual computation as a subgraph.

Example 2. We now present a computation and its graph (Fig. 3). In this graph r, r_1 abbreviate some resumptions

$$\rho = \rho_0 = \langle read(y) : (fork(v)) : E), \{\}, \alpha\beta, \gamma \rangle = \langle r, \{\}, \alpha\gamma, \gamma \rangle \longrightarrow$$
$$\rho_1 = \langle read(y) : (fork(v) : E), \{\}, \alpha\beta\gamma, \epsilon \rangle = \langle r, \{\}, \alpha\beta\gamma, \epsilon \rangle \longrightarrow$$
$$\rho_2 = \langle fork(v) : E, \{y = \alpha\}, \beta\gamma, \epsilon \rangle = \langle r1, \{y = \alpha\}, \beta\gamma, \epsilon \rangle \longrightarrow$$
$$\rho_3 = \langle E, \{y = \alpha, v = 1\}, \epsilon, \langle E, \{y = \alpha, v = 0\}, \beta\gamma, \epsilon \rangle \rangle (\longrightarrow \otimes)$$

In this graph, a subprocess is denoted by a rectangle (node) containing a resumption and a state. For every rectangle there is an identifier representing it. Node G symbolizes the input generator and node R symbolizes the output receiver. The I/O channels are represented by arrows from right to left. The label on such an arrow denotes the stream on the channel at a certain stage. We use the notation $\langle B, A \rangle$ to denote the channel from process A to process B and we use $\langle B, A \rangle = \alpha\beta$ to denote that its label is $\alpha\beta$. The top-down edges represent transitions. The computation can be represented in the following way:

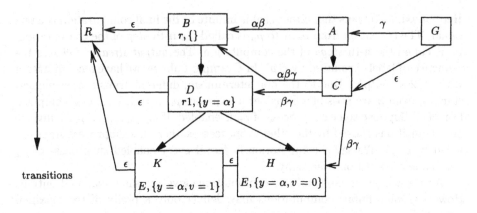

Fig. 3. The graph of a computation

- $\rho = \rho_0$ is represented by the following path (right to left) from G to R:
  $\langle R, B \rangle (= \epsilon), \langle B, A \rangle (= \alpha\beta), \langle A, G \rangle (= \gamma)$
- ρ_1 is represented by the path
  $\langle R, B \rangle (= \epsilon), \langle B, C \rangle (= \alpha\beta\gamma), \langle C, G \rangle (= \epsilon)$.
  The transition leading to the new stage is caused by the 0-th subprocess. It moves γ from the input stream to the buffer of the first subprocess. The edge from A to C represents this transition.
- ρ_2 is represented by the path
  $\langle R, D \rangle (= \epsilon), \langle D, C \rangle (= \beta\gamma), \langle C, G \rangle (= \epsilon)$.
  The transition edge from B to D is caused by the read-statement in B.
- ρ_3 is represented by the path
  $\langle R, K \rangle (= \epsilon), \langle K, H \rangle (= \epsilon), \langle H, C \rangle (= \beta\gamma), \langle C, G \rangle (= \epsilon)$.
  This stage is caused by the fork-statement in D, so there are two transition edges from D, to H and to K. H is the mother, K is the daughter created by the fork.

Notice that a path given above represents more than just a stage ρ_i in a computation, because the label of the last arrow of a path is in fact the global output stream until this stage. This stream can be considered as the 'buffer stream' for the output receiver R, but this stream will never decrease in length. Thus ρ_i and the global output stream together constitute a generalized stage in this computation. Notice that we now use a single graph for different stages in one computation, so one channel may correspond to several arrows. For instance, different arrows to R always denote the global output channel at different points in time. On the other hand, one arrow can correspond to the contents of a channel in more than one stage, e.g. $\langle C, G \rangle$ occurs in both ρ_1 and ρ_2. This computation transforms the initial input stream $\gamma = \langle A, G \rangle$ to the final output stream that is empty.

Now we give a general definition of the notion of the graph of one computation. In the graph of Example 2 we used A, B, C, etc. to identify nodes. In the general case we will use a Dewey like numbering system for this purpose. Given a configuration $\rho = \langle r, \sigma, \epsilon, \zeta \rangle$ and a computation $c(\rho)$ we always have that the 0-th stage ρ_0 equals ρ.

Beginning of the graph-the 0-th step: The rightmost node denotes the input gen-

$$\boxed{2} \overset{\epsilon}{\longleftarrow} \boxed{1, r, \sigma} \overset{\epsilon}{\longleftarrow} \boxed{0} \overset{\zeta}{\longleftarrow} \boxed{-1}$$

Fig. 4.

erator and it has identification number -1. The leftmost node denotes the output receiver and it has identification number 2. The interior nodes with identification number 0 and 1 are the roots of transition trees to be constructed. In node 1 we add r and σ, yielding $[1, r, \sigma]$. The horizontal arrows (*horizontal edges*) in this stage, having $\epsilon, \epsilon, \zeta$ as labels respectively, are $\langle 2, 1 \rangle, \langle 1, 0 \rangle$ and $\langle 0, -1 \rangle$. These horizontal arrows together form a *horizontal path* of ρ_0.

Extension of the graph-the (n+1)-th step : Suppose the graph is already drawn up to the n-th stage $\rho_n = \langle r_m, \rho_m, \eta_m, \ldots \langle r_i, \sigma_i, \eta_i, \ldots, \langle r_1, \sigma_1, \eta_1, \zeta \rangle \ldots \rangle \rangle$. We define here the extension of the graph only for the case that $\rho_n \longrightarrow \rho_{n+1}$ is caused by the i-th subprocess where ρ_i does not begin with a fork-statement. Let $\rho_{n+1} = \langle \ldots, r_{i+1}, \sigma_{i+1}, \eta_{i+1}, \langle r_i', \sigma_i', \eta_i', \ldots \langle \ldots \rangle \ldots \rangle \rangle$. If r_i begins with a *read*-statement, then $\eta_i' = rest(\eta_i)$ (i.e. η_i without the first element). If ρ_i writes a value α, then $\eta_{i+1}' = \eta_{i+1}\alpha$. In the other situations, η_i and η_{i+1} do not change.

Let $[p, \rho_i, \sigma_i]$ be the corresponding node in ρ_n and let $\langle q_2, p \rangle = \eta_{i+1}$ and $\langle p, q_1 \rangle = \eta_i$ be the outgoing and incoming arrows in the horizontal path of ρ_n. We extend the graph by adding a new node $[p0, r_i', \sigma_i']$ and vertical and horizontal connections in the following way : The *horizontal path* of ρ_{n+1} is obtained by

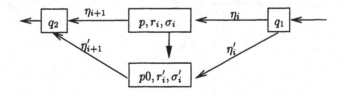

Fig. 5.

replacing $\langle q_2, p \rangle$ and $\langle p, q_1 \rangle$ in the horizontal path of ρ_n by $\langle q_2, p0 \rangle$ and $\langle p0, q_1 \rangle$. It is easy to see from the construction of the graph of a computation that every stage of this computation corresponds to a unique path from the input generator to the output receiver. The labels on the horizontal arrows are just the buffer streams.

4 The Graph Of All Computations

In this section we first define a special computation $C(\rho)$. Then we use $C(\rho)$ as the backbone to construct the graph of all computations $\mathbf{C}$.

The special computation $C(\rho)$ begins with the transition caused by the 0-th process if it is enabled. In the next steps, the transition are successively caused by the subprocesses from right to left (if they are enabled). After the transition caused by the leftmost subprocess we start anew with the 0-th subprocess. This procedure is repeated as often as possible. Example 2 described such a computation.

The graph of all computations $\mathbf{C}$ is defined from $C(\rho)$ as follows. Whenever the graph of $C(\rho)$ is extended by a transition, in $\mathbf{C}$ many horizontal arrows will be added (including the arrows needed for the new stage of $C(\rho)$). For example, suppose $p0$ is the newly constructed node. Then for every arrow $\langle q, p \rangle = \eta$, already existing in $\mathbf{C}$, a new $\langle q, p0 \rangle$ is added. In general this arrow will contain the stream η again. However, if node p begins with a statement of writing α then we will have $\langle q, p0 \rangle = \eta\alpha$ (see the leftmost picture of Fig. 6). Now suppose that $q0$ is the newly constructed node. Then, if q does not begin with a read-statement, for every arrow $\langle q, p \rangle = \eta$, a new arrow $\langle q0, p \rangle = \eta$ is added. If, on the other hand, q begins with a read statement, then arrows $\langle q0, p \rangle$ are only added if $\langle q, p \rangle = \eta \neq \epsilon$,and we define $\langle q0, p \rangle = rest(\eta)$ (see the middle picture of Fig. 6). The rightmost picture of Fig. 6 gives an impression how the graph $\mathbf{C}$ usually looks like after a few nodes of $C(\rho)$ are constructed.

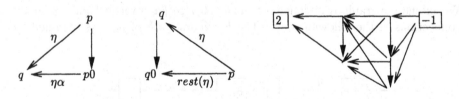

Fig. 6.

Example 3. Let $\rho = \langle \text{write}(1) : (\text{write}(2) : E), \sigma, \epsilon, 3 \rangle$. Then $C(\rho) : \rho_0 \longrightarrow \rho_1 \longrightarrow \rho_2 \longrightarrow \rho_3$ defined in the sequel can be used to construct the graph $\mathbf{C}$ of all computations:

In Fig. 7 we have $\rho_0 = \rho : \langle 2, 1 \rangle, \langle 1, 0 \rangle, \langle 0, -1 \rangle$. The transition from node 0 to node 00 generates the path of $\rho_1 : \langle 2, 1 \rangle, \langle 1, 00 \rangle, \langle 00, -1 \rangle$. After the transition from node 1 to node 10 the path $\rho_2 : \langle 2, 10 \rangle, \langle 10, 00 \rangle, \langle 00, -1 \rangle$ is added. However, in $\mathbf{C}$ we also add an arrow $\langle 10, 0 \rangle = \epsilon$ based on the existing arrow $\langle 1, 0 \rangle = \epsilon$. The transition $\rho_0 \longrightarrow \rho_1' : \langle 2, 10 \rangle, \langle 10, 0 \rangle, \langle 0, -1 \rangle$ will occur in computations other than $C(\rho)$. The next step in $C(\rho)$ corresponds with the transition from node 10 to node 100, and yields the path of $\rho_3 : \langle 2, 100 \rangle, \langle 100, 00 \rangle, \langle 00, -1 \rangle$. In $\mathbf{C}$ we also add the arrow $\langle 100, 0 \rangle = \epsilon$. Thus the path $\rho_2' : \langle 2, 100 \rangle, \langle 100, 0 \rangle, \langle 0, -1 \rangle$ can again be used for other computations, for example, in $\rho_0 \longrightarrow \rho_1' \longrightarrow \rho_2' \longrightarrow \rho_3 \longrightarrow \otimes$.

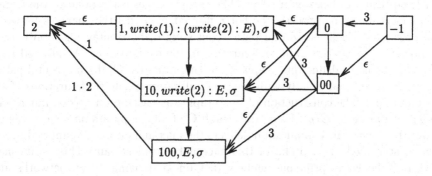

Fig. 7. A graph of all computations

The graph of all computations enables us to set up the following results, thus establishing the first part of the Kahn principle. The proofs of the following results can be found in [BN94].

Lemma 2. *If $\langle p, q \rangle$ and $\langle p, q' \rangle$ exist in C, then one is a prefix of the other. If $\langle q, p \rangle$ and $\langle q', p \rangle$ exist in C, then one is a postfix of the other.*

Theorem 3. *Every horizontal path in C denotes a stage of some computation.*

Theorem 4. *Every stage of a computation can be found as a horizontal path in C.*

Theorem 5. *For any stage ρ_n of any computation $c(\rho)$, the label of $\langle 2, p \rangle$ in the horizontal path of ρ_n is the output stream of this computation to this stage.*

Theorem 6. *Every fair computation delivers the maximal output stream and every computation delivers a prefix of this maximal output stream.*

Notice that the arrows in $\mathbf{C}$ are well defined by the construction of $\mathbf{C}$, i.e. we have at most one arrow between two nodes and we have only one label on an arrow. We have to prove that the horizontal paths in $\mathbf{C}$ correspond exactly with the horizontal paths in the graphs of individual computations. The main technical complication in the proof is the case that in a computation an arrow $\langle q0, p0 \rangle$ is derived from $\langle q0, p \rangle$ because p takes a step in this computation, though in $\mathbf{C}$ $\langle q0, p0 \rangle$ has been derived from $\langle q, p0 \rangle$. We then have to prove that in both cases the labels on $\langle p0, q0 \rangle$ agree. Only then we can say that $\mathbf{C}$ contains the graphs of all computations as subgraphs. The above results can then be established.

5 Conclusions and Future Work

In this paper we have defined a nondeterministic transition system for a language with which dynamic linear arrays of processes can be specified. For each computation defined by the transition system, we can define a computation graph, in which the nodes are (snapshots of) processes, characterized by a state and the remainder of program still to be executed, and in which two kinds of connections exist, vertical and horizontal edges. The vertical edges correspond to the transitions caused by one process, the horizontal edges model the channels between the processes. Each 'horizontal path', consisting of horizontal edges only, corresponds to an intermediate configuration in the computation. We showed that it was possible to define the 'graph of all computations' C, the horizontal paths of which correspond exactly with the set of all intermediate configurations of all computations. The construction of this graph was based on a special fair right-to-left computation $C(\rho)$. Using this graph C of all computations we are able to prove that every fair computation delivers the maximal output stream and every computation delivers a prefix of this maximal output stream. This establishes part I of the Kahn principle, namely that linear dynamic Kahn networks are deterministic.

It is natural to extend this research to part II of the Kahn principle. In order to prove the second half of the Kahn principle, we have to show that this output stream equals the smallest solution of a system of equations to be derived from the initial configuration. This smallest solution can be obtained from a suitably defined denotational semantics. We have done this [NB96] in the framework of metric spaces by applying the Banach fixed point theorem.

Several extensions of the results derived here come to mind. A natural idea is to study the general case introduced in [K74], i.e. to allow processes to expand into arbitrary networks, and to allow feedback loops, (sequences of) channels starting from and arriving at the same node. Apart from notational inconveniences, we do not see many problems. The advantage of using linear arrays of processes is that the graph of all computations can be depicted as a two dimensional structure. In the general case, a configuration, a snapshot of the system, will be a two dimensional graph in itself. This means that the new 'vertical edges' should now be drawn in the third dimension, and therefore computation graphs will be three dimensional. However, we expect that all our results will carry over to this general case.

Acknowledgements. Thanks go to Joost Kok, who offered a few essential pointers into the literature on dataflow nets. We are also indebted to the other members of the Amsterdam Concurrency Group, headed by Jaco de Bakker, for their stimulating remarks during a presentation of our results.

References

[AG78] Arvind and K.P. Gostelow, Some relationships between asynchronous interpreters of a data flow language, in: Formal description of programming concepts, W.J. Neuhold (ed.), pp. 95- 119, North-Holland, 1978.

[A81] A. Arnold, Sémantique des processus communicants, RAIRO Theor.Inf. 15,2, pp.103-139, 1981.

[B86] A. de Bruin, Experiments with continuation semantics: jumps, backtracking, dynamic networks, Ph.D. Thesis, Free University Amsterdam, 1986.

[B88] M. Broy, Nondeterministic data flow programs: how to avoid the merge anomaly, Science of Computer Programming, 10, pp.65-85, 1988.

[BA81] J. Brock and W. Ackerman, Scenarios: a model of non-determinate computation, in: Formalization of programming concepts, LNCS 107,pp.252-259, Springer, 1981.

[BBB93] J. W. de Bakker, F. van Breugel and A. de Bruin, Comparative semantics for linear arrays of communicating processes, a study of the UNIX fork and pipe commands, in: A.M. Borzyszkowski and S. Sokolowski (eds.) Proc. Math. Foundations of Computer Science, LNCS 711, pp.252-261, Springer, 1993.

[BB85] A. P. W. Böhm and A. de Bruin, The denotational semantics of dynamic networks of processes, ACM Trans. Prog. Lang. Syst, 7, pp.656-679,1985.

[BN94] A. de Bruin, S. H. Nienhuys-Cheng, Towards a proof of the Kahn principle for linear dynamic networks. Technical report cs-eur-94-06, Erasmus University of Rotterdam. Obtainable by anonymous ftp from ftp.cs.few.eur.nl from directory pub/doc/techreports/1994 as eur-cs-94-06.

[C72] J.M. Cadiou, Recursive definitions of partial functions and their computations, Ph.D. thesis, Stanford Univ., 1972.

[F82] A.A. Faustini, An operational semantics for pure data flow, in: Automata, languages and programming, 9th. Coll., LNCS 140, M.Nielsen and E.M. Schmidt (eds.), pp.212-224, 1982.

[JK90] B. Jonsson and J.N. Kok, Towards a complete hierarchy of compositional dataflow networks, in: Proc. theoretical aspects of computer software, LNCS 526, pp.204-225, Springer, 1990.

[K74] G. Kahn, The semantics of a simple language for parallel programming,in: Proc. IFIP74, J.L. Rosenfeld (ed.), North-Holland, pp.471-475, 1974.

[K86] J. Kok, Denotational semantics of nets with nondeterminism, in: European Symp. Programming, Saarbrücken, LNCS 206, pp.237-249, Springer, 1986.

[K78] P. Kosinski, A straight-forward denotational semantics for nondeterminate data flow programs, in: Proc. 5th ACM PoPL, pp.214-219, 1978.

[KM77] G. Kahn and D.B. MacQueen, Coroutines and networks of parallel processes, in: Proc. IFIP77, B. Gilchrist (ed.), North-Holland, pp.993-998, 1977.

[LS89] N. Lynch and E. Stark, A proof of the Kahn principle for input/output automata, Information and Computation, 82, 1, pp.81-92, 1989.

[SN85] J. Staples and V. Nguyen, A fixpoint semantics for nondeterministic data flow, JACM, 32 (2), pp.411-444, 1985.

[NB96] S. H. Nienhuys-Cheng and A. de Bruin, A proof of Kahn's fixed-point characterization for linear dynamic networks (preprint, also to appear as technical report)

Shortest Path Problems with Time Constraints

X. Cai[1], T. Kloks[2] * and C. K. Wong[3]**

[1] Department of Systems Engineering and Engineering Management
The Chinese University of Hong Kong
Shatin, Hong Kong
[2] Department of Mathematics and Computing Science
Eindhoven University of Technology
P.O.Box 513, 5600 MB Eindhoven
The Netherlands
[3] Department of Computer Science and Engineering
The Chinese University of Hong Kong
Shatin, Hong Kong

Abstract. We study a new version of the shortest path problem. Let $G = (V, E)$ be a directed graph. Each arc $e \in E$ has two numbers attached to it: a transit time $b(e, u)$ and a cost $c(e, u)$, which are functions of the departure time u at the beginning vertex of the arc. Moreover, postponement of departure (i.e., waiting) at a vertex may be allowed. The problem is to find the shortest path, i.e., the path with the least possible cost, subject to the constraint that the total traverse time is at most some number T. Three variants of the problem are examined. In the first one we assume arbitrary waiting times, where it is allowed to wait at a vertex without any restriction. In the second variant we assume zero waiting times, namely, waiting at any vertex is strictly prohibited. Finally, we consider the general case where there is a vertex-dependent upper bound on the waiting time at each vertex. Several algorithms with pseudopolynomial time complexity are proposed to solve the problems. First we assume that all transit times $b(e, u)$ are positive integers. In the last section, we show how to include zero transit times.

1 Introduction

One of the most studied problems in graph algorithms is the shortest path problem. A natural extension is the problem where one would like to find the shortest path subject to some constraint, such as the total time required to traverse the path being at most some number T. This kind of problem is also often studied (see, e.g., [5, 6]), and is known to be NP-complete ([1, 4]).

In this article we study yet a new extension of the problem. We address the situations where the transit time and the cost to traverse an arc are varying over time, which depend upon the departure time at the beginning vertex of the arc. Moreover, waiting times at vertices are considered being decision variables. Our problem is to determine an optimal path as well as the optimal waiting times at the

---

* Email: ton@win.tue.nl
** On leave from IBM T. J. Watson Research Center, P.O.Box 218, Yorktown Heights, NY 10598, U.S.A.

vertices along the path, subject to the constraint that the total traverse time of the path is at most T, where the total traverse time is the sum of all waiting times and all transit times. The practical motivation lies in the fact, that, that, for example due to traffic jams, the cost and time to get from one place to another depends on the time when one takes a certain route.

We address three variants of the problem. The first one assumes no constraints on the waiting times, namely, it is allowed to wait at any vertex arbitrarily. The second variant assumes zero waiting time, i.e., waiting at any vertex is strictly prohibited. Finally, we consider the situation where each vertex has a waiting time limit. Several algorithms are derived to compute the solutions. We first examine algorithms under the assumption that transit times on all arcs are strictly positive. Three algorithms are proposed to solve the three variants of the problem under this assumption, with time complexity of $O(T(m+n))$, $O(T(m+n))$, and $O(T(m+n\log T))$, respectively, where m and n are respectively the numbers of arcs and vertices. We then generalize the algorithms to cases with nonnegative transit times. We show that the three variants of the problem can be solved by algorithms with running time $O(T(m + n\log n))$, $O(T(m+n\log n))$, and $O(T(m+n\log n+n\log T))$ pseudopolynomial time complexity, indicating that they are NP-complete in the ordinary sense only.

2 Problem formulation

Let $G = (V, E)$ be a directed graph, without loops or multiple arcs. Let $b(e, u)$ be the *transit time* needed to traverse an arc e. This transit time is dependent on the discrete values of the time $u = 0, 1, \ldots, T$. $T \geq 0$ is a given number, which is the maximum time that is allowed to traverse a whole path (see below). Let $c(e, u)$ be the *length*, or *cost*, of an arc $e \in E$. (For consistency, we shall use the term *length* for the rest of the article.) This length is also dependent on the discrete value of the time $u = 0, 1, \ldots, T$. More specifically, both the transit time $b(e, u)$ and the length $c(e, u)$ are functions of the departure time u at the beginning vertex of e. We assume that all transit times $b(e, u)$ are nonnegative integers and all lengths $c(e, u)$ are nonnegative. Throughout the article, we let $n = |V|$ and $m = |E|$. We also write $c(x, y, u)$ and $b(x, y, u)$ if $e = (x, y)$.

Definition 1 *A waiting time $w(x)$ at a vertex x is a nonnegative integer, and $U_x \geq 0$ is its upper bound.*

Definition 2 *Let $P = (s = x_1, \ldots, x_r = x)$ be a path from s to x. Let $w(x_i)$ $(i = 1, \ldots, r)$ be waiting times at vertices $x_1, \ldots, x_r$. Let $T(x_1) = w(x_1)$ and define recursively*

$$T(x_i) = w(x_i) + T(x_{i-1}) + b(x_{i-1}, x_i, T(x_{i-1})) \text{ for } i = 2, \ldots, r.$$

The departure time of a vertex x_i on P is defined as $T(x_i)$ if $1 \leq i < r$.

Definition 3 *Let $P = (s = x_1, \ldots, x_r = x)$ be a path from s to x. The time of P, given the waiting times, is defined as $T(x_r)$. A path has time at most t, if there are waiting times for the vertices on the path such that the time of the path with these waiting times is at most t.*

Definition 4 *Let $P = (s = x_1, \ldots, x_r = x)$ be a path from s to x. Let $T(x_i)$ be the departure time at x_i for $1 \leq i < r$, given the waiting times of the vertices on the path. Let $\mathcal{L}(x_1) = 0$ and define recursively*

$$\mathcal{L}(x_i) = \mathcal{L}(x_{i-1}) + c(x_{i-1}, x_i, T(x_{i-1})) \text{ for } i = 2, \ldots, r.$$

The length of P, with the given waiting times, is defined as $\mathcal{L}(x_r)$.

We are interested in the problem of finding a path P^* from a given vertex s to another vertex y, $y \in V \setminus \{s\}$, and determining the optimal values of the waiting times at the vertices on the path, so that the length of the path is minimal subject to the following constraints:

C1. The time of P^* is at most T.
C2. The waiting time at each vertex x along P^* is at most $\mathcal{U}_x$.

For completeness, we adopt the following convention:

Definition 5 *For a vertex $x \in V \setminus \{s\}$ and a given number $t \leq T$, the length of a shortest path from s to x with time t is said to be ∞ if*

(i) there does not exist any path from s to x, or
(ii) all paths from s to x either have times greater than t or violate the constraint C2 above.

3 Arbitrary waiting times

Starting from this section we consider the time-varying problem as formulated in Section 2. We examine, in this section, the problem when there is no constraint imposed on the waiting times. In other words, in the Constraint C2 (see Section 2) we assume that $\mathcal{U}_x = \infty$ for all $x \in V$. Again, in this section we assume that all $b(x, y, u)$ are positive integers, and that all $c(x, y, u)$ are nonnegative.

The problem is to find a path from s to another vertex of time at most T and, meanwhile, to determine the optimal waiting times at each vertex along the path, so that the length of the path is minimal. We refer to this problem as the *Time-Varying Constrained Shortest Path with Arbitrary Waiting Times problem* (TCSP-AWT problem).

Definition 6 *$d_A(x, t)$ is the length of a shortest path from s to x of time at most t, where waiting at any vertex is not restricted.*

The following Lemma gives us a recursive relation to compute $d_A(x, t)$. Note that the optimal waiting times can be obtained implicitly by the recursive computations. This will be elaborated later in Remark ??.

Lemma 1 *$d_A(s, t) = 0$ for all t and $d_A(y, 0) = \infty$ for all $y \neq s$. For $t > 0$ and $y \neq s$, we have:*

$$d_A(y, t) = \min\left\{d_A(y, t - 1), \min_{\{x | (x, y) \in E\}} \min_{\{u | u + b(x, y, u) = t\}} \{d_A(x, u) + c(x, y, u)\}\right\}$$

The proof of this lemma is not very complicated, and, for reasons of space linitations, we decided to omit it. It will be included in the full version of the paper.

Definition 7 *For every arc* $(x,y) \in E$ *and for* $t = 0, \ldots, T$, *let*

$$\Upsilon_A(x,y,t) = \min_{\{u \mid u+b(x,y,u)=t\}} \{d_A(x,u) + c(x,y,u)\}.$$

We adopt the convention that $\Upsilon_A(x,y,t) = \infty$ *whenever* $\{u \mid u + b(x,y,u) = t\} = \emptyset$.

The result below follows directly from Lemma 1.

Corollary 1

$$d_A(y,t) = \min\left\{d_A(y,t-1), \min_{\{x \mid (x,y) \in E\}} \Upsilon_A(x,y,t)\right\}.$$

It can be seen from Definition 7 that $\Upsilon_A(x,y,t)$ is the solution of an optimization problem which determines the optimal departure time u at the beginning vertex x of the arc (x,y) when one wants to arrive at the vertex y at time t. Corollary 1 indicates that whenever $d_A(y,t)$ is to be updated for a given vertex y and a time t, we have to solve a number of optimization problems to obtain $\Upsilon_A(x,y,t)$ for all $(x,y) \in E$.

Given a time t and a vertex y, $\min_{\{x \mid (x,y) \in E\}} \Upsilon_A(x,y,t)$ could be performed by a naive approach of enumerating $0 \leq u \leq T$ to find those satisfying $u+b(x,y,u) = t$ for all $(x,y) \in E$. This would however require a worst-case running time of $O(Tm)$. Consequently, an algorithm with this approach would require a worst-case running time of the order of T^2m. So we need some mechanism to make the evaluation of $\Upsilon_A(x,y,t)$ more efficiently. Our idea in the algorithm below is to firstly sort the values of $u + b(x,y,u)$ for all $u = 1, 2, \cdots, T$ and all arcs $(x,y) \in E$, before the recursive relation as given in Lemma 1 is applied to compute $d_A(y,t)$ for all $y \in V$ and $t = 1, 2, \cdots, T$.

We describe the algorithm in Figure 1. Then we show its correctness in Lemma 2, and its worst-case running time in Lemma 3. Some implementation details which guarantee its worst-case running time are given in the proof of Lemma 3.

Algorithm TCSP-AWT
begin
Initialize $d_A(s,t) = 0$ and $\forall_{x \neq s} d_A(x,t) = \infty$ for $t = 0, \ldots, T$
Sort all values $u + b(x,y,u)$ for $u = 1, \ldots, T$ and for all arcs $(x,y) \in E$
For $t = 1, \ldots, T$ do
    For every arc $(x,y) \in E$ do $\Upsilon_A(x,y,t) := \infty$
    If $u + b(x,y,u) = t$ for some arc (x,y) and some u then
        $\Upsilon_A(x,y,t) := \min\{\Upsilon_A(x,y,t), d_A(x,u) + c(x,y,u)\}$
    For every vertex y do $d_A(y,t) := \min\{d_A(y,t-1), \min_{(x,y) \in E} \Upsilon_A(x,y,t)\}$
end.

Fig. 1. Algorithm TCSP-AWT

Lemma 2 *After the termination of the algorithm TCSP-AWT, each computed value $d_A(x,t)$ is the length of a shortest path from s to x of time at most t.*

Proof. With induction on t we show that upon completion of the tth iteration of the algorithm, $d_A(x,t)$ is the length of a shortest path of time at most t from s to x.

For $t = 0$, all $d_A(x,0)$ are given the correct values by the initialization.

Assume for all $t' < t$, all values $d_A(x,t')$ are correct.

First consider the update of $\Upsilon_A(x,y,t)$. If there is no u with $u + b(x,y,u) = t$, then $\Upsilon_A(x,y,t) = \infty$, implying that there is no path from s to y via the arc (x,y) and of time exactly t. If there are values u for which $u+b(x,y,u) = t$, then $\Upsilon_A(x,y,t)$ is correctly updated by the induction, since $b(x,y,u) > 0$ and thus $u < t$.

Finally, in the last line of the algorithm, the correct values of $d_A(y,t)$ are computed, by Corollary 1. This proves the lemma. $\qquad\qquad\qquad\qquad\quad$ □

Lemma 3 *The algorithm described above can be implemented such that it runs in $O(T(n+m))$ time.*

From Lemma 2 and Lemma 3 we now obtain

Theorem 1 *The TCSP-AWT problem with positive transit times can be solved in $O(T(n+m))$ time.*

4 Zero waiting times

In this section we consider the case in which no waiting times are allowed at any vertices. We refer to this problem as the *Time-Varying Constrained Shortest Path with Zero Waiting Times problem* (TCSP-ZWT problem). We shall see that the results on the TCSP-ZWT problem are somewhat similar to those described in section 3 on the TCSP-AWT problem. We examine them separately as TCSP-AWT and TCSP-ZWT are virtually two different variants of the time-varying problem we are studying. This section provides a complete treatment to the TCSP-ZWT problem.

Let $G = (V, E)$ be a directed graph. Again, associated with each arc $(x,y) \in E$ there are two numbers $b(x,y,u)$ and $c(x,y,u)$, which are functions of the departure time u at the vertex x. Also in this section we assume that all times $b(x,y,u)$ are positive integers, and all values $c(x,y,u)$ are nonnegative.

Recall Definitions 2, 3 and 4 on departure times at vertices, time of path, and length of path. Note that the waiting times in these definitions should be set to zero for the TCSP-ZWT problem.

Definition 8 $d_Z(x,t)$ *is the length of a shortest path from s to x of time exactly t. If such a path does not exist, then $d_Z(x,t) = \infty$.*

Note that the definition of $d_Z(x,t)$ is different from that of $d_A(x,t)$. This is because of the constraint that no waiting is allowed at any vertex in the present problem. This constraint also causes the result in the following Lemma to be different from that in Lemma 1.

Lemma 4 $d_Z(s,0) = 0$ and $d_Z(y,0) = \infty$ for all $y \neq s$. For $t > 0$ and $y \neq s$, we have:

$$d_Z(y,t) = \min_{\{x|(x,y)\in E\}} \min_{\{u|u+b(x,y,u)=t\}} \Big\{ d_Z(x,u) + c(x,y,u) \Big\}.$$

Because of space limitations, we decided to omit this proof. It will be included in the full version of the paper.

Definition 9 For each arc (x,y) and each $1 \leq t \leq T$, define

$$\Upsilon_Z(x,y,t) = \min_{\{u|u+b(x,y,u)=t\}} \{d_Z(x,u) + c(x,y,u)\},$$

and adopt the convention that $\Upsilon_Z(x,y,t) = \infty$ whenever the set $\{u|u+b(x,y,u) = t\}$ is empty.

The result below follows directly from Lemma 4.

Corollary 2 For $1 \leq t \leq T$, and for each vertex y,

$$d_Z(y,t) = \min_{\{x|(x,y)\in E\}} \Upsilon_Z(x,y,t).$$

We see that we have to evaluate $\Upsilon_Z(x,y,t)$ for all $(x,y) \in E$ when $d_Z(y,t)$ is to be updated. Again, to compute $\Upsilon_Z(x,y,t)$ as efficiently as possible, our idea in the algorithm presented below is to use sorting. We sort in advance the values of $u + b(x,y,u) = t$ for all $u = 1,2,\cdots,T$ and all arcs $(x,y) \in E$.

Our algorithm is described in Figure 2. Then, its correctness and its worst-case running time are given in Lemma 5 and Lemma 6, respectively.

Algorithm TCSP-ZWT
begin
Initialize $d_Z(s,0) = 0$, and $d_Z(x,0) = \infty$ for all $x \neq s$ and $t = 0,1,...,T$
Sort all values $u + b(x,y,u)$ for all $u = 1,...,T$ and for all arcs $(x,y) \in E$
For $t = 1,...,T$ do
    For every arc $(x,y) \in E$ do $\Upsilon_Z(x,y,t) := \infty$
    For all arcs $(x,y) \in E$ and all u such that $u + b(x,y,u) = t$ do
        $\Upsilon_Z(x,y,t) := \min\{\Upsilon_Z(x,y,t), d_Z(x,u) + c(x,y,u)\}$
    For every vertex y do $d_Z(y,t) := \min_{(x,y)\in E} \Upsilon_Z(x,y,t)$
For every vertex y do $d_Z^*(y) = \min_{0 \leq t \leq T} d_Z(y,t)$
end.

Fig. 2. Algorithm TCSP-ZWT

Lemma 5 After the termination of the algorithm TCSP-ZWT, each computed value $d_Z^*(x)$ is the length of a shortest path from s to x of time at most T.

Lemma 6 *The algorithm described above can be implemented such that it runs in* $O(T(n+m))$ *time.*

The proof is similar to that for Lemma 3.
From Lemma 5 and Lemma 6 we now obtain

Theorem 2 *The TCSP-ZWT problem with positive transit times can be solved in* $O(T(n+m))$ *time.*

5   Bounded waiting times

In this section we consider the case where waiting at a vertex is allowed, but there is an upper bound $\mathcal{U}_x$ on the waiting time at a vertex x.

Again, we assume that all $b(x, y, u)$ are positive integers, and that all $c(x, y, u)$ are nonnegative. Recall the definition of $\mathcal{U}_x$ (Definition 1). Clearly we may assume that $\mathcal{U}_x \leq T$ for all $x \in V$. Recall also Definitions 2, 3, 4, and 5 on departure times at vertices, time of path, and length of path.

The problem is to find a shortest path from s to a vertex $x \in V \setminus \{s\}$ of time at most T, subject to the constraint that the waiting time at any vertex x on the path is not greater than $\mathcal{U}_x$. We refer to this problem as the *Time-Varying Constrained Shortest Path with Constrained Waiting Times problem* (TCSP-CWT problem).

Definition 10 $d_C(x, t)$ *is the length of a shortest feasible path from s to x of time exactly t and with waiting time zero at x, subject to the constraint that the waiting time at any other vertex y on the path is not greater than $\mathcal{U}_y$. If such a feasible path does not exist, then $d_C(x, t) = \infty$.*

Definition 11 $d_C^*(x)$ *is the length of a shortest feasible path from s to x of time at most T.*

Lemma 7

$$d_C^*(x) = \min_{0 \leq t \leq T} d_C(x, t)$$

Proof. Consider a shortest feasible path P of time at most T. Let $t \leq T$ be the time of P, with waiting time zero at x. Then $d_C^*(x) = d_C(x, t)$. □

Lemma 8 $d_C(s, 0) = 0$ *and* $d_C(x, 0) = \infty$ *for all* $x \neq s$. *For* $t > 0$ *and* $y \neq s$, *we have:*

$$d_C(y, t) = \min_{\{x | (x, y) \in E\}} \min_{(u_A, u_D) \in \mathcal{F}(x, y, t)} \left\{ d_C(x, u_A) + c(x, y, u_D) \right\},$$

where $\mathcal{F}(x, y, t) = \{(u_A, u_D) \mid u_D + b(x, y, u_D) = t \wedge 0 \leq u_D - u_A \leq \mathcal{U}_x\}$.

Because of space limitations, we had to omiy this proof. The proof will appear in the full version of the paper.

Definition 12 *For each arc $(x, y) \in E$ and each $1 \leq t \leq T$, define*

$$\Upsilon_C(x, y, t) = \min_{(u_A, u_D) \in \mathcal{F}(x, y, t)} \{d_C(x, u_A) + c(x, y, u_D)\},$$

and adopt the convention that $\Upsilon_C(x, y, t) = \infty$ whenever the set $\mathcal{F}(x, y, t)$ is empty.

From Lemma 8, we have:

Corollary 3 *For $1 \leq t \leq T$, and for each vertex y*

$$d_C(y, t) = \min_{\{x | (x, y) \in E\}} \Upsilon_C(x, y, t).$$

In addition to the idea of sorting the values of $u + b(x, y, u)$ as discussed previously, our key idea in the algorithm to be presented below is the use of a binary heap. For every vertex x, we maintain a binary heap, which contains the values of $d_C(x, u_A)$ for all $\max\{0, t - \mathcal{U}_x\} \leq u_A \leq t$. Using this data structure, initialization and finding the minimum take constant time. Each insertion and each deletion take $O(\log \mathcal{U}_x) = O(\log T)$ time [2]. For convenience, we introduce the following notation.

Definition 13 $d_C^m(x, t)$ *is the minimum in the heap.*

We need $d_C^m(x, t)$ when evaluating $\Upsilon_C(x, y, t)$. We see from Definition 12 that we have to solve an optimization problem of minimizing $\{d_C(x, u_A) + c(x, y, u_D)\}$ subject to $(u_A, u_D) \in \mathcal{F}(x, y, t)$, to obtain $\Upsilon_C(x, y, t)$. Clearly, given (x, y) and t, a value of u_D that satisfies $u_D + b(x, y, u_D) = t$ is known and, consequently, the corresponding value of $c(x, y, u_D)$ is known. Thus, solving the optimization problem reduces to solving a problem of minimizing $\{d_C(x, u_A)\}$ subject to $\max\{0, u_D - \mathcal{U}_x\} \leq u_A \leq u_D$ (recall the definition of $\mathcal{F}(x, y, t)$ in Lemma 8). Therefore, if $d_C^m(x, t)$ is known for $t = u_D$, we can obtain $\Upsilon_C(x, y, t)$, which is equal to the minimum of $d_C^m(x, u_D) + c(x, y, u_D)$ for all $(x, y) \in E$ and all u_D satisfying $u_D + b(x, y, u_D) = t$.

We describe our algorithm for the TCSP-CWT problem in Figure 3. Note that Algorithm TCSP-CWT computes iteratively $d_C(y, t)$ for all y at $t = 0, 1, \cdots, T$. At any time t, the algorithm keeps all $d_C^m(y, u)$ for all vertices y and all $u \leq t - 1$. Nevertheless, for each vertex y, the algorithm only maintains one heap $Heap_y$. After $d_C(y, t)$ is obtained, the new $Heap_y$ at time t is obtained by deleting $d_C(y, t - \mathcal{U}_y - 1)$ from the heap (if $t - \mathcal{U}_y - 1 \geq 0$) and inserting $d_C(y, t)$.

Remark 1 The algorithm TCSP-CWT computes the length $d_C^*(y)$ of the shortest feasible path from the source s to each vertex y with time at most T (the correctness of the algorithm is shown in Lemma 9 below). By a backtracking procedure, one may obtain the path as well as the optimal departure time at each vertex on the path. Then, the optimal waiting times can be computed by using the departure times. See discussions in Remark ??.

We now show the correctness of the algorithm.

Lemma 9 *After the termination of the algorithm TCSP-CWT, $d_C(y, t)$ is the length of a shortest feasible path from s to y of time exactly t and with waiting time zero at the vertex y.*

Algorithm TCSP-CWT
begin
Initialize $d_C(s,0) := 0$ and $\forall_{x \neq s} d_C(x,0) = \infty$; $\forall_x \forall_{t>0}$ Initialize $d_C(x,t) := \infty$;
      $\forall_x$ Initialize $Heap_x := \{d_C(x,0)\}$ and $d_C^m(x,0) := d_C(x,0)$
Sort all values $u + b(x,y,u)$ for all $u = 1,\ldots,T$ and for all arcs (x,y)
For $t = 1,\ldots,T$ do
    For every arc (x,y) do $\Upsilon_C(x,y,t) := \infty$
    For all arcs (x,y) and all u_D such that $u_D + b(x,y,u_D) = t$ do
      $\Upsilon_C(x,y,t) := \min\{\Upsilon_C(x,y,t), d_C^m(x,u_D) + c(x,y,u_D)\}$
    For every vertex y do $d_C(y,t) = \min_{\{x|(x,y) \in E\}} \Upsilon_C(x,y,t)$
    For every vertex y update the heap as follows
      Insert-heap$_{(y)}$ $d_C(y,t)$
      If $t > \mathcal{U}_y$ then **delete-heap**$_{(y)}$ $d_C(y, t - \mathcal{U}_y - 1)$
    For every vertex y do
      $u_A :=$ **Minimum-heap**$_{(y)}$
      $d_C^m(y,t) := d_C(y, u_A)$
For every vertex y do $d_C^*(y) = \min_{0 \leq t \leq T} d_C(y,t)$
end.

Fig. 3. Algorithm TCSP-CWT

Lemma 10 *The algorithm TCSP-CWT can be implemented such that it runs in* $O(T(m + n \log T))$ *time.*

Combining Lemma 9 with Lemma 10, we obtain

Theorem 3 *The TCSP-CWT problem with positive transit times can be solved in* $O(T(m + n \log T))$ *time.*

6 Taking care of the zeros

In this section we show that we can extend our methods to allow zero transit times. Notice that we cannot expect the same running-time bounds. (Otherwise, taking all transit times zero, the algorithm TCSP-AWT would solve the shortest path problem in linear time. This can obviously not be true.) We propose an approach to handle zero transit times. The approach holds for all three problems TCSP-AWT, TCSP-ZWT, and TCSP-CWT. In the following we describe our approach in details for the TCSP-AWT problem. The particulars of the approach for the other problems can be similarly derived following the same idea.

Consider a network $G = (V, E)$. At the tth step of the algorithm TCSP-AWT, we firstly apply, as usual, the algorithm to a subgraph $G' = (V, E')$. This subgraph G' has the same vertex set V as G, but its edge set $E' = \{e\}$ are those for which $b(e,t) > 0$. Then, after the values of $d_A(y,t)$, namely, the lengths of the shortest paths from s to each vertex y, $y \in V$, are obtained by the algorithm TCSP-AWT, we create, for each $y \in V$, an artificial arc from s to y. Call this arc $e'_y = (s,y)$. The length of $e'_y = (s,y)$ is set to $d_A(y,t)$, namely, $c(e'_y, t) = d_A(y,t)$, and the transit time on $e'_y = (s, y)$ is assumed to be t. Then, we construct a new subgraph $G'' = (V, E'')$. The vertex set V of the subgraph G'' is the same as that of G. The edge set $E'' = \{e\}$ consists of two groups of elements, one being those $\{e\}$ for which $b(e, t) = 0$, and the other being those $\{e'_y\}$ for all $y \in V \setminus \{s\}$. If there are double arcs from s to y, delete that arc from E'' which has the larger length (or break up a tie arbitrarily if they have equal lengths).

When the subgraph G'' is created, we apply a 'common' shortest path algorithm (say, Dijkstra's algorithm, see [3, 2]) to G'' to find the shortest path from s to each $y \in V$. In applying such an algorithm, we ignore the transit times and the problem is thus a classical shortest path problem. The 'common' shortest path algorithm works as follows.

It maintains two sets S and S'. The set S contains vertices for which the final shortest path lengths have been determined, while the set S' contains vertices for which upper bounds on the final shortest path lengths have been known. Initially, S contains only the source s, and the lengths of the vertices in S' are set to $d_A(y, t)$. Repeatedly, select the vertex $x \notin S$, for which the distance from s is the shortest. Put x in S, and for all outgoing arcs $e = (x, y) \in E''$, update $d_A(y, t) := \min\{d_A(y, t), d_A(x, t) + c(e, t)\}$. The algorithm terminates if all vertices are in S.

Notice that the above describes the well-known shortest path algorithm of Dijkstra. We refer to this procedure as the shortest path procedure, **SP**. It is well-known that this procedure can find the length of a shortest path from s to each $y \in V$ (see, for example, [1]).

We are going to show that our approach is correct in terms of finding an optimal solution at each time t for the original problem TCSP-AWT. For any vertex $y \in V$ and time $t > 0$, we can see that any path from s to y with time at most t must be one of the paths of the following type:

1. a path from s to y of time at most $t - 1$,
2. a path from s to y of time exactly t, which must pass an arc $(x, y) \in E'$ with $b(x, y, t) > 0$, or
3. a path from s to y of time exactly t, which must pass an arc $(x, y) \in E''$ with $b(x, y, t) = 0$.

In fact, for each vertex $y \in V$, our approach firstly uses the algorithm TCSP-AWT to determine the shortest path among those of types 1 and 2. After this is done, it creates an artificial arc $e'_y = (s, y)$ to represent this shortest path. Then, it uses the procedure SP to further determine the shortest length of all possible paths. The shortest path can be one with only the artificial arc e'_y (in this case the algorithm TCSP-AWT had in fact found the optimum) or a path of type 3 (in this case, the procedure SP has found a shorter length than that obtained by the algorithm TCSP-AWT). Since any vertex y can only be reached by one of the paths of the three types, the approach has considered all possible paths and is thus optimal. Formally, we have

Lemma 11 *Consider the approach: at each $t = 0, 1, \cdots, T$, apply the algorithm TCSP-AWT to G', then apply the procedure SP to G'' to update $d_A(y, t)$ for all*

$y \in V$. After the tth iteration, $d_A(y,t)$ is the length of a shortest path from s to y of time at most t.

To prove this lemma, let us introduce the following result first. Notice that we do not change the definitions of $d_A(x,t)$ and $\Upsilon(x,y,t)$ (recall Definitions 6 and 7).

Lemma 12 *Assume that $d_A(x,t')$ are correct for all $x \in V$ and all $0 \le t' < t$. Then, for each vertex $y \in V \setminus \{s\}$, the tth iteration of the algorithm TCSP-AWT obtains the length of a shortest path among all the paths of types 1 and 2.*

Proof. Let $d'_A(y,t) = \min_{\{x|(x,y)\in E'\}}\{\Upsilon(x,y,t)\}$, where

$$\Upsilon(x,y,t) = \min_{\{u|u+b(x,y,u)=t\}}\{d_A(x,u) + c(x,y,u)\}$$

as defined by Definition 7.

If $d_A(x,t')$ are correct for $t' < t$, then by definition, $d_A(x,t-1)$ is the length of a shortest path among those from s to x that have time at most $t - 1$. Moreover, notice that $d'_A(y,t)$ is the length of the shortest path among those from s to x that must include an arc $(x,y) \in E'$ and of time exactly t, provided that $d_A(x,t')$ are correct for all $x \in V$ and all $t' < t$. Since the algorithm TCSP-AWT computes $d_A(y,t) = \min\{d_A(x,t-1), d'_A(y,t)\}$, it is not hard to see that the lemma is correct. □

Proof of Lemma 11. With induction on t, we now show that $d_A(x,t)$ is the length of a shortest path from s to x of time at most t.

When $t = 0$, the algorithm TCSP-AWT firstly initializes $d_A(s,0) = 0$ and $d_A(x,0) = \infty$ for all $x \ne s$. Then, a subgraph G'' is created, and the procedure SP is applied to this graph. Clearly, this procedure can correctly obtain, for each $x \in V$, the length of a shortest path from s to x in the graph G''. Hence the values for $d_A(x,t)$ are correct for $t = 0$.

Now assume that the values of $d_A(x,t')$ are correct for all $x \in V$ and $t' < t$. Under this assumption, it follows from Lemma 12 that the values for $d_A(x,t)$ obtained by the algorithm TCSP-AWT are the lengths of shortest paths of types 1 and 2. Now consider the subgraph G'' created with artificial arcs $e'_y = (s,y)$ associated with these lengths. As the procedure SP is in fact the algorithm of Dijkstra, it can find the length of a shortest path P from s to y in the graph G'', for each $y \in V$. Moreover, the time of this path is at most t, since all arcs except those artificial arcs in G'' have zero transit times. The artificial arcs in G'' have a transit time t, but all of them originates from s and thus any path from s to y in G'' can contain at most one such arc. By the notation of our approach, $d_A(y,t)$ (updated by the procedure SP) is the length of this shortest path P.

We now claim that $d_A(y,t)$ is also the length of the shortest path from s to y of time at most t in the original graph G. Suppose this is not true, namely, there exists another path $\hat{P}$ from s to y, which has a length $\mathcal{L}_{\hat{P}} < d_A(y,t)$. Clearly, this cannot be a path of type 1, 2, or 3, otherwise such a path would have implied that the procedure SP, namely, Dijkstra's algorithm, is not optimal. The only possibility is that $\hat{P}$ is a path with time greater than t, which is however infeasible for the given t. This proves the claim, and therefore the lemma. □

For each $t = 0, 1, 2, \cdots, T$, the subgraphs G' and G'' can be constructed in $O(m + n)$ time, and the procedure SP can be implemented such that it runs in $O(m + n \log n)$ time (see [1]). This is the additional running time to update the solutions obtained by the algorithms TCSP-AWT, TCSP-ZWT, or TCSP-CWT. In summary, we have

Theorem 4 *The problems TCSP-AWT, TCSP-ZWT, and TCSP-CWT with non-negative transit times can be solved in times $O(T(m + n \log n))$, $O(T(m + n \log n))$ and $O(T(m + n \log T + n \log n))$, respectively.*

7 Conclusions

In this article we considered a new extension of the ordinary shortest path problem. Three variants have been examined in details. Algorithms with pseudopolynomial time complexity were proposed, which solve these problems.

Of course, there are many variations of the problems discussed in this paper. For example, one could study the problem where a *speedup* is also allowed. The transit time of an arc may be shortened, which however may incur an additional cost. In such a case, both the transit times on arcs and the waiting times at vertices become decision variables. One should determine not only an optimal path, but also the optimal transit times on the arcs and the waiting times at the vertices along the path.

References

1. Ahuja, R. K., T. L. Magnanti and J. B. Orlin, *Network flows: theory, algorithms, and applications*, Prentice Hall, Englewood Cliffs, New Jersey. (1993)
2. Cormen, T. H., C. E. Leiserson and R. L. Rivest, *Introduction to algorithms*, The MIT press, Cambridge, Massachusetts. (1990)
3. Dijkstra, E. W., A note on two problems in connection with graphs, *Numer. Math.*, 1, (1959), pp. 269–271.
4. Handler, G. Y. and I. Zang, A dual algorithm for the constrained shortest path problem, *Networks*, 10, (1980), pp. 293–310.
5. Hassan, M. M. D, Network reduction for the acyclic constrained shortest path problem, *European Journal of Operational Research*, 63, (1992), pp. 121–132.
6. Skiscim, C. C. and B. L. Golden, Solving k-shortest and constrained shortest path problems efficiently, *Annals of Operations Research*, 20, (1989), pp. 249–282.

Parallel Alternating-Direction Access Machine*

Bogdan S. Chlebus[1], Artur Czumaj[2]**, Leszek Gąsieniec[1,3],
Mirosław Kowaluk[1], and Wojciech Plandowski[1]

[1] Instytut Informatyki, Uniwersytet Warszawski, Banacha 2, 02-097 Warszawa, Poland.
[2] Heinz Nixdorf Institute and Department of Mathematics & Computer Science, Universität-GH Paderborn, D-33095 Paderborn, Germany.
[3] Max-Planck-Institut für Informatik, Im Stadtwald, D-66123 Saarbrücken, Germany.

Abstract. This paper presents a theoretical study of a model of parallel computations called Parallel Alternating-Direction Access Machine (PADAM). PADAM is an abstraction of the multiprocessor computers ADENA /ADENART and a prototype architecture USC/OMP. The main feature of PADAM is the organization of access to the global memory:
(1) the memory modules are arranged as a 2-dimensional array,
(2) each processor is assigned to a row and a column,
(3) the processors switch synchronously between row and column access modes, and can access any of the assigned modules in each mode without conflicts.
Since the PADAM processors have such a restricted access to the partially shared memory, developing tools to enhance flexibility of access to the memory is important. The paper concentrates on these issues.

1 Introduction

An important goal in the study of parallel computation is to develop models which are close to real machines but abstract from technical details and provide a vehicle to design and analyze algorithms (see [2, 4, 7, 9, 10, 12]). In this paper we present theoretical investigations of a model of a multiprocessor computer with a (partially shared) global memory: Parallel Alternating-Direction Access Machine (PADAM). The model is an abstraction of the architecture of a recently built computer ADENA/ADENART [5, 6], and has been previously considered by Hwang, Tseng, and Kim [3] under the name of orthogonal multiprocessor (OMP). The organization of the global memory of ADENA/OMP (and their abstraction PADAM) is a characteristic feature of this architecture. The memory modules are arranged as an array and each processor has the assigned row and column that

---

* Work partially supported by EC Cooperative Action IC-1000 (project ALTEC: *Algorithms for Future Technologies*) and a research grant from Matsushita Electric Industrial Company Ltd.
** Partially supported by DFG-Graduiertenkolleg "Parallele Rechnernetzwerke in der Produktionstechnik", ME 872/4-1 and by DFG-Sonderforschungsbereich 376 "Massive Parallelität: Algorithmen, Entwurfsmethoden, Anwendungen".

it is able to access. The processors switch simultaneously between the column and row access modes and then access the assigned groups of memory modules without conflicts.

The restricted access of processors to memory is an advantage in terms of hardware complexity and costs, but it hinders mapping algorithms on this model. Hence enhancing flexibility of memory access is of primary importance. What is not provided by hardware could be done by software. Our goal is to enhance the architecture by algorithmic means, and we aim at developing algorithms that could be applied on computers like ADENA/OMP and become useful tools. We develop a deterministic and optimal routing scheme. Next we show how to implement access of each processor to all the memory modules. Finally, we present general PRAM simulations. All these algorithms are variations on the same theme: communication and memory access.

Models of computation. The *Parallel Alternating-Direction Access Machine (PADAM)* consists of a set of n processors and a global memory. The global memory is organized as an $n \times n$ array of *memory units* (MUs). In one step a processor can either execute an instruction on its local data or access an MU for reading or writing. Let $P(i)$ denote the ith processor, and $M[i, j]$ denote the MU in the ith row and the jth column. Processor $P(i)$ can access the MUs in the ith row and in the ith column. The system is always in one of two states: either the *row-access mode* or the *column-access mode*. While in the row-access mode, all the processors may access their rows of MUs, that is, processor $P(i)$ can read or write to MU $M[i, k]$, for all k, $1 \leq k \leq n$. Similarly, if the computer is in the column-access mode, processor $P(i)$ can access $M[k, i]$, for all $1 \leq k \leq n$. Switching modes is programmable, that is, there is an instruction available, say switch, to change between the row and column access modes. All the processors execute the same program. The computations are globally synchronized. In particular, if the processors execute a loop then the next instruction is being started only after all the processors have already exited the loop; the same holds for the instruction switch.

A *Distributed Memory Machine* (DMM) consists of a number of synchronized processors and memory modules. Each module can be accessed by any processor in one step, modulo possible conflicts with other processors. The DMM has been studied under various names in the literature, usually the authors assumed a specific way to resolve conflicts for access to modules, examples are *Direct Connection Machine* in [8] and *S*PRAM* in [12].

A *Parallel Random Access Machine* (PRAM) has a number of synchronized processors and the global shared memory. Each processor has a random access to any memory word, modulo possible conflicts with other processors. So PRAM is like a DMM in which each module has the capacity of one word. The PRAM model has proven to be particularly appealing and popular among parallel models (see [2, 4, 7]). The reason is that its key feature, the random access to memory, greatly facilitates the design and analysis of algorithms.

The DMM is more realistic than the PRAM, it captures the phenomena occurring if the global memory is partitioned into modules. Simulating PRAM on

DMM has been a subject of extensive study (see [11]). PADAM is more restricted than DMM, because each MU can be accessed directly by at most two processors. We show that the PADAM is strictly weaker than both the DMM and PRAM (see Theorems 8 and 13).

Miscellany. If some property depending on n holds with the probability $1 - n^{-a}$, for any constant a, then the property is said to hold *with high probability*, abbreviated w.h.p.. All the presented randomized algorithms are *Las Vegas*, in the sense that they always output the correct solution.

A PADAM algorithm is *optimal* if its time-processor product is equal within a constant factor to the time of the fastest sequential algorithm solving the same problem.

We often refer to certain indexings of the MUs determined by linear orderings. Define the *row-major order* as follows: the MUs in row i precede the MUs in row j, if $i < j$, and all the rows are ordered from left to right. The *column-major order* is defined similarly.

Due to space limitations some proofs and details are omitted and will appear in the full version of the paper.

2 Routing

Suppose that each MU stores at most h packets, and the packets need to be rearranged among the MUs in such a way that each MU receives at most h new packets. This is the *h-h routing* problem.

We first describe an optimal deterministic 1-1 routing algorithm operating in time $\mathcal{O}(n)$. Let us suppose that initially the packets are stored in the MUs, at most one packet at each $M[i, j]$. The array M of MUs is partitioned into $\sqrt{n}$ *horizontal strips*, which are segments of $\sqrt{n}$ rows, and similarly into $\sqrt{n}$ *vertical strips* comprised of $\sqrt{n}$ columns. The strips are numbered from left to right and from the bottom up, respectively. The $\sqrt{n}$ MUs in a row that are in the same vertical strip make a *horizontal segment*, the *vertical segments* are defined similarly. An intersection of a horizontal and vertical strip is called a *block*. There are n blocks, each having n elements organized as an $\sqrt{n} \times \sqrt{n}$ array. The algorithm can be conceptually divided into three phases:

PADAM ROUTING
Step 1. Route the packets to their vertical strips.
Step 2. Route the packets to their blocks.
Step 3. Finish routing in the blocks.

We describe the phases in the reversed order.

Step 3. Routing packets inside the blocks:
Let us suppose that the packets have already been moved to their destination blocks. To complete routing, we convert blocks into rows, then route packets in the rows, and convert rows back to blocks. The details follow.

Enumerate the blocks in the (regular) row-major order, which is analogous to the ordering of the array of MUs. We move all the packets from the ith block to the ith row. To this end, the kth row inside a row of blocks, for $1 \leq k \leq \sqrt{n}$, is shifted cyclically $(k-1) \cdot \sqrt{n}$ columns to the right. More precisely, a packet in the row i and column j, for $1 \leq i, j \leq n$, is moved to the column number $(j + ((i-1) \bmod \sqrt{n}) \cdot \sqrt{n}) \bmod n$ in the ith row. Now all the packets from a block are in different columns. Permute the packets in the columns so that all the packets from the ith block are moved to the ith row. This transformation of blocks to rows has the property that the packet from the lth column and mth row (enumeration inside the block) is the lth element in the mth horizontal segment. This determines a translation of addresses in the block to addresses in the row. Permute the packets in each row, translating block addresses to row addresses, to achieve the desired routing in the block. Finally convert back rows to blocks in a reversed way, the ith row to the ith block.

Step 2. Routing packets inside the vertical strips:
Suppose that the packets have already been moved to their destination vertical strips. We show how to move the packets to their destination blocks. Consider a vertical strip. The first stage is to sort the packets in each column on their block addresses. This can be done in time $\mathcal{O}(n)$, one processor per column, as follows. The processor scans the column and finds the number a_i of packets with destination address in the ith block. The number a_i is stored at the top MU of the ith vertical segment of the column. Next the processor computes the partial sums of these numbers: $f_1 = 0$, and $f_k = a_1 + \ldots + a_{k-1}$, for $k > 1$. Now the ith packet among those going to the jth block is moved to the location $i + f_j$ in the column. This can be done by one additional scan of the column, and completes sorting.

The sorted packets in each column are partitioned into two categories: $\mathcal{A}$-*packets* and $\mathcal{B}$-*packets*, as follows. For each block, if there are i packets in the column with this destination address, then $i - (i \bmod \sqrt{n})$ packets are designated as $\mathcal{A}$-packets and the remaining ones as the $\mathcal{B}$-packets. Next the columns are sorted in a stable way such that the $\mathcal{A}$-packets precede the $\mathcal{B}$-packets. This can be done in time $\mathcal{O}(n)$ similarly as sorting on the block addresses. The $\mathcal{A}$-packets are dealt with first. Afterwards we add some dummy $\mathcal{B}$-packets to make exactly $\sqrt{n}$ packets with each block address, and the $\mathcal{B}$-packets are sorted on their block addresses. This operation moves them to their blocks. The dummy packets are discarded then.

It remains to consider the $\mathcal{A}$-packets in detail. At this moment, every vertical segment contains only $\mathcal{A}$-packets with the same block address. Rearrange the $\mathcal{A}$-packets in such a way that these in the rows congruent to i modulo $\sqrt{n}$ go to the ith block (we refer to this permutation of rows in the vertical strips as σ). Next solve the routing problem in each block treating the original block addresses as row addresses inside the block. Afterwards the packets are rearranged according to the permutation σ^{-1}.

Lemma 1. *At this stage of the algorithm the packets are in their proper destination blocks.*

Step 1. Routing the packets to their vertical strips:
First the packets in each row are sorted on their destination vertical-strip addresses. This is done in time $\mathcal{O}(n)$ similarly as sorting columns on their block

addresses. Next, for each vertical-strip address, if there are some i packets with this destination address, then $i - (i \bmod \sqrt{n})$ are designated as $\mathcal{A}$-packets and the remaining ones as $\mathcal{B}$-packets. When the $\mathcal{A}$-packets will have been routed, the $\mathcal{B}$-packets are padded with the dummy packets to make exactly $\sqrt{n}$ packets with each vertical strip address in each column. The packets are sorted in each row and this operation places them in the proper vertical strip. It remains to show how to handle the $\mathcal{A}$-packets. First the $\mathcal{A}$-packets and $\mathcal{B}$-packets are sorted in a stable way in the rows preserving that the $\mathcal{A}$-packets precede the $\mathcal{B}$-packets. Now we permute the $\mathcal{A}$-packets so that those in the columns congruent to i modulo $\sqrt{n}$ are moved to the ith vertical strip (we refer to this permutation of columns as γ). When inside the vertical strips, treat the original vertical-strip addresses as block addresses inside the strip, and move the packets to the respective blocks. Next move the packets from the ith block to the ith column inside vertical strips, and finish the routing by rearranging the packets according to the permutation γ^{-1}.

Lemma 2. *Routing to the vertical strips is correct and can be accomplished in time $\mathcal{O}(n)$.*

Adaptation to the h-h case is as follows. While routing inside the blocks, all the packets from a block are reallocated to a row, then the routing in that row is performed. While routing inside vertical strips, we need to define indexing of the locations of packets in a column to determine the sorting order, and then the $\mathcal{A}$-packets and $\mathcal{B}$-packets. The indexing of packet locations is such that all the packets from the MU in a row i are to precede the packets in the MU in the row j, if $i < j$. For each vertical-strip address, if there are some i packets with this destination address, then some $i - (i \bmod (\sqrt{n} \cdot h))$ of them are designated as $\mathcal{A}$-packets and the remaining ones as $\mathcal{B}$-packets. The phase of routing packets to their destination vertical strips is modified accordingly.

Traditionally, all the packets stored in one unit of storage are referred to as the *queue*.

Theorem 3. *The algorithm PADAM ROUTING solves an instance of the h-h routing problem on the PADAM with n processors in time $\mathcal{O}(h \cdot n)$, while the queues are of size $\mathcal{O}(h)$.*

3   Emulating DMM

In this section we show how to implement on the PADAM an access of each processor to any memory unit. This is the same as emulating a DMM with the same processors and MUs. Two emulations are presented. One is deterministic and uses the original memory addressing of the PADAM, the other is randomized, the randomization is used in shifting cyclically the addresses in rows. Each processor creates its memory request as a packet storing the address of a MU, a memory word x in it, and the stipulation that either x is to be read or that some value y is to be written to x. The set of packets created by all the processors, that are to be realized concurrently, is called a *batch of memory requests*. We do not assume any specific semantics regarding conflicts for access of MUs, but all the popular ones (like EREW, CREW, arbitrary CRCW, priority CRCW) can be used.

3.1 Deterministic Emulation

The emulation is performed in a step-by-step manner. Each processor creates a packet and places it in its row in the first column. The packets are first organized into groups determined by the MU addresses. For each group, only one packet is designated to be routed, its selection depends on the conflict-of-access-resolution protocol. In the case of a writing step, this packet will carry the value which is to be written. In the case of a reading step, this packet will be also routed back by a reversed route and will bring the value retrieved from the accessed MU. This value will be distributed among the packets in the group, if necessary, and finally these packets will be sent back to their original rows.

EMULATION E1

Step 1. The packets have been placed in the first column. Sort them on their destination-column addresses by emulating the Batcher's sorting network (see [9]). Now the groups of packets with the same addresses are in contiguous block of locations in the first column.

Step 2. Select one packet (if any) from each group. The selection is made depending on the conflict-resolution protocol.

Step 3. The selected packets are routed up to fill a contiguous block of MUs in the first column, including the top row, one packet per MU. This is accomplished by first computing the addresses for the packets by the parallel prefix algorithm executed on a simulated tree, and then by routing on a simulated butterfly by applying a monotone-routing algorithm (see [9]). After this the selected packets remain sorted on their destination-column addresses. We consider only the selected packets from now on.

Step 4. For each group of the packets with the same destination-column address, if there are some i packets in the group then designate $i - (i \bmod \sqrt{n})$ packets as A-packets, the remaining being the B-packets. This can be done on a simulated tree. The A-packets are then routed to a contiguous block of locations, preserving their relative order. This is accomplished similarly as in Step 3.

We first deal with the A-packets.

Step 5. The ith vertical segment of $\sqrt{n}$ A-packets is routed $i - 1$ columns to the right. Then the processors switch to the column-access mode. The ith processor, for $i \leq \sqrt{n}$, looks for $\sqrt{n}$ packets in its column in the consecutive rows starting from the row $(i - 1)\sqrt{n} + 1$. If there are any packets there then they are moved to their destination rows.

Step 6. The processors switch to the row-access mode. Each processor reads the first $\sqrt{n}$ MUs in its column looking for packets, and if it finds any then they are moved to their destination columns.

Lemma 4. *The A-packets are routed in time* $\mathcal{O}(\sqrt{n})$.

This completes routing of the $\mathcal{A}$-packets. The next steps take care of the $\mathcal{B}$-packets. They are already in a contiguous block of memory locations, and the packets with the same column-destination addresses also make contiguous blocks. These blocks will be now moved to their destination columns. Before that we need to inform the processors where they are to look for packets in their columns. **Step 7.** For each packet which is the first in a group of packets with the same column address j, and which is in the row i, create a *special* packet to carry the number i and send it to the jth processor, say to $M[j, j]$. To this end, the special packet from the ith row is first moved to the column $\lfloor (i-1)/\sqrt{n} \rfloor + 1$. Then the processors switch to the column-access mode and the kth processor, for $k \leq \sqrt{n}$, scans $\sqrt{n}$ MUs, starting from the row $(k-1)\sqrt{n} + 1$. If a packet is found then it is routed to its destination row. Finally the processors switch to the row-access mode and scan the first $\sqrt{n}$ columns and move the packets (if any) to their column addresses. The special packets are handled in time $\mathcal{O}(\sqrt{n})$, because there is never more than a single packet in each MU, what follows from the fact that they have distinct row addresses.

Step 8. Now the $\mathcal{B}$-packets are moved to their destination columns in one step. The processors switch to the column-access mode and go to the rows which numbers have just been distributed by the special packets. Then they move the packets waiting there to their destination rows.

This completes the emulation and proves:

Theorem 5. *The algorithm* E1 *on a* PADAM *with* n *processors performs one batch of memory requests in time* $\mathcal{O}(\sqrt{n})$.

The following matching lower bound holds.

Theorem 6. *Any emulation of the random access to memory units on the* n-*processor* PADAM *requires* $\Omega(\sqrt{n})$ *operations per batch of memory requests in the worst case.*

3.2 Randomized Emulation

The emulation is based on an *offset addressing* mechanism. It is determined by a sequence a_i of *offsets*, where $1 \leq i \leq n$ and $1 \leq a_i \leq n$. Instead of $M[i, j]$, the access to the ith row and the jth column is meant to be to $M'[i, j] = M[i, (j + a_i) \bmod n]$. Just before the computation starts, each $P(i)$ generates a random offset a_i and stores it in $M[i, 1]$. Then the input is read and stored in the MUs. The following is an emulation of a single memory reference.

EMULATION E2
Step 1. Each processor $P(i)$ generates a memory access request and places the respective (ordinary) packet p_i in $M[i, 1]$. Assume that packets addresses are all distinct, otherwise proceed as in E1.
Step 2. The packets are sorted on their row addresses. The processor $P(i)$ such that the packet at $M[i, 1]$ is the first in a block of packets with the same row address, say r_i, generates a packet to inquire about the offset a_{r_i}. The inquiring

packets are routed to their respective rows in the first column on a simulated butterfly, and then routed back (these are instances of monotone routing). The offsets they bring are disseminated among other packets with the same row address on a simulated tree.

Step 3. The packets p_i are sorted on their offset column addresses k_i and stored again in the column $M[*, 1]$.

Step 4. Each processor $P(i)$ checks if the packet residing now in $M[i, 1]$ is the first in the block of packets with the same key, say k_i. If this is the case then it generates a special packet storing i and the address $M[k_i, 1]$.

Step 5. The special packets are routed to their destination addresses on a simulated butterfly. Then each processor $P(i)$ checks $M[i, 1]$ for such a packet.

Step 6. Each processor $P(i)$ moves the (ordinary) packet from $M[i, 1]$ to $M[i, k_i]$, where k_i is the key of the packet.

Step 7. The processors switch to the column access mode and processor $P(j)$ scans consecutive locations in the column j, starting from the address delivered by the special packet (if any) and moves the packets waiting there (if any) to their final destinations.

Theorem 7. *The algorithm* E2 *performs one batch of memory requests on a* PADAM *with n processors in time $\mathcal{O}(\log n)$ w.h.p..*

The following matching lower bound holds.

Theorem 8. *Any randomized emulation of random memory access on the n-processor* PADAM *has the expected number of $\Omega(\log n)$ steps per batch of memory requests.*

4 PRAM Simulations

In this section we consider simulations of a PRAM on the PADAM. First we describe a deterministic one of a PRAM with some restrictions on the number of processors and the size of memory. Next we show a general randomized simulation of a p-processor CRCW PRAM with unbounded memory on the n-processor PADAM. The simulation has an optimal delay for $p = \mathcal{O}(n)$ and for $p = \Omega(n^{1+\epsilon})$, for any constant $\epsilon > 0$. (If t is the running time of a CRCW PRAM algorithm and $pt > n^2$, then it is assumed that each MU of the PADAM can store $\Omega(\frac{pt}{n^2})$ words.)

4.1 Deterministic Simulation

We present a deterministic simulation of a PRAM with memory $\mathcal{O}(n^2)$ and p processors, for $p \leq n^2$. The algorithm uses $\mathcal{O}(1)$ memory of each MU, but alternatively could be easily modified to an emulation of a DMM with p processors and the MUs of the PADAM. It is possible to use the algorithm E1 directly to obtain a similar simulation (or the respective DMM emulation) as follows. Divide the PRAM processors into $\lceil p/n \rceil$ groups of at most n processors. To simulate one step, perform $\lceil p/n \rceil$ simulation phases, in each one taking care of n processors. Each phase takes $\mathcal{O}(\sqrt{n})$ steps, and the delay is $\mathcal{O}(p/\sqrt{n})$. We present an alternative simulation with delay $\mathcal{O}(\sqrt{p})$, what is better if $p > n$.

The simulation is an extension of emulation E1, so we just sketch it. Initially the PRAM processors are divided into p/n groups and the groups assigned to the PADAM processors. Let $\mathcal{K}$ be the set of MUs $M[i,j]$ with $i,j \leq \lceil \sqrt{p} \rceil$.

SIMULATION S1
Step 1. The processors create packets storing the memory requests, and place them in their rows in the first p/n columns. Next the packets are allocated to the square $\mathcal{K}$, one packet per MU. This can be accomplished in a straightforward way in time $p/n + \sqrt{p} \leq 2 \cdot \sqrt{p}$.
Step 2. The packets in $\mathcal{K}$ are sorted in time $\mathcal{O}(\sqrt{p})$ on their destination-column addresses.

A packet is defined to be an $\mathcal{A}$-packet if all the packets in its column in $\mathcal{K}$ have the same destination-column address. The remaining packets are $\mathcal{B}$-packets.
Step 3. The $\mathcal{A}$-packets are moved to their destination rows. Then the processors switch to the row access mode, scan the first $\sqrt{p}$ MUs in their rows and move the $\mathcal{A}$-packets residing there to their final destinations.
Step 4. The $\mathcal{B}$-packets with the same column address are in contiguous blocks in at most two columns. For each such block(s) of packets going to the column i create a special packet p_i with the row addresses of the blocks. Then the special packets are routed to the diagonal MUs, p_i to $M[i,i]$, and retrieved by the processors.
Step 5. The $\mathcal{B}$-packets are routed along the rows to their destination columns. Then the processors switch to the column access mode and deliver the $\mathcal{B}$-packets to their final destinations.

This completes the description of the simulation and proves:

Theorem 9. *A p-processor PRAM with $\mathcal{O}(n^2)$ memory words can be simulated on the n-processor PADAM with delay $\mathcal{O}(\sqrt{p})$, provided $p \leq n^2$.*

4.2 Randomized Simulation

In this subsection a randomized simulation of a CRCW PRAM is presented. The algorithm hashes the memory of the PRAM into the (usually smaller) address space of the PADAM. First we give an optimal simulation of a n-processor CRCW PRAM on the n-processor PADAM. Next we extend it to the case of a p-processor CRCW PRAM, for any p. Our algorithm achieves the optimal delay $\mathcal{O}(\lceil p/n \rceil)$ for $p \geq n^{1+\varepsilon}$. Because our simulation hashes the memory of the PRAM we do not assume any bounds on the size of the memory of the PRAM.

A PRAM consists of p processors $\bar{P}(1), \bar{P}(2), \ldots, \bar{P}(p)$ and a shared memory with cells $U = \{0, \ldots, u-1\}$. The PADAM has n processors $P(1), P(2), \ldots, P(n)$. In our simulations the shared memory cells of the PRAM are distributed among the rows of the PADAM using a hash function $h : U \to \{1, \ldots, n\}$ drawn at random from the universal class of hash functions $\mathcal{R}(n,n,d)$ introduced by Dietzfelbinger and Meyer auf der Heide [1]. $\mathcal{R}(n,n,d)$ consists of functions $h =$

$h(f, a_0, \ldots, a_{n-1})$, where f is a polynomial of degree $d-1$ in the field of residues of some large prime taken modulo n^2, $a_0, \ldots, a_{n-1} \in \{0, \ldots, n-1\}$, and $h(x) = (f(x) \bmod n + a_{f(x) \operatorname{div} n}) \bmod n$ for $x \in U$.

Lemma 10. *Any n requests of hash values of $h \in \mathcal{R}(n, n, d)$ can be obtained in time $\mathcal{O}(\log n)$ on the n-processor* PADAM.

The family of hash functions $\mathcal{R}(n, n, d)$ has the following useful property:

Fact 1 [1] *If h is a random hash function from class $\mathcal{R}(n, n, d)$, then, for a constant d sufficiently large, and for any $X \subseteq U$, the following estimation holds w.h.p. :* $\max_{1 \le i \le n} \{|h^{-1}(i) \cap X|\} = \mathcal{O}\left(\frac{|X|}{n} + \log n\right)$.

A dictionary is a data structure supporting two operations: LOOKUP(x) - which returns the information associated with x that is currently stored in the dictionary; and INSERT(x, c) - which stores x together with the information c associated with x in the dictionary; if x was already stored then it is first removed from the dictionary.

Fact 2 [1] *There is a randomized sequential implementation of a dictionary which uses $\mathcal{O}(n)$ space and that performs each of a sequence of $\mathcal{O}(n)$ dictionary operations in constant time w.h.p..*

Let $\mathcal{M} = \{m_1, \ldots, m_p\} \subseteq U$. In a given PRAM step, $\bar{P}(i)$ wants to access cell m_i and either reads its content as x_i (in the reading phase) or writes y_i to m_i (in the writing phase). We refer to the respective pair (m_i, x_i) or (m_i, y_i) as the packet μ_i. At the beginning of the simulation we choose randomly a hash function $h : U \to \{1, \ldots, n\}$ from $\mathcal{R}(n, n, d)$. Function h defines the row $h(x)$ of the PADAM in which cell x of the PRAM will be stored.

SIMULATION S2
Step 1. Each processor $P(i)$ creates a packet μ_i, places it in M$[i, 1]$, and evaluates $h(m_i)$.
Step 2. All the packets are sorted on their PRAM addresses from $\mathcal{M}$. Now the groups of packets with the same addresses are in contiguous block of locations in the first column. Select one packet from each group according to the writing-conflict-resolution protocol.
Step 3. The selected packets are routed up to fill a contiguous block of MUs in the first column, including the top row, one packet per MU. We consider only these packets from now on.
Step 4. These packets are sorted on their destination row addresses $h(m_i)$. Using the obtained sorted sequence, each packet computes the number of packets with the same destination row and the number of packets with the same destination row which precede it in the sorted sequence. This is accomplished by emulating a binary tree.
Step 5. The first packet destined for each row is moved to that row, as on a binary tree.

Step 6. The left packets are handled as follows. If the ith packet μ_i (with respect to the order from Step 4) was moved in Step 5 to row k and there are j packets to be delivered to row k, then in the next $j-1$ moves: processor $P(i+l)$ writes μ_{i+l} to $M[i+l,k]$ in move l and $P(k)$ reads μ_{i+l} from that cell. Afterwards $P(k)$ stores all packets destined for row k.

Step 7. Each processor $P(i)$ keeps in its row i a local dictionary. In the writing phase, $P(i)$ executes operation INSERT(m_j, y_j) for every packet μ_j with $h(m_j) = i$. In the reading phase, $P(i)$ executes operation $x_j = $ LOOKUP(m_j) for every $h(m_j) = i$.

Step 8. The values read are distributed among the read packets left in Step 2.

Theorem 11. *Simulation* S2 *performs one step of the n-processor* CRCW PRAM *on the n-processor* PADAM *in time* $\mathcal{O}(\log n)$ *w.h.p..*

The algorithm above can be extended to simulate the p-processor PRAM on the n-processor PADAM for any p. We use the following simple class of hash functions. Let q be a prime and $U = \{0, \ldots, u-1\}$ for $u \leq q$. For $s \geq 1$ define the set $\mathcal{H}_s^1 = \{h_{a,b} : 0 \leq a, b < q\}$ where $h_{a,b}(x) = (a + bx \bmod q) \bmod s$, for $x \in U$ and $0 \leq a, b < q$.

Fact 3 ([1]) *Let $h : U \to \{0, \ldots, s-1\}$ be a random function from class $\mathcal{H}_s^1$, then for any $X \subseteq U$, if $2|X|^{c+2} \leq s$ then $\Pr(\max_{1 \leq i \leq s}\{|h^{-1}(i) \cap X|\} = 1) \geq 1 - |X|^{-c}$*

Observe that a function $h \in \mathcal{H}_s^1$ can be stored in $O(1)$ cells, and can be generated and evaluated in constant time by one processor. Hence it can be generated by $P(1)$ and broadcast by emulating a binary tree to all the processors of the PADAM in time $\mathcal{O}(\log n)$. Afterwards each processor can evaluate h in constant time.

Theorem 12. *One step of the p-processor* CRCW PRAM *can be simulated on the n-processor* PADAM *in the optimal time* $\mathcal{O}(\lceil p/n \rceil)$, *provided $p = \Omega(n^{1+\epsilon})$ for any constant $\epsilon > 0$. For smaller values of p the simulation has delay $\mathcal{O}(\log p \cdot \lceil p/n \rceil)$.*

Emulation E2 and simulation S2 are optimal. This follows from the following lower bound on randomized PRAM simulations.

Theorem 13. *Any randomized simulation of the n-processor* EREW PRAM *on the* PADAM, *that operates correctly with the probability larger than $\frac{1}{2} + \varepsilon$, for some constant $\varepsilon > 0$, requires time $\Omega(\log n)$.*

5   Conclusion and Open Problems

We considered PADAM, an abstraction of the architecture OMP and ADENA. General tools to enhance programming on such a computer were developed, among them h-h routing algorithm, emulations of access to all the global memory, and PRAM simulations. The obtained results show that ADENA/OMP/PADAM is a parallel-computer architecture with a potential for general-purpose applications.

Sometimes we presented more than one solution to the same algorithmic problem, say, a deterministic and randomized one, because often algorithms that look modest and simple turn out to be best in terms of actual performance. Which of these algorithms could be of practical value depends on the technical parameters of physical machines and could be verified only by tests.

New versions of ADENA have the memory modules organized as a 3-dimensional array. All the results presented in this report can be adapted to the analogous variant of PADAM with a 3-dimensional array of memory modules.

It would be interesting to compare PADAM with the mesh with buses by designing mutual simulations, and compare optimal algorithms for various problems on both models.

References

1. M. Dietzfelbinger and F. Meyer auf der Heide, Dynamic Hashing in Real Time, in *Informatik: Festschrift zum 60. Geburtstag von Günter Hotz*, ed. J. Buchmann, H. Ganzinger, and W. J. Paul, B. G. Teuber Verlagsgesellschaft, Leipzig, 1992, pp. 95–119.
2. A. Gibbons and W. Rytter, *"Efficient Parallel Algorithms"*, Cambridge University Press, 1988.
3. K. Hwang, P.-S. Tseng, and D. Kim, On Orthogonal Multiprocessor for Parallel Scientific Computations, *IEEE Transactions on Computers* 38 (1989) 47–61.
4. J. JáJá, *"An Introduction to Parallel Algorithms"*, Addison Wesley, Reading, MA, 1992.
5. H. Kadota, K. Kaneko, I. Okabayashi, T. Okamoto, T. Mimura, Y. Nakakura, A. Wakatani, M. Nakajima, J. Nishikawa, K. Zaiki, and T. Nogi, Parallel Computer ADENART - its Architecture and Application, in *Proceedings of the 5th ACM International Conference on Supercomputing*, 1991, pp. 1–8.
6. H. Kadota, K. Kaneko, Y. Tanikawa, and T. Nogi, VLSI Parallel Computer with Data Transfer Network: ADENA, in *Proceedings of the International Conference on Parallel Processing*, Vol. I, 1989, pp. 319–322.
7. R. M. Karp and V. Ramachandran, Parallel Algorithms for Shared-Memory Machines, in *"Handbook of Theoretical Computer Science, Vol. A: Algorithms and Complexity"*, ed. J. van Leeuwen, Elsevier Science Publishers, 1990, pp. 870–941.
8. C. P. Kruskal, L. Rudolph, and M. Snir, A Complexity Theory of Efficient Parallel Algorithms, *Theoretical Computer Science* 71 (1990) 95–132.
9. F. T. Leighton, *"Introduction to Parallel Algorithms and Architectures: Arrays, Trees, Hypercubes"*, Morgan Kaufman Publishers, San Mateo, California, 1991.
10. W. F. McColl, General Purpose Parallel Computing, in *"Lectures on Parallel Computation"*, ed. A. Gibbons and P. Spirakis, Cambridge University Press, 1993, pp. 337–391.
11. F. Meyer auf der Heide, Hashing Strategies for Simulating Shared Memory on Distributed Memory Machines, in *Proceedings of the 1st Heinz Nixdorf Symposium "Parallel Architectures and their Efficient Use,"* ed. F. Meyer auf der Heide, B. Monien, A.L. Rosenberg, 1992, Lecture Notes on Computer Science 678, pp. 20–29.
12. L. G. Valiant, General Purpose Parallel Architectures, in *"Handbook of Theoretical Computer Science, Vol. A: Algorithms and Complexity"*, ed. J. van Leeuwen, Elsevier Science Publishers, 1990, pp. 943–972.

Specification and Verification of Timed Lazy Systems[*]

Flavio Corradini[1] and Marco Pistore[2]

[1] School of Cognitive and Computing Sciences, University of Sussex
Brighton BN1 9QH, England
e-mail: flavioc@cogs.sussex.ac.uk

[2] Dipartimento di Informatica, Università di Pisa
Corso Italia 40, I-56100 Pisa, Italy
e-mail: pistore@di.unipi.it

Abstract. In this paper *CCS* is equipped with a simple operational semantics that allows us to describe and reason about the performance of systems which proceed by reacting to external stimuli. Based on the new operational semantics, *lazy performance equivalence* is introduced as a natural extension of the standard interleaving bisimulation. It turns out to be preserved by all *CCS* contexts.

The problem of automatically checking lazy performance equivalence is also tackled. Because of the lazy character of our calculus, an infinite transition graph is associated with every non-trivial *CCS* term. Nevertheless, lazy performance equivalence can be provided with an alternative finite characterization and existing algorithms for checking bisimulation based semantics can be applied to the latter.

1 Introduction

Recently, in the field of semantics for process calculi, a great deal of interest has been stirred up by equivalence notions which incorporate some measure of efficiency (see, e.g., [15, 3, 1, 12]).

Some of them [1, 2, 10, 11, 12, 6, 5] rely on the basic idea that systems are discriminated not only according to what actions the processes can do, but also considering the time consumed for their execution. Systems are distributed over the space and each sequential component is equipped with a local clock, whose elapsing is set dynamically during the execution of the actions by the corresponding sequential component. Every action a has a fixed duration $\delta(a)$, which represents the time needed for its execution, and there is no time-passing between the execution of actions from the same sequential component, i.e., the actions are *eager*, or urgent. Eagerness of actions favour an "observational" point of view: systems proceed autonomously (without any cooperation with the environment) and an external observer simply watches what is going on.

---

[*] Research supported by CEE HCM "EXPRESS", by Esprit Basic Research project CONFER and by CNR.

In contrast, in this paper, we concentrate on systems whose actions are *lazy*, that is, actions can be delayed arbitrarily long before firing. Consider, for instance, a semaphore used to serialize the accesses to a shared resource. Since the times of request and release of the resource are not predictable *a priori*, the semaphore has to be ready to perform "p" and "v" actions at every time. Laziness of actions favours an "interacting" point of view. The observer interacts with the system to figure out what actions are possible. In our opinion, this point of view fits well with the intuition behind concurrency models such as process algebras (see [8] for a clear formulation of this idea).

Within the same framework of [1, 2, 10, 11, 12, 6, 5], lazy actions permit us to define a bisimulation based equivalence, called *lazy performance equivalence*, relating systems that can perform the same actions at the same time. Lazy performance equivalence holds several pleasant properties:

- it gives rise to a "well-timed" semantics in the sense that the action observation time coincides with the action execution time (differently from [1, 2, 10, 11, 12, 6, 5] where *ill-timed* traces but *well-caused* are allowed in order to develop a smooth theory of timed systems, see [1, 2] for details);
- it is a natural extension of Milner's *observational equivalence*;
- it is a congruence with respect to parallel composition with synchronization (differently from [11, 12]).

However, lazy actions introduce a non trivial difficulty to be managed. The transition system associated with every system (also trivial) is infinite branching. Indeed, a system has to be ready to perform its actions at every time later than the actual one. This results in an infinity of transitions in its operational description. Moreover, since the states of our transition systems are composed of sequential processes with associated local clocks that increase as the system computation proceeds, the transition systems we are facing are, in fact, acyclic and they can have infinite paths. These two different kinds of infinity give rise to problems when trying to implement the proposed semantics, since existing algorithms for checking bisimulation based semantics only apply to finite state transition systems. Nevertheless, we show that lazy performance equivalence is decidable in the class of systems with associated a finite standard untimed transition system. To establish this result we provide lazy performance equivalence with an alternative finite characterization, so that the two well known algorithms for checking bisimulation equivalence, i.e, the *on-the-fly* algorithm in [9] and the *partitioning* algorithm in [13], can be applied to this alternative characterization.

The paper is organized as follows. In Section 2 we introduce *CCS*, its timed operational semantics and lazy performance equivalence. A comparison with observational equivalence is also provided. In Section 3 we present the finite alternative characterization of lazy performance equivalence. Finally in Section 4 we give concluding remarks and further work.

2 The language and its timed behavioural semantics

The language used in the paper as a case study is Milner's *CCS* [14]. Below we report its syntax. As usual, the set of atomic actions is denoted by A, its complementation by $\bar{A}$ and τ is the invisible action. $Act = A \cup \bar{A}$ (ranged over by $a, b, \ldots$) is the set of visible actions and $Act_\tau = Act \cup \{\tau\}$ (ranged over by μ) is the set of all actions. Complementation is extended to Act by $\bar{\bar{a}} = a$. A *durational function* $\delta : Act \to \mathbf{N}^+$ assigns to each action the time needed for its execution. We assume that $\delta(\bar{a}) = \delta(a)$. Process *variables*, used for recursive definitions, are ranged over by x.

CCS is the set (ranged over by $p, p', \ldots$) of closed (i.e., without free variables) and guarded (i.e., variable x in a rec $x.p$ term, can appear only within $a._-$ and wait $n._-$ contexts) terms generated by the grammar below:

$$p ::= \text{nil} \mid a.p \mid \text{wait } n.p \mid p + p \mid p \mid p \mid p \backslash \{a\} \mid x \mid \text{rec } x.p$$

with $n \in \mathbf{N}^+$. *Sequential* processes (ranged over by $s, s', \ldots$) are CCS terms of the form nil, $a.p$, wait $n.p$, $p + p$ or rec $x.p$.

Process nil denotes a terminated process. By prefixing a process p with an action a, we get a process $a.p$ which can do an action a and then behaves like p. Process wait $n.p$ is a timed version of standard CCS's $\tau.p$; it denotes an invisible action of duration n. The alternative composition of p_1 and p_2 is denoted by $p_1 + p_2$, while $p_1 \mid p_2$ denotes the parallel composition of p_1 and p_2 that can perform any interleaving of the actions of p_1 and p_2 or a synchronization whenever p_1 and p_2 can perform complementary actions. Finally, $p \backslash \{a\}$ is a process which behaves like p except for actions $a, \bar{a}$ that are forbidden, and rec $x.p$ is used for recursive definitions.

We could extend the language with a relabelling operator $_-[\Phi]$; $p[\Phi]$ stands for a process which can perform action $\Phi(a)$ whenever p can perform action a. As usual in the durational setting, we should take into account duration-preserving relabelling functions in order to develop a smooth theory of processes with durational actions. See [12] for a detailed discussion on this point.

2.1 The operational semantics

In this subsection we provide CCS with a timed operational semantics via labelled transition systems. A *labelled transition system* is a triple (D, L, T) where D is a set of *states*, L is a set of *labels* and $T = \{ \xrightarrow{\mu} \subseteq D \times D \mid \mu \in L \}$ is the *transition relation*. We will write $d \xrightarrow{\mu} d'$ instead of $(d, d') \in \xrightarrow{\mu}$.

In our case, the states, called *timed states*, are terms obtained by extending the syntax of CCS with a *local clock prefixing* operator, $n \triangleright _-$, with $n \in \mathbf{Z}$. Set D of timed states is the set of terms generated by the following syntax:

$$d ::= n \triangleright s \mid d \mid d \mid d \backslash \{a\}$$

with s a sequential process and $n \in \mathbf{Z}$. Note that we allow also negative clock values. This is just for technical reasons and has no semantic content; it allow us to give an easier and comprehensive coincidence proof between the performance sensitive equivalence we are going to present and its finite characterization.

Intuitively, $n \triangleright p$ denotes a *CCS* process p with associated local clock $n \in \mathbf{Z}$. When p is not a sequential process, $n \triangleright p$ can be (canonically) reduced to a timed state by applying the following *clock distribution equations* as rewriting rules from left to right:

$$n \triangleright (p \mid p') = (n \triangleright p) \mid (n \triangleright p')$$
$$n \triangleright (p \backslash \{a\}) = (n \triangleright p) \backslash \{a\}$$

Given a timed state d, $\mathtt{maxclock}(d)$ denotes the maximum clock value within d. More precisely, function $\mathtt{maxclock} : D \to \mathbf{Z}$ is defined by the following axioms and inference rules:

$$\mathtt{maxclock}(n \triangleright s) = n$$
$$\mathtt{maxclock}(d_1 \mid d_2) = \max(\mathtt{maxclock}(d_1), \mathtt{maxclock}(d_2))$$
$$\mathtt{maxclock}(d \backslash \{a\}) = \mathtt{maxclock}(d)$$

Function $\mathtt{minclock}$, that returns the minimum clock value within a timed state, can be defined similarly.

Each transition of our transition systems is labelled by a pair of the form $\langle \mu, n \rangle$, where $\mu \in Act_\tau$ is the actual performed action and $n \in \mathbf{Z}$ is its completing time. Formally, the transition relation is defined through a set of axioms and inference rules, listed in Table 1[3]; symmetric rules for PAR and ALT have been omitted. The rule for action prefixing PREF states that process $a.p$ with local clock n, can perform an action a at any time $n + d + \delta(a)$, where $d \geq 0$ is the delay before the execution of action a is started. Rule WAIT deals similarly with wait $n. \_$ prefixes; also in this case, the execution can be delayed arbitrarily long. Rule PAR states that if component d_1 of a parallel composition $d_1 \mid d_2$ can perform an action μ at a time $n \geq \mathtt{maxclock}(d_2)$, then $d_1 \mid d_2$ can perform the same action at the same time. Condition $n \geq \mathtt{maxclock}(d_2)$ guarantees that the time increases as the computation proceeds and, hence, that only well-timed executions are taken into account. Rule SYNCH deals with synchronization: d_1 and d_2 in $d_1 \mid d_2$ can synchronize, if they are able to perform complementary actions at the same time. Rules ALT for nondeterministic composition, RES for restriction and REC for recursive definition are as usual. In this latter case $p[\mathbf{rec}\ v.\ p/v]$ denotes process p with every free occurrence of v substituted by $\mathbf{rec}\ v.\ p$.

Some properties of our operational semantics immediately follow. First of all we show that the time increases as the computation proceeds, so that only well-timed traces are considered. Then, we show that the system as a whole can idle. This states the "lazy" character of the systems we are dealing with.

---

[3] It is worthwhile observing that these rules are parametric *w.r.t.* the chosen duration function δ. Hence, we should write $\to_\delta$. For sake of simplicity, the subscript will always be omitted whenever clear from the context.

$$\text{PREF}\frac{}{(n \triangleright a.p) \overset{\langle a, n+d+\delta(a)\rangle}{\longrightarrow} (n+d+\delta(a)) \triangleright p} \, d \geq 0$$

$$\text{WAIT}\frac{}{(n \triangleright \texttt{wait } m.p) \overset{\langle \tau, n+d+m \rangle}{\longrightarrow} (n+d+m) \triangleright p} \, d \geq 0$$

$$\text{ALT}\frac{n \triangleright p \overset{\langle \mu, n' \rangle}{\longrightarrow} d}{n \triangleright (p+p') \overset{\langle \mu, n' \rangle}{\longrightarrow} d} \qquad \text{PAR}\frac{d_1 \overset{\langle \mu, n \rangle}{\longrightarrow} d_1'}{d_1 \mid d_2 \overset{\langle \mu, n \rangle}{\longrightarrow} d_1' \mid d_2} \, n \geq \text{maxclock}(d_2)$$

$$\text{SYNCH}\frac{d_1 \overset{\langle a, n \rangle}{\longrightarrow} d_1' \text{ and } d_2 \overset{\langle \bar{a}, n \rangle}{\longrightarrow} d_2'}{d_1 \mid d_2 \overset{\langle \tau, n \rangle}{\longrightarrow} d_1' \mid d_2'}$$

$$\text{RES}\frac{d \overset{\langle \mu, n \rangle}{\longrightarrow} d'}{d \backslash \{a\} \overset{\langle \mu, n \rangle}{\longrightarrow} d' \backslash \{a\}} \, \mu \notin \{a, \bar{a}\} \qquad \text{REC}\frac{n \triangleright p[\texttt{rec } v.\, p/v] \overset{\langle \mu, n' \rangle}{\longrightarrow} d}{n \triangleright \texttt{rec } v.\, p \overset{\langle \mu, n' \rangle}{\longrightarrow} d}$$

Table 1. The structural rules for the operational semantics.

Proposition 1. *Let $d \overset{\langle \mu, n \rangle}{\longrightarrow} d'$ be derivable from the rules in Table 1. Then:*

 $-$ $n \geq \text{maxclock}(d)$ and $n = \text{maxclock}(d')$ (no ill-timed); and
 $-$ $n \geq \text{maxclock}(d) + \text{duration}(\mu)$ (idling, as a whole)

where $\text{duration}(\mu) = \delta(a)$ if $\mu = a$, $\text{duration}(\mu) = \delta(a)$ if $\mu = \tau$ and an $(a, \bar{a})$-communication has been performed, or $\text{duration}(\mu) = n'$ if a $\texttt{wait } n'.$ _ action has been performed.

The following examples show two properties (infinite branching and acyclicity) that distinguish timed processes from untimed ones.

Example 1 (infinite branching). The transition system associated with a timed *CCS* process is, in general, infinite branching. Consider, for instance, timed state $0 \triangleright a.\texttt{nil}$ that, in virtue of rule PREF in Table 1, can perform action a at every time $t \geq \delta(a)$. The transition system generated by $a.\texttt{nil}$ has infinite transitions, labelled by $\langle a, t \rangle$, and infinite states of the form $t \triangleright \texttt{nil}$. On the contrary, ordinary untimed *CCS* processes can perform only a finite number of (initial) transitions (if they are guarded and have finite summation).

Example 2 (acyclicity). In each sequence of transitions a state can appear at most once. This is because the values of local clocks within a timed state increase as the computation proceeds. Process $p = \texttt{rec } v.\, a.v$ which, intuitively, can perform an action a and then becomes itself (and hence it is cyclic in a untimed context), performs computations of the form $0 \triangleright p \overset{\langle a, t_1 \rangle}{\longrightarrow} t_1 \triangleright p \overset{\langle a, t_2 \rangle}{\longrightarrow} t_2 \triangleright p \overset{\langle a, t_3 \rangle}{\longrightarrow} \cdots$, with $t_{i+1} > t_i$, so that no cycles are created.

2.2 Observational equivalence

The behavioural equivalence we are going to present is called *lazy performance equivalence* and is based on the classical notion of bisimulation. It relates *CCS* processes whenever they can perform the same actions at the same times. It turns out to be a congruence with respect to all *CCS* operators.

Definition 2 (lazy performance equivalence).

1. A binary relation $\mathcal{R}$ over D is a *lazy performance bisimulation*[4] if and only if for each $(d_1, d_2) \in \mathcal{R}$ and for each $\mu \in Act_\tau$:
   - $d_1 \xrightarrow{\langle \mu, n \rangle} d_1'$ implies $d_2 \xrightarrow{\langle \mu, n \rangle} d_2'$ and $(d_1', d_2') \in \mathcal{R}$;
   - $d_2 \xrightarrow{\langle \mu, n \rangle} d_2'$ implies $d_1 \xrightarrow{\langle \mu, n \rangle} d_1'$ and $(d_1', d_2') \in \mathcal{R}$.
2. We say that two states d_1 and d_2 are *lazy performance equivalent*, denoted $d_1 \sim d_2$, if and only if there exists a performance bisimulation $\mathcal{R}$ such that $(d_1, d_2) \in \mathcal{R}$.
3. We say that two processes p_1, p_2 are *lazy performance equivalent*, denoted $p_1 \sim_p p_2$, if and only if $0 \triangleright p_1 \sim 0 \triangleright p_2$.

Lazy performance equivalence is preserved by the consistent updating of local clocks: given two lazy performance equivalent states d_1 and d_2 and an integer m, then $\mathtt{addclock}(d_1, m)$ is lazy performance equivalent to $\mathtt{addclock}(d_2, m)$, where $\mathtt{addclock}(d, m)$ denotes the state obtained by adding m to all local clocks within d. Formally function $\mathtt{addclock}(d, m)$ is defined by the following inference rules:

$$\mathtt{addclock}(n \triangleright s, m) = (n + m) \triangleright s$$
$$\mathtt{addclock}(d_1 \mid d_2, m) = \mathtt{addclock}(d_1, m) \mid \mathtt{addclock}(d_2, m)$$
$$\mathtt{addclock}(d \backslash \{a\}, m) = \mathtt{addclock}(d, m) \backslash \{a\}$$

Lemma 3. *Let d be a timed state and $m \in \mathbf{Z}$. Then:*
$d \xrightarrow{\langle \mu, n \rangle} d'$ *iff* $\mathtt{addclock}(d, m) \xrightarrow{\langle \mu, n+m \rangle} \mathtt{addclock}(d', m)$.

Proposition 4. *Let d_1 and d_2 be two timed states and $m \in \mathbf{Z}$. Then:*
$d_1 \sim d_2$ *iff* $\mathtt{addclock}(d_1, m) \sim \mathtt{addclock}(d_2, m)$.

The following proposition shows that lazy performance equivalence is a congruence with respect to prefixing operators, nondeterministic composition, parallel composition and restriction. If we consider also open terms, it can be proved that $\sim_p$ is a congruence also for recursion following the techniques of [14].

Proposition 5. *Let $n \in \mathbf{N}^+$, and p_1, p_2, p be processes such that $p_1 \sim_p p_2$. Then $a.p_1 \sim_p a.p_2$, $\mathtt{wait}\ n.\ p_1 \sim_p \mathtt{wait}\ n.\ p_2$, $p_1 + p \sim_p p_2 + p$, $p_1 \mid p \sim_p p_2 \mid p$ and $p_1 \backslash \{a\} \sim_p p_2 \backslash \{a\}$.*

---

[4] We should define lazy performance bisimulation and the induced equivalence *w.r.t.* a duration function δ. As usual, we omit it.

Proof. Cases $a.p_1 \sim_p a.p_2$ and $\mathtt{wait}\ n.\,p_1 \sim_p \mathtt{wait}\ n.\,p_2$ immediately follow by Proposition 4. We only prove $p_1 \,|\, p \sim_p p_2 \,|\, p$; the proof for the other cases follows standard techniques [14]. We know that $p_1 \,|\, p \sim_p p_2 \,|\, p$ iff $0 \triangleright (p_1 \,|\, p) \sim 0 \triangleright (p_2 \,|\, p)$, i.e., by the clock distribution equations, iff $0 \triangleright p_1 \,|\, 0 \triangleright p \sim 0 \triangleright p_2 \,|\, 0 \triangleright p$. This is immediate by the statement:

$$d_1 \sim d_2 \ \text{and}\ \mathtt{maxclock}(d_1) = \mathtt{maxclock}(d_2)\ \text{imply}\ d_1 \,|\, d \sim d_2 \,|\, d$$

which follows from the fact that

$$\mathcal{R} = \{(d_1 \,|\, d, d_2 \,|\, d) \mid d_1 \sim d_2\ \text{and}\ \mathtt{maxclock}(d_1) = \mathtt{maxclock}(d_2)\}$$

is a lazy performance bisimulation. $\qquad\qquad\qquad\qquad\qquad\qquad\qquad\qquad\quad\square$

2.3 Relationships between $\sim_p$ and observational equivalence

Lazy performance equivalence is strictly finer than standard Milner's *(strong) observational equivalence* $(\sim_i)$. The proof of this statement relies on a lemma that establishes a correspondence between our timed transition system and Milner's untimed one. We will use $\overset{\mu}{\longmapsto}$ to denote the standard interleaving operational semantics. A function $\mathtt{forget}(\_)$ will be used that, given a state d, returns the untimed *CCS* process p obtained by forgetting all time prefixing $n \triangleright (\_)$ within d. It is defined by the following rules[5]:

$$\mathtt{forget}(n \triangleright s) = s$$
$$\mathtt{forget}(d_1 \,|\, d_2) = \mathtt{forget}(d_1) \,|\, \mathtt{forget}(d_2)$$
$$\mathtt{forget}(d \backslash \{a\}) = \mathtt{forget}(d) \backslash \{a\}$$

Lemma 6. *Let d be a state and $p = \mathtt{forget}(d)$. Then:*

1. *$p \overset{\mu}{\longmapsto} p'$ implies $d \overset{\langle \mu, n \rangle}{\longrightarrow} d'$ with $p' = \mathtt{forget}(d')$;*
2. *$d \overset{\langle \mu, n \rangle}{\longrightarrow} d'$ implies $p \overset{\mu}{\longmapsto} p'$ with $p' = \mathtt{forget}(d')$.*

Proposition 7. *Let p_1 and p_2 be processes. Then: $p_1 \sim_p p_2$ implies $p_1 \sim_i p_2$.*

Proof. By $p_1 \sim_p p_2$, follows that $0 \triangleright p_1 \sim 0 \triangleright p_2$. By Lemma 6 it is easy to show that relation $\mathcal{R} = \{(\mathtt{forget}(d_1), \mathtt{forget}(d_2)) \mid d_1 \sim d_2\}$ is an observational bisimulation in the sense of [14]. $\qquad\qquad\qquad\qquad\qquad\qquad\qquad\quad\square$

The converse does not hold in general: consider processes $a.\mathtt{nil} \,|\, b.\mathtt{nil}$ and $a.b.\mathtt{nil} + b.a.\mathtt{nil}$ that are observational bisimilar but not lazy performance bisimilar because $0 \triangleright a.\mathtt{nil} | 0 \triangleright b.\mathtt{nil}$ can perform an action a at time $t = \max\{\delta(a), \delta(b)\}$ followed by an action b at the same time while $0 \triangleright (a.b.\mathtt{nil} + b.a.\mathtt{nil})$ cannot.

---

[5] Clearly $\mathtt{wait}\ n.\_$ prefixes have to be intended now as standard $\tau.\_$ prefixes.

3 Decidability results

Within the setting of ordinary untimed CCS, it has been shown that strong observational equivalence [14] is decidable on the class of finite state processes. A process p is finite state if it can reach, via transitions, a finite set of processes, i.e., p is finite state iff the set $\{p' \mid p \overset{\mu_1}{\longmapsto} \overset{\mu_2}{\longmapsto} \cdots \overset{\mu_n}{\longmapsto} p'\}$ is finite (where, $\overset{\mu}{\longmapsto}$, denotes the standard interleaving operational semantics) .

In this section we show that lazy performance equivalence is also decidable in the class of processes which are finite state according to the standard interleaving operational semantics[6]. Unfortunately, as we have seen in Examples 1 and 2, simple finite state processes give rise to infinite transition systems. In order to use the standard algorithms for checking bisimulation based semantics we give a finite alternative characterization of lazy performance equivalence.

3.1 Dealing with acyclicity

Consider $p = a.\texttt{nil} \mid \texttt{rec } x.\, b.x$ (with $\delta(b) = 5$ and $\delta(a) = 7$) and computation:

$$0 \triangleright p \overset{\langle b,7\rangle}{\longrightarrow} 0 \triangleright a.\texttt{nil} \mid 7 \triangleright \texttt{rec } x.\, b.x \overset{\langle b,18\rangle}{\longrightarrow} 0 \triangleright a.\texttt{nil} \mid 18 \triangleright \texttt{rec } x.\, b.x$$
$$\overset{\langle b,50\rangle}{\longrightarrow} 0 \triangleright a.\texttt{nil} \mid 50 \triangleright \texttt{rec } x.\, b.x \cdots$$

Note that an infinite number of different states are reached during this computation and that the difference between maximal clock and minimal clock grows unboundedly. However, in state $0 \triangleright a.\texttt{nil} \mid 50 \triangleright \texttt{rec } x.b.x$, action a can only happen at a time greater than (or equal to) 50, or, equivalently, can only start at a time greater than (or equal to) 43. This means that we can update the clock value of the slower subprocess $a.\texttt{nil}$ to 43 without affecting the transitional semantics of the state: clearly, $43 \triangleright a.\texttt{nil} \mid 50 \triangleright \texttt{rec } x.\, b.x \sim_p 0 \triangleright a.\texttt{nil} \mid 50 \triangleright \texttt{rec } x.\, b.x$. Now both local clocks in $43 \triangleright a.\texttt{nil} \mid 50 \triangleright \texttt{rec } x.b.x$ can be decreased by 43 time units, obtaining state $0 \triangleright a.\texttt{nil} \mid 7 \triangleright \texttt{rec } x.\, b.x$ which has been investigated. To show how the above transformations can be used to decide lazy performance equivalence, we need some preliminary definitions and results. With $\texttt{maxduration}(p)$ we denote the time of the (initial) action of process p with maximal duration. It is defined by:

$$\texttt{maxduration}(\texttt{nil}) = 0 \qquad \texttt{maxduration}(a.p) = \delta(a)$$
$$\texttt{maxduration}(\texttt{wait } n.\, p) = n \qquad \texttt{maxduration}(p\backslash\{a\}) = \texttt{maxduration}(p)$$
$$\texttt{maxduration}(p_1 + p_2) = \max\{\texttt{maxduration}(p_1), \texttt{maxduration}(p_2)\}$$
$$\texttt{maxduration}(p_1 \mid p_2) = \max\{\texttt{maxduration}(p_1), \texttt{maxduration}(p_2)\}$$
$$\texttt{maxduration}(\texttt{rec } x.\, p) = \texttt{maxduration}(p)$$

Clearly, $\texttt{maxduration}(p)$ is finite for every CCS process p. Given a timed state d, $\texttt{maxduration}(d)$ is defined by $\texttt{maxduration}(\texttt{forget}(d))$.

---

[6] In the following, for *finite state* processes we mean finite state processes according to the standard interleaving operational semantics. Indeed, the timed operational semantics of a CCS process is finite state if and only if it is deadlocked.

Let $C[\_]$ denote the context of a timed subcomponent within a global timed state, i.e., $C[\_]$ is a term of the following syntax:

$$C[\_] ::= \_ \mid d \mid C[\_] \mid C[\_] \mid d \mid C[\_]\backslash\{a\}.$$

With $C[n \triangleright s]$ we denote the state obtained from $C[\_]$ by replacing $\_$ with $n \triangleright s$.

The following lemma shows how local clocks of slower sub-systems can be updated. In order to be faithful with the (operational) semantics, however, the local clock of a slower sub-system can be updated to a time less or equal than the actual time (the maximum clock value of the global system) minus the maximal duration of the actions the sub-system can perform.

Lemma 8. *Assume $d = C[n \triangleright p]$ and $m = \text{maxclock}(d) - \text{maxduration}(p)$. If $n \leq m$ then $d \sim C[m \triangleright p]$.*

The next step is to show how a timed state d can be transformed, preserving lazy performance equivalence, into a timed state $\hat{d}$, such that the difference between the minimal and the maximal clocks of $\hat{d}$ is less or equal than the maximal duration of the actions that d can (initially) perform.

Proposition 9. *For each timed state d there exists a state $\hat{d}$ such that $d \sim \hat{d}$ and $\text{maxclock}(\hat{d}) - \text{minclock}(\hat{d}) \leq \text{maxduration}(\hat{d})$.*

Proof. Define $\text{update}(d, t)$ as follows:

$$\text{update}(n \triangleright s, t) = \max\{n, t - \text{maxduration}(s)\} \triangleright s$$
$$\text{update}(d_1 \mid d_2, t) = \text{update}(d_1, t) \mid \text{update}(d_2, t)$$
$$\text{update}(d\backslash\{a\}, t) = \text{update}(d, t)\backslash\{a\}$$

Then $\hat{d} = \text{update}(d, \text{maxclock}(d))$ obviously satisfies the property $\text{maxclock}(\hat{d}) - \text{minclock}(\hat{d}) \leq \text{maxduration}(\hat{d})$.

Moreover $\text{update}(d, \text{maxclock}(d))$ can be seen as an iterate application of Lemma 8, so that $d \sim \hat{d}$: indeed, $\text{update}(n \triangleright p, t) = n \triangleright p$ if $n \geq t - \text{maxduration}(p)$ and $\text{update}(n \triangleright p, t) = (t - \text{maxduration}(p)) \triangleright p$ if $n \leq t - \text{maxduration}(p)$. The latter equivalence corresponds to the updating described in Lemma 8. $\square$

In the following $\hat{d}$ denotes state $\text{update}(d, \text{maxclock}(d))$. Proposition 9 shows how the difference between minimal and maximal clock can be bounded. This does not solve the problem of acyclicity because the maximal clock of a state is not affected; in fact $\text{maxclock}(\hat{d}) = \text{maxclock}(d)$ and $\text{maxduration}(\hat{d}) = \text{maxduration}(d)$. It is possible, however, to update all the clocks of a state, so that the maximal clock is re-set to 0.

Definition 10 (normalized states). Given a timed state d, the *normalized state* corresponding to d is $[d] = \text{addclock}(\hat{d}, -\text{maxclock}(\hat{d}))$.

By Proposition 4 and Proposition 9 follows that the normalization procedure does not affect the equivalence of states with the same maximal time.

Corollary 11. *Let d_1 and d_2 be two timed states such that* $\texttt{maxclock}(d_1) = \texttt{maxclock}(d_2)$. *Then:* $d_1 \sim d_2$ *iff* $[d_1] \sim [d_2']$.

The normalization procedure guarantees that only a finite number of different normalized states are generated from a finite state process.

Proposition 12. *Let p be a finite state process. Then following set is finite:*

$$S = \{[d_k] \mid 0 \triangleright p \xrightarrow{\langle \mu_1, n_1 \rangle} d_1 \cdots d_{k-1} \xrightarrow{\langle \mu_k, n_k \rangle} d_k\}.$$

Proof. Since p is finite state, by Lemma 6, also set

$$\{\texttt{forget}(d_k) \mid 0 \triangleright p \xrightarrow{\langle \mu_1, n_1 \rangle} d_1 \cdots d_{k-1} \xrightarrow{\langle \mu_k, n_k \rangle} d_k\}$$

is finite. Since $\texttt{forget}(d) = \texttt{forget}([d])$, then set $F = \{\texttt{forget}(d) \mid d \in S\}$ is finite. If S were infinite then there should be a process $p' \in F$ such that set $\{d \in S \mid \texttt{forget}(d) = p'\}$ is infinite. This is clearly impossible because states in this set differ only for the value of (a finite number of) local clocks and, since the states are normalized, all their clocks are in $[-\texttt{maxduration}(p'), 0]$. □

3.2 Dealing with infinite-branching

Example 1 shows that very simple processes can give rise to infinite branching transition graphs. In this subsection we show how, given a pair of timed states, it is possible to decide their (lazy performance) equivalence by checking a finite number of (initial) transitions. In order to do this, the notion of *efficient bisimulation* is given. It is a lazy performance bisimulation where, however, the number of transitions that timed states can perform is bounded.

Definition 13 (efficient bisimulation).
A binary relation $\mathcal{R}$ over D is an *efficient bisimulation* if and only if for each $(d_1, d_2) \in \mathcal{R}$ and $t = \max\{\texttt{maxduration}(d_i) + \texttt{maxclock}(d_i) \mid i = 1, 2\}$ we have:

1. $d_1 \xrightarrow{\langle \mu, n \rangle} d_1'$ and $n \leq t$ imply $d_2 \xrightarrow{\langle \mu, n \rangle} d_2'$ and $([d_1'], [d_2']) \in \mathcal{R}$;
2. $d_2 \xrightarrow{\langle \mu, n \rangle} d_2'$ and $n \leq t$ imply $d_1 \xrightarrow{\langle \mu, n \rangle} d_1'$ and $([d_1'], [d_2']) \in \mathcal{R}$.

We now establish the coincidence between lazy performance bisimulation and efficient bisimulation. A preliminary proposition is needed.

Proposition 14. *Let d be a state, $t = \texttt{maxduration}(d) + \texttt{maxclock}(d)$ and $n, n' \geq t$. If $d \xrightarrow{\langle \mu, n \rangle} d'$ then $d \xrightarrow{\langle \mu, n' \rangle} d''$ for some d'' s.t. $\widehat{d'} = \texttt{addclock}(\widehat{d''}, n - n')$.*

Theorem 15. *Let p_1 and p_2 be two processes; $p_1 \sim_p p_2$ if and only if there exists an efficient bisimulation $\mathcal{R}$ such that $(0 \triangleright p_1, 0 \triangleright p_2) \in \mathcal{R}$.*

Corollary 16. *Let p_1 and p_2 be two finite state agents. Then it is decidable whether $p_1 \sim_p p_2$.*

Proof. By Theorem 15, to decide whether $p_1 \sim_p p_2$ it is sufficient to exploit an efficient bisimulation $\mathcal{R}$ such that $(0 \triangleright p_1, 0 \triangleright p_2) \in \mathcal{R}$. By Proposition 12 both p_1 and p_2 have a finite set of reachable normalized states. Let S_1 and S_2 be the sets of normalized states reachable by p_1 and p_2 respectively. Since S_1 and S_2 are finite, there is a finite number of possible efficient bisimulations $\mathcal{R}$ such that $\mathcal{R} \subseteq S_1 \times S_2$. Moreover, given a relation $\mathcal{R} \subseteq S_1 \times S_2$, it is decidable whether $\mathcal{R}$ is an efficient bisimulation or not. Indeed, by Definition 13, only a finite number of transitions have to be checked for each pair of states in $\mathcal{R}$. □

More efficiently, the algorithms in [9] and [13] can be applied to decide efficient bisimulation of finite state processes. We refer to [7] for a deeper explanation.

4 Concluding remarks, related work and further research

In this paper we have proposed a new semantics for comparing functional behaviour and performance of systems that evolve via external stimuli. These systems are characterized by the fact that their actions are lazy, that is, they can be delayed arbitrarily long before firing. The resulting equivalence has nice mathematical properties and is decidable in the class of systems that can be described operationally by a finite standard untimed transition system.

In [7] we show how the theory extends in a natural way to deal with time dependents operators. These permit to introduce urgency in our calculus. An example of application of our theory is also presented. Moreover, the reader can also find full details of all proofs omitted in this paper and an extended section of related work within the durational setting.

In [1, 2], in [11, 12] and in [5, 6] three different equivalences have been defined for the language we investigated. These equivalences, however, rely on eager actions and ill–timed computations are allowed. Due to this lazy performance equivalence is unrelated with all of them.

The class of systems we considered has some analogies with that in [15]. There, a preorder for relating systems with respect to speed has been studied in the so-called abstract time approach. It has been shown that a precongruence can be properly defined by using lazy actions. However, because the different technical assumptions between our work and their (actions are atomic and time "passes between" them in [15], while the elapsing of the global clock is set only by the execution of actions in our setting), a formal comparison is not immediate. Further work will be devoted to study the relationships between these two different points of views. This would permit to transfer techniques and analytic concepts from one theory to the other.

In [4] a similar equivalence of lazy performance equivalence is presented, where ill-timed traces (even if well-caused) are allowed. It is worth noting that these traces give more discriminating power to the semantics, indeed, lazy performance equivalence turns out to be weaker than that proposed in [4].

We would like to conclude this section by mentioning another interesting line of further research. It is concerned with the "strong" character of our new theory. We assume that every action a concurrent systems can perform is observable to an external environment. This leads to a quite discriminating semantics. Thus, the study of "weak" versions of our lazy performance equivalence, would be interesting.

Acknowledgement: We thank Gian Luigi Ferrari for useful discussions on the topic of this paper. The anonymous referees are also thanked for helpful comments and suggestions.

References

1. L. Aceto and D. Murphy. On the ill–timed but well–caused. In *CONCUR'93, LNCS 715*, pages 97–111. Springer-Verlag, 1993.
2. L. Aceto and D. Murphy. Timing and causality in process algebra. Technical Report 9/93, University of Sussex, 1993. To appear in *Acta Informatica*.
3. S. Arun-Kumar and M. Hennessy. An efficient preorder for processes. *Acta Informatica*, 29:737–760, 1992.
4. F. Corradini. Compositionality for processes with durational actions. In *ICTCS'95*, World Scientific, 1995.
5. F. Corradini, R. Gorrieri and M. Roccetti. Performance preorder and competitive equivalence. Technical Report 95/01, University of Bologna, 1995.
6. F. Corradini, R. Gorrieri and M. Roccetti. Performance preorder: Ordering processes with respect to speed. In *MFCS'95, LNCS 969*, pages 444–453. Springer Verlag 1995.
7. F. Corradini and M. Pistore. Specification and Verification of Timed Systems. Technical Report 96/107, University of L'Aquila, 1996.
8. R. De Nicola and M. Hennessy: Testing Equivalences for Processes. *Theoretical Computer Science*, 34, pages 83–133, 1984.
9. G-C. Fernandez and L. Mounier. "On-the-fly" verification of behavioural equivalences and preorders. In *CAV'91, LNCS 575*, pages 181–191. Springer Verlag, 1992.
10. G-L. Ferrari and U. Montanari. Dynamic matrices and the cost analysis of concurrent programs. In *AMAST'95, LNCS 936*. Springer Verlag, 1995.
11. R. Gorrieri and M. Roccetti. Towards performance evaluation in process algebras. In *AMAST'93*, pages 289–296. Springer-Verlag, 1993.
12. R. Gorrieri, M. Roccetti and E. Stancampiano. A theory of processes with durational actions. *Theoretical Computer Science*, 140(1):73–94, 1995.
13. P.C. Kanellakis and S.C. Smolka. CCS expressions, finite state processes and three problem of equivalence. In *Second ACM Symposium on Principles of Distributed Computing*, 1983.
14. R. Milner. *Communication and Concurrency*. Prentice Hall International, 1989. International Series on Computer Science.
15. F. Moller and C. Tofts. Relating processes with respect to speed. In *CONCUR'91, LNCS 527*, pages 424–438. Springer Verlag, 1991.

A Class of Information Logics with a Decidable Validity Problem*

Stéphane Demri

LEIBNIZ Laboratory - IMAG Institute
46, Avenue Félix Viallet
38031 Grenoble Cedex, France
Email: Stephane.Demri@imag.fr

Abstract. For a class of propositional information logics defined from Pawlak's information systems, the validity problem is proved to be decidable using a significant variant of the standard filtration technique. Actually the decidability is proved by showing that each logic has the strong finite model property and by bounding the size of the models. The logics in the scope of this paper are characterized by classes of Kripke-style structures with interdependent equivalence relations and closed by the so-called restriction operation. They include Gargov's data analysis logic with local agreement and Nakamura's logic of graded modalities.

1 Introduction

The information logics derived from Pawlak's *information systems* [Paw81] have been the object of very active research (see for example [Orło84, Orło85, FdCO85, Kon87, Vak91b, Nak93]). Indeed the information systems provide relevant solutions to numerous problems in knowledge representation which is a central problem in Artificial Intelligence. The information systems have been proposed for the representation of knowledge by introducing the concept of *rough sets* which leads to the notion of approximation for sets of objects by means of equivalence relations. This contributes to model the uncertainty of the knowledge. The rough sets are based on the notion of *indiscernibility relations* that are binary relations identifying objects having the same description with respect to a given set of attributes. The indiscernibility relations are equivalence relations and so the logics of indiscernibility relations can be viewed as multimodal logics for which the modal operators behave as in the modal logic S5, *except* that they are not necessarily independent. Numerous logics of this type have been studied in the past (see e.g., [Orło85, FdCO85]).

An *information system* can be seen as a structure $(OB, AT, \{Val_a : a \in AT\}, f)$ such that OB is the non-empty set of *objects*, AT is the non-empty set of *attributes*, for all $a \in AT$, Val_a is the non-empty set of *values* of a and f is a mapping $OB \times AT \rightarrow \bigcup_{a \in AT} Val_a$ such that for all $(x, a) \in OB \times AT$,

---

* This work has been supported by the Centre National de la Recherche Scientifique (C.N.R.S.), France.

$f(x, a) \in Val_a$. In that setting, two objects o_1, o_2 are said to be *indiscernible* with respect to a set of attributes $A \subseteq AT$ (in short $o_1 \ ind(A) \ o_2$) iff for all $a \in A$, $f(o_1, a) = f(o_2, a)$. Different generalizations of the notion of information system (for instance by changing the profile of f with $\emptyset \neq f(o, a) \subseteq Val_a$) and various other relations between the objects (similarity, weak indiscernibility, ...) can be found for instance in [Vak91a]. The modal logics obtained from the information systems are multimodal logics such that the relations in the Kripke-style semantical structures correspond to relations between objects in the underlying information systems. Hence the relations are interdependent; for instance if $B \subseteq A$ then $ind(A) \subseteq ind(B)$. The decidability of the validity problem for various information logics has been an issue of interest in the past (see for example the precious Vakarelov's contributions in [Vak91b, Vak91a]). The aim of this paper is to prove that various information logics derived from Pawlak's information systems have a decidable validity problem by defining an original filtration construction (see e.g. [Gab72, Seg71]). The decidability is proved by showing that each logic has the strong finite model property and by bounding the size of the models. Logics defined in [Nak93, Gar86] shall illustrate the general construction.

The paper is structured as follows. In section 2, a class of information logics (LA-logics) is defined by refining the *local agreement condition* defined in [Gar86]. In Section 3, an original filtration construction is presented in order to show that every LA-logic has the strong finite model property and that the validity problem is decidable. In Section 4, we show how to apply the results of the previous section to logics defined in [Gar86, Nak93]. As a side-effect of our work, a sound and complete axiomatization is defined for the logic introduced in [Nak93]. In section 5, possible extensions of the present work are discussed.

2 Preliminaries

A (propositional) *modal language* L is determined by four sets which are supposed to be pairwise disjoint, a set Φ_0 of *propositional variables*, a set m of *modal constants*, a set of *propositional operators*, and a set of *modal operators*. The set M of *modal expressions* is the smallest set that satisfies the following conditions: $m \subseteq M$ and if $\oplus$ is any n-ary modal operator and $a_0, \ldots, a_{n-1} \in M$ then $\oplus(a_0, \ldots, a_{n-1}) \in M$. The set Σ of L-*formulae* is the smallest set that satisfies the following conditions: $\Phi_0 \subseteq \Sigma$, if $\oplus$ is any n-ary propositional operator and $A_1, \ldots, A_n \in \Sigma$ then $\oplus(A_1, \ldots, A_n) \in \Sigma$ and if $a \in M$ and $A \in \Sigma$ then $\{\Box_a A, \Diamond_a A\} \subseteq \Sigma$ -for the sake of simplicity $\Box_a$ and $\Diamond_a$ are also called *modal operators*. We assume throughout the paper that any modal language used in the sequel satisfies the following conditions: ϕ_0 is a fixed countable set of propositional variables and the propositional operators are the unary $\neg$ and the binary $\Leftrightarrow$, $\Rightarrow$, $\vee$, $\wedge$. Let L be a modal language. We write $sub(A)$ (resp. $mw(A)$) to denote the set of *subformulae* of the formula A (resp. the *modal weight* of A, i.e. the number of occurrences of modal operators in A -of the form $\Box_a$ or $\Diamond_a$). We also write $sub_M(A)$ to denote the set of modal expressions occurring in the formula A. As usual, by a L-*model* we understand a triple $(W, (R_a)_{a \in M}, V)$ such that W is a non-empty set -set of *objects* or *worlds*-, for all $a \in M$, R_a is a binary relation on W, and V is a mapping $\phi_0 \to \mathcal{P}(W)$, the power set of W. The set of L-models is written mod_L. Let $M = (W, (R_a)_{a \in M}, V)$ be a L-model. We say that a formula A is *satisfied by the object* $u \in W$ *in* M (written $M, u \models A$) when the following conditions are satisfied.

- $M, u \models P$ iff $u \in V(P)$, for all $P \in \phi_0$, $M, u \models \neg A$ iff not $M, u \models A$,
- $M, u \models A \wedge B$ iff $M, u \models A$ and $M, u \models B$,
- $M, u \models \Box_a A$ iff for all $v \in R_a(u)$, $M, v \models A$,
- $M, u \models \Diamond_a A$ iff there is $v \in R_a(u)$ such that $M, v \models A$.

We omit the conditions for the other logical operators. Since the interpretations of $\Box_a$ and $\Diamond_a$ are not independent, in the sequel only the operators of the form $\Box_a$ are used. A formula A is *true* in a L-model M (written $M \models A$) iff for all $u \in W$, $M, u \models A$.

In the sequel, by a *logic* $\mathcal{L}$, we understand a triple[2] $(L, \mathcal{S}, \models_{\mathcal{L}})$ such that L is a modal language, $\mathcal{S} \subseteq mod_L$ and $\models_{\mathcal{L}}$ is the restriction of $\models$ to the sets $\mathcal{S}$ and L (satisfiability relation). For all models $M \in \mathcal{S}$, M is said to be a *model for* $\mathcal{L}$. A L-formula A is said to be $\mathcal{L}$-*valid* iff A is true in all L-models of $\mathcal{S}$. A L-formula A is said to be $\mathcal{L}$-*satisfiable* iff there is $M = (W, (R_a)_{a \in M}, V) \in \mathcal{S}$, $u \in W$ such that $M, u \models_{\mathcal{L}} A$. A logic $\mathcal{L} = (L, \mathcal{S}, \models_{\mathcal{L}})$ has the *strong finite model property* iff for every $\mathcal{L}$-satisfiable formula A, there exist $M = (W, (R_a)_{a \in M}, V) \in \mathcal{S}$ and $w \in W$ such that W is finite and $M, w \models_{\mathcal{L}} A$. As usual, an *instance of the validity (resp. satisfiability) problem for* $\mathcal{L}$ consists in: is the L-formula A $\mathcal{L}$-valid (resp. $\mathcal{L}$-satisfiable)? It is immediate that the validity problem for $\mathcal{L}$ is decidable iff the satisfiability problem for $\mathcal{L}$ is decidable.

A logic $\mathcal{L} = (L, \mathcal{S}, \models_{\mathcal{L}})$ is said to be a *LA-logic* iff there is a set of linear[3] orders over M, say $\mathcal{LO}_{\mathcal{L}}$, such that for all L-models $M = (W, (R_a)_{a \in M}, V)$, $M \in \mathcal{S}$ iff for all $a \in M$, R_a is an equivalence relation and for all $u \in W$, there is $\rhd \in \mathcal{LO}_{\mathcal{L}}$ such that for all $a, b \in M$, if $a \rhd b$ then $R_a(u) \subseteq R_b(u)$. Take for instance $\mathcal{L}' = (L', \mathcal{S}', \models_{\mathcal{L}'})$ with $M' = \{1, 2\}$ and for all L'-models $M = (W, (R_a)_{a \in M'}, V)$, $M \in \mathcal{S}'$ iff R_1 and R_2 are equivalence relations and $R_1 \subseteq R_2$. $\mathcal{L}'$ is an example of LA-logic where $\mathcal{LO}_{\mathcal{L}'}$ is a singleton. The set $\mathcal{S}'$ can be related to the set of information systems as follows. Let $(OB, AT, \{Val_a : a \in AT\}, f)$ be an information system and $\emptyset \neq AT' \subseteq AT$. The L'-model $(OB, (R_a)_{a \in M'}, V)$ with $(\star)$ $R_1 = ind(AT)$ and $R_2 = ind(AT')$ belongs to $\mathcal{S}'$. Moreover for all $M = (W, (R_a)_{a \in M'}, V) \in \mathcal{S}'$, there exist an information system $(OB, AT, \{Val_a : a \in AT\}, f)$ and $\emptyset \neq AT' \subseteq AT$ such that $(\star)$. Actually, take $OB = W$, $AT = \{a, a'\}$, $Val_a = \{R_1(x) : x \in W\}$, $Val_{a'} = \{R_2(x) : x \in W\}$, for all $x \in W$, $f(x, a) = R_1(x)$, $f(x, a') = R_2(x)$ and $AT' = \{a'\}$.

---

[2] It is possible to define a logic in terms of *frames* (structures of the form $(W, (R_a)_{a \in M})$) but the definition of logic used in the paper is sufficient for our needs.

[3] A binary relation $\rhd$ over U is said to be *linear* iff $\rhd$ is reflexive, transitive, totally connected (for all $x, y \in U$ either $(x, y) \in \rhd$ or $(y, x) \in \rhd$) and antisymmetric (for all $x, y \in U$ if $(x, y) \in \rhd$ and $(y, x) \in \rhd$ then $x = y$).

The term of LA-logic refers to the local agreement condition defined in [Gar86]. Equivalence relations R and S on a set W are said to be in *local agreement* (LA) iff for all $u \in W$ either $R(u) \subseteq S(u)$ or $S(u) \subseteq R(u)$. It is easy to show that for any LA-logic $\mathcal{L} = (\mathsf{L}, \mathcal{S}, \models_{\mathcal{L}})$, for any model $\mathcal{M} = (W, (R_a)_{a \in \mathsf{M}}, V) \in \mathcal{S}$, for all $a, b \in \mathsf{M}$, R_a and R_b are in local agreement. The property stated in Proposition 1 might explain why the local agreement condition has been introduced in [Gar86].

Proposition 1. *Let R and S be two equivalence relations on a set W. (1) R and S are in local agreement iff (2) $R \cup S$ is transitive.*

Proof. By way of example we show (1) $\rightarrow$ (2). Assume $(x, y) \in R$ and $(y, z) \in S$. If $R(x) \subseteq S(x)$ then by transitivity of S, $(x, z) \in S$. Now assume $S(x) \subseteq R(x)$. If $S(y) \subseteq R(y)$ then by transitivity of R, $(x, z) \in R$. Now assume $R(y) \subseteq S(y)$. Since R and S are equivalence relations, $R(x) = R(y)$, $S(y) = S(z)$ and therefore $S(x) \subseteq R(x) = R(y) \subseteq S(y) = S(z)$. So $(x, z) \in S$. The case $(x, y) \in S$ and $(y, z) \in R$ is similar. Since R and S are transitive, $R \cup S$ is therefore transitive.

The property stated in Proposition 2 below shall be needed in the sequel.

Proposition 2. *Let $(R_i)_{i \in \{1, \ldots, n\}}$ be a finite family of equivalence relations on the set W such that the relations are pairwise in local agreement. For all $x \in W$, there is a permutation s of $\{1, \ldots, n\}$ such that $R_{s(1)}(x) \subseteq \ldots \subseteq R_{s(n)}(x)$.*

The proof is by an easy induction on n.

3   A Restriction Construction for the LA-logics

The aim of this section is to show that every LA-logic has the strong finite model property. Since the size of the models is shown to be bounded, it shall entail that the validity problem for every LA-logic is decidable. Although the modal operators for each LA-logic behave as modal operators for S5, the usual filtration construction for the multimodal logics $S5_k$ cannot be used straightforwardly for the LA-logics. Indeed, in the model obtained by filtration, the accessibility relations have to be equivalence relations *and* the relationships between the relations have to be preserved. That is why, a variant of the traditional construction is proposed in this section. Instead of defining equivalence classes of worlds (as done in the standard filtration constructions), restrictions[4] of models are used. In the rest of this section $\mathcal{L}$ denotes a LA-logic $(\mathsf{L}, \mathcal{S}, \models_{\mathcal{L}})$ unless otherwise stated. Let $\mathcal{M} = (W, (R_a)_{a \in \mathsf{M}}, V) \in \mathcal{S}$ and $W' \subseteq W$. The *restriction* of $\mathcal{M}$ to W', noted $\mathcal{M}_{|W'}$, is the L-model $(W', (R'_a)_{a \in \mathsf{M}}, V')$ such that for all $a \in \mathsf{M}$, $R'_a = R_a \cap W' \times W'$ and for all $P \in \phi_0$, $V'(P) = V(P) \cap W'$. Proposition 3 below states that the class of models for a LA-logic is closed under the restriction operation.

---

[4] In the literature restrictions are defined for instance in [Cer94, HM92].

Proposition 3. *For all* $\mathcal{M} = (W, (R_a)_{a \in M}, V) \in \mathcal{S}$ *and* $W' \subseteq W$, $\mathcal{M}_{|W'} \in \mathcal{S}$.

Proof. It is easy to check that for all $a \in M$, R'_a is an equivalence relation on W'. Now assume that for some $a, b \in M$ and $w \in W'$, $R_a(w) \subseteq R_b(w)$. So $R_a(w) \cap W' \subseteq R_b(w) \cap W'$ and therefore $R'_a(w) \subseteq R'_b(w)$. Since there is $\rhd \in \mathcal{LO}_{\mathcal{L}}$ such that for all $a, b \in M$, $w \in W$, if $a \rhd b$ then $R_a(w) \subseteq R_b(w)$, then for all $a, b \in M$, $w \in W'$, if $a \rhd b$ then $R'_a(w) \subseteq R'_b(w)$.

It is not difficult to show that for all L-formulae A, $w \in W'$, $a \in M$, if for all $b \in sub_M(A)$, $R_b(w) \subseteq R_a(w)$ then $\mathcal{M}, w \models A$ iff $\mathcal{M}_{|R_a(w)}, w \models A$.

Proposition 4. *Let* $\{a_1, \ldots, a_n\} \subseteq M$, $\mathcal{M} = (W, (R_a)_{a \in M}, V) \in \mathcal{S}$, $x \in W$ *such that* $R_{a_1}(x) \subseteq \ldots \subseteq R_{a_n}(x)$. *Assume* $(x, y) \in R_{a_k}$ *for some* $k \in \{1, \ldots, n\}$. *Then,* *(1) for all* $k' \in \{k, \ldots, n\}$, $R_{a_{k'}}(x) = R_{a_{k'}}(y)$ *and (2) for all* $k' \in \{1, \ldots, k-1\}$, $R_{a_{k'}}(y) \subseteq R_{a_k}(y)$.

Proof. (1) Since $R_{a_k}(x) \subseteq R_{a_{k'}}(x)$ then $(x, y) \in R_{a_{k'}}$ and $R_{a_{k'}}(x) = R_{a_{k'}}(y)$ since $R_{a_{k'}}$ is an equivalence relation. (2) Assume $z \in R_{a_{k'}}(y)$. Either $(x, z) \in R_{a_{k'}}$ or $(x, z) \in R_{a_k}$ since $R_{a_k} \cup R_{a_{k'}}$ is transitive (see Proposition 1). If $(x, z) \in R_{a_k}$ then $R_{a_k}(y) = R_{a_k}(x) = R_{a_k}(z)$ and therefore $z \in R_{a_k}(y)$. Now assume $(x, z) \in R_{a_{k'}}$. Since $R_{a_k}(x) = R_{a_k}(y)$ and $z \in R_{a_{k'}}(x) \subseteq R_{a_k}(x)$ then $z \in R_{a_k}(y)$.

For any finite sequence of natural numbers σ, we write $set(\sigma)$ to denote the set of elements occurring in σ. For example $set((1, 2, 3, 3, 4)) = \{1, 3, 2, 4\}$.

3.1 The Construction

Let A be a L-formula, $\mathcal{M} = (W, (R_a)_{a \in M}, V) \in \mathcal{S}$, $w \in W$ such that $\mathcal{M}, w \models A$. Assume that the modal expressions occurring in A are exactly $a_1, \ldots, a_n$ with $R_{a_1}(w) \subseteq \ldots \subseteq R_{a_n}(w)$ (see Proposition 2). The set F is defined by $F = \{\Box_b A' : \exists \Box_a A' \in sub(A), \exists b \in sub_M(A)\}$. For all $x \in W$, for all sequences $\sigma = (j_1, \ldots, j_k)$ such that $set(\sigma) \subseteq \{1, \ldots, n\}$ and $R_{a_{j_1}}(x) \subseteq \ldots \subseteq R_{a_{j_k}}(x)$, the set F^σ_x is defined as follows,

$$F^\sigma_x = \{\Box_{a_i} A' \in F : \exists k' \in \{1, \ldots, k\} \text{ with } i = j_{k'}, \mathcal{M}, x \models \neg \Box_{a_i} A', \text{ and}$$
$$\text{if } k' \geq 2 \text{ then } \mathcal{M}, x \models \Box_{a_{j_{k'-1}}} A'\}$$

Observe that[5] $card(F) \leq n \times mw(A)$, $card(F^\sigma_x) \leq mw(A)$ and $F^\sigma_x = \emptyset$ when σ is the empty sequence -written Λ. We write $W'(\mathcal{M}, w, A)$ to denote the new set of objects (also abbreviated by W') and each element in W' shall be written[6] $w(\sigma, i, \Box_{a_p} A')$ where σ is a sequence of elements of $\{1, \ldots, n\}$ without repetition, $i \in \{1, \ldots, n\}$, $\Box_{a_p} A' \in F$ and $p \notin set(\sigma)$. There is only one exception, the object w is written $w((1, \ldots, n), 0, \Lambda)$. Since all the involved sets are finite, the set W'

---

[5] For any finite set U, $card(U)$ denotes the cardinal of U.

[6] Although a more economical writing is possible, we believe that the present one insures a good readability of the forthcoming proofs.

shall be finite. The set $W_0 = \{w((1,\ldots,n),0,\Lambda)\}$ is first defined and then W_{i+1} from W_i. For all $w' = w(\sigma,i,?) \in W_i$ ('?' is either Λ or a formula of the form $\square_{a_p} B$), for all $\square_{a_j} A' \in F^{\sigma}_{w'}$, there is $u \in W$ such that $(w',u) \in R_{a_j}$ and $\mathcal{M}, u \models \neg A'$. If $\sigma = (j_1,\ldots,j_k)$ then we write k' to denote the element of $\{1,\ldots,k\}$ such that $j_{k'} = j$. The existence of k' is guaranteed by the definition of $F^{\sigma}_{w'}$. The set W_{i+1} is augmented with the element $u = w((j'_1,\ldots,j'_{k'-1}),i+1,\square_{a_j} A')$ with $set((j'_1,\ldots,j'_{k'-1})) = set((j_1,\ldots,j_{k'-1}))$ and $R_{a_{j'_1}}(u) \subseteq \ldots \subseteq R_{a_{j'_{k'-1}}}(u)$ -whenever $k' = 1$ the sequence $(j'_1,\ldots,j'_{k'-1})$ is empty.

There exists $\alpha \in \{0,\ldots,n\}$ such that $W_\alpha \neq \emptyset$ and $W_{\alpha+1} = \emptyset$ since the length of the sequences of natural numbers strictly decreases. Moreover if $w(\sigma,i,?) \in W_i$ then the length of σ is at most $n-i$. The set W' is defined as $\bigcup_{i \in \{0,\ldots,\alpha\}} W_i$ and the restricted model is $\mathcal{M}' = \mathcal{M}_{|W'}$. For all $i \in \{0,\ldots,\alpha-1\}$, $card(W_{i+1}) \leq card(W_i) \times mw(A)$ and therefore $card(W') \leq 1 + n \times mw(A)^n$. This construction is more general than the construction defined in [HM92] to prove the NP-completeness of the satisfiability problem for the propositional modal logic S5. Indeed for $n = 1$ the construction in [HM92] and ours are identical.

3.2 How the Construction Captures Enough Worlds

Now assume that $\mathcal{M}, w \models A$ for some L-formula A, $\mathcal{M} = (W,(R_a)_{a \in \mathbb{M}},V) \in \mathcal{S}$, $w \in W$ and the distinct modal expressions occurring in A are exactly $a_1,\ldots,a_n$ with $R_{a_1}(w) \subseteq \ldots \subseteq R_{a_n}(w)$. The rest of the section is mainly devoted to show that $\mathcal{M}, w \models A$ iff $\mathcal{M}_{|W'(\mathcal{M},w,A)}, w \models A$. For a sequence $\sigma = (j_1,\ldots,j_k)$ without repetition such that $set(\sigma) \subseteq \{1,\ldots,n\}$ we write $\sigma' = v(j_1,\ldots,j_k)$ to denote a sequence of elements of $\{1,\ldots,n\}$ without repetition such that $set(\sigma) = set(\sigma')$. This notation is used when it is not necessary to know the exact value of σ'. Moreover $\sigma_1.\sigma_2$ denotes the concatenation of two sequences.

Proposition 5. *For all $i \in \{1,\ldots,\alpha\}$, $w' = w((j_1,\ldots,j_k),i,\square_{a_p} A') \in W_i$,*

1. *$R_{a_{j_1}}(w') \subseteq \ldots \subseteq R_{a_{j_k}}(w')$, $R_{a_{j_k}}(w') \subseteq R_{a_p}(w')$, $\mathcal{M}, w' \models \neg A'$, and*
2. *for all $j \in \{1,\ldots,n\} \setminus set((j_1,\ldots,j_k))$, $R_{a_{j_k}}(w') \subseteq R_{a_j}(w')$.*

Proof. (1) Obvious from the construction of $w((j_1,\ldots,j_k),i,\square_{a_p} A')$.
(2) By induction on i.
Base case $(i = 1)$: assume $(w,w') \in R_{a_p}$ with $w' = w((j_1,\ldots,j_k),1,\square_{a_p} A') \in W_1$. It follows that $p = k + 1$ and $set((j_1,\ldots,j_k)) = \{1,\ldots,k\}$. From (1), $R_{a_{j_k}}(w') \subseteq R_{a_p}(w')$. Since $R_{a_p}(w) = R_{a_p}(w')$ and from Proposition 4, for all $j \in \{k+1,\ldots,n\} = (\{1,\ldots,n\} \setminus set((j_1,\ldots,j_k)))$, $R_{a_j}(w) = R_{a_j}(w')$ then for all $j \in \{1,\ldots,n\} \setminus set((j_1,\ldots,j_k))$, $R_{a_{j_k}}(w') \subseteq R_{a_j}(w')$.
Induction step: let $w'' = w(v(j_1,\ldots,j_{k'-1}).(j_{k'},\ldots,j_k),i,\square_a A'')$ with $(w'',w') \in R_{a_{j_{k'}}}$ and $w' = w((j_1,\ldots,j_{k'-1}),i+1,\square_{a_{j_{k'}}} A')$. By the induction hypothesis, for all $j \in \{1,\ldots,n\} \setminus set((j_1,\ldots,j_k))$, $R_{a_{j_k}}(w'') \subseteq R_{a_j}(w'')$. Since $R_{a_{j_{k'}}}(w'') \subseteq \ldots \subseteq R_{a_{j_k}}(w'')$, then for all $j \in \{1,\ldots,n\} \setminus set((j_1,\ldots,j_{k'-1}))$, $R_{a_{j_{k'}}}(w'') \subseteq R_{a_j}(w'')$. Since $R_{a_{j_{k'}}}(w'') = R_{a_{j_{k'}}}(w')$ and considering Proposition 4, we get that for all $j \in \{1,\ldots,n\} \setminus set((j_1,\ldots,j_{k'-1}))$, $R_{a_{j_{k'-1}}}(w') \subseteq R_{a_j}(w')$.

Proposition 6 below states that the set W' contains *enough* worlds to check whether $\mathcal{M}, w \models A$ holds.

Proposition 6. *For all* $w' \in W'$ *and* $\Box_{a_j} A' \in F$, *if* $\mathcal{M}, w' \models \neg\Box_{a_j} A'$ *then there is* $w'' \in W'$ *such that* $(w', w'') \in R_{a_j}$ *and* $\mathcal{M}, w'' \not\models A'$.

Proof. By induction on i when $w' \in W_i$.

Base case ($i = 0$): assume that $\mathcal{M}, w \models \neg\Box_{a_j} A'$. There exists $\Box_{a_k} A' \in F_w^{(1,\ldots,n)}$ and there is $u = w((j_1, \ldots, j_{k-1}), 1, \Box_{a_k} A') \in W_1$ such that $(w, u) \in R_{a_k}$. Since $R_{a_k}(w) \subseteq R_{a_j}(w)$ (by definition of $F_w^{(1,\ldots,n)}$) we have $(w, u) \in R_{a_j}$.

Induction step: assume $\mathcal{M}, w' \models \neg\Box_{a_j} A''$, $w' = w((j_1, \ldots, j_{k'-1}), i+1, \Box_{a_{j_{k'}}} A')$. If $j \in set((j_1, \ldots, j_{k'-1}))$ then a new object in W_{i+2} is built. There exists $\Box_{a_k} A'' \in F_{w'}^{(j_1, \ldots, j_{k'-1})}$ and there is $u_1 = w((j'_1, \ldots, j'_{k''-1}), i+2, \Box_{a_k} A'') \in W_{i+2}$ such that $(w', u_1) \in R_{a_k}$ and $\mathcal{M}, u_1 \models \neg A''$. By definition of $F_{w'}^{(j_1, \ldots, j_{k'-1})}$ and by Proposition 5(1) we have $R_{a_k}(w') \subseteq R_{a_j}(w')$ and therefore $(w', u_1) \in R_{a_j}$. Now assume $j \notin set((j_1, \ldots, j_{k'-1}))$. So $w \neq w'$. There exists

$$u_1 = w(v(j_1, \ldots, j_{k'-1}).(j_{k'}, \ldots, j_k), i, ?) \in W_i$$

such that $(u_1, w') \in R_{a_{j_{k'}}}$. By Proposition 5, $R_{a_{j_{k'}}}(u_1) \subseteq R_{a_j}(u_1)$. It follows that $(u_1, w') \in R_{a_j}$ and therefore $\mathcal{M}, u_1 \models \neg\Box_{a_j} A''$ since R_{a_j} is an equivalence relation. By the induction hypothesis, there is $u_2 \in W'$ such that $(u_1, u_2) \in R_{a_j}$, $\mathcal{M}, u_2 \not\models A''$ and therefore $(w', u_2) \in R_{a_j}$ since $R_{a_j}(u_1) = R_{a_j}(w')$. ∎

Proposition 7. *A* L-*formula* A *is* $\mathcal{L}$-*satisfiable iff it is satisfiable in a model for* $\mathcal{L}$ *with at most* $1 + n \times mw(A)^n$ *objects where* $n = card(sub_M(A))$.

The proof of Proposition 7 follows the lines of the proof of Lemma 6.1 in [Lad77].

Proof. Assume there exist $\mathcal{M} = (W, (R_a)_{a \in \mathcal{M}}, V) \in \mathcal{S}$ and $w \in W$ such that $\mathcal{M}, w \models A$. Let $\mathcal{M}'$ be $\mathcal{M}_{|W'(\mathcal{M},w,A)}$. For all objects $u' \in W'$ and for all $B \in sub(A)$, $\mathcal{M}, u' \models B$ iff $\mathcal{M}', u' \models B$ (including A). We proceed by induction on the structure of B. The only nontrivial case is when B has the form $\Box_a B'$. Assume $u' \in W'$. If $\mathcal{M}, u' \models \Box_a B'$ then for all $v \in W$ such that $(u', v) \in R_a$ we have $\mathcal{M}, v \models B'$. In particular for all $v \in W'$ such that $(u', v) \in R'_a$ we have $\mathcal{M}, v \models B'$. By the induction hypothesis, for all $v \in W'$, $\mathcal{M}', v \models B'$. So $\mathcal{M}', u' \models \Box_a B'$. Now assume $\mathcal{M}, u' \not\models \Box_a B'$. From Proposition 6, there exists $v \in W'$ such that $(u', v) \in R_a$ and $\mathcal{M}, v \not\models B'$. It follows that $(u', v) \in R'_a$ and by the induction hypothesis $\mathcal{M}', v \not\models B'$. Hence $\mathcal{M}', u' \not\models \Box_a B'$. ∎

Corollary 8. $\mathcal{L}$ *has the strong finite model property and the validity problem for* $\mathcal{L}$ *is decidable.*

The construction in this section generalizes the technique used in [Dem96] to the set of LA-logics. Corollary 8 takes advantage of the fact that for any LA-logic $\mathcal{L} = (L, \mathcal{S}, \models_\mathcal{L})$, $\mathcal{S}$ is closed by the restriction operation and any binary relations R, S of a model for $\mathcal{L}$ are in local agreement. Actually it can be easily shown (following the lines of this section) that for all R-*logics* $\mathcal{L}$, $\mathcal{L}$ has the strong finite model property and the validity problem for $\mathcal{L}$ is decidable where a logic $\mathcal{L} = (L, S, \models_{\mathcal{L}})$ is said to be a R-*logic* iff S is closed by the restriction operation, for all $\mathcal{M} = (W, (R_a)_{a \in M}, V) \in S$, $a, b \in M$, R_a is an equivalence relation and R_a and R_b are in local agreement. Every LA-logic is a R-logic and the converse does not hold. For instance, take some $P \in \phi_0$ and define S_P as the set of L-models $(W, (R_a)_{a \in M}, V)$ such that $V(P)$ is finite, and for all $a, b \in M$, R_a is an equivalence relation and, R_a and R_b are in local agreement. For a given modal language L, $\mathcal{L} = (L, S_P, \models_{\mathcal{L}})$ is a R-logic but $\mathcal{L}$ *is not* a LA-logic. Although every R-logic has a decidable validity problem, up to now there is no general method to define a sound and complete axiomatization for each R-logic. To characterize the complexity class of the satisfiability problem for a given R-logic remains an open problem. However, for all $k \geq 0$, the satisfiability problem for a R-logic $\mathcal{L}$ restricted to the L-formulae A such that the number of distinct modal expressions occurring in A is less than k, is **NP-complete**. Actually any $\mathcal{L}$-satisfiable formula A of such kind has a model of polynomial *size* with respect to the *size* of A (at most $1 + k \times mw(A)^k$ worlds). So if the language of a given R-logic $\mathcal{L}$ has a finite set of modal constants and no modal operator then the satisfiability problem for $\mathcal{L}$ is **NP-complete**. For the sake of comparison, for all $k \geq 2$, the satisfiability problem for the multimodal logics $S5_k$ is **PSPACE-complete** [HM92].

4 Applications to Logics from the Literature

The LA-logics can be related to logics from the literature as it is presented below.

4.1 Gargov's Data Analysis Logic with Local Agreement

The logic DALLA defined in [Gar86] (originally called DAL), restricts the class of models of the logic DAL [FdCO85] by requiring that any two indiscernibility relations of a model are in (LA). A complete axiomatization of DALLA is given in [Gar86]. The decidability of the validity problem for the logic DALLA is open, as mentioned in [Gar86]. The logic DALLA $= (L_D, S_D, \models_{DALLA})$ is defined as follows. L_D has a countable set of modal constants i and the set of modal operators is $\{\cap, \cup^*\}$ -intersection[7] and transitive closure of union. The set of modal expressions (resp. L_D-formulae) is denoted by **eac** (resp. **for**). For all $\mathcal{M} = (W, (R_a)_{a \in \mathbf{eac}}, V) \in S_D$, for all $a, b \in \mathbf{eac}$, $\oplus \in \{\cap, \cup^*\}$, R_a is an equivalence relation, $R_{a \oplus b} = R_a \oplus R_b$ and R_a and R_b are in local agreement. Consider the Hilbert-style system **dalla** containing the following axiom schemes and inference rules ($A, B \in \mathbf{for}$, $a, a_1, a_2 \in \mathbf{eac}$):

P. All formulae having the form of a classical propositional tautology
K. $\Box_a(A \Rightarrow B) \Rightarrow (\Box_a A \Rightarrow \Box_a B)$, **T.** $\Box_a A \Rightarrow A$, **5.** $\Diamond_a A \Rightarrow \Box_a \Diamond_a A$

---

[7] For the sake of simplicity, the modal operators and the relational operations are denoted by the same symbols.

U. $\Box_{a_1 \cup^* a_2} A \Leftrightarrow \Box_{a_1} A \wedge \Box_{a_2} A$, **I.** $\Box_{a_1 \cap a_2} A \Leftrightarrow \Box_{a_1} A \vee \Box_{a_2} A$
MP. From A and $A \Rightarrow B$ infer B, **NR.** From A infer $\Box_a A$.

For all L_D-formulae A, A is DALLA-valid iff A is a theorem of `dalla` [Gar86]. Although DALLA is not *stricto sensu* a LA-logic (because of the condition $R_{a \oplus b} = R_a \oplus R_b$), there exists a simple translation between DALLA and the LA-logic[8] DALLA' defined below. Consider the LA-logic[8] DALLA' $= (L_{D'}, S_{D'}, \models_{D'})$) where $L_{D'}$ is the subset of L_D without the modal operators $\{\cap, \cup^*\}$ and $\mathcal{LO}_{DALLA'}$ is the set of all the linear orders over i. Consider the mapping of formulae T from L_D into $L_{D'}$ defined as follows: $T(P) = P$ for all $P \in \phi_0$, $T(\Box_a A_1) = \Box_a T(A_1)$ for all $a \in$ i, $T(A_1 \wedge A_2) = T(A_1) \wedge T(A_2)$, $T(\neg A_1) = \neg T(A_1)$, $T(\Box_{a_1 \cap a_2} A_1) = T(\Box_{a_1} A_1) \vee T(\Box_{a_2} A_1)$ and $T(\Box_{a_1 \cup^* a_2} A_1) = T(\Box_{a_1} A_1) \wedge T(\Box_{a_2} A_1)$.

Proposition 9. *For all L_D-formulae A, A is DALLA-valid iff $T(A)$ is DALLA-valid iff $T(A)$ is DALLA'-valid.*

The proof is immediate considering the replacement of equivalents in `dalla`, the completeness of `dalla` with respect to the DALLA-validity and the fact that for all $A \in L_{D'}$, A is DALLA-valid iff A is DALLA'-valid. DALLA has therefore the strong finite model property and the validity problem for DALLA is decidable.

4.2 Nakamura's Logic of Graded Modalities

The logic of graded modalities (LGM $= (L_{LGM}, S_{LGM}, \models_{LGM})$ for short) introduced in [Nak93] (see also [Nak92]) is based on the graded equivalence relations, i.e. the graded similarity in Zadeh's meaning [Zad71]. Although the decidability of LGM is proved in [Nak93] using the rectangle method developed in [HC68], we prove that LGM has the strong finite model property and we give an upper bound for the size of the models (therefore we prove the decidability). The set of modal constants of L_{LGM} is the set of real numbers $[0, \dots, 1]$ and L_{LGM} has no modal operator. For all L_{LGM}-models $\mathcal{M} = (W, (R_\lambda)_{\lambda \in [0, \dots, 1]}, V)$, $\mathcal{M} \in S_{LGM}$ iff there is $\mu : W \times W \to [0, \dots, 1]$ such that (1) for all $x \in W$, $\mu(x, x) = 1$, (2) for all $x, y \in W$, $\mu(x, y) = \mu(y, x)$, (3) for all $x, z \in W$, $\mu(x, z) \geq \bigvee\{min(\mu(x, y), \mu(y, z)) : y \in W\}$ ($\bigvee S$: least upper bound of S) and (4) for all $\lambda \in [0, \dots, 1]$, $R_\lambda = \{(x, y) \in W \times W : \mu(x, y) \geq \lambda\}$. The binary relations in a model for LGM are equivalence relations. Now consider the LA-logic LGM' $= (L_{LGM}, S_{LGM'}, \models_{LGM'})$ such that $\mathcal{LO}_{LGM'} = \{\geq\}$ where $\geq$ is the usual linear order on $[0, \dots, 1]$. So for all $\mathcal{M} = (W, (R_\lambda)_{\lambda \in [0, \dots, 1]}, V) \in S_{LGM'}$, $\lambda, \lambda' \in [0, \dots, 1]$, if $\lambda \geq \lambda'$ then $R_\lambda \subseteq R_{\lambda'}$. By Corollary 8, LGM' has the strong finite property and the validity problem for LGM' is decidable. In the sequel we show that for all $A \in L_{LGM}$, A is LGM-valid iff A is LGM'-valid although $S_{LGM} \neq S_{LGM'}$.

---

[8] In [DO96] a complete axiomatization of DALLA' is given: it corresponds to `dalla` where the axiom schemes U. and I. are deleted and the axiom schema $(\Box_{a_1} \Box_{a_2} A \wedge \Box_{a_2} \Box_{a_1} A) \Leftrightarrow (\Box_{a_1} A \wedge \Box_{a_2} A)$ for all $a_1, a_2 \in$ **eac** is added.

Proposition 10. Let $M = (W, (R_\lambda)_{\lambda \in [0,...,1]}, V) \in S_{LGM}$. Then $M \in S_{LGM'}$ and for all $A \in L_{LGM}$, $w \in W$, $M, w \models_{LGM} A$ iff $M, w \models_{LGM'} A$.

The proof is by an easy verification. Proposition 11 below states a converse result.

Proposition 11. Let $M = (W, (R_\lambda)_{\lambda \in [0,...,1]}, V) \in S_{LGM'}$, $w \in W$ and A be a L_{LGM}-formula. The set of real numbers occurring in A is written $\{\lambda_1, \ldots, \lambda_n\}$ (by increasing order). Let M' be $M_{|R_{\lambda_1}(w)} = (R_{\lambda_1}(w), (R'_\lambda)_{\lambda \in [0,...,1]}, V')$ and $M'' = (R_{\lambda_1}(w), (R''_\lambda)_{\lambda \in [0,...,1]}, V')$ be the model such that,

1. for all $\lambda \in [0, \ldots, \lambda_1]$ $R''_\lambda = R'_{\lambda_1}$,
2. for all $i \in \{1, \ldots, n-1\}$, $\lambda \in]\lambda_i, \ldots, \lambda_{i+1}]$ $R''_\lambda = R'_{\lambda_{i+1}}$ and
3. for all $\lambda \in]\lambda_n, \ldots, 1]$, $R''_\lambda = R'_1$.

Then $M'' \in S_{LGM}$ and for all $w' \in R_{\lambda_1}(w)$, $M, w' \models_{LGM'} A$ iff $M', w' \models_{LGM'} A$ iff $M'', w' \models_{LGM} A$.

Proof. For all $\lambda \in \{\lambda_1, \ldots, \lambda_n\}$, $R'_\lambda = R''_\lambda$ so for all $w' \in R_{\lambda_1}(w)$, $M', w' \models A$ iff $M'', w' \models A$. Moreover, considering the remark after the proof of Proposition 3, $M, w' \models A$ iff $M', w' \models A$ (remember $R_{\lambda_n} \subseteq \ldots \subseteq R_{\lambda_1}$). Now we prove that $M'' \in S_{LGM}$. It is obvious that $M'' \in S_{LGM'}$. Consider the function $\mu : R_{\lambda_1}(w) \times R_{\lambda_1}(w) \to [0, \ldots, 1]$ such that for all $(x, y) \in R_{\lambda_1}(w) \times R_{\lambda_1}(w)$, $\mu(x, y) = max\{\lambda : (x, y) \in R''_\lambda\}$. This definition is correct since $\{\lambda : (x, y) \in R''_\lambda\}$ is not empty ($\lambda_1 \in \{\lambda : (x, y) \in R''_\lambda\}$, justifying the construction of M') and the maximum of $\{\lambda : (x, y) \in R''_\lambda\}$ always exists (justifying the construction of M''). The possible values for $max\{\lambda : (x, y) \in R''_\lambda\}$ are in the set $\{\lambda_1, \ldots, \lambda_n, 1\}$. (1) Since R''_1 is reflexive then for all $x \in R_{\lambda_1}(w)$, $(x, x) \in R''_1$ and therefore $\mu(x, x) = 1$. (2) For all $x, y \in R_{\lambda_1}(w)$, $\mu(x, y) = max\{\lambda : (x, y) \in R''_\lambda\} = max\{\lambda : (y, x) \in R''_\lambda\}$ (by symmetry of the relations R''_λ) and therefore $\mu(x, y) = \mu(y, x)$. (3) For any $x, y, z \in R_{\lambda_1}(w)$, we write κ to denote $min(\mu(x, y), \mu(y, z))$. By definition of μ, $\kappa = min(max\{\lambda : (x, y) \in R''_\lambda\}, max\{\lambda : (y, z) \in R''_\lambda\})$. It follows that $max\{\lambda : (x, y) \in R''_\lambda\} \geq \kappa$. There is $\kappa' \geq \kappa$ such that $(x, y) \in R''_{\kappa'}$ and therefore $(x, y) \in R''_\kappa$ since $R''_{\kappa'} \subseteq R''_\kappa$. In a similar way it can be shown that $(y, z) \in R''_\kappa$. By transitivity, $(x, z) \in R''_\kappa$ and therefore by definition of μ, $\mu(x, z) \geq \kappa$. So for all $x, z \in R_{\lambda_1}(w)$, for all $y \in R_{\lambda_1}(w)$, $\mu(x, z) \geq min(\mu(x, y), \mu(y, z))$. It follows that for all $x, z \in R_{\lambda_1}(w)$, $\mu(x, z) \geq \bigvee\{min(\mu(x, y), \mu(y, z)) : y \in R_{\lambda_1}(w)\}$. (4) By construction of M'', $(x, y) \in R''_{\lambda'}$ iff $\lambda' \in \{\lambda : (x, y) \in R''_\lambda\}$ iff $\lambda' \leq max\{\lambda : (x, y) \in R''_\lambda\}$ iff $\lambda' \leq \mu(x, y)$. So for all $x, y \in R_{\lambda_1}(w)$, $\lambda \in [0, \ldots, 1]$, $R''_\lambda = \{(x, y) : \mu(x, y) \geq \lambda\}$. This terminates the proof.

Proposition 10 entails that $S_{LGM} \subseteq S_{LGM'}$ but $S_{LGM'} \not\subseteq S_{LGM}$. For instance in a model for LGM', R_0 may not be the universal relation although in each model for LGM R_0 is the universal relation. Proposition 10 and 11 entail that for all $A \in L_{LGM}$, A is LGM-valid iff A is LGM'-valid. Hence LGM has the strong finite model property and the validity problem for LGM is decidable[9]. As

---

[9] One of the anonymous referees has noticed that the decidability of some extensions of logics with graded modalities is proved in [MMN95].

mentioned in [Nak92], the axiomatization of LGM is an interesting open problem. As a side-effect of our work, we define a sound and complete axiomatization of LGM using standard techniques for modal logics. Consider the Hilbert-style system $1gm'$ containing the following axiom schemes and inference rules ($A, B \in L_{LGM}$, $\lambda_1, \lambda_2 \in [0, \ldots, 1]$):

P. All formulae having the form of a classical propositional tautology,
K'. $\Box_{\lambda_1}(A \Rightarrow B) \Rightarrow (\Box_{\lambda_1} A \Rightarrow \Box_{\lambda_1} B)$, **T'.** $\Box_{\lambda_1} A \Rightarrow A$, **5'.** $\Diamond_{\lambda_1} A \Rightarrow \Box_{\lambda_1} \Diamond_{\lambda_1} A$
<. $\Box_{\lambda_1} A \Rightarrow \Box_{\lambda_2} A$ when $\lambda_1 < \lambda_2$, **MP. + RN.**

By using the canonical model construction (see e.g [Mak66]), it is standard to prove that for all $A \in L_{LGM}$, A is a theorem of $1gm'$ iff A is LGM'-valid. As a consequence, $1gm'$ is a sound and complete system for the logic LGM.

The logics LGM' and DALLA' correspond to extreme situations for the set of LA-logics: $\mathcal{LO}_{LGM'}$ is a singleton whereas $\mathcal{LO}_{DALLA'}$ contains all the linear orders on i. Although the concepts of rough sets and fuzzy sets are different (see e.g. a discussion in [Paw85]), the technique developed in Section 3 can be applied to logics derived either from the notion of rough sets (DALLA for instance) or from the notion of fuzzy sets (LGM for instance). However our technique does not seem to be applicable to the modal fuzzy logics presented in [Yin88] (for instance) since the valuation of any formula in a model is a fuzzy set which is not the case with LGM.

5   Concluding Remarks

In the paper we have shown that the logics DALLA and LGM respectively defined in [Gar86, Nak93] have the strong finite model property and that the validity problem is decidable. Actually, we have shown more than that since this result has been generalized to any R-logic (see the end of Section 3). As a side-effect of our work, we have defined a simple complete axiomatization for LGM which was until now an open problem stated in [Nak92]. Although the filtration technique introduced in Section 3.1 seems to be limited to the set of R-logics, it provides an elegant construction strongly guided by the properties of relations satisfying the local agreement condition. This technique cannot be applied in a straightforward way to the family of logics $DALD_i^{i'}$ defined in [DO96]. This could be a direction for further investigations.

Acknowledgments: The author thanks Professor Ewa Orłowska for the stimulating discussions and suggestions about this work and the anonymous referees for their valuable remarks and suggestions.

References

[Cer94]   Claudio Cerrato. Decidability by filtrations for graded normal logics (graded modalities V). *Studia Logica*, 53(1):61–73, 1994.

302

[Dem96]   Stéphane Demri. The validity problem for the logic DALLA is decidable, 1996. To appear in the Bulletin of the Polish Academy of Sciences.

[DO96]   Stéphane Demri and Ewa Orłowska. Logical analysis of indiscernibility, 1996. In: Orłowska, Ewa (ed.), Reasoning with Incomplete Information. To appear.

[FdCO85] Luis Fariñas del Cerro and Ewa Orłowska. DAL - A logic for data analysis. *Theoretical Computer Science*, 36:251–264, 1985. Corrigendum in T.C.S., 47:345, 1986.

[Gab72]   Dov Gabbay. A general filtration method for modal logics. *Journal of Philosophical Logic*, 1:29–34, 1972.

[Gar86]   George Gargov. Two completeness theorems in the logic for data analysis. Technical Report 581, Institute of Computer Science, Polish Academy of Sciences, Warsaw, 1986.

[HC68]   G. Hughes and Max Cresswell. *An introduction to modal logic*. Methuen and Co., 1968.

[HM92]   Joseph Halpern and Yoram Moses. A guide to completeness and complexity for modal logics of knowledge and belief. *Artificial Intelligence*, 54:319–379, 1992.

[Kon87]   Beata Konikowska. A formal language for reasoning about indiscernibility. *Bulletin of the Polish Academy of Sciences*, 35:239–249, 1987.

[Lad77]   Richard Ladner. The computational complexity of provability in systems of modal propositional logic. *SIAM Journal of Computing*, 6(3):467–480, September 1977.

[Mak66]   David Makinson. On some completeness theorems in modal logic. *The Journal of Symbolic Logic*, 12:379–384, 1966.

[MMN95]   Maarten Marx, Szabolos Mikulás, and Istvan Németi. Taming Logic. *Journal of Logic, Language and Information*, 4:207–226, 1995.

[Nak92]   Akira Nakamura. On a logic based on fuzzy modalities. In *22nd International Symposium on Multiple-Valued Logic, Sendai*, pages 460–466. IEEE Computer Society Press, May 1992.

[Nak93]   Akira Nakamura. On a logic based on graded modalities. *IEICE Transactions*, E76-D(5):527–532, May 1993.

[Orło84] Ewa Orłowska. Logic of indiscernibility relations. In Andrzej Skowron, editor, *Computation Theory, 5th Symposium, Zaborów, Poland*, pages 177–186. LNCS 208, Springer-Verlag, 1984.

[Orło85] Ewa Orłowska. Logic approach to information systems. *Fundamenta Informaticae*, 8:361–379, 1985.

[Paw81]   Zdzislaw Pawlak. Information systems theoretical foundations. *Information Systems*, 6(3):205–218, 1981.

[Paw85]   Zdzislaw Pawlak. Rough sets and fuzzy sets. *Fuzzy sets and systems*, 17:99–102, 1985.

[Seg71]   Krister Segerberg. An essay in classical modal logic (three vols.). Technical Report Filosofiska Studier nr 13, Uppsala Universitet, 1971.

[Vak91a] Dimiter Vakarelov. A modal logic for similarity relations in Pawlak knowledge representation systems. *Fundamenta Informaticae*, 15:61–79, 1991.

[Vak91b] Dimiter Vakarelov. Modal logics for knowledge representation systems. *Theoretical Computer Science*, 90:433–456, 1991.

[Yin88]   Ming-Sheng Ying. On standard models of fuzzy modal logics. *Fuzzy sets and systems*, 26:357–363, 1988.

[Zad71]   Lofti Zadeh. Similarity relations and fuzzy orderings. *Information Systems*, 3:177–200, 1971.

On the Power of Nonconservative PRAM

Anders Dessmark and Andrzej Lingas

Department of Computer Science, Lund University, Box 117, S-221 00 Lund, Sweden;
email {Anders.Dessmark, Andrzej.Lingas}@dna.lth.se

Abstract. An alternative simple method of exploiting word parallelism in a nonconservative RAM and PRAM model is considered. In effect, improved bounds for parallel integer sorting in the nonconservative and conservative EREW PRAM models are obtained.

1 Introduction

In the analysis of algorithms for RAM's (especially for PRAM's), the machine word is often assumed to be of length logarithmic in the input size. In this way, the machine random address space is sufficiently large to store the input. Although this assumption is reasonable it still seems to be a bit arbitrary. With the exception of sorting [4, 6, 8], not too much is known about problem complexity in the so called nonconservative RAM model where machine word is of substantially larger size.

In the introduction to [4], Albers and Hagerup argue that "algorithms using a nonstandard word length should not hastily be discarded as unfeasible and beyond practical relevance". The truth of these words has been fully confirmed recently by Andersson *et al.* in [2]. First, they iterated the $O(n)$-time reduction of sorting n integers of value $\leq 2^m$ to sorting n integers of value $\leq 2^{m/2}$ due to Kirkpatrick and Reisch[8] $2\lceil \log \log n \rceil$ times. In this way, they reduced the input problem to that of sorting n integers of bit length $\leq m/\log^2 n$ in time $O(n \log \log n)$. To solve the latter problem, they simply applied the nonconservative linear-time sorting algorithm due to Albers and Hagerup [4] that uses words of length $O(m \log n \log \log n)$ [4]. In result, an $O(n \log \log n)$-time algorithm for sorting n integers $\leq 2^m$ using the conservative (for sorting) word length $O(\log n + m)$ has been obtained.

The known efficient nonconservative algorithms succeed to use the restricted parallelism hidden in unit-cost operations on longer words [4, 6, 8]. A straightforward approach is to pack consecutive groups of input items into consecutive long words and then solve the subproblems induced by the groups faster taking advantage of the restricted parallelism. For instance, in the nonconservative sorting algorithm due to Albers and Hagerup [4], $\Omega(k)$ integers are packed in a single word and then sorted using only $O(\log k)$ operations (k is here the multiplicative factor of nonconservativeness).

We propose an alternative general approach of using the restricted parallelism of nonconservative RAM. Simply, we divide if possible the input problem into $\Omega(k)$ analogous subproblems. Next, we place the i-th input item of the l-th subproblem into the l-th field of the i-th machine word. Further, we run if possible an oblivious algorithm on such a subproblem vector performing the number of operations proportional to that taken by a single subproblem with a standard word length. Now, it depends on the problem nature whether some substantial merging step is needed (e.g., sorting), or the original problem is already solved.

We provide a general framework for efficient word-parallel simulation of k computations of an oblivious PRAM or RAM (see Section 3, [10]), or of an oblivious Turing machine (see [12, 13]), by a PRAM or RAM, respectively.

Our general approach yields in particular an improved bound on the time and the nonconservativeness factor for linear-work parallel integer sorting in the EREW PRAM model. We show that n integers of size $\leq 2^m$ can be sorted in time

$O(\log n \log \log n)$ using an EREW PRAM with $n/\log n \log \log n$ processors and word length $O((\log n + m) \log n)$.

In [4], Albers and Hagerup also presented a method of reduction of parallel conservative integer sorting to parallel nonconservative integer sorting. Combining their method with our new bounds on nonconservative parallel integer sorting, we also obtain improved bounds on conservative parallel integer sorting in the EREW PRAM model. In particular, we show that n integers of size $n^{O(1)}$ can be sorted in time $O(\log^{3/2} n)$ on a EREW PRAM with $n/\log n$ processors. Previously, such a result was known only in the CREW PRAM model [4].

We also shortly discuss an application of our approach to scalar product computation.

2 Simulation of k-wise RAM Instructions

A PRAM is a synchronous parallel machine consisting of a finite number of processors and a finite number of memory cells accessible to all processors. A RAM is a PRAM with a single processor. Following [4], the PRAM instruction set is assumed to include addition, subtraction, comparison, shift and the bitwise operations of AND and OR. The shift operation $shift(x,i)$ is given by $\lfloor x2^i \rfloor$.

In an EREW PRAM, a memory cell can be accessed only by a single processor. In a CREW PRAM several processors can simultaneously read the same cell but concurrent writing is still disallowed. In a CRCW PRAM, both concurrent reading and concurrent writing are allowed.

Usually, one assumes that a PRAM memory cell stores a bit-word of a length logarithmic in the input size. In the case of sorting n integers in the range 1 through 2^m, the word length is assumed to be $O(\log n + m)$.

We shall also use the so called nonconservative model of RAM and PRAM with substantially larger word length [9, 4]. To exploit the larger word length, we shall assume like in [4] that a word consists of at least k fields of equal length. The leftmost bit of each field serves as a test bit and the remaining are used to store strings over binary alphabet, in particular integers. Usually, k strings are stored in the k rightmost fields of a word, the remaining fields serve as a temporary storage. A k-tuple of string sequences $(x_1^i, x_2^i, ...)$ $i = 1, ..., k$, is given in *orthogonal-word representation* if there are consecutive machine words such that for $i = 1, ..., k$, $j = 1, 2, ...$, the string x_j^i is in the i-th field of the j-th cell.

Throughout the paper, we shall measure the time complexity of RAM's and PRAM's according to the unit-cost criterion with bounded word length.

Assuming the above notation and conventions, we have the following useful lemma.

Lemma 1. *Let M be a nonconservative EREW PRAM whose word consists of at least k fields of length l. Let I be the set of the basic PRAM instructions different from the shift instruction.*
After an $O(\log k)$-time and $O(1)$-processor preprocessing, for any instruction in I and any k-tuple of pairs of arguments given in orthogonal-word representation, M can produce the corresponding k-tuple of results of the instruction in orthogonal-word representation in time $O(1)$ using $O(1)$ processors (in case of comparison, the control bit of the i-th field of the resulting word, $1 \leq i \leq k$, is set to one iff the contents of the corresponding field of the first argument word is greater than that of the second one).
Also, after an $O(\log k)$-time and $O(l)$-processor preprocessing, for any k-tuple of arguments in orthogonal-word representation and $0 \leq q < l$, M can produce the corresponding k-tuple of the arguments shifted by q, in orthogonal-word representation, in time $O(1)$ using $O(1)$ processors.

Proof. The lemma is obvious for the bitwise operations of AND and OR. For the remaining instructions in I we need to use a constant number of masks. In particular, we need the mask C_1 where control bits are set to 1 and the remaining bits to 0 and the complementary mask C_0. It is not difficult to see that these two masks can be produced in time $O(\log k)$ on an EREW PRAM with $O(1)$ processors (cf. p. 466 in [4]). Due to them and the separating control bits, we easily obtain the lemma for addition by applying single addition and AND operations of M.

For comparison, the lemma follows from the method due to Paul and Simon [11]. For two argument words A_1 and A_2, we can obtain the result on control bits simply by $(A_1 \vee C_1) - (A_2 \wedge C_0)$.

A function copying the control bits in A to the remaining positions of the corresponding fields in A has been considered in [4]. It is given by $A \wedge C_1 - shift(A \wedge C_1, -l + 1)$. By applying the k-wise comparison and the function copying the control bits, we can reduce the k-wise subtraction to to its restricted case where the first argument is never smaller than the second. The so restricted k-wise subtraction can be directly implemented by a single subtraction of M.

Finally, the k-wise shift by q can be implemented in a straightforward way by a single shift of M by q and the bitwise AND operation with a special mask. The mask has 0's on the q leftmost positions in each field and 1's on the remaining ones. Clearly, we can create such masks for $0 \leq q < l$ in time $O(\log k)$ using $O(l)$ processors of M.

Using the result of k-wise comparison for two k-tuples in orthogonal-word representation and the function copying the control bits, we can also easily produce the corresponding k-tuples of maxima and minima in orthogonal-word representation, i.e., to implement a k-wise comparator. Hence, assuming the preprocessing of Lemma 1, we obtain the following corollary.

Corollary 2. *The functions of copying control bits and of k-wise comparator can be implemented in constant time using a single processor of M.*

3 k-wise Simulation of an Oblivious PRAM

By naturally extending the definition of an oblivious RAM (see [10]), we say that a PRAM is *oblivious* if the sequence of addresses of memory cells accessed in its i-th step by its k-th processor (including the direct and indirect addresses) is uniquely determined by the input size, k and i. However, for our purposes, we need a stronger version of the above definition where even the type of the instruction in the i-th step is uniquely determined by the three parameters.

We shall say that a PRAM is *strongly oblivious* if the instruction (given by its type and memory addresses of the arguments and the result) performed in its i-th step by its k-th processor is uniquely determined by the input size, k, and i, and if the instruction is *shift(x,j)* then the argument j is also uniquely determined by the input size, k, and i.

In the following lemma and further, by an *on-line simulation* of a PRAM computation we mean simulations of the single instructions that comprise the computation in the partial order given by the computation. Importantly, printing the output in its original form is also included in the simulation.

The following lemma shows that if an oblivious PRAM doesn't use shift then it can be simulated by a strongly oblivious PRAM within the same asymptotic resource bounds.

Lemma 3. *An oblivious PRAM which doesn't use shift, runs in time $t(n)$, uses $p(n)$ processors and word length $l(n)$, can be on-line simulated by a strongly oblivious PRAM running in time $O(t(n))$, using $p(n)$ processors and word length $l(n)$.*

Proof. sketch. Let M be the oblivious *PRAM* to be simulated. The main idea is to use a PRAM with stored program (see [1]) encoding M. The PRAM assigns a processor to each processor of M. In order to perform the instructions of the corresponding processor of M, the assigned processor tests the code of the type of the current instruction for equality with the codes of all the possible instruction types u. Next, using the copy bit function, it produces a mask C_u consisting entirely of 1's or 0's, depending whether the test is positive or negative. Finally, it runs a loop for all instruction types u, trying to perform u on arguments masked with the help of C_u. In result, only the encoded instruction is performed. Since M is oblivious, the addresses of the arguments depend only on the input size, the number of the processor and the number of the parallel step. The shift used for the copy bit function has always the same second argument. We conclude that the simulating PRAM is strongly oblivious.

It remains to observe that the generation of the program encodings for the $p(n)$ simulated processors takes $O(1)$-time and $p(n)$ processors.

The concept of an oblivious PRAM enables us to exploit the word parallelism according to the following theorem.

Theorem 4. *Let M be a strongly oblivious PRAM, or an oblivious PRAM which doesn't use shift, running in time $t(n)$, using $p(n)$ processors and word length $l(n)$. Suppose that M produces output composed of $r(n)$ words. Let $p'(n)$ denote $O(p(n) + r(n) + l(n))$ in the strongly oblivious case and $O(p(n) + r(n))$ in the shift-free case, respectively. For k-tuple inputs of size n, the k computations of M on the k-tuple can be on-line simulated by a single strongly oblivious PRAM of the same type of read-write access running in time $O(t(n) + k)$, using $p'(n)$ processors and word length $O(kl(n))$.*

Proof. By Lemma 3 we may assume w.l.o.g. that M is strongly oblivious. Divide the first n memory cells of a PRAM with word length $O(kl(n))$ into consecutive fields of length $\theta(l(n))$. Pack the i-th item of the q-input into the q-th field of the i-th cell. It is clear that the packing step can be done easily in time $O(k)$ using n EREW processors. Since M is strongly oblivious it remains to simulate each instruction performed by M simultaneously on k-tuples hold in common memory cells. After the preprocessing specified in Lemma 1, it takes $O(1)$ time and $O(1)$ processors by this lemma. In this way, in parallel for $q = 1, ..., k$, the q-th computation is on-line simulated on the q-th track corresponding to the q-th fields of the memory cells within the resource bounds specified in the lemma. To print the k outputs, we first use $r(n)$ processors to extract the output contents of on the first track, then that on the second *etc.* which totally takes $O(k)$ time. It remains to observe that the simulating PRAM is strongly oblivious.

4 k-wise RAM Simulation of an Oblivious TM

A multitape deterministic Turing Machine (TM for short) is oblivious if the positions of its heads in each step are uniquely determined by the input word size. A TM running in time $T(n)$ can be easily simulated by a RAM running in time $O(T(n))$ and using memory cells of constant length with the exception of $O(1)$ memory cells keeping the head positions (see [1]). Hence, we obtain the following theorem similarly as Theorem 4.

Theorem 5. *Let M be an oblivious multitape TM running in time $t(n)$. For k-tuple input words of length n, the k computations of M on the k-tuple can be on-line simulated by a RAM using word length $O(k + \log t(n))$ and running in time $O(t(n))$.*

Proof. sketch. Suppose first that $k = 1$. Let m be the number of tapes of M. The transition function of M is specified by a finite set of $m + 1 + 2m + m + 1$-tuples whose first m positions give the contents of the cells under the m heads, the $m+1$-st position correspond to the current state, the next $2m$ positions specify the head movements, the further m positions give the new contents of the cells under the heads, and the last position gives the new state. For each such a tuple $a = (a_1, ..., a_{4m+2})$, the simulating RAM tests the contents of registers holding the contents of the tape cells under the heads and the current state for equality with $a_1, a_2, ..., a_{m+1}$, respectively. The results of the equality tests are conjuncted by the AND operation, and then copied to produce the vector C^a consisting entirely of either 1's or 0's. Next, C^a is composed with the remaining components of a by the AND operation, i.e., $a_{m+2} \wedge C^a_{m+2}$ through $a_{4m+2} \wedge C^a_{4m+2}$ are produced, in order to update the head positions and the current state in case all the tests were positive. For example, we can interpret $a_{m+2} = 1, a_{m+3} = 0$, and $a_{m+2} = 0, a_{m+3} = 1$, as the move of the first head to the left or right, respectively. Hence, the position of the first head should be increased by $-a_{m+2} \wedge C^a_{m+2} + a_{m+3} \wedge C^a_{m+3}$ etc. Since the transition function of M is finite, the PRAM needs $O(1)$-time to test and simulate all the tuples a specifying it. Also, the memory cells keeping the contents of the tape cells and the current state of M need to store $O(1)$ bits, and only the cells keeping the positions of the heads need to store a logarithmic in $T(n)$ number of bits.

Now, to obtain this theorem for an arbitrary k, we proceed analogously as in the proof of Theorem 4, observing that it is sufficient to keep track of the positions of M's heads only, say, in the first simulated computation since M is oblivious.

In [7], Hopcroft, Paul and Valiant used the four Russian method to simulate a multitape TM by a RAM with a logarithmic speed up. By inspecting their simulation in [7], we obtain the following 'oblivious' version of their result.

Lemma 6. *Let M be an oblivious multitape TM running in time $t(n) \geq n \log n$. If $\log t(n)$ is computable on a RAM in time $t(n)/\log t(n)$ then M can be simulated by a strongly oblivious RAM using word length $O(\log t(n))$ in time $O(t(n)/\log t(n))$.*

By combining the above lemma with the sequential variant of Theorem 4, we obtain the following theorem (trading time for word length in comparison with Theorem 5).

Theorem 7. *Let M be an oblivious multitape TM running in time $t(n) \geq n \log n$. For k-tuple input words of length n, the k computations of M on the k-tuple can be simulated by a strongly oblivious RAM using word length $O(k \log t(n))$ and running in time $O(t(n)/\log t(n))$.*

Pippinger and Fisher [12], and Schnorr [13] have shown that a multitape TM running in time $t(n)$ can be simulated by an oblivious two-tape TM running in time $O(t(n) \log t(n))$. This yields the following corollary from Theorems 7, 5.

Corollary 8. *Let M be a multitape TM running in time $t(n)$ and let u be a k-tuple of input words of length n. The k computations of M on u can be simulated* by a strongly oblivious RAM using word length $O(k \log t(n))$ and running in time $O(t(n))$. Also, these k computations of M can be on-line simulated by a strongly oblivious RAM using word length $O(\max\{k, \log n\})$ and running in time $O(t(n) \log t(n))$.

5 Applications

5.1 Nonconservative Sorting

The acyclic AKS sorting network of constant degree and $O(\log n)$ depth is built of $O(n \log n)$ comparator gates [3]. It gives rise to an $O(\log n)$-time, work optimal strongly oblivious EREW PRAM for sorting provided that the expander needed for AKS network is given. The expander can be constructed in logarithmic time by an EREW PRAM on $n/\log n$ processors [5]. Hence, we obtain the following theorem.

Theorem 9. *After $O(\log n)$-time, $n/\log n$- processor preprocessing on an EREW PRAM, n integers $\leq 2^m$ can be sorted by a strongly oblivious EREW PRAM running in time $O(\log n)$ and using n processors and word length $O(\log n + m)$.*

Combining the above theorem with Theorem 4, we obtain the following useful lemma.

Lemma 10. *After $O(\log n)$-time, $n/\log n$-processor preprocessing on an EREW PRAM, $O(\log n)$ groups of $n/\log n$ integers $\leq 2^m$ can be sorted by an EREW PRAM running in time $O(\log n)$ and using $n/\log n$ processors and word length $O((\log n + m) \log n)$.*

By Theorem 1(a) in [4], two sorted sequences of at most n integers $\leq 2^m$ can be merged in time $O(\log n)$ using an EREW PRAM with $O(n \log \log n / \log^2 n)$ processors and word length $O((\log n + m) \log n)$ (under the assumption that the input integers are already packed into consecutive fields of consecutive machine words). Hence, by Lemma 10 and Brent's principle [8] we obtain the following result (cf. Theorem 1 in [4] and Theorem 3 in [2]).

Theorem 11. *A sequence of n integers $\leq 2^m$ can be sorted by an EREW PRAM running in time $O(\log n \log \log n)$ and using $n/\log n \log \log n$ processors with machine words of length $O((\log n + m) \log n)$.*

Proof. To combine Lemma 10 with Theorem 1(a) in [4], it is sufficient to note that a sequence of at most n integers $\leq 2^m$ can be easily packed into consecutive $\Theta(\log n) \times \Theta(n/\log n)$ fields of consecutive machine words in time $O(\log n)$ using an EREW PRAM with $n/\log n$ processors.

5.2 Conservative Sorting

Albers and Hagerup derived their new upper bounds on integer sorting in the EREW PRAM model from their results on nonconservative integer sorting in this model (respectively Theorems 1, 2 in [4]). By using our results on nonconservative parallel integer sorting (i.e., Theorem 11) we can analogously derive stronger upper bounds on integer sorting in the EREW PRAM model.

Following [4], we reduce the problem of sorting integers $\leq 2^m$ to that of sorting integers $\leq 2^s$ where $s < m$ by using the method of radix sorting. By taking s bits at a time, we need $O(m/s)$ phases to fulfill the reduction. We shall choose s as the largest integer satisfying $s \leq \log n$ and $s^2 \leq \log n + m$. Analogously as in [4], we implement each phase in two stages.

First, we split the input integers into $r = \lceil n/2^s \rceil$ groups of $\leq 2^s$ integers and apply the algorithm of Theorem 11 to each of the groups independently (stabilizing it as described in Section 2 in [4]). The splitting can be easily done in time $O(\log n)$ on an EREW PRAM with $O(n/\log n)$ processors and word length $O(\log n + m)$. Let $t = \max\{\log n, s \log \log s\}$. By Brent's principle, the first stage takes time $O(t)$ and $O(n/t)$ processors in the EREW PRAM model with word length $O(\log n + m)$. In the second stage, we merge the sorted groups analogously as on page 469 in [4], using optimal prefix sums [8]. It takes time $O(\log n)$ on an EREW PRAM with $n/\log n$ processors [4]. We conclude that a single phase takes time $O(t)$ and $O(n/t)$ processors in the EREW PRAM model with word length $O(\log n + m)$.

Hence, by the definition of s, we obtain the following upper bounds on the conservative integer sorting in the EREW PRAM model improving the results from [4].

Theorem 12. *Let $t = \max\{\log n, \sqrt{\log n + m} \log \log(\log n + m)\}$, and let s be the largest integer satisfying $s \leq \log n$ and $s^2 \leq \log n + m$. A sequence of n integers $\leq 2^m$ can be sorted by an EREW PRAM running in time $O(tm/s)$, using n/t processors and word length $O(\log n + m)$.*
In particular, if $2^m = n^{O(1)}$ the time and processor bounds are respectively $O(\log^{3/2} n)$ and $n/\log n$.

5.3 Scalar Product

For the sake of simplicity, as an example of application of the k-wise RAM simulation of an oblivious TM we have chosen the problem of computing the scalar product of two integer vectors.

The Schönhage-Strassen algorithm for the product of two n-bit integers [14] can be implemented by an oblivious TM running in time $O(n \log n \log \log n)$. Hence, we obtain the following corollary from Theorems 7, 5.

Corollary 13. *The scalar product of two vectors, each composed of k n-bit integers, can be computed in time $O(n \log \log n)$ on a RAM with word length $O(k \log n)$ or in time $O(n \log n \log \log n)$ on a RAM with word length $O(\max\{k, \log n\})$.*

References

1. A. Aho, J.E. Hopcroft and J.D. Ullman. The Design and Analysis of Computer Algorithms. Addison-Wesley, Reading, MA, 1974.
2. A. Andersson, T. Hagerup, S. Nilsson and R. Raman. Sorting in Linear Time ? Proc. 27-th Annual ACM Symposium on Theory of Computing, 1995, pp. 427-436.
3. M. Ajtai, J. Komlos and E. Szemeredi. An $O(n \log n)$ sorting network. Proc. 15-th Annual ACM Symposium on Theory of Computing, 1983, pp. 1-9.
4. S. Albers and T. Hagerup. Improved Parallel Integer Sorting without Concurrent Writing. Proceedings of Symposium on Discrete Algorithms, 1992, pp. 463-472.
5. R. Cole and U. Vishkin. Approximate Parallel Scheduling. Part 1: The Basic Technique with Applications to Optimal Parallel List Ranking in Logarithmic Time. SIAM J. Comput. 17(1), 1988, pp. 128-142.
6. T. Hagerup and H. Shen. Improved nonconservative sequential and parallel integer sorting. Information Processing Letters 36, 1990, pp. 57-63.
7. J. Hopcroft, W. Paul and L. Valiant. On time versus space and related problems. In Proc. 16-th Annual IEEE Symposium on the Foundations of Computer Science, 1975, pp. 57-64.
8. R. M. Karp and V. Ramachandran. Parallel algorithms for shared memory machines. In: J. van Leeuwen, ed., Handbook of Theoretical Computer Science, Vol. A (Elsevier Science Publishers, 1990), pp. 869-941.
9. D.G. Kirkpatrick and S. Reisch. Upper bounds for sorting integers on random access machines. Theoretical Computer Science 28, 1984, pp. 263-276.
10. R. Ostrovsky. Efficient Computation on Oblivious RAMs. In Proc. 22-nd Annual ACM Symposium on Theory of Computing, 1990, pp. 514-523.
11. W.J. Paul and J. Simon. Decision trees and random access machines. Proc. International Symposium on Logic and Algorithmic, Zürich, 1980, pp. 331-340.
12. N. Pippenger and M. Fischer. Relations among complexity measures. JACM 26, pp. 361-381.
13. C. Schnorr. The network complexity and the Turing machine complexity of finite functions. Acta Inform. 7, 1976, pp. 95-107.
14. A. Schönhage and V. Strassen. Schnelle Multiplikation grosser Zahlen. Computing 7, 1971, pp. 281-322.

Self-Similarity Viewed as a Local Property via Tile Sets

Bruno Durand

LIP, ENS-Lyon CNRS, 46 Allée d'Italie, 69364 Lyon Cedex 07, France.

Abstract. A self-similar image is defined by its global invariance through a finite number of contractive transformations. We show that, in the plane, a self-similar figure can also be constructed according to *local* rules: given a finite number of similarities (also called an Iterated Function System, IFS), we construct a tile set with the following properties:
- any nth iterate of the IFS can be represented by a finite tiling of the plane,
- any finite non-trivial tiling of the plane is formed only by iterates of the IFS.

Furthermore, we construct another tile set such that all "correctly initialized" tilings of the whole plane represent the IFS attractor.

1 Introduction

Matching rule in tilings are local constraints. Thus, tile sets have been used to model atomic positions in materials defined by short-range interactions. A traditional approach is then to study the periodicity or quasi-periodicity properties of tilings that can be formed. This study has been revived by the study of quasicrystals (see [4] for an overview on the subject and pertinent references such as [6]).

In another hand, global properties of sets such as self-similarity, have been intensively studied using the theory of fractal sets. But if this theory models quite well global properties of objects such as crystals, self-similarity does not correspond directly to physical interactions.

We prove that, if a figure is defined by a finite number of similarities, then it can also be defined by local rules. To do that, we construct tile sets, the tilings of which represent the figure in certain senses. In the first of these tile sets, we are interested on "finite" tilings, for the second one, we observe tilings of the whole plane.

The paper is organized as follows. We first recall briefly usual definitions and the background for both the tiling theory and the IFS' theory. We then present the desired properties of our tile sets: we explain in what sense our tilings verify the self-similarity properties of an IFS. The next Section contains the construction of both tile sets and a proof of our basic result. The proof is given in full details in the easy case where similarities defining the IFS are only homotheties, then we indicate some improvements of our construction for other more complicated cases. The last section is devoted to open problems.

2 Preliminaries

2.1 Tilings

A tile is a square the sides of which are colored. Colors belong to a finite set C called the *color set*. A set of tiles $\mathcal{T}$ is a subset of C'^4. All tiles have the same (unit) size. A tiling of the plane is *valid* if and only if all pairs of adjacent sides have the same color. Notice that it is not allowed to turn tiles. The following well-known theorem is due to Berger [2] in 1966 and a simplified proof was given in 1971 by Robinson [8].

Theorem 1. *Given a tile set, it is undecidable whether this tile set can be used to tile the plane.*

We can also define *finite tilings*. We assume that the set of colors contains a special "blank" color and that the set of tiles contains a "blank" tile *i.e.* a tile whose sides are blank. A finite tiling is an *almost everywhere blank* tiling of the plane. If there is at least one non-blank tile, then the tiling is called *non-trivial*.

Another undecidability result can be proved simply by using a construction presented by Robinson in [8] which reduces the undecidability of the halting problem for Turing Machines into it:

Theorem 2. *Given a tile set with a blank tile, it is undecidable whether this tile set can be used to form a valid finite non-trivial tiling of the plane.*

It is often convenient to use other notions of tiles that differ slightly from above:

- One can use arrows on tiles; a tiling is considered as valid if and only if all pairs of adjacent sides have the same color, and for each arrow of the plane, its head points out on the tail of an arrow in the adjacent cell.
- One can replace squares by polygons of the plane and ask that two adjacent polygons neither overlap nor create holes.
- One can put a color not only on the sides of the squares put also on their corners; four corners in contact should have the same color.
- One could just assign a state (out of a finite set) to each considered cell and fix a neighborhood. The matching condition is replaced by a relation between states that should be verified in the neighborhood of each cell.

It is folklore that all these notions are equivalent: there exist transformations of tile set from one notion into another that preserve existence of valid tilings, periodicity or non-periodicity, etc. In order to clarify our proofs, we use these notions together in our constructions.

Another useful technical construction is the super-imposition: given two tile sets A and B, one can create a "product" tile set $A \times B$. Tiles of $A \times B$ are composed by one tile of B superimposed on a tile of A. A-tiles (resp. B-tiles) should match other A-tiles (resp. B-tiles) and there is no matching requirement between A-tiles and B-tiles.

2.2 Iterated Function Systems

This theory has been initiated by Hutchinson [3]; we use notations of Barnsley's book [1].

In the following, we shall consider $\mathcal{H}(\mathbb{R}^n)$ the set of all compact nonempty subsets of $\mathbb{R}^n$, endowed with the Hausdorff metric. The Hausdorff distance h between the compacts A and B is defined by

$$h(A, B) = \max \left(\max_{x \in A} d(x, B), \max_{y \in B} d(A, y) \right)$$

where d is the usual distance between a point and a compact set of $\mathbb{R}^n$.

An *Iterated Function Systems* (IFS) consists of a finite number k of strictly contractive transformations $w_i : \mathbb{R}^n \to \mathbb{R}^n$. Its attractor Ω is the unique fixed point of the strictly contracting mapping

$$W : \mathcal{H}(\mathbb{R}^n) \longrightarrow \mathcal{H}(\mathbb{R}^n)$$
$$A \longmapsto \bigcup_{1 \leq i \leq k} w_i(A).$$

This attractor is often qualified as a *self-similar* set since $\Omega = W(\Omega)$. This self-similarity is more "visual" if the transformations are affine transformations of $\mathbb{R}^n$. Thus people often use to consider IFS defined by similarities (translations composed with homotheties and rotations); the whole theory is unchanged and they can generate "visually self-similar" images in the case of image analysis, or a self-similarity without deformation in the field of quasicrystals theory.

Our results deal with planar figures ($n = 2$). In the following, we call nth iterate of an IFS, the result of its iteration n times on a unit square.

3 Our result

We explain below what properties should have the tile sets we construct. They will be more precisely formalized in Theorems 3 and 5.

3.1 Finite figures

Given as inputs a set of homotheties I that defines an IFS, we effectively construct in Section 4 a tile set τ_I with a blank tile having the following properties:

Property 1 *Every nth iterate of the IFS can be represented by the tile set.*

This property means that if you consider the nth iterate of the IFS, it can be scaled up to fit a valid finite tiling of the plane with τ_I. Thus if you look only to one of the color-components, you see a discretization of the nth iterate of the IFS (see Fig. 1).

Property 2 *Every finite non-trivial tiling of the plane by τ_I represents a disjointed union of iterates of I.*

This property is much more difficult to obtain: the only finite tilings that can be formed represent nth iterates of the IFS. You cannot obtain a figure such that one part is developed at rank n, and another part at rank $k \neq n$ (see Fig. 2).

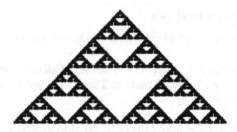

Fig. 1. Example: a discretization of a 5th iterate of the Sierpinski triangle

Fig. 2. Example: this discretization cannot be obtained

3.2 Infinite figures

In order to deal with infinite tilings of the plane, we construct in Section 4 a tile set τ'_I with the property that "correctly initialized" tilings of the plane "correspond" to the attractor of the IFS. "Correctly initialized" means that there exists somewhere in the tiling, the representation of a finite segment, side of one of the iterated squares. Thus the representation of I either covers the plane, or only covers a cone. But in this last case, by a compactness property, there exists a tiling of the plane by τ'_I such that the representation of I covers the whole plane. These tilings "correspond" to the attractor of the IFS because if one cut out of the tilings a finite square, then then this square contains an approximation of the attractor. The bigger is the cuted square, the better is the approximation.

4 The construction

In this section, we first construct an adequate tile set corresponding to an IFS (subsection 4.1), and then we prove a Theorem describing its properties (subsection 4.2). In the last subsection (4.3), we present some improvements of our result, by explaining how the construction should be modified to get them.

4.1 The basic construction

Let us consider first an IFS being a union of homotheties composed only by translations, both of rational parameters. We assume that the unit square contains the IFS's attractor (otherwise we scale down parameters) and that the images of the unit square by all the transformations are disjointed; thus these images are strictly contained in the unit square (see Fig. 5). The construction below is still correct if two image squares have at most one point in common.

Observe now that if in a rectangle, bisectors are equal to diagonals, the the rectangle is a square. This property can be represented by special tiles: consider the tile set of Fig. 3. It is not difficult to see that the only valid finite non-trivial tilings of the plane that can be done are squares (of even size). One bisector-diagonal would have been enough to ensure this result, but with these two diagonals, we can locate the center of the square.

[htb]

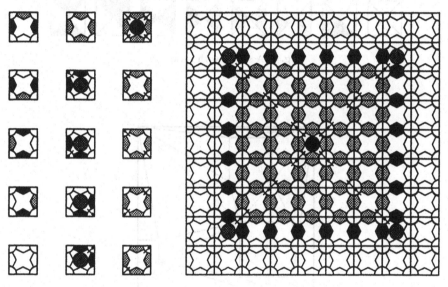

Fig. 3. A tile set to construct squares

By analogy to the cellular automata theory, one can generate "signals" of any rational ratio, and then produce a tile set that represent only a given signal (see [7] and Fig. 4). Let us consider now the first iteration of the IFS on a unit square. Then we can set up a triangulation on all these points (vertices of squares see Fig. 5). Note that all edges of the triangulation are chosen outside the images of the square. As parameters are rational, this figure can be scaled up so that all points have integer positions. All these edges can be generated with different signals. The number of edges is finite, hence we need only a finite number of tiles to set up the triangulation. In each vertex, we have the following information:

- all directions of edges that start at this vertex,
- which vertex will be found at the other end of this edge.

We do not have the information of the length of each edge, which will allow our tile set to draw the figure at different scales.

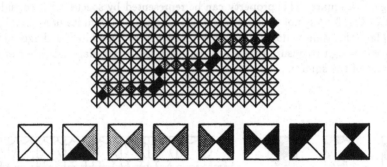

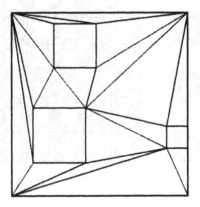

Fig. 4. A signal with ratio 2/5

Fig. 5. The construction at step 1

The triangulation of our figure is not sufficient for all cases; our aim is to reconstruct unambiguously the figure when an edge is given, assuming that directions of all edges starting from all vertices are known. Measuring lengths of edges is not permitted. Thus we have to cope with a few other particular cases. The two first ones are when to edges are superimposed. Note that there cannot be other cases of intersection since images of the square by the transformations are supposed to be strictly contained in it, and disjointed. The solution for these cases is depicted in Fig. 6a. and b. In case b. we could not have chosen the other diagonal because diagonals of all squares would have superimposed one on the other. In case a., one can see that diagonals enter the smaller self-similar figures. We can nevertheless construct our tile set since the number of crossed images is bounded. The last problem arises when two adjacent corners of the square are in case of Fig. 6b. We have to add one or several other points that can set up the sizes of the squares (the new point is M in Fig. 7). This addition is always possible since there is a space between the two considered squares (they have no common point by hypothesis).

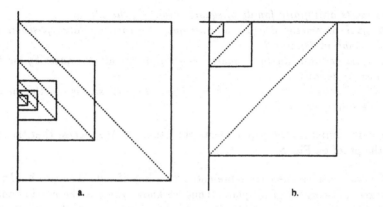

Fig. 6. The two particular cases

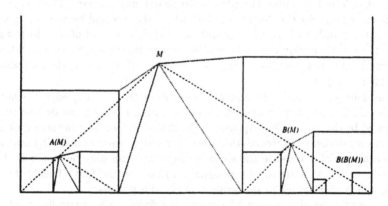

Fig. 7. The last problem

Note that all vertices of the figure are represented by different tiles in the tile set. These tiles code (in their colour scheme) the angles at wich the signals have to travel to other vertices. Now, we only need to add a set of "initialization" tiles. They correspond to the smallest size the squares can be drawn. As this size is known for each image square, we just have to count the number of cells, adding k new tiles for each edge of length k. We obtain the tile set τ'_I.

4.2 Our Theorem

Theorem 3. *Consider a valid tiling of the plane by τ'_I containing an edge. This tiling represents the attractor of I:*

- *there exists an infinite family of squares drawn by the tiling,*
- *each square is embedded inside a larger one, at a correct relative position* wrt. *a transformation defining I,*
- *if a square contains another square, then it contains all its images by the transformations defining I,*
- *if a square does not contain any other square, then its size is exactly one of the possible initialization sizes.*

This theorem formalizes the requirements of Section 3.2. We suggest that the reader follows the proof on Fig. 8.

Proof. Let us consider one edge represented by a line of cells. The two border cells are vertexes and corners of a square. Thus we know where is the outside and the inside. Let us start by the outside. Our edge belongs to a triangle and ,the tiling being correct, its two other edges are obtained from our vertices by signals. A their intersection, we must find a cell representing another vertex of the figure. We can now iterate this process and obtain *all* vertices of a figure representing one iteration such as depicted in 5. in particular, the edge is completed into a square. There is not any choice of tile except for the "father square" where there would be as many possible choices as the number of transformations defining I. The kind of tiles forming the edges of the father determines what will be its own position *wrt.* its own father. If one square is in contact with one edge or one vertex of its father, then we also use edges added in Fig. 6.

Let us look now inside the square drawn from the starting edge. Either the square is small, and then composed by initialization tiles. Let us denote its side length by γ. In this case, other squares at the same level are also initialization since there is no room inside to iterate the process of signal construction. Otherwise, its size is a multiple of γ because the signal construction can only be scaled up by an integer factor. We can start the inside signal construction and find a family of square children, which correspond to an iteration of the IFS process.

Thus the number of ancestors of a square is infinite, each square has a father, a correct relative position *wrt.* his father, and either it has children, or it has a small initializing size as well as his brothers.

Now let us add a few other tiles in order to transform our tile set τ'_I into a tile set τ_I which will have all properties required in Section 3.1.

We duplicate tiles representing edges and vertices of squares into another set, and we delete the outside color of each side to replace it by a "blank" color. we add a color that must propagate along edges and force that the outside part is blank. With this addition, there are two type of squares that can be obtained: normal squares (with a father and some children similarly to τ'_I), and special "outside" squares with the same family of normal children (exactly as in τ'_I) but without father and surrounded by blanks. We manage so that the blank color never appears inside a square.

With these tiles, we obtain a tile set τ''_I such that the following theorem holds. It does not yet corresponds to Section 3.1's requirements, because we are not sure

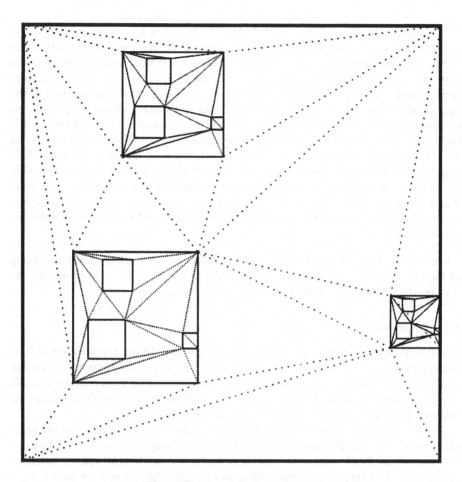

Fig. 8. The construction at step 2

that nth iterates are represented since several parts of the iterates may develop at different speeds. Thus our tile sets represent the attractor viewed at a certain precision which can be different from the nth iterate.

Theorem 4. *Any approximation of the attractor of I can be represented by a finite non-trivial tiling of the plane by τ_I''. Any finite non-trivial tiling of the plane by τ_I'' is formed by the juxtaposition of a finite number of figures representing approximations of the attractor of I.*

Proof. The proof is exactly done as in Theorem 3, excepted for the initialization process of the converse which does not start from an edge: we assume that there exists somewhere a non-blank tile in a finite tiling of the plane by τ_I. On its row, a blank tile appears somewhere in the left since the tiling is finite. Consider the tile of the row just on the right side of this blank tile. This tile must represent a left side of an outside square. By the same method, we go up and down and find a complete left edge of an outside square. The following of the proof is the same as in Theorem 3.

Our goal is now to obtain nth iterates of I and not an approximation of the attractor. Thus we have to impose that the levels of development of all parts of the tilings are the same. We change slightly "initialization tiles": we impose that the smallest embedded squares send two superimposed signals to define the position of their fathers, let's say a blue one and a red one. The blue will set the position of the squares as it was done above, the red ones count the number of embedded squares: a square that receive a red signal from its children will send it to its father one tile lower if it is a signal defining an upper corner (one tile upper if it is a signal defining an lower corner). Red signals must coincide on edges of intermediate squares. It imposes that embedded figures are developed to the same level. It is always possible to define these signals since we are sure that a father of a given square is at least 2 tiles bigger: it can be imposed for small squares and, as the growth of squares is exponential, it is still the case for bigger squares. We thus obtain τ_I and the following theorem holds:

Theorem 5. *Any nth iterate of I can be represented by a finite non-trivial tiling of the plane by τ_I. Any finite non-trivial tiling of the plane by τ_I is formed by the juxtaposition of a finite number of figures representing iterates of I.*

4.3 Improvements

Rotations We assume now that transformations of the IFS are similarities and thus include rotations. We only present below a solution where the rotation angles are multiples of $\pi/4$, and when vertices of image squares have rational coordinates. We claim that a tile set construction is possible for all angles that divide a multiple of 2π, even if vertices of image squares have irrational coordinates.

In our case, it is enough to color the inside square by a color representing their rotation. Then for each angle of rotation, we draw a triangulation as before. As there is at most 8 angles of rotation, we just scale up the 8 possible unit squares (oriented accordingly) and their triangulation diagrams in order that all vertices have integer coordinates.

We think that we can allow all rotation angles that divide a multiple of π. This is much more complicated since vertices of image square are not at rational positions. We do not prove this improvement here. We also think that it is possible to generalize the construction to the case where vertices of image squares have recursive irrational coordinates, if these irrational numbers can be computed fast enough by a cellular automaton. We do not know what is the exact necessary complexity bound on these real numbers.

Overlapping A the beginning of the section. we have assumed that images of the unit square by all the transformations of the IFS are disjointed. Let us assume now that it is not the case. The idea is to add a few edges in the triangulation *inside* the image squares, as shown in Fig. 9. In this figure, is drawn in dotted lines the triangulation that positions the two image squares. These dotted lines cross the second order image squares but not the third order. Thus we can product colors and force that all these edges are in the same "slide". Thus two or more diagrams can be superimposed without interfering. The product must be finite hence the necessary and sufficient condition for the correction of our construction is that it is possible to draw a triangulation without entering an infinite number of iterated squares at different relative positions. Alas, this condition is not, *a priori*, recursive. It is true even in the case of Fig. 10 where an infinite number of edges which do not belong to its descendants, enter the interior of a square. We do not know any recursive criterium equivalent to this condition.

We have just now to prove that our modified tiles set still verify Theorems 3 and 5. There is only one case which is different from the previous proofs: when two image squares cover two adjacent corners and are in contact, there is no room to add an intermediate point as we did in Fig. 7. The solution can be found in [5]. In this paper, Jarkko Kari constructs a tile set that has exactly the properties we impose, and that is based on the Hilbert path.

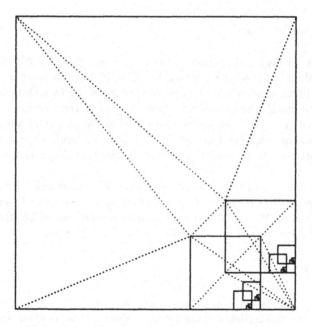

Fig. 9. The triangulation in an overlapping diagram

This condition does not always hold, For instance, when the attractor contains a continuous path that goes from an edge to another one, we cannot construct our triangulation.

Other frames of tiles In all previous sections, we have used square tiles. Our construction is still valid with triangular, or hexagonal tiles. It is only needed that the grid is itself self-similar in order to do the proper scalings. Remark that the IFS's attractor is always the same (it does not depend on the compact we start with).

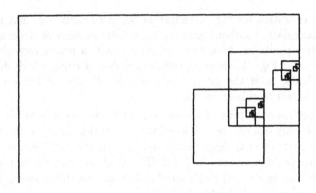

Fig. 10. A more difficult case

5 Open problems

The extension in higher dimensional spaces (in particular to $\mathbb{R}^3$) does not look difficult. We think that our construction is still valid, but the proof seems becoming much more difficult especially in the case of overlapping images. Our result is clearly not true in the line $\mathbb{R}$: the Cantor set cannot be locally generated.

We have not yet solved the case of rotations of any angle (which allow to generate figures such as the maple leaf or the black spleenwort fern), we just claim that it is possible, but the construction is much more complicated and not yet completely proved.

We think that it should be interesting to find all possible self-similar grids using only a finite number of polygons. It would allow us to construct tile sets with more complicated frames. We think that our simulation result would be still correct. An other open problem can be found in Section 4.3.

References

1. M. Barsnley. *Fractals everywhere.* Academic press, 1988.
2. R. Berger. The undecidability of the domino problem. *Memoirs of the American Mathematical Society,* 66, 1966.
3. J. Hutchinson. Fractal and self-similarity. *Indiana Univ. Journal of Mathematics,* 30(5):713–747, 1981.
4. K. Ingersent. *Matching rules for quasicrystalline tilings,* pages 185–212. World Scientific, 1991.
5. J. Kari. Reversibility and surjectivity problems of cellular automata. *Journal of Computer and System Sciences,* 48:149–182, 1994.
6. L. S. Levitov. *Commun. Math. Phys.,* 119(627), 1988.
7. J. Mazoyer. Cellular automata and tilings. In D. Beauquier, editor, *Actes des journées "Polyominos et Pavages",* Université Paris XII - Val de Marne., 1991.
8. R.M. Robinson. Undecidability and nonperiodicity for tilings of the plane. *Inventiones Mathematicae,* 12:177–209, 1971.

Simulation of Specification Statements in Hoare Logic

Kai Engelhardt and Willem-Paul de Roever*

Christian-Albrechts-Universität zu Kiel, Preußerstraße 1–9, 24105 Kiel, Germany

Abstract. Data refinement is a powerful technique to derive implementations in terms of low-level data structures like bytes from specification in terms of high-level data structures like queues. The higher level operations need not be coded as ordinary programs; it is more convenient to introduce specification statements to the programming language and use them instead of actual code. Specification statements represent the maximal program satisfying a given Hoare-triple. Sound and (relatively) complete simulation techniques allow for proving data refinement by local arguments. A major challenge for simulation consists of expressing the weakest lower level specification simulating a given higher level specification w.r.t. a given relation between these two levels of abstraction. We present solutions to this challenge for upward and downward simulation in both partial and total correctness frameworks, thus reducing the task of proving data refinement to proving validity of certain Hoare-triples.

In this paper we use specification statements $\phi \rightsquigarrow \psi$ that are defined such that program S refines $\phi \rightsquigarrow \psi$ iff Hoare-triple $\{\phi\} S \{\psi\}$ is valid.

We address a fundamental problem of data refinement. Given a specification statement $\phi \rightsquigarrow \psi$ on higher abstraction level A and an abstraction relation ρ between A and a lower level C, what is the least refined concrete level program S such that it simulates the abstract level specification statement w.r.t. ρ (see Fig. 1 in which $\subseteq$ denotes some form of weak commutativity)?

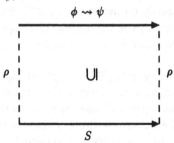

Fig. 1. The main question of this paper: How to express S?

Answers are developed for the two standard kinds of simulation: downward and upward simulation [10]. For the remaining two kinds answers can be found using similar techniques [5]. The remainder of this paper consists of three sections. In Sect. 1 we briefly sketch a framework for expressing and reasoning about data refinement. The

* Authors' email: {ke|wpr}@informatik.uni-kiel.de. Supported by ESPRIT BRA project REACT (no. 6021).

rather syntactic nature of this framework is forced by the need to integrate syntax-based data refinement methods à la VDM, Z, and Reynolds's, see [5]. Sect. 2 presents simulation theorems that are new for our Hoare-style framework. The calculation of data refinements of specifications is a well known technique in refinement calculi [1, 12, 7]. Finally we critically review what we achieved in the conclusion section.

Only a few selected proofs are provided in this paper; details are given in [5].

1 Technical Background

Validity of a partial correctness Hoare-triple $\{\phi\}\ S\ \{\psi\}$ informally means: If precondition ϕ holds, and if program S terminates, then postcondition ψ holds after termination of S [9].

Assertions (formally introduced in Sect. 1.3) are first order predicate logic expressions ranging over two different categories of variables: program variables whose values are subject to change during program execution, and *logical variables* whose appearance is restricted to Hoare-style assertions and whose values are not accessible to, and therefore not changed by, programs. The sole use of logical variables is to increase the expressiveness of assertions [13, 3]. Without such variables, a statement as simple as $x := x + 1$ cannot be characterized by a single Hoare-triple.

Operations at the most concrete level are pieces of program text in some sequential, imperative programming language like Guarded Commands [4]. For this purpose we introduce a toy programming language with recursion in Sect. 1.4.

This restriction to a programming language makes no sense on higher levels since there the specification of input/output behaviour is not necessarily tied to one particular implementation. Instead one has to decouple specifications from implementations and work with pure specifications [1, 12, 11]

To provide a common ground for reasoning about operations on all levels we add *specification statements* to our toy programming language. Remember, the meaning of Hoare-triple $\{\phi\}\ S\ \{\psi\}$ depends on the actual input/output behaviour of program S. However, ϕ and ψ together characterize the maximal input/output behaviour that S is allowed to expose. We use specification statement $\phi \rightsquigarrow \psi$ to denote this maximal behaviour.

Our formal justification of reducing the task of proving data refinement to proving validity of Hoare-style partial correctness formulae is carried out in Sect. 2 mainly using a calculus of binary relations on states. These relations are represented by syntactic objects called *relational terms* which are constructed from assertions, abstraction relations, and toy language programs (inc. specification statements).

Inclusion between input/output behaviour coincides with refinement. Hence, the obvious choice for correctness formulae on top of relational terms are inclusions between relational terms. Hoare-triples are just a different notation for particular inclusions and so are proof obligations for data refinement as given, e.g., in [10, 5].

1.1 Expressions and Predicates

The standard syntax and semantics of first order languages can be found in standard textbooks on mathematical logic. We use carrier set *Val* and the following syntactic categories: *variables Var $\ni$ x, arithmetic expressions Exp $\ni$ e*, and first order *predicates Pred $\ni$ p* (with equality between arithmetic expressions).

Instead of taking the satisfaction-based view we prefer to speak about *evaluating* predicates in interpretations obtaining a truth value in $\mathbb{B} = \{f\!f, tt\}$, where tt stands for true and $f\!f$ for false. State space Σ consists of all total functions $\sigma : Var \longrightarrow Val$, a fact that we write $\Sigma \stackrel{\text{def}}{=} [Var \longrightarrow Val]$ for short.

The *variant state* $(\sigma : x \mapsto v)$ denotes a state similar to σ except for its value in variable x which is v. This notation extends straightforwardly to lists of values and lists of variables.

We omit the definitions of semantics for arithmetic expressions $\mathcal{E}[\![.]\!] : Exp \longrightarrow (\Sigma \longrightarrow Val)$ and for first order predicates $\mathcal{I}[\![.]\!] : Pred \longrightarrow (\Sigma \longrightarrow \mathbb{B})$ since they are standard.

We write $t[^e/_x]$ for the term obtained from term t by substituting all free occurrences of variable x by expression e possibly requiring a renaming of bound variables in t such that free occurrences of variables in e are not bound by quantifiers in t. States $\sigma, \tau \in \Sigma$ *coincide* on $V \subseteq Var$ (denoted by $\sigma \equiv_V \tau$) iff $\sigma(x) = \tau(x)$, for all $x \in V$. We let $fvar(t)$ denote the free variables of term t. The lion's share of predicates in this paper will be used to characterize binary relations over states instead of sets of states. How shall we do that, given the above semantics?

In the presentation above all variables are similar in nature, for they all belong to Var and there is one uniform state space whose elements provide values for all variables in Var. Such states are rather inconvenient when dealing with several levels of abstraction like we do in data refinement. When discussing a particular data refinement step, say, from level A to level C we distinguish normal variables, i.e., program variables common to both levels and not subject to replacement during this step, from representation variables occurring on only one of the levels. The logical variables occurring in assertions are yet another kind of variables.

These separations between designated subsets of variables lead to semantics of predicates relative to *two* nonvoid subsets of Var. We develop this relational semantics of predicates via an intermediate step of predicates and semantics relative to *one* set of variables $V \subseteq Var$.

Definition 1. Let $V \subseteq Var$, and define state space Σ^V by $\Sigma^V \stackrel{\text{def}}{=} [V \longrightarrow Val]$, *projection* $\sigma|_V \in \Sigma^V$ of state $\sigma \in \Sigma$ onto variables V by $\sigma|_V \equiv_V \sigma$, syntactic category $Pred^V \stackrel{\text{def}}{=} \{ p \in Pred \mid fvar(p) \in V \}$ of *predicates over* V, and semantic function $\mathcal{I}[\![.]\!]^V : Pred^V \longrightarrow 2^{\Sigma^V}$ by $\mathcal{I}[\![p]\!]^V \stackrel{\text{def}}{=} \{ (\tau|_V) \mid \tau \in \mathcal{I}[\![p]\!] \}$.

Next we consider two sets $V, W \subseteq Var$ of variables instead of single set V. A notion of combining two states from different state spaces to form a bigger state ranging over the union of the two corresponding domains is helpful for defining the new semantics in terms of the one given in the previous definition.

Definition 2. Let $V, W \subseteq Var$. The *composition* $\sigma^V \dagger \sigma^W \in \Sigma^{V \cup W}$ of $\sigma^V \in \Sigma^V$

with $\sigma^W \in \Sigma^W$ is defined by: $(\sigma^V \dagger \sigma^W)(x) \stackrel{\text{def}}{=} \begin{cases} \sigma^W(x) \text{ if } x \in W \\ \sigma^V(x) \text{ if } x \in V \setminus W \end{cases}$

The syntactic category of *predicates over V and W* is given by $Pred^{V,W} \overset{\text{def}}{=}$ $Pred^{V \cup W}$. Its semantics $\mathcal{I}[\![.]\!]^{V,W} : Pred^{V,W} \longrightarrow 2^{\Sigma^V \times \Sigma^W}$ is derived straightforwardly from Def. 1.

$$\mathcal{I}[\![p]\!]^{V,W} \overset{\text{def}}{=} \left\{ (\sigma^V, \sigma^W) \mid (\sigma^V \dagger \sigma^W) \in \mathcal{I}[\![p]\!]^{V \cup W} \right\}$$

1.2 Abstraction Relations

To formalize reasoning about programs on different levels of abstraction we shall introduce separate instances of the above defined syntactic and semantic categories as well as semantic functions.

On abstraction level A objects are decorated with superscript A, e.g., Var^A, whenever confusion is likely to occur. By Σ^A we refer to Σ^{Var^A}.

Now consider two abstraction levels, say, A and C, for abstract and concrete, respectively. We assume w.l.o.g. that the normal variables are collected in some list $\mathbf{x} = (x_1, \ldots, x_n)$, whose elements are $Var^C \cap Var^A$. Usually we are somewhat sloppy about the distinction between lists of variables and their elements, as we shall see in the declaration of a for instance. Concatenation of lists a and c is denoted by juxtaposition ac. We assume that names of representation variables on different levels are disjoint; those on the abstract level are $\mathbf{a} = (a_1, \ldots, a_{n_A}) = Var^A \setminus Var^C$ and those on the concrete level are $\mathbf{c} = (c_1, \ldots, c_{n_C}) = Var^C \setminus Var^A$.

An abstraction relation α expresses a relation between two abstraction levels C and A, more precisely, between the values of representation variables on C and those on A. This motivates that the basic elements of the syntactic category Abs^{CA} of abstraction relations are first order predicates with free occurrences of representation variables from levels C and A only, i.e., $fvar(\alpha) \subseteq \mathbf{ac}$. The interpretation function $\mathcal{A}[\![.]\!]^{CA} :$ $Abs^{CA} \longrightarrow 2^{\Sigma^C \times \Sigma^A}$ is defined independently from the actual values of normal variables $x \in Var^C \cap Var^A$ as long as these values coincide on the two levels under consideration.

$$\mathcal{A}[\![\alpha]\!]^{CA} \overset{\text{def}}{=} \left\{ (\sigma, \tau) \mid (\sigma, \tau) \in \mathcal{I}[\![\alpha]\!]^{\mathbf{xc}, \mathbf{xa}} \wedge \sigma \equiv_{\mathbf{x}} \tau \right\} \tag{1}$$

1.3 Hoare-style Assertions

On each level of abstraction A Hoare-style assertions $\phi^A \in Pred^A$ are just first order logic predicates over two disjoint sets of variables: logical variables $Lvar^A$ and program variables Var^A, i.e., $Pred^A \overset{\text{def}}{=} Pred^{Lvar^A, Var^A}$. Similarly we refer to semantics function $\mathcal{I}[\![.]\!]^{Lvar^A, Var^A}$ by $\mathcal{I}[\![.]\!]^A$. Because of the different role logical variables play in our calculus we give their state space another name: $\Gamma^A \overset{\text{def}}{=} \Sigma^{Lvar^A}$.

1.4 Programs

Define the syntactic category $Prog^C \ni S$ of *programs*, where $x \in Var^C, \phi, \psi \in Pred^C$, recursion variable X, and $e, be \in Pred^{Var^C}$ by:

$$S ::= \mathbf{abort} \mid x := e \mid be \rightarrow \mid \phi \rightsquigarrow \psi \mid X \mid S_1 ; S_2 \mid S_1 \,[\!]\, S_2 \mid \mu X . S(X)$$

In the sequel we will consider only well-formed programs, i.e., ones that do not contain free occurrences of recursion variables like X, to avoid difficulties arising from free occurrences of recursion variables. This restriction allows us to define the semantics of programs without giving an explicit meaning to program X. (See [3] for a more formal set-up.)

$$\mathcal{P}[\![.]\!]^{C} : \mathit{Prog}^{C} \longrightarrow 2^{\Sigma^{C} \times \Sigma^{C}}$$

$$\mathbf{abort} \mapsto \emptyset$$

$$x := e \mapsto \{ (\sigma, \tau) \mid \tau = (\sigma : x \mapsto \mathcal{E}[\![e]\!]\sigma) \}$$

$$be \rightarrow \mapsto \left\{ (\sigma, \sigma) \mid \sigma \in \mathcal{I}[\![be]\!]^{\mathit{Var}^{C}} \right\}$$

$$\phi \rightsquigarrow \psi \mapsto \bigcap_{\gamma} \left\{ (\sigma, \tau) \mid (\gamma, \sigma) \in \mathcal{I}[\![\phi]\!]^{C} \Rightarrow (\gamma, \tau) \in \mathcal{I}[\![\psi]\!]^{C} \right\} \qquad (2)$$

$$S_1 ; S_2 \mapsto [\![S_1]\!] ; [\![S_2]\!]$$

$$S_1 \,[\!]\, S_2 \mapsto [\![S_1]\!] \cup [\![S_2]\!]$$

$$\mu X . S(X) \mapsto \bigcup_{i \in \mathbb{N}} [\![S^i(\mathbf{abort})]\!]$$

As usual, $S^0(\mathbf{abort}) \stackrel{\mathrm{def}}{=} \mathbf{abort}$ and $S^{i+1}(\mathbf{abort}) \stackrel{\mathrm{def}}{=} S(S^i(\mathbf{abort}))$ for $i \in \mathbb{N}$.

1.5 Relational Terms

Syntactic objects denoting binary relations occurred so far in the shape of predicates, abstraction relations, and programs. These objects constitute the basic building blocks of the syntactic category Rel of *relational terms*.

Operators on relational terms enable us to express maximal solutions for each of the relations on the LHS of $r_1 ; r_2 \subseteq r$ in terms of the remaining two relations (see Fig. 2).

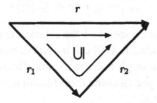

Fig. 2. Motivation for operators on relational terms.

For each pair (D, R) of sets of variables we introduce a syntactic category $\mathit{Rel}^{D,R} \ni r$ of relational terms with domain D and range R, where $r_1 \in \mathit{Rel}^{M,D}$, $r_2 \in \mathit{Rel}^{M,R}$, for some $M \subseteq \mathit{Var}$, and b is some basic building block: a predicate $\phi \in \mathit{Pred}^{A}$ if $(D, R) = (\mathit{Lvar}^{A}, \mathit{Var}^{A})$, an abstraction relation $\alpha \in \mathit{Abs}^{CA}$ if $(D, R) = (\mathit{Var}^{C}, \mathit{Var}^{A})$, or a program $S \in \mathit{Prog}^{A}$ if both D and R are Val^{A}.

$$r ::= b \mid \neg r \mid r_1 \rightsquigarrow r_2$$

The semantics $[.]^{D,R} : Rel^{D,R} \longrightarrow 2^{\Sigma^D \times \Sigma^R}$ of relational terms is defined next. $[\alpha]^{Var^C,Var^A} \stackrel{\text{def}}{=} A[\alpha]^{CA}$, $[\phi]^{Lvar^A,Var^A} \stackrel{\text{def}}{=} \mathcal{I}[\phi]^A$, $[S]^{Var^A,Var^A} \stackrel{\text{def}}{=} \mathcal{P}[S]^A$,

$$[\neg r]^{D,R} \stackrel{\text{def}}{=} \left\{ (\sigma,\tau) \in \Sigma^D \times \Sigma^R \mid (\sigma,\tau) \notin [r]^{D,R} \right\} \text{, and}$$

$$[r_1 \rightsquigarrow r_2]^{D,R} \stackrel{\text{def}}{=} \left\{ (\sigma,\tau) \mid \forall \zeta \left((\zeta,\sigma) \in [r_1]^{M,D} \Rightarrow (\zeta,\tau) \in [r_2]^{M,R} \right) \right\} \quad (3)$$

Note that $\mathcal{P}[\phi \rightsquigarrow \psi]^C = [\phi \rightsquigarrow \psi]^{Var^C,Var^C}$, i.e., (3) is consistent with (2). We introduce the following, more common relational terms as abbreviations in Rel: $r^{-1} \stackrel{\text{def}}{=} \neg(\neg r \rightsquigarrow \neg(\text{true} \rightarrow))$, $r_1 ; r_2 \stackrel{\text{def}}{=} \neg(r_1^{-1} \rightsquigarrow \neg r_2)$, $\langle r_3 \rangle r_2 \stackrel{\text{def}}{=} r_2 ; r_3^{-1}$, and $[r_3] r_2 \stackrel{\text{def}}{=} \neg \langle r_3 \rangle \neg r_2$. For $\alpha \in Abs^{CA}$ and $\phi \in Pred^A$ with $\mathbf{a} = Var^A \setminus Var^C$ we observe that the relational terms $\langle \alpha \rangle \phi$ and $[\alpha] \phi \in Rel^{Lvar^A,Var^C}$ are expressed by $\exists \mathbf{a} (\alpha \wedge \phi)$ and $\forall \mathbf{a} (\alpha \Rightarrow \phi) \in Pred^{Lvar^A,Var^C}$, respectively. Keeping this correspondence in mind we may use $\langle \alpha \rangle \phi$ and $[\alpha] \phi$ as assertions in $Pred^{Lvar^A,Var^C}$.

1.6 Correctness Formulae

We introduce the syntactic category $Cform \ni cf$ of *correctness formulae*, defined by: $cf ::= r \subseteq r'$, where $r, r' \in Rel^{D,R}$. Symbol $\subseteq$ is pronounced as "refines" and interpreted as relational inclusion. It is a partial order on relational terms.

For $G : A \longrightarrow B$ and $F : B \longrightarrow A$ monotone functions between partially ordered classes $(A, \leq)$ and $(B, \preceq)$ the pair (F, G) is called a *Galois connection* iff, for all $a \in A$ and $b \in B$: $F(b) \leq a \Leftrightarrow b \preceq G(a)$. We have the following two well-known Galois connections that are reformulations of the Schröder equivalences and go back to de Morgan's work in 1860.

$$r_2 \subseteq r_1 \rightsquigarrow r \Leftrightarrow r_1 ; r_2 \subseteq r \quad (4)$$

$$r_1 ; r_2 \subseteq r \Leftrightarrow r_1 \subseteq [r_2] r \quad (5)$$

Recall Fig. 2. In case r_2 is to be specified the solution is expressed using the leads-to operator $\rightsquigarrow$. Hoare, He, and Sanders [10] call it *weakest postspecification*. Similarly, but this time using the box operator [.], one can express Hoare, He, and Sander's *weakest prespecification*. Hoare-style partial correctness formulae are just abbreviations for some particular correctness formulae.

$$\{\phi\} S \{\psi\} \stackrel{\text{def}}{=} \phi ; S \subseteq \psi \quad (6)$$

The most general program S satisfying Hoare-triple $\{\phi\} S \{\psi\}$ is $\phi \rightsquigarrow \psi$ by (4).

1.7 Simulation

Next we introduce two concepts of simulation for single operations. When combined with adequate treatment of program initialization and finalization—a topic left out because of space limitations—simulation provides sound and complete methods for proving data refinement [10, 8, 5].

Definition 3. Let $\rho \in Abs^{AC}$, $\alpha \in Abs^{CA}$, $S^A \in Prog^A$, and $S^C \in Prog^C$:

- S^C *downward simulates* S^A w.r.t. ρ iff $\rho \, ; S^C \subseteq S^A \, ; \rho$.

- S^C *upward simulates* S^A w.r.t. α iff $S^C \, ; \alpha \subseteq \alpha \, ; S^A$.

2 Simulation Theorems

Throughout this section we assume that we reason about two levels of abstraction, abstract level A and concrete level C with representation variables a and c, respectively, and with normal variables x. Let assertions $\phi, \psi \in Pred^A$, abstraction relations $\alpha \in Abs^{CA}$ and $\rho \in Abs^{AC}$. The logical variables of a level may change in the course of action but we always start with x_0 as logical variables on level A. Such a change is harmless because the logical variables of two specifications do not interfere with each other opposite to what is the case for with program variables. In those cases in which the logical variables of the specification under consideration differ from x_0 they are $x_0' a_0'$, a list of fresh logical variables similar in length to the list of abstract level program variables xa.

2.1 Downward Simulation of Specifications

For downward simulation Fig. 1 more precisely looks like Fig. 3.

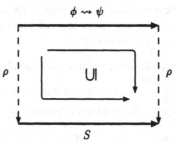

Fig. 3. What is the least refined S s.t. S downward simulates $\phi \leadsto \psi$ w.r.t. ρ?

The following theorem provides a specification statement that expresses least refined S that downward simulates $\phi \leadsto \psi$ w.r.t. ρ.

Theorem 4. $max_{pc} \stackrel{\text{def}}{=} \exists a \left(\rho \wedge xa = x_0' a_0' \right) \leadsto \exists a \left(\rho \wedge \forall x_0 \left(\phi[{}^{x_0' a_0'}/_{xa}] \Rightarrow \psi \right) \right)$
specifies the least refined program that downward simulates $\phi \leadsto \psi$ *w.r.t.* ρ.

Proof. In order to express the least refined S that downward simulates $\phi \leadsto \psi$ w.r.t. ρ we use Galois connection (4)

$$\rho \, ; S \subseteq \phi \leadsto \psi \, ; \rho \tag{7}$$

$$\Leftrightarrow \qquad S \subseteq \rho \leadsto (\phi \leadsto \psi \, ; \rho) \tag{8}$$

Up to here, nothing new has been achieved. This is all known at least since 1987 [10]. Unfortunately, the RHS $\rho \leadsto (\phi \leadsto \psi \, ; \rho)$ is not yet a concrete level program (e.g., a specification statement). Expanding the meaning of this correctness formula according to the definitions from Sect. 1 we arrive at a semantic characterization of the concrete level program: $(\sigma, \tau) \in \mathcal{P}[S]^C$ implies

$$\forall \zeta \left((\zeta, \sigma) \in \mathcal{A}[\rho]^{AC} \Rightarrow \exists \xi \left((\zeta, \xi) \in \mathcal{P}[\phi \rightsquigarrow \psi]^A \wedge (\xi, \tau) \in \mathcal{A}[\rho]^{AC} \right) \right) \quad (9)$$

How to express the relation characterized in (9) as a specification statement? To match the pattern of specification statements' semantics we have to replace the universally quantified abstract level state ζ by some new logical state γ' coinciding with ζ. Therefore we introduce fresh logical variables x_0' and a_0' for normal variables x and abstract level representation variables a, respectively. Now define a new logical state space $\Gamma' \stackrel{\text{def}}{=} [x_0'a_0' \longrightarrow Val]$. The main task is to substitute all occurrences of x and a evaluated in ζ by x_0' and a_0' evaluated in γ'. Let $\zeta \in \Sigma^A$ and choose $\gamma' \in \Gamma'$ such that $\zeta(xa) = \gamma'(x_0'a_0')$. The two sides of the universally quantified implication (9) are treated separately, starting with the LHS. By the substitution and coincidence lemmas $(\zeta, \sigma) \in \mathcal{A}[\rho]^{AC}$ holds iff $(\gamma', \sigma) \in \mathcal{I}[\rho[^{a_0'}/_a] \wedge x = x_0']^C$ does. The predicate logic term in semantic brackets can be simplified further to $\exists a \left(\rho \wedge xa = x_0'a_0' \right)$. We proceed with the RHS in (9), $(\zeta, \xi) \in \mathcal{P}[\phi \rightsquigarrow \psi]^A$, which is equivalent to $\forall \gamma \left((\gamma, \zeta) \in \mathcal{I}[\phi]^A \Rightarrow (\gamma, \xi) \in \mathcal{I}[\psi]^A \right)$. By the assumption made about ζ and γ' this is equivalent to $\forall \gamma \left((\tilde{\gamma}, \zeta) \in \mathcal{I}[\phi[^{x_0'a_0'}/_{xa}]]^{x_0'a_0'x_0, xa} \Rightarrow (\tilde{\gamma}, \xi) \in \mathcal{I}[\psi]^A \right)$, where $\tilde{\gamma} = \gamma \dagger \gamma'$, i.e., an element of a bigger logical state space. Observe that ζ is now rendered superfluous since there are no free occurrences of elements of x or a in $\phi[^{x_0'a_0'}/_{xa}]$ left. Hence ζ can be replaced by any other abstract level program state, e.g., ξ, i.e., $\forall \gamma \left((\tilde{\gamma}, \xi) \in \mathcal{I}[\phi[^{x_0'a_0'}/_{xa}] \Rightarrow \psi]^{x_0'a_0'x_0, xa} \right)$.

Next we replace the semantic universal quantification by a syntactic one: $(\gamma', \xi) \in \mathcal{I} \left[\left[\forall x_0 \left(\phi[^{x_0'a_0'}/_{xa}] \Rightarrow \psi \right) \right] \right]^{x_0'a_0', xa}$, exploiting that the γ part of $\tilde{\gamma}$ is not needed anymore. In the last step we express the existential quantification over ξ syntactically.

$$\exists \xi \left((\xi, \tau) \in \mathcal{A}[\rho]^{AC} \wedge (\gamma', \xi) \in \mathcal{I} \left[\left[\forall x_0 \left(\phi[^{x_0'a_0'}/_{xa}] \Rightarrow \psi \right) \right] \right]^{x_0'a_0', xa} \right)$$

$$\Leftrightarrow (\gamma', \tau) \in \mathcal{I} \left[\left[\exists a \left(\rho \wedge \forall x_0 \left(\phi[^{x_0'a_0'}/_{xa}] \Rightarrow \psi \right) \right) \right] \right]^C$$

$$\square$$

Using the downward simulation theorem together with the Galois connection (4) and the definition of Hoare-triples (6) the task of proving a downward simulation correct is reduced to proving validity of one Hoare-triple per operation.

Next we investigate under which conditions this rather complicated solution can be trimmed down to a simpler one. The otherwise almost unreadable solutions may be simplified substantially. Fortunately, the lion's share of practical cases belongs to these simplifiable ones.

Lemma 5. *If the inverse of the relation denoted by ρ is total and functional, or if ϕ is the freeze predicate $xa = x_0$, then $\exists a (\rho \wedge \phi) \rightsquigarrow \exists a (\rho \wedge \psi)$ specifies the least refined program that downward simulates $\phi \rightsquigarrow \psi$ w.r.t. ρ.*

By considering the identity relation as abstraction relation the downward simulation theorem suggests a *normal form* for specification statements in the partial correctness interpretation (just drop all ρ and $\exists a$ from max_{pc}).

2.2 Upward Simulation of Specifications

For upward simulation Fig. 1 more precisely looks like Fig. 4.

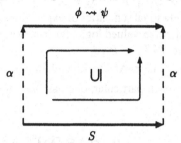

Fig. 4. What is the least refined S s.t. S upward simulates $\phi \rightsquigarrow \psi$ w.r.t. α?

Theorem 6. $[\alpha] \exists x_0 \left(\neg \psi[^{x_0' a_0'}/_{xa}] \wedge \phi \right) \rightsquigarrow [\alpha] (xa \neq x_0' a_0')$ *specifies the least refined program that upward simulates* $\phi \rightsquigarrow \psi$ *w.r.t.* α.

Proof. By reading Fig. 4 from right to left and α from A to C we discover an instance of the downward simulation problem.   □

The complexity of this last specification statement, especially the use of negations, usually strains one's capabilities of making up an intuition about the meaning of this solution. However, for total and functional α the simplification stated in Lemma 5 applies since then upward and downward simulation are equivalent.

2.3 Total Correctness

The program semantics employed so far neglects termination of programs. As a consequence every nonterminating program, denoted by the empty relation, is considered a correct implementation of every other program or specification; in other words, we have been concerned with development steps preserving partial correctness only. Clearly, this is an undesirable property of the theory developed so far.

We introduce a special state $\perp$, which is added to Σ, to express nontermination on the semantic level. That termination indeed refines nontermination while keeping $\subseteq$ as refinement relation is ensured by adding all pairs (σ, τ) to a relation r once $(\sigma, \perp) \in r$.

Simulation, our main technique, also works for proving refinement of total correctness between suitably adjusted data types. However, certain technical difficulties arise from the complexity added to the semantic model:

– The semantic domain for relations is not closed under relational inversion.

- Intersection instead of union models recursion, but intersection does not distribute over sequential composition.

- Upward simulation is unsound in case abstract operations expose infinite nondeterminism, which severely restricts the use of specification statements.

For each pair of sets (or lists) of variables (D, R) we define: $K^{D,R} \stackrel{\text{def}}{=} \{\perp\} \times \Sigma_{\perp}^{R}$ and $\mathcal{R}_{\perp}^{D,R} \stackrel{\text{def}}{=} \{ r \subseteq \Sigma_{\perp}^{D} \times \Sigma_{\perp}^{R} \mid K^{D,R} \subseteq r \supseteq r ; K^{R,R} \}$. This characterization is due to Paul Gardiner. The relational semantics of expressions and predicates is replaced by standard semantics using three-valued logic, for instance, the three-valued predicate semantics has the signature $T[\![.]\!]_{\perp} : Pred \longrightarrow (\Sigma_{\perp} \longrightarrow \mathbb{B}_{\perp})$. This construction is standard and therefore omitted here. Abstraction relation $\alpha \in Abs^{CA}$ is given a strictly bigger semantics than before. In particular, one has that $\mathcal{A}[\![\alpha]\!]_{\perp}^{CA} \in \mathcal{R}_{\perp}^{Var^{C}, Var^{A}}$ and $\mathcal{A}[\![\alpha]\!]^{CA} \subseteq \mathcal{A}[\![\alpha]\!]_{\perp}^{CA}$.

$$\mathcal{A}[\![\alpha]\!]_{\perp}^{CA} \stackrel{\text{def}}{=} \left\{ (\sigma, \tau) \ \middle| \ \begin{array}{l} (\sigma|_{\mathbf{c}}, \tau|_{\mathbf{a}}) \in \mathcal{I}[\![\alpha]\!]^{\mathbf{c,a}} \wedge \sigma \equiv_{\mathbf{x}} \tau \\ \vee \exists \zeta \in \Sigma^{\mathbf{a}} (T[\![\alpha]\!]_{\perp}^{\mathbf{c,a}}(\sigma|_{\mathbf{c}}, \zeta) = \perp_{\mathbb{B}}) \end{array} \right\} \tag{10}$$

By allowing also $\perp$ as σ and τ in (2) we define a total correctness semantics of specification statements $\mathcal{P}[\![\phi \rightsquigarrow \psi]\!]_{\perp}^{C} \in \mathcal{R}_{\perp}^{C}$. Relational terms and correctness formulae are defined relative to the semantics of expressions, predicates, abstraction relations, and programs. With the total, instead of partial, correctness interpretations of these lower level terms we can stick to the old definitions of relational terms and correctness formulae. Refinement still coincides with relational containment, and validity of Hoare-triple $\{\phi\}\, S\, \{\psi\}$ now means: If precondition ϕ holds, then program S terminates and postcondition ψ holds after termination of S.

A simulation theorem for termination preserving downward simulation is derived similar to the partial correctness version in Sect. 2.1 up to the point where we leave the syntactic ground and descend into the semantic reasoning. This turning point is (9). So we have to redo the remaining, bigger part of the problem, i.e., to express the RHS of (8) as a specification statement.

The clue to this part of the problem is the observation that the partial correctness solution is almost the total correctness solution, too. Modulo possible differences in the evaluation of the predicates ϕ, ψ, and ρ only state pairs $(\sigma, \perp)$ must be added to the total correctness solution exactly for those σ for which nontermination shall be possible.

Relation $r \in Rel^{D,R}$ *terminates* when started in state $\sigma \in \Sigma_{\perp}^{D}$ if the state pair $(\sigma, \perp)$ does not occur in the total correctness semantics of r. Then we say σ belongs to the *domain of termination* (denoted by $dt(r)$) of r.

$$dt(r) \stackrel{\text{def}}{=} \left\{ \sigma \in \Sigma_{\perp}^{D} \ \middle| \ (\sigma, \perp) \notin [\![r]\!]_{\perp}^{D,R} \right\} \tag{11}$$

We are interested in a syntactic characterization of the domain of termination of the RHS of (8), that is, we are looking for a predicate $\pi \in Pred^{Var^{C}}$ such that (12) holds:

$$dt(\rho \rightsquigarrow (\phi \rightsquigarrow \psi ; \rho)) = \mathcal{I}[\![\pi]\!]_{\perp}^{Var^{C}} \tag{12}$$

Corollary 7. *Predicate* $\exists a\,(\rho \wedge \exists x_0\,(\phi)) \in Pred^{Var^C}$ *is a solution to* π *in* (12).

Lemma 8. *A pair* $(\sigma, \tau) \in (\Sigma^C)^2$ *of proper states is in the total correctness semantics of the partial correctness solution to the downward simulation problem iff it belongs to the total correctness solution.* $(\sigma, \tau) \in \mathcal{P}[\![max_{pc}]\!]_\perp^C \Leftrightarrow \mathcal{A}[\![\rho]\!]_\perp^{AC} ; \{(\sigma, \tau)\} \subseteq \mathcal{P}[\![\phi \rightsquigarrow \psi]\!]_\perp^A ; \mathcal{A}[\![\rho]\!]_\perp^{AC}$.

What exactly is wrong about the total correctness semantics $\mathcal{P}[\![max_{pc}]\!]_\perp^C$ of the partial correctness solution? It may be sub-optimal since its domain of termination, syntactically characterized by $\exists a\,(\rho)$, is possibly bigger than the domain of termination of the total correctness solution, i.e., $\exists a\,(\rho \wedge \exists x_0\,(\phi)) \not\supseteq \exists a\,(\rho)$. The next lemma explains how to strengthen the precondition of max_{pc} such that those pairs $(\sigma, \perp)$ are no longer missing.

Lemma 9. *Let* $\phi, \psi \in Pred^A$ *and let* $\pi \in Pred^{Var^A}$.

$$\mathcal{P}[\![(\pi \wedge \phi) \rightsquigarrow \psi]\!]_\perp^A = \left\{ (\sigma, \tau) \in \Sigma_\perp^A \;\middle|\; \sigma \in \mathcal{I}[\![\pi]\!]_\perp^{Var^A} \Rightarrow (\sigma, \tau) \in \mathcal{P}[\![\phi \rightsquigarrow \psi]\!]_\perp^A \right\}$$

Finally the total correctness solution can be characterized syntactically by adding its domain of termination as a conjunct to the precondition of the partial correctness solution.

Theorem 10. *The following specification statement expresses the maximal solution of* (7) *in* S.

$$(\exists a\,(\rho \wedge \exists x_0\,(\phi))) \wedge \exists a\,(\rho \wedge xa = x_0'a_0') \rightsquigarrow \exists a\,\left(\rho \wedge \forall x_0\,\left(\phi[^{x_0'a_0'}/_{xa}] \Rightarrow \psi\right)\right)$$

Proof. By Corollary 7, Lemma 9, and Lemma 8. $\qquad\qquad\Box$

A similar theorem can be formulated for upward simulation using that the domain of termination of $[\alpha]\,(\alpha ; (\phi \rightsquigarrow \psi))$ is expressed by $\forall a\,(\alpha \Rightarrow \exists x_0\,(\phi))$.

3 Conclusion, Comparison, and Acknowledgment

The simulation theorems presented in the previous section appear rather complicated at first sight since the expressions obtained for the general simulation problems are too complex to be of much practical relevance. However, they are not over-complicated since one can generate examples demonstrating the need for this complexity (cf. [5]). Seemingly, this complexity can be avoided by switching to predicate transformers as semantical domain instead of binary relations on states [1, 12, 6]. One is tempted to ask: Why does all this look so simple in refinement calculi? The predicate transformer semantics of specification statements need not be conjunctive. This freedom allows for much simpler looking solutions to the problems addressed in this paper. If one insists on working with universally conjunctive (read: implementable without backtracking) specifications in the refinement calculus, then one obtains exactly the same, ugly expressions in general. This can be seen as follows. The least refined, (w.r.t. *wlp* semantics) conjunctive predicate transformer refining $\phi \leadsto \psi$ is exactly our normal form. Thus, any implementation of $\phi \leadsto \psi$ also refines our normal form. The same holds for the concrete level refinement of $\phi \leadsto \psi$: Our solution is the least refined conjunctive refinement of the refinement calculus solution.

An earlier but rather faulty attempt to solve the simulation problems appeared as [2]. Our first correct solution to the downward simulation problem for partial correctness was possible due to a suggestion by Ralph Back bridging a gap in our proof. However, the new proofs presented here do not depend anymore on his clue.

We thank Rudolf Berghammer for providing the historical information concerning Galois connections (4) and (5).

The present paper summarizes about 100 pages of our forthcoming book on comparing methods for data refinement and their theoretical foundations [5].

References

1. R. J. Back and J. von Wright. Refinement calculus, part I: Sequential nondeterministic programs. In *Stepwise Refinement of Distributed Systems*, volume 430 of *LNCS*, pages 43–66, 1990. Springer-Verlag.

2. J. Coenen, J. Zwiers, and W.-P. de Roever. Assertional data reification proofs: Survey and perspective. In J. Morris and R. Shaw, editors, *Proceedings of the 4th Refinement Workshop*, Workshops in Computing, pages 91–114. Springer-Verlag, 1991.

3. P. Cousot. *Handbook of Theoretical Computer Science: Formal Models and Semantics*, volume B, chapter 15, pages 841–993. Elsevier/MIT Press, 1990.

4. E. W. Dijkstra. Guarded commands, nondeterminacy and formal derivation of programs. *Communications of the ACM*, 18(8):453–457, 1975.

5. K. Engelhardt, W.-P. de Roever, et al. Data refinement. submitted to Prentice-Hall, an annotated table of contents is available at http://www.informatik.uni-kiel.de/~ke/proposal.ps.

6. P. Gardiner. Algebraic proofs of consistency and completeness. *FACS*, 1994. submitted.

7. P. H. Gardiner and C. Morgan. Data refinement of predicate tranformers. *Theoretical Computer Science*, (87):143–162, 1991.

8. P. H. Gardiner and C. Morgan. A single complete rule for data refinement. *Formal Aspects of Computing*, 5:367–382, 1993.

9. C. Hoare. Proofs of correctness of data representation. *Acta Informatica*, 1:271–281, 1972.

10. C. Hoare, He J., and J. Sanders. Prespecification in data refinement. *Information Processing Letters*, 25:71–76, 1987.

11. C. Morgan and T. Vickers. *On the Refinement Calculus*. Springer-Verlag, 1994.

12. J. M. Morris. Laws of data refinement. *Acta Informatica*, 26:287–308, 1989.

13. E.-R. Olderog. On the notion of expressiveness and the rule of adaptation. *Theoretical Computer Science*, 24:337–347, 1983.

Equational Properties of Iteration in Algebraically Complete Categories

Zoltán Ésik*
Dept. of Computer Science
A. József University
Árpád tér 2
6720 Szeged
Hungary
esik@inf.u-szeged.hu

Anna Labella**
Dept. of Computer Science
University of Rome "La Sapienza"
Via Salaria 113
00198 Rome
Italy
labella@dsi.uniroma1.it

Abstract. The main result is the following completeness theorem: If the fixed point operation over a category is defined by initiality, then the equations satisfied by the fixed point operation are exactly those of iteration theories. Thus, in such categories, the equational axioms of iteration theories provide a sound and complete axiomatization of the equational properties of the fixed point operation.

1 Introduction

Iteration theories provide an axiomatic treatment of the equational properties of the fixed point (or dagger or iteration) operation in cartesian categories, and in Lawvere theories in particular. The book [11] contains convincing evidence that all of the valid equations which hold for a "constructive" fixed point operation are captured by the axioms of iteration theories. But there are models in which the fixed point operation is not constructive. Let $\mathcal{C}$ be a category with a given collection $\mathcal{F}$ of functors $\mathcal{C}^{n+p} \to \mathcal{C}^n$, $n, p \geq 0$ containing the projections and closed under composition and target tupling. Suppose that for each $F : \mathcal{C}^{n+p} \to \mathcal{C}^n$ in $\mathcal{F}$ and each $\mathcal{C}^p$-object y, there is an initial F_y-algebra $(F^\dagger y, \mu_{F,y})$, where F_y denotes the endofunctor $F(-, y) : \mathcal{C}^n \to \mathcal{C}^n$. It is well-known, see e.g. [11], that the assignment $y \mapsto F^\dagger y$ is the object map of a unique functor $F^\dagger : \mathcal{C}^p \to \mathcal{C}^n$ such that $\mu_F = (\mu_{F,y})_{y \in \mathcal{C}^p}$ is a natural transformation (in fact isomorphism, see [24]) $F \cdot \langle F^\dagger, \mathbf{1}_p \rangle \to F^\dagger$. (Here, $\mathbf{1}_p$ stands for the identity functor $\mathcal{C}^p \to \mathcal{C}^p$.) Borrowing terminology from [20], we call the pair $(\mathcal{C}, \mathcal{F})$ an algebraically complete category if $F^\dagger$ is in $\mathcal{F}$ whenever F is. When $\mathcal{C}$ is an ω-category and the functor F preserves colimits of ω-chains, $F^\dagger$ can be constructed by the well-known initial algebra construction, cf. [1, 6, 2, 35, 34], or the book [5]. But unless

---

* Partially supported by grant no. T16344 of the National Foundation of Hungary for Scientific Research, and by the US–Hungarian Joint Fund under grant no. 351.
** Partially supported by the EEC–HCM project EXPRESS.

one has some additional assumptions on the category C and the functors $\mathcal{F}$, the dagger operation in algebraically complete categories is not constructive.

In this paper our concern is the logic of initiality. What are all of the equations that hold (up to isomorphism) for the dagger operation in algebraically complete categories? Some questions related to this topic have been studied in several papers, see e.g. [28, 32, 10, 20, 21, 12, 7, 13]. In the case that C is an ω-category [34, 35] and the functors preserve all ω-colimits, the dagger operation is constructive. Hence the valid equations are exactly those of iteration theories. See [10], or [16] for the case that C is an ω-cpo. In [12], it has been shown that the dagger operation in algebraically complete categories satisfies *Conway's classical identities* [14] for the regular sets, except for the equation $A^{**} = A^{*}$. Then in [18], it is shown that in the particular case that the category C is a poset, so that the initial algebras are least pre-fixed points and hence the Park induction principle holds, the valid equations satisfied by the dagger operation are again those of iteration theories. (Kozen's axiomatization [23] of the regular sets may be seen as an instance of the completeness of the Park induction principle.) In this paper, we generalize this result for the (non-constructive) fixed point operation in algebraically complete categories. This general result seems to indicate that the iteration theory identities also capture the equational properties of non-constructive fixed point operations. Our argument is based on recent advances on the axiomatization of iteration theories: A complete set of axioms consists of a small set of equations and an identity associated with each finite group, see [19].

In programming languages, one may define new data types by taking initial algebras of functors corresponding to data type constructors. (The choice of the right category is a non-trivial task, see [35, 34, 28, 3, 29], to mention only a few references.) Our main result shows that the calculus of iteration theories is a useful formal tool for establishing the equivalence of two specifications. See Section 7.

All proofs are omitted from this extended abstract.

2 Preliminaries

We refer to [4, 25] for basic notions of categories, and [22] for 2-categories. In any 2-category C, we will denote horizontal composition by $\cdot$ and vertical composition by $\star$. Thus, for any horizontal morphisms $f, f' : A \to B$, $g, g' : B \to C$, and for any vertical morphisms $u : f \to f'$ and $v : g \to g'$, both $f \cdot g$ and $f' \cdot g'$ are horizontal morphisms $A \to C$, and $u \cdot v$ is a vertical morphism $f \cdot g \to f' \cdot g'$. And if f, g, h are given horizontal morphisms $A \to B$ and $u : f \to g$ and $v : g \to h$, then $u \star v$ is a vertical morphism $f \to h$. As usual, we write also f for the identity vertical morphism corresponding to a horizontal morphism $f : A \to B$, and we use the notation 1_A for the horizontal identity $A \to A$ as well as for the vertical identity $1_A \to 1_A$.

A 2-*cell* is specified by two horizontal morphisms $f, g : A \to B$ and a vertical morphism $u : f \to g$. When $f = g$ and u is the vertical identity $f \to f$, we write just f for the corresponding 2-cell. Horizontal and vertical composition of 2-cells is defined as usual. For any 2-category $\mathcal{C}$, the 2-cells form a category $\mathbf{Cell}_{\mathcal{C}}$, which is in fact a 2-category with a suitable notion of a morphism between 2-cells.

A 2-theory is a 2-category T whose objects are the natural numbers $n \geq 0$ such that there are *distinguished horizontal morphisms* $i_n : 1 \to n$, $i \in [n] = \{1, \ldots, n\}$ with the following coproduct property: For any 2-cells $\alpha_i = (u_i : f_i \to g_i) : 1 \to p$, $i \in [n]$, there is a unique 2-cell

$$\alpha : n \to p$$

such that $i_n \cdot \alpha = \alpha_i$, for all $i \in [n]$. We denote this α as $\langle \alpha_1, \ldots, \alpha_n \rangle$. Thus, writing

$$\alpha = (\langle u_1, \ldots, u_n \rangle : \langle f_1, \ldots, f_n \rangle \to \langle g_1, \ldots, g_n \rangle),$$

we have

$$i_n \cdot \langle f_1, \ldots, f_n \rangle = f_i$$
$$i_n \cdot \langle g_1, \ldots, g_n \rangle = g_i$$
$$i_n \cdot \langle u_1, \ldots, u_n \rangle = u_i,$$

for all $i \in [n]$. The operation defined by the above coproduct conditions is called *tupling*. In the case that $n = 0$, it follows that there is a unique 2-cell $0 \to n$ determined by a (unique) horizontal morphism 0_n. Further, it follows that each identity 2-cell $\mathbf{1}_n$ is determined by the distinguished morphisms i_n:

$$\mathbf{1}_n = \langle 1_n, \ldots, n_n \rangle.$$

As an additional assumption we require that $1_1 = \mathbf{1}_1$, so that $\langle \alpha \rangle = \alpha$, for any 2-cell $\alpha : 1 \to p$. Note that the underlying category of a 2-theory T determined by the horizontal morphisms is a Lawvere theory, cf. [27].

Morphisms of 2-theories are 2-functors that preserve the distinguished morphisms i_n. It follows that each 2-theory morphism preserves the tupling operation.

Example 1. The 2-theory T_0 has horizontal morphisms $n \to p$ all functions $[n] \to [p]$. Each horizontal morphism has a vertical identity, and there is no other vertical morphism. The composite $f \cdot g$ of the horizontal morphisms f and g is their function composite. The 2-theory T_0 is initial in the category of 2-theories.

Example 2. Suppose that $\mathcal{C}$ is a category. The 2-theory $\mathbf{Th}(\mathcal{C})$ has horizontal morphisms $n \to p$ all functors $\mathcal{C}^p \to \mathcal{C}^n$. (Note the reversal of the arrow). For $f, g : n \to p$, a vertical morphism $f \to g$ is a natural transformation. The definition of horizontal and vertical composition is standard. For each $i \in [n]$, $n \geq 0$, the distinguished morphism i_n is the ith projection $\mathcal{C}^n \to \mathcal{C}$.

Example 3. Milner's synchronization trees [30] and their morphisms form a 2-theory. See [11].

Each 2-theory T contains a least sub 2-theory T_0' determined by the images of the 2-cells in T_0 under the unique morphism $T_0 \to T$. The horizontal morphisms $n \to p$ in T_0' are those of the form

$$\langle (1\rho)_p, \ldots, (n\rho)_p \rangle, \tag{1}$$

where ρ is a function $[n] \to [p]$. Each vertical morphism in T_0' is also of this form, since each vertical morphism is a vertical identity. The morphisms (1) are called *base* and are usually identified with the function ρ. (In non-trivial 2-theories, this identification is completely legal, since T_0' is isomorphic to T_0.) We call a base morphism surjective, injective, or bijective, if the corresponding function has the appropriate property. A bijective base morphism is sometimes called a *base permutation*.

It follows from the definition that each object $n + m$ of a 2-theory is the coproduct of the objects n and m. Indeed, let κ denote the base morphism corresponding to the inclusion $[n] \to [n + m]$, and let λ correspond to the translated inclusion $[m] \to [n + m]$. Then for any 2-cells $\alpha = (u : f \to f')$: $n \to p$ and $\beta = (v : g \to g') : m \to p$ there is a unique 2-cell $\langle \alpha, \beta \rangle = (\langle u, v \rangle :$ $\langle f, g \rangle \to \langle f', g' \rangle) : n + m \to p$ with

$$\kappa \cdot \langle \alpha, \beta \rangle = \alpha$$
$$\lambda \cdot \langle \alpha, \beta \rangle = \beta.$$

The operation defined by these conditions is called *pairing* and is an extension of the tupling operation. Another useful operation is that of *separated sum*. Let κ and λ be the base morphisms defined above, and let $\kappa' : p \to p + q$ and $\lambda' : q \to p + q$ be defined in the same way. Then the separated sum of the 2-cells $\alpha = (u : f \to f') : n \to p$ and $\beta = (v : g \to g') : m \to q$ is the unique 2-cell $\alpha \oplus \beta = (u \oplus v : f \oplus g \to f' \oplus g') : n + m \to p + q$ such that

$$\kappa \cdot (\alpha \oplus \beta) = \alpha \cdot \kappa'$$
$$\lambda \cdot (\alpha \oplus \beta) = \beta \cdot \lambda'.$$

These operations have several useful properties:

$$\langle \alpha, \langle \beta, \gamma \rangle \rangle = \langle \langle \alpha, \beta \rangle, \gamma \rangle$$
$$\alpha \oplus (\beta \oplus \gamma) = (\alpha \oplus \beta) \oplus \gamma$$
$$\langle \alpha, 0_p \rangle = \alpha = \langle 0_p, \alpha \rangle$$
$$\alpha \oplus 0_0 = \alpha = 0_0 \oplus \alpha$$
$$(\alpha \oplus \beta) \cdot \langle \gamma, \delta \rangle = \langle \alpha \cdot \gamma, \beta \cdot \delta \rangle$$
$$(\alpha \oplus \beta) \cdot (\gamma \oplus \delta) = \alpha \cdot \gamma \oplus \beta \cdot \delta,$$

whenever the 2-cells $\alpha, \beta, \gamma, \delta$ have appropriate source and target.

3 Algebras in 2-theories

In this section we define f-algebras in a 2-theory.

Definition 1. Suppose that $f : n \to n + p$ is a horizontal morphism in a 2-theory T. An **f-algebra** (g, u) consists of a horizontal morphism $g : n \to p$ and a vertical morphism $u : f \cdot \langle g, 1_p \rangle \to g$. Suppose that (g, u) and (h, v) are f-algebras. An f-algebra morphism $(g, u) \to (h, v)$ is a vertical morphism $w : g \to h$ such that

$$u \star w = (f \cdot \langle w, 1_p \rangle) \star v,$$

i.e., the following diagram commutes:

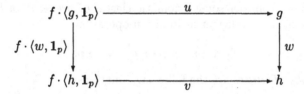

Definition 2. Suppose that $f : n \to n + p$ is a horizontal morphism in the 2-theory T. The f-algebra (g, u) is an **initial f-algebra**, if for each f-algebra (h, v) there exists a unique morphism $(g, u) \to (h, v)$.

It is well-known that if (g, u) is an initial f-algebra then u is a vertical isomorphism.

Definition 3. An **algebraically complete 2-theory** is a 2-theory T such that each horizontal morphism $f : n \to n+p$ has a specified initial f-algebra $(f^\dagger, \mu_f)$.

Example 4. Suppose that $(\mathcal{C}, \mathcal{F})$ is an algebraically complete category. Then the functors $\mathcal{F}$ determine a sub 2-theory of $\mathbf{Th}(\mathcal{C})$ that we denote by $\mathbf{Th}(\mathcal{C}, \mathcal{F})$. This 2-theory has all initial f-algebras. Indeed, if $f : n \to n + p$ in $\mathbf{Th}(\mathcal{C}, \mathcal{F})$, then $(f^\dagger, \mu_f)$ is an initial f-algebra, since for each $\mathcal{C}^p$-object y, $(f^\dagger y, \mu_{f,y})$ is an initial f_y-algebra in the usual sense. See the Introduction. It follows that $\mathbf{Th}(\mathcal{C}, \mathcal{F})$ is an algebraically complete 2-theory.

Example 5. Suppose that $\mathcal{C}$ is an ω-category, i.e., $\mathcal{C}$ has an initial object and colimits of all ω-chains. Then $\mathcal{C}^n$ is also an ω-category, for each $n \geq 0$. Further, if $\mathcal{F}$ denotes the collection of all functors $\mathcal{C}^{n+p} \to \mathcal{C}^n$ which preserve all colimits of ω-chains, $(\mathcal{C}, \mathcal{F})$ is an algebraically complete category. Hence $\mathbf{Th}_\omega(\mathcal{C}) = \mathbf{Th}(\mathcal{C}, \mathcal{F})$ is an algebraically complete 2-theory.

More generally, an ω-continuous 2-theory is a 2-theory T such that each vertical category $T(n, p)$ is an ω-category. Moreover, composition preserves initial objects in the first argument and colimits of ω-chains in either argument. Each ω-continuous 2-theory is algebraically complete.

Example 6. In [21], several examples of a category $\mathcal{C}$ are given such that each functor $\mathcal{C} \to \mathcal{C}$ has an initial algebra. These examples include the category of countable sets and the category of vector spaces of dimension at most countable. If $\mathcal{C}$ is such and $\mathcal{F}$ denotes the collection of *all* functors $\mathcal{C}^{n+p} \to \mathcal{C}^n$, then $(\mathcal{C}, \mathcal{F})$ is algebraically complete. (The proof of this fact uses the pairing identity, see below.) Thus $\mathbf{Th}(\mathcal{C})$ is also algebraically complete.

By Definition 3, each algebraically complete 2-theory comes with a dagger operation defined on the horizontal morphisms $n \to n + p$. In the next section, we will study some equational properties of the dagger operation.

4 Conway theories and iteration theories

In this section we consider some axiomatic classes of *preiteration theories*, i.e., theories T enriched by a dagger or iteration operation

$$f : n \to n + p \mapsto f^\dagger : n \to p.$$

The *Conway identities* are the following equations:

- LEFT ZERO IDENTITY

$$(0_n \oplus f)^\dagger = f,$$

all $f : n \to p$. The particular case that $n = 1$ is called the scalar left zero identity.
- RIGHT ZERO IDENTITY

$$(f \oplus 0_q)^\dagger = f^\dagger \oplus 0_q,$$

all $f : n \to n + p$. The particular case that $n = q = 1$ is called the scalar right zero identity.
- PAIRING IDENTITY

$$\langle f, g \rangle^\dagger = \langle f^\dagger \cdot \langle h^\dagger, 1_p \rangle, \ h^\dagger \rangle,$$

for all $f : n \to n + m + p$ and $g : m \to n + m + p$, where

$$h = g \cdot \langle f^\dagger, 1_{m+p} \rangle : m \to m + p.$$

The subcase that $m = 1$ is called the scalar pairing identity.
- PERMUTATION IDENTITY

$$(\pi \cdot f \cdot (\pi^{-1} \oplus 1_p))^\dagger = \pi \cdot f^\dagger,$$

for all $f : n \to n + p$ and for all base permutations $\pi : n \to n$. (Here, π^{-1} denotes the inverse of π.)

Definition 4. A **Conway theory** is a preiteration theory satisfying the Conway identities.

It is known that each Conway theory also satisfies the following identities:

- PARAMETER IDENTITY

$$(f \cdot (\mathbf{1}_n \oplus g))^\dagger = f^\dagger \cdot g,$$

all $f : n \to n + p$, $g : p \to q$. The particular case that $n = 1$ is called the scalar parameter identity.

- FIXED POINT IDENTITY

$$f^\dagger = f \cdot \langle f^\dagger, \mathbf{1}_p \rangle,$$

for all $f : n \to n + p$. When $n = 1$, this equation is the scalar fixed point identity.

- COMPOSITION IDENTITY

$$(f \cdot \langle g, 0_n \oplus \mathbf{1}_p \rangle)^\dagger = f \cdot \langle (g \cdot \langle f, 0_m \oplus \mathbf{1}_p \rangle)^\dagger, \mathbf{1}_p \rangle,$$

for all $f : n \to m + p$, $g : m \to n + p$. When $n = m = 1$, this identity is called the scalar composition identity.

- DOUBLE DAGGER IDENTITY

$$f^{\dagger\dagger} = (f \cdot (\langle \mathbf{1}_n, \mathbf{1}_n \rangle \oplus \mathbf{1}_p))^\dagger,$$

for all $f : n \to n + n + p$. When $n = 1$, this equation is the scalar double dagger identity.

Theorem 5. *Each of the following groups of identities is a complete axiomatization of the class of Conway theories:*

1. *The parameter, composition and double dagger identities.*
2. *The scalar versions of the parameter, composition, double dagger and pairing identities.*

Proof. See [11]. □

Suppose that G is a finite group on the set $[n]$, for some $n \geq 1$. For each $i, j \in G$, let us write ij for the product of i and j in the group G. In any theory T, we associate with G the base morphisms $\rho_i^G : n \to n$ defined by:

$$j_n \cdot \rho_i^G = (ij)_n$$

for all $i, j \in [n]$. We let τ_n denote the unique base morphism $n \to 1$.

Definition 6. The **group identity associated with the group G** is the equation

$$\mathbf{1}_n \cdot \langle f \cdot (\rho_1^G \oplus \mathbf{1}_p), \ldots, f \cdot (\rho_n^G \oplus \mathbf{1}_p) \rangle^\dagger = (f \cdot (\tau_n \oplus \mathbf{1}_p))^\dagger, \quad f : 1 \to n + p.$$

Definition 7. An **iteration theory** is a Conway theory satisfying the group identities.

The axioms of iteration theories give a sound and complete axiomatization of the equational properties of iteration in the following models:

1. Theories of ω-continuous functions on ω-cpo's with a bottom element, or more generally, continuous theories. In these theories, the dagger operation is the least (pre-)fixed point operation.
2. Theories of contraction functions on complete metric spaces, or Elgot's iterative theories. In these theories, the dagger operation is essentially defined by unique fixed points.
3. Theories $\mathbf{Th}_\omega(\mathcal{C})$ of ω-functors on ω-categories $\mathcal{C}$, or ω-continuous 2-theories. In these theories, the dagger operation is defined by initiality and the iteration theory equations hold up to isomorphism.
4. Matrix theories over completely or countably additive semirings. In these theories the dagger operation is defined by a star operation which involves infinite geometric sums.

For proofs of the above facts, see [11], where original references may be found. In most of the above theories, the dagger operation is constructive in some sense.

Remark 8. Iteration theories were introduced in [8, 9] and axiomatically in [16]. The axioms in [16] involve the Conway identities and a complicated equation scheme, the commutative identity. The completeness of the group identities in conjunction with the Conway identities is the main result of [19]. In fact, it is shown in [19] that the Conway identities and a subcollection of the group identities associated with the groups G_i, $i \in I$ is complete iff each (simple) finite group divides one of the groups G_i. See [26] for the definition of the divisibility relation. Thus, iteration theories do not have a "finite" axiomatization, see also [17].

5 The main result

Suppose that T is an algebraically complete 2-theory. When $f : n \to n + p$ and $g : p \to q$ in T, we define $f_g = f \cdot (1_n \oplus g) : n \to n + q$. Note that when (h, u) is an f-algebra, $(h \cdot g, u \cdot g)$ is an f_g-algebra.

We say that the *parameter identity holds* in T if for any horizontal morphisms f and g, the f_g-algebra $(f^\dagger \cdot g, \mu_f \cdot g)$ is initial, so that the equation

$$(f \cdot (1_n \oplus g))^\dagger = f^\dagger \cdot g$$

holds up to isomorphism.

Since the dagger operation in algebraically complete categories is defined pointwise, we have:

Proposition 9. *If $(\mathcal{C}, \mathcal{F})$ is an algebraically complete category, then $\mathbf{Th}(\mathcal{C}, \mathcal{F})$ satisfies the parameter identity.*

Theorem 10. *Suppose that T is an algebraically complete 2-theory satisfying the parameter identity. Then all of the Conway identities hold in T up to isomorphism.*

Proof. See [11]. The fact that the pairing identity holds in all theories $\mathbf{Th}(\mathcal{C}, \mathcal{F})$ where $(\mathcal{C}, \mathcal{F})$ is an algebraically complete category is an extension of a result of Bekič, see [32], and is proved in [28]. See also the product theorem in [21]. □
The main result of this paper is the following theorem:

Theorem 11. *Any algebraically complete 2-theory satisfying the parameter identity satisfies all of the iteration theory identities up to isomorphism.*

Corollary 12. *An equation involving the dagger operation holds up to isomorphism in all algebraically complete 2-theories satisfying the parameter identity iff it holds in all iteration theories.*

Proof. By Theorem 11, any such theory satisfies at least the iteration theory identities. But any ω-continuous 2-theory is algebraically complete, and an equation holds in all ω-continuous 2-theories or in all theories $\mathbf{Th}_\omega(\mathcal{C})$ on ω-categories, iff it holds in all iteration theories. See [10]. □

Corollary 13. *An equation involving the dagger operation holds up to isomorphism in all theories $\mathbf{Th}(\mathcal{C}, \mathcal{F})$ where $(\mathcal{C}, \mathcal{F})$ is an algebraically complete category iff it holds in all iteration theories.*

Since the equational theory of iteration theories is in the complexity class **P**, we have:

Corollary 14. *There is a polynomial time algorithm to decide whether an equation involving dagger holds in all algebraically complete categories.*

In order to prove Theorem 11, by Theorem 10 we only need to show that each algebraically complete 2-theory satisfies any group identity. The details may be found in the full version of the paper.

6 Iterating vertical morphisms

Suppose that $f : n \to n + p$ and $g : m \to m + p$ in a 2-theory T such that the initial algebras $(f^\dagger, \mu_f)$ and $(g^\dagger, \mu_g)$ exist. Suppose further that u is a vertical morphism $f \to g$, giving an interpretation of the "data type constructor" f in g. By the next lemma, it is possible to give a canonical interpretation of $f^\dagger$ in $g^\dagger$.

LEMMA 6.1 *If $(f^\dagger, \mu_f)$ is initial and $(\overline{g}, v)$ is a g-algebra, and if $u : f \to g$, then there is a unique vertical morphism $u^\dagger : f^\dagger \to \overline{g}$ such that the following square commutes:*

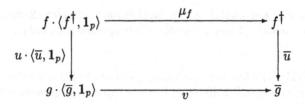

In particular, if $(\overline{g}, v)$ is $(g^{\dagger}, \mu_g)$, it follows that there is a unique $u^{\dagger} : f^{\dagger} \to g^{\dagger}$ such that

$$\mu_f \star u^{\dagger} = (u \cdot \langle u^{\dagger}, 1_p \rangle) \star \mu_g.$$

This defines the dagger operation $u \mapsto u^{\dagger}$ on vertical morphisms in any algebraically complete 2-theory. It follows that for each $n, p \geq 0$, $\dagger$ is a functor $T(n, n + p) \to T(n, p)$. Below we will show that this operation also satisfies the iteration theory identities. In fact, we reduce the proof of this fact to Theorem 11.

It is shown in [11] that when T is a 2-theory, then so is the 2-category $\mathbf{Cell}_T$, whose horizontal morphisms are the 2-cells of T.

THEOREM 6.2 *If T is an algebraically complete 2-theory, then $\mathbf{Cell}_T$ is also algebraically complete. Moreover, if the parameter identity holds in T then it holds in $\mathbf{Cell}_T$ as well.*

7 An example

We use equational reasoning to show that two data type specifications are equivalent. We will make use of the *scalar commutative identity* [11] which holds in all iteration theories and which may be seen as a generalization of the group identities:

$$(f \cdot (\tau_n \oplus 1_p))^{\dagger} = \langle f \cdot (\rho_1 \oplus 1_p), \ldots, f \cdot (\rho_n \oplus 1_p) \rangle^{\dagger},$$

for all $f : 1 \to n + p$ and for all base morphisms $\rho_i : n \to n$, $i \in [n]$.

Let $\mathbf{Set}$ denote the ω-category of sets, and let F and G denote the following functors, specified by their object maps:

$$F : \mathbf{Set} \to \mathbf{Set}$$
$$X \mapsto A + X \times X \times X$$

and

$$G : \mathbf{Set}^3 \to \mathbf{Set}$$
$$(X, Y, Z) \mapsto A + X \times Y \times Z,$$

where A is a given a set. Thus F defines the data type $F^{\dagger}$ of all ternary trees with leaves labeled in the set A.

Let H denote the functor $\mathbf{Set}^2 \to \mathbf{Set}$:

$$H = G \cdot \langle 1_2, 2_2, G^\dagger \cdot \langle 2_2, 1_2 \rangle \rangle.$$

Moreover, define

$$K = G \cdot \langle 1_1, H^\dagger, G^\dagger \cdot \langle 1_1, H^\dagger \rangle \rangle,$$

so that K is functor $\mathbf{Set} \to \mathbf{Set}$. In fact, all of the functors F, G, H, K are ω-functors. It is not clear what data type the functor K defines. But using the Conway identities, one may show that

$$K^\dagger = 1_3 \cdot \langle G, G \cdot \langle 2_3, 1_3, 3_3 \rangle, G \cdot \langle 3_3, 1_3, 2_3 \rangle \rangle^\dagger,$$

i.e., $K^\dagger$ is the first component of the initial solution of the system of equations:

$$X = G(X, Y, Z)$$
$$Y = G(Y, X, Z)$$
$$Z = G(Z, X, Y)$$

By the scalar commutative identity, $K^\dagger$ is isomorphic to $F^\dagger$, so that K also defines the data type of ternary trees with leaves labeled in A.

References

1. J. Adamek. Free algebras and automata realizations in the language of category theory. *Comm. Math. Unive. Carolinae*, 15(1974), 589–602.
2. J. Adamek and V. Koubek. Least fixed point of a functor. *J. of Computer and System Sciences*, 19(1979), 163–178.
3. P. America and J.J.M.M. Rutten. Solving reflexive domain equations in a category of complete metric spaces. *J. of Computer and System Sciences*, 35(1989), 343–375.
4. A. Asperti and G. Longo. *Categories, Types, and Structures*. Foundations of Computing Series. MIT Press, 1991.
5. J. Adamek and V. Trnkova. *Automata and Algebras in Categories*. Kluwer Academic Publishers, 1990.
6. M.A. Arbib. Free dynamics and algebraic semantics. in: *Mathematical Foundations of Computer Science '76*, LNCS 56, 1977, 212–227.
7. R. Backhouse, M. Birjsterveld, R. van Geldrop and J. van der Woude. Categorical fixed point calculus. in: proc. conf. *Category Theory and Computer Science '95*, LNCS 953, Springer-Verlag, 1995, 159–179.
8. S.L. Bloom, C.C. Elgot and J.B. Wright. Solutions of the iteration equation and extension of the scalar iteration operation. *SIAM J. Computing*, 9(1980), 26–45.
9. S.L. Bloom, C.C. Elgot and J.B. Wright. Vector iteration in pointed iterative theories. *SIAM J. Computing*, 9(1980), 525–540.
10. S.L. Bloom and Z. Ésik. Equational logic of circular data type specification. *Theoretical Computer Science*, 63(1989), 303–331.
11. S.L. Bloom and Z. Ésik. *Iteration Theories: The Equational Logic of Iterative Processes*. EATCS Monographs on Theoretical Computer Science. Springer–Verlag, 1993.

12. S.L. Bloom and Z. Ésik. Some equational laws of initiality in 2CCC's. *Int. J. on Foundations of Computer Science*, 6(1995), 95–118.

13. S.L. Bloom and Z. Ésik. Fixed-point operations on CCC's. Part 1. *Theoretical Computer Science*, 155(1996), 1–38.

14. J.C. Conway. *Regular Algebra and Finite Machines.* Chapman and Hall, 1971.

15. C.C. Elgot. Monadic computation and iterative algebraic theories. In J. C. Shepherdson, editor, *Logic Colloquium 1973, Studies in Logic*, volume 80. North Holland, Amsterdam, 1975.

16. Z. Ésik. Identities in iterative and rational algebraic theories. *Computational Linguistics and Computer Languages*, 14(1980), 183-207.

17. Z. Ésik. The independence of the equational axioms of iteration theories. *J. Computer and System Sciences*, 36(1988), 66–76.

18. Z. Ésik. Completeness of Park induction. *Theoretical Computer Science*, to appear.

19. Z. Ésik. Group axioms for iteration. To appear.

20. P. Freyd. Algebraically complete categories. in: proc. conf. *Category Theory, Como 1990*, LNM vol. 1488, Springer–Verlag, 1991, 95–104.

21. P. Freyd. Remarks on algebraically compact categories. in: *Applications of Categories in Computer Science*, London Math. Society Lecture Notes Series, vol. 77, Cambridge University Press, 1992, 95–106.

22. G.M. Kelly and R. Street. Review of the elements of 2-categories. in: *LNM* 420, Springer–Verlag, 1974, 76–103.

23. D. Kozen. A completeness theorem for Kleene algebras and the algebra of regular events. *Information and Computation*, 110(1994), 366–390.

24. J. Lambek. A fixpoint theorem for complete categories. *Mathematische Zeitschrift*, 103(1968), 151–161.

25. S. Mac Lane. *Categories for the Working Mathematician.* Graduate Texts in Mathematics. Springer–Verlag, 1971.

26. G. Lallement. *Semigroups and Combinatorial Applications.* Wiley-Interscience, 1979.

27. F.L. Lawvere. Functorial semantics of algebraic theories. *Proceedings of the National Academy of Sciences USA*, 50(1963), 869-873.

28. D. Lehmann and M.B. Smyth. Algebraic specification of data types: a synthetic approach. *Mathematical Systems Theory*, 14(1981), 97–139.

29. M.E. Majster-Cederbaum and F. Zetzsche. Towards a foundation for semantics in complete metric spaces. *Information and Computation*, 90(1991), 217-243.

30. R. Milner. *A Calculus for Communicating Systems.* LNCS 92, Springer–Verlag, 1980.

31. P.S. Mulry. Categorical fixed point semantics. *Theoretical Computer Science*, 70(1990), 85-97.

32. G.D. Plotkin. *Domains.* Lecture Notes. Department of Computer Science, University of Edinburgh, 1983.

33. A.K. Simpson. A characterisation of the least fixed point operator by dinaturality. *Theoretical Computer Science*, 118(1993), 301-314.

34. M.B. Smyth and G.D. Plotkin. The category theoretic solution of recursive domain equations. *SIAM J. Computing*, 11(1982), 761-783.

35. M. Wand. Fixed point constructions in order enriched categories. *Theoretical Computer Science*, 8(1979), 13-30.

36. J.B. Wright, J. Thatcher, J. Goguen, and E.G. Wagner. Rational algebraic theories and fixed-point solutions. In *Proc. 17th IEEE Symposium on Foundations of Computing*, 1976, 147-158.

On Unconditional Transfer

Henning Fernau*

Wilhelm-Schickard-Institut für Informatik, Universität Tübingen
Sand 13, D-72076 Tübingen, Germany
email: fernau@informatik.uni-tuebingen.de

Abstract. In this paper, we investigate the concept of unconditional transfer within various forms of regulated grammars like programmed grammars, matrix grammars, grammars with regular control, grammars controlled by bicoloured digraphs, and periodically time-variant grammars, especially regarding their descriptive capacity.

Moreover, we sketch the relations to restricted parallel mechanisms like k-, uniformly k-, partition-, and function-limited ET0L systems, as well as (unordered) scattered context grammars.

In this way, we solve a number of open problems from the literature.

1 Introduction

Regulated rewriting is one of the main and classic topics of formal language theory [7, 28], since there, basically context-free rewriting mechanisms are enriched by different kinds of regulations (we restrict ourselves to context-free core rules in this paper), hence generally enhancing the generative power of such devices compared to the context-free languages. In this way, it is possible to describe more natural phenomena using context-independent derivation rules, see [7].

One of the main problems investigated in the theory of formal languages is the question of the relation between language classes. We will concentrate on this question in the present paper.

More recently, investigation of limited forms of parallel rewriting became popular, see, e.g., the works of Wätjen [30, 32]. Such investigations add to the understanding of parallelism in rewriting and are, at first glance, a rather different form of context-free rewriting. In our opinion, one of the most beautiful facts of this theory is the close connection between these two variations of context-freeness, first observed by Dassow [6], cf. also [8] and [15, Theorem 3.1].

Surprisingly enough, although grammars with unconditional transfer were already introduced in the very first papers on regulated rewriting, lots of problems remained unsolved. In this paper, we (at least partially) solve open problems stated in [6, 8, 9, 17, 25, 30, 32].

For reasons of space, we cannot introduce every concept formally, nor it is possible to prove every assertion. We restrict ourselves to some key notions and techniques. A long version of this paper can be obtained from the author.

* Supported by Deutsche Forschungsgemeinschaft grant DFG La 618/3-1.

Conventions: $\subseteq$ denotes inclusion, $\subsetneq$ denotes strict inclusion, $|M|$ is the number of elements in the set M. $\emptyset$ denotes the empty set. $\mathbb{N}$ respectively $\mathbb{N}_0$ are the sets of natural numbers excluding respectively including zero. $\multimap\!\!\rightarrow$ indicates a partial function. The empty word is denoted by λ. The length of a word x is denoted by $|x|$. If $x \in V^*$, where V is some alphabet, and if $W \subseteq V$, then $|x|_W$ denotes the number of occurrences of letters from W in x. Sometimes, we use bracket notations like $\mathcal{L}(1\mathrm{lE}[\mathrm{P}]\mathrm{T0L}) = \mathcal{L}(\mathrm{P},\mathrm{CF}[-\lambda],\mathrm{ut})$ in order to say that the equation holds both in the case of excluding erasing productions (as indicated by the P and $-\lambda$ enclosed in brackets) and in the case of admitting erasing productions (neglecting the bracket contents). We consider two languages L_1, L_2 to be equal iff $L_1 \setminus \{\lambda\} = L_2 \setminus \{\lambda\}$, and we simply write $L_1 = L_2$ in this case. We term two devices describing languages equivalent if the two described languages are equal. We have a similar interpretation of the inclusion relation of languages.

We presuppose some knowledge of formal language theory on side of the reader. Especially, the Chomsky hierarchy $\mathcal{L}(\mathrm{CF}) \subsetneq \mathcal{L}(\mathrm{CS}) \subsetneq \mathcal{L}(\mathrm{REC}) \subsetneq \mathcal{L}(\mathrm{RE})$ should be known.

2 Regulated rewriting

A *grammar controlled by a bicoloured digraph* [7, 23, 25, 33] or *G-grammar* is an 8-tuple $G = (V_N, V_T, P, S, \Gamma, \Sigma, \Phi, h)$ where

- V_N, V_T, P, S define, as in a phrase structure grammar, the set of nonterminals, terminals, context-free core rules, and the start symbol, respectively;
- Γ is a bicoloured digraph, i.e., $\Gamma = (U, E)$, where U is a finite set of nodes and $E \subseteq U \times \{g, r\} \times U$ is a finite set of directed edges (arcs) coloured by g or r ("green" or "red");
- $\Sigma \subseteq U$ are the initial nodes;
- $\Phi \subseteq U$ are the final nodes;
- $h : U \to (2^P \setminus \{\emptyset\})$ relates nodes with rule sets.

There are two different definitions of the appearance checking mode in the literature: We say that $(x, u) \Rightarrow (y, v)$ ($(x, u) \Rightarrow_c (y, v)$, respectively) holds in G with $(x, u), (y, v) \in (V_N \cup V_T)^* \times U$, if either $x = x_1 \alpha x_2$, $y = x_1 \beta x_2$, $\alpha \to \beta \in h(u)$, $(u, g, v) \in E$ or every (one, respectively) rule of $h(u)$ is not applicable to x, $y = x$, $(u, r, v) \in E$. The reflexive transitive closure of $\Rightarrow$ ($\Rightarrow_c$, respectively) is denoted by $\stackrel{*}{\Rightarrow}$ ($\stackrel{*}{\Rightarrow}_c$, respectively). The corresponding languages generated by G are defined by

$$L(G_{[c]}) = \{x \in V_T^* \mid \exists u \in \Sigma \exists v \in \Phi(S, u) \stackrel{*}{\Rightarrow}_{[c]} (x, v)\}.$$

G is said to be with *unconditional transfer* [25] (there, the notion of "full checking" is used instead) iff $(u, g, v) \in E \iff (u, r, v) \in E$. If $E \cap U \times \{r\} \times U = \emptyset$, G is *without appearance check*. The corresponding language families are denoted (this time deviating from [7]) by $\mathcal{L}(G_{[c]}, \mathrm{CF},\mathrm{ac})$, $\mathcal{L}(G_{[c]}, \mathrm{CF},\mathrm{ut})$, and $\mathcal{L}(G_{[c]}, \mathrm{CF})$. Here and in the following, we write $\mathrm{CF}-\lambda$ instead of CF if we do not allow erasing core rules of the form $A \to \lambda$.

By definition, we have the following relations:

Lemma 2.1 Let $X \in \{\mathrm{CF}, \mathrm{CF} - \lambda\}$. Then, we have:

- $\mathcal{L}(\mathrm{G}_c, X) = \mathcal{L}(\mathrm{G}, X) \subseteq \mathcal{L}(\mathrm{G}_{[c]}, X, \mathrm{ac})$, and
- $\mathcal{L}(\mathrm{G}_{[c]}, X, \mathrm{ut}) \subseteq \mathcal{L}(\mathrm{G}_{[c]}, X, \mathrm{ac})$. □

We consider three special cases of G-grammars in the following.

- A *programmed grammar* [7, 24, 28, 29] has the following features:
  - o Every node contains exactly one rule. Therefore, both modes of appearance checks coincide in this case.
  - o There are no designated initial or final nodes. It is possible to start a derivation in each node containing a rule whose left-hand side equals to the start symbol S, and it is possible to stop anywhere when a terminal string has been derived.

  Usual notation of rules: $(r : A \to w, \sigma, \phi)$, where r – called label – is the name of the particular node, and σ – called success field – (or ϕ – called failure field –, respectively) is the set of nodes green arcs (or red arcs, respectively), starting from r, are pointing to.

  As language families, we obtain, e.g., $\mathcal{L}(\mathrm{P}, \mathrm{CF}, \mathrm{ac})$, $\mathcal{L}(\mathrm{P}, \mathrm{CF}, \mathrm{ut})$, and $\mathcal{L}(\mathrm{P}, \mathrm{CF})$.
- A *matrix grammar* [1, 7, 28] has the following features:
  - o Every node contains exactly one rule. Therefore, both modes of appearance checks coincide in this case.
  - o If there is a red arc from node u to node v, then there is also a green arc from node u to node v. Only the initial nodes (not necessarily containing rules with left-hand side S) are allowed to have more than one ingoing green arc, while only the final nodes are allowed to have more than one outgoing green arc. Moreover, between every final node and every initial node, there is a green arc.

  As language families, we obtain, e.g., $\mathcal{L}(\mathrm{M}, \mathrm{CF}, \mathrm{ac})$, $\mathcal{L}(\mathrm{M}, \mathrm{CF}, \mathrm{ut})$, and $\mathcal{L}(\mathrm{M}, \mathrm{CF})$.
- A *(periodically) time-variant grammar* [7, 26, 27, 28] has the following features:
  - o If there is a red arc from node n to node n', then there is also a green arc from node n to node n'. Every node has exactly one ingoing green arc and one outgoing green arc. In other words, the graph of green arcs has a simple ring structure.
  - o There is one designated initial node, and every node can be a final node.

  Usual notation: there is a periodic function $\phi : \mathbb{N} \to 2^P \setminus \{\emptyset\}$ prescribing the rule set $\phi(n)$ of the nth derivation step.

  Now, it makes sense to consider both the $\Rightarrow$ and the $\Rightarrow_c$ derivations, leading to the families of languages $\mathcal{L}(\mathrm{TV}_{[c]}, \mathrm{CF}, \mathrm{ac})$, $\mathcal{L}(\mathrm{TV}_{[c]}, \mathrm{CF}, \mathrm{ut})$, and $\mathcal{L}(\mathrm{TV}_{[c]}, \mathrm{CF})$. For reasons unknown to us, only the c-mode has been considered in the literature [7, 26, 27, 28].

Another way of viewing at such regulations of the derivation is the consideration of control lanuages. Without introducing them formally, we mention the class of regularly controlled grammars leading to the families of languages $\mathcal{L}(\mathrm{rC}, \mathrm{CF}, \mathrm{ac})$, $\mathcal{L}(\mathrm{rC}, \mathrm{CF}, \mathrm{ut})$, and $\mathcal{L}(\mathrm{rC}, \mathrm{CF})$ [7, 18, 28].

From the monograph [7], the following relations are known:

Theorem 2.2 Let $X \in \{G_c, P, M, rC, TV_c\}$, $Y \in \{CF, CF - \lambda\}$, $Z \in \{ac, \lambda\}$. Then $\mathcal{L}(X, Y, Z) = \mathcal{L}(G, Y, Z)$.

From [7, 10, 12, 20, 21], we know the following:

Theorem 2.3 • $\mathcal{L}(G, CF) \subsetneq \mathcal{L}(G, CF, ac) = \mathcal{L}(RE)$.
• $\mathcal{L}(G, CF - \lambda) \subsetneq \mathcal{L}(G, CF - \lambda, ac) \subsetneq \mathcal{L}(CS)$.

What about unconditional transfer? Although already Rosenkrantz treated this case in his famous starting paper [24], very little was known in the literature.

Theorem 2.4 Let L be some recursively enumerable language over the alphabet V_T. Then, there is a (P,CF,ut) grammar $\tilde{G}$ such that $w \in L$ iff $w\$\# \in L(\tilde{G})$, where $\{\$, \#\} \cap V_T = \emptyset$.

Proof. By [24, Theorem 5] and [7, Theorem 1.2.5], we know that any recursively enumerable language $L \subseteq V_T^*$ can be generated by a (P,CF,ac) grammar $G = (V_N, V_T, P, S)$ with $V_N = \{B_1, \ldots, B_n\}$, $P = \{p_1, \ldots, p_m\}$, and productions $(p_i : A_i \to w_i, \sigma_i, \phi_i)$. We introduce a (P,CF,ut) grammar $\tilde{G} = (\tilde{V}_N, \tilde{V}_T, \tilde{P}, \tilde{S})$ with $\tilde{V}_N = V_N \cup \{\tilde{S}, F, \sigma\}$ and $\tilde{V}_T = V_T \cup \{\$, \#\}$. $\tilde{P} = \{\text{init}\} \cup \{p_i^+, p_i^-, p_i' \mid 1 \le i \le m\} \cup \{t_1, \ldots, t_{n+1}\}$ contains (1) one initialization production (init : $\tilde{S} \to S\$\sigma, \{p_i^+, p_i^- \mid A_i = S\}$), (2) termination productions $(t_i : B_i \to F, \{t_{i+1}\})$ for $1 \le i \le n$ and $(t_{n+1} : \sigma \to \#, \{t_{n+1}\})$, and (3), for every p_i, $1 \le i \le m$, simulating rules

$$(p_i^- : A_i \to F, \{q^+; q^- \mid q \in \phi_i\}) \quad ,$$
$$(p_i^+ : A_i \to w_i\sigma, \{p_i'\}) \quad , \text{ and}$$
$$(p_i' : \sigma \to \lambda, \{q^+, q^- \mid q \in \sigma_i\} \cup \{t_1\}) .$$

When simulating a production p_i of G applied to a sentential form w, nondeterministically it has to be guessed whether A_i is contained in w or not. Then, the correctness of this guess has to be checked. In the negative case, $|w|_{A_i} = 0$ is checked by $A_i \to F$. Otherwise, $|w|_{A_i} > 0$ is tested in the following way: first, we try to apply $A_i \to w_i\sigma$; if anything has been correct so far, the sentential form $w' \ne w$ obtained in this way contains two occurrences of σ. The production p_i' erases the superfluous one. If our guess was incorrect, the further absence of σ testifies this error. If the rightmost σ has been erased by p_i', this error is signalled by the fact that the rightmost letter of w' (and of all words obtained from w') is $\$$. □

Corollary 2.5 $\mathcal{L}(P, CF, ut)$ contains non-recursive languages. □

Together with Theorem 2.3, we get:

Corollary 2.6 $\mathcal{L}(P, CF - \lambda, ut) \subsetneq \mathcal{L}(P, CF, ut)$ □

We say that a language class $\mathcal{L}$ is closed under *endmarking* iff $L \in \mathcal{L}$, $L \subseteq V^*$ and $\# \notin V$ implies $L\{\#\} \in \mathcal{L}$. Using standard constructions, one obtains the following closure properties.

Lemma 2.7 $\mathcal{L}$(P,CF,ut) is closed under homomorphism, union and endmarking; $\mathcal{L}$(P,CF$-\lambda$,ut) is closed under non-erasing homomorphism (but not closed under arbitrary erasing homomorphisms), union and endmarking. □

Theorem 2.8 The following statements are equivalent:

- $\mathcal{L}$(P,CF,ut) = $\mathcal{L}$(RE);
- $\mathcal{L}$(P,CF,ut) is closed under intersection with regular sets;
- $\mathcal{L}$(P,CF,ut) is closed under intersection;
- $\mathcal{L}$(P,CF,ut) is closed under left or right derivatives;
- $\mathcal{L}$(P,CF,ut) is closed under general sequential machine (gsm) mappings;
- $\mathcal{L}$(P,CF,ut) is closed under inverse homomorphisms.

Proof. If $\mathcal{L}$(P,CF,ut) were closed under intersection with regular sets, then, by Theorem 2.4, every recursively enumerable language would lie in $\mathcal{L}$(P,CF,ut) due to its closure under homomorphisms. A similar argument holds for left or right derivatives and gsm mappings. By [22], every gsm mapping τ as a special case of a rational transduction can be represented as the composition of an endmarking operation followed by some homomorphisms and inverse homomorphisms. If $\mathcal{L}$(P,CF,ut) were closed under inverse homomorphisms, $\mathcal{L}$(P,CF,ut) would be closed under gsm mappings, hence $\mathcal{L}$(P,CF,ut) = $\mathcal{L}$(RE). □

Note that none of the closure properties required in the theorem above can be effective, since for (P,CF,ut) grammars, the emptiness problem is decidable, while it is undecidable for (P,CF,ac) (= type-0-) grammars [24].

We encounter a similar situation in the λ-free case. We omit proofs here. We just remark that the constant m in the theorem below equals the number of production labels of the given (P,CF$-\lambda$,ac) grammar.

Theorem 2.9 Let L be some (P,CF$-\lambda$,ac) language over the alphabet V_T. Then, there is a constant $m > 0$ and a (P,CF$-\lambda$,ut) grammar $\tilde{G}$ such that $w \in L$ iff $w\$\#^m \in L(\tilde{G})$, where $\{\$, \#\} \cap V_T = \emptyset$. □

Theorem 2.10 The following statements are equivalent:

- $\mathcal{L}$(P,CF$-\lambda$,ut) = $\mathcal{L}$(P,CF$-\lambda$,ac);
- $\mathcal{L}$(P,CF$-\lambda$,ut) is closed under intersection with regular sets and restricted homomorphisms;
- $\mathcal{L}$(P,CF$-\lambda$,ut) is closed under left or right derivatives;
- $\mathcal{L}$(P,CF$-\lambda$,ut) is closed under λ-free gsm mappings;
- $\mathcal{L}$(P,CF$-\lambda$,ut) is closed under inverse homomorphisms and restricted homomorphisms. □

Corollary 2.11 If for some language family $\mathcal{L}$ we know that

$$\mathcal{L}(P,CF[-\lambda],ut) \subseteq \mathcal{L} \subseteq \mathcal{L}(P,CF[-\lambda],ac),$$

and $\mathcal{L}$ is closed under intersection with regular sets [and restricted homomorphisms], then $\mathcal{L} = \mathcal{L}$(P,CF[$-\lambda$],ac). □

What happens in case of unconditional transfer in connection with other regulation mechanisms? We can show the following equalities, some of them were already known in the literature. Note that the case of TV_c-grammars is somewhat tricky, but we had to omit the proof for reasons of space.

Theorem 2.12 $\mathcal{L}(\mathrm{P,CF}[-\lambda],\mathrm{ut}) = \mathcal{L}(\mathrm{M,CF}[-\lambda],\mathrm{ut}) = \mathcal{L}(\mathrm{rC,CF}[-\lambda],\mathrm{ut}) = \mathcal{L}(\mathrm{TV}_c,\mathrm{CF}[-\lambda],\mathrm{ut}) = \mathcal{L}(\mathrm{G}_c,\mathrm{CF}[-\lambda],\mathrm{ut})$ □

Can there be something more interesting within the case of unconditional transfer? This is the case, as it is shown in the next theorem which solves an open problem raised in [25], where the corresponding statement in the case of left derivations is shown.

Theorem 2.13 For $X \in \{\mathrm{CF}, \mathrm{CF} - \lambda\}$, $\mathcal{L}(\mathrm{G},X,\mathrm{ut}) = \mathcal{L}(\mathrm{G},X,\mathrm{ac})$ holds.

Proof. We only prove the inclusion $\mathcal{L}(\mathrm{P},X,\mathrm{ac}) \subseteq \mathcal{L}(\mathrm{G},X,\mathrm{ut})$. Let $L \in \mathcal{L}(\mathrm{P},X,\mathrm{ac})$, $L \subseteq V_T^*$, $L = \bigcup_{a \in V_T}(\{a\}V_T^+ \cap L) \cup ((V_T \cup \{\lambda\}) \cap L)$. We construct a $(\mathrm{G,CF,ut})$ grammar $G' = (V_N \cup \{\#, F, S'\}, V_T, P', S', \Gamma, \Sigma, \Phi, h)$ with $\Gamma = (U, E)$ such that $L(G') = \{a\}V_T^+ \cap L$. By the easily seen closure of $\mathcal{L}(\mathrm{G},X,\mathrm{ut})$ under union, one obtains the required grammar for L. Let $G = (V_N, V_T, P, S)$ be a $(\mathrm{P},X,\mathrm{ac})$ grammar generating $\{w \in V_T^+ \mid aw \in L\}$ which exists due to the closure properties of $\mathcal{L}(\mathrm{P}, X, \mathrm{ac})$. Any production $(p_i : A_i \to w_i, \sigma_i, \phi_i)$ of G is converted into two production nodes F_i^+, F_i^- with $h(F_i^+) = \{A_i \to w_i, \# \to F\}$, $E \cap \{F_i^+\} \times \{g,r\} \times U = \{F_i^+\} \times \{g,r\} \times (\{F_j^+, F_j^- \mid p_j \in \sigma_i\} \cup \{t\})$, and $h(F_i^-) = \{A_i \to F\}$, $E \cap \{F_i^-\} \times \{g,r\} \times U = \{F_i^-\} \times \{g,r\} \times \{F_j^+, F_j^- \mid p_j \in \phi_i\}$. There is one initial node I containing the production $S' \to \#S$ with green and red arcs leading to $\{F_i^+, F_i^- \mid (p_i : S \to w_i, \sigma_i, \phi_i) \in P\}$, and one final node T containing the production $\# \to a$ with green and red arcs leading to itself. □

Especially, $\mathcal{L}(\mathrm{G,CF}-\lambda,\mathrm{ut})$ is an abstract family of languages.

Theorem 2.14 • $\mathcal{L}(\mathrm{TV,CF,ut}) = \mathcal{L}(\mathrm{G,CF,ut}) = \mathcal{L}(\mathrm{RE})$

• $\mathcal{L}(\mathrm{TV,CF}-\lambda,\mathrm{ut}) = \mathcal{L}(\mathrm{G,CF}-\lambda,\mathrm{ut}) \subsetneq \mathcal{L}(\mathrm{CS})$

Proof. We show how to simulate a $(\mathrm{G,CF,ut})$-grammar by a time-variant one. Let $G = (V_N, V_T, P, S, \Gamma, \Sigma, \Phi, h)$ be a $(\mathrm{G,CF,ut})$ grammar with $\Gamma = (U, E)$; $U = \{U_1, \ldots, U_n\}$, $U_i = \{A_{i1} \to w_{i1}, \ldots, A_{ir_i} \to w_{ir_i}\}$. We give a simulating $(\mathrm{TV,CF,ut})$-grammar $(V_N \cup \{X^{(i)} \mid X \in V_N, 1 \leq i \leq n\} \cup \{S', F\} \cup U, P', S', \phi)$, where ϕ is a function with period $2n + 3$ defined as follows: for $i = 1, \ldots, n$, let

$$\phi(i) = \{A_{i\rho} \to A_{i\rho}^{(i)} \mid 1 \leq \rho \leq r_i\} \cup \{U_j \to U_j \mid j \neq i\},$$

$$\phi(n+i) = \{A_{j\rho}^{(j)} \to F \mid j \neq i, 1 \leq \rho \leq r_k\} \cup \{U_j \to U_j \mid j \neq i\};$$

$$\phi(2n+1) = \{A_{i\rho}^{(i)} \to w_{i\rho} \mid 1 \leq i \leq n, 1 \leq \rho \leq r_i\},$$

$$\phi(2n+2) = \{S' \to SU_i \mid U_i \in \Sigma\} \cup \{U_i \to U_j \mid (U_i, g, U_j) \in E\},$$

$$\phi(2n+3) = \{U_i \to U_i \mid U_i \in U\} \cup \{U_i \to \lambda \mid U_i \in \Phi\}.$$

Due to the closure of $\mathcal{L}(G,CF-\lambda,ut)$ under derivatives and catenation, and of the family $\mathcal{L}(TV,CF-\lambda,ut)$ under union, the λ-free case follows immediately. □

3 Parallel rewriting

A *partition-k-limited ETOL system* is a sixtuple $G = (V, V', \{P_1, \ldots, P_r\}, \omega, \pi, k)$ where V' is a non-empty subset (terminal alphabet) of the alphabet V, $\omega \in V^+$, and each so-called table P_i is a finite subset of $V \times V^*$ which satisfies the condition that, for each $a \in V$, there is a word $w_a \in V^*$ such that $a \to w_a \in P_i$, such that each P_i defines a finite substitution $\sigma_i : V^* \to 2^{V^*}$. π is a partition of V, i.e., $\pi = \{\pi_1, \ldots, \pi_s\}$ with $\bigcup_{i=1}^s \pi_i = V$ and $1 \le i < j \le s$ implies $\pi_i \cap \pi_j = \emptyset$, and k is some natural number. G is called propagating if no table contains an erasing production $a \to \lambda$. Similarly, the notion of deterministic system is inherited from the theory of Lindenmayer systems, i.e., G is called deterministic if no table contains two different productions with the same left-hand side. According to G, $x \Rightarrow y$ (for $x, y \in V^*$) iff there is a table P_i and $x = x_0\alpha_1x_1 \cdots \alpha_nx_n$, $y = x_0\beta_1x_1 \cdots \beta_nx_n$ such that $\alpha_\nu \to \beta_\nu \in P_i$ for each $1 \le \nu \le n$, such that, for each part π_j of the partition of π, either (1) $|\alpha_1 \cdots \alpha_n|_{\pi_j} = k$ or (2) $|\alpha_1 \cdots \alpha_n|_{\pi_j} < k$ and $|x_0x_1 \cdots x_n|_{\pi_j} = 0$.

As usual, the language generated by G is given by $L(G) = \{w \in V'^* \mid \omega \overset{*}{\Rightarrow} w\}$, where $\overset{*}{\Rightarrow}$ denotes the reflexive transitive closure of $\Rightarrow$.

As special cases, we define *uniformly k-limited ETOL systems* (introduced by Wätjen and Unruh in [31, 32]) restricting ourselves to partitions of the form $\pi = \{V\}$, and *k-limited ETOL systems* (introduced by Wätjen in [30]) via the restriction to partitions of the form $\pi = \{\{a\} \mid a \in V\}$. This way, we obtain the language families $\mathcal{L}(k\ell_pE[P][D]T0L)$, $\mathcal{L}(ku\ell E[P][D]T0L)$, and $\mathcal{L}(k\ell E[P][D]T0L)$.

The motivation to consider these partially parallel rewriting systems in this place is the nice correspondence between limited rewriting and unconditional transfer (in programmmed grammars), between partition-limited rewriting and apperance checking, as well as between uniformly limited rewriting and programmed grammars without appearance checks.

In [9], we showed that 1ℓET0L languages can be non-recursive, introducing so-called 1ℓET0L machines. By a similar technique, it is possible to show that $\mathcal{L}(k\ell$ET0L$) \not\subseteq \mathcal{L}(REC)$ for any $k \ge 1$, which solves the mentioned problem. We sketch the proof idea in the following. The idea is to simulate an *r-register machine* or *r-RM* which consists of r registers $R_1, \ldots R_r$, each of them capable of storing one natural number $\rho_1, \ldots, \rho_r$. It can be supplied with a program (*r-RMP*). Its syntax can be described as follows.

- a_i $(1 \le i \le r)$ is an r-RMP. (Increment the contents ρ_i of register i by one.)
- s_i $(1 \le i \le r)$ is an r-RMP. (If $\rho_i > 0$, decrement ρ_i by one.)
- If P_1 and P_2 are r-RMPs, then so is P_1P_2. (First, follow the instructions of P_1 and then the instructions of P_2.)
- If P is an r-RMP, then so is $(P)_i$ $(1 \le i \le r)$. (While $\rho_i > 0$ do P.)
- Nothing else is an r-RMP.

r-RMPs (while-programs) are a well-known formalization of computability. Hence, for every recursively enumerable set M of natural numbers, there exists an r-RMP P such that M is the range of f_P.

A $k\ell ET0L$ machine is given by $M = (V, V', \{P_1, \ldots, P_t\}, \{\sigma, x, y, R\})$, where $V, V', \{P_1, \ldots, P_t\}$ are just the first three components of a $k\ell ET0L$ system, and σ, x, y, R are special symbols in V. Furthermore, the terminal alphabet V' equals $\{\sigma, y\}$. We say that M computes the function $f : \mathbb{N}_0 \multimap \mathbb{N}_0$ iff the corresponding $k\ell ET0L$ system $G_{M,n} = (V, V', \{P_1, \ldots, P_t\}, x^{kn}R\sigma, k)$ generates a word of the form $y^{km}\sigma$ if and only if $m = f(n)$. Especially, there is at most one word in $\{y\}^*\{\sigma\} \cap L(G_{M,n})$.

Theorem 3.1 Any r-RMP can be simulated by a $k\ell ET0L$ machine. □

The main problem is to test the occurrence of a symbol, corresponding to a non-zero-test in a while-loop. This is overcome introducing a special symbol σ (the success witness, as it was called by Stotskii [29]), see the proof of Theorem 2.4.

Corollary 3.2 $\mathcal{L}(k\ell ET0L)$ contains non-recursive languages. □

In the following theorem, we solve a number of open problems from the literature. We list these problems together with their sources below.

- [6] Is the inclusion $\mathcal{L}(1\ell EPT0L) \subseteq \mathcal{L}(P,CF - \lambda, ut)$ strict or not? Confer [8] for a similar question on so-called periodically funtion-limited EPT0L systems which is solved in passing.

- [17] Gärtner left open the propagating and deterministic case of partition-limited ET0L systems.

- [30] Wätjen asked the relation between $\mathcal{L}(k\ell EPT0L)$ and $\mathcal{L}(k'\ell EPT0L)$. Our theorem solves this question partially for $k' = 1$.

- [32] What is the relation between $\mathcal{L}(ku\ell E[P]T0L)$ and $\mathcal{L}(k\ell E[P]T0L)$?

Theorem 3.3 For every $k \geq 1$, we have:

(1) $\mathcal{L}(ku\ell E[P]T0L) \subseteq \mathcal{L}(P,CF[-\lambda]) \subsetneq \mathcal{L}(P,CF[-\lambda], ac)$,

(2) $\mathcal{L}(k\ell E[P]T0L) \subseteq \mathcal{L}(1\ell E[P]T0L) = \mathcal{L}(P,CF[-\lambda], ut)$,

(3) $\mathcal{L}(k\ell_p E[P]T0L) = \mathcal{L}(k\ell_p E[P]DT0L) = \mathcal{L}(1\ell_p E[P]DT0L) = \mathcal{L}(P,CF[-\lambda], ac)$,

(4) $\mathcal{L}(ku\ell E[P]T0L) \neq \mathcal{L}(k\ell E[P]T0L)$.

Proof. The relation (1) is proved in [15] and further discussed in [11, 16]. As regards (2), we sketch the proof of $\mathcal{L}(1\ell EPT0L) \supseteq \mathcal{L}(P,CF - \lambda, ut)$ in the following. Assertion (3) either follows by a direct simulation similar to Theorem 2.13, see [14], or indirectly using Corollary 2.11. For the last assertion (4), see [12].

Let $G = (V_N, V_T, P, S)$ be a (P,CF$-\lambda$,ut) grammar. Let the set of labelled productions of G be $P = \{(p_j : A_j \to w_j, \sigma_j, \sigma_j) \mid 1 \leq j \leq m\}$. The set of labels (nodes) is denoted by Lab(P). We construct a $1\ell EPT0L$ system $G' = (V, V_T, H, S', 1)$ simulating G as follows. Let $V = V_G \cup \{S', F\} \cup \{[a, p] \mid a \in$

$V_G, p \in \mathrm{Lab}(P)\} \cup \{a', a'', a''' \mid a \in V_G\} \cup \{\tilde{A}, \hat{A}, \bar{A} \mid A \in V_N\}$, $H = \{h_I, h_T\} \cup$ $\{h_p, h_{p,1}, \ldots, h_{p,4} \mid p \in \mathrm{Lab}(P)\}$. The tables are defined as follows. (We include only such productions which do not lead to the failure symbol F; they have to be supplemented otherwise because of the completeness condition inherited from L systems.)

The initialization table h_I embraces $\{S' \rightarrow v[a, p] \mid (S, q) \overset{*}{\Rightarrow}_G (va, p)$ and $a \in$ $V_G, |v| \geq 2; q, p \in \mathrm{Lab}(P)\} \cup \{S' \rightarrow w \mid (S, q) \overset{*}{\Rightarrow}_G (w, p)$ and $w \in V_T^2 \cup V_T; q, p \in$ $\mathrm{Lab}(P)\}$. The termination table h_T contains $\{[x, p] \rightarrow x, x \rightarrow x \mid x \in V_T, p \in$ $\mathrm{Lab}(P)\}$. Note that a premature attempt to apply the termination table to a string still containing "real" nonterminal symbols would inevitably introduce the failure symbol F.

During a simulation, we face a string of the form $w_1[x, p_j]w_2$. This testifies that in G there is a derivation $(S, q) \overset{*}{\Rightarrow} (w_1 x w_2, p_j)$. Since the other case has already been treated in the initialization table, we assume $|w_1 w_2| \geq 2$.

The first and simple case we deal with is $A_j \neq x$. Now, the marker can stay at the symbol x in its place, and the actual simulation takes place elsewhere. This is accomplished with the table h_{p_j} which contains $\{[x, p_j] \rightarrow [x, q] \mid x \in V_G \setminus \{A_j\} \wedge q \in \sigma_j\} \cup \{A_j \rightarrow w_j\} \cup \{Y \rightarrow Y \mid Y \in V_G \setminus \{A_j\}\}$. Observe that the 'unconditional transfer' is done automatically via the definition of a derivation step in 1lEPT0L systems.

The second case, $A_j = x$, is more complicated. Why? It is possible that there is another A_j in the string $w_1 w_2$ not hidden in the disguise $[A_j, p_j]$. Both forms of A_j should have a chance to be chosen to take part in the rewriting $A_j \rightarrow w_j$ which has to be simulated. This is done by the following four tables.

- $h_{p_j,1}$ contains productions $[A_j, p_j] \rightarrow \hat{A}_j$, $A_j \rightarrow \tilde{A}_j$, and, for every $Y \in V_G \setminus \{A_j\}$, marking productions $Y \rightarrow Y'$.

- $h_{p_j,2}$ contains $\hat{A}_j \rightarrow \bar{A}_j$, $\tilde{A}_j \rightarrow \bar{A}_j$, $A_j \rightarrow A_j''$ and, for every $Y \in V_G \setminus \{A_j\}$, marking productions $Y' \rightarrow Y''$ and $Y \rightarrow Y$ [in order to conserve unmarked symbols]. Note that after a successful application of this table, at least one and at most two occurrences of $\bar{A}_j$ are present. Moreover, there is at least one occurrence of the form a'' for some $a \in V_G$, since we presume $|w_1[x, p_j]w_2| \geq 3$, and we have monotone productions only.

- $h_{p_j,3}$ contains $\bar{A}_j \rightarrow A_j'$ and, for every $Y \in V_G$, $Y \rightarrow Y$ and $Y'' \rightarrow Y'''$.

- $h_{p_j,4}$ embraces $\{A_j' \rightarrow v[a, q] \mid w_j = va, a \in V_G, q \in \sigma_j\} \cup \{Y''' \rightarrow Y, Y \rightarrow Y \mid Y \in V_G\} \cup \{\bar{A}_j \rightarrow A_j\}$. $\qquad \square$

In some sense, also the family of (unordered) scattered context [(u)SC] languages is defined by some grammatical mechanism with restricted parallelism. In [7], an appearance checking feature is defined for such grammars, too. For example, a *scattered context grammar (with appearance checking)* is a construct $G = (V_N, V_T, P, S, F)$, where V_N, V_T, S have the usual meanings, and P is a finite set of finite sequences of context-free productions, $P = \{p_1, \ldots, p_n\}$, written as $p_i : (\alpha_{i1}, \ldots, \alpha_{ir_i}) \rightarrow (\beta_{i1}, \ldots, \beta_{ir_i})$, $1 \leq i \leq n$.

Let $\mathrm{Lab}(P)$ be the set of labels p_{ij} of occurrences of productions $\alpha_{ij} \rightarrow \beta_{ij}$ in P; then, $F \subseteq \mathrm{Lab}(P)$. For $x, y \in V_G^*$, we say that x directly derives y, written as

$x \Rightarrow y$, iff, for some $1 \leq i \leq n$,

$x = x_1 \alpha_{i,j_1} x_2 \alpha_{i,j_2} \cdots x_m \alpha_{i,j_m} x_{m+1}, \quad x_j \in V_G^*$ for $1 \leq j \leq m$,

$y = x_1 \beta_{i,j_1} x_2 \beta_{i,j_2} \cdots x_m \beta_{i,j_m} x_{m+1}, \quad p_{ij} \in F$ for $j \neq j_k, 1 \leq k \leq m$, and x_j does not contain as subwords $\alpha_{i,k}$ for any $k \in \{j_{k-1}+1, \ldots, j_k - 1\}$ [with $j_0 = 0$ and $j_{m+1} = r_i + 1$] for all $k \in \{1, \ldots, m+1\}$. As usual, $\overset{*}{\Rightarrow}$ denotes the reflexive transitive closure of $\Rightarrow$, and $L(G) = \{w \in V_T^* \mid S \overset{*}{\Rightarrow} w\}$. Unconditional transfer means $F = \text{Lab}(P)$. The generated language families are denoted, e.g., by $\mathcal{L}(\text{SC,ac})$ and $\mathcal{L}(\text{SC,ut})$.

We summarise some old results in the following theorem, giving references to literature in case of previously known results. Then, we give our new results. Detailed definitions and proofs can be found in [13].

Theorem 3.4 • $\mathcal{L}(\text{uSC}[-\lambda]) = \mathcal{L}(\text{P,CF}[-\lambda]) \subsetneq \mathcal{L}(\text{SC}[-\lambda])$ [12, 19];

• $\mathcal{L}(\text{uSC}[-\lambda]) \subsetneq \mathcal{L}(\text{uSC}[-\lambda],\text{ac}) = \mathcal{L}(\text{P,CF}[-\lambda],\text{ac})$ [10, Section 4.2] [20, 21];

• $\mathcal{L}(\text{SC}-\lambda,\text{ac}) = \mathcal{L}(\text{CS}) \subsetneq \mathcal{L}(\text{SC}) = \mathcal{L}(\text{SC,ac}) = \mathcal{L}(\text{RE})$ [7] [19, Section 3] [12].

Theorem 3.5 • $\mathcal{L}(\text{uSC}[-\lambda],\text{ut}) = \mathcal{L}(\text{P,CF}[-\lambda],\text{ut})$;

• $\mathcal{L}(\text{CS}) = \mathcal{L}(\text{SC}-\lambda,\text{ut}) \subsetneq \mathcal{L}(\text{SC,ut}) = \mathcal{L}(\text{RE})$.

Proof. The relation $\mathcal{L}(\text{uSC}[-\lambda],\text{ut}) = \mathcal{L}(\text{P,CF}[-\lambda],\text{ut})$ can be proved very similar to Theorem 3.3(2).

We show the inclusion $\mathcal{L}(\text{CS}) \subseteq \mathcal{L}(\text{SC}-\lambda, \text{ut})$ in the following. Let $L \in \mathcal{L}(\text{CS})$, $L \subseteq V_T^*$. Then, $L_{ab} = \{w \in V_T^+ \mid awb \in L\} \in \mathcal{L}(\text{CS})$. Let $G = (V_N, V_T, P, S)$ be a context-sensitive grammar in Kuroda normal form generating L_{ab} with $V_N = \{A_1, \ldots, A_n\}$. We construct an SC grammar with unconditional transfer $G' = (V_N', V_T, P', S')$ generating $\{a\}L_{ab}\{b\}$. Let $V_N' = V_N \cup \{S', F, L, R\}$. There is a start production $(S') \to (LSR)$. For every production of the form $AB \to CD$, we take a simulating production $(A, A_1, \ldots, A_n, L, R, B) \to (C, F, \ldots, F, F, F, D)$. Observe that, in case for example A is not contained in the present sentential form, then $L \to F$ would introduce the failure symbol. For every production of the form $C \to c$, we take $(A_1, \ldots, A_n, C) \to (F, \ldots, F, c)$. These productions allow turning placeholders of terminal symbols into terminal symbols in a left-to-right manner. Finally, we derive the left and right markers: $(L, A_1, \ldots, A_n, R) \to (a, F, \ldots, F, b)$. By the easily proved closure of $\mathcal{L}(\text{SC}-\lambda, \text{ut})$ under union and since $\mathcal{L}(\text{CF}) \subseteq \mathcal{L}(\text{SC}-\lambda, \text{ut})$, we see that $L \in \mathcal{L}(\text{SC}-\lambda, \text{ut})$. $\square$

4 Summary, Open Problems and Discussion

We considered various regulation mechanisms in connection with unconditional transfer. Especially, TV- and G-grammars are in some sense very interesting cases. They may also be looked upon from another point of view: both of them yield a simple (in case of time-variant, very simple) regime regulating the application of production sets, which can be seen as a form of regulation in cooperating distributive grammar systems, a very modern topic in formal language theory, cf. the monograph [5]. This is also emphasized in one of the very first papers on grammar systems [2]. Moreover, observe that the close connections between TV- and G-grammars obtained in this paper also provide another proof for the fact that in case of G-grammars, without loss of generality one can assume planar control graphs (more precisely, we see that we can even assume rings as control graphs), also in case of unconditional transfer, cf. [23, 33].

Let us mention the main open question in this area, namely the question whether the restriction of the appearance checking mechanism to unconditional transfer within programmed grammars entails a restricted class of languages or not. In the literature [29], such a separation is claimed, but we are not convinced the argument given there. This question seems to possess an invincibility similar to the question whether the appearance checking feature itself enhances the descriptional power or not, a problem only recently and independently solved in [21] (cf. also [10, 12]) and [20]. Interestingly, such separation result exists in case of leftmost derivations without erasing rules [29].

We close the discussion with some remarks on accepting grammars. In case of regulated and restricted parallel rewriting, taking the original idea and motivation of the grammar type as basis for the definition of accepting grammar, the descriptive power of such mechanisms has only recently been investigated [3, 4, 15]. Since checking the applicability of an accepting context-free rule means testing words instead of single letters, it is possible to simulate, e.g., context-sensitive grammars in Kuroda normal form by accepting programmed grammars with unconditional transfer without λ-productions (of the form $\lambda \to A$). We obtained similar results for the other types of regulated rewriting with unconditional transfer and and for parallel rewriting. In particular, this means that $(P,CF-\lambda,ut)$ grammars are more powerful as language acceptors than as language generators.

Acknowledgements: We are grateful for discussions with our colleagues Henning Bordihn, Volker Diekert, Rudi Freund, Markus Holzer, Anca Muscholl and Klaus Reinhardt on this topic.

References

1. S. Ábrahám. Some questions of phrase-structure grammars. *Comput. Linguistics*, 4:61–70, 1965.
2. A. Atanasiu and V. Mitrana. The modular grammars. *International Journal of Computer Mathematics*, 30:101–122, 1989.
3. H. Bordihn and H. Fernau. Accepting grammars with regulation. *International Journal of Computer Mathematics*, 53:1–18, 1994.
4. H. Bordihn and H. Fernau. Accepting grammars and systems via context condition grammars. *Journal of Automata, Languages and Combinatorics*, 1(2), 1996.
5. E. Csuhaj-Varjú et al. *Grammar Systems: A Grammatical Approach to Distribution and Cooperation*. London: Gordon and Breach, 1994.
6. J. Dassow. A remark on limited 0L systems. *J. Inf. Process. Cybern. EIK*, 24(6):287–291, 1988.
7. J. Dassow and Gh. Păun. *Regulated Rewriting in Formal Language Theory*. Berlin: Springer, 1989.
8. H. Fernau. On function-limited Lindenmayer systems. *J. Inf. Process. Cybern. EIK*, 27(1):21–53, 1991.

9. H. Fernau. Membership for 1-limited ETOL languages is not decidable. *J. Inf. Process. Cybern. EIK*, 30(4):191–211, 1994.

10. H. Fernau. Observations on grammar and language families. Technical Report 22/94, Universität Karlsruhe (Germany), Fakultät für Informatik, August 1994. Most of this report will appear in Fundamenta Informaticae in 1996.

11. H. Fernau. A note on uniformly limited ETOL systems with unique interpretation. *Information Processing Letters*, 54:199–204, 1995.

12. H. Fernau. A predicate for separating language classes. *EATCS Bulletin*, 56:96–97, June 1995.

13. H. Fernau. Scattered context grammars with regulation. Submitted.

14. H .Fernau. Remarks on propagating partition-limited ETOL systems. Submitted.

15. H. Fernau and H. Bordihn. Remarks on accepting parallel systems. *International Journal of Computer Mathematics*, 56:51–67, 1995.

16. H. Fernau and D. Wätjen. Remarks on regulated limited ETOL systems and regulated context-free grammars. In preparation.

17. S. Gärtner. *Partitions-limitierte Lindenmayer-Systeme*. Aachen: Shaker-Verlag, 1995. (Dissertation Technische Universität Braunschweig, Germany).

18. S. Ginsburg and E. H. Spanier. Control sets on grammars. *Mathematical Systems Theory*, 2:159–177, 1968.

19. J. Gonczarowski and M. K. Warmuth. Scattered versus context-sensitive rewriting. *Acta Informatica*, 27:81–95, 1989.

20. D. Hauschildt and M. Jantzen. Petri net algorithms in the theory of matrix grammars. *Acta Informatica*, 31:719–728, 1994.

21. F. Hinz and J. Dassow. An undecidability result for regular languages and its application to regulated rewriting. *EATCS Bulletin*, 38:168–173, 1989.

22. J. Karhumäki and M. Linna. A note on morphic characterization of languages. *Discrete Applied Mathematics*, 5:243–246, 1983.

23. A. Pascu and Gh. Păun. On the planarity of bicolored digraph grammar systems. *Discrete Mathematics*, 25:195–197, 1979.

24. D. J. Rosenkrantz. Programmed grammars and classes of formal languages. *Journal of the Association for Computing Machinery*, 16(1):107–131, 1969.

25. G. Rozenberg and A. K. Salomaa. Context-free grammars with graph-controlled tables. Technical Report DAIMI PB-43, Institute of Mathematics at the University of Aarhus (Denmark), January 1975.

26. A. K. Salomaa. On grammars with restricted use of productions. *Annales academiae scientarum Fennicae*, Serie A-454:1–32, 1969.

27. A. K. Salomaa. Periodically time-variant context-free grammars. *Information and Control*, 17:294–311, 1970.

28. A. K. Salomaa. *Formal Languages*. Academic Press, 1973.

29. E. D. Stotskii. Управление выводом в формальных грамматиках. Проблемы передачи информации, VII(3):87–102, 1971.

30. D. Wätjen. k-limited OL systems and languages. *J. Inf. Process. Cybern. EIK*, 24(6):267–285, 1988.

31. D. Wätjen. On k-uniformly-limited TOL systems and languages. *J. Inf. Process. Cybern. EIK*, 26(4):229–238, 1990.

32. D. Wätjen and E. Unruh. On extended k-uniformly-limited TOL systems and languages. *J. Inf. Process. Cybern. EIK*, 26(5/6):283–299, 1990.

33. D. Wood. Bicolored digraph grammar systems. *RAIRO Informatique théorique et Applications/Theoretical Informatics and Applications*, 1:45–50, 1973.

(poly(log log n), poly(log log n))–Restricted Verifiers are Unlikely to Exist for Languages in $\mathcal{NP}$[*]

Dimitris Fotakis[1] and Paul Spirakis[1,2]

[1] Computer Engineering and Informatics Dept.
Patras Univ., 265 00 Rion, Patras, Greece. Email: fotakis@cti.gr

[2] Computer Technology Institute
Kolokotroni 3, 262 21 Patras, Greece. Email: spirakis@cti.gr

Abstract. The aim of this paper is to present a proof of the equivalence of the equalities $\mathcal{NP} = \mathcal{PCP}(\log \log n, 1)$ and $\mathcal{P} = \mathcal{NP}$. The proof is based on producing long pseudo-random bit strings through random walks on expander graphs. This technique also implies that for any language in $\mathcal{NP}$ there exists a restricted verifier using $\log n + c$, c is a constant, random bits. Furthermore, we prove that the equality of classes $\mathcal{NP}$ and $\mathcal{PCP}(\mathrm{poly}(\log \log n), \mathrm{poly}(\log \log n))$ implies the inclusion of $\mathcal{NP}$ in $\mathcal{DTIME}\left(n^{\mathrm{poly}(\log \log n)}\right)$. Also, some technical details of the proof of $\mathcal{NP} = \mathcal{PCP}(\log n, 1)$ are used for showing that a certain class of (poly(log log n), poly(log log n))–restricted verifiers does not exist for languages in $\mathcal{NP}$ unless $\mathcal{P} = \mathcal{NP}$.

1 Introduction

A language (decision problem) is in the class $\mathcal{NP}$ if, for every input x which belongs to this language, there exists a membership proof, say π_x, which can be checked in polynomial time by some Turing machine. The majority of typical desicion problems in $\mathcal{NP}$ have concise and simple membership proofs. For example a membership proof for the 3-coloring problem consists of a 3-coloring assignment. Surely, to distinguish a proper 3-coloring from a coloring that is proper on all but one of the vertices, one really has to read the color of every vertex, i.e., the whole proof.

This is not the case with *probabilistically checkable proofs*. Probabilistically checkable proofs are inspected by *verifiers* (polynomial time Turing machines) which proceed as follows: After reading the input x and a random string r, they decide which positions of the proof they want to read. Subsequently, they either accept the input x or reject it — only on the knowledge of (few) queried bits. A language is said to have a probabilistically checkable proof if, for all x in the language, there exists a proof π_x which the verifier accepts for all random strings

---

[*] This work was supported by the Information Technology Programme of EU under the project ALCOM-IT.

r, while, for all x not in the language, the verifier rejects all proofs for a majority of the random strings.

Feige et al. [FGLSS91] were the first who used results in the theory of interactive proofs to obtain some non-approximability results for the clique number. Feige et al. exploited a technique of Ajtai, Komlós and Szemerédi [AKS87] and Impagliazzo and Zuckerman [IZ89] for generating long pseudo-random bit strings in order to prove that if $\mathcal{NP} = \mathcal{PCP}(\log n, q(n))$, then the clique number of a graph cannot be approximated within a factor of $2^{\Omega(\log n/q(n))}$ unless $\mathcal{P} = \mathcal{NP}$. Since Arora and Safra [AS92] characterized $\mathcal{NP}$ as $\mathcal{PCP}(\log n, \operatorname{poly}(\log\log n))$, they also proved that it is $\mathcal{NP}$–hard to approximate the clique number within a factor of $2^{\log n/\operatorname{poly}(\log\log n)}$.

Arora et al. [ALMSS92] improved on the work of Arora and Safra by characterized $\mathcal{NP}$ as $\mathcal{PCP}(\log n, 1)$. This result implies [ALMSS92] that MAXSNP–hard problems do not have Polynomial Time Approximation Schemes (PTAS) unless $\mathcal{P} = \mathcal{NP}$, and that for some $\epsilon > 0$ the clique number of a graph cannot be approximated in polynomial time within a factor of n^ϵ unless $\mathcal{P} = \mathcal{NP}$.

Hougardy, Prömel and Steger [HPS94] provide an excellent survey on the recent work concerning probabilistically checkable proofs and their concequences for approximation algorithms.

Relying on [AKS87] and [IZ89] for generating long pseudo-random bit strings by random walks on expander graphs, we further develop the ideas of [AS92] and [ALMSS92] in order to prove that $\mathcal{PCP}(\log\log n, 1) = \mathcal{NP}$ is equivalent to $\mathcal{P} = \mathcal{NP}$. Our techniques also imply that any language in $\mathcal{NP}$ has a restricted verifier that uses exactly $\log n + c$ (c is a constant) random bits and reads $\mathcal{O}(1)$ positions from a membership proof of polynomial length. Furthermore, we prove that $(\operatorname{poly}(\log\log n), \operatorname{poly}(\log\log n))$–restricted verifiers do not exist for languages in $\mathcal{NP}$, unless $\mathcal{NP} \subseteq \mathcal{DTIME}\left(n^{\operatorname{poly}(\log\log n)}\right)$.

The rest of this paper is organized as follows: After some notation and techniques, we prove the fundamental technical lemmas. Then we apply these lemmas to different classes of restricted verifiers. We conclude with some open questions.

1.1 Notation and Techniques

A *verifier* V is a polynomial time Turing machine with access to an input x and a string r of random bits. Furthermore, the verifier has access to a proof π via an oracle, which takes as input a position of the proof the verifier wants to query and outputs the corresponding bit of the proof π.

The result of V's computation, usually denoted by $V(x, r, \pi)$, is either AC-CEPT or REJECT. We always assume verifiers to be non-adaptive, that is we assume that the bits a verifier queries solely depend on the random bits r, but not on the outcome of any previously queried bits.

An $(r(n), q(n))$–restricted verifier is a verifier that for inputs x of length n uses at most $\hat{r}(n)$ random bits and inspects at most $\hat{q}(n)$ bits from π, where $\hat{r}(n)$ and $\hat{q}(n)$ are integral functions such that $\hat{r}(n) = \mathcal{O}(r(n))$ and $\hat{q}(n) = \mathcal{O}(q(n))$.

Definition 1 (Arora, Safra [AS92]). A language L is in $\mathcal{PCP}(r(n), q(n))$ iff there exists an $(r(n), q(n))$–restricted verifier V such that:

(a) For all $x \in L$ there exists a proof π_x such that

$$\mathrm{Prob}_r\,[V(x,r,\pi_x) = \text{ACCEPT}] = 1 \ ,$$

(b) while for all $x \notin L$ every proof π satisfies

$$\mathrm{Prob}_r\,[V(x,r,\pi) = \text{ACCEPT}] < \tfrac{1}{4} \ .$$

$\square$

The probability $\mathrm{Prob}_r\,[\ldots]$ in Definition 1 is computed with respect to the uniform distribution on $\{0,1\}^{f(|x|)}$. Also, the constant $1/4$ may be replaced by any constant α between 0 and 1. Let $\mathcal{PCP}_\alpha(.\,,.)$ denote the class of languages defined in the same way as $\mathcal{PCP}(.\,,.)$ except that the constant $1/4$ in Definition 1 is replaced by α. Since the probability of getting a wrong answer can be made arbitrarily (but constantly) small by repeating the run of the restricted verifier a (suitable) constant number of times, we have $\mathcal{PCP}_\alpha(.\,,.) = \mathcal{PCP}(.\,,.)$ for any constant α.

A method presented in [AKS87] and [IZ89] for producing long pseudo-random bit strings implies that $\mathcal{NP} = \mathcal{PCP}_{n-\epsilon}(\log n, \log n)$ for some constant $\epsilon > 0$. The following lemma gives the theoretical background for this method:

Lemma 2 (Ajtai, Komlós, Szemerédi [AKS87]). *Let $\mathcal{G}_\varrho$ be an infinite family of d-regular graphs with the following property: If $G = (V, E)$ is a member of $\mathcal{G}_\varrho$ and A denotes its adjacency matrix multiplied by $1/d$ then all but the largest eigenvalue of A are less than $1 - \varrho$ and positive.*
Then for every subset C of V with $|C| \le |V|/16$ there exists a constant c such that the probability that a random walk on G of length $\kappa \cdot c$ arrives in every c-th step in a vertex of C is at most $2^{-\kappa}$.

The application of this method is explained in the proof of Lemma 3. The existence of families of graphs $\mathcal{G}_\varrho$ satisfying the requirements of Lemma 2 is based on the existence of *constant degree expanders*. An explicit construction of constant degree expanders is given by Gabber and Galil [GG81]. The so-called Gabber–Galil expander has the advantage that we do not need to explicitly construct the entire graph. In particular, for any vertex in the expander, it is possible to compute the neighbouring vertices in time polynomial in $\log |V|$. Random walks on Gabber–Galil expanders are used (e.g. [MR95]) for obtaining probability amplification results for randomized algorithms.

2 $\mathcal{P} = \mathcal{NP}$ and $\mathcal{NP} = \mathcal{PCP}(\log\log n, 1)$ are Equivalent

Then, we are ready to prove some fundamental technical lemmas.

Lemma 3. *Let L be a language in $\mathcal{PCP}_{1/16}(r(n), q(n))$. Then for any $0 < \delta < 1$, L has a restricted verifier R' achieving an error rate δ, using $\hat{r}(n) + f\left(\log \frac{1}{\delta}\right)$ random bits and inspecting $\hat{q}(n) \log \frac{1}{\delta}$ proof bits.*

Proof. Since $L \in \mathcal{PCP}_{1/16}(r(n), q(n))$, there exists a restricted verifier R for L such that achieves an error rate smaller than $1/16$, uses $\hat{r}(n)$ random bits and inspects $\hat{q}(n)$ proof bits. The restricted verifier R' with the desired properties will be the result of multiple invocations of R. Lemma 2 implies that the random bits for multiple invocations of R can be provided by a random walk on a Gabber–Galil expander, under the assumption that $r(n)$ can be bounded from above by a polynomial of n. The restricted verifier R' should proceed as follows:

(1) It should read a random string r of length $\hat{r}(n) + f\left(\log \frac{1}{\delta}\right)$
(2) It should construct [GG81] a regular graph $G(V, E)$ that should be a member of an infinite family of graphs $\mathcal{G}_\varrho$ for which Lemma 2 holds. The graph G should have a vertex for every possible bit string of length $\hat{r}(n)$. Let d be the degree of G and c be the constant of Lemma 2.
(3) R' should perform a random walk (as required by Lemma 2) on the graph G. Let $\kappa \cdot c$ be the length of the random walk. Let r_i be the bit string that is associated with the vertex reached by the random walk at the $(i \cdot c)$-th, $i = 1, \ldots, \kappa$, step. If R' has been invoked as $R'(x, r, \pi)$, then R' should invoke R for each bit string r_i (κ times) as $R(x, r_i, \pi)$. R' should produce ACCEPT iff all invocations of R produce ACCEPT too.

Since error probability of R is less than $1/16$, there exists a $C \subseteq V(G), |C| \leq |V(G)|/16$ such that a vertex $u \in C$ iff it is associated with a bit string r_i that causes invocation $R(x, r_i, \pi)$ to result in wrong outcome. Let κ (the number of times that R is invoked by R') be $\kappa = \log \frac{1}{\delta}$. Then Lemma 2 implies that the probability P_{error} of *all* bit strings r_i causing invocations $R(x, r_i, \pi)$ to result in wrong outcomes will be $P_{error} \leq 2^{-\log(1/\delta)} = 2^{\log \delta} = \delta$.

Furthermore, let $f\left(\log \frac{1}{\delta}\right) = c\left(\log \frac{1}{\delta}\right)(\log d)$. Then a bit string r of length $\hat{r}(n) + f\left(\log \frac{1}{\delta}\right)$ can completely specify the steps of the random walk on G ($\hat{r}(n)$ bits are used for choosing the initial vertex uniformly at random and $c\left(\log \frac{1}{\delta}\right)(\log d)$ bits are used for specifying the $c \log \frac{1}{\delta}$ steps of the walk).

Eventually, since R is invoked by R' $\log \frac{1}{\delta}$ times, the number of proof bits inspected by R' is $\hat{q}(n) \log \frac{1}{\delta}$. $\qquad\square$

Remark. Obviously, Lemma 3 can be applied for δ not being constant. So, the characterization of $\mathcal{NP}$ as $\mathcal{PCP}_{n^{-\epsilon}}(\log n, \log n)$ can be obtained [HPS94] by Lemma 3 for $\delta = n^{-\epsilon}$.

Lemma 4. *Let L be any language in $\mathcal{PCP}(r(n), q(n))$. Then there exists a $(r(n), q(n))$-restricted verifier $\hat{R}$ for L that uses exactly $r(n) + t$ random bits, where t does not depend on n.*

Proof. Since $L \in \mathcal{PCP}(r(n), q(n))$, there exists a restricted verifier R achieving error rate smaller than $1/16$, using no more than $\alpha r(n)$, $\alpha > 1$, random bits and inspecting $\hat{q}(n)$ proof bits. Under the assumption that $r(n)$ is bounded from above by a polynomial of n, one can apply Lemma 3 for $\delta = 2^{-(4\alpha+1)}$ in order to obtain a verifier R' that achieves error rate smaller than $\frac{1}{16\alpha}$, uses $\alpha r(n) + f(4\alpha+1)$ random bits and inspects $(4\alpha+1)\hat{q}(n)$ proof bits. The verifier $\hat{R}$

with the desired properties will be the result of multiple invocations of R'. Under the assumption on the order of $r(n)$, Lemma 2 implies that the random bits for multiple invocations of R' can be provided by a random walk on a Gabber–Galil expander.

Let $\hat{R}(x, r, \pi)$ be an invocation of $\hat{R}$, where r is a random string of length $r(n) + t$. Let κ be the number of times the verifier $\hat{R}$ should invoke the verifier R' and let $t = \kappa \cdot f(4\alpha + 1) + t'$. The verifier $\hat{R}$ interprets the random string r as consisting of the concatenation of the random substrings $r = r^0 \circ r^1 \circ r^2 \circ \cdots \circ r^\kappa \circ r^{\kappa+1}$, where $|r^0| = r(n)$, $|r^i| = f(4\alpha + 1)$, $i = 1, \ldots, \kappa$, and $|r^{\kappa+1}| = t'$. The restricted verifier $\hat{R}$ should proceed as follows:

(1) It should construct [GG81] a regular graph $\hat{G}(\hat{V}, \hat{E})$ that should be a member of an infinite family of graphs $\hat{\mathcal{G}}_\varrho$ for which Lemma 2 holds. The graph $\hat{G}$ should have a vertex for every possible bit string of length $r(n)$. Let $\hat{d}$ be the degree of $\hat{G}$ and $\hat{c}$ be the constant of Lemma 2.

(2) $\hat{R}$ should perform a random walk (as required by Lemma 2) on the graph $\hat{G}$. Let $\alpha \cdot \kappa \cdot \hat{c}$ be the length of the random walk and let m_i be the bit string associated with the vertex reached by the $(i \cdot \hat{c})$-th, $i = 1, \ldots, \alpha \cdot \kappa$, step of the walk. Let r_j be the bit string that consists of the concatenation of
(a) α consecutive results of the random walk; and
(b) a substring r^{j+1}, $|r^{j+1}| = f(4\alpha + 1)$, provided by the random string r.
More precisely $r_j = m_{j\alpha+1} \circ m_{j\alpha+2} \circ \cdots \circ m_{j\alpha+\alpha} \circ r^{j+1}$, $j = 0, \ldots, \kappa - 1$.
Additionally, since $|r_j| = \alpha r(n) + f(4\alpha + 1)$, r_j can be used as a random string for the verifier R'.

(3) $\hat{R}$ should invoke $R'(x, r_j, \pi)$, $j = 0, \ldots, \kappa - 1$ (κ times). $\hat{R}$ should produce ACCEPT iff all invocations of R' produce ACCEPT too.

Any bit string r_j consists of the concatenation of two substrings, $r_j = r_j^1 \circ r_j^2$, of length $|r_j^1| = \alpha r(n)$ and $|r_j^2| = f(4\alpha + 1)$. The bit strings r_j^1 are used for choosing the initial vertex of the random walk that the verifier R' performs on the graph G (proof of Lemma 3). By [AKS87] and [MR95], we know that random walks on Gabber–Galil expanders are rapidly mixing. In other words, given any starting vertex, after a small number of steps we expect the random walk to be at a uniformly distributed vertex independent of choice of the initial vertex. The random walk that $\hat{R}$ performs on $\hat{G}$ can be viewed as the composition of $\alpha \cdot \kappa$ different random walks, each generating a different random string m_i. Each of these smaller random walks has constant length $\hat{c}$, whereas we would require $\Theta(\log |\hat{V}|)$ steps to get close to stationary distribution. On the other hand, we choose the initial vertex according to stationary distribution, and this works in our favour. The previous remarks provide the intuition behind the fact that the bit strings r_j^1 are almost uniformly distributed. Further, the bit strings r_j^2 specify the steps of the random walk that the verifier R' performs on graph G. The strings r_j^2 are distributed uniformly at random because they are provided by the random string r. Let $2^{r(n)} = N$ and $2^{\alpha r(n)} = N^\alpha$. Since the error rate of R' is less than $\frac{1}{16^\alpha}$, less than $\left(\frac{N}{16}\right)^\alpha$ bit strings r_j^1 cause the verifier R' to result in wrong outcome.

Let $\hat{C}$ be any set of vertices of $\hat{G}$, $\hat{C} \subseteq \hat{V}(\hat{G})$, such that $|\hat{C}| = |\hat{V}(\hat{G})|/16$. Let w be a random walk of length $\alpha \cdot \hat{c}$ on the graph $\hat{G}$ such that every $\hat{c}$-th step of w reaches only vertices of $\hat{C}$. Let r_w be the bit string of length $\alpha r(n)$ that is produced by the concatenation of the bit strings of length $r(n)$ associated with the vertices (of $\hat{C}$) reached by the $(i \cdot \hat{c})$-th, $i = 1, \ldots, \alpha$, steps of w. The number of different bit strings r_w that can be produced this way is $|\hat{C}|^\alpha = \left(\frac{|\hat{V}(\hat{G})|}{16} \right)^\alpha = \left(\frac{N}{16} \right)^\alpha$. Therefore, since r_j^2 have been specified by the random string r, there exists a $\hat{C} \subseteq \hat{V}(\hat{G}), |\hat{C}| \le |\hat{V}(\hat{G})|/16$ such that the following proposition holds:

Proposition 5. *There exists an one-to-one correspondence between*

(a) *the bit strings r_j^1, $|r_j^1| = \alpha r(n)$, which are the initial substrings of the bit strings r_j, $|r_j| = \alpha r(n) + f(4\alpha + 1)$, that causes the verifier R' to result in wrong outcome; and*

(b) *the bit strings that can be produced by the concatenation of the bit strings m_i, $i = 1, \ldots, \alpha$, associated with the vertices reached by a random walk w which has length $\alpha \cdot \hat{c}$ and only reaches vertices of $\hat{C}$ at every $\hat{c}$-th step.*

Remark. The explicit construction of the set $\hat{C}$ may be computationally intractable because we have no knowledge about the distribution of the "misleading" strings for the verifier R'. The same holds for the set C in the proof of Lemma 3. However, a one-to-one correspondence between the "misleading" strings and the vertices of $\hat{C}$ or C suffices in both cases. So, explicit constructions are required neither in the proof of Lemma 3 nor in the proof of Lemma 4.

Proposition 5 implies that there exists a $\hat{C} \subseteq \hat{V}(\hat{G})$ such that the verifier $\hat{R}$ results in wrong outcome iff the random walk of (2) only reaches vertices of $\hat{C}$. Since Lemma 2 holds for $\hat{G}$ and $|\hat{C}| \le |\hat{V}(\hat{G})|/16$, the probability that a random walk of length $\alpha \cdot \kappa \cdot \hat{c}$ reaches at every $\hat{c}$-th step a vertex of $\hat{C}$ is at most $2^{-\alpha \cdot \kappa}$. We have already assumed that $\alpha > 1$. Henceforth, the verifier $\hat{R}$ should invoke R' twice ($\kappa = 2$) in order to achieve error rate $1/4$.

The complete specification of the random walk requires $r(n)$ random bits for choosing the initial vertex uniformly at random and $2 \cdot \alpha \cdot \hat{c} \cdot \log \hat{d}$ random bits for the specification of the $2 \cdot \alpha \cdot \hat{c}$ steps of the walk. Thus, the verifier $\hat{R}$ requires $r(n) + 2 \cdot f(4\alpha + 1) + 2 \cdot \alpha \cdot \hat{c} \cdot \log \hat{d}$ random bits. If we let $t' = 2 \cdot \alpha \cdot \hat{c} \cdot \log \hat{d}$ and $t = 2 \cdot f(4\alpha + 1) + t'$, then $\hat{R}$ requires exactly $r(n) + t$ random bits, where t does not depend on n. Finally, the bits that $\hat{R}$ should read from the proof π remains $\mathcal{O}(q(n))$ since α is a constant and R' is only invoked twice. $\qquad \square$

By applying Lemma 4 for $r(n) = \log n$ and for $r(n) = \log \log n$ we obtain the following corollaries:

Corollary 6. *For any language in $\mathcal{NP}$ there exists a $(\log n, 1)$-restricted verifier that uses exactly $\log n + t$ random bits, where t does not depend on n.*

Corollary 7. *For any language in $\mathcal{PCP}(\log \log n, 1)$ there exists a $(\log \log n, 1)$-restricted verifier that uses exactly $\log \log n + t$ random bits, where t does not depend on n.*

Remark. The proof of Lemma 4 says that random walks of length M in Gabber–Galil expanders with nodes representing $\alpha r(n)$ bits can be emulated by random walks of length αM in Gabber–Galil expanders with nodes representing $r(n)$ bits. Moreover, this holds for α not being constant.

Lemma 8. *Let L be any language over Σ^* and R a $(r(n), q(n))$-restricted verifier for L. Then for any $x \in \Sigma^*$ there exists a SAT instance S_x of size $\mathcal{O}(2^{\hat{r}(n)}\hat{q}(n))$ that is satisfiable iff $x \in L$.*

Proof. For any $x \in \Sigma^*$, $|x| = n$, we will exploit the verifier R in order to construct a SAT instance S_x such that S_x is satisfiable iff x is an element of L.

For every position of a membership proof π we introduce a variable whose values TRUE and FALSE correspond to the values 1 and 0 of the bit at this position. By using these variables the SAT instance S_x is obtained as follows:

- For any possible random string r let S_r denote the boolean formula that expresses which proofs π are accepted by R on input x. Since R queries only $\hat{q}(n)$ bits from the proof, the size of the formulas S_r is $\mathcal{O}(\hat{q}(n))$. Furthermore, since $|r| = \hat{r}(n)$, the number of formulas S_r is $2^{\hat{r}(n)}$.
- Let S_r' be the formula S_r written in conjuctive normal form. Note that each S_r' still contains $\mathcal{O}(\hat{q}(n))$ boolean variables.
- Let S_x be the conjuction of all the formulas S_r'. Since formulas S_r' contain $\mathcal{O}(\hat{q}(n))$ variables and the number of S_r' is $2^{\hat{r}(n)}$, the size of the SAT instance S_x is $\mathcal{O}(2^{\hat{r}(n)}\hat{q}(n))$.

$\square$

Since $\mathcal{P} = \mathcal{PCP}(0,0)$, Lemmas 4 and 8 imply the following:

Theorem 9. $\mathcal{NP} = \mathcal{PCP}(\log\log n, 1) \Leftrightarrow \mathcal{P} = \mathcal{NP}$.

Proof. If $\mathcal{NP} = \mathcal{PCP}(\log\log n, 1)$, then Lemma 4 guarantees that for any language $L \in \mathcal{NP}$ there exists a $(\log\log n, 1)$-restricted verifier R such that $\hat{r}(n) = \log\log n + t$, t is a constant. Also, Lemma 8 implies that the size of the corresponding SAT instance S_x is $\mathcal{O}(\log n)$. Thus, for any $L \in \mathcal{NP}$ and for any $x \in \Sigma^*$ we can decide whether x is in L in polynomial time. $\square$

3 On Verifiers Using poly($\log\log n$) Random Bits

Obviously, Lemma 4 can be applied with α not being constant. Let us consider a $((\log\log n)^\gamma, 1)$-restricted verifier R that uses $\hat{r}(n) = \beta \cdot (\log\log n)^\gamma$ random bits, where β, γ are integral constants, $\beta, \gamma > 1$. Since $r(n)$ can be bounded from above by any polynomial of n, we can apply Lemma 4 to R with parameter $\alpha = \beta \cdot \log\log n$. Consequently, $f(4\alpha + 1) = \mathcal{O}(\log\log n)$. The verifier $\hat{R}$, that is constructed as described by the proof of Lemma 4, uses $\mathcal{O}((\log\log n)^{\gamma-1})$ random bits and invokes the verifier R $\mathcal{O}(\log\log n)$ times. Thus we can apply Lemma 4 to a $((\log\log n)^\gamma, 1)$-restricted verifier in order to obtain a $((\log\log n)^{\gamma-1}, \log\log n)$-restricted verifier. Furthermore, $(\gamma - 1)$ repeated applications of Lemma 4 to a $((\log \log n)^\gamma, 1)$–restricted verifier R result in the construction of a $(\log \log n, (\log \log n)^{\gamma-1})$–restricted verifier $\tilde{R}$. A final application of Lemma 4 to $\tilde{R}$ results in a $(\log \log n, (\log \log n)^{\gamma-1})$–restricted verifier $\hat{R}$ that uses only $\log \log n + t$ random bits, where t does not depend on n. Similarly, γ repeated applications of Lemma 4 to a $((\log \log n)^\gamma, \text{poly}(\log \log n))$–restricted verifier R result in a $(\log \log n, \text{poly}(\log \log n))$–restricted verifier $\hat{R}$ that uses only $\log \log n + t$ random bits.

The previous observations are summarized by the following corollaries:

Corollary 10. *For any language in* $\mathcal{PCP}((\log \log n)^\gamma, 1)$ *where* γ *is an integral constant, there exists a* $(\log \log n, (\log \log n)^{\gamma-1})$*–restricted verifier that uses* $\log \log n + t$ *random bits, where* t *does not depend on* n.

Corollary 11. *For any language in* $\mathcal{PCP}(\text{poly}(\log \log n), \text{poly}(\log \log n))$ *there exists a* $(\log \log n, \text{poly}(\log \log n))$*–restricted verifier that uses* $\log \log n + t$ *random bits, where* t *does not depend on* n.

Corollary 10 implies that under the assumption that $\mathcal{NP}$ is included either in $\mathcal{PCP}((\log \log n)^2, 1)$ or in $\mathcal{PCP}(\log \log n, \log \log n)$, the size of the SAT instance S_x, which is constructed according to the proof of Lemma 8, is $\mathcal{O}(\log n \cdot \log \log n)$. Thus, we have:

Theorem 12. *Unless* $\mathcal{NP} \subseteq \mathcal{DTIME}\left(n^{\log \log n}\right)$*, the class* $\mathcal{NP}$ *is included neither in* $\mathcal{PCP}((\log \log n)^2, 1)$ *nor in* $\mathcal{PCP}(\log \log n, \log \log n)$.

Further, Corollary 11 and the proof of Lemma 8 imply that if the class $\mathcal{NP}$ is included in $\mathcal{PCP}(\text{poly}(\log \log n), \text{poly}(\log \log n))$, then the size of the corresponding SAT instance S_x is $\mathcal{O}(\log n \cdot \text{poly}(\log \log n))$. Thus, we can state the following:

Theorem 13. *Unless* $\mathcal{NP} \subseteq \mathcal{DTIME}\left(n^{\text{poly}(\log \log n)}\right)$*, the class* $\mathcal{NP}$ *is not included in* $\mathcal{PCP}(\text{poly}(\log \log n), \text{poly}(\log \log n))$.

3.1 A Stronger Result for a Certain Class of Verifiers

Then, we use some technical prerequisites in order to prove a slightly stronger negative result about a certain class of $(\log \log n, \text{poly}(\log \log n))$–restricted verifiers for languages in $\mathcal{NP}$.

Let $x, y \in \Sigma^*$ be two bit strings such that $|x| = |y|$. Then x and y are called δ-close iff the fraction of bits on which they differ is less than δ. An *encoding scheme* is a function $E : \Sigma^* \mapsto \Sigma^*$ such that for all $x, x' \in \Sigma^*$ with $|x| = |x'|$ the encodings $E(x)$ and $E(x')$ have the same length and coincide in at most $\frac{1}{2}$ of their bits, that is $E(x)$ and $E(x')$ are not $\frac{1}{2}$-close. For any arbitrary but fixed encoding scheme and for any constant $d \in \mathbb{N}$ we can define a new encoding scheme E' with respect to d by

$$E'(y) = (E(y_1), \ldots, E(y_d)) , \tag{1}$$

where $y = y_1 \cdots y_d$ and $|y_1| = \ldots = |y_d|$.

A *solution verifier* V is a verifier which in addition has access to a solution s, which it can query via an oracle in the same way as the membership proof π. An $(r(n), q(n), b(n))$–*restricted solution verifier* is a solution verifier which for inputs x of length n uses at most $\hat{r}(n)$ random bits and queries at most $\hat{q}(n)$ blocks of length $\hat{b}(n)$ from s and π, where the starting positions of such blocks are all congruent on modulo $\hat{b}(n)$. Further, $\hat{r}(n)$, $\hat{q}(n)$ and $\hat{b}(n)$ are integral functions such that $\hat{r}(n) = \mathcal{O}(r(n))$, $\hat{q}(n) = \mathcal{O}(q(n))$ and $\hat{b}(n) = \mathcal{O}(b(n))$.

Definition 14 (Arora, Safra [AS92]). Let R be a p-relation and E be an encoding scheme. Then (R, E) is in $\mathcal{PCS}(r(n), q(n), b(n))$ iff there exists an $(r(n), q(n), b(n))$–restricted solution verifier V such that:

(a) For all $x, y \in \Sigma^*$ with $(x, y) \in R$ there exists a proof $\pi_{x,y}$ such that

$$\text{Prob}_r[V(x, r, E(y), \pi_{x,y}) = \text{ACCEPT}] = 1 \ ,$$

(b) For all $x, s \in \Sigma^*$ such that s is not $\frac{1}{4}$-close to the encoding $E(y)$ of a solution $y \in R(x)$ every proof π satisfies

$$\text{Prob}_r[V(x, r, s, \pi) = \text{ACCEPT}] < \frac{1}{4} \ .$$

$\square$

The following technical lemma introduces the idea of recursive proof checking. This lemma shows how two solution verifiers can be composed to form a new verifier that queries fewer bits.

Lemma 15 (Arora, Safra [AS92]). *Let $r_i(n)$, $b_i(n)$ be positive functions. If $(R_{3SAT}, \hat{E}) \in \mathcal{PCS}(r_1(n), 1, b_1(n))$ for some encoding scheme $\hat{E}$ and $\tilde{E}$ is an encoding scheme such that for any constant $d \in \mathbb{N}$ the encoding scheme $\tilde{E}'$ given by equation (1) satisfies $(R_{3SAT}, \tilde{E}') \in \mathcal{PCS}(r_2(n), 1, b_2(n))$, then*

$$(R_{3SAT}, \hat{E}) \in \mathcal{PCS}(r_1(n) + r_2(\text{poly}(b_1(n))), 1, b_2(\text{poly}(b_1(n)))) \ .$$

The proof of Lemma 15 is based on techniques first described by Cook in [Coo71]. Let us make the following assumption in order to apply Lemma 15:

Assumption 16. *There exists an encoding scheme E such that*

(a) $(R_{3SAT}, E) \in \mathcal{PCS}(\log \log n, \text{poly}(\log \log n), \text{poly}(\log \log n))$,

(b) *there exists a* $(\log \log n, \text{poly}(\log \log n), \text{poly}(\log \log n))$–*restricted solution verifier V for* (R_{3SAT}, E) *which queries the solution string only $\mathcal{O}(1)$ times, and*

(c) *if* $(R_{3SAT}, E) \in \mathcal{PCS}(\log \log n, 1, \text{poly}(\log \log n))$, *then for any constant $d \in \mathbb{N}$ the encoding scheme E' given by Equation (1) satisfies $(R_{3SAT}, E') \in \mathcal{PCS}(\log \log n, 1, \text{poly}(\log \log n))$.*

Since the solution verifier V uses only $\mathcal{O}(\log\log n)$ random bits, we can assume, without loss of generality, that the length of any proof which can be used by V will be $\mathcal{O}(\text{poly}(\log n))$. Then, we can use V for constructing a new $(\log\log n, 1, \text{poly}(\log\log n))$–restricted solution verifier $\hat{V}$. The idea behind the construction of $\hat{V}$ is simple: Using the proof π that V would read, we construct $\hat{V}$ that uses a proof $\hat{\pi}$ which contains for every possible random string r the sequence of $\text{poly}(\log\log n)$ bits that V would read from π on input x and random string r. For the verification process the verifier $\hat{V}$ reads a consecutive sequence of $\text{poly}(\log\log n)$ bits from $\hat{\pi}$ that depend on r and uses these bits to determine what the verifier V would have answered if it had received these bits as answers for its queries to π.

The new proof $\hat{\pi}$ for $\hat{V}$ contains the encoding of π under the E_1 encoding scheme that is defined as follows:

Definition 17 (Babai et al. [BFLS91]). Let x be a bit vector of length l, $x \in \mathbb{F}_2^l$. Let p be the smallest prime larger than $\log^2 l$. We may assume that $\log l$ and $\log\log l$ are integer (otherwise elongate x by a suitable number of bits). Set $m = \lceil \log l / \log\log l \rceil$. Let H be a subset of $\mathbb{F}_p$ of size $\log l$. Note that $|\text{H}|^m \geq l$. Thus, we may interpret x as a function $f_x : \text{H}^m \mapsto \{0,1\}$. Then $E_1(x) : \mathbb{F}_p^m \mapsto \mathbb{F}_p$ is the low degree extension of f_x. $E_1(x)$ is a polynomial of degree at most $|\text{H}|$ in each variable and agrees with f_x on H^m. $\qquad\square$

The use of the encoding scheme E_1 provides a restricted solution verifier with efficient polynomial testers like LOW DEGREE TEST [RS92] and EXTENDED LFKN-TEST [LFKN92]. The construction of the verifier $\hat{V}$ from V is similar to the construction found in the proof of Theorem 18 ([ALMSS92, HPS94]). Theorem 18 is used in the proof of $\mathcal{NP} = \mathcal{PCP}(\log n, 1)$ and exploits the encoding scheme E_1 and the LOW DEGREE TEST.

Theorem 18 (Arora et al. [ALMSS92]).
If a $(\log n, \text{poly}(\log n), \text{poly}(\log n))$–restricted solution verifier that queries the solution string a constant number of times exists for (R_{3SAT}, E_1), then $(R_{3SAT}, E_1) \in \mathcal{PCS}(\log n, 1, \text{poly}(\log n))$.

The verifier $\hat{V}$ implies that $(R_{3SAT}, E) \in \mathcal{PCS}(\log\log n, 1, \text{poly}(\log\log n))$. Furthermore, we have assumed the encoding scheme E to satisfy the requirements of Lemma 15. So, we can apply Lemma 15 with respect to the encoding scheme $\hat{E} = \tilde{E} = E$ and the functions $r_1(n) = r_2(n) = \log\log n$ and $b_1(n) = b_2(n) = \text{poly}(\log\log n)$. We therefore deduce that $(R_{3SAT}, E) \in \mathcal{PCS}(\log\log n, 1, \text{poly}(\log\log\log\log n))$.

The following lemmas are also proved in [ALMSS92]:

Lemma 19 (Arora et al. [ALMSS92]). *Let E_0 denote the encoding scheme given by $E_0 : y \mapsto \{y^T z\}_{z \in \mathbb{F}_2^{|y|}}$. Then $(R_{3SAT}, E_0) \in \mathcal{PCS}(n^3, 1, 1)$.*

Lemma 20 (Arora et al. [ALMSS92]). *Let E_0' be the encoding scheme defined by Equation (1). Then for all constants $d \in \mathbb{N}$ one has $(R_{3SAT}, E_0') \in \mathcal{PCS}(n^3, 1, 1)$.*

Lemmas 19 and 20 imply that we can apply Lemma 15 with respect to the encoding schemes $\hat{E} = E$ and $\tilde{E} = E_0$ and to the functions $r_1(n) = \log\log n$, $r_2(n) = n^3$ and $b_1(n) = \text{poly}(\log\log\log\log n)$, $b_2(n) = 1$. We therefore deduce that $(R_{3SAT}, E) \in \mathcal{PCS}(\log\log n, 1, 1)$. Obviously, this implies that 3SAT is in $\mathcal{PCP}(\log\log n, 1)$. Since 3SAT is an $\mathcal{NP}$-complete language [GJ79], we have $\mathcal{NP} = \mathcal{PCP}(\log\log n, 1)$ and subsequently $\mathcal{P} = \mathcal{NP}$. So, we have proved that:

Theorem 21. *If there exists an encoding scheme E that satisfies Assumption 16, then $\mathcal{P} = \mathcal{NP}$.*

Under Assumption 16, Corollary 11 implies that Theorem 18 can also be applied to a $(\text{poly}(\log\log n), \text{poly}(\log\log n), \text{poly}(\log\log n))$–restricted solution verifier that uses membership proofs of length $\mathcal{O}(\text{poly}(\log n))$.

Moreover, Assumption 16.b can be removed if we handle the solution string s (that is of length $\mathcal{O}(\text{poly}(\log n))$) as the proof π in the construction of the verifier $\hat{V}$. Furthermore, it is plausible to assume that a properly encoded solution can be distinguished in any proof used by a restricted verifier for a language in $\mathcal{NP}$. The combination of Theorem 18 with the previous observations provides strong evidence that unless $\mathcal{P} = \mathcal{NP}$, the class $\mathcal{NP}$ is not included in $\mathcal{PCP}(\log\log n, \text{poly}(\log\log n))$. The same holds for the classes $\mathcal{NP}$ and $\mathcal{PCP}(\text{poly}(\log\log n), \text{poly}(\log\log n))$ if there exists a $(\text{poly}(\log\log n), \text{poly}(\log\log n))$–restricted verifier that uses membership proofs of length $\mathcal{O}(\text{poly}(\log n))$.

4 Conclusions – Open Problems

Corollaries 10 and 11 combined with the proof of Lemma 8 are used for showing that unless the class $\mathcal{NP}$ is contained in $\mathcal{DTIME}\left(n^{\text{poly}(\log\log n)}\right)$, $\mathcal{NP}$ is not included in $\mathcal{PCP}(\text{poly}(\log\log n), \text{poly}(\log\log n))$. The question that naturaly arises is whether the equality of classes $\mathcal{P}$ and $\mathcal{NP}$ is implied by the existence of $(\text{poly}(\log\log n), \text{poly}(\log\log n))$–restricted verifiers for languages in $\mathcal{NP}$. We prove that the answer is positive for restricted solution verifiers using membership proofs of length $\mathcal{O}(\text{poly}(\log n))$ and satisfying Assumptions 16.b and 16.c. These assumptions are not of major importance and may be removed. Concerning languages in the class $\mathcal{NP}$, restricted solution verifiers are not expected to be more powerful than ordinary restricted verifiers. However, there is no proof about the equivalence of the equalities $\mathcal{NP} = \mathcal{PCP}(\text{poly}(\log\log n), \text{poly}(\log\log n))$ and $\mathcal{P} = \mathcal{NP}$.

The existence of a restricted verifier that uses membership proofs of length $\mathcal{O}(\text{poly}(\log n))$ for a language in $\mathcal{NP}$ implies the inclusion of $\mathcal{NP}$ in $\mathcal{DTIME}\left(n^{\text{poly}(\log n)}\right)$. Another open question is whether the existence of such a restricted verifier implies the inclusion of $\mathcal{NP}$ in $\mathcal{P}$.

Lemma 4 can be applied to a $(\log n, 1)$–restricted verifier in order to construct a $\left(\frac{\log n}{\log\log n}, \log\log n\right)$–restricted verifier for any language in $\mathcal{NP}$. However, the existence of $\left(\frac{\log n}{\log\log n}, 1\right)$ and $\left(\frac{\log n}{\text{poly}(\log\log n)}, 1\right)$–restricted verifiers for languages in $\mathcal{NP}$ are open problems.

References

[AKS87] M. Ajtai, J. Komlós, and E. Szemerédi. Deterministic simulation in logspace. *Proc. of the 19th ACM Symposium on Theory of Computing*, pp. 132–140, 1987.

[ALMSS92] S. Arora, C. Lund, R. Motwani, M. Sudan, and M. Szegedy. Proof verification and hardness of approximation problems. *Proc. of the 33th Annual IEEE Symposium on Foundations of Computer Science*, pp. 14–23, 1992.

[AS92] S. Arora and S. Safra. Probabilistic checking of proofs: A new characterization of $\mathcal{NP}$. *Proc. of the 33th Annual IEEE Symposium on Foundations of Computer Science*, pp. 2–13, 1992.

[BFLS91] L. Babai, L. Fortnow, L. Levin, and M. Szegedy. Checking Computations in Polylogarithmic Time. *Proc. of the 23th ACM Symposium on Theory of Computing*, pp. 21–31, 1991.

[Coo71] S.A. Cook. The complexity of theorem-proving procedures. *Proc. of the 3rd ACM Symposium on Theory of Computing*, pp. 151–158, 1971.

[FGLSS91] U. Feige, S. Goldwasser, L. Lovász, S. Safra, and M. Szegedy. Approximating clique is almost $\mathcal{NP}$-complete. *Proc. of the 32th Annual IEEE Symposium on Foundations of Computer Science*, pp. 2–12, 1991.

[GG81] O. Gabber and Z. Galil. Explicit constructions of linear–sized superconcentrators. *Journal of Computer and System Sciences* 22, pp. 407–420, 1981.

[GJ79] M.R. Garey and D.S. Johnson. *Computers and Intractability: A Guide to the Theory of $\mathcal{NP}$-Completeness*. Freeman, San Francisco, 1979.

[HPS94] S. Hougardy, H.J. Proömel, and A. Steger. Probabilistically Checkable Proofs and their Consequenses for Approximation Algorithms. *Discrete Mathematics* 136, pp. 175–223, 1994.

[IZ89] T. Impagliazzo and D. Zuckerman. How to recycle random bits. *Proc. of the 30th Annual IEEE Symposium on Foundations of Computer Science*, pp. 248–253, 1989.

[LFKN92] C. Lund, L. Fortnow, H. Karloff, and N. Nisan. Algebraic methods for interactive proof systems. *Journal of the Assosiation for Computing Machinery* 39 (4), pp. 859–868, 1992.

[MR95] R. Motwani and P. Raghavan. *Randomized Algorithms*. Cambridge University Press, New York, 1995.

[RS92] R. Rubinfeld and M. Sudan. Testing Polynomial Functions Efficiently and over Rational Domains. *Proc. 3rd Annual ACM-SIAM Symposium on Discrete Algorithms*, pp. 23–32, 1992.

Minimizing Congestion of Layouts for ATM Networks with Faulty Links

Leszek Gąsieniec[1]* Evangelos Kranakis[2]** Danny Krizanc[2]*** Andrzej Pelc[3]†

[1] Max-Planck Institut für Informatik, Im Stadtwald, Saarbrücken D-66123, Germany.
[2] Carleton University, School of Computer Science, Ottawa, ON, K1A 5B6, Canada.
[3] Département d'Informatique, Université du Québec à Hull, Hull, Québec J8X 3X7, Canada.

Abstract. We consider the problem of constructing virtual path layouts for an ATM network consisting of a complete network K_n of n processors in which a certain number of links may fail. Our main goal is to construct layouts which tolerate any configuration of up to f layouts and have a least possible congestion. First, we study the minimal congestion of 1-hop f-tolerant layouts in K_n. For any positive integer f we give upper and lower bounds on this minimal congestion and construct f-tolerant layouts with congestion corresponding to the upper bounds. Our results are based on a precise analysis of the diameter of the network $K_n[\mathcal{F}]$ which results from K_n by deleting links from a set $\mathcal{F}$ of bounded size. Next we study the minimal congestion of h-hop f-tolerant layouts in K_n, for larger values of the number h of hops. We give upper and lower bounds on the order of magnitude of this congestion, based on results for 1-hop layouts. Finally, we consider a random, rather than worst case, fault distribution. Links fail independently with constant probability $p < 1$. Our goal now is to construct layouts with low congestion that tolerate the existing faults with high probability. For any $p < 1$, we show such layouts in K_n, with congestion $O(\log n)$.

1 Introduction

Broadband Integrated Digital Services (or BISDNs, for short) are meant to accomodate various kinds of data traffic, including voice, video, image, file transfer, as well as interactive. Such systems are being built with standard 150 Mbps optical fiber local access and provide true high rate file transfer, video conferencing, on demand HDTV, interface with high-speed LANs, etc.

---

* On leave from Institute of Informatics, Warsaw University, ul. Banacha 2, 02-097, Warszawa, Poland. WWW: http://www.mpi-sb.mpg.de/~leszek/, Email: leszek@mpi-sb.mpg.de.

** Research supported in part by NSERC (Natural Sciences and Engineering Research Council of Canada) grants. Email: kranakis@scs.carleton.ca.

*** Research supported in part by NSERC (Natural Sciences and Engineering Research Council of Canada) grants. Email: krizanc@scs.carleton.ca.

† Research supported in part by NSERC grant OGP 0008136. Email: pelc@uqah.uquebec.ca.

Requirements in new and emerging information services require a new transfer mode for BISDN. ATM (Asynchronous Transfer Mode) was developed [2]
as a packet structure for BISDNs. ATM is a new multiplexing and switching
technology that results in more cost effective solutions of greater flexibility than
several separate individually optimized technologies [2]. Because of this, it has
significant commercial as well as public service applications. This technology is
thoroughly described in the literature [7, 6].

For standard networks, routing is based on variable size data units, and has
been the topic of extensive studies in the literature, e.g., see [8, 1]. However,
this is not suitable for present day multimedia environments. By contrast, in
ATM networks routing is based on relatively small fixed-sized packets. Such
packets are routed through a layout of "virtual" paths, as well as sequences of
such virtual paths, also called "virtual" channels. Although such layouts may be
time-expensive to set-up from scratch by the network user, they remain fixed for
relatively long time. In addition, in a network where links may fail it is important
to establish a virtual path layout which guarantees fault-free transmission of the
packets. The construction has to take into account the available capacity of the
existing links, i.e. the congestion bounds of the links cannot be exceeded. Hence
it is important to construct fault-tolerant virtual path layouts with the least
possible congestion.

The model we use in this paper is based on the Virtual Path Layout model
introduced by Gerstel and Zaks [4, 5]. Messages may be transmitted through
arbitrarily long virtual paths. Packets are routed along those paths by maintaining a routing field whose subfields determine intermediate destinations of
the packet, i.e. end-points of virtual paths on its way to the final destination.
In such a network it is important to construct path layouts that minimize the
hop number (i.e. the number of virtual paths used to travel between any two
nodes) as a function of edge-congestion (i.e. the number of virtual paths passing
through a link).

1.1 Notation and definitions

In this paper we consider the problem of constructing virtual path layouts for a
complete network of n processors in which a certain number, say f, of links may
fail. We also assume that the number of faults may be arbitrary but the network
resulting after removing the faulty links remains connected. Before proceeding
with an outline of the main results of the paper we give the following definitions.

- K_n is the complete network on a set X of n nodes.
- For any set $\mathcal{F}$ of links of K_n, $K_n[\mathcal{F}]$ is the network resulting from K_n by
  deleting all links from $\mathcal{F}$.
- A virtual path (VP) in a network is a simple chain in this network (i.e. a
  non-repetitive sequence $(v_1, ..., v_k)$ of nodes).
- A virtual channel (VC) of length k, joining nodes u and v, is a sequence
  $p_1, p_2, ..., p_k$ of VP's such that p_1 begins at node u, p_k ends at node v and
  the beginning of p_{i+1} coincides with the end of p_i, for $i < k$.

- A h-hop (virtual path) layout $\mathcal{P}$ in K_n is a collection of virtual paths, such that every pair of nodes u and v is joined by a VC of length at most h, composed of VP's from $\mathcal{P}$.
- The congestion of a layout $\mathcal{P}$ in K_n is the maximum number of VP's from $\mathcal{P}$ passing through any link of K_n.
- A layout $\mathcal{P}$ in K_n is h-hop f-tolerant, for positive integers h and f, if, for any set $\mathcal{F}$ of links such that $|\mathcal{F}| \leq f$ and $K_n[\mathcal{F}]$ is connected, any pair of nodes u and v is joined by a VC of length at most h, composed of VP's from $\mathcal{P}$ not containing links from $\mathcal{F}$.
- The length of a simple path in a network is the number of links in this path.
- The diameter of a connected network is the maximum over shortest paths between all pairs of distinct nodes.
- By $[n]_k$, for positive integers n and k, we denote the descending factorial $n(n-1)\cdots(n-k+1)$. For $k \leq 0$, we define $[n]_k = 1$.

1.2 Results of the paper

First, we study the minimal congestion of 1-hop f-tolerant layouts in K_n. For any positive integer f we give upper and lower bounds on this minimal congestion and construct f-tolerant layouts with congestion corresponding to the upper bounds. Our results are based on a precise analysis of the diameter of the network $K_n[\mathcal{F}]$ when the number $|\mathcal{F}|$ of faulty links is bounded by $F(n,k) = k(n - \frac{k+3}{2})$. The bounds $F(n,k)$ on the number of faults play an important role in our considerations, as they yield thresholds for the diameter of $K_n[\mathcal{F}]$. We show that

- if the number of faults is $f = F(n,k)$ then the minimal congestion of a 1-hop f-tolerant layout is $\Theta(k[n-2]_k)$.

Next we study the minimal congestion of h-hop f-tolerant layouts in K_n, for larger values of the number h of hops. We give upper and lower bounds on the order of magnitude of this congestion. More precisely, assuming that $F(n,k) < f \leq F(n,k+1)$ and $0 < h < n$, we show that the congestion depends on the ratio $\frac{k}{h}$, in the following way:

- there exists a h-hop f-tolerant layout in K_n with congestion

$$O\left(\frac{k}{h}[n-2]_{\lceil \frac{k+2}{h} \rceil - 1}\right),$$

- every h-hop f-tolerant layout in K_n has congestion

$$\Omega\left(\frac{k}{h^2}[n-2]_{\lfloor \frac{k}{h} \rfloor - 1}\right).$$

In the last section, we consider an alternative assumption on fault distribution. Instead of imposing an upper bound f on the number of faulty links and assuming their worst case location, we adopt a random approach. Assume that links fail independently with constant probability $p < 1$. Our goal is to construct layouts with low congestion that tolerate the existing faults with high probability. For any $p < 1$, we show such layouts in K_n, with congestion $O(\log n)$.

2 One hop

In this section we study the minimal congestion of 1-hop f-tolerant layouts in K_n. For any positive integer f we give upper and lower bounds on this minimal congestion and construct f-tolerant layouts with congestion corresponding to the upper bounds.

We first define a sequence of integers which will play an important role in our considerations. These integers are called *key-points*. For natural numbers n and $k \leq n-2$ define $F(n,k) = k(n-\frac{k+3}{2})$. Notice that $F(n,k+1) = F(n,k)+n-k-2$. Any natural number $f \leq \frac{(n-1)(n-2)}{2}$ can be uniquely represented as $F(n,k)+r$, where $k \leq n-2$ and $0 < r \leq n-k-2$.

The following result gives a lower bound on the congestion of any 1-hop f-tolerant layout in K_n.

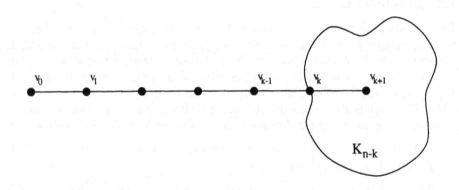

Fig. 1. The network G

Theorem 1. *Let $f = F(n,k)+r$, where $k \leq n-2$ and $0 < r \leq n-k-2$. Every 1-hop f-tolerant layout has congestion $\Omega(rk[n-2]_k)$.*

Proof. Let $(v_0, v_1, ..., v_k, v_{k+1})$ be any simple path of length $k+1$ in K_n. Consider the network G on the set X of all nodes, whose links are those from the above path and those of the complete graph on $X \setminus \{v_0, v_1, ..., v_{k-1}\}$ (see figure 1). Let $\mathcal{F}$ be the set of all links in K_n which are not links of G. Thus $G = K_n[\mathcal{F}]$ and

$$|\mathcal{F}| = \frac{n(n-1)}{2} - \frac{(n-k)(n-k-1)}{2} - k = F(n,k).$$

Consider any 1-hop f-tolerant layout $\mathcal{P}$ in K_n. We claim that there are at least $r+1$ VP's in $\mathcal{P}$ joining v_0 and v_{k+1}, which have the prefix $(v_0, v_1, ..., v_k)$. Suppose not and let $p_1, ..., p_t$, for $t \leq r$, be all such VP's in $\mathcal{P}$ with this prefix. Let Z be the set of links in all those VP's following link (v_{k-1}, v_k). Clearly $|Z| \leq r$ and hence $|\mathcal{F} \cup Z| \leq f$. However, all VP's in $\mathcal{P}$ joining v_0 and v_{k+1} must contain links from $\mathcal{F} \cup Z$, which contradicts 1-hop f-tolerance of $\mathcal{P}$.

It follows that for any pair u and v of distinct nodes and for any simple path $(u, v_1, ..., v_k, v)$ of length $k+1$ joining them, there are at least $r+1$ VP's in $\mathcal{P}$ joining u and v, which have the prefix $(u, v_1, ..., v_k)$. There are $\frac{n(n-1)}{2}$ pairs of nodes and $[n-2]_k$ simple paths of length $k+1$ joining each pair. For each such path, the $r+1$ VP's (of length at least $k+1$) contribute at least $(r+1)(k+1)$ links. Thus the sum of numbers of links in all VP's of the layout $\mathcal{P}$ is at least $(r+1)(k+1)\frac{n(n-1)}{2}[n-2]_k$. Consequently the congestion of $\mathcal{P}$ is at least the average of this total per link, i.e. $(r+1)(k+1)[n-2]_k = \Omega(rk[n-2]_k)$. □

Since $F(n, k) = F(n, k-1) + n - (k-1) - 2$, we obtain the following corollary concerning congestion of 1-hop f-tolerant layouts for f being a key-point.

Corollary 2. *If $f = F(n, k)$, where $k \leq n-2$ then every 1-hop f-tolerant layout has congestion $\Omega(k[n-2]_k)$.*

Next we focus attention on the upper bound on congestion of 1-hop f-tolerant layouts. We first prove the following lemma.

Lemma 3. *If $|\mathcal{F}| \leq F(n, k)$ then the diameter of $K_n[\mathcal{F}]$ is at most $k+1$.*

Proof. Let $\mathcal{F}$ be any set of links in K_n such that $K_n[\mathcal{F}]$ is connected. Suppose that the diameter of $K_n[\mathcal{F}]$ exceeds $k+1$. There exists a pair of nodes u and v with the shortest path P in $K_n[\mathcal{F}]$ joining them. Suppose that P has length $k+2$. Let Y be the set of all nodes not belonging to this path. Thus $|Y| = n - k - 3$.

For any node w in Y, the distance between nodes v_1 and v_2 in the path, such that links (w, v_1) and (w, v_2) do not belong to $\mathcal{F}$, is at most 2: otherwise using the detour v_1, w, v_2 instead of the segment between v_1 and v_2 would create a shortcut, thus contradicting the minimality of P. It follows that for any node $w \in Y$ and all but three consecutive nodes v_i from P, the link (w, v_i) must belong to $\mathcal{F}$. On the other hand, all links joining nodes from the path P, except links in this path, must be in $\mathcal{F}$: otherwise a shortcut would again be possible. Thus the size of $\mathcal{F}$ is at least

$$(n - k - 3)k + \frac{(k+3)(k+2)}{2} - (k+2) = F(n, k) + 1.$$

Consequently, if the shortest path in $K_n[\mathcal{F}]$ joining u and v has length $k+2$, then $|\mathcal{F}| \geq F(n, k) + 1$. The same argument shows that if this shortest path has length $k + 2 + x$, for a positive integer x, then $|\mathcal{F}| \geq F(n, k + x) + 1$. Since $F(n, k) < F(n, k+1)$, this implies that if the diameter of $K_n[\mathcal{F}]$ exceeds $k+1$ then $|\mathcal{F}| \geq F(n, k) + 1$. This concludes the proof. □

The proofs of theorem 1 and lemma 3 provide the following characterization of the key-point $F(n, k)$: this is the maximum number f such that, whenever f links are deleted from the complete network K_n, the resulting network has diameter at most $k+1$.

Lemma 3 implies an upper bound on congestion of 1-hop f-tolerant layouts which matches the lower bound from theorem 1 when f is a key-point (cf. corollary 2).

Theorem 4. *Let $f \leq F(n, k)$, where $k \leq n - 2$. Then the layout consisting of all VP's of length at most $k + 1$ is 1-hop f-tolerant and has congestion $O(k[n-2]_k)$.*

Proof. Lemma 3 implies that the layout is 1-hop f-tolerant. In order to estimate its congestion, notice that, due to symmetry, the number of VP's containing link l is the same, for any link l. Each VP of length i contributes i to the congestion count. There are $[n]_{i+1}$ such VP's, hence the total contribution is $\sum_{i=1}^{k+1} i[n]_{i+1}$. Since all links are equally loaded, in order to obtain congestion, this sum should be divided by $\frac{n(n-1)}{2}$, thus giving congestion equal to $2\sum_{i=1}^{k+1} i[n-2]_{i-1}$. Let $a_i = i[n-2]_{i+1}$. We have $\frac{a_{i+1}}{a_i} \geq 2$ for $i \leq n - 3$, hence the series in question grows faster than geometric and consequently the order of magnitude of the sum of its initial segment is the same as that of the last term. It follows that the congestion of our layout is $O(k[n-2]_k)$. □

The upper bound given in the above theorem is not tight for values of f which are not key-points. In fact we conjecture that it is the lower bound from theorem 1 which is tight for such values, up to a multiplicative constant.

Conjecture. Let $f = F(n, k) + r$, where $k \leq n - 2$ and $0 < r \leq n - k - 2$. There exists a 1-hop f-tolerant layout with congestion $O(r(k + 1)[n - 2]_k)$.

Notice that we formulated the conjecture putting $O(r(k+1)[n-2]_k)$ instead of $O(rk[n-2]_k)$ to include the case $k = 0$. Although we cannot prove this conjecture in general, we construct an appropriate layout for $k = 0$, i.e. when $f \leq n - 2$. Let $v_0, ..., v_{n-1}$ be a labeling of all nodes of the complete network K_n. In the sequel all operations on node indices are performed modulo n. We say that two nodes v_i and v_j are k-neighbors if $k = \min(j - i, i - j)$. Note that $k \leq \lfloor \frac{n}{2} \rfloor$.

Consider the following 1-hop f-tolerant layout $\mathcal{P}_f$ in K_n, for $f \leq n - 2$. If f is equal to $n - 3$ or $n - 2$ then $\mathcal{P}_f$ consists of all VP's of length 1 and 2. For $f \leq n - 4$ the layout $\mathcal{P}_f$ is the union of $f + 3$ disjoint layouts $\mathcal{P}_f^0, ..., \mathcal{P}_f^{f+2}$, defined as follows. The layout $\mathcal{P}_f^0$ consists of all VP's of length 1 in K_n. All other layouts $\mathcal{P}_f^i$, for $i = 1, ..., f + 2$, consist of the following VP's of length 2. Any pair of k-neighbors in X, for $k = 1, ..., \lfloor \frac{n}{2} \rfloor$, except pairs (v_{i-1}, v_{k+i-1}) and (v_{k+i-1}, v_{2k+i-1}), is joined by a VP of length 2 with middle node v_{k+i-1}. This node is called *central* for fixed i and k. (see figure 2). Notice that, for fixed i and k, every link is in at most two VP's of $\mathcal{P}_f^i$ (cf. figure 2).

Lemma 5. *For any $f \leq n - 2$, the layout $\mathcal{P}_f$ is 1-hop f-tolerant and has congestion $O(f)$.*

Proof. For any pair of nodes u and v and all $i = 1, ..., f + 2$, VP's in layouts $\mathcal{P}_f^i$ joining u and v have distinct middle nodes and hence are link disjoint. For at most two values of i the layout $\mathcal{P}_f^i$ does not contain a VP joining these nodes. Hence there are at least f link disjoint VP's of length 2 in $\mathcal{P}_f$ joining u and v.

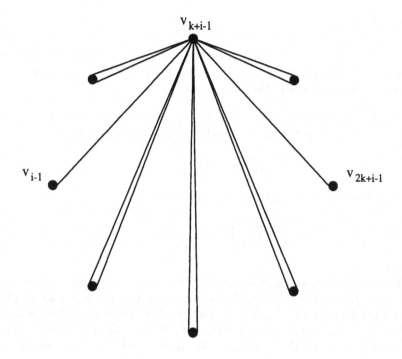

Fig. 2. VP's joining k-neighbors in $\mathcal{P}_f{}^i$

Together with the VP of length 1 this gives at least $f + 1$ link disjoint VP's joining these nodes. This proves that the layout $\mathcal{P}_f$ is 1-hop f-tolerant.

It remains to estimate the congestion of layout $\mathcal{P}_f$. Fix a link $l = (u, v)$ and $i \in \{1, .., f+2\}$. Each of nodes u and v can be central for at most one value of k. For a fixed k, the link l is in at most two VP's of $\mathcal{P}_f{}^i$. Hence there are at most four VP's in $\mathcal{P}_f{}^i$ containing link l. This link is also in one VP from $\mathcal{P}_f{}^0$. This gives the upper bound $4(f + 2) + 1 = 4f + 9$ on congestion of the layout $\mathcal{P}_f$.

$\square$

Notice that the gap between the upper and the lower bounds obtained in theorems 1 and 4 for general values of $f = F(n, k) + r$, is a factor of $\frac{n-k}{r}$. For k close to n, as well as for f close to the nearest larger key-point, the orders of magnitude of the upper and lower bounds meet. In some cases, however, e.g. for $r = 1$ and $k = \frac{n}{2}$, this gap becomes a factor of $\Theta(n)$. The exact order of magnitude of the minimal congestion of 1-hop f-tolerant layouts for such values f remains open (cf. the conjecture).

3   Many hops

In this section we study the minimal congestion of h-hop f-tolerant layouts in K_n, for larger values of the number h of hops. We give upper and lower bounds on the order of magnitude of this congestion.

Theorem 6. *Assume that $F(n,k) < f \leq F(n, k+1)$ and $0 < h < n$.*
1. There exists a h-hop f-tolerant layout in K_n with congestion

$$O(\frac{k}{h}[n-2]_{\lceil \frac{k+2}{h} \rceil - 1}).$$

2. Every h-hop f-tolerant layout in K_n has congestion

$$\Omega(\frac{k}{h^2}[n-2]_{\lfloor \frac{k}{h} \rfloor - 1}).$$

Proof.
1. Let f be as assumed and let $|\mathcal{F}| \leq f$. According to lemma 3, there exists a path P of length $k+2$ in $K_n[\mathcal{F}]$, between any two nodes u and v. There exists a virtual channel $C = (P_1, ..., P_h)$ of length h, joining u and v, such that all VP's P_i are segments of P and every path P_i has length at most $\lceil \frac{k+2}{h} \rceil$. It follows that the layout Q consisting of all virtual paths of length not greater than $\lceil \frac{k+2}{h} \rceil$ is h-hop f-tolerant. In view of theorem 4, the congestion of layout Q is $O(\frac{k}{h}[n-2]_{\lceil \frac{k+2}{h} \rceil - 1})$.

2. Fix a h-hop f-tolerant layout $\mathcal{P}$. Let $P = (v_0, v_1, ..., v_k)$ be any simple path of length k in K_n. Consider the unique virtual channel of length h joining v_0 and v_k whose VP's are consecutive segments $S_1, ..., S_h$ of P such that $S_1, ..., S_i$ have length $\lceil \frac{k}{h} \rceil$ and $S_{i+1}, ..., S_h$ have length $\lfloor \frac{k}{h} \rfloor$, for some $i \leq s$.

Consider the set $\mathcal{F}$ of links of size $F(n,k)$, defined in the proof of theorem 1. In the network $G = K_n[\mathcal{F}]$ the path P is the only simple path joining v_0 and v_k. Since $f \geq F(n,k)$, there must exist a VP in the layout $\mathcal{P}$ which contains one of the segments S_i. The value i must be the same, $i = i_0$, for at least $\frac{[n]_{k+1}}{h}$ paths P. For a given simple path S of length L, where $L = \lceil \frac{k}{h} \rceil$ or $L = \lfloor \frac{k}{h} \rfloor$, there exist $[n - L - 1]_{k+1-L-1}$ simple paths of length k in which S is the i_0th segment. It follows that there are at least

$$N = \frac{[n]_{k+1}}{h[n - \lfloor \frac{k}{h} \rfloor - 1]_{k+1 - \lfloor \frac{k}{h} \rfloor - 1}}$$

VP's of length $\Omega(\frac{k}{h})$ in the layout $\mathcal{P}$. Consequently, the congestion of $\mathcal{P}$ is at least

$$\Omega(\frac{k}{h} N \frac{2}{n(n-1)}) = \Omega(\frac{k}{h^2}[n-2]_{\lfloor \frac{k}{h} \rfloor - 1}).$$

$\square$

If $h \geq k + 2$, both the upper and the lower bound in the above theorem become constant. Indeed, in this case the layout consisting of VP's of length 1 (i.e. individual links only) is h-hop f-tolerant and has congestion 1.

4 Random faults

In this section we consider an alternative assumption on fault distribution. Instead of imposing an upper bound f on the number of faulty links and assume their worst case location, we adopt a random approach. Assume that links fail independently with constant probability $p < 1$. Our goal now is to construct layouts with low congestion that tolerate the existing faults with high probability. More precisely, a layout $\mathcal{P}$ is called h-hop p-safe, for a given positive real $p < 1$, if whenever links in the set $\mathcal{F}$ are chosen randomly with probability p, the probability that there exists a VC of length h in $K_n[\mathcal{F}]$, composed of VP's from $\mathcal{P}$, between any pair of nodes, is at least $1 - \frac{1}{n}$.

Theorem 7. *For any $p < 1$, there exists a 1-hop p-safe layout in K_n, with congestion $O(\log n)$.*

Proof. Let $q = 1 - (1 - p)^2$, $c = \frac{-3}{\log q}$ and $f = \lceil c \log n \rceil$. In view of lemma 5, the layout $\mathcal{P}_f$ defined in section 2 has congestion $O(f) = O(\log n)$. It remains to show that this layout is p-safe.

Let $\mathcal{F}$ be a set of links chosen according to the definition of a p-safe layout. Consider any pair of distinct nodes u and v. There are f (link disjoint) VP's of length 2 joining u and v, in the layout $\mathcal{P}_f$. Consider any of those VP's, call it P. The probability that at least one of the two links of P is in $\mathcal{F}$ is equal to q. For distinct link disjoint paths P the above events are independent. Hence the probability that for every VP in $\mathcal{P}_f$, joining u and v, at least one of its links is in $\mathcal{F}$, is at most $q^f \le q^{c \log n}$. Since there are less than n^2 pairs (u, v), the probability that this happens for at least one such pair is at most $n^2 q^{c \log n}$. We have

$$n^2 q^{c \log n} = n^2 n^{c \log q} = \frac{1}{n},$$

in view of $c = \frac{-3}{\log q}$. This proves that the layout $\mathcal{P}_f$ is p-safe. $\square$

Since layout $\mathcal{P}_f$ considered in the above proof consists of VP's of length at most two, the following corollary holds.

Corollary 8. *For any $p < 1$, the layout consisting of all VP's of length one is 2-hop p-safe.*

In fact, in the 2-hop case we have $n - 2$ VC's for each pair of nodes. The probability that all of them are faulty for some pair is not only at most $\frac{1}{n}$ but indeed decreases exponentially in n.

The comparison of theorems 1 and 7 shows a dramatic difference between the worst case and random scenarios. Suppose that a fixed fraction, say $\frac{1}{4}$, of all links are faulty, and let $h = 1$. If we require worst case fault-tolerance from a layout, theorem 1, applied for a k linear in n, implies that the layout must have congestion exponential in n. However, if we assume that this fraction of faulty links is distributed randomly in the complete network K_n, theorem 7 shows that we can construct a layout fault-tolerant with high probability, which has only logarithmic congestion.

5 Conclusion

We presented upper and lower bounds on the minimal congestion of fault-tolerant layouts in complete networks, assuming the worst case fault distribution among links. We also showed that congestion can dramatically decrease if faults are distributed randomly and almost certain, rather than worst case, fault tolerance is required.

In the worst case scenario our bounds on congestion of 1-hop layouts are tight, up to a multiplicative constant, for some values of the number of faults, called key-points. In between these values there remain gaps between upper and lower bounds: we conjecture that the lower bounds are tight. Another interesting problem is to generalize the study of fault-tolerant layouts from the complete network to other graphs.

References

1. B. Awerbuch, A. Bar-Noy, N. Linial and D. Peleg, "Improved Routing with Succinct Tables", Journal of Algorithms 11, 307-341 (1990).
2. J. Y. Le Boudec, "The Asynchronous Transfer Mode: A Tutorial", Computer Networks and ISDN Systems, 24:279-309, 1992.
3. I. Cidon, O. Gerstel and S. Zaks, "A Scalable Approach to Routing in ATM Networks", WDAG 94.
4. O. Gerstel and S. Zaks, "The Virtual Path Layout Problem in Fast Networks", PODC 94.
5. O. Gerstel and S. Zaks, "The Virtual Path Layout Problem in ATM Ring and Mesh Networks", in Proceedings of SIROCCO 94.
6. D. E. McDysan and D. L. Spohn, "ATM: Theory and Applications", McGraw-Hill Series on Computer Communication, 1995.
7. M. de Prycker, "Asynchronous Transfer Mode: Solutions for Broadband ISDN", Ellis Horwood Limited, 1993 (2nd edition).
8. N. Santoro and R. Khatib, "Labeling and Implicit Routing in Networks", Comput. J. 28(1985), 5-8.

Polynomial Automaticity, Context-Free Languages, and Fixed Points of Morphisms (Extended Abstract)

Ian Glaister* and Jeffrey Shallit**

Department of Computer Science
University of Waterloo
Waterloo, Ontario, Canada N2L 3G1
shallit@graceland.uwaterloo.ca

Abstract. If L is a formal language, we define $A_L(n)$ to be the number of states in the smallest deterministic finite automaton that accepts a language that agrees with L on all inputs of length $\leq n$. This measure is called *automaticity*. In this paper, we first study the closure properties of the class DPA of languages of deterministic polynomial automaticity, i.e., those languages L for which there exists k such that $A_L(n) = O(n^k)$. Next, we discuss similar results for a nondeterministic analogue of automaticity, introducing the classes NPA (languages of nondeterministic polynomial automaticity) and NPLA (languages of nondeterministic poly-log automaticity). We then show how to construct a context-free language with automaticity arbitrarily close to the maximum possible. Finally, we conclude with some remarks about the automaticity of sequences, focusing on fixed points of homomorphisms.

1 Introduction.

In two previous papers [11, 7], the second author and co-authors studied the concept of *automaticity*: roughly speaking, how closely a formal language L can be approximated by regular languages L'; also see [10]. In this paper, the third of a series, we introduce three new complexity classes based on this concept, and study their properties.

As usual, we define a finite automaton M to be a 5-tuple, $(Q, \Sigma, \delta, q_0, F)$, where Q is a finite set of *states*, Σ is a finite *input alphabet*, q_0 is the *start state*, and F is a set of *final states*. The map δ is called the *transition function*. If M is *deterministic*, then δ maps $Q \times \Sigma$ to Q. If M is *nondeterministic*, then δ maps $Q \times \Sigma$ to 2^Q. The map δ is extended to $Q \times \Sigma^*$ in the obvious manner. In either case, we define $L(M)$, the *language accepted by* M, to be the set $\{x \in \Sigma^* : \delta(q_0, x) \cap F \neq \emptyset\}$. For more on these concepts, see, for example, [5].

---

* Author's current address: Array Systems Computing Inc., 1120 Finch Avenue West, 8th Floor, North York, ON M3J 3H7 Canada. E-mail: ian@array.ca

** Research supported in part by a grant from NSERC. Please direct all correspondence to this author.

We denote the class of all deterministic finite automata by DFA, and the class of all nondeterministic finite automata by NFA. By $|M|$ we mean $|Q|$, the number of states in the machine M.

Let ϵ denote the empty string, and let $\Sigma^{\leq n} = \epsilon + \Sigma + \Sigma^2 + \cdots + \Sigma^n$, the set of all strings over Σ of length at most n. Let L, L' be languages with $L, L' \subseteq \Sigma^*$. If $L \cap \Sigma^{\leq n} = L' \cap \Sigma^{\leq n}$, we say that L' is an nth order approximation to L.

Given a language $L \subseteq \Sigma^*$, we define the function $A_L(n)$, the deterministic automaticity of L, as follows:

$$A_L(n) = \min\{|M| \ : \ M \in \text{DFA and } L(M) \cap \Sigma^{\leq n} = L \cap \Sigma^{\leq n}\}.$$

Informally, $A_L(n)$ counts the number of states in the smallest finite automaton that accepts some nth order approximation to L.

Similarly, we define $N_L(n)$, the nondeterministic automaticity of L, as follows:

$$N_L(n) = \min\{|M| \ : \ M \in \text{NFA and } L(M) \cap \Sigma^{\leq n} = L \cap \Sigma^{\leq n}\}.$$

We now introduce three new complexity classes:

1. deterministic polynomial automaticity, or DPA:

$$\text{DPA} = \{L \subseteq \Sigma^* \ : \ \exists k \text{ such that } A_L(n) = O(n^k)\}.$$

2. nondeterministic polynomial automaticity, or NPA:

$$\text{NPA} = \{L \subseteq \Sigma^* \ : \ \exists k \text{ such that } N_L(n) = O(n^k)\}.$$

3. nondeterministic poly-log automaticity, or NPLA:

$$\text{NPLA} = \{L \subseteq \Sigma^* \ : \ \exists k \text{ such that } N_L(n) = O((\log n)^k)\}.$$

There are clear analogies of these classes with more traditional ones such as P, NP, and NC. In this paper, we first discuss the closure properties of these new classes.

It is perhaps worth pointing out that, unlike the classes P and NP, and as a consequence of the non-uniformity of the model, these classes contain uncountably many languages. If this troubles the reader, one can restrict one's attention to the recursive languages in these classes without altering any of the results in this paper.

Next, we examine the relationship between context-free languages and automaticity, showing in particular how to construct a CFL with automaticity arbitrarily close to the maximum possible.

Finally, we discuss the automaticity of sequences, focusing in particular on the automaticity of fixed points of homomorphisms.

2 Results on DPA.

In this section, we discuss the properties of the class DPA. First, we show that DPA consists of a strict hierarchy of complexity classes. Second, we study the closure properties of DPA.

In [11] it is shown that if L is a unary language (i.e., L is defined over an alphabet with one letter), then $A_L(n) = O(n)$. In this case, DPA is trivially closed under every operation. Hence, for the remainder of this section and the next one, we assume that $|\Sigma| \geq 2$.

First, we state three useful lemmas. Proofs can be found, for example, in [6, 11]. Let $L \subseteq \Sigma^*$, and let S be a finite set of strings over Σ. Suppose for all $x, y \in S$, there exists a $w \in \Sigma^*$ such that $|xw|, |yw| \leq n$, and exactly one of xw, yw is in L. Then we call S an n-dissimilar set of strings for L.

Lemma 1 (Kaneps and Freivalds) $A_L(n)$ is equal to the cardinality of the largest n-dissimilar set of strings for L.

Lemma 2 For $n \geq 0$ we have

$$A_L(n) \leq 2 + \sum_{w \in L \cap \Sigma^{\leq n}} |w| \leq 2 + n|L \cap \Sigma^{\leq n}|.$$

Lemma 3 The language L is regular if and only if $A_L(n) = O(1)$. The same statement holds for $N_L(n)$.

We now sketch the proof of:

Theorem 4 For all integers $k \geq 0$, there is a language L_k such that $A_{L_k}(n) = \Theta(n^k)$.

Proof. For L_0 we may take any regular language, by Lemma 3. Now let k be an integer ≥ 1, and define

$$L_k = \{0^{a_1} 1 0^{a_2} 1 \cdots 0^{a_k} 1 0^{a_1} 1 0^{a_2} 1 \cdots 0^{a_k} 1 : a_1, \ldots, a_k \geq 0\}.$$

Let $n' = \lfloor n/2 \rfloor$. It is now not hard to show, using Lemmas 1 and 2, that $A_{L_k}(n) = \Theta(n^k)$, where the constant implied in the Θ depends on k. Details will appear in the final paper. ∎

Now we move on to the closure properties of DPA.

Theorem 5 The class DPA is closed under union, intersection, complement, and inverse homomorphism.

Proof. Left to the reader. ∎

Now we turn to properties under which DPA is not closed. First, we prove the following lemma:

Lemma 6 Let $(r_k)_{k \geq 0}$ be any sequence of integers satisfying $r_0 = 1$ and $r_{k+1} \geq 2r_k$. Define $L = \{(0+1)^* 1(0+1)^{r_k} : k \geq 0\}$. Put $n = r_k + r_{k-1}$ for $k \geq 1$. Then $A_L(n) \geq 2^{r_{k-1}}$.

Proof. Let w, x be two distinct strings in $(0+1)^{r_{k-1}}$. Without loss of generality, we may write

$$w = w_1 w_2 \cdots w_{t-1} 1 w_{t+1} \cdots w_{r_{k-1}}$$
$$x = x_1 x_2 \cdots x_{t-1} 0 w_{t+1} \cdots x_{r_{k-1}}$$

for some t with $1 \leq t \leq r_{k-1}$. Let $y = 0^{r_k + t - r_{k-1}}$. We have $wy \in (0+1)^{t-1}1(0+1)^{r_k}$, so $wy \in L \cap \Sigma^{\leq n}$. On the other hand, we claim that $xy \notin L$. If it were, then there would be a suffix of xy of the form $1(0+1)^\ell$, for some $\ell \in R$. The first 1 in this suffix must come from x. Without loss of generality, let it be x_s, so $x = x_1 x_2 \cdots x_{s-1} 1 x_{s+1} \cdots x_{r_{k-1}}$, for some $s \neq t$. In this case we have $\ell = r_k + t - s \in R$. But either $1 \leq s \leq t-1$, or $t+1 \leq s \leq i$. Hence either $r_k + 1 \leq \ell \leq r_k + t - 1 < r_{k+1}$, or (if $t \neq r_{k-1}$) $r_{k-1} < r_k - r_{k-1} + t \leq \ell \leq r_k - 1$. Both cases contradict $\ell \in R$.

It follows that $(0+1)^{r_{k-1}}$ forms a set of n-dissimilar strings. Hence $A_L(n) \geq 2^{r_{k-1}}$. ∎

Theorem 7 *The class DPA is not closed under concatenation.*

Proof. Let $L_1 = (0+1)^*$, and $L_2 = \{1(0+1)^{2^k} : k \geq 0\}$. Then it is easy to see that $A_{L_1}(n) = O(1)$, and $A_{L_2}(n) = O(n)$.

Now define $L = L_1 L_2$, and put $r_k = 2^k$ in Lemma 6. Then $n = 2^k + 2^{k-1} = 3 \cdot 2^{k-1}$. It follows that $A_L(n) \geq 2^{k-1}$. Thus, for infinitely many n, we have $A_L(n) \geq 2^{n/3}$, and so $L \notin$ DPA. ∎

Theorem 8 *The class DPA is not closed under reversal.*

Proof. Let $L' = \{(0+1)^{2^k}1(0+1)^* : k \geq 0\}$. It is easy to see that $A_{L'}(n) = O(n)$. On the other hand, $L'^R = L$, where $L = \{(0+1)^*1(0+1)^{2^k} : k \geq 0\}$. From the previous theorem, $L \notin$ DPA. ∎

Theorem 9 *The class DPA is not closed under quotient by regular sets.*

Proof. Let $L = \{ww10^{2^k} : w \in (0+1)^k, k \geq 0\}$, and let $R = 10^*$. By Lemma 6, we know that

$$A_L(n) \leq 2 + \sum_{w \in L \cap \Sigma^{\leq n}} |w| \leq 2 + \sum_{0 \leq j \leq n} j|L \cap \Sigma^j| = O(n^2). \qquad (1)$$

Now $L/R = \{ww : w \in (0+1)^*\}$. Let $S = S_n = \{w \in (0+1)^* : |w| = \lfloor n/2 \rfloor\}$. Then it is easy to see that S_n is an n-dissimilar string set for L/R, and so $A_{L/R}(n) = \Omega(2^{n/2})$. ∎

Theorem 10 *The class DPA is not closed under Kleene closure.*

Proof. Define $L = \{b(a+b)^{k^2-1} : k \geq 2\} \subseteq \Sigma^* = (a+b)^*$. We will show that $L \in$ DPA and $L^* \notin$ DPA. To see that $L \in$ DPA, we use Lemma 2. We have $|L \cap \Sigma^{\leq n}| = O(\sqrt{n})$, so $A_L(n) = O(n^{3/2})$.

We now show that $A_{L^*}(n) \geq n^{n^{1/8}/8}$ for all n sufficiently large. First, we introduce some definitions. We say that a string $w \in (a+b)^*$ is *valid for position* j if there exists a way to write $w = w_1 w_2 \cdots w_r$, where $r \geq 1$, $w_i \in (a+b)^+$ for $1 \leq i \leq r$, $w_i \in L$ for $1 \leq i < r$, the first symbol of w_r is b, and $|w_r| = j$. Note that a word may be valid for no positions, or for several. As an example, the string $baaabaaaaaaaaabaaab$ is valid for positions $1, 5, 14, 18$. Then it is easy to see that $w \in L^+$ if and only if there exists a $k > 1$ such that w is valid for position k^2.

Next, let $S = \{s_1, s_2, \ldots, s_k\}$ be a nonempty set of positive integers. If S satisfies the following two conditions, then we call it *good*:

(a) $s_1 = 1$;

(b) for all i with $2 \leq i \leq k$, there exists an integer $t_i > 1$ such that $s_i - s_{i-1} = t_i^2$.

We call s_k the *weight* of the set S, k the *size* of the set S, and $\max_{2 \leq i \leq k}(s_i - s_{i-1})$ the *span* of the set S.

If S is good, then there is a word $w = w(S)$ of length equal to the weight of S, such that w is valid exactly for the positions specified by S. Namely, we can take

$$w = ba^{s_k - s_{k-1} - 1} ba^{s_{k-1} - s_{k-2} - 1} b \cdots ba^{s_2 - s_1 - 1} ba^{s_1 - 1}.$$

Note that the map that sends S to $w(S)$ is injective.

Now suppose S and T are different good sets of weight at most m. Then we claim there exists a word y with $|y| \leq \lceil (m+2)/2 \rceil^2$ such that $w(S)y \in L^+$ and $w(T) \notin L^+$. Since S and T are different, without loss of generality there exists c such that $c \in S$ and $c \notin T$. Choose $y = a^{\lceil (m+2)/2 \rceil^2 - c}$. Then $w(S)$ is valid for position c, so $w(S)y$ is valid for position $\lceil (m+2)/2 \rceil^2$. Hence $w(S)y \in L^+$.

On the other hand, suppose $w(T)$ is valid for a position p. Then $p \neq c$ and $1 \leq p \leq m$. Thus if $w(T)y$ is valid for a position q, we must have

$$\lceil (m+2)/2 \rceil^2 - c + 1 \leq q \leq \lceil (m+2)/2 \rceil^2 - c + m.$$

This implies that

$$\lceil (m+2)/2 \rceil^2 + 1 \leq q \leq \lceil (m+2)/2 \rceil^2 + m. \tag{2}$$

If m is even, say $m = 2t$, then (2) implies that $t^2 < q < (t+2)^2$. If $w(T)y \in L^+$, q must be a square, and so $q = (t+1)^2$. But then $w(T)$ is valid for position c, a contradiction. Similar reasoning handles the case where m is odd. It follows that $w(T)y \notin L^+$.

Now fix an n, and define the collection C_n to be those good sets S of span at most $(n^{1/8} + 2)^2$ and size $k = \lceil n^{1/8} \rceil + 1$. Each set has weight at most $1 + (n^{1/8} + 2)^2(k-1) = O(n^{3/8})$.

Now define $U = \{w(S) : S \in C_n\}$. The cardinality of U is the cardinality of C_n, which corresponds to the number of possible choices of S with the given span and size. Each of the $k - 1$ possible differences $s_i - s_{i-1}$ can be any one of at least $n^{1/8}$ possible squares, and so there are $(n^{1/8})^{n^{1/8}}$ possibilities for S. We claim that, for all n sufficiently large, U is an n-dissimilar string set for L^+ (and hence, for L^*). This is clear, since by the reasoning above, if S and T are different elements of C_n, then $w(S)$ and $w(T)$ are distinguishable by a string y whose length is $O((n^{3/8})^2) = O(n^{3/4})$. Hence $|w(S)y| = O(n^{3/4} + n^{3/8})$, and the same length bound holds for $w(T)y$. It therefore follows that $A_{L^*}(n) \geq |U| \geq (n^{1/8})^{n^{1/8}}$, and so $L^* \notin \text{DPA}$. $\blacksquare$

Theorem 11 *If $|\Sigma| \geq 3$, then DPA is not closed under homomorphism.*

Proof. Let $L = \{(0+1)^* 2 (0+1)^{2^k} : k \geq 0\}$, and define $h(0) = 0$, $h(1) = h(2) = 1$. Then it is easy to see that $A_L(n) = O(n)$, but $h(L)$ is not in DPA by Theorem 7. $\blacksquare$

3   Results on NPA and NPLA.

In this section, we obtain results on languages with nondeterministic polynomial automaticity: the class NPA.

We start with a new lower bound technique for nondeterministic automaticity. Let U be a finite set of strings. Then we say that U is a set of *uniformly n-dissimilar* strings if for each string $u \in U$ there exists a string w such that

(i) $|uw| \leq n$ and $uw \in L$; and

(ii) for every string $v \in U$ such that $u \neq v$, we have $|vw| \leq n$ and $vw \notin L$.

We sometimes call the string w a *witness* for u.

Then we have the following

Lemma 12 *If U is a set of uniformly n-dissimilar strings for L, then $N_L(n) \geq |U|$.*

Proof. Consider a string $u \in U$. By the definition, there exists a witness string w satisfying conditions (i) and (ii). Let $M = (Q, \Sigma, \delta, q_0, F)$ be any nondeterministic finite automaton that accepts an nth order approximation to L. Now $uw \in L$, and since M accepts all strings in L of length $\leq n$, we have $\delta(q_0, uw) \cap F \neq \emptyset$. Hence there exists at least one state $q \in \delta(q_0, u)$ such that $p \in \delta(q, w)$, where $p \in F$.

However, for every other string $v \in U$, with $v \neq u$, we must have $q \notin \delta(q_0, v)$. For if $q \in \delta(q_0, v)$, we would have $p \in \delta(q_0, vw)$ and so $vw \in L$, a contradiction (since $|vw| \leq n$).

Hence every set $\delta(q_0, u)$ contains a state q which does not appear in any other set $\delta(q_0, v)$ for $u \neq v$. It follows that there must be at least $|U|$ different states in Q. $\blacksquare$

This simple, but powerful, lemma will allow us to estimate the nondeterministic automaticity for a wide variety of languages; see below. However, unlike the case of deterministic automaticity, the lower bound provided by Lemma 12 is not tight. An example of this is the following: consider the set $L = \{0^i 1^j \ : \ i \neq j\}$. Then a simple argument shows that a set of uniformly n-dissimilar strings for L can contain no more than 2 strings. Yet, we know from Lemma 3 that $N_L(n) \neq O(1)$.

Proposition 13 *Let* $L = \{0^i 1^i \ : \ i \geq 0\}$. *Then* $N_L(n) = \Omega(n)$.

Proof. The set $\{\epsilon, 0, 00, \dots, 0^{\lfloor n/2 \rfloor}\}$ forms a set of uniformly n-dissimilar strings for L; the witness for 0^i is 1^i. It follows that $N_L(n) \geq \lfloor n/2 \rfloor + 1$. ∎

Using Lemma 12, we can also prove a theorem analogous to Theorem 4:

Proposition 14 *For all integers* $k \geq 0$, *there is a language* L_k *such that* $N_{L_k}(n) = \Theta(n^k)$.

Proof. Consider the languages L_k introduced in the proof of Theorem 4. The set S there is actually a uniformly n-dissimilar string set for L_k, and so exactly the same upper and lower bounds follow. ∎

Here is another application of Lemma 12:

Proposition 15 *Let* $L = \{ww \ : \ w \in (0+1)^*\}$. *Then* $N_L(n) = \Omega(2^{n/2})$.

Proof. The set $S = S_n = (0+1)^{\lfloor n/2 \rfloor}$ forms a uniformly n-dissimilar string set for L; the witness for w is w itself. It follows that $N_L(n) \geq 2^{\lfloor n/2 \rfloor}$. ∎

Next, we prove a simple result on some operation under which the class NPA is closed:

Proposition 16 *The class* NPA *is closed under the operations of union, intersection, concatenation, Kleene closure, and inverse homomorphism.*

Proof. Let M_1 be an NFA accepting an nth order approximation to L_1, and let M_2 be an NFA accepting an nth order approximation to L_2. Then we can make an NFA accepting an nth order approximation to $L_1 \cup L_2$ by using the usual construction, as given, for example, in [5, p. 31]. The construction gives an automaton with $|M_1| + |M_2| + 2$ states. The other properties can be proved similarly. ∎

Proposition 17 *The class* NPA *is not closed under complement.*

Proof. See [11, §5, Example 4]. ∎

Proposition 18 *The class* NPA *is not closed under quotient by regular sets.*

Proof. Consider the language $L = \{ww\,1\,0^{2^k} : w \in (0+1)^k, k \geq 0\}$ introduced in the proof of Theorem 9. By the same argument given there, $L \in$ NPA. Let $R = 1\,0^*$. Then $L/R = \{ww : w \in (0+1)^*\}$. But by Proposition 15, we have $N_{L/R} = \Omega(2^{n/2})$. ∎

Finally, we examine the languages with nondeterministic poly-log automaticity: the class NPLA.

Theorem 19 *The class NPLA is closed under the operations of union, intersection, concatenation, Kleene closure, and inverse homomorphism.*

Proof. Left to the reader. ∎

Theorem 20 *The class NPLA is not closed under complement.*

Proof. Let $L = \{w \in (0+1)^* : |w|_0 \neq |w|_1\}$. By [11, Theorem 17], we know that $L \in$ NPLA. Now $\overline{L} = \{w \in (0+1)^* : |w|_0 = |w|_1\}$. If this language were in NPLA, then so would $\overline{L} \cap 0^*1^*$, by Theorem 16. But $\overline{L} \cap 0^*1^* = \{0^i 1^i : i \geq 0\}$, which by Proposition 13 is not in NPLA. Hence $\overline{L} \notin$ NPLA. ∎

4 Automaticity and context-free languages

In this section we briefly discuss the automaticity of context-free languages. As shown in [11, Theorem 5], there exists a CFL L_s such that $A_{L_s}(n) = \lfloor (n + 3)/2 \rfloor$. This is the language with essentially the lowest-possible deterministic automaticity. On the other hand, in [11, §5; Example 4], it is shown that the CFL $L_4 = \{w \in (0+1)^* : w \neq w^R\}$ has automaticity $A_{L_4}(n) = \Omega(2^{n/2})$. This raises the question, what is the maximum possible deterministic automaticity for a context-free language over $\{0, 1\}$?

We know from [11, Theorem 9] that if $L \subseteq (0+1)^*$, then $A_L(n) = O(2^n/n)$. We have not been able to find a CFL with deterministic automaticity $\Omega(2^n/n)$, but in the following theorem we construct a sequence of languages with deterministic automaticity arbitrarily close to 2^n:

Theorem 21 *For all real $\epsilon > 0$, there exists a CFL of deterministic automaticity $\Omega(2^{n(1-\epsilon)})$.*

Proof. First, we introduce the following notation. If w is a string with $|w| = n$, then by w_{-i}, $(1 \leq i \leq n)$ we mean the symbol w_{n-i+1}.

To prove the result, we will show that for all integers $r \geq 1$, there exists a language L_r of automaticity $\Omega(2^{\lfloor rn/(r+1) \rfloor - r})$. Let

$$L_r = \{w\,0\,1^a\,0^b : w \in (0+1)^*, 1 \leq a \leq r, b \geq 0, w_{-(rb+a)} = 1\}.$$

It is easy to see that that L_r is a CFL for all $r \geq 1$. We now prove that $A_{L_r}(n) \geq 2^{\lfloor rn/(r+1) \rfloor - r}$. To do this, we exhibit an n-dissimilar string set $S = S_{n,r}$ of cardinality $2^{\lfloor rn/(r+1) \rfloor - r}$.

Assume $n \geq r + 5$, so that $\lfloor rn/(r+1) \rfloor - r \geq 1$, and define

$$S = S_{n,r} = \{w \; : \; |w| = \lfloor rn/(r+1) \rfloor - r\}.$$

Pick two distinct strings from S, say x and y. There must be some position k at which x and y differ, say $x_{-k} \neq y_{-k}$. Clearly $1 \leq k \leq \lfloor rn/(r+1) \rfloor - r$ by construction. Without loss of generality, assume $x_{-k} = 1$ and $y_{-k} = 0$. Write $k = rb + a$ with $1 \leq a \leq r$ and $0 \leq b \leq n/(r+1) - 1$.

Consider $z = 0 \, 1^a \, 0^b$. Then clearly $xz \in L$, but $yz \notin L$. Also,

$$|xz| = |yz| = \left\lfloor \frac{rn}{r+1} \right\rfloor - r + a + b + 1 = \left\lfloor \frac{rn}{r+1} \right\rfloor + \frac{n}{r+1} \leq n, \qquad (3)$$

so x and y are n-dissimilar.

To complete the proof, take $r = \lceil 2/\epsilon \rceil$. Then for $n \geq (r+1)^2$, we have $A_{L_r}(n) \geq 2^{n(1-\epsilon)}$. $\blacksquare$

5 Automaticity of Sequences and Fixed Points of Homomorphisms

In the previous section, we were interested in the automaticity of formal languages. We now turn to discussing the automaticity of sequences.

First, some notation. Let k be an integer ≥ 2 and define $\Sigma_k = \{0, 1, \ldots, k-1\}$. If $w \in \Sigma_k^*$, then by $[w]_k$ we mean w evaluated as a base-k integer, that is, if $w = w_1 w_2 \cdots w_r$, then $[w]_k = \sum_{1 \leq i \leq r} w_{-i} k^{i-1}$. (By w_{-i} we mean w_{r-i+1} — a notation we introduced in §6.)

Let M be a DFA with output, $M = (Q, \Sigma, \delta, q_0, \Delta, \tau)$. The output alphabet is Δ and the output function τ maps Q to Δ. On input w, the machine M outputs the single symbol $\tau(\delta(q_0, w))$.

Suppose $(s_i)_{i \geq 0}$ is a sequence over the finite alphabet Δ. If there exists a DFA with output M such that for all $i \geq 0$, we have $s_i = \tau(\delta(q_0, w^R))$ for all $w \in \Sigma_k^*$ such that $[w]_k = i$, then the sequence $(s_i)_{i \geq 0}$ is said to be k-automatic. (Note that the slightly awkward definition results from the problem of "leading zeroes" input, and that the machine reads the input number starting with the least significant digit.)

Let φ be a homomorphism from Δ^* to Δ^*. If there is a symbol $a \in \Delta$ such that $\varphi(a) = ax$ for some $x \in \Delta^*$, then

$$y = ax\varphi(x)\varphi^2(x)\varphi^3(x)\cdots = \lim_{j \to \infty} \varphi^j(a)$$

is a fixed point of φ; that is, $\varphi(y) = y$. If further φ is nonerasing (i.e., $\varphi(b) \neq \epsilon$ for all $b \in \Delta$), then y is infinite. If $|\varphi(b)| = k$ for all $b \in \Delta$, then φ is said to be k-uniform. A 1-uniform homomorphism is called a coding. A well-known theorem of Cobham [2] states that $(s_i)_{i \geq 0}$ is the image (under a coding) of a fixed point of a k-uniform homomorphism if and only if $(s_i)_{i \geq 0}$ is k-automatic.

Given a sequence (s_i), we can define its automaticity $A_s(n)$ as follows: $A_s(n)$ is the number of states in the smallest DFA with output M for which $s_i = \tau(\delta(q_0, w^R))$ for all $w \in \Sigma_k^*$ with $[w]_k = i$ and $i \leq n$. Clearly (s_i) is an automatic sequence if and only if $A_s(n) = O(1)$. If $A_s(n) = O(\log n)$, then we say (s_i) is *k-quasiautomatic*.

An alternative definition of automaticity of sequences, which is easily seen to be equivalent, is as follows: we look at the *k*-kernel of $(s_i)_{i \geq 0}$:

$$\{(s_{k^a n + b})_{n \geq 0} \; : \; a \geq 0, \; 0 \leq b < k^a\},$$

but only compute the subsequences up to the n'th term of the initial sequence. Hence, if one is looking at, say, $(s_{2i})_{i \geq 0}$, one actually only knows this sequence up to $i = \lfloor n/2 \rfloor$. Now one has a finite collection of sequences, known to various lengths, and one identifies two sequences as the same if they agree on the terms to which they are known. The size of the minimum set of different sequences is then the *k*-automaticity of the sequence at n.

A natural question then is to determine the automaticity of fixed points of non-uniform homomorphisms. We conclude this paper by using some theorems of Diophantine approximation [1] to exhibit a homomorphism whose fixed point is not *k*-quasiautomatic for any $k \geq 2$.

If α is a real irrational number, we can expand it uniquely as an infinite continued fraction, $\alpha = [a_0, a_1, a_2, \ldots]$. The a_i are called the *partial quotients* of α. We say the partial quotients of α are *bounded by B* if $a_i \leq B$ for all $i \geq 1$ [9]. We define $p_n/q_n = [a_0, a_1, \ldots, a_n]$, and call p_n/q_n the *n*th *convergent*. We define a'_n, the *n*th *complete quotient*, to be $[a_n, a_{n+1}, \ldots]$. We define $\{\alpha\} = \alpha - \lfloor \alpha \rfloor$, the *fractional part* of α, and $\|\alpha\| = \min(\alpha - \lfloor \alpha \rfloor, \lceil \alpha \rceil - \alpha)$, the *distance to the nearest integer*.

We then have

Lemma 22 *Let α be an irrational real number, $0 < \alpha < 1$, with partial quotients bounded by B. Let the numbers $0, \{\alpha\}, \{2\alpha\}, \ldots, \{n\alpha\}, 1$ be arranged in ascending order and let them be labeled $p_0, p_1, p_2, \ldots, p_{n+1}$. Then*

$$\min_{0 \leq i \leq n} (p_{i+1} - p_i) \geq \frac{1}{(B+2)n}.$$

Proof. Omitted for space considerations. ∎

Our next lemma is a result on inhomogeneous approximation. Unlike the traditional versions of this theorem, the requirement that α has bounded partial quotients allows us to bound the size of the integers that effect the desired approximation.

Lemma 23 *Let α be an irrational real number, $0 < \alpha < 1$, with partial quotients bounded by B. Let $0 \leq \beta < 1$ be a real number. Then for all $N \geq 1$ there exist integers p, q with $0 \leq p, |q| \leq (B+2)N^2$ such that $|p\alpha - \beta - q| \leq \frac{1}{N}$.*

Proof. Omitted for space considerations. ∎

The next lemma shows that Sturmian sequences have the property that for all pairs of subsequences of the form $(s_{rn+c})_{n\geq 0}$, $(s_{rn+d})_{n\geq 0}$, there is a small witness n that shows that these subsequences are different.

Lemma 24 *Let $0 < \alpha < 1$ be an irrational real number with partial quotients bounded by B. Define the Sturmian word $s_1 s_2 s_3 \cdots$ by $s_n = \lfloor (n+1)\alpha \rfloor - \lfloor n\alpha \rfloor$ for $n \geq 1$. Let $r \geq 2$ be an integer. Then for all integers c, d with $0 \leq c, d < r$, $c \neq d$, there exists an integer n with $0 \leq n \leq 4(B+2)^3 r^3$ such that $s_{rn+c} \neq s_{rn+d}$.*

Proof. We use the "circular representation" for intervals in $[0, 1)$, identifying the point 0 with the point 1, and considering each point modulo 1. See, for example, [4, §3.8, §23.2].

It is easy to see that $s_n = 1 \iff \{n\alpha\} \in [1 - \alpha, 1)$. Hence if we could find n such that

$$\{(rn + c)\alpha\} \in [1 - \alpha, 1); \quad \text{and} \quad \{(rn + d)\alpha\} \in [0, 1 - \alpha);$$

it would follow that $s_{rn+c} \neq s_{rn+d}$.

Now

$$\{(rn + c)\alpha\} \in [1 - \alpha, 1) \iff \{rn\alpha\} \in I_c := [-(c + 1)\alpha, -c\alpha);$$
$$\{(rn + d)\alpha\} \in [0, 1 - \alpha) \iff \{rn\alpha\} \in I_d := [-d\alpha, -(d + 1)\alpha).$$

We have $\mu(I_c) + \mu(I_d) = 1$; hence these intervals have nontrivial intersection whenever $c \neq d$. In fact, the endpoints of these intervals are precisely of the form $\{-i\alpha\}$ for some i with $0 \leq i \leq r$. Let $p_0, p_1, \ldots, p_{n+1}$ denote the points 0, $\{\alpha\}$, $\{2\alpha\}$, $\ldots$, $\{r\alpha\}$, 1 arranged in increasing order. It follows that $\mu(I_c \cap I_d) \geq \min_{0 \leq i \leq r}(p_{i+1} - p_i)$, and by Theorem 22, we know this quantity is bounded below by $\frac{1}{(B+2)r}$.

Now let m be the midpoint of the interval $I_c \cap I_d$. To find n with $s_{rn+c} \neq s_{rn+d}$, it suffices to find integers n, t with

$$|rn\alpha - m - t| < \frac{\mu(I_c \cap I_d)}{2} < \frac{1}{2(B + 2)r}.$$

By a folklore result (see, e.g., [8]), since α has partial quotients bounded by B, we know that $r\alpha$ has partial quotients bounded by $r(B + 2)$. By Lemma 23, it follows that such an n exists with $n \leq r(B + 2)(2(B + 2)r)^2 = 4(B + 2)^2 r^3$. ∎

Theorem 25 *Let $0 < \alpha < 1$ be an irrational real number with bounded partial quotients. Let $s_n = \lfloor (n + 1)\alpha \rfloor - \lfloor n\alpha \rfloor$ for $n \geq 1$. Then for all $k \geq 2$, the k-automaticity of the sequence $(s_n)_{n\geq 1}$ is $\Omega(n^{1/5})$.*

Proof. Set $i = \lceil (\log_k n)/5 \rceil$, and take $r = k^i$. Then by Lemma 24 for $1 \le c, d < k^i$ there exists an $m \le 4(B+2)^3 k^{3i}$ such that $s_{k^i m + c} \ne s_{k^i m + d}$. For this value of m, we have $k^i m + c = O(n^{4/5})$, and the same bound holds for $k^i m + d$. It follows that for n sufficiently large, all of the $k^i - 1$ sequences $(s_{k^i m + c})_{m \ge 0}$ are distinguishable if one knows the sequence up to the first n terms. Hence the k-automaticity of the sequence is $\Omega(n^{1/5})$. ∎

It now follows from this result, for example, that the fixed point of the homomorphism $1 \to 10$, $0 \to 1$ is not k-quasiautomatic. This follows because this fixed point can be obtained as a Sturmian sequence by setting $\alpha = (\sqrt{5} - 1)/2$.

6 Acknowledgments.

This paper is based in part on the M. Math. thesis of the first author [3]. Jean-Paul Allouche and Jonathan Buss read a draft of this paper and offered several corrections.

References

1. J. W. S. Cassels. *An Introduction to Diophantine Approximation.* Cambridge University Press, 1957.

2. A. Cobham. Uniform tag sequences. *Math. Systems Theory* **6** (1972), 164–192.

3. I. Glaister. Automaticity and closure properties. Master's thesis, University of Waterloo, April 1995.

4. G. H. Hardy and E. M. Wright. *An Introduction to the Theory of Numbers.* Oxford University Press, 1989.

5. J. E. Hopcroft and J. D. Ullman. *Introduction to Automata Theory, Languages, and Computation.* Addison-Wesley, 1979.

6. J. Kaneps and R. Freivalds. Minimal nontrivial space complexity of probabilistic one-way Turing machines. In B. Rovan, editor, *MFCS '90 (Mathematical Foundations of Computer Science)*, Vol. 452 of *Lecture Notes in Computer Science*, pages 355–361. Springer-Verlag, 1990.

7. C. Pomerance, J. M. Robson, and J. O. Shallit. Automaticity II: Descriptional complexity in the unary case. Submitted, 1994.

8. J. Shallit. Some facts about continued fractions that should be better known. Technical Report CS-91-30, University of Waterloo, Department of Computer Science, July 1991.

9. J. Shallit. Real numbers with bounded partial quotients: a survey. *Enseign. Math.* **38** (1992), 151–187.

10. J. Shallit and Y. Breitbart. Automaticity: Properties of a measure of descriptional complexity. In P. Enjalbert, E. W. Mayr, and K. W. Wagner, editors, *STACS 94: 11th Annual Symposium on Theoretical Aspects of Computer Science*, Vol. 775 of *Lecture Notes in Computer Science*, pages 619–630. Springer-Verlag, 1994.

11. J. Shallit and Y. Breitbart. Automaticity I: Properties of a measure of descriptional complexity. To appear, *J. Comput. System Sci.*

Causal Testing*

Ursula Goltz and Heike Wehrheim

Institut für Informatik, University of Hildesheim
Postfach 101363, D–31113 Hildesheim, Germany
Fax: (+49)(05121)883-768
{goltz,wehrheim}@informatik.uni-hildesheim.de

Abstract

We suggest an equivalence notion for event structures as a semantic model of concurrent systems. It combines the notion of testing (or failure) equivalence with respect to the timing of choices between different executions with a precise account of causalities between action occurrences as in causal semantics. This fills an open gap in the lattice of equivalences considered in comparative concurrency semantics. We show that our notion coincides with a "canonical" equivalence obtained as the usual testing performed on causal trees. Furthermore, we show that it is invariant under action refinement, thus fulfilling a standard criterion for non-interleaving equivalences.

1   Introduction

For systematic investigations concerning semantics of concurrent systems, it is useful to investigate all possibilities to consider systems as semantically equivalent. This leads to a better understanding of the crucial features of systems. This line of research is sometimes referred to as *comparative concurrency semantics*. On the other hand, for practical purposes of specifying and verifying system properties, it is necessary to provide a number of suitable equivalence notions in order to be able to choose always the simplest possible view of the system.

The most comprehensive investigation of the possible varieties so far has been undertaken in [18]. Two main lines which have been followed there can be sketched as follows. The first aspect which is most dominant in the classical concurrency approaches (interleaving semantics) is the so-called *linear time - branching time spectrum* [17]. Here different possibilities are discussed to what extent the points of choice between different executions of systems are taken into account. In the linear time approach, a system is equated with its set of

---

*The research reported in this paper was partially supported by the Human Capital and Mobility Cooperation Network "EXPRESS" (Expressiveness of Languages for Concurrency).

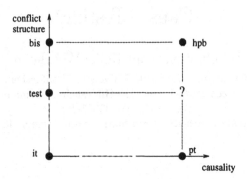

possible executions (*interleaving trace equivalence*, it), that is points of choice are neglected. At the other end of the spectrum, *bisimulation equivalence* (bis) considers choices very precisely (even the timing of internal choices). Between these two extremes, for example the notion of (interleaving) *testing* (or *failure*) *equivalence* (test) is located. The other aspect to follow is whether causalities between action occurrences are taken into account. In the so-called interleaving approaches these are neglected. Using more expressive system models like Petri nets or event structures, equivalences can be defined which take causality into account. The "linear time" variant, where causalities are taken into account but points of choice are fully neglected is ofter referred to as *pomset trace equivalence* (pt).

For getting a complete picture, at least these two aspects need to be considered also in combined form (there are still other aspects like the effects of internal actions which we do not discuss here). However, it has turned out that this is not trivial, as soon as choices are taken into account. One criterion which has been used to evaluate semantic equivalences is invariance under action refinement: If two systems are considered equivalent then one expects that after refining actions in both systems in the same way the resulting systems are still equivalent. Using this criterion it has been shown that *pomset bisimulation* as a simple idea to combine both aspects fails [19]. *History preserving bisimulation* (hpb) has then been suggested as an equivalence which takes both causalities and points of choice fully into account (and is invariant under action refinement).

What has not been investigated in detail until now is how to extend other equivalences of the linear time - branching time spectrum to respect causality, i.e. to respect pomset trace equivalence. In particular testing (or failure) equivalence seems, for many applications, to provide exactly the suitable view of a system with respect to the choices. A first attempt to generalise testing to take care of causality has been undertaken in [2] . However here the same problem occurs as for pomset bisimulation: The subtle interplay between conflict and causality is not captured by this notion as we will show here using again the refinement criterion.

In this paper, we suggest an equivalence which respects both interleaving testing and pomset trace equivalence, and we show that it is invariant under action refinement. In particular, it coincides with interleaving testing for sequential systems. We first define this equivalence for prime event structures with binary conflict as the most basic model which represents causalities. Since we want to use this simple model, we consider only conflict-free refinement, however our results could easily be generalised to arbitrary refinements using a more expressive version of event structures. In order to verify that our equivalence is indeed the natural one, we show that it coincides with the equivalence obtained as the canonical version of testing on causal trees (Section 3), similarly as history preserving bisimulation is obtained as bisimulation on causal trees [16]. It would be straightforward to define also a "failure formulation" of our equivalence.

We see this piece of work as a contribution to complete the lattice of semantic notions for concurrent systems; we do not want to discuss here whether it is possible to see causalities via some form of test. At least, we would need more than a sequential test. For a discussion of this problem for the case of pomset trace equivalence see [14].

Concerning other related work, there are a number of approaches to define testing equivalences which are invariant under action refinement [21, 11, 12, 3]. However, all these approaches use the idea of split or ST equivalences: It is taken into account that actions have a duration, however they do not take causalities precisely into account (they will in general not respect pomset trace equivalence).

Due to lack of space, all proofs have been omitted. They can be found in the full version of this paper [10].

2 Basic Definitions

In this section we will introduce the basic definitions we use throughout the paper. These are mainly event structures and their configurations, causal trees and the transformation of event structures into causal trees.

2.1 Event Structures

Our new testing equivalence requires the use of a model of systems which represents causal relationships among its entities. For this purpose we have chosen prime event structures [22] here.

Let *Act* be a set of actions, $\mathbf{E}$ a set of events.

2.1 Definition. A *(labelled) prime event structure* over $\mathbf{E}$ is a tuple $\mathcal{E} = \langle E, \leq, \#, l \rangle$ where

- $E \subseteq \mathbf{E}$ is a set of *events*,

- $\leq\ \subseteq E \times E$ is a partial order (the *causality* relation) satisfying the *principle of finite causes*:

$$\forall e \in E : \{d \in E \mid d \leq e\} \text{ is finite,}$$

- $\# \subseteq E \times E$ is an irreflexive, symmetric relation (the *conflict* relation) satisfying the *principle of conflict heredity*:

$$\forall d, e, f \in E : d \leq e, d \# f \Rightarrow e \# f,$$

- $l : E \to Act$ is a *labelling function*.

A prime event structure represents a system in the following way: the action names are activities the system may perform, an event labelled $a \in Act$ stands for a particular occurrence of an action, $d \leq e$ denotes that e cannot occur before d has and $d \# e$ denotes that d and e can never occur together in one run. From the causality relation we can also derive a notion of *causal independence*: $d \operatorname{co} e \Leftrightarrow \neg(d < e \vee e < d \vee d \# e)$.

The set of all prime event structures is denoted $\mathbb{E}_{prim}$, $\mathbf{0}$ stands for the empty event structure $\langle \varnothing, \varnothing, \varnothing, \varnothing \rangle$, the components of an event structure $\mathcal{E}$ are denoted $E_{\mathcal{E}}, \leq_{\mathcal{E}}, \#_{\mathcal{E}}$ and $l_{\mathcal{E}}$. The index will be omitted if clear from the context. $E \in \mathbb{E}_{prim}$ is *conflict-free* iff $\#_{\mathcal{E}} = \varnothing$.

For $X \subseteq E_{\mathcal{E}}$, the *restriction of $\mathcal{E}$ to X* is defined as

$$\mathcal{E}|_X := \langle X, \leq \cap (X \times X), \# \cap (X \times X), l|_X \rangle.$$

The behaviour of an event structure is described by its configurations which are sets of events with certain properties.

2.2 Definition.

A subset $X \subseteq E$ of events of a prime event structure $\mathcal{E}$ is *left-closed* iff, for all $d, e \in E$, $e \in X \wedge d \leq e \Rightarrow d \in X$.

X is *conflict-free* iff $\mathcal{E}|_X$ is conflict-free.

X is called a *configuration* iff it is left-closed and conflict-free. $\mathcal{C}(\mathcal{E})$ denotes the set of all configurations of $\mathcal{E}$.

A configuration $X \in \mathcal{C}(\mathcal{E})$ is called *complete* iff $\forall d \in E : d \notin X \Rightarrow \exists e \in X$ with $e \# d$.

For all $X \subseteq E$ being configurations, $\mathcal{E}|_X$ can also be seen as a poset (partially ordered set) since the conflict relation is empty, and therefore we will directly use the symbol X to denote the poset $\langle E_X, \leq_X, l_X \rangle = \langle X, \leq_{\mathcal{E}} \cap (X \times X), l_{\mathcal{E}}|_X \rangle$.

Two posets X, X' are said to be *isomorphic* ($X \simeq X'$) iff there exists a bijective function $f : E_X \to E_{X'}$ such that $e \leq_X e' \Leftrightarrow f(e) \leq_{X'} f(e')$ and $l_X(e) = l_{X'}(f(e))$. The isomorphism class of a poset is called a *pomset* [15], for a poset X this will be denoted by $[X]_{\simeq}$. We write $X \in \mathbf{p}$ if the poset/configuration X is in the pomset $\mathbf{p}$.

For the definition of causal testing we will use posets to represent partial executions and extensions of posets by one event to represent possible continuations. For two posets p, q, p is a *direct prefix* of q ($p \prec q$) if $E_p \subseteq E_q$, $q|_{E_p} = p$, $E_q \smallsetminus E_p = \{e\}, e \in \mathbf{E}$, and e is a maximal event of q.

A particular property of labelled event structures will play an important role in the definition of causal testing, namely autoconcurrency. $\mathcal{E} \in \mathbb{E}_{prim}$ is *without autoconcurrency* iff

$$\forall X \in \mathcal{C}(\mathcal{E}), \forall d, e \in X : d \text{ co } e \text{ and } l(d) = l(e) \Rightarrow d = e.$$

Event structures are often graphically represented. In the pictures the event names are mostly omitted — the event structure is only shown up to isomorphism — and the causality relation is depicted by arrows.

2.3 Example.

$$\mathcal{E}: \quad \begin{matrix} a \rightarrow b \\ \# \\ c \end{matrix}$$

The figure shows an event structure with three events labelled a, b and c. The action a is causal for b and c is in conflict with a and by conflict inheritance also with b.

The possible configurations of $\mathcal{E}$ are $\varnothing, \{a\}, \{a, b\}$ and $\{c\}$, $\{a, b\}$ and $\{c\}$ are complete.

2.2 Causal Trees

The second causality-based model we use for our definition of causal testing are causal trees [6, 7]. Causal trees are essentially synchronisation trees [13] which carry in their labels additional information about the causes of actions. A label in a causal tree consists of an action name and a set of backward pointers being simply numbers which point to the arcs which are causes for the action.

The advantage of causal trees especially for our purpose of finding a causal variant of an interleaving equivalence relation is their interleaving representation as a tree, however carrying all information about causality. Thus every equivalence on sychronisation trees can in a natural way be lifted to causal trees by replacing action names by the new labels.

2.4 Definition. A *causal tree* over *Act* is a tree $\langle N, A, \varphi \rangle$ where

- N is the set of *nodes*,
- $A \subseteq N \times N$ is the set of *arcs* and

- $\varphi: A \to Act \times 2^{\mathbb{N}}$ is the *labelling function*.

The labelling function is extended to paths in a causal tree in the standard way.

For the comparison of the two causal testing notions we use the standard construction of Darondeau and Degano [7] to transform event structures into causal trees. First the *traces* of an event structure are derived.

2.5 Definition. Let $\mathcal{E} = \langle E, \leq, \#, l \rangle$ be a prime event structure. A *trace* of $\mathcal{E}$ is a word $\sigma = e_1 \dots e_n$ such that

- $X_\sigma = \{e_1, \dots, e_n\}$ is a configuration of $\mathcal{E}$,
- each event $e \in X_\sigma$ occurs exactly once in σ and
- $e_i <_{X_\sigma} e_j$ implies $i < j$ for all i, j.

Hence a trace is just a linearisation of a configuration. The set of traces of $\mathcal{E}$ is denoted $Tr(\mathcal{E})$. The length of a trace σ is denoted $|\sigma|$, $\sigma[i]$ denotes the i-th component of the trace σ. Analogously the traces of a configuration X, denoted $tr(X)$, can be defined.

In the causal tree of an event structure $\mathcal{E}$, the nodes are now simply the traces of $\mathcal{E}$ and an arc exists between two traces if the second one is an extension of the first one. The causes in the labels of the arc have to be computed from the causality relation of $\mathcal{E}$.

2.6 Definition. The *causal tree of an event structure* $\mathcal{E}$, $CT(\mathcal{E})$, is the tree $\langle Tr(\mathcal{E}), A, \varphi \rangle$ such that

- $A = \{(\sigma, \sigma e) \mid \sigma, \sigma e \in Tr(\mathcal{E})\}$ and
- $\varphi((\sigma, \sigma e)) = \langle l_\mathcal{E}(e), K \rangle$ where

$$K = \{|\sigma_2| + 1 \mid \exists e' <_{X_{\sigma e}} e \text{ and } \sigma_1, \sigma_2 \text{ such that } \sigma = \sigma_1 e' \sigma_2\} .$$

For an event structure $\mathcal{E}$ and a trace $\sigma = e_1 \dots e_n \in Tr(\mathcal{E})$ we define $\varphi_\mathcal{E}(\sigma)$ to be $\langle a_1, K_1 \rangle \dots \langle a_n, K_n \rangle$ such that $l_\mathcal{E}(e_i) = a_i$ and $K_i = \{i - j \mid \exists j : e_j <_\mathcal{E} e_i\}$. In the causal tree of $\mathcal{E}$, $\varphi_\mathcal{E}(\sigma)$ is the labelling of the path leading from the root to the node σ. A similar definition can be made for posets: $\varphi_p(\sigma)$ stands for the (causal) labelling of a trace $\sigma \in tr(p)$.

For an isomorphism $f: p \to q$ between posets p and q two traces $\sigma \in tr(p), \sigma' \in tr(q)$ are called f-*isomorphic* iff $f(\sigma[i]) = \sigma'[i]$ holds for all $1 \leq i \leq |\sigma|$.

3 Testing

In this section we will develop a testing equivalence on event structures which takes causality into account. As a crucial property of this equivalence we require that it is *invariant* under refinement of actions. The testing notion on event structures will be compared — via the transformation of event structures into causal trees — with a testing definition on causal trees. Thus we can ensure that our causal testing relation is the "canonical" one. Furthermore, we classify our equivalence with respect to the related notions.

3.1 Interleaving Testing

Testing in general is based on the idea that systems can be distinguished by executing tests on them and comparing their results. For process algebras De Nicola and Hennessy [9] suggested a notion of (interleaving) testing where the system to be tested is set in parallel with a test and the outcome of a test is signalled by a specific success symbol. Two systems are equivalent if they pass exactly the same set of tests. We will use an alternative formulation (also from [9]) of this testing notion here: A test consists of a string of actions s and a set of actions L. A process passes this test if after every execution of s an action from the set L is possible next.

We define this notion for event structures as follows.

3.1 Definition. Let $\mathcal{E} \in \mathbb{E}_{prim}$.

For X, X' configurations of $\mathcal{E}$, $X \xrightarrow{a}_{\mathcal{E}} X'$ iff $X \subseteq X'$, $X' \smallsetminus X = \{e\}$ and $l_{\mathcal{E}}(e) = a$.

For $s = a_1 \ldots a_n$ write $\varnothing \xrightarrow{s}_{\mathcal{E}} X_n$, $X_n \in \mathcal{C}(\mathcal{E})$ iff there exist configurations $X_1 \ldots X_{n-1}$ such that $\varnothing \xrightarrow{a_1}_{\mathcal{E}} X_1 \xrightarrow{a_2}_{\mathcal{E}} \ldots \xrightarrow{a_n}_{\mathcal{E}} X_n$.

3.2 Definition. Let $\mathcal{E}, \mathcal{F} \in \mathbb{E}_{prim}$, $s \in Act^*$ and $L \subseteq Act$.

$\mathcal{E}$ *after* s *MUST* L iff for all X such that $\varnothing \xrightarrow{s}_{\mathcal{E}} X$ there exists an $a \in L$, $X' \in \mathcal{C}(\mathcal{E})$ such that $X \xrightarrow{a}_{\mathcal{E}} X'$.

$\mathcal{E}$ and $\mathcal{F}$ are interleaving testing equivalent ($\mathcal{E} \approx_{test} \mathcal{F}$) iff for all $s \in Act^*, L \subseteq Act$, $\mathcal{E}$ *after* s *MUST* L iff $\mathcal{F}$ *after* s *MUST* L.

This equivalence will now be extended to capture causalities.

3.2 Causal Testing

A kind of causal testing on event structures has already been defined by Aceto, De Nicola and Fantechi [2]. Their idea is that the experiments on event structures are pomsets instead of words and the behaviour which is tested for after the experiment consists of a set of actions. More precisely, an event structure $\mathcal{E}$ fulfills the test $\mathcal{E}$ *after* **p** *MUST* L if for every configuration in **p** there is at least one $a \in L$ which is possible to execute next. However, this equivalence is not invariant under refinement, as can be shown by the following example. Under this definition $\mathcal{E}$ and $\mathcal{F}$ are equivalent.

$$\mathcal{E}: \quad a \qquad\qquad \mathcal{F}: \quad a \qquad\qquad a$$
$$\downarrow \qquad\qquad\qquad \downarrow \qquad\quad + \quad \downarrow$$
$$b \;\#\; b \qquad\qquad b \;\#\; b \qquad\quad b$$

After refining the action a into $a_1 \to a_2$ (for a formal definition of refinement see Section 4) the event structures are not equivalent any more since $r(\mathcal{E})$ after a_1 MUST $\{b\}$ whereas $r(\mathcal{F})$ fails this test (if the action a_1 marked with a star is chosen no action b can be performed next).

$$r(\mathcal{E}): \quad a_1 \qquad\qquad r(\mathcal{F}): \quad a_1 \qquad\qquad a_1^*$$
$$\downarrow \qquad\qquad\qquad \downarrow \qquad\qquad \downarrow$$
$$a_2 \qquad\qquad\qquad a_2 \qquad + \quad a_2$$
$$\downarrow \qquad\qquad\qquad \downarrow \qquad\qquad \downarrow$$
$$b \;\#\; b \qquad\qquad b \;\#\; b \qquad\qquad b$$

As soon as actions are assumed to have some duration, $\mathcal{E}$ and $\mathcal{F}$ should be distinguished. They can be distinguished if the actions which are required to take place after the experiments are being causally related to the experiments themselves and thus to the actions which have happened so far. Instead of sets of actions L we therefore could use sets of direct extensions of the executed pomsets as tests. The first idea would be to define $\mathcal{E}$ *after* $\mathbf{p}$ *MUST* $\mathbf{Q}$ as: after all configurations in $\mathbf{p}$ there must be at least one $\mathbf{q} \in \mathbf{Q}$ such that a configuration isomorphic to $\mathbf{q}$ is possible next. This equivalence can indeed distinguish $\mathcal{E}$ and $\mathcal{F}$ since $\mathcal{E}$ *after* a *MUST* $\left\{ \begin{matrix} a \\ b \end{matrix} \right\}$ but $\mathcal{F}$ *after* a *MUST* $\left\{ \begin{matrix} a \\ b \end{matrix} \right\}$.

3.3 Definition.

Let $\mathcal{E}, \mathcal{F} \in \mathbb{E}_{prim}$ and $\mathbf{p}$ a pomset, $\mathbf{Q}$ a set of pomsets such that $\forall \mathbf{q} \in \mathbf{Q}$: $\mathbf{p} \prec \mathbf{q}$.

$\mathcal{E}$ *after* $\mathbf{p}$ *MUST*$_w$ $\mathbf{Q}$ iff for all $X \in \mathcal{C}(\mathcal{E}), X \in \mathbf{p}$ there exists $\mathbf{q} \in \mathbf{Q}$ and $X' \in \mathcal{C}(\mathcal{E})$ such that $X' \in \mathbf{q}$.

$\mathcal{E}$ and $\mathcal{F}$ are *weak causal testing equivalent* ($\mathcal{E} \approx_{wct} \mathcal{F}$) iff for all $\mathbf{p}, \mathbf{Q}$, $\mathcal{E}$ *after* $\mathbf{p}$ *MUST*$_w$ $\mathbf{Q} \Leftrightarrow \mathcal{F}$ *after* $\mathbf{p}$ *MUST*$_w$ $\mathbf{Q}$.

However this equivalence is still not invariant under refinement which can be seen by the following example:

$$\mathcal{E}: a \;\#\; a \qquad a \qquad\qquad \mathcal{F}: a \;\#\; a \qquad\qquad a$$
$$\downarrow \quad\; \downarrow \qquad \downarrow \qquad\qquad\qquad \downarrow \quad\; \downarrow \quad\#\quad \downarrow$$
$$b \quad\; b \;\#\; b \qquad\qquad\qquad b \quad\; b \qquad\qquad b$$

$\mathcal{E}$ and $\mathcal{F}$ are weak causal testing equivalent. After refining a into $a_1 \to a_2$ they are however not equivalent anymore since $r(\mathcal{F})$ *after* $\begin{matrix} a_1 \to a_2 \\ a_1 \end{matrix}$ *MUST*$_w$ $\left\{ \begin{matrix} a_1 \to a_2 \to b \\ a_1 \end{matrix} \right\}$ (after the configuration containing the actions marked with star no action b is possible) but $r(\mathcal{E})$ fulfills the test.

$$r(\mathcal{E}): a_1 \ \# \ a_1 \qquad a_1$$
$$\downarrow \qquad \downarrow \qquad \downarrow$$
$$a_2 \qquad a_2 \qquad a_2$$
$$\downarrow \qquad \downarrow \qquad \downarrow$$
$$b \qquad b \ \# \ b$$

$$r(\mathcal{F}): a_1 \ \# \ a_1^* \qquad\qquad a_1^*$$
$$\downarrow \qquad \downarrow \ \ \backslash \qquad \downarrow$$
$$a_2 \qquad a_2 \quad \# \quad a_2^*$$
$$\downarrow \qquad \downarrow \ \ \backslash \qquad \downarrow$$
$$b \qquad b \qquad\qquad b$$

This phenomenon is due to the fact that the equivalence does not treat autoconcurrency properly: Both $\mathcal{E}$ and $\mathcal{F}$ fulfill the test $after \ {}^a_a \ MUST_w \ \left\{ {a \to b \atop a} \right\}$. But whereas in $\mathcal{E}$ the action b can be causally related to either of the two concurrent a's, in $\mathcal{F}$ there is a configuration where the action b can only causally follow one of the actions a. What we need is a definition of causal testing where $\mathcal{F}$ already does not fulfill the test and thus $\mathcal{E}$ and $\mathcal{F}$ are not equivalent. For this we now move from pomsets in the tests to posets.

3.4 Definition.

Let $\mathcal{E}, \mathcal{F}$ be prime event structures, p a poset and Q a set of posets such that $\forall q \in Q : p \prec q$.

$\mathcal{E}$ after p MUST Q iff forall $X \in C(\mathcal{E})$ and for all isomorphisms $f: X \to p$ there exist $q \in Q$, $X' \in C(\mathcal{E})$, f' such that $X \subseteq X'$, $f': X' \to q$ is an isomorphism and $f'|_X = f$.

$\mathcal{E}$ and $\mathcal{F}$ are *causal testing equivalent* ($\mathcal{E} \approx_{ct} \mathcal{F}$) iff for all p, Q:
$\mathcal{E}$ after p MUST $Q \Leftrightarrow \mathcal{F}$ after p MUST Q.

Now the above two event structures can be distinguished: For $p = {}^{1a}_{2a}$ and $Q = \left\{ {1a \atop 2a \to 3b} \right\}$, where $1a$ stands for the event 1 labelled a, we get: $\mathcal{E}$ after p MUST Q but $\mathcal{F}$ after p MŲST Q.

For event structures without autoconcurrency the two equivalence notions coincide:

3.5 Theorem. Let $\mathcal{E}, \mathcal{F}$ be without autoconcurrency.

Then $\mathcal{E} \approx_{wct} \mathcal{F} \Longleftrightarrow \mathcal{E} \approx_{ct} \mathcal{F}$.

The proof follows immediately from the following lemma.

3.6 Lemma. Let X, Y be two finite posets without autoconcurrency, $X \simeq Y$, and let $f, g: X \to Y$ be two isomorphisms between X and Y. Then $f = g$.

Next the definition of testing on causal trees is developed. For this we adapt the testing definition of De Nicola and Hennessy [9] to causal trees, that is, the experiments and tests are constructed over the alphabet $Act \times 2^{\mathbb{N}}$ instead of over Act. Since causal trees are an augmentation of synchronisation trees with causality information this directly gives us a canonical notion of causal testing, following the idea from Vaandrager [16] and Aceto [1] where it is shown that causal bisimulation (bisimulation on causal trees) equals history preserving bisimulation.

3.7 Definition.

Let T_1, T_2 be causal trees, w a word over $Act \times 2^{\mathbb{N}}$ and $L \subseteq Act \times 2^{\mathbb{N}}$ a set of labels.

T_1 *after* w *MUST* L iff for all paths u in T_1 from the root of T_1 to a node n such that $\varphi_1(u) = w$ there exists a label $\langle a, K \rangle \in L$ and an arc r starting from n such that $\varphi_1(r) = \langle a, K \rangle$.

T_1 and T_2 are *causal tree testing equivalent*, $T_1 \approx_{ctt} T_2$, iff for all words w and sets of labels L, T_1 *after* w *MUST* $L \Leftrightarrow T_2$ *after* w *MUST* L.

The following theorem shows that the two test notions coincide. This suggests that our notion of causal testing is indeed the natural extension of interleaving testing for causality.

3.8 Theorem. Let $\mathcal{E}, \mathcal{F}$ be prime event structures.
$$\mathcal{E} \approx_{ct} \mathcal{F} \Longleftrightarrow CT(\mathcal{E}) \approx_{ctt} CT(\mathcal{F}) \ .$$

3.3 Classification

We now classify the new equivalence according to the scheme sketched in Section 1. In particular we will compare causal testing with three equivalences: with pomset trace equivalence which does not take the branching structure but the causality between action occurrences into account, with history preserving bisimulation which completely takes both branching structure and causality into account and with interleaving testing which takes branching to the same degree into account as causal testing but does not treat causality. It turns out that causal testing lies strictly between pomset trace equivalence and history preserving bisimulation and that it implies interleaving testing, hence fills exactly the gap indicated in Section 1.

We start by defining pomset trace equivalence and history preserving bisimulation.

3.9 Definition. Let $\mathcal{E}, \mathcal{F} \in \mathbb{E}_{prim}$. Let $Pomsets(\mathcal{E}) := \{[X]_{\simeq} \mid X \in \mathcal{C}(\mathcal{E})\}$.

$\mathcal{E}$ and $\mathcal{F}$ are *pomset trace equivalent* ($\mathcal{E} \approx_{pt} \mathcal{F}$) iff $Pomsets(\mathcal{E}) = Pomsets(\mathcal{F})$.

3.10 Definition. Let $\mathcal{E}, \mathcal{F} \in \mathbb{E}_{prim}$. A relation $\mathcal{R} \subseteq \mathcal{C}(\mathcal{E}) \times \mathcal{C}(\mathcal{F}) \times 2^{E_{\mathcal{E}} \times E_{\mathcal{F}}}$ is called a *history preserving bisimulation between $\mathcal{E}$ and $\mathcal{F}$* if $(\varnothing, \varnothing, \varnothing) \in \mathcal{R}$ and whenever $(X, Y, f) \in \mathcal{R}$ then

- f is an isomorphism between X and Y,

- $X \xrightarrow{a}_{\mathcal{E}} X' \Rightarrow \exists Y', f'$ with $Y \xrightarrow{a}_{\mathcal{F}} Y', (X', Y', f') \in \mathcal{R}$ and $f'|_X = f$,
- $Y \xrightarrow{a}_{\mathcal{F}} Y' \Rightarrow \exists X', f'$ with $X \xrightarrow{a}_{\mathcal{E}} X', (X', Y', f') \in \mathcal{R}$ and $f'|_X = f$.

$\mathcal{E}$ and $\mathcal{F}$ are *history preserving bisimilar* ($\mathcal{E} \approx_{hpb} \mathcal{F}$) iff there exists a history preserving bisimulation between $\mathcal{E}$ and $\mathcal{F}$.

The following theorem gives the envisaged classification. For completeness, we include also some other known inclusions.

3.11 Theorem.

$$
\begin{array}{ccc}
\approx_{bis} & \supset & \approx_{hpb} \\
\cap & & \cap \\
\approx_{test} & \supset & \approx_{ct} \\
\cap & & \cap \\
\approx_{it} & \supset & \approx_{pt}
\end{array}
$$

(all inclusions being strict.)

In particular, $\approx_{ct}$ coincides with $\approx_{test}$ for sequential systems ($co = \varnothing$).

4 Refinement

Next we will define the operation of action refinement and show that our notion of causal testing is invariant under refinement. In event structures, action refinement means substituting single events by complex event structures. A refinement function describes this substitution: it maps actions (and thereby all events labelled with this action) to finite non-empty conflict-free event structures. The restriction to conflict-free refinements is due to the chosen model of prime event structures, if one moves to for instance flow event structures [4] it can be dropped.

The operation itself is defined as follows (see [20]): in the refinement of an event structure $\mathcal{E}$ by a refinement function r every event e labelled a is replaced by a disjoint copy of $r(a)$, the causality and the conflict relation are inherited from $\mathcal{E}$.

4.1 Definition.

- A function $r: Act \rightarrow \mathbb{E}_{prim} - \{0\}$ is called a *refinement function* for prime event structures if $\forall a \in Act : r(a)$ is finite and conflict-free.
- Let $\mathcal{E} \in \mathbb{E}_{prim}$ and r a refinement function. Then $r(\mathcal{E})$ is defined as
  - $E_{r(\mathcal{E})} = \{(e, e') \mid e \in E_{\mathcal{E}}, e' \in E_{r(l_{\mathcal{E}}(e))}\},$

$$- (d, d') \leq_{r(\mathcal{E})} (e, e') \text{ iff } d <_{\mathcal{E}} e \text{ or } (d = e \wedge d' \leq_{r(l_{\mathcal{E}}(d))} e'),$$
$$- (d, d') \#_{r(\mathcal{E})} (e, e') \text{ iff } d \#_{\mathcal{E}} e \text{ and}$$
$$- l_{r(\mathcal{E})}(e, e') = l_{r(l_{\mathcal{E}}(e))}(e').$$

The behaviour of an event structure is described by the set of its configurations. The behaviour of the refined event structure $r(\mathcal{E})$ can immediately be derived from the behaviour of $\mathcal{E}$ and the behaviour of the substituted event structures [19].

We can now prove that causal testing indeed possesses the property which we required to hold: Causal testing equivalence is invariant under refinement. If two systems are causal testing equivalent then after refining actions in both systems in the same way the resulting systems are still causal testing equivalent.

4.2 Theorem. Let $\mathcal{E}, \mathcal{F} \in \mathbb{E}_{prim}$ and r a refinement function for prime event structures.
$$\mathcal{E} \approx_{ct} \mathcal{F} \Rightarrow r(\mathcal{E}) \approx_{ct} r(\mathcal{F}).$$

As a corolloray we get that for event structures without autoconcurrency weak causal testing is also invariant under refinement.

Acknowledgements. Thanks to Rob van Glabbeek, Arend Rensink and Walter Vogler for inspiring discussions and in particular to Rob van Glabbeek for suggesting to compare any kind of causal variant of interleaving equivalences with the corresponding causal tree version. Many thanks to Thomas Gehrke for his help in preparing the final version of this paper.

References

[1] Luca Aceto. History preserving, causal and mixed-ordering equivalences over stable event structures. *Fundamenta Informaticae*, 17, 1992.

[2] Luca Aceto, Rocco De Nicola, and A. Fantechi. Testing equivalences for event structures. In M. Venturini Zilli, editor, *Mathematical Models for the Semantics of Parallelism*, volume 280 of *Lecture Notes in Computer Science*, pages 1–20. Springer-Verlag, 1987.

[3] Luca Aceto and Uffe Engberg. Failures semantics for a simple process language with refinement. In S. Biswas and K. V. Nori, editors, *Foundations of Software Technology and Theoretical Computer Science*, volume 590 of *Lecture Notes in Computer Science*, pages 89–108. Springer-Verlag, 1991.

[4] Gérard Boudol and Ilaria Castellani. Permutations of transitions: An event structure semantics for CCS and SCCS. In de Bakker et al. [8], pages 411–427.

[5] W. R. Cleaveland, editor. *Concur '92*, volume 630 of *Lecture Notes in Computer Science*. Springer-Verlag, 1992.

[6] Philippe Darondeau and Pierpaolo Degano. Causal trees. In G. Ausiello, M. Dezani-Ciancaglini, and S. Ronchi Della Rocca, editors, *Automata, Languages and Programming*, volume 372 of *Lecture Notes in Computer Science*, pages 234–248. Springer-Verlag, 1989.

[7] Philippe Darondeau and Pierpaolo Degano. Causal trees = interleaving + causality. In I. Guessarian, editor, *Semantics of Systems of Concurrent Processes*, volume 469 of *Lecture Notes in Computer Science*, pages 239–255. Springer-Verlag, 1990.

[8] J. W. de Bakker, W.-P. de Roever, and Grzegorz Rozenberg, editors. *Linear Time, Branching Time and Partial Order in Logics and Models for Concurrency*, volume 354 of *Lecture Notes in Computer Science*. Springer-Verlag, 1989.

[9] Rocco De Nicola and Matthew Hennessy. Testing equivalences for processes. *Theoretical Computer Science*, 34:83–133, 1984.

[10] Ursula Goltz and Heike Wehrheim. Causal Testing. Technical Report 5/96, Institut für Informatik, Universität Hildesheim, 1996.

[11] M. Hennessy. Concurrent testing of processes. In Cleaveland [5], pages 94–107.

[12] Lalita Jategaonkar and Albert Meyer. Testing equivalences for Petri nets with action refinement. In Cleaveland [5], pages 17–31.

[13] Robin Milner. *Communication and Concurrency*. Prentice-Hall, 1989.

[14] Gordon Plotkin and Vaughan Pratt. Teams can see pomsets. extended abstract; available by ftp, August 1990.

[15] Vaughan R. Pratt. Modeling concurrency with partial orders. *International Journal of Parallel Programming*, 15(1):33–71, 1986.

[16] Frits Vaandrager. An explicit representation of equivalence classes of history preserving bisimulation. Unpublished manuscript, 1989.

[17] R. J. van Glabbeek. The linear time – branching time spectrum. In J. C. M. Baeten and J. W. Klop, editors, *Concur '90*, volume 458 of *Lecture Notes in Computer Science*, pages 278–297. Springer-Verlag, 1990.

[18] Rob van Glabbeek. *Comparative Concurrency Semantics and Refinement of Actions*. PhD thesis, Free University of Amsterdam, 1990.

[19] Rob van Glabbeek and Ursula Goltz. Equivalences and refinement. In I. Guessarian, editor, *18éme Ecole de Printemps d'Informatique Théorique Semantique du Parallelisme*, volume 469 of *Lecture Notes in Computer Science*, 1990.

[20] Rob van Glabbeek and Ursula Goltz. Refinement of actions in causality based models. In J. W. de Bakker, W.-P. de Roever, and Grzegorz Rozenberg, editors, *Stepwise Refinement of Distributed Systems — Models, Formalisms, Correctness*, volume 430 of *Lecture Notes in Computer Science*, pages 267–300. Springer-Verlag, 1990. Report version: Arbeitspapiere der GMD 428, Gesellschaft für Mathematik und Datenverarbeitung.

[21] Walter Vogler. Failure semantics based on interval semiwords is a congruence for refinement. *Distributed Computing*, (4):139–162, 1991.

[22] Glynn Winskel. An introduction to event structures. In de Bakker et al. [8], pages 364–397.

Construction of List Homomorphisms by Tupling and Fusion

Zhenjiang Hu[1], Hideya Iwasaki[2], Masato Takeichi[1]

[1] Department of Information Engineering, The University of Tokyo,
Hongo 7–3–1, Bunkyo-ku, Tokyo 113, Japan
Email: hu@ipl.t.u-tokyo.ac.jp and takeichi@u-tokyo.ac.jp
[2] Educational Computer Centre, The University of Tokyo,
Yayoi 2–11–16, Bunkyo-ku, Tokyo 113, Japan
Email: iwasaki@rds.ecc.u-tokyo.ac.jp

Abstract. List homomorphisms are functions which can be efficiently computed in parallel since they ideally suit the divide-and-conquer paradigm. However, some interesting functions, e.g., the maximum segment sum problem, are not list homomorphisms. In this paper, we propose a systematic way of embedding them into list homomorphisms so that parallel programs are derived. We show, with an example, how a simple, and "obviously" correct, but possibly inefficient solution to the problem can be successfully turned into a semantically equivalent almost homomorphism by means of two transformations: tupling and fusion.

1 Introduction

List homomorphisms [Bir87] are those functions on finite lists that *promote* through list concatenation – that is, functions h for which there exists a binary operator $\oplus$ such that, for all finite lists xs and ys,

$$h\,(xs \mathbin{+\!\!+} ys) = h\,xs \oplus h\,ys$$

where $+\!\!+$ denotes list concatenation. Examples of list homomorphisms are simple functions, such as *sum* and *max* which return the sum and the largest of the elements of a list respectively.

Intuitively, the definition of list homomorphisms implies that the computations of $h\,xs$ and $h\,ys$ are independent each other and can be carried out in parallel, which can be viewed as expressing the well-known divide-and-conquer paradigm of parallel programming. Therefore, the implications for parallel program derivation are clear; if the problem is a list homomorphism, then it only remains to define an efficient $\oplus$ in order to produce a highly parallel solution.

However, there are some useful and interesting list functions that are not list homomorphisms. One example is the function mss, which finds the sum of contiguous segment within a list whose members have the largest sum among all segments. For example, we have

$$mss\,[3, -4, 2, -1, 6, -3] = 7$$

where the result is contributed by the segment $[2, -1, 6]$. The *mss* is not a list homomorphism, since knowing *mss xs* and *mss ys* is not enough to allow computation of *mss* $(xs + ys)$.

To solve this problem, Cole [Col93] proposed an informal approach showing how to embed these functions into list homomorphisms. His method consists of constructing a homomorphism as a tuple of functions where the original function is one of the components. The main difficulty is to guess which functions must be included in a tuple in addition to the original function and to prove that the constructed tuple is indeed a list homomorphism. The examples given by Cole show that this usually requires a lot of ingenuity from the program developer. The purpose of this paper is to give a formal derivation of such list homomorphisms containing the original non-homomorphic function as its component. Our main contributions are as follows.

First, unlike Cole's informal study, we propose a *systematic* way of discovering the extra functions which are to be tupled with the original function to construct a list homomorphism. We give two main theorems, the Tupling Theorem and the Almost Fusion Theorem, showing how to derive a "true" list homomorphism from recursively defined functions by means of tupling and how to calculate a new homomorphism from the old by means of fusion.

Second, our main theorems for tupling and fusion are given in a *calculational* style [MFP91, TM95, HIT96] rather than being based on the fold/unfold transformation [Chi92, Chi93]. Therefore, infinite unfoldings, once inherited in the fold/unfold transformation, can be definitely avoided by the theorems themselves. Furthermore, although we restrict ourselves to list homomorphisms, our theorems could be extended naturally for homomorphisms of arbitrary data structures (e.g., trees) with the theory of *constructive algorithmics* [Fok92].

Third, our derivation of parallel program proceeds in a *formal* way, leading to a *correct* solution with respect to the initial specification. We start with a simple, and "obviously" correct, but possibly inefficient solution to the problem, and then transform it based on our rules and algebraic identities into a semantically equivalent list homomorphism. We demonstrate our method through a non-trivial example: maximum segment sum problem, once informally discussed by Cole [Col93].

This paper is organized as follows. In Sect.2, we review the notational conventions and basic concepts used in this paper. In Sect.3, we explain the concept of Cole's almost homomorphism and show the difficulty in deriving almost homomorphisms. After showing our specifications in Sect.4, we focus ourselves on the derivation of parallel programs from specifications with two important theorems, namely the tupling and the fusion theorems, in Sect.5. Finally, we give the concluding remarks with related work in Sect.6.

2 Preliminary

In this section, we briefly review notational conventions and the basic concepts in [Bir87], known as Bird-Meertens Formalism, used in the rest of this paper.

2.1 Functions

Functional application is denoted by a space and the argument which may be written without brackets. Thus $f\,a$ means $f\,(a)$. Functions are curried and application associates to the left. Thus $f\,a\,b$ means $(f\,a)\,b$. Functional application is regarded as more binding than any other operator, so $f\,a \oplus b$ means $(f\,a) \oplus b$ but not $f\,(a \oplus b)$. Functional composition is denoted by a centralized circle $\circ$. By definition, $(f \circ g)\,a = f\,(g\,a)$. Functional composition is an associative operator, and the identity function is denoted by id.

The projection function π_i will be used to select the i-th component of tuples, e.g., $\pi_1\,(a,b) = a$. The $\vartriangle$ and $\times$ are two important operators related to tuples, defined by $(f \vartriangle g)\,a = (f\,a,\ g\,a), \ \ (f \times g)\,(a,b) = (f\,a,\ g\,b)$.

Infix binary operators will often be denoted by $\oplus, \otimes$. The $\vartriangle$ can be naturally extended to functions with two arguments, i.e., $a\,(\oplus \vartriangle \otimes)\,b = (a \oplus b,\ a \otimes b)$.

2.2 Lists

Lists are finite sequences of values of the same type. A list is either empty, a singleton, or the concatenation of two other lists. We write $[\,]$ for the empty list, $[a]$ for the singleton list with element a (and $[\cdot]$ for the function taking a to $[a]$), and $xs \mathbin{+\!\!+} ys$ for the concatenation of xs and ys. Concatenation is associative, and $[\,]$ is its unit. For example, the term $[1] \mathbin{+\!\!+} [2] \mathbin{+\!\!+} [3]$ denotes a list with three elements, often abbreviated to $[1, 2, 3]$.

2.3 List Homomorphisms

A function h satisfying the following three equations will be called a *list homomorphism*.

$$
\begin{aligned}
h\,[\,] &= \iota_\oplus \\
h\,[x] &= f\,x \\
h\,(xs \mathbin{+\!\!+} ys) &= h\,xs \oplus h\,ys
\end{aligned}
$$

It soon follows from this definition that $\oplus$ must be an associative binary operator with unit $\iota_\oplus$. We usually use $(\!| f, \oplus |\!)$ to denote the unique function h[3]. For example, both the functions sum and max are list homomorphisms defined by $sum = (\!| id, + |\!)$, $max = (\!| id, \uparrow |\!)$. where $\uparrow$ denotes the binary maximum function with the unit of $-\infty$.

2.4 Parallelism: Map and Reduction

Map is the operator which applies another function to every item in a list. It is written as an infix $*$, which is informally defined by

$$
f * [x_1, x_2, \cdots, x_n] = [f\,x_1, f\,x_2, \cdots, f\,x_n].
$$

---

[3] Strictly speaking, we should write $(\!| \iota_\oplus, f, \oplus |\!)$ to denote the unique function h. We can omit the $\iota_\oplus$ because it is the unit of $\oplus$.

Reduction (also known as *fold*) is the operator which collapses a list into a single value by repeated application of some binary operator. It is written as an infix /. Informally, for an associative binary operator $\oplus$ with unit $\iota_{\oplus}$, we have

$$\oplus/\,[x_1, x_2, \cdots, x_n] = x_1 \oplus x_2 \cdots \oplus x_n.$$

It is not difficult to see that $*$ and / have simple massively parallel implementations on many architectures [Ski90]. For example, $\oplus/$ can be computed in parallel on a tree-like structure with the combining operator $\oplus$ applied in the nodes, whereas $f*$ is totally parallel.

The relevance of list homomorphisms to parallel programming can be seen clearly from the Homomorphism Lemma [Bir87]: $([f, \oplus]) = (\oplus/) \circ (f*)$. Every list homomorphism can be written as the composition of a reduction and a map. The implications for parallel program derivation become clear: if a problem is a list homomorphism, then it only remains to define $\oplus$ and f in order to produce a highly parallel solution. The performance of this program is governed by the complexities of $\oplus$ and f. The major problem remained is that many useful functions are not list homomorphisms themselves.

3 Almost Homomorphisms

As stated in the previous section, quite a few useful functions are not list homomorphisms. Cole argued informally that some of them can be converted into so-called *almost homomorphisms* [Col93] by tupling them with some extra functions. An almost homomorphism is a composition of a projection function and a list homomorphism.

In fact, it may be surprising to see that every function can be represented in terms of an almost homomorphism[Gor95]. Let k be a non-homomorphic function. Consider a new function g such that $g\,x = (x, k\,x)$. The tuple-function g is homomorphic, i.e., $g\,(xs + \!\!+ \,ys) = (xs + \!\!+ \,ys, k\,(xs + \!\!+ \,ys)) = g\,xs \oplus g\,ys$, where $(xs, a) \oplus (ys, b) = (xs + \!\!+ \,ys, k\,(xs + \!\!+ \,ys))$, and we have the almost homomorphism for k defined by $k = \pi_2 \circ g = \pi_2 \circ ([g \circ [\cdot], \oplus])$. However, it is quite expensive and meaningless in that it does not make use of the previously computed values a and b) and computes k from scratch! In this sense, we say it is not an expected "true" almost homomorphism.

In order to derive a "true" almost homomorphism, a suitable tuple-function should be carefully defined, making full use of previously computed values. Cole reported several case studies of such derivation with parallel algorithms as a result, and stressed that in each case, the derivation requires a lot of intuition [Col93]. In this paper, we shall propose a systematic approach to this derivation.

4 Specification

We aim at a formal derivation of parallel programs by constructing list homomorphisms including the original problems as its component (i.e., almost homomorphisms). To talk about parallel program derivation, we should be clear about specifications. It is strongly advocated by Bird [Bir87] that specifications should be direct solutions to problems. Therefore, our specification for a problem p will be a simple, and "obviously" correct, but possibly inefficient solution with the form of

$$p = p_n \circ \cdots \circ p_2 \circ p_1 \qquad (1)$$

where each p_i is a (recursively defined) function. This reflects our way of solving problems; a (big) problem p may be solved through multiple passes in which a simpler problem p_i is solved by a recursion.

We shall use the maximum segment sum problem mss as our running example. This problem is of interest because there are efficient but non-obvious algorithms to compute it, both in sequential [Bir87] and in parallel[CS92, Col93]. An obviously correct solution to the problem is:

$$mss = max^s \circ (sum *^s) \circ segs \qquad (2)$$

which is implemented by three passes: computing the set including all of the contiguous segments by $segs$, summing the elements of each by $sum *^s$, and selecting from the set the largest value of the sums by max^s. Clearly, max^s and $sum *^s$ should be functions over *sets*. Similar to lists, a set is either empty $\{ \}$, s singleton $\{x\}$, or the union of two sets $s_1 \cup s_2$. The difference lies in that the union operation is not only associative but also commutative and idempotent. The definition style of functions over sets is quite similar to those over lists. For instance, max^s can be defined directly as follows.

$$
\begin{aligned}
max^s \{ \} &= -\infty \\
max^s \{x\} &= x \\
max^s (s_1 \cup s_2) &= max^s s_1 \uparrow max^s s_2
\end{aligned}
$$

Like the list map operator, the *map* operator over sets, denoted by $*^s$, applies another function to every items in a set (e.g., $sum *^s$ in our case). In the following, we shall focus on defining $segs$.

The $segs$, the function computing the set of all (contiguous) segments of a list, can be recursively defined by:

$$
\begin{aligned}
segs\ [] &= \{ \} \\
segs\ [x] &= \{[x]\} \\
segs\ (xs ++ ys) &= segs\ xs \cup segs\ ys \cup (tails\ xs\ \mathcal{X}_{++}\ inits\ ys).
\end{aligned}
$$

The last equation says that all segments of the concatenation of two lists xs and ys can be obtained from segments in both xs and ys and the segments produced crosswisely by concatenating all tail segments of xs (i.e., the segments in xs ending with the last element) with initial segments of ys (i.e., the segments in ys starting with the the first element). Note that the *inits*, *tails*, and $\mathcal{X}_{++}$ are considered as standard functions in [Bir87] (though our definitions are slightly different). The *inits* is a function returning all initial segments of a list, while the *tails* is a function returning all tail segments. They can be defined directly by:

$$
\begin{aligned}
inits\ [] &= [] \\
inits\ [x] &= [[x]] \\
inits\ (xs ++ ys) &= inits\ xs ++ (xs ++) * (inits\ ys)
\end{aligned}
$$

and

$$\begin{aligned}
tails\ [] &= [] \\
tails\ [x] &= [[x]] \\
tails\ (xs \mathbin{+\!\!+} ys) &= (\mathbin{+\!\!+} ys) * (tails\ xs) \mathbin{+\!\!+} tails\ ys.
\end{aligned}$$

The operator $\mathcal{X}_\oplus$ is usually called *cross* operator, defined informally by

$$[x_1, \cdots, x_n]\ \mathcal{X}_\oplus\ [y_1, \cdots, y_m] = \{x_1 \oplus y_1, \cdots, x_1 \oplus y_m, \cdots, x_n \oplus y_1, \cdots, x_n \oplus y_m\},$$

which crosswisely combines elements in two lists with operator $\oplus$. An obvious property with cross operator is

$$(f *^s) \circ \mathcal{X}_\oplus = \mathcal{X}_{f \circ \oplus}. \tag{3}$$

So much for the specification of the *mss* problem. It is a naive correct solution to the problem without concerning efficiency and parallelism at all. It will be shown that our method can derive a correct $O(\log n)$ parallel time algorithm by constructing a "true" almost homomorphism.

5 Derivation

Our derivation of a "true" almost homomorphism from the specification (1) is carried out in the following way: we derive an almost homomorphism for p_1 (a recursion) first, then fuse p_2 with the derived almost homomorphism to obtain another almost homomorphism, and repeat this fusion until p_n is fused. We are confronted with two problems here: (a) How can a "true" almost homomorphism be derived from a recursive definition? (b) How can a new almost homomorphism be calculated from a composition of another function and an old one?

5.1 Deriving Almost Homomorphisms

Although some functions cannot be described directly as list homomorphisms, they may be easily described by (mutual) recursive definitions while some other functions might be used (see *segs* in Sect.4 for an example) [Fok92]. In this section, we propose a way of deriving almost homomorphisms from such (mutual) recursive definitions, systematically discovering extra functions that should be tupled with the original function to turn it into a "true" list homomorphism. The "true" list homomorphism must fully reuse the previously computed values, as discussed in Sect.3.

Our approach is based on the following theorem. For notational convenience, we define

$$\Delta_1^n f_i = f_1 \vartriangle f_2 \vartriangle \cdots \vartriangle f_n.$$

Theorem 1 Tupling. Let $h_1, \cdots, h_n$ be mutually defined as follows.

$$\begin{aligned}
h_i\ [] &= \iota_{\oplus_i} \\
h_i\ [x] &= f_i\ x \\
h_i\ (xs \mathbin{+\!\!+} ys) &= ((\Delta_1^n h_i)\ xs) \oplus_i ((\Delta_1^n h_i)\ ys)
\end{aligned} \tag{4}$$

Then,
$$\Delta_1^n h_i = (\![\Delta_1^n f_i, \ \Delta_1^n \oplus_i]\!)$$
and $(\iota_{\oplus_1}, \cdots, \iota_{\oplus_n})$ is the unit of $\Delta_1^n \oplus_i$.

Proof. According to the definition of list homomorphisms, it is sufficient to prove that

$$
\begin{aligned}
(\Delta_1^n h_i)\,[\,] &= (\iota_{\oplus_1}, \cdots, \iota_{\oplus_n}) \\
(\Delta_1^n h_i)\,[x] &= (\Delta_1^n f_i)\,x \\
(\Delta_1^n h_i)\,(xs \!+\!\!+ ys) &= ((\Delta_1^n h_i)\,xs)\,(\Delta_1^n \oplus_i)\,((\Delta_1^n h_i)\,ys).
\end{aligned}
$$

The first two equations are trivial. The last can be proved by the following calculation.

$$
\begin{aligned}
&LHS \\
=\quad &\{ \text{ Def. of } \Delta \text{ and } \vartriangle \} \\
&(h_1(xs \!+\!\!+ ys), \cdots, h_n(xs \!+\!\!+ ys)) \\
=\quad &\{ \text{ Def. of } h_i \} \\
&(((\Delta_1^n h_i)\,xs) \oplus_1 ((\Delta_1^n h_i)\,ys), \cdots, ((\Delta_1^n h_i)\,xs) \oplus_n ((\Delta_1^n h_i)\,ys)) \\
=\quad &\{ \text{ Def. of } \vartriangle \text{ and } \Delta \} \\
&RHS \qquad\qquad\qquad\qquad\qquad\qquad\qquad\qquad\qquad\qquad\qquad\quad \square
\end{aligned}
$$

Theorem 1 says that if h_1 is mutually defined with other functions (i.e., $h_2, \cdots h_n$) which *traverse over the same lists* in the *specific form* of (4), then tupling $h_1, \cdots, h_n$ will definitely give a list homomorphism. It follows that h_1 is an almost homomorphism: the projection function π_1 composed with the list homomorphism for the tuple-function. It is worth noting that this style of tupling can avoid repeatedly redundant computations of $h_1, \cdots, h_n$ in the computation of the list homomorphism of $\Delta_1^n h_i$ [Tak87]. That is, all previous computed results by $h_1, \cdots, h_n$ can be fully reused, as expected in "true" almost homomorphisms.

Practically, not all recursive definitions are in the form of (4). They, however, can be turned into such form by a simple transformation. Let's see how the tupling theorem works in deriving a "true" almost homomorphism from the definition of *segs* given in Sect.4.

First, we determine what functions are to be tupled, i.e., $h_1, \cdots, h_n$. As explained above, the functions to be tupled are those which traverse over the same lists in the definitions. So, from the definition of *segs*:

$$segs\ (xs \!+\!\!+ ys) = segs\ xs \cup segs\ ys \cup (\underline{tails\ xs}\ \mathcal{X}_{\!+\!\!+}\ inits\ ys)$$

we know that *segs* needs to be tupled with *tails* and *inits*, because *segs* and *inits* traverse the same list xs whereas *segs* and *tails* traverse the same list ys as underlined. Going to the definition of *inits*,

$$inits\ (xs \!+\!\!+ ys) = \underline{inits\ xs} \!+\!\!+ (\underline{xs} \!+\!\!+)\ast(inits\ ys)$$

we find that the *inits* needs to be tupled with *id*, the identity function, since $xs = id\ xs$. Similarly, The *tails* needs to be tupled with *id*. Note that *id* is the identity function over lists defined by

$$
\begin{aligned}
id\,[\,] &= [\,] \\
id\,[x] &= [x] \\
id\,(xs \!+\!\!+ ys) &= id\ xs \!+\!\!+ id\ ys
\end{aligned}
$$

To summarize the above, the functions to be tupled are *segs, inits, tails,* and *id,* i.e., our tuple function will be *segs* △ *inits* △ *tails* △ *id.*

Next, we rewrite the definitions of the functions in the above tuple to the form of (4), i.e., deriving $f_1, \oplus_1$ for *segs,* $f_2, \oplus_2$ for *inits,* $f_3, \oplus_3$ for *tails,* and $f_4, \oplus_4$ for *id.* In fact, this is straightforward: just selecting the corresponding recursive calls from the tuples. From the definition of *segs,* we have

$$
\begin{aligned}
f_1\, x &= \{[x]\} \\
(s_1, i_1, t_1, d_1) \oplus_1 (s_2, i_2, t_2, d_2) &= s_1 \cup s_2 \cup (t_1 \mathcal{X}_{+\!\!+} i_2).
\end{aligned}
$$

It would be helpful for understanding the above derivation if we notice the following correspondences: s_1 to *segs* xs, i_1 to *inits* xs, t_1 to *tails* xs, d_1 to *id* xs, s_2 to *segs* ys, i_2 to *inits* ys, t_2 to *tails* ys, d_2 to *id* ys. Similarly, for *inits, tails* and *id,* we have

$$
\begin{aligned}
f_2\, x &= [[x]] \\
(s_1, i_1, t_1, d_1) \oplus_2 (s_2, i_2, t_2, d_2) &= i_1 +\!\!+ (d_1 +\!\!+) * i_2 \\
f_3\, x &= [[x]] \\
(s_1, i_1, t_1, d_1) \oplus_3 (s_2, i_2, t_2, d_2) &= (+\!\!+ d_2) * t_1 +\!\!+ t_2 \\
f_4\, x &= [x] \\
(s_1, i_1, t_1, d_1) \oplus_4 (s_2, i_2, t_2, d_2) &= d_1 +\!\!+ d_2
\end{aligned}
$$

Finally, we apply Theorem 1 and get the following list homomorphism.

$$
segs \vartriangle inits \vartriangle tails \vartriangle id = (\!|\Delta_1^4 f_i, \Delta_1^4 \oplus_i|\!)
$$

And our almost homomorphism for *segs* is thus obtained:

$$
segs = \pi_1 \circ (\!|\Delta_1^4 f_i, \Delta_1^4 \oplus_i|\!). \tag{5}
$$

It would be intersting to see that the above derivation is practically *mechanical.* Note that the derivation of the unit of the new binary operator (e.g., $\Delta_1^4 \oplus_i$) is omitted because this is trivial; the new tuple function applying to empty list will give exactly this unit (e.g., $(segs \vartriangle inits \vartriangle tails \vartriangle id)\,[\,])$. The derivation of units will be omitted in the rest of the paper as well.

5.2 Fusion with Almost Homomorphisms

In this section, we show how to fuse a function with an almost homomorphism, the second problem (b) listed at the beginning of Sect. 5.

It is well known that list homomorphisms are suitable for program transformation in that there is a general rule called *Fusion Theorem* [Bir87], showing how to fuse a function with a list homomorphism to get another list homomorphism.

Theorem 2 Fusion. Let h and $(\!|f, \oplus|\!)$ be given. If there exists $\otimes$ such that $\forall x, y.\ h\,(x \oplus y) = h\,x \otimes h\,y$, then $h \circ (\!|f, \oplus|\!) = (\!|h \circ f, \otimes|\!)$. $\qquad\square$

This fusion theorem, however, cannot be used directly for our purpose. As seen in (5), we usually derive an almost homomorphism and we hope to know how to fuse functions with almost homomorphisms, namely we want to deal with the following case:

$$h \circ (\pi_1 \circ (\![\Delta_1^n f_i, \Delta_1^n \oplus_i]\!)).$$

We'd like to shift π_1 left and promote h into the list homomorphism. Our fusion theorem for this purpose is given below.

Theorem 3 Almost Fusion. Let h and $(\![\Delta_1^n f_i, \Delta_1^n \oplus_i]\!)$ be given. If there exist $\otimes_i$ $(i = 1, \cdots, n)$ and a map $H = h_1 \times \cdots \times h_n$ where $h_1 = h$ such that for all j,

$$\forall x, y.\ h_i\,(x \oplus_i y) = H\,x \otimes_i H\,y \qquad (6)$$

then

$$h \circ (\pi_1 \circ (\![\Delta_1^n f_i, \Delta_1^n \oplus_i]\!)) = \pi_1 \circ (\![\Delta_1^n (h_i \circ f_i), \Delta_1^n \otimes_i]\!)$$

Proof. We prove it by the following calculation.

$$
\begin{aligned}
& h \circ (\pi_1 \circ (\![\Delta_1^n f_i, \Delta_1^n \oplus_i]\!)) \\
=\ & \quad \{ \text{ By } \pi_1 \text{ and } H \ \} \\
& \pi_1 \circ H \circ (\![\Delta_1^n f_i, \Delta_1^n \oplus_i]\!) \\
& \quad \{ \text{ Theorem 2, and the proves below } \} \\
& \pi_1 \circ (\![\Delta_1^n (h_i \circ f_i), \Delta_1^n \otimes_i]\!)
\end{aligned}
$$

To complete the above proof, we need to show that for any x and y,

$$
\begin{aligned}
H\,(x\,(\Delta_1^n \oplus_i)\,y) &= (H\,x)\,(\Delta_1^n \otimes_i)\,(H\,y) \\
H \circ (\Delta_1^n f_i) \quad &= \Delta_1^n (h_i \circ f_i)
\end{aligned}
$$

The second equation is easy to prove. For the first, we argue that

$$
\begin{aligned}
& LHS \\
=\ & \quad \{ \text{ Expanding } \Delta, \text{ Def. of } \vartriangle \ \} \\
& H\,(x \oplus_1 y, \cdots, x \oplus_n y) \\
=\ & \quad \{ \text{ Expanding } H, \text{ Def. of } \times \ \} \\
& (h_1\,(x \oplus_1 y), \cdots, h_n\,(x \oplus_n y)) \\
=\ & \quad \{ \text{ Assumption } \} \\
& (H\,x \otimes_1 H\,y, \cdots, H\,x \otimes_n H\,y) \\
=\ & \quad \{ \text{ Def. of } \vartriangle, \Delta \ \} \\
& RHS
\end{aligned}
$$

$\square$

Theorem 3 suggests a way of fusing a function h with the almost homomorphism $\pi_1 \circ (\![\Delta_1^n f_i, \Delta_1^n \oplus_i]\!)$ in order to get another almost homomorphism; trying to find $h_2, \cdots, h_n$ together with $\oplus_1, \cdots, \oplus_n$ that meet the equation (6). Note that, without lose of generality we restrict the projection function of our almost homomorphisms to π_1 in the theorem.

Returning to our running example, recall that we have reached the point:

$$mss = max^s \circ (sum *^s) \circ (\pi_1 \circ (\![\Delta_1^4 f_i, \Delta_1^4 \oplus_i]\!)).$$

We demonstrate how to fuse $sum *^s$ with $\pi_1 \circ (\![\Delta_1^4 f_i, \Delta_1^4 \oplus_i]\!)$ by Theorem 3. Let $H = (sum *^s) \times h_2 \times h_3 \times h_4$ where h_2, h_3, h_4 await to be determined. In addition, we need to derive $\otimes_1, \otimes_2, \otimes_3$, and $\otimes_4$ based on the following equations according to Theorem 3:

$$sum *^s ((s_1, i_1, t_1, d_1) \oplus_i (s_2, i_2, t_2, d_2)) =$$
$$(sum *^s s_1, h_2 i_1, h_3 t_1, h_4 d_1) \otimes_i (sum *^s s_2, h_2 i_2, h_3 t_2, h_4 d_2) \quad (i = 1, \cdots, 4).$$

Now the derivation procedure becomes clear; calculating each LHS of the above equations to promote $sum *^s$ into s_1 and s_2, and determining the unknown functions (h_i and $\otimes_i$) by matching with its RHS. As an example, consider the following calculation of the LHS of the the equation for $i = 1$.

$$\begin{aligned}
& (sum *^s) ((s_1, i_1, t_1, d_1) \oplus_1 (s_2, i_2, t_2, d_2)) \\
= \quad & \{ \text{ Def. of } \oplus_1 \ \} \\
& (sum *^s) (s_1 \cup s_2 \cup (t_1 \mathcal{X}_{+\!\!+} i_2)) \\
= \quad & \{ \ f *^s (s1 \cup s2) = f *^s s1 \cup f *^s s2 \ \} \\
& sum *^s s_1 \cup sum *^s s_2 \cup sum *^s (t_1 \mathcal{X}_{+\!\!+} i_2) \\
= \quad & \{ \ (3) \ \} \\
& (sum *^s s_1 \cup sum *^s s_2 \cup (t_1 \mathcal{X}_{sum \circ +\!\!+} i_2)) \\
= \quad & \{ \text{ cross operator, } sum \ \} \\
& (sum * s1 \cup sum * s_2 \cup ((sum * t_1) \mathcal{X}_+ (sum * i_2)))
\end{aligned}$$

Matching the last expression with its corresponding RHS:

$$(sum *^s s_1, h_2 i_1, h_3 t_1, h_4 d_1) \otimes_1 (sum *^s s_2, h_2 i_2, h_3 t_2, h_4 d_2)$$

will give

$$h_2 = h_3 = sum*$$
$$(s_1, i_1, t_1, d_1) \otimes_1 (s_2, i_2, t_2, d_2) = s_1 \cup s_2 \cup (t_1 \mathcal{X}_+ i_2).$$

The gl others can be similarly derived.

$$\begin{aligned}
h_4 &= sum \\
(s_1, i_1, t_1, d_1) \otimes_2 (s_2, i_2, t_2, d_2) &= i_1 +\!\!+ (d_1+) * i_2 \\
(s_1, i_1, t_1, d_1) \otimes_3 (s_2, i_2, t_2, d_2) &= (+d_2) * t_1 +\!\!+ t_2 \\
(s_1, i_1, t_1, d_1) \otimes_4 (s_2, i_2, t_2, d_2) &= d_1 + d_2
\end{aligned}$$

To use Theorem 3, we also need to consider the f part whose results are as follows.

$$\begin{aligned}
f_1' \, x &= ((sum *^s) \circ f_1) \, x = \{x\} \\
f_2' \, x &= ((sum*) \circ f_2) \, x \ = [x] \\
f_3' \, x &= ((sum*) \circ f_3) \, x \ = [x] \\
f_4' \, x &= (sum \circ f_1) \, x \ \ \ = x
\end{aligned}$$

According to Theorem 3, we soon have

$$(sum *^s) \circ segs = \pi_1 \circ (\![\Delta_1^4 f_i', \Delta_1^4 \otimes_i]\!).$$

Again, we can fuse max^s with the above almost homomorphism (in this case, $H = max^s \times max \times max \times id$) and get the following almost homomorphism for mss, the final result for mss.

$$mss = \pi_1 \circ (\![id, \Delta_1^4 \otimes_i']\!)$$

where
$$(s_1,i_1,t_1,d_1) \otimes'_1 (s_2,i_2,t_2,d_2) = s1 \uparrow s_2 \uparrow (t_1 + i_2)$$
$$(s_1,i_1,t_1,d_1) \otimes'_2 (s_2,i_2,t_2,d_2) = i_1 \uparrow (d_1 + i_2)$$
$$(s_1,i_1,t_1,d_1) \otimes'_3 (s_2,i_2,t_2,d_2) = (t_1 + d_2) \uparrow t_2$$
$$(s_1,i_1,t_1,d_1) \otimes'_4 (s_2,i_2,t_2,d_2) = d_1 + d_2$$

Thus we got the same result as informally given by Cole [Col93]. In practical terms, the algorithm looks so promising that on many architectures we can expect an $O(\log n)$ algorithm with $O(n/(\log n))$ processors.

6 Concluding Remarks and Related Work

In this paper, we propose a formal and systematic approach to the derivation of efficient parallel programs from specifications of problems via *manipulation* of almost homomorphisms, namely the construction of almost homomorphisms from recursive definitions (Theorem 1) and the fusion of a function with almost homomorphisms (Theorem 3). It is different from Cole's informal way[Col93].

Tupling and fusion are two well-known techniques for improving programs. Chin [Chi92, Chi93] gave an intensive study on it. His method tries to fuse and/or tuple *arbitrary* functions by *fold-unfold* transformations while keeping track of function calls and using clever control to avoid infinite unfolding. In contrast to his costly and complicated algorithm to keep out of non-termination, our approach makes use of structural knowledge of list homomorphisms and constructs our tupling and fusion rules in a calculational style without worrying about infinite unfoldings.

Our approach to the tupling of mutual recursive definitions is much influenced by the *generalization algorithm* [Tak87, Fok92]. Takeichi showed how to define a higher order function common to all functions mutually defined so that multiple traversals of the same data structures in the mutual recursive definition can be eliminated. Because higher order functions are suitable for partial evaluation but not good for program derivation, we employ tuple-functions and develop the corresponding fusion theorem.

Construction of list homomorphisms has gained great interest because of its importance in parallel programming. Barnard et.al. [BSS91] applied the Third Homomorphism Theorem [Gib94] for the language recognition problem. The Third Homomorphism Theorem says that an algorithm h which can be formally described by two specific sequential algorithms (*leftwards* and *rightwards reduction* algorithms) is a list homomorphism. Although the existence of an associative binary operator is guaranteed, the theorem does not address the question of the existence – let alone the construction – of a direct and efficient way of calculating it. To solve this problem, Gorlatch [Gor95] imposed additional restrictions, left associativity and right associativity, on the leftwards and rightwards reduction functions so that an associative binary operator $\oplus$ could be derived in a systematic way. However, finding left-associative binary operators is usually not easier than finding associative operators. In comparison, our derivation is more constructive: we derive list homomorphism directly from mutual recursive representations and then fuse it with other functions.

418

Acknowledgement

Many thanks are to Akihiko Takano, Fer-Jan de Vries and other CACA members for many enjoyble discussions.

References

[Bir87] R. Bird. An introduction to the theory of lists. In M. Broy, editor, *Logic of Programming and Calculi of Discrete Design*, pages 5–42. Springer-Verlag, 1987.

[BSS91] D. Barnard, J. Schmeiser, and D. Skillicorn. Seriving associative operators for language recognition. In *Bulletin of* EATCS (43), pages 131–139, 1991.

[Chi92] W. Chin. Safe fusion of functional expressions. In *Proc. Conference on Lisp and Functional Programming*, San Francisco, California, June 1992.

[Chi93] W. Chin. Towards an automated tupling strategy. In *Proc. Conference on Partial Evaluation and Program Manipulation*, pages 119–132, Copenhagen, June 1993. ACM Press.

[Col93] M. Cole. Parallel programming, list homomorphisms and the maximum segment sum problems. Report CSR-25-93, Department of Computing Science, The University of Edinburgh, May 1993.

[CS92] W. Cai and D.B. Skillicorn. Calculating recurrences using the Bird-Meertens Formalism. Technical report, Department of Computing and Information Science, Queen's University, 1992.

[Fok92] M. Fokkinga. A gentle introduction to category theory — the calculational approach —. Technical Report Lecture Notes, Dept. INF, University of Twente, The Netherlands, September 1992.

[Gib94] J. Gibbons. The third homomorphism theorem. Technical report, University of Auckland, 1994.

[Gor95] S. Gorlatch. Constructing list homomorphisms. Technical Report MIP-9512, Fakultät für Mathematik und Informatik, Universität Passau, August 1995.

[HIT96] Z. Hu, H. Iwasaki, and M. Takeichi. Deriving structural hylomorphisms from recursive definitions. In *ACM SIGPLAN International Conference on Functional Programming* (ICFP '96), Philadelphia, Pennsylvania, May 1996. ACM Press.

[MFP91] E. Meijer, M. Fokkinga, and R. Paterson. Functional programming with bananas, lenses, envelopes and barbed wire. In *Proc. Conference on Functional Programming Languages and Computer Architecture* (LNCS 523), pages 124–144, Cambridge, Massachuetts, August 1991.

[Ski90] D.B. Skillicorn. Architecture-independent parallel computation. *IEEE Computer*, 23(12):38–51, December 1990.

[Tak87] M. Takeichi. Partial parametrization eliminates multiple traversals of data structures. *Acta Informatica*, 24:57–77, 1987.

[TM95] A. Takano and E. Meijer. Shortcut deforestation in calculational form. In *Proc. Conference on Functional Programming Languages and Computer Architecture*, pages 306–313, La Jolla, California, June 1995.

Probabilistic Metric Semantics for a Simple Language with Recursion

Marta Kwiatkowska and Gethin Norman

School of Computer Science University of Birmingham,
Edgbaston, Birmingham B15 2TT, UK

Abstract. We consider a simple divergence-free language RP for reactive processes which includes prefixing, deterministic choice, action-
guarded probabilistic choice, synchronous parallel and recursion. We
show that the probabilistic bisimulation of Larsen & Skou is a congruence
for this language. Following the methodology introduced by de Bakker
& Zucker we give denotational semantics to this language by means of
a complete metric space of (deterministic) probabilistic trees defined in
terms of the powerdomain of closed sets. This new metric, although not
an ultra-metric, nevertheless specialises to the metric of de Bakker &
Zucker. Our semantic domain admits a full abstraction result with respect to probabilistic bisimulation.

1 Introduction

Probabilistic and stochastic phenomena are important in many areas of computing, for example, distributed systems, fault tolerance, communication protocols
and performance analysis, and thus formal and automated tools for reasoning
about such systems are needed. This paper makes a contribution towards the
foundations of languages for specifying probabilistic systems, and thus furthers
understanding of the probabilistic phenomena which have so far proved trouble-
some to handle by conventional techniques, see e.g. the probabilistic powerdomain construction [9].

The recent trend in the semantics of programming languages has been to
supply a language with three pairwise "equivalent" semantics: operational, denotational and logical. Each semantics gives a different view of the language –
the operational focuses on the transition system, denotational on compositionality, while the logical on the properties and satisfaction – and a statement of their
"equivalence" states how closely they are related. The work of Kozen [13] for a
while language with random assignment is a pre-cursor of this approach in the
area of probabilistic languages, but so far no framework encompassing the three
semantics has been proposed for a probabilistic extension of a process algebra.

In this paper we consider a probabilistic variant of a process algebra (a "reactive" language in the terminology of [20]) based on CCS [17] and CSP [8].
The calculus contains recursion, deterministic choice and concurrency, but instead of non-deterministic choice it has (action guarded) probabilistic choice.
The operational semantics of this language is given in terms of the probabilistic

transition systems and probabilistic bisimulation of Larsen & Skou [14]. The calculus is provided with a denotational, metric-space semantics derived following the techniques introduced by de Bakker & Zucker [5] for the non-probabilistic case. We show the semantics to be fully abstract with respect to the probabilistic bisimulation. Our result can be seen as complementing the framework of Larsen & Skou who (without considering a calculus) give a logical characterization of probabilistic bisimulation in terms of probabilistic modal logic.

Existing research in this area has focussed mainly on the operational side, see e.g [2, 4, 6, 10, 12, 15, 19, 21]. In [2, 12, 15, 19] complete axiomatizations of the constructed probabilistic process calculi are given, with [15] dealing with a reactive model and [2, 12] generative models in the terminology of [20]. The probabilistic powerdomain construction [9] has been applied to give domain-theoretic semantics to certain languages, but as yet no fully abstract metric model has been proposed. Fully abstract characterizations for testing equivalences are included in [11, 4, 21]; denotational semantics is given in [11], but recursion is not considered. [18] introduces denotational semantics for probabilistic CSP in terms of conditional probability measures on the space of infinite traces. A "metric" for ϵ-bisimulation can be found in [6]; in contrast to ours, it does not satisfy the axioms of a metric.

We omit most details of the proofs from this version of the paper.

2 Probabilistic Transition Systems and Bisimulation

We assume the reader has some knowledge of metric spaces and the methodology for metric denotational semantics (see e.g. [5]).

Let D be a set. A *probability distribution with countable support* on D is a function $f : D \longrightarrow [0,1]$ such that the set $s(f) = \{d \in D \mid f(d) > 0\}$ is countable and $\sum_{d \in D} f(d) = 1$. Unless otherwise stated, by a *probability distribution* we shall mean a probability distribution with countable support. Let D be a set, and let $\mu(D)$ denote the family of probability distributions on D. Given any probability distribution f, and a set D such that $s(f) \subseteq D$, f can be extended to f_D such that $f_D \in \mu(D)$. When it is clear from the context what the set D is we write f instead of f_D.

Proposition 1. *The family $\mu(D)$ of probability distributions on D is a metric space with respect to the metric:*

$$d_\mu(f,g) = \frac{1}{2} \sum_{p \in s(f) \cup s(g)} |f(p) - g(p)| \ .$$

Furthermore, for all f and $g \in \mu(D)$, $0 \le d_\mu(f,g) \le 1$.

We recall the notions of probabilistic transition systems and probabilistic bisimulation introduced originally by Larsen & Skou [14]. A *probabilistic transition system* is a tuple $S = (P, Act, Can, \mu)$ where P is a set of processes (states),

Act is a set of observable actions, Can is an Act-indexed family of sets of processes where Can_a is the set of processes capable of performing the action a as their initial move, μ is a family of probabilistic distributions, $\mu_{p,a} : P \longrightarrow [0,1]$, for $a \in Act$, $p \in Can_a$, indicating the possible next states and their probabilities after p has performed a, i.e. $\mu_{p,a}(q) = \lambda$ means that the probability that p becomes q after performing a is λ.

Note that it is required that $\sum_{p' \in P} \mu_{a,p}(p') = 1$ since $\mu_{p,a}$ is a probability distribution. A probabilistic transition can, for a given state p and action a, be thought of as yielding a probabilistic distribution on the set of all processes P. The notation for probabilistic transitions is as follows:

$$p \xrightarrow{a}_\lambda p' \text{ whenever } p \in Can_a \text{ and } \mu_{p,a}(p') = \lambda$$
$$p \xrightarrow{a} p' \text{ whenever } p \xrightarrow{a}_\lambda p' \text{ for some } \lambda > 0 \ .$$

It should be noted that Larsen and Skou's definition models *reactive* systems in the terminology of van Glabbeek et al [20]. In the reactive model for probabilistic processes, a button-pressing experiment suceeds with probability 1, or else fails. When successful, the process makes an internal state transition according to a probability distribution associated with the depressed button. More formally, in the reactive model the probabilities are *action guarded*, meaning there is a (single) probability distribution for each process and action it can perform, thus imposing *determinism* at the language level.

Definition 2. Let (P, Act, Can, μ) be a probabilistic transition system. A *probabilistic bisimulation* R on P is an equivalence relation $R \subseteq P \times P$ such that whenever pRq the following holds:

$$\forall a \in Act \ \forall S \in P/R \ . \ p \xrightarrow{a}_\lambda S \iff q \xrightarrow{a}_\lambda S$$

where P/R denotes the quotient of P by R and for any $p \in P$ and $S \in P/R$ $p \xrightarrow{a}_\lambda S$ if and only if $\lambda = \sum_{s \in S} \mu_{p,a}(s)$. Two processes p and q are *probabilistic bisimilar* (notation $p \sim q$) if they are contained in a probabilistic bisimulation. The largest probabilistic bisimulation is denoted by $\sim$.

3 Language RP and its Operational Semantics

We consider a divergence-free probabilistic process algebra based on CCS [17] and CSP [8], referred to as RP, which includes recursion and deterministic choice, but instead of the usual non-determinism has action-guarded probabilistic choice. The language derives from the need to model reactive systems. We choose RP instead of PCCS [6] as RP is more intuitive for reactive processes and avoids the need for two types of transitions.

Let Act denote the set of actions (ranged over by $a, b \ldots$), and $\mathcal{X}$ the set of process variables (ranged over by $x, y \ldots$), both sets being countable. The syntax of all expressions is as follows:

$$q ::= \underline{0} \mid x \mid \sum_{i \in I} a_{\mu_i}.q_i \mid q_1 \oplus q_2 \text{ where } \mathcal{A}(q_1) \cap \mathcal{A}(q_2) = \emptyset \mid q_1 \| q_2 \mid fix \ x.q$$

where $a \in Act$, $x \in \mathcal{X}$, $\underline{0}$ denotes the inactive process, $\sum_{i \in I} a_{\mu_i} . q_i$ (where $a \in Act$, I is an index set, and $\mu_i \in (0, 1]$ is a countable set of real numbers such that $\sum_{i \in I} \mu_i = 1$) denotes probabilistic choice, $q_1 \oplus q_2$ denotes deterministic choice (we require that the sets of initial actions of q_1 and q_2 are disjoint), $q_1 \| q_2$ denotes synchronous parallel, and $fix\, x.q$ denotes recursion. In the case of I finite we also write $a_{\mu_1} . q_1 + a_{\mu_2} . q_2 + ... a_{\mu_n} . q_n$. Formally, the set $\mathcal{A}(q)$ of initial actions of q is defined inductively by setting $\mathcal{A}(\underline{0}) = \mathcal{A}(x) = \emptyset$, $\mathcal{A}(\sum_{i \in I} a_{\mu_i} . q_i) = \{a\}$, $\mathcal{A}(fix\, x.q) = \mathcal{A}(q)$, and $\mathcal{A}(q_1 \oplus q_2) = \mathcal{A}(q_1 \| q_2) = \mathcal{A}(q_1) \cup \mathcal{A}(q_2)$. We only consider the subset of the *guarded* expressions $\mathcal{E}$ defined over syntax:

$$p ::= \underline{0} \mid \sum_{i \in I} a_{\mu_i} . q_i \mid p_1 \oplus p_2 \text{ where } \mathcal{A}(p_1) \cap \mathcal{A}(p_2) = \emptyset \mid p_1 \| p_2 \mid fix\, x.p .$$

Observe that prefixing is a special case of probabilistic choice: $a.p$ is equivalent to $a_1.p$, meaning that after a is performed the process becomes p with probability 1. The syntactic restriction on the choice operator is necessary to draw comparisons with Larsen and Skou's formalism [14]. The operational semantics is as follows:

$$\textbf{Act} \quad \frac{}{\sum_{i \in I} a_{\mu_i} . p_i \xrightarrow{a}_\mu p_j} \quad j \in I \text{ and } \mu = \sum_{p_i = p_j} \mu_i$$

$$\textbf{Sum}_1 \quad \frac{p_1 \xrightarrow{a}_\mu q}{p_1 \oplus p_2 \xrightarrow{a}_\mu q} \qquad \textbf{Sum}_2 \quad \frac{p_2 \xrightarrow{a}_\mu q}{p_1 \oplus p_2 \xrightarrow{a}_\mu q}$$

$$\textbf{Par} \quad \frac{p_1 \xrightarrow{a}_{\mu_1} q_1 \text{ and } p_2 \xrightarrow{a}_{\mu_2} q_2}{p_1 \| p_2 \xrightarrow{a}_{\mu_1 \mu_2} q_1 \| q_2} \qquad \textbf{Rec} \quad \frac{p\{fix\, x.p/x\} \xrightarrow{a}_\mu q}{fix\, x.p \xrightarrow{a}_\mu q}$$

where $q\{p/x\}$ denotes the result of changing all free occurences of x in q by p, with change of bound variables to avoid clashes. Following the usual convention, we define the set RP of *(guarded) processes* of the language as the set of expressions in $\mathcal{E}$ with no free variables.

Proposition 3. *Probabilistic bisimulation is a congruence for the language RP, i.e. it is preserved by all contexts of the language.*

Furthermore, the equational laws below, derived following Milner [16], characterise RP:

$$
\begin{array}{llll}
(\oplus 1) & p_1 \oplus p_2 & = p_2 \oplus p_1 \\
(\oplus 2) & p_1 \oplus (p_2 \oplus p_3) & = (p_1 \oplus p_2) \oplus p_3 \\
(\oplus 3) & p \oplus \underline{0} & = p \\
(\textbf{Par1}) & p_1 \| p_2 & = p_2 \| p_1 \\
(\textbf{Par2}) & p \| \underline{0} & = p \\
(\textbf{Rec1}) & fix\, x.p & = p\{fix\, x.p/x\} \\
(\textbf{Rec2}) & q = p\{q/x\} & \Rightarrow q = fix\, x.p \\
(\textbf{Act}) & \sum_{i \in I} a_{\mu_i} . p_i & = \sum_{i \in I \setminus J} a_{\mu_i} . p_i + a_\mu . p
\end{array}
$$

where $p \in RP$ and $J \subseteq I$ such that for all $j \in J$ we have $p_j = p$ and $\mu = \sum_{j \in J} \mu_j$. It follows from the **Rec** laws that fixed points are unique up to bisimulation.

4 A Metric for Probabilistic Computations

We first turn our attention to probabilistic computations, which should be thought of as suitable generalizations of sequential computations (= sequences of steps) of de Bakker & Zucker [5]. As in the non-probabilistic case, a probabilistic process will be represented by a certain set of such computations.

Intuitively, a probabilistic computation step will be represented by a pair consisting of an action and a probabilistic distribution, i.e. an element of the set $A \times \mu(D)$, where D is assumed to be the set of all probabilistic computations. Thus, each such step $p = (a, f)$, for some $f \in \mu(D)$ and $a \in A$, can be viewed as the process which can perform the action a and become a process $q \in D$ with probability $f(q)$. To allow for termination we also require a distinguished element p_0 to model the inactive process. This gives:

$$D \cong \{p_0\} \cup A \times \mu(D)$$

as the candidate for a domain equation for probabilistic computations.

We proceed by applying the techniques of [5] to derive an inductively defined collection of metric spaces $(D_n, d), n = 0, 1, \ldots$, where the elements of the spaces model *finite* probabilistic computations. Informally, $D_0 \subseteq D_1 \subseteq \ldots \subseteq D_n \ldots$ form a sequence of sets, where as n increases the number of probabilistic processes which are modelled increases, with D_n modelling the processes capable of performing one probabilistic action at a time up to *the depth n*. Formally:

Definition 4 (Finite probabilistic computations). Let $D_n, n = 0, 1, \ldots$, be a collection of carrier sets defined inductively by:

$$D_0 = \{p_0\} \quad \text{and} \quad D_{n+1} = \{p_0\} \cup A \times \mu(D_n)$$

where A is a set of actions. Let $D_\omega = \bigcup_n D_n$ denote bounded computations.

For simplicity, we consider any $f \in \mu(D_n)$ as the extension of f to D_ω, i.e. $f_{D_\omega} \in \mu(D_\omega)$, with the subscript often dropped. We now explain the intuition behind our metric on probabilistic computations D_ω: the distance is set to 1 if the computations differ on the initial action, and to a (possibly infinite) sum derived from the distances between the resulting distributions otherwise. The latter involves the notion of a *truncation* on distributions, which we now define.

Definition 5. Let $f \in \mu(D_\omega)$. For $k \in \mathbb{N}$ define the kth *truncation* of f, $f[k] \in \mu(D_k)$, as follows. The support of $f[k]$ is given by $s(f[k]) \stackrel{def}{=} \bigcup \{ p[k] \mid p \in s(f) \}$, and for any $q \in D_k$,

$$f[k](q) = \begin{cases} 0 & \text{if } q \notin s(f[k]) \\ \sum \{ f(p) \mid p \in s(f) \text{ and } p[k] = q \} & \text{otherwise} \end{cases}$$

where for $p \in D_\omega$ the auxiliary truncation on probabilistic computations, $p[k] \in D_k$, is defined inductively on $k \in \mathbb{N}$ by putting $p[0] = p_0$ for all p and

$$p[k+1] = \begin{cases} p_0 & \text{if } p = p_0 \\ (a, f[k]) & \text{if } p = (a, f) \text{ for some } a \in A \text{ and } f \in \mu(D_\omega) \end{cases} .$$

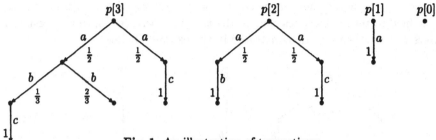

Fig. 1. An illustration of truncations.

The truncation of probabilistic distributions (and respectively of probabilistic computations, which we omit for reasons of space) satisfies the following properties useful in proofs of properties of our metric, as truncations are an integral part of its definition. These properties are, moreover, reminiscent of the properties of projection spaces of Große-Rhode and Ehrig [7].

Proposition 6.

(a) If $f \in \mu(D_n)$ then $f[k] \in \mu(D_k)$ when $0 \le k \le n$ and $f[k] = f_{D_k}$ when $k \ge n$
(b) If $f \in \mu(D_\omega)$ then for all k, m $(f[m])[k] = f[\min\{m, k\}]$
(c) For all $f, g \in \mu(D_\omega)$ and $k \in \mathbb{N}$ $d_\mu(f[k], g[k]) \le d_\mu(f, g)$
(d) For all $f, g \in \mu(D_\omega)$ if $f[m] = g[m]$ then $f[k] = g[k]$ for all $0 \le k \le m$.

We now define a metric on probabilistic computations. In the non-trivial case of computations starting with the same action, the distance is set to an infinite sum of distances between the truncations of the two distributions, with each summand weighted by the depth of the truncation in inverse proportion.

Definition 7. Let $(D_n)_{n \in \mathbb{N}}$, D_ω be the carrier sets defined in Definition 4. Define the metric d by induction on the structure of elements of D_n by putting $d(p_0, p_0) = 0$, $d(p_0, (a, f)) = 1$, $d((a, f), p_0) = 1$, and

$$
d((a, f), (\tilde{a}, g)) = \begin{cases} 1 & \text{if } a \ne \tilde{a} \\ \sum_{k=0}^{\infty} 2^{-(k+1)} d_\mu(f[k], g[k]) & \text{otherwise} . \end{cases}
$$

Lemma 8. Let (D_ω, d) be as above, then $0 \le d(p, q) \le 1$ for all $p, q \in D_\omega$.

We now prove the following for D_ω, and simultaneously for each D_n.

Theorem 9. (D_ω, d) is a metric space.

Proof. 1. We show $d(p, q) = 0$ if and only if $p = q$. In the non-trivial case of $p \ne q$ the result follows by definition of d except when $p = (a, f)$ and $q = (a, g)$: since $p \ne q$ we must have $f \ne g$, and thus from d_μ being a metric and Proposition 6(a) we have that $d_\mu(f[m], g[m]) = d_\mu(f, g) \ne 0$ for $m = \min_n\{s(f), s(g) \subseteq D_n\}$, and thus $d(p, q) \ne 0$ as required.
2. $d(p, q) = d(q, p)$ by definition of d and d_μ a metric on all D_n, $n \in \mathbb{N}$.
3. The inequality $d(p, q) + d(q, r) \ge d(p, r)$ follows from Lemma 8 in all cases except $p = (a, f)$, $q = (a, g)$ and $r = (a, h)$, in which case it holds since d_μ is a metric. $\qquad \square$

It should be noted that our metric is not an ultra-metric. An ultrametric can be defined in terms of truncations in the standard way, see [3], but it results in different convergence as demonstrated in the example below.

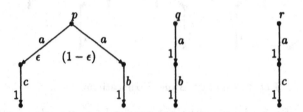

Fig. 2. 'Smooth' metric d (this paper) vs 'discrete' ultrametric of [3]

Example 1. Consider the processes in Figure 2. We have that:

$$d(p,q) = \frac{\epsilon}{2} \ , \quad d(p,r) = \frac{(1-\epsilon)}{2} \quad \text{and} \quad d(q,r) = \frac{1}{2}$$

and hence as $\epsilon \to 0$ the distance $d(p,q) \to 0$ while $d(p,r) \to \frac{1}{2}$. On the other hand, in the metric of [3], the distances between p and q, and between p and r, are $\frac{1}{2}$ for any $\epsilon \in (0,1)$.

Our metric nevertheless specialises to the metric of de Bakker & Zucker [5]. To see this consider a restriction, for each $n \in \mathbb{N}$, of the set $\mu(D_n)$ to the set of *point distributions* of D_n, i.e. the set $\{ \eta_p \,|\, p \in D_n \}$ where

$$\eta_p(q) = \begin{cases} 1 \text{ if } p = q \\ 0 \text{ otherwise} \end{cases}$$

and inductively we denote $\{p_0\} \cup A \times \{\eta_p \,|\, p \in D_n^\eta\}$ by D_{n+1}^η. Intuitively, if $p = (a, \eta_q) \in D_n^\eta$ then the probability of p performing the action a and becoming q is 1, and the probability of p becoming any other process is 0. This can be compared with de Bakker and Zucker's construction of simple processes, where the elements are of the form $p = p_0$ or $p = (a, q)$, for a action and q process. We have the following.

Proposition 10. *The metric d coincides with the metric of de Bakker & Zucker on the subspace D_ω^η of D_ω, i.e. for all p, q in D_ω^η:*

$$d(p,q) = \begin{cases} 0 & \text{if } p = q \\ \\ 2^{1-m} & \text{otherwise where } m = \min_k\{ p[k] \neq q[k] \} \end{cases}.$$

We now apply the standard completion technique to derive the domain D of probabilistic computations as consisting of D_ω together with all limit points $p = \lim_{n \to \infty} p_n$, with $\langle p_n \rangle_n$ a Cauchy sequence in D_ω, such that $p_n \in D_n \ \forall n \in \mathbb{N}$.

Definition 11. Define the space (D, d) of *probabilistic computations* as the metric completion of (D_ω, d).

We show that d satisfies the required domain equation by constructing isometric embeddings. Categorical techniques of [1] have not been used as it is unclear how to define a functor to represent this construction; this is due to the fact that our metric is not defined inductively in correspondence with the inductively defined metric spaces.

Theorem 12. D *satisfies the domain equation* $D \cong \{p_0\} \cup A \times \mu(D)$.

Proof. Let $\grave{D} \overset{def}{=} \{p_0\} \cup A \times \mu(D)$.

1. First define $\psi : \grave{D} \to D$ by

$$\psi(\grave{p}) = \begin{cases} p_0 & \text{if } \grave{p} = p_0 \\ \lim_{n \to \infty} p_n & \text{otherwise} \end{cases}$$

   where, assuming $\grave{p} = (a, g)$ for some $a \in A$ and $g \in \mu(D)$, $p_n = (a, f_n)$ with $f_{n+1} = g[n]$ for $n \in \mathbb{N}$. This is well-defined as $p_n \in D_n$ and the sequence $(p_n)_n$ can be shown to be Cauchy with respect to d. Finally, we demonstrate that that ψ is an isometry.

2. For the opposite direction, we define the map $\phi : D \to \grave{D}$ by

$$\phi(p) = \begin{cases} p_0 & \text{if } p = p_0 \\ (a, g) & \text{otherwise} \end{cases}$$

   where, assuming wlog $p = \lim_{n \to \infty} p_n$ with $\langle p_n \rangle_n$ Cauchy, $p_n = (a, f_n)$ for some $a \in A$ and $f_n \in \mu(D_{n-1})$ for all $n \geq 1$, $g : D \to [0, 1]$ is defined by $g(q) = \lim_{n \to \infty} f_n(q)$ for $q \in D$. To show that this is well-defined, i.e. $\grave{p} \in \grave{D}$, we show $\lim_{n \to \infty} f_n(q)$ exists for all $q \in D$ and $g \in \mu(D)$, i.e. $\sum_{q \in D} g(q) = 1$ and g has countable support. Finally, we show that ϕ is an isometry. $\square$

5 Domain Equation for Reactive Processes

Observe that the probabilistic computations (the elements of D) are represented either by p_0 (termination), or are limits $\lim_{n \to \infty} p_n$ of Cauchy sequences of (finite) computations, where the limit is of the form $(a, \lim_n f_n)$, and thus initially can only perform the action a. To allow choice it is necessary to use *sets* of elements of D as denotations for probabilitic processes. As we wish to maintain consistency with Larsen & Skou's approach, we mimic the syntactic restrictions in the semantic domain by requiring that such sets must satisfy the following *reactiveness* condition.

Definition 13. Let $X \subseteq D_\omega$. X is said to satisfy the *reactiveness* condition if, for any $p, q \in D_\omega$ where $p = (a, f)$ and $q = (\tilde{a}, g)$, if $p, q \in X$ then it must be the case that either $a \neq \tilde{a}$ or $p = q$.

The above guarantees, for any $a \in A$, the existence of at most one element of the form (a, f) in the set X, and so the probability of performing an a transition for any one of these sets is either 1 or 0.

To extend our metric to sets of probabilistic computations we use the Hausdorff distance. As before, we introduce a sequence of metric spaces $(P_n, d)_n$ $n \in \mathbb{N}$.

Definition 14. Let (P_n, d) $n = 0, 1, \ldots$ be a collection of metric spaces defined inductively by

$$P_0 = \{p_0\} \quad \text{and} \quad P_{n+1} = \{p_0\} \cup \mathcal{P}_r (A \times \mu(P_n))$$

where A is a set of actions and $\mathcal{P}_r$ denotes the powerset operator restricted to the subsets which satisfy the reactiveness condition. Put $P_\omega = \bigcup_n P_n$ and then define d on P_ω (or on any P_n where $n \in \mathbb{N}$) to be the Hausdorff distance with respect to d as defined on D_ω. Let (P, d) denote the completion of (P_ω, d).

Observe that for any $X \in P_\omega$ we have that $X \in \mathcal{P}_r(A \times \mu(P_n))$ for some $n \in \mathbb{N}$. Then for any distinct elements $p, q \in X$ such that $p = (a, f)$ and $q = (\hat{a}, g)$ with $a \neq \hat{a}$, we have by definiton of the metric d that $d(p, q) = 1$. It follows that X is closed, since the only Cauchy sequences included in X are the trivial ones. Thus, $P_\omega \subseteq \mathcal{P}_c(D_\omega)$, and from the completion techniques it follows that $P \subseteq \mathcal{P}_c(D)$, and hence d is indeed a metric on P_ω.

Moreover, Hahn's Theorem can be used in the proof of the following.

Theorem 15. Let $\mathcal{P}_{rc}(A \times \mu(P))$ denote the closed subsets of $(A \times \mu(P))$ satisfying the reactiveness condition. Then

$$P \cong \{p_0\} \cup \mathcal{P}_{rc} (A \times \mu(P)) \ .$$

The theorem is proved by an adaptation of a similar result in [5] for the non-probabilistic case. We note that truncations on $f \in \mu(P_\omega)$ are defined as for D, and we define the truncation function on P inductively by putting $X[n] = \{p[n] \mid p \in X\}$ for any $X \subseteq A \times \mu(P)$.

6 Denotational Semantics

We have obtained P as a solution of a domain equation (assuming $A = Act$), and can now give denotational semantics for our language RP. The next step is to define the semantic operators on P.

Definition 16. The *degree* of a process $p \in P$ is defined inductively by putting $deg(p_0) = 0$, $deg(p) = n$ if $p \in P_n \setminus P_{n-1}$ for some $n \geq 1$, and $deg(p) = \infty$ otherwise. We then say a process p is *finite* if $deg(p) = n$ for some $n \in \mathbb{N}$ and *infinite* otherwise.

Thus, each $p \in P$ is either finite, or it is infinite, in which case $p = \lim p_n$, $(p_n)_n$ Cauchy, with each p_n of degree n. We now define the operators "$\cup$" and "$\|$" on P to model deterministic choice and synchronous parallel; this is achieved by first defining the operators on finite processes and then extending the definition to limits of Cauchy sequences.

Definition 17. Let $p \in P$, $X, Y \in \mathcal{P}_{rc}(A \times \mu(P))$ with finite degree, $(p_i)_i$, $(q_i)_i$ Cauchy sequences of finite processes.
(a) (union) Put $p \cup p_0 = p_0 \cup p = p$, $X \cup Y$ is the set theoretic union of X and Y, and define $(\lim_i p_i) \cup (\lim_j q_j) = \lim_k (p_k \cup q_k)$.
(b) (parallel) Put $p\|p_0 = p_0\|p = p$, and define

$$X\|Y = \begin{cases} \{x\|y \mid x \in X,\ y \in Y\ \&\ d(x,y) < 1\} & \text{if there exists } x \in X \text{ and } y \in Y \\ & \text{such that } d(x,y) < 1 \\ p_0 & \text{otherwise} \end{cases}$$

where for $x = (a, f)$ and $y = (a, g)$ put $x\|y \stackrel{def}{=} (a, f\|g)$ with

$$(f\|g)(p) \stackrel{def}{=} \begin{cases} f(p_1)g(p_2) & \text{if } p = p_1\|p_2 \\ 0 & \text{otherwise} \end{cases}$$

for any $p \in P$, and define $(\lim p_i)\|(\lim q_j) \stackrel{def}{=} \lim_k (p_k\|q_k)$.

Lemma 18. *For all X, $\tilde{X}$ and $Y \in P$ with finite degree*

$$d(X \cup Y, \tilde{X} \cup Y) \le d(X, \tilde{X}) \quad \text{and} \quad d(X\|Y, \tilde{X}\|Y) \le d(X, \tilde{X})\ .$$

Theorem 19. *$\cup$ and $\|$ are well defined and continuous operators on P subject to the restriction that $X \cup Y$ satisfies the reactiveness condition.*

Recall that $\mathcal{E}$ denotes the (guarded) expression with free variables, while RP is the set of closed (guarded) expressions. As usual, in order to handle the variables x of $\mathcal{E}$, we introduce the semantic map $\mathcal{M} : \mathcal{E} \to (E \to P)$ parametrised by environments E, ranged over by ρ, defined by $E = \mathcal{X} \to P$. In addition, we shall require an auxiliary function $\Phi : ([0,1] \times P)^\infty \to (P \to [0, \infty))$, defined as follows: for any $p = \langle (\mu_i)_i, (p_i)_i \rangle_{i \in I} \in ([0,1] \times P)^\infty$, $\Phi(p) = f_p$ where for any $q \in P$

$$f_p(q) = \begin{cases} 0 & \text{if } q \ne p_i \text{ for all } i \in I \\ \sum_{j \in J} \mu_j & \text{otherwise where } J = \{j \mid j \in I \text{ and } q = p_j\}\ . \end{cases}$$

We now define denotational metric semantics for RP expressions $\mathcal{E}$. Recursive processes are defined as limits of Cauchy chains of unfoldings of the map $\mathcal{M}$.

Definition 20 (Denotational semantics). Define $\mathcal{M} : \mathcal{E} \to (E \to P)$ inductively on the structure of elements of $\mathcal{E}$ as follows:

$$\mathcal{M}(\underline{0})(\rho) = \{p_0\}$$
$$\mathcal{M}(\textstyle\sum_{i \in I} a_{\mu_i} . p_i)(\rho) = \{(a, \Phi(\langle(\mu_i)_i, (\mathcal{M}(p_i)(\rho))_i\rangle_{i \in I}))\}$$
$$\mathcal{M}(p_1 \oplus p_2)(\rho) = \mathcal{M}(p_1)(\rho) \cup \mathcal{M}(p_2)(\rho)$$
$$\mathcal{M}(p_1 \| p_2)(\rho) = \mathcal{M}(p_1)(\rho) \| \mathcal{M}(p_2)(\rho)$$
$$\mathcal{M}(fix\ x.p)(\rho) = \lim_{k \to \infty} \mathcal{M}^k(p)(\rho)$$

where $\mathcal{M}^0(p)(\rho) = p_0$ and $\mathcal{M}^{k+1}(p)(\rho) = \mathcal{M}(p)(\rho\{\mathcal{M}^k(p)(\rho)/x\})[k+1]$.

The well-definedness of the semantic map follows from the lemma below.

Lemma 21. *Let $fix\ x.p \in \mathcal{E}$, and let the sequence q_k denote $\mathcal{M}^k(p)(\rho)$, $k \in \mathbb{N}$. Then $q_{k+1}[k] = q_k$ for all $k \in \mathbb{N}$.*

7 Full Abstraction

Finally, we obtain that P is a fully abstract model of the language RP with respect to probabilistic bisimulation. The result follows from Lemma 22 below.

Lemma 22. *For all $a \in A$ and $p \in$ RP:*

1. *$p \overset{a}{\longrightarrow}$ if and only if there exists $(a, f) \in \mathcal{M}(p)$.*
2. *For any $q \in$ RP if $S_{\mathcal{M}(q)} = \{\tilde{q} \mid \tilde{q} \in$ RP and $\mathcal{M}(\tilde{q}) = \mathcal{M}(q)\}$ then we have $p \overset{a}{\longrightarrow}_\mu S_{\mathcal{M}(q)}$ if and only if $f(\mathcal{M}(q)) \geq \mu$.*
3. *If $f(r) > 0$ for some $r \in P$ then if $S_r = \{q \mid q \in$ RP and $\mathcal{M}(q) = r\}$ then $p \overset{a}{\longrightarrow}_\mu S_q$ if and only if $f(q) \geq \mu$.*

Theorem 23. *Let $\mathcal{M} : \mathcal{E} \to (E \to P)$ be the semantic map of Definition 20. Then for all $p, q \in$ RP,*

$$p \sim q \quad \text{if and only if} \quad \mathcal{M}(p) = \mathcal{M}(q) .$$

8 Conclusions and Further Work

We have derived a metric space model for a probabilistic extension of a process calculus, which can be further extended with an asynchronous concurrency operator by following, to a large extent, the techniques introduced by de Bakker & Zucker [5].

Although the continuity of prefixing (and also of the asynchronous concurrency operator) fails in our model, our metric is 'smooth' (as apposed to the 'discrete' metric of [3]), and hence closer in spirit to the probabilistic powerdomain construction. It remains to be seen if a suitable combination of our metric and the standard metric which yields continuity can be found. Finally, we intend to consider the addition of non-deterministic choice and apply our results to existing probabilistic process calculi, e.g. PCCS [6].

Acknowledgements: We would like to thank Achim Jung, Michael Huth, Christel Baier and Reinhold Heckmann for discussions and suggestions.

References

1. P.H.M.America and J.J.M.M.Rutten. Solving reflexive domain equations in a category of complete metric spaces, *JCSS, 39, no.3, 1989.*
2. J.C.M.Baeten, J.A.Bergstra and S.A.Smolka. Axiomatising probabilistic processes: ACP with generative probability, *Proc. Concur'92, LNCS, 630, Springer, 1992.*
3. C.Baier and M.Kwiatkowska. Domain equations for probabilistic processes, *preprint.*
4. I.Christoff. Testing equivalences and fully abstract models for probabilistic processes, *Proc. Concur'90, LNCS, 458, Springer, 1990.*
5. J.W.de Bakker and J.I.Zucker. Processes and the denotational semantics of concurrency, *Information and Control, 1/2, 1984.*
6. A.Giacalone, C.-C.Jou and S.A.Smolka. Algebraic reasoning for probabilistic concurrent systems, *In Proc. Programming Concepts and Methods, IFIP, 1990.*
7. M.Große-Rhode and H.Ehrig. Transformation of combined data type and process specifications using projection algebras, *LNCS, 430, Springer, 1989.*
8. C.A.Hoare. Communicating sequential processes, *Prentice Hall, 1985.*
9. C.Jones. Probabilistic non-determinism, *PhD Thesis, University of Edinburgh, 1990.*
10. B.Jonsson and K.G.Larsen. Specification and refinement of probabilistic processes, *Proc. IEEE Logic in Computer Science (LICS), 1991.*
11. B.Jonsson and Wang Yi. Compositional testing preorders for probabilistic processes, *Proc. IEEE Logic in Computer Science (LICS), 1995.*
12. C.-C.Jou and S.Smolka. Equivalences, congruences and complete axiomatizations for probabilistic processes, *Proc. Concur'90, LNCS, 458, Springer, 1990.*
13. D.Kozen. Semantics of probabilistic programs, *Proc. IEEE Symposium on Foundations of Computer Science (FOCS), 1979.*
14. K.G.Larsen and A.Skou. Bisimulation through probabilistic testing, *Information and Computation, 94, 1991.*
15. K.G.Larsen and A.Skou. Compositional verification of probabilistic processes, *Proc. Concur'92, LNCS, 630, Springer, 1992.*
16. R.Milner. Calculi for synchrony and asynchrony, *TCS, 25(3), 1983.*
17. R.Milner. Communication and concurrency, *Prentice Hall, 1989.*
18. K.Seidel. Probabilistic communicating processes, *TCS, 152, 1995.*
19. C.Tofts. A synchronous calculus of relative frequency, *Proc. Concur'90, LNCS, 458, Springer, 1990.*
20. R.J.van Glabbeek, S.A.Smolka, B.Steffen and C.Tofts. Reactive, generative and stratified models of probabilistic processes, *Proc. Concur'92, LNCS, 630, Springer, 1992.*
21. S.Yuen, R.Cleaveland, Z.Dayar and S.A.Smolka. Fully abstract characterizations of testing preorders for probabilistic processes, *Proc. Concur'94, LNCS, 836, Springer, 1994.*

Dynamic Graphs [1]

Andrea Maggiolo-Schettini
Dipartimento di Informatica, Università di Pisa
56125 Pisa, Corso Italia 40, Italy

Józef Winkowski
Instytut Podstaw Informatyki PAN
01-237 Warszawa, ul. Ordona 21, Poland

Abstract

A model of processes of rewriting graphs is proposed in which concurrency and branching can be represented. Operations on structures representing processes of rewriting graphs are defined that allow one to construct such structures from simple components and to characterize sets of processes of rewriting graphs, including sets generated by graph grammars.

1 Motivation and introduction

In this paper we extend a model of processes of rewriting graphs which has been developed in the theory of graph grammars.

A graph grammar has rules called graph productions, where a rule p says that a certain pattern L_p can be replaced by another pattern R_p if it occurs in a graph. A concrete form of graph productions, and principles stating how a production can be applied to a graph and what graph is the result of application depend on the specific approach. In the so called algebraic approach the patterns in productions are graphs, and an application of a production p to a graph G giving a graph H as the result, $G \xrightarrow{p} H$, is defined as a replacement of an occurrence G' of the pattern L_p in G by a copy H' of the pattern R_p such that the part of G which remains unchanged, that is $I = (G - G') \cup (G' \cap H')$, is a subgraph of G and $H = (G - G') \cup H'$. The common subgraph $G' \cap H'$ of G' and H' plays the role of a context in the presence of which the part $G' - H'$ of G' is removed from G resulting in graph I, and next, the part $H' - G'$ of H' is added to I resulting in graph H. Processes of rewriting graphs are represented by sequences of applications of productions of the form $G_0 \xrightarrow{p_1} G_1 \xrightarrow{p_2} ... \xrightarrow{p_n} G_n$, called derivations. It is assumed that applications which are independent in the sense that they do not intersect except in the contexts can be performed concurrently. The potential concurrency of applications of productions in a process of rewriting graphs is reflected in the so called shift-equivalence of representing

[1] This work has been supported by the Italian National Council for Research (CNR-GNIM), by the Polish Academy of Sciences (IPI PAN), and by COMPUGRAPH Basic Research Esprit Working Group n. 7183.

derivations, that is in the possibility of obtaining such derivations one from another by repeatedly exchanging contiguous independent steps. Consequently, a process of rewriting graphs can be defined as an equivalence class of derivations with respect to shift-equivalence, that is as a derivation trace in the sense of [CELMR 94].

In [CELMR 94] it is shown that derivation traces of so called safe graph grammars can be ordered such that they form prime event structures, and in [CMR 96] it is shown that derivation traces of safe graph grammars correspond exactly to isomorphism classes of partially ordered structures called graph processes.

Both derivation traces and graph processes are defined for concrete graph grammars with the idea of associating with each grammar a semantical meaning in the form of the set of its derivation traces or graph processes.

In this paper we introduce models of processes of rewriting graphs independently of graph grammars, and we define operations allowing us to combine such models and, consequently, to generate sets of processes of concrete graph grammars. As with graph processes, our models, called dynamic graphs, are enriched variants of contextual occurrence nets in the sense of [MR 95].

A dynamic graph consists of data elements which represent nodes and edges of graphs taking part in a process of rewriting or in a collection of such processes, and of events which represent rewriting steps. When representing a single process a dynamic graph may be equipped with some canonical representations of the initial graph and of the resulting graph, and then, by analogy with concatenable processes of [DMM 89], called a concatenable dynamic graph.

Dynamic graphs may be considered to be either concrete or abstract, that is up to isomorphism.

An important feature which distinguishes dynamic graphs from derivation traces and graph processes is that abstract concatenable dynamic graphs admit natural universal operations with the aid of which one can represent derivation traces of arbitrary graph grammars, not only safe ones.

The paper is organized as follows. Section 2 recalls the concept of graph and defines canonical representations of graphs. Section 3 describes the underlying contextual occurrence nets of dynamic graphs, called occurrence structures. Section 4 describes concrete dynamic graphs and constructions on such graphs. Section 5 describes concrete and abstract concatenable dynamic graphs and operations on abstract concatenable dynamic graphs. Section 6 describes how finite abstract dynamic graphs can be represented as sequential compositions of indecomposable components, and how such representations can be transformed. Finally, section 7 gives an idea of how dynamic graphs can represent derivation traces of arbitrary graph grammars. For lack of space, all the proofs are omitted and will be given in the full paper.

2 Graphs

Let L be a fixed set of labels. In the sequel we assume that L is well-ordered.

2.1. Definition. A *labelled graph* over L (or a *graph*) is $G = (V, E, s, t, l)$, where V and E are disjoint sets (the set of *nodes* and the set of *edges*), $s : E \to V$ and $t : E \to V$ are functions assigning to each $e \in E$ some $s(e) \in V$ (the *source* of e) and some $t(e) \in V$ (the *target* of e), and $l : V \cup E \to L$ is a function (a *labelling*) assigning to each $x \in V \cup E$ a label $l(x) \in L$. $\square$

We denote $V \cup E$ by X and use subscripts, $X_G, V_G, E_G, s_G, t_G, l_G$, when necessary to avoid a confusion.

Each $x \in X$ is called an *element* of G (a *graph element*). We say that G is *finite* if so is X.

2.2. Definition. A *morphism* from a graph G to another graph G' is a mapping $f : X_G \to X_{G'}$ such that $f(V_G) \subseteq V_{G'}$, $f(E_G) \subseteq E_{G'}$, $s_{G'}(f(e)) = f(s_G(e))$ and $t_{G'}(f(e)) = f(t_G(e))$ for all $e \in E_G$, and $l_{G'}(f(x)) = l_G(x)$ for all $x \in X_G$. $\square$

Graphs over L and their morphisms form a category $GR(L)$. It is known that each pair $(G \xleftarrow{f} G^0 \xrightarrow{f'} G')$ of graph morphisms has a pushout $(G \xrightarrow{g} G'' \xleftarrow{g'} G')$, where $(X_G \xrightarrow{g} X_{G''} \xleftarrow{g'} X_{G'})$ is a pushout of $(X_G \xleftarrow{f} X_{G^0} \xrightarrow{f'} X_{G'})$ in the category of sets, $s_{G''}(g(e)) = g(s_G(e))$ and $t_{G''}(g(e)) = g(t_G(e))$ for $e \in E_G$, and $s_{G''}(g'(e')) = g'(s_{G'}(e'))$ and $t_{G''}(g'(e')) = g'(t_{G'}(e'))$ for $e' \in E_{G'}$.

In the sequel we exploit graphs of a particular form.

2.3. Definition. A *arranged graph* is a graph G such that:

(1) for each label λ nodes with this label are arranged into a (possibly transfinite) sequence
$$(\lambda, 0), ..., (\lambda, m), ...,$$

(2) for each label ν edges with this label, and with a source node (λ, i) and a target node (μ, j) are arranged into a (possibly transfinite) sequence
$$(\nu, (\lambda, i), (\mu, j), 0), ..., (\nu, (\lambda, i), (\mu, j), n), ...$$

Such a graph G has a *code* of the form uv, written as $code(G)$, where u is the sequence of lexicographically ordered nodes and v is a sequence of lexicographically ordered edges. $\square$

Due to the fact that each set can be well-ordered we can arrange nodes and edges into sequences as in (1) and (2) of the above definition. Consequently, for each graph G there exists an arranged graph G' and an isomorphism $f : G' \to G$, called a *representation* of G by G'. In particular, each graph has a nonempty set of codes.

Taking into account the fact that the set consisting of the well-ordered set of natural numbers followed by L is well-ordered and that codes of a graph are ordered lexicographically, we conclude that each graph G has a representation by an arranged graph G' with the least code. We say that such a representation is *canonical*, call G' the *canonical form* of G, and write it as $form(G)$.

From the fact that the only automorphism of a well-ordered set is the identity we obtain the following result which will play a role in defining operations on our models of processes of rewriting graphs.

2.4. Proposition. Two graphs are isomorphic iff their canonical forms are identical. □

3 Occurrence structures

The way in which a process of rewriting graphs develops can be represented by what we call an occurrence structure. Such a structure can be defined as a branching version of a contextual occurrence net in the sense of [MR 95] similar to a branching process in the sense of [Eng 91].

3.1. Definition. An *occurrence structure* is $N = (X, Y, F, C)$, where X and Y are disjoint sets (the set of *data elements* and the set of *events*) and $F \subseteq X \times Y \cup Y \times X$ and $C \subseteq X \times Y$ are disjoint relations (the *flow relation* and the *context relation*) such that:

(1) $(F \cup FC \cup C^{-1}F)^*$ is a partial order $\leq$ on $X \cup Y$ (the *precedence relation*),

(2) there are no $x \in X$ and $y, y' \in Y$ such that yFx, $y'Fx$, and $y \neq y'$ (no backward branching at data elements),

(3) the relation defined by:

   $u \# v$ iff there exist $x \in X$ and $y, y' \in Y$ such that xFy, xFy', $y \neq y'$, $y \leq u$, and $y' \leq v$

   (the *conflict relation*) is irreflexive,

(4) $u \# v$ excludes uCv,

(5) for each $y \in Y$ there exists some $x \in X$ such that xFy or some $x' \in X$ such that yFx'. □

We denote $X \cup Y$ by U and use subscripts, $U_N, X_N, Y_N, F_N, C_N, \leq_N, \#_N$, when necessary.

For each event $y \in Y$ the sets $Cy = \{x \in X : xCy\}$, $Fy = \{x \in X : xFy\}$, $yF = \{x \in X : yFx\}$ represent respectively the context in which y occurs and the data y consumes and produces. The relation in (1) is defined such that the context of each event is present unchanged during entire execution of this event.

The set of data elements which are both minimal and maximal is called the *static part* of N and it is written as $static(N)$.

Each maximal conflict-free antichain of $(U, \leq)$, where conflict-freeness means the lack of #-related elements, is called a *cut* of N. Such a cut is said to be *proper* if it contains only such events which are minimal or maximal.

It can be shown that each cut which contains an event y contains also all the data elements belonging to the context in which y occurs, that is all $x \in Cy$.

Given a cut Z, we define

$$head_Z(N) = \{u \in U : x \leq u \leq z \text{ for some } x \in X \text{ and } z \in Z\},$$

$$tail_Z(N) = \{u \in U : z \leq u \leq x \text{ for some } x \in X \text{ and } z \in Z\}.$$

We call $head_Z(N)$ the *configuration* corresponding to Z, call the restriction of N to the set of data elements of Z the *state* corresponding to Z, and write such a state as $state_Z(N)$. If the cut Z is proper then the state corresponding to Z is said to be *complete*. Otherwise the state corresponding to Z is said to be *partial*.

From (3) of 3.1 it follows that each maximal antichain of a configuration is a cut.

Given a cut Z and an event $y \in Y$, we say that y is *enabled* at Z if $Fy \cup Cy \subseteq Z$ and $Fy \cap Cy' = \emptyset$ for all $y' \in tail_Z(N)$.

It can be shown that if an event y is enabled at a cut Z then the set

$$Z' = \begin{cases} (Z - F_N y) \cup yF_N & \text{if } yF_N \neq \emptyset \\ (Z - F_N y) \cup \{y\} & \text{otherwise} \end{cases}$$

is a cut of N. We call it the *result* of executing y at Z.

Cuts of N are ordered by the relation: $Z \sqsubseteq Z'$ iff $head_Z(N) \subseteq head_{Z'}(N)$.

We say that N is *finitary* if for all cuts Z', Z'' such that $Z' \sqsubseteq Z''$ the set of cuts Z such that $Z' \sqsubseteq Z \sqsubseteq Z''$ is finite. We say that N is *lower bounded* if it has a least cut and if this cut coincides with the set of minimal elements of U. The state corresponding to such a least cut is called the *initial state* of N and it is written as $initial(N)$. We say that N is *bounded* if it lower bounded and has a greatest cut and if this greatest cut coincides with the set of maximal elements of U. The state corresponding such a greatest cut is called the *final state* of N and it is written as $final(N)$.

If the set Y of events is empty then $N = initial(N) = final(N)$ and we call N a *state*.

Finally, if the conflict relation $\#$ is empty then we say that N is *deterministic*. In this case the set of cuts of N is a lattice with $Z \sqcap Z'$ defined as the set of maximal elements of the set $head_Z(N) \cap head_{Z'}(N)$ and $Z \sqcup Z'$ defined as the set of minimal elements of the set $tail_Z(N) \cap tail_{Z'}(N)$.

3.2. Proposition. Given a finitary occurrence structure N, for each pair of cuts Z and Z' of this structure such that $Z \sqsubseteq Z'$, there exists a finite sequence $y_1, ..., y_n$ of events of N and a finite chain $Z = Z_0 \sqsubseteq Z_1 \sqsubseteq ... \sqsubseteq Z_n = Z'$ of cuts such that each y_i is enabled at Z_{i-1} and it has the result Z_i. $\square$

3.3. Definition. A *morphism* from an occurrence structure N to another such a structure N' is a mapping $f : U_N \to U_{N'}$ such that:

(1) $f(X_N) \subseteq X_{N'}$ and $f(Y_N) \subseteq Y_{N'}$,

(2) f is injective on Y_N,

(3) $f(F_N y) = F_{N'} f(y)$, $f(C_N y) = C_{N'} f(y)$, and $f(y F_N) = f(y) F_{N'}$, for all $y \in Y_N$,

(4) $f(F_N y \cap y' F_N) = f(F_N y) \cap f(y' F_N)$, $f(y F_N \cap C_N y') = f(y F_N) \cap f(C_N y')$, $f(F_N y \cap F_N y') = f(F_N y) \cap f(F_N y')$, and $f(F_N y \cap C_N y') = f(F_N y) \cap f(C_N y')$, for all $y, y' \in Y_N$ □.

Condition (3) of this definition says that the context of an event $f(y)$ and the data this event consumes and produces are images under f of the respective context and data of y. Together with (4) this condition implies that f preserves incomparability of events w.r. to the precedence relation.

Occurrence structures and their morphisms form a category OS.

A morphism as in in 3.3 is written as $f : N \to N'$. The following proposition says that such a morphism induces a simulation of the process represented by N in the process represented by N'.

3.4. Proposition. Given a morphism $f : N \to N'$, for each cut Z of N the set $f(Z)$ is an antichain of $(U_{N'}, \leq_{N'})$ such that if an event $y \in Y_N$ is enabled at Z then the event $f(y)$ is enabled at each cut Z' of N' such that $f(Z) \subseteq Z'$ and the result of executing $f(y)$ at Z' contains the image under f of the result of executing y at Z. □

4 Dynamic graphs

Processes of rewriting graphs can be represented by occurrence structures whose data elements form graphs. We call such structures concrete dynamic graphs and define them as follows.

Let L be a set of labels as in section 2.

4.1. Definition. A *concrete dynamic graph* over L is $D = (V, E, Y, F, C, s, t, l)$, where $N_D = (X, Y, F, C)$ with $X = V \cup E$ is an occurrence structure and $G_D = (V, E, s, t, l)$ is a graph over L such that:

(1) the restrictions of D to to the static part of N_D and to the states of N_D are subgraphs of G_D,

(2) the restrictions of D to Cy, $Fy \cup Cy$, and $yF \cup Cy$ are subgraphs of G_D for all $y \in Y$. □

We use subscripts, U_D, X_D, V_D, E_D, Y_D, F_D, C_D, $\leq_D$, s_D, t_D, l_D, when necessary.

We say that D is *finitary* (resp.: *lower bounded, bounded*) if such is N_D. By *cuts* of D we mean cuts of N_D. By the *static part* of D, *static(D)*, we mean the restriction of D to the static part of N_D. By *states* of D we mean the restrictions of D to states of N_D. For each cut Z by $state_Z(D)$, $head_Z(D)$, $tail_Z(D)$ we denote the restrictions of D to $state_Z(N_D)$, $head_Z(N_D)$, $tail_Z(N_D)$, respectively. The initial state of D, $initial(D)$, and the final state of D, $final(D)$, are defined as the restrictions of D to $initial(N_D)$ and $final(N_D)$, respectively. Enabling and effect of execution of an event $y \in Y$ are understood as corresponding to the occurrence structure N_D.

If the underlying occurrence structure N_D coincides with its static part then we call D a *static graph*. Finally, if N_D is deterministic then we say that D is *deterministic*.

4.2. Example. The concrete dynamic graph D with $V_D = \{p, q, x, y\}$, $E_D = \{a, b, c, d, u, v, w\}$, $Y_D = \{e, f, g\}$, $F_D e = \{a, b, p\}$, $C_D e = \{x, y\}$, $eF_D = \{u\}$, $F_D f = \{c, d, q\}$, $C_D f = \{x, y\}$, $fF_D = \{v\}$, $F_D g = \{u, v\}$, $C_D g = \{x, y\}$, $gF_D = \{w\}$, $s_D(a) = s_D(c) = s_D(u) = s_D(v) = s_D(w) = x$, $s_D(b) = p$, $s_D(d) = q$, $t_D(b) = t_D(d) = t_D(u) = t_D(v) = t_D(w) = y$, $t_D(a) = p$, $t_D(c) = q$ represents a process which consists of two potentially concurrent events e and f followed by an event g, where e stands for the replacement of the edge a from x to p and the edge b from p to y by the edge u from x to y, f stands for the replacement of the edge c from x to q and the edge d from q to y by the edge v from x to y, and g stands for the replacement of the edges u and v from x to y by the edge w from x to y. $\square$

4.3. Definition. A *morphism* from a concrete dynamic graph D to another such a graph D' is a morphism f from the underlying occurrence structure N_D to $N_{D'}$, where f is a graph morphism from the underlying graph G_D to $G_{D'}$. $\square$

Concrete dynamic graphs over L and their morphisms form a category $DG(L)$. This category has some pushouts corresponding to natural operations on processes.

4.4. Proposition. For each static graph D and each lower bounded concrete dynamic graph D' with an injective morphism $f : initial(D') \to D$ such that each edge in $X_D - f(X_{initial(D')})$ has the source and the target nodes in $X_D - f(X_{initial(D')})$ or in $f(X_{static(D')})$, the pair $(D \xleftarrow{f} initial(D') \xhookrightarrow{\subseteq} D')$ has a pushout $(D \xrightarrow{g} D'' \xleftarrow{g'} D')$ with g being inclusion and D'' having a proper cut Z and a subset Z' of this cut such that the restriction of D'' to $head_Z(D'')$ coincides with D and that to the set

$$\{u \in U_{D''} : z \leq u \leq x \text{ for some } x \in X_{D''} \text{ and } z \in Z'\}$$

is isomorphic to D'. If D' is bounded then D'' is also bounded and there exists a unique morphism $h : final(D') \to final(D'')$ such that, for all u in $final(D')$, $h(u) = g'(u)$ whenever either side is defined. $\square$

4.5. Proposition. For each finitary bounded concrete dynamic graph D and each finitary lower bounded concrete dynamic graph D' such that there exists an isomorphism $f : initial(D') \to final(D)$ from the initial state of D' to the final state of D, the pair of morphisms given by $(D \xrightarrow{\supseteq} final(D) \xleftarrow{\perp} initial(D') \xrightarrow{\subseteq} D')$ has a pushout $(D \xrightarrow{g} D'' \xleftarrow{g'} D')$ with g being inclusion and D'' having a proper cut Z such that the restriction of D'' to $head_Z(D'')$ coincides with D and that to $tail_Z(D'')$ is isomorphic to D'. If D' is bounded then D'' is also bounded and there exists a unique morphism $h : final(D') \to final(D'')$ such that, for all u in $final(D')$, $h(u) = g'(u)$ whenever either side is defined. $\square$

4.6. Proposition. For each lower bounded concrete dynamic graph D and D' such that there exists an isomorphism $f : initial(D') \to initial(D)$ from the initial state of D' to the initial state of D, the pair of morphisms given by $(D \xrightarrow{\supseteq} initial(D) \xleftarrow{\perp} initial(D') \xrightarrow{\subseteq} D')$ has a pushout $(D \xrightarrow{g} D'' \xleftarrow{g'} D')$. $\square$

5 Operations on dynamic graphs

In order to construct models of complex processes by combining models of components of such processes, we define operations on dynamic graphs. To this end we restrict ourselves to finitary lower bounded and bounded dynamic graphs, we equip such graphs with some specific representations of their initial and final states, and we consider structures thus obtained independently of inessential details of concrete representations. The respective formalization is as follows.

5.1. Definition. A concrete *left concatenable* (resp.: *concatenable*) dynamic graph is $P = (V, E, Y, F, C, s, t, l, i)$ (resp.: $P = (V, E, Y, F, C, s, t, l, i, j)$), where $D_P = (V, E, Y, F, C, s, t, l)$ is a finitary lower bounded (resp.: bounded) concrete dynamic graph with the underlying occurrence structure $N_P = (X = V \cup E, Y, F, C)$ and the underlying graph $G_P = (V, E, s, t, l)$, $i_P = i$ is a representation of the initial state of D_P by its canonical form, and $j_P = j$ is a representation of the final state of D_P by its canonical form. $\square$

We use subscripts, U_P, X_P, V_P, E_P, Y_P, F_P, C_P, $\leq_P$, s_P, t_P, l_P, when necessary, and we apply to P the terminology introduced for the usual concrete dynamic graphs.

5.2. Definition. A *morphism* from a concrete left concatenable (resp.: concatenable) dynamic graph P to another such a graph P' is a morphism f from the underlying concrete dynamic graph D_P to $D_{P'}$ such that $i_{P'} = i_P f$ (resp.: such that $i_{P'} = i_P f$ and $j_{P'} = j_P f$). $\square$

5.3. Definition. An *abstract* left concatenable (resp.: concatenable) dynamic graph is an isomorphism class π of concrete left concatenable (resp.: concatenable) dynamic graphs. Each member P of such a class is called an *instance* of π and we write π as $[P]$. $\square$

If $\pi = [P]$ with a concatenable static graph P then the initial state and the final state of P are equal and P reduces to the only state thus obtained. In this case i_P and j_P are two representations of this state by its canonical form and they determine a unique automorphism of this form. In such a case we call π a *symmetry*. If in addition i_P and j_P are equal then they can be replaced by any other equal representations without leaving the equivalence class π. In such a case we can identify π with the class of graphs isomorphic to G_P (that is with the abstract graph $[G_P]$), and we call it an *identity*.

In the sequel we do not say explicitly whether a dynamic graph is concrete or abstract if it is clear from the context.

Operations are defined for abstract left concatenable and concatenable dynamic graphs.

We start with the observation that all initial states of instances of an abstract left concatenable dynamic graph π are isomorphic and they are instances of an identity. We call such an identity the *source* of π and write it as $\partial_0(\pi)$. If π is concatenable the the same holds true for final states of instances and we call the respective identity the *target* of π and write it as $\partial_1(\pi)$. Thus for left concatenable dynamic graphs we obtain the operation of taking the source, $\pi \mapsto \partial_0(\pi)$, and for concatenable dynamic graphs we obtain the operations of taking the source and the target, $\pi \mapsto \partial_0(\pi)$ and $\pi \mapsto \partial_1(\pi)$.

Other operations on dynamic graphs can be introduced as follows taking into account 4.4 - 4.6.

5.4. Proposition. Given an identity σ, a left concatenable dynamic graph π, and an injective morphism g from the canonical form of the initial state of an instance P' of π to the canonical form of an instance S' of σ such that each edge in $X_{form(initial(D_{S'}))} - g(i_{P'}^{-1}(X_{initial(D_{P'})}))$ has the source and the target nodes in $X_{form(initial(D_{S'}))} - g(i_{P'}^{-1}(X_{initial(D_{P'})}))$ or in $g(i_{P'}^{-1}(X_{static(D_{P'})}))$, there exists a unique left concatenable dynamic graph α, called the *application* of π to σ at g and written as $\sigma \overset{\pi}{\Rightarrow}_g$, such that the following conditions are satisfied for an instance A of α, an instance S of σ, an instance P of π, and a morphism $a : D_P \to D_A$ between the respective underlying concrete dynamic graphs D_P and D_A:

(1) $(D_S \overset{\subseteq}{\to} D_A \overset{a}{\leftarrow} D_P)$ is a pushout of the pair of morphisms given by $(D_S = initial(D_S) \overset{is}{\leftarrow} form(initial(D_S)) \overset{g}{\leftarrow} form(initial(D_P)) \overset{i_P}{\to}$

$initial(D_P) \overset{\subseteq}{\to} D_P)$,

(2) the representation i_A of the initial state of A is equal to i_S.

Such an application α has the source $\partial_0(\alpha) = \partial_0(\sigma) = \sigma$ and we say that it is *realized* by the instances S, P, A, of σ, π, α, respectively, and by the morphism $a : D_P \to D_A$. If π is concatenable then the same holds true and, in addition, there exist: a unique injective morphism $r : final(D_P) \to final(D_A)$ such that $r(u) = a(u)$ for all u in $final(D_P)$, and a unique injective morphism $h : form(final(D_P)) \to form(final(D_A))$ such that $hj_A = j_P r$. In this case

α has the target $\tau = \partial_1(\alpha)$, called the *result* of α, with an instance T such that $D_T = final(D_A)$ and the injective morphism $h : form(final(D_P)) \rightarrow form(final(D_T))$, we write it as $\sigma \Rightarrow_{g,h} \tau$, and say that it is *realized* by S, P, T, A, and $a : D_P \rightarrow D_A$. $\square$

5.5. Proposition. Given a concatenable dynamic graph α and a left concatenable dynamic graph β such that $\partial_0(\beta) = \partial_1(\alpha)$, there exists a unique left concatenable dynamic graph ϱ, called the *sequential composition* of α and β and written as $\alpha; \beta$, such that the following conditions are satisfied for an instance A of α, an instance B of β, and an instance R of ϱ:

(1) the underlying concrete dynamic graph D_R is equal to the pushout object D of a pushout $(D_A \xrightarrow{a} D \xleftarrow{b} D_B)$ of the pair of morphisms given by $(D_A \xleftarrow{\supseteq} final(D_A) \xrightarrow{i_A} form(final(D_A)) = form(initial(D_B)) \xrightarrow{i_B} initial(D_B)) \xrightarrow{\subseteq} D_B)$,

(2) the representation i_R of the initial state of R is equal to the composition of i_A, inclusion, and a.

Such a dynamic graph ϱ has the source $\partial_0(\varrho) = \partial_0(\alpha)$. If β is concatenable then the same holds true and, in addition, there exists a unique isomorphism $r : final(D_B) \rightarrow final(D_R)$ defined by $r(u) = b(u)$ for all u in $final(D_B)$, $j_R = j_B r$, and $\partial_1(\varrho) = \partial_1(\beta)$. $\square$

5.6. Proposition. Given a left concatenable dynamic graph α and a left concatenable dynamic graph β such that $\partial_0(\beta) = \partial_0(\alpha)$, there exists a unique left concatenable dynamic graph ϱ, called the *indeterministic sum* of α and β and written as $\alpha + \beta$, such that the following conditions are satisfied for an instance A of α, an instance B of β, and an instance R of ϱ:

(1) the underlying concrete dynamic graph D_R is equal to the pushout object D of a pushout $(D_A \xrightarrow{a} D \xleftarrow{b} D_B)$ of the pair of morphisms given by $(D_A \xleftarrow{\supseteq} initial(D_A) \xrightarrow{i_A} form(initial(D_A)) = form(initial(D_B)) \xrightarrow{i_B} initial(D_B)) \xrightarrow{\subseteq} D_B)$,

(2) the representation i_R of the initial state of R is equal to the composition of i_A, inclusion, and a.

Such a dynamic graph ϱ has the source $\partial_0(\varrho) = \partial_0(\alpha) = \partial_0(\beta)$. $\square$

6 Sequential representations of concatenable dynamic graphs

In this section we concentrate on concatenable dynamic graphs. Such graphs can be represented in the form of a sequential composition of steps, each step corresponding to an application of a *prime* dynamic graph, where a prime dynamic graph is a concatenable dynamic graph whose instance has exactly one event and its context and elements the event consumes and produces.

6.1. Theorem. Each concatenable dynamic graph with instances having events has a *sequential representstion* of the form

$$(\sigma_1 \overset{\pi_1}{\Rightarrow}_{g_1,h_1} \tau_1); (\sigma_2 \overset{\pi_2}{\Rightarrow}_{g_2,h_2} \tau_2); ...; (\sigma_n \overset{\pi_n}{\Rightarrow}_{g_n,h_n} \tau_n)$$

where $\pi_1, \pi_2,..., \pi_n$ are prime concatenable dynamic graphs. $\square$

A concatenable dynamic graph may have many different sequential representations. Such representations can be obtained one from another by "permuting" components which are independent in the following sense (cf. [MR 92]).

6.2. Definition. Given an application $\alpha = (\sigma \overset{\pi}{\Rightarrow}_{g,h} \tau)$ of π to σ at g, and an application $\alpha' = (\sigma' \overset{\pi'}{\Rightarrow}_{g',h'} \tau')$ of π' to $\sigma' = \tau$ at g', we say that α and α' are *sequentially independent* if each instance A'' of $\alpha'' = (\sigma \overset{\pi}{\Rightarrow}_{g,h} \tau); (\sigma' \overset{\pi'}{\Rightarrow}_{g',h'} \tau')$ has a proper cut Z such that:

(1) α is realized by instances S, P, T, A of σ. π, τ, α, respectively, and by a morphism $a : D_P \rightarrow D_A$, such that D_A is the restriction of $D_{A''}$ to $head_Z(A'')$,

(2) α' is realized by instances S', P', T', A' of $\sigma', \pi', \tau', \alpha'$, respectively, and by a morphism $a' : D_{P'} \rightarrow D_{A'}$, such that $D_{A'}$ is the restriction of $D_{A''}$ to $tail_Z(A'')$,

(3) $h(j_P^{-1}(X_{final(D_P)})) \cap g'(i_{P'}^{-1}(X_{initial(D_{P'})})) \subseteq$
$h(j_P^{-1}(X_{static(D_P)})) \cap g'(i_{P'}^{-1}(X_{static(D_{P'})}))$,

(4) there is no $x' \in X_{P'}$ such that $a'(x')C_{A''}a(y)$ for some $y \in Y_P$ and $a'(x')F_{A''}a'(y')$ for some $y' \in Y_{P'}$. $\square$

6.3. Theorem. If an application $\alpha = (\sigma \overset{\pi}{\Rightarrow}_{g,h} \tau)$ of π to σ at g and an application $\alpha' = (\sigma' \overset{\pi'}{\Rightarrow}_{g',h'} \tau')$ of π' to $\sigma' = \tau$ at g' are sequentially independent then there exist θ, e, f, e', f' such that $\beta = (\sigma \overset{\pi'}{\Rightarrow}_{e,f} \theta)$ and $\beta' = (\theta \overset{\pi}{\Rightarrow}_{e',f'} \tau')$ are sequentially independent and $\alpha; \alpha' = \beta; \beta'$. $\square$

7 Dynamic graphs as semantics of graph grammars

Each left concatenable dynamic graph μ can be interpreted as a semantics of a graph grammar in the sense that each prefix of μ (that is a concatenable dynamic graph α such that $\mu = a ; \beta$ for some β) is a derivation of this grammar. To this end it suffices to consider the grammar with the source of μ playing the role of starting graph and the prime dynamic graphs occurring in the sequential representations of prefixes of μ playing the role of productions. Indeed, in the framework of the algebraic theory of graph grammars, each prime dynamic graph π can be regarded as a production $p = (L \stackrel{\supseteq}{=} L \cap R \stackrel{\subseteq}{=} R)$, where $L = initial(D_P)$ and $R = final(D_P)$ for an instance P of π, and each application $\sigma \stackrel{\pi}{\Rightarrow}_{g,h} \tau$ of π to σ at g with the result τ can be regarded as a direct derivation from S to T by applying p at the image of L under the morphism

$$L = initial(D_P) \stackrel{i_P^{-1}}{\to} form(initial(D_P)) \stackrel{g}{\to} form(D_S) \stackrel{i_S}{\to} D_S,$$

where S and T are the respective instances of σ and τ. Consequently, each prefix α of μ can be regarded as a sequence of direct derivations corresponding to the subsequent components of a sequential representation of α, that is as a derivation. With such an interpretation in mind each prefix of μ plays a role similar to that of a derivation trace in the sense of [CELMR 94], and μ itself plays the role of a semantics of the respective graph grammar.

A precise description of the relationships between concatenable dynamic graphs and derivation traces would require recalling the respective notions and it is not a subject of the present paper. We would like only to point out that our formalism applies to grammars which need not be safe as in [CELMR 94]. On the other hand, we restrict a little the concept of graph grammar by considering only injective applications of productions.

References

[CELMR 94] Corradini, A., Ehrig, H., Löwe, M., Montanari, U., Rossi, F.: *An Event Structure Semantics for Safe Graph Grammars*, in Proc. IFIP Working Conference PROCOMET, 1994, 417-439

[CMR 96] Corradini, A., Montanari, U., Rossi, F.,: *Graph Processes*, to appear in Fundamenta Informaticae 26 (1996)

[DMM 89] Degano, P., Meseguer, J., Montanari, U., *Axiomatizing Net Computations and Processes*, in the Proceedings of 4th LICS Symposium, IEEE, 1989, 175-185

[Eng 91] Engelfriet, J., *Branching Processes of Petri Nets*, Acta Informatica 28 (1991), 575-591

[MR 92] Montanari, U., Rossi, F., *Graph Grammars as Context- Dependent Rewriting Systems*, in Raoult, J.C. (Editor): Proc. of CAAP'92, Springer LNCS 581 (1992) 232-247

[MR 95] Montanari, U., Rossi, F., *Contextual Nets*, Acta Informatica 32 (1995), 545-596

Equations on Trees

Sabrina Mantaci and Antonio Restivo

Dipartimento di Matematica ed Applicazioni, Università di Palermo
via Archirafi, 34 - 90123 Palermo - ITALY
(e-mail: {sabrina,restivo}@altair.math.unipa.it)

Abstract. We introduce the notion of equation on trees, generalizing the corresponding notion for words, and we develop the first steps of a theory of tree equations. The main result of the paper states that, if a pair of trees is the solution of a tree equation with two indeterminates, then the two trees are both powers of the same tree. As an application, we show that a tree can be expressed in a unique way as a power of a primitive tree. This extends a basic result of combinatorics on words to trees. Some open problems are finally proposed.

1 Introduction

In this paper we are mainly concerned with k-ary trees whose vertices are labeled by letters of an alphabet A. We look at a labeled k-ary tree as a generalization of a word, in the sense that words correspond to the particular case of $k = 1$, i.e. to unary trees.

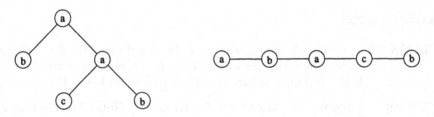

A binary tree   A word

Given that combinatorics on words is a well developed theory, with several applications to computer science, it appears to be a natural task to generalize results and methods of this theory to trees. This approach was initiated by M. Nivat in [10], where in particular he extended the notion of code to trees. It can be considered a special part of the theory of tree automata and languages (cf. [11], [12]). Recent contributions in this direction have been given by the authors and al. in [3] and in [4].

In section 2 of this paper we introduce the basic notions of combinatorics on trees: subtree, prefix, suffix, concatenation, power, primitiveness, morphism.

The theory of word equations constitutes an important chapter of combinatorics on words, that appears in several branches of mathematics and theoretical computer science. In section 3 we introduce the notion of equation on trees, generalizing the corresponding notion for words, and we give the first elements of a theory of tree equations. The point of view here developed is very different from that considered, for instance, in [2].

A very general problem in the theory of tree equations is to decide the existence of a solution of a given equation. In [9] Makanin solved this problem in the special case of word equations, i.e. in the case of unary trees. We do not know whether this result can be extended to general k-ary trees.

In section 4 we investigate the combinatorial structure of the set of solutions of a tree equation with two indeterminates. Our main result states that, if a pair of trees is solution of such an equation, then the two trees are both powers of the same tree. In the particular case of unary trees this corresponds to a classical result in the theory of word equations [8].

As an application, we show in section 5 that a tree can be expressed in a unique way as a power of a primitive tree. This extends a basic result of combinatorics on words. Some open problems and suggestions for further researches are finally proposed.

2 Basic definitions

We recall here few notions and notations concerning words and free monoids; we refer for unexplained ones to [1] and to [7]. Some of the basic definitions and notations concerning trees are from [10].

Free monoids and *semigroups* generated by a (not necessairely finite) alphabet A are denoted by A^* and $A^+ = A^* \setminus \{\epsilon\}$ respectively. The elements of A^* are called *words*. A word $u \in A^*$ is a *prefix* (*suffix* resp.) of a word $v \in A^*$ if $v = uw$ ($v = wu$ resp.) for some $w \in A^*$. A subset L of A^* is *prefix closed* if, for any $u, v \in A^*$, $uv \in L$ implies $u \in L$. Let $I\!N$ denote the set of natural numbers.

We now introduce a very general notion of labeled tree, in which the arity of each node depends on its label. This notion will be useful in the definition of morphism for trees, given at the end of this section. We first need the following definition.

Definition 1. A *graded alphabet* (X, α) is a set of letters X endowed with an arity function

$$\alpha : X \longrightarrow I\!N$$

assigning a natural number to each letter in X.

In the following definition, $I\!N$ is regarded as a numerable set of symbols, and $I\!N^*$ is the set of words over the alphabet $I\!N$.

Definition 2. Given a graded alphabet (X, α) we define a labeled tree over X with arity function α a partial mapping

$$\tau : I\!N^* \longrightarrow X$$

where the domain $dom(\tau)$ is a finite and prefix closed subset of $I\!N^*$ and, for all $v \in dom(\tau)$ and $i \in I\!N$, if $v \cdot i \in dom(\tau)$, then $i \leq \alpha(\tau(v))$. We say that τ is an α-ary tree over X. The elements in $dom(\tau)$ are called *nodes*.

Let τ be an α-ary tree over X and let u be a node of τ (i.e. $u \in dom(\tau)$). If v is the node of τ such that $u = v \cdot i$ for some $i \leq \alpha(\tau(v))$, we say that v is the *father* of u and that u is the *i-th son* of v. A node without sons is called *leaf*. The only node without father is called the *root* of the tree. The set $B(\tau) = \{u \cdot i \mid u \in dom(\tau), u \cdot i \notin dom(\tau)\}$ is called the *border* of τ. We will call the *size* of a tree τ the number of its nodes, and we will denote it by $|\tau|$. We will denote by $(X, \alpha)^{\#}$ the set of all labeled trees over a graded alphabet (X, α).

Example 1. If $X = \{x, y\}$ and $\alpha(x) = 2$, $\alpha(y) = 3$, then the tree

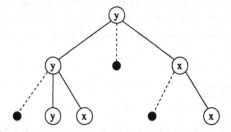

is an α-ary tree over X, whose domain is the set $\{\epsilon, 1, 3, 12, 13, 32\}$ and its border is the set $B(\tau) = \{2, 11, 31, 121, 122, 123, 131, 132, 321, 322\}$.

Note that in the figures in this paper, when we want to give evidence to the arity of a node, we draw all the edges outgoing from the node, denoting with dotted edges and black points the lacking sons. If we are given a graded alphabet (X, α) where the arity function is a constant function, that is, $\alpha(x) = k$, $\forall x \in X$, then we have the traditional notion of k-ary tree. We now give the formal definition:

Definition 3. Let $\Sigma = \{1, 2, \cdots, k\}$ and let A be a finite alphabet. A *k-ary tree* over the label-alphabet A is a partial mapping $\tau : \Sigma^* \to A$ whose domain $dom(\tau)$ is a finite and a prefix closed subset of Σ^*.

We will denote by $A_k^{\#}$ the set of all finite k-ary trees over A. When k is implicit we will write $A^{\#}$ instead of $A_k^{\#}$. In the examples in this paper $A^{\#}$ will denote the set of all binary trees over A. Particular elements of $A^{\#}$ are the *empty tree*, denoted by Ω, whose domain is the empty set, and the *punctual trees*, that is, trees whose domain is $\{\epsilon\}$. A punctual tree with label a will be simply denoted by a.

If τ and τ' are two k-ary trees, τ is a *subtree* of τ' if there exists a node $v \in dom(\tau')$ such that:

i) $S(v, \tau) = \{vu \mid u \in dom(\tau)\} \subseteq dom(\tau')$, and
ii) $\tau(u) = \tau'(vu)$

In this case we say that τ is a subtree of τ' *rooted at* the node v. If $v = \epsilon$, then $S(\epsilon, \tau) = dom(\tau) \subseteq dom(\tau')$ and τ coincides with the restriction of τ' to $dom(\tau)$. In this case we write $\tau \sqsubseteq \tau'$ and we say that τ is an *initial subtree* or a *prefix* of τ'. If $S(v, \tau) = v\Sigma^* \cap dom(\tau')$, then we say that τ is a *terminal subtree* or a *suffix* of τ'.

Remark. It can be easily proved by induction that a k-ary tree with n nodes has a border with cardinality equal to $(k-1)n + 1$. Then, if $k \neq 1$, given the cardinality of the border, one can obtain the size of the tree. Notice that in the case of words (1-ary trees) the border contains one element indipendently from the size of the word.

Definition 4. Given a finite alphabet X, an *ordered labeled tree* over X is a tree in which the sons of each node are ordered.

When we draw an ordered tree, we assume that the sons of each vertex are ordered from left to right. We will denote by $O(X)$ the set of all ordered labeled trees over X. By convention, we will denote ordered trees by Latin letters, while α-ary (k-ary) trees will be denoted by Greek letters.

Notice that an α-ary (k-ary) tree is a particular ordered tree in which the sons of each node are distinguished and they are ordered by means of their lexicographic order. Then, if (X, α) is a graded alphabet, it is naturally defined the application

$$\omega : (X, \alpha)^\# \longrightarrow O(X)$$

that associates to an α-ary tree the corresponding ordered tree. Notice that this application is not injective. In fact, for instance, both trees τ_1 and τ_2 over (X, α) with $X = \{x, y\}$ and $\alpha(x) = 3$ and $\alpha(y) = 2$, drawn in the figure below

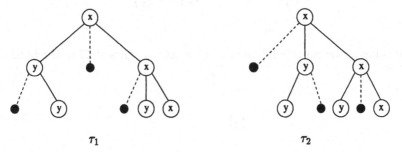

τ_1 $\qquad\qquad\qquad\qquad$ τ_2

have as image the same ordered tree

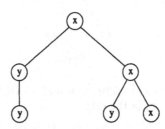

Remark. The definition of k-ary trees given here is consistent with the definition of word. In fact a word $w \in A^*$ can be described as a partial mapping from the set of natural numbers $I\!N$ (that is isomorphic to Σ^* when $\Sigma = \{1\}$) to a finite alphabet A, whose domain is the set $\{0, 1, \ldots, |w| - 1\}$ (isomorphic in Σ^* to the set $\{\epsilon, 1, 1^2, \ldots, 1^{|w|-1}\}$, that is a prefix closed subset of $\{1\}^*$). Thus we can always consider a word as a labeled 1-ary tree over A. Moreover notice that in the case of words the border contains only one element $1^{|w|}$.

We now define the *concatenation* between trees. Intuitively the concatenation between two trees τ_1 and τ_2 is obtained by "attaching" the root of τ_2 to one of the elements of the border of τ_1. It is clear that this operation cannot have as result a unique tree, since in general the border of a tree contains more than one element. Then the concatenation between two trees will be a set of trees containing as many trees as the number of elements in $B(\tau_1)$. First we define the concatenation of two trees at a given element of the border.

Definition 5. Let τ_1 and τ_2 be two k-ary trees and let $B(\tau_1)$ denote the border of τ_1. The concatenation of τ_2 to τ_1 at $b \in B(\tau_1)$ is the tree $\tau_1 (b) \tau_2$ defined by:

$$dom(\tau_1 (b)\tau_2) = dom(\tau_1) \cup b \cdot dom(\tau_2)$$

$$\forall u \in dom(\tau_1 (b) \tau_2), \ (\tau_1 (b) \tau_2)(u) = \begin{cases} \tau_1(u) & \text{if } u \in dom(\tau_1) \\ \tau_2(v) & \text{if } bv = u \text{ and } v \in dom(\tau_2) \end{cases}$$

For example, given the binary trees:

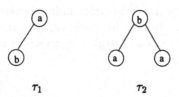

$$\tau_1 \qquad\qquad \tau_2$$

the following is the concatenation of τ_2 to τ_1 at the element 12 of the border:

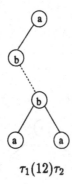

$$\tau_1(12)\tau_2$$

Definition 6. Let τ_1 and τ_2 be two trees and let $B(\tau_1)$ be the border of τ_1. The *concatenation* of τ_1 with τ_2 is the set:

$$\tau_1 \cdot \tau_2 = \{\tau_1 (b) \tau_2 \mid b \in B(\tau_1)\}$$

Remark that, according to the definitions of prefix and suffix given above, τ_1 (τ_2 resp.) is prefix (suffix resp.) of any tree in $\tau_1 \cdot \tau_2$, as it is easy to verify. The notion of concatenation of two trees can be extended in the usual way to the concatenation of two sets of trees. If T_1 and T_2 are two families of trees, the concatenation of T_1 and T_2 is defined as the union of the sets $\tau_1 \cdot \tau_2$ where $\tau_1 \in T_1$ and $\tau_2 \in T_2$. It is denoted by $T_1 \cdot T_2$. Notice that this operation is not associative since, for example, the tree:

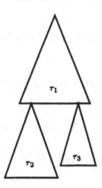

belongs to the set $(\tau_1 \cdot \tau_2) \cdot \tau_3$ but not to the set $\tau_1 \cdot (\tau_2 \cdot \tau_3)$.

Given a set of trees T we can recursively define T^n by: $T^0 = \{\Omega\}$, $T^1 = T$ and $T^n = T^{n-1} \cdot T$. Finally we will denote by $T^\# = \bigcup_n T^n$.

A particular case is when $T = \{\tau\}$. In this case we can define the sets τ^n and $\tau^\# = \bigcup_n \tau^n$. This allows us to give the following definition:

Definition 7. Given two trees τ, τ_0, we say that τ is a *power* of τ_0, if $\tau \in \tau_0^\#$.

The notion of primitiveness plays an important role in the theory of words. This notion can be extended to trees:

Definition 8. Given a tree τ, we say that τ is a *primitive tree* if the condition $\tau \in \tau_0^n$ for some $\tau_0 \in A^\#$ implies $\tau = \tau_0$ and $n = 1$.

Example 2. Consider the binary trees τ_1 and τ_2 in the figure below:

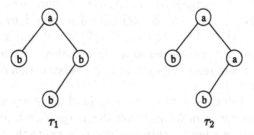

τ_1 is a primitive tree since it is not a power of another tree, whereas τ_2 is not primitive since it is power of the (primitive) tree τ in the following figure:

$$\tau$$

We remark that in the case of unary trees, the notion of primitive tree coincides with the one of primitive word.

We now introduce the notion of morphism for trees, which is fundamental in our work.

Definition 9. Given an alphabet A and a graded alphabet (X, α), a *morphism* of $(X, \alpha)^{\#}$ into $A^{\#}$ is an application

$$\varphi : (X, \alpha)^{\#} \longrightarrow A^{\#}$$

such that, for any $x \in (X, \alpha)$, $|B(\varphi(x))| = \alpha(x)$ and

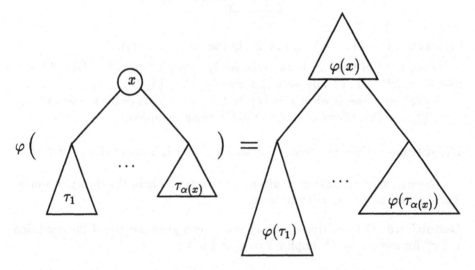

A morphism is said to be *nonerasing* if the only tree having as image the empty tree (in $A^{\#}$), is the empty tree in $(X, \alpha)^{\#}$. A morphism $\varphi : (X, \alpha)^{\#} \longrightarrow A^{\#}$, where $X = \{x_1, \cdots, x_k\}$, can be uniquely determined by the set of trees $T = \{\varphi(x_1), \cdots, \varphi(x_k)\} \subseteq A^{\#}$. Then $\varphi((X, \alpha)^{\#}) = T^{\#}$ and any tree τ in the image of the morphism φ is obtained as a concatenation of trees in T. We call *T-decomposition* of τ any representation of τ as concatenation of elements of T.

Remark. The need of dealing with α-ary trees (and not k-ary trees) as domain of the morphism depends from the fact that the image of each node labeled, say, x, must be a tree having a border with cardinality equal to the arity of x. Then, if we allow that two different elements of X, say x and y, have as image trees with different size, we have to suppose that x and y have different arities.

3 Tree equations

Definition 10. Let $X = \{x_1, \cdots, x_k\}$ be a finite alphabet. A *tree equation* is a pair of ordered trees $t_1, t_2 \in O(X)$. We denote it by (t_1, t_2) and we call $\{x_1, \cdots, x_k\}$ the set of indeterminates.

We say that a tree equation (t_1, t_2) admits a solution in $A^\#$ if there exists an arity function $\alpha : X \longrightarrow \mathbb{N}$, two trees $\tau_1, \tau_2 \in (X, \alpha)^\#$ such that $\omega(\tau_1) = t_1$ and $\omega(\tau_2) = t_2$, and a nonerasing morphism $\varphi : (X, \alpha)^\# \longrightarrow A^\#$ such that $\varphi(\tau_1) = \varphi(\tau_2)$. The multiset of trees $\{\chi_1, \cdots, \chi_k\} = \{\varphi(x_1), \cdots, \varphi(x_k)\}$ is called a *solution* of (t_1, t_2).

Example 3. Consider the tree equation (t_1, t_2) in the set of indeterminates $\{x, y, z\}$, expressed in the form:

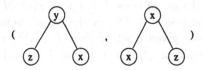

By taking α such that $\alpha(x) = \alpha(y) = 3$ and $\alpha(z) = 4$, and the pair of trees τ_1 and τ_2 over (X, α)

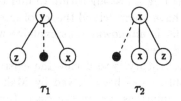

one obtains as solution the set of binary trees:

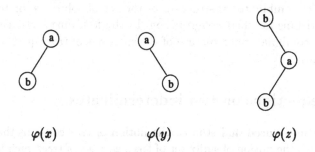

$$\varphi(x) \qquad \varphi(y) \qquad \varphi(z)$$

Indeed one can easily verify that $\tau = \varphi(\tau_1) = \varphi(\tau_2)$ is the following binary tree:

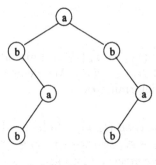

τ

Remark. The notion of tree equation is in a sort of a dual correspondence with the notion of *tree code* introduced by Nivat in [10]. If the multiset $T = \{\chi_1, \cdots, \chi_k\}$ is a solution of an equation (t_1, t_2), then there exists a tree $\tau \in T^{\#}$ that admits two different T-decompositions, i.e. T is not a tree code in the sense of Nivat [10]. For instance, the set of trees $T = \{\varphi(x), \varphi(y), \varphi(z)\}$ in example 3 is not a tree code, since the tree $\tau \in T^{\#}$, in the figure, admits two different T-decompositions. Conversely, given a set of trees $T \subset A^{\#}$, if T is not a tree code, then there exists an equation (t_1, t_2) such that T is a solution of (t_1, t_2).

Notice that if the arity function is fixed, the problem of solvability of a tree equation is trivially decidable. In fact, when we know the arity of a variable, we know the cardinality of the border of its image by φ and, consequentely, the size of this image. Then we have to test only a finite number of cases. We are only faced to the problem of designing a fast algorithm to find the solution of a tree equation. Thus, in order to have a non trivial theory of tree equations, we need that the arity of the variables is unknown. This explains why we define a tree equation as a pair of ordered trees.

A general problem in this theory is to decide the existence of a solution of a given equation. This problem has been solved by Makanin [9] (cf. also [5]) in the special case of word equations, i.e. in the case of unary trees. It is not known whether this result can be extended to general k-ary trees. Another related research line is to develop methods to describe the set (in general infinite) of all solutions of a given tree equation. A step in this direction is either to investigate the combinatorial structure of the set of solutions, or to search for explicit algorithms for their computation. In the following section we shall investigate the combinatorial structure of the solutions of tree equations with two indeterminates.

4   Tree equations on two indeterminates

In [10] Nivat introduced and studied the notion of tree code. In that paper he introduced also the notion of suffix set of trees as a set of trees such that none of its elements is suffix of another one. In particular he proved the following lemma for tree codes:

Lemma 11. *Every suffix set of trees is a tree code.*

This lemma, interpreted in terms of tree equation, gives the following:

Corollary 12. *If $\{\sigma, \zeta\}$ is a solution of a tree equation in two indeterminates, then either σ is suffix of ζ, or ζ is suffix of σ.*

Proof. The multiset $\{\sigma, \zeta\}$ such that $\sigma = \zeta$ is trivially solution of tree equations (it is sufficient to take any ordered tree t over the set of indeterminates $\{x, y\}$, consider the ordered tree t' obtained by interchanging x with y and y with x, and take the tree equation (t, t')). It is obvious that in this case σ is suffix of ζ (and vice-versa). If $\sigma \neq \zeta$ then the existence of a tree equation implies the existence of a tree having a double $\{\sigma, \zeta\}$-decomposition, that is, $\{\sigma, \zeta\}$ is not a tree code, then, by lemma 11 either σ is suffix of ζ or ζ is suffix of σ

We now prove that all solutions of tree equations in two indeterminates have a particular structure. This result generalizes the analogous result for equations on words with two indeterminates (cf. [6], [7], [8])

Theorem 13. *Let (σ, ζ) be the solution of a tree equation on two indeterminates. Then there exists a tree χ such that $\sigma, \zeta \in \chi^\#$, that is, σ and ζ are powers of the same tree χ.*

Proof. The proof is by induction on $|\sigma| + |\zeta|$.

Initialization step: if $|\sigma| + |\zeta| = 2$ then the only possible case is that $|\sigma| = |\zeta| = 1$, that is, σ and ζ are punctual trees. But by Corollary 12 we have that if $\{\sigma, \zeta\}$ is solution of a tree equation, then one of the two trees must be suffix of the other one. Since $|\sigma| = |\zeta| = 1$, the only possibility is that $\sigma = \zeta$ and the statement is trivially verified.

Induction step: Let $|\sigma| + |\zeta| = n$. Let us suppose that the theorem is true for all solutions $\{\sigma', \zeta'\}$ of tree equations such that $|\sigma'| + |\zeta'| < n$.

If $\sigma = \zeta$ then we are done. Otherwise one tree is suffix of the other one (cf. Corollary 12). We can suppose that σ is suffix of ζ. Then there exists a tree $\eta \subset \zeta$ and an element $b \in B(\eta)$ such that $\zeta = \eta(b)\sigma$. By a previous remark, if $\{\sigma, \zeta\}$ is solution of a tree equation, then there exists a tree $\tau \in \{\sigma, \zeta\}^\#$ having two different $\{\sigma, \zeta\}$-decompositions. If, in turn, we decompose σ into $\eta(b)\sigma$, we obtain two different $\{\sigma, \eta\}$-decompositions of τ, i.e. $\{\sigma, \eta\}$ is solution of a tree equation, with $|\sigma| + |\zeta| < n$. By the inductive hypothesis, there exists a tree χ such that $\sigma, \eta \in \chi^\#$. Since $\zeta \in \eta \cdot \sigma \subseteq \chi^\#$, it follows that $\sigma, \zeta \in \chi^\#$. This concludes the proof.

In terms of tree codes the Theorem 13 can be restated as follows: *if a set $\{\sigma, \zeta\}$ of trees is not a code, then σ and ζ are both power of the same tree χ.*

Example 4. Consider the following pair of binary trees:

$$\sigma \qquad\qquad \zeta$$

$\{\sigma,\zeta\}$ is not a tree code. Indeed the tree:

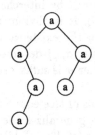

admits two different $\{\sigma,\zeta\}$-decompositions. Remark that σ and ζ are both powers of the same punctual tree a. In terms of equations, the pair $\{\sigma,\zeta\}$ is a solution of the equation

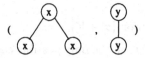

Indeed, by taking $\alpha(x) = 3$, $\alpha(y) = 4$, and the pair of trees over $(\{x,y\},\alpha)$

one easily verify that the morphism that maps x in σ and y in ζ gives rise to two identical trees.

Let us remark that, unlike the case of words, the converse of the previous result does not hold: there exist pairs of trees that are powers of the same tree and that are tree codes. An example is given by the following pair of binary trees:

It remains open the problem to give a complete characterization of tree codes with two elements.

5 Applications and conclusions

As an application of the result in section 4, we extend a well known combinatorial property of words (cf. [6] and [7]) to trees.

Theorem 14. *Let τ be a labeled tree. Then there exists a unique primitive tree τ_0 such that $\tau \in \tau_0^\#$.*

Proof. By induction on the size of the tree.

Initialization step: If τ is a punctual tree a, then τ is a primitive tree and we are done.

Induction step: Let us suppose that the theorem is true for every tree τ' such that $|\tau'| < |\tau|$. If τ is primitive, the statement is verified. Otherwise, there exists a tree τ_0 with $|\tau_0| < |\tau|$ such that $\tau \in \tau_0^\#$. By the inductive hypothesis the statement is true for τ_0, then there exists a primitive tree τ_1 such that $\tau_0 \in \tau_1^\#$, then $\tau \in \tau_0^\# \subseteq \tau_1^\#$.

This primitive tree is unique. In fact if there were two different primitive trees τ_1 and τ_2 such that $\tau \in \tau_1^n$ and $\tau \in \tau_2^m$ then τ would have a double $\{\tau_1, \tau_2\}$-decomposition, that is, there exists a tree equation in two variables such that (τ_1, τ_2) is a solution. By Theorem 13, τ_1 and τ_2 must be powers of the same tree, and this contradicts the primitiveness of τ_1 and τ_2.

In the previous section we observed that, unlike the case of words, there exist pairs of trees that are powers of the same tree and that, at the same time, are codes, i.e. are not solution of a tree equation. In order to go deeply in this problem, let us stress the differences and the similarities between the case of words and the case of trees.

Given two words u and v in a free monoid A^*, the following three conditions are equivalent (cf. [6]):

1) $u^+ \cap v^+ \neq \emptyset$;
2) $\{u, v\}$ is not a code;
3) $u, v \in w^+$, for some $w \in A^*$.

Given two labeled trees σ and τ, consider the following conditions:

1') $\sigma^\# \cap \tau^\# \neq \{\Omega\}$;
2') $\{\sigma, \tau\}$ is not a tree code;
3') $\sigma, \tau \in \chi^\#$, for some $\chi \in A^\#$.

We proved in Section 4 that *1')* $\Rightarrow$ *2')* $\Rightarrow$ *3')*. However unlike the case of words, *3')* $\not\Rightarrow$ *2')* and *3')* $\not\Rightarrow$ *1')* as shown by an example in section 4. Moreover *2'* $\not\Rightarrow$ *1')*, that is, there exits pair of trees that are not tree codes, but that do not have a common power. For example the following pair of trees

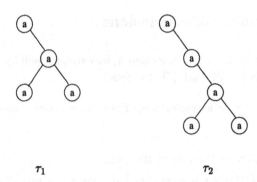

$$\tau_1 \qquad\qquad\qquad \tau_2$$

is not a tree code. In fact it can be easily seen that the following tree have two different $\{\tau_1, \tau_2\}$-decompositions:

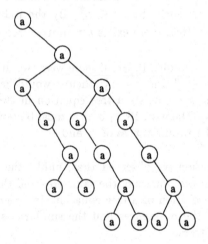

In terms of equations, $\{\tau_1, \tau_2\}$ is solution of the following tree equation:

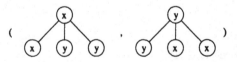

Nevertheless it can be proved that τ_1 and τ_2 cannot have a common power.

These remarks show that the situation is more complicated for trees than for words. In particular it remains open the problem to characterize those pairs of trees that are tree codes.

A more general problem is to extend the result of Makanin (see [9]) from words to trees, i.e. to prove the decidability of the solution of a general tree equation. Apart from the answer to this general problem, there remains plenty of space for investigations on the structure of the solutions of tree equations of some specific form.

456

6   Aknowledgements

We are very grateful to Sergio Salemi for his useful suggestions, and, in particular, for providing the last example of this paper.

References

1. S. Eilenberg. *Automata, Languages and Machines*. Vol. A, Academic Press, 1974.
2. C. C. Elgot, S. L. Bloom, R. Tindell. "On the algebraic structure of rooted trees". *J. Comput. System Sci.* 16, pagg. 362–399 , 1978.
3. D. Giammarresi, S. Mantaci, F. Mignosi, A. Restivo. "Periodicities on trees". *Preprint*, 1995.
4. D. Giammarresi, S. Mantaci, F. Mignosi, A. Restivo. "A periodicity theorem for trees". *Proc. 13th World Computer Congress - IFIP '94*, Amburgo, Germania, 1994; vol. A-51 pagg. 473–478. *Elsevier Science B.V. (North-Holland)*, 1994.
5. J. Jaffar. "Minimal and complete word unification". *J. ACM vol. 37* ; pagg. 47–85, 1990.
6. A. Lentin, M. P. Schützenberger. "A combinatorial problem in the theory of free monoids". *Proc. of the University of North Carolina* ; pagg. 67–85, 1967.
7. M. Lothaire. *Combinatorics on Words*. Addison-Wesley, Reading, MA 1983.
8. R. C. Lyndon, M. P. Schützenberger. "The equation $a^m = b^n c^p$ in a free group". *Michigan Math. J. vol. 9* ; pagg. 289–298, 1962.
9. G. S. Makanin. "The problem of solvability of equations in a free semigroup". *Math, USSR Sbornik vol. 32* ; pagg. 129–198, 1977. (in AMS, 1979).
10. M. Nivat. "Binary tree codes". *Tree automata and languages*; pagg. 1–19 *Elsevier Science Publishers B.V. (North-Holland)*, 1992.
11. J. Thatcher. "Tree automata,an informal survey" in A. Aho (ed.), *Currents in the theory of computing, Pretence Hall*, 1973.
12. W. Thomas. "Logical aspects in the theory of tree languages" in B. Courcelle (ed.), *Proceedings ninth colloquium on Trees in Algebra and Programming, Cambridge University Press*, 1984.

On the Equivalence Problem for E-Pattern Languages

(Extended Abstract)

Enno Ohlebusch[1] and Esko Ukkonen[2]

[1] Technische Fakultät, University of Bielefeld,
P.O. Box 100131, 33501 Bielefeld, Germany
email: enno@TechFak.Uni-Bielefeld.DE
[2] Department of Computer Science, University of Helsinki,
P.O. Box 26, FIN-00014 Helsinki, Finland
email: Esko.Ukkonen@cs.Helsinki.FI

Abstract. On the one hand, the inclusion problem for nonerasing and erasing pattern languages is undecidable; see [JSSY95]. On the other hand, the language equivalence problem for NE-pattern languages is trivially decidable (see [Ang80a]) but the question of whether the same holds for E-pattern languages is still open. It has been conjectured by Jiang et al. [JSSY95] that the language equivalence problem for E-pattern languages is also decidable. In this paper, we introduce a new normal form for patterns and show, using the normal form, that the language equivalence problem for E-pattern languages is decidable in many special cases. We conjecture that our normal form procedure decides the problem in the general case, too. If the conjecture holds true, then the normal form is the shortest pattern generating a given E-pattern language.

1 Introduction

A pattern α is a word over two disjoint alphabets $\Sigma \cup V$, where Σ is an alphabet of terminals and V is an alphabet of variables. The language $L_{E,\Sigma}(\alpha)$ generated by the pattern α is obtained in the obvious way: One takes all words obtained by substituting words over Σ for the variables in α. In the same way, one gets the language $L_{NE,\Sigma}(\alpha)$ – only in this case the replacement of a variable with the empty word λ is forbidden. Here E stands for erasing, whereas NE stands for nonerasing pattern language.

It is known that the inclusion problem (is the language generated by a pattern included in the language generated by another pattern?) for nonerasing and erasing pattern languages is undecidable; see [JSSY95]. However, the language equivalence problem (do two patterns generate the same language?) for nonerasing patterns is trivially decidable: Two patterns α and β generate the same NE-language if and only if they are identical up to variable renaming. It is very easy to see that this does not hold for E-patterns. Jiang et al. [JSSY95] conjectured that the language equivalence problem for erasing pattern languages is decidable (cf. also [Sal94]). However, they did not state a decision procedure.

In this paper, we investigate the E-language equivalence problem and show that the conjecture is true in many cases. Our "decision procedure" is based on a decidable and length-decreasing reduction relation $\rightarrow \subset (\Sigma \cup V)^+ \times (\Sigma \cup V)^+$. Roughly speaking, pattern α reduces to another pattern α' (which will be denoted by $\alpha \rightarrow \alpha'$) if α' can be obtained from the pattern α by deleting certain variables from α and $\alpha = \sigma(\alpha')$, where σ is a (linear) substitution. Since $\rightarrow$ is length-decreasing, every reduction sequence starting from a pattern α must end in an irreducible pattern, a so-called normal form. Moreover, $\rightarrow$ is language-preserving, that is, if $\alpha \rightarrow \alpha'$, then $L_{E,\Sigma}(\alpha) = L_{E,\Sigma}(\alpha')$. We conjecture that reduction to normal form yields a decision procedure:

Conjecture 1: Two E-patterns define the same language over an alphabet Σ with $card(\Sigma) \geq 3$ if and only if their normal forms with respect to $\rightarrow$ are identical, up to renaming of variables.

Note that in Conjecture 1, the restriction $card(\Sigma) \geq 3$ is essential. This is witnessed by the following example.

Example 1. Let $\Sigma = \{0,1\}$, $\alpha = x01y0z$, and $\beta = x0y10z$. Both α and β are in normal form w.r.t. $\rightarrow$; see Section 4. Moreover, $L_{E,\Sigma}(\alpha) = L_{E,\Sigma}(\beta)$; see [JKS+94]. However, α and β are not identical up to renaming of variables.

We will prove in Section 4 that the conjecture can be paraphrased as follows: If $\alpha, \beta \in (\Sigma \cup V)^+$ are patterns and $card(\Sigma) \geq 3$, then $L_{E,\Sigma}(\alpha) = L_{E,\Sigma}(\beta)$ if and only if there are substitutions $f : var(\alpha) \rightarrow var(\beta)^*$ and $g : var(\beta) \rightarrow var(\alpha)^*$ such that $f(\alpha) = \beta$ and $g(\beta) = \alpha$.

Also note that if Conjecture 1 holds true, then the normal form is the *shortest* pattern that generates a given E-pattern language. This is because $\rightarrow$ is a length-decreasing reduction relation. Hence we conjecture, in fact, that our normal form is a minimal description of the corresponding language.

We will prove in Sections 3 and 5 that the equivalence problem for E-pattern languages can in many cases be decided by this procedure, for instance whenever:

- the underlying alphabet Σ contains two constants that do not occur in the patterns (this is always true if Σ is infinite),
- $card(\Sigma)$, the cardinality of Σ, exceeds the number of terminal segments in one pattern by at least two,
- the normal form of one pattern has independent variable segments,
- the normal form of one pattern is linear,
- the normal form of one pattern is a one-variable pattern,
- the normal form of one pattern does not contain consecutive variables.

It is interesting to note that the first two special cases can already be proved by further developing a proof technique used in [JSSY95]; see Section 3. However, we will indicate why this technique is not sufficient to prove the general case. Furthermore, we will pose two open questions. An affirmative answer to one of them would prove decidability of E-language equivalence. Due to space limitations, some of the proofs had to be omitted in this extended abstract. The full version of the paper can be found in [OU95]. For a nice introduction to patterns and reasons why they are of interest in formal language theory, the reader is referred to [Sal94, Sal95]. Aspects of inductive inference and the theory of learning – although closely related to patterns – will not be discussed in this paper, details can be found in [Ang80b, IJ91, KMU95], for instance.

The paper is organized as follows. In Section 2, basic definitions are recalled and in Section 3 our first results are presented. Section 4 explains the normal form approach, while Section 5 contains decidability results based on normal forms. Finally, in Section 6 it is shown that in contrast to what has been claimed in [DF96], the recent approach of Dányi and Fülöp does not solve the problem.

2 Basic Definitions

In the sequel, we use standard terminology and assume that Σ and V are two disjoint alphabets. Elements of Σ are called *terminals* (or *constants*) and elements of V are called *variables*. Usually, $0, 1, a, b, c$ will denote constants, whereas $x, y, z, x_1, \ldots$ will stand for variables. Any word α over the union $\Sigma \cup V$ is said to be a *pattern*. The set of variables appearing in α will be denoted by $var(\alpha)$. A *substitution* σ is a mapping from V to $(\Sigma \cup V)^*$. Every substitution σ extends uniquely to a *morphism* $h : (\Sigma \cup V)^* \to (\Sigma \cup V)^*$, where $h(a) = a$ for every $a \in \Sigma$ and $h(x) = \sigma(x)$ for every $x \in V$. Conversely, every morphism $h : (\Sigma \cup V)^* \to (\Sigma \cup V)^*$ with $h(a) = a$ for every $a \in \Sigma$ can be viewed as a substitution. In the sequel, we will identify a substitution with its corresponding morphism that keeps terminals fixed. Henceforth σ, ν, f, g, h will denote substitutions. A bijective substitution $\sigma : V \to V$ is called a *variable renaming*. Two patterns α and β are said to be *identical up to variable renaming* if and only if there is a variable renaming σ such that $\alpha = \sigma(\beta)$. A pattern α with $card(var(\alpha)) = k + 1$ is in *canonical form* if the variables occurring in α are precisely $\{x_0, x_1, x_2, \ldots, x_k\}$ and for every i with $0 \le i < k$, the leftmost occurrence of x_i in α is to the left of the leftmost occurrence of x_{i+1} in α. Clearly, every pattern α has a canonical form α', and furthermore α and α' are identical up to variable renaming. Moreover, the canonical form of a pattern can be computed in linear time. Thus, it can be tested in linear time whether or not two patterns are identical up to variable renaming. Given two E-patterns α and β, it is not difficult to prove that if there is a substitution h with $h(\beta) = \alpha$, then $L_{E,\Sigma}(\alpha) \subseteq L_{E,\Sigma}(\beta)$.

Let α be a pattern. α has a unique representation $\alpha_0 u_1 \alpha_1 u_2 \ldots \alpha_{m-1} u_m \alpha_m$, where $\alpha_0, \alpha_m \in V^*$, $\alpha_i \in V^+$ for $0 < i < m$, and $u_i \in \Sigma^+$ for $1 \le i \le m$, $m \ge 0$. We call the subwords u_i the *terminal segments* of α. Furthermore, the α_i are called the *variable segments* of α. By imposing syntatic restrictions, we obtain special patterns. We say that α has *independent variable segments* if for all $1 \le i < j \le m$ we have $var(\alpha_i) \cap var(\alpha_j) = \emptyset$. Pattern α is *linear* if every variable $x \in var(\alpha)$ occurs exactly once in α. Pattern α is said to be *terminal free* if no terminal occurs in it. Pattern α is a *one-variable pattern* if it contains occurrences of one variable only.

3 Preliminary Results

On the one hand, Jiang *et al.* [JSSY95] have shown that the inclusion problem for E-pattern languages is in general undecidable. On the other hand, this problem is decidable for terminal free patterns; see [JSSY95]. Consequently, the E-language equivalence problem for terminal free patterns is decidable. Despite the fact that the inclusion problem for general E-pattern languages is undecidable, it might be the case that the decidability of the equivalence problem for general E-pattern languages can be proved by showing the decidability of the inclusion problem for certain E-patterns. This follows from the following theorem owing to Jiang *et al.* [JKS+94, JSSY95].

Theorem 1. *Let* $\alpha, \beta \in (\Sigma \cup V)^+$. *Moreover, let* $\alpha = \alpha_0 u_1 \alpha_1 u_2 \ldots \alpha_{m-1} u_m \alpha_m$ *and* $\beta = \beta_0 v_1 \beta_1 v_2 \ldots \beta_{n-1} v_n \beta_n$ *be their unique representations. If* $L_{E,\Sigma}(\alpha) = L_{E,\Sigma}(\beta)$ *and* $card(\Sigma) \geq 3$, *then* $m = n$ *and* $u_i = v_i$ *for* $1 \leq i \leq m$. *If* $card(\Sigma) \geq 4$, *then it further follows* $L_{E,\Sigma}(\alpha_i) = L_{E,\Sigma}(\beta_i)$ *for* $0 \leq i \leq m$.

Again, the assumption $card(\Sigma) \geq 3$ is crucial; see Example 1. In view of the above theorem, we introduce the following notion.

Definition 2. Two patterns α and β are called *similar* if and only if in their unique representations the respective terminal segments coincide, that is to say, $\alpha = \alpha_0 u_1 \alpha_1 u_2 \ldots \alpha_{m-1} u_m \alpha_m$ and $\beta = \beta_0 u_1 \beta_1 u_2 \ldots \beta_{m-1} u_m \beta_m$.

Open question 1: Is the inclusion problem decidable for similar E-patterns?

If the answer was positive in case $card(\Sigma) \geq 3$, then the decidability of the equivalence problem for general E-pattern languages would follow: Given patterns $\alpha = \alpha_0 u_1 \alpha_1 u_2 \ldots \alpha_{m-1} u_m \alpha_m$ and $\beta = \beta_0 v_1 \beta_1 v_2 \ldots \beta_{n-1} v_n \beta_n$, first check whether $m = n$ and $u_i = v_i$ for $1 \leq i \leq m$. If not, then $L_{E,\Sigma}(\alpha) \neq L_{E,\Sigma}(\beta)$. If so, then decide $L_{E,\Sigma}(\alpha) = L_{E,\Sigma}(\beta)$ by testing inclusion in both directions.

It is important to note that the undecidability of the inclusion problem for general E-pattern languages does, *a priori*, not imply the undecidability of the inclusion problem for similar E-patterns. This is because the patterns used in the undecidability proof of [JSSY95] are not similar. And in fact, we can prove that the inclusion problem for similar E-patterns is decidable provided that the alphabet Σ contains two constants not occurring in the patterns.

Definition 3. Let $V = \{x_1, \ldots, x_n\}$ be a set of variables and Σ be an alphabet with $card(\Sigma) \geq 2$. For every pair of letters a, b in Σ, $a \neq b$, and an integer $k > 0$, we define a substitution $\tau_{k,a,b} : V \to \{a, b\}^*$ by

$$\tau_{k,a,b}(x_i) = ab^{k*i+1}aab^{k*i+2}a \ldots ab^{k*(i+1)}a, \quad 1 \leq i \leq n.$$

Lemma 4. *Let* Σ *be an alphabet and let* $\alpha, \beta \in (\Sigma \cup V)^+$ *be similar patterns of the form* $\alpha = \alpha_0 u_1 \alpha_1 u_2 \ldots \alpha_{m-1} u_m \alpha_m$ *and* $\beta = \beta_0 u_1 \beta_1 u_2 \ldots \beta_{m-1} u_m \beta_m$. *If* Σ *contains two distinct letters* a *and* b *which do not occur in* α *and* β, *then* $\eta_{|\beta|,a,b}(\alpha) \in L_{E,\Sigma}(\beta)$ *if and only if there exists a substitution* $h : var(\beta) \to var(\alpha)^*$ *such that* $h(\beta) = \alpha$.

Proof. The *if* part is trivially true. The *only if* part can be proved similar to Lemma 7.1 of [JSSY95] as follows. Let $var(\alpha) = \{x_1, \ldots, x_n\}$ and $k = |\beta|$. Since $\tau_{k,a,b}(\alpha) \in L_{E,\Sigma}(\beta)$, there is a $\nu : var(\beta) \to \Sigma^*$ such that $\nu(\beta) = \tau_{k,a,b}(\alpha)$. For every $x \in var(\alpha)$, $\tau_{k,a,b}(x)$ consists of k segments of the form $ab^j a$. Thus, for every $x_i \in var(\alpha)$, there is at least one segment $ab^{j_i} a$, $k * i + 1 \leq j_i \leq k * (i+1)$, such that none of the appearance(s) of this segment in $\nu(\beta)$ is split by any partition of $\nu(\beta)$ into $\nu(\beta')$ and $\nu(\beta'')$ where $\beta = \beta'\beta''$. For every $x \in var(\alpha)$, we choose one such segment to be the anchor segment of x in $\tau_{k,a,b}(\alpha)$ w.r.t. β. We also say that this segment anchors x.

Suppose there is a variable $y \in var(\beta)$ such that $\nu(y)$ contains a terminal different from a and b, say c, occurring in β. Let c occur l-times in β. Then c appears more than l-times in $\nu(\beta)$. Apparently, c also occurs l-times in $\tau_{k,a,b}(\alpha)$. This, however, contradicts the equality $\nu(\beta) = \tau_{k,a,b}(\alpha)$. So for any variable $y \in var(\beta)$, it follows $\nu(y) \in \{a,b\}^*$. Suppose that y occurs in β_i, $0 < i < m$ (the cases $i = 0$ and $i = m$ are similar and simpler). Let c be the last letter of u_i and d the first letter of u_{i+1} ($c = d$ is possible). Let the last letter of u_i be the pth appearance of c and the first letter of u_{i+1} be the qth appearance of d in β. Since $\nu(y)$ does not contain any letter from a terminal segment, $\nu(y)$ is in between the pth appearance of c and the qth appearance of d in $\nu(\beta)$. This observation plays a key rôle in the proof.

We define a substitution $h : var(\beta) \to var(\alpha)^*$ as follows. For each $y \in var(\beta)$, let

$$ab^{j_{i_1}} a a b^{j_{i_2}} a \ldots ab^{j_{i_r}} a, \quad r \geq 0,$$

be the word obtained from $\nu(y)$ by deleting all the incomplete segments and segments that are not anchor segments. Note that the indices $i_1, i_2, \ldots, i_r$ are not necessarily distinct. Hereby a segment $ab^{j_{i_s}} a$ anchors a variable $x_{i_s} \in var(\alpha)$, $1 \leq i_s \leq n$. Then we define $h(y) = x_{i_1} x_{i_2} \ldots x_{i_r}$. Clearly, $h(\beta) = \alpha$ because each appearance of x in α has exactly one anchor segment w.r.t. β.

Theorem 5. *Let Σ be an alphabet and let $\alpha, \beta \in (\Sigma \cup V)^+$ be similar patterns. If Σ contains two distinct constants a and b not occurring in α and β, then the following statements are equivalent:*

1. $L_{E,\Sigma}(\alpha) \subseteq L_{E,\Sigma}(\beta)$.
2. $\tau_{|\beta|,a,b}(\alpha) \in L_{E,\Sigma}(\beta)$.
3. *There exists a substitution $h : var(\beta) \to var(\alpha)^*$ such that $h(\beta) = \alpha$.*

Proof. The implications $(1) \Rightarrow (2)$ and $(3) \Rightarrow (1)$ are trivially true. $(2) \Rightarrow (3)$ has been proved in Lemma 4.

Since the membership problem is decidable (see [Ang80a, JKS+94]), it is decidable whether $L_{E,\Sigma}(\alpha) \subseteq L_{E,\Sigma}(\beta)$ for two similar patterns a and b provided there are $a, b \in \Sigma$ which do not appear in α and β.

Corollary 6. *If $card(\Sigma) \geq 3$, then the equivalence problem for E-pattern languages is decidable whenever the underlying alphabet Σ contains two terminals that do not occur in the patterns.*

Notice that the inclusion problem (and hence the equivalence problem) for E-pattern languages is particularly decidable if Σ is infinite.

We next discuss the limits of the above proof technique. Clearly, we would like to get rid of the annoying condition "Σ has to contain two distinct constants a and b that do not occur in α and β". The crucial point is that this condition enabled us to infer the second implication below (cf. Lemma 4)

$$L_{E,\Sigma}(\alpha) \subseteq L_{E,\Sigma}(\beta) \Rightarrow \eta_{|\beta|,a,b}(\alpha) \in L_{E,\Sigma}(\beta) \Rightarrow \exists h : h(\beta) = \alpha.$$

However, if there are no extra constants, then it is not clear at all which encoding $\eta_{|\beta|,l_1,l_2}(\alpha)$ should be taken, i.e., which constants $l_1, l_2 \in \Sigma$, $l_1 \neq l_2$ should be chosen to encode the variables in α. A possible solution would be to generalize the above method: Find a finite test set $F \subset L_{E,\Sigma}(\alpha)$ such that $F \subset L_{E,\Sigma}(\beta)$ implies the existence of a substitution h such that $h(\beta) = \alpha$. A very natural candidate would be the set of all encodings, i.e., the set

$$F = \{\eta_{|\beta|,l_1,l_2}(\alpha) \mid l_1, l_2 \in \Sigma, l_1 \neq l_2\}.$$

But the next example shows that this set is not sufficient when $card(\Sigma) = 3$ (the situation might be different for $card(\Sigma) \geq 4$).

Example 2. Let $\Sigma = \{a, b, c\}$, $\alpha = y_1\, a\, y_2\, b\, y_3 y_2 y_4$ and $\beta = z_1\, a\, z_2\, b\, z_3 z_2$. There is a substitution $f : var(\alpha) \to var(\beta)^*$ such that $f(\alpha) = \beta$. Hence $L_{E,\Sigma}(\beta) \subseteq L_{E,\Sigma}(\alpha)$. The opposite inclusion does not hold because $cacbcb \in L_{E,\Sigma}(\alpha)$ but $cacbcb \notin L_{E,\Sigma}(\beta)$. And indeed, there is no substitution $g : var(\beta) \to var(\alpha)^*$ such that $g(\beta) = \alpha$. Despite this fact, we have $\eta_{|\beta|,l_1,l_2}(\alpha) \in L_{E,\Sigma}(\beta)$ for all $l_1, l_2 \in \Sigma$, $l_1 \neq l_2$. We exemplify this by showing (i) $\eta_{|\beta|,a,b}(\alpha) \in L_{E,\Sigma}(\beta)$ and (ii) $\eta_{|\beta|,c,a}(\alpha) \in L_{E,\Sigma}(\beta)$. First of all note that $\eta_{|\beta|,l_1,l_2}(y_i)$ starts with a prefix $l_1 l_2$ and ends with a suffix $l_2 l_1$. Hence $\eta_{|\beta|,l_1,l_2}(y_i) = l_1 l_2 w_{y_i} l_2 l_1$ for some subword w_{y_i} of $\eta_{|\beta|,l_1,l_2}(y_i)$.

$$(i) \quad \eta_{|\beta|,a,b}(\alpha) = abw_{y_1} ba\, a\, abw_{y_2} ba\, b\, abw_{y_3} ba\, abw_{y_2} ba\, abw_{y_4} ba.$$

Define the substitution ν by $\nu(z_1) = \nu(z_2) = \lambda$ and

$$\nu(z_3) = w_{y_1} ba\, a\, abw_{y_2} ba\, b\, abw_{y_3} ba\, abw_{y_2} ba\, abw_{y_4} ba.$$

Then clearly $\nu(\beta) = \eta_{|\beta|,a,b}(\alpha)$, hence $\eta_{|\beta|,a,b}(\alpha) \in L_{E,\Sigma}(\beta)$.

$$(ii) \quad \eta_{|\beta|,c,a}(\alpha) = caw_{y_1} ac\, a\, caw_{y_2} ac\, b\, caw_{y_3} ac\, caw_{y_2} ac\, caw_{y_4} ac.$$

Now define the substitution ν by $\nu(z_1) = caw_{y_1} ac\, a\, caw_{y_2}$, $\nu(z_2) = c$ and $\nu(z_3) = caw_{y_3} ac\, caw_{y_2} ac\, caw_{y_4} a$.

Note that the condition of having two extra constants could be dropped in Lemma 4, if E-language equivalence was preserved under alphabet extensions, more precisely, if the question below would have an answer in the affirmative.

Open question 2: Does the equivalence:

$$L_{E,\Sigma}(\alpha) = L_{E,\Sigma}(\beta) \quad \Leftrightarrow \quad L_{E,\Sigma'}(\alpha) = L_{E,\Sigma'}(\beta)$$

hold for $card(\Sigma) \geq 3$, $\alpha, \beta \in (\Sigma \cup V)^+$ and $\Sigma' = \Sigma \cup \{a\}$, where $a \notin \Sigma$?

Again, the condition $card(\Sigma) \geq 3$ is essential (cf. Example 1). By variations of the above proof technique, we are able to show that E-language inclusion (hence E-language equivalence) is also decidable in other special cases.

Proposition 7. *Let Σ be an alphabet and let $\alpha, \beta \in (\Sigma \cup V)^+$ be similar patterns of the form $\alpha = \alpha_0 u_1 \alpha_1 u_2 \ldots \alpha_{m-1} u_m \alpha_m$ and $\beta = \beta_0 u_1 \beta_1 u_2 \ldots \beta_{m-1} u_m \beta_m$. If $card(\Sigma) \geq m + 2$, $m \geq 0$, then it is decidable whether $L_{E,\Sigma}(\alpha) \subseteq L_{E,\Sigma}(\beta)$.*

Proposition 8. *Let $\alpha, \beta \in (\Sigma \cup V)^+$ be similar patterns. If there are $a, b \in \Sigma$, $a \neq b$, such that every terminal segment of α (hence of β) contains at least one letter different from a and b, then it is decidable whether $L_{E,\Sigma}(\alpha) \subseteq L_{E,\Sigma}(\beta)$.*

4 Variable Elimination and Normal Forms

Solving the equivalence problem for NE-pattern languages is easy: Two patterns generate the same language if and only if they are identical up to renaming of variables; see [Ang80a]. For E-patterns, however, this is not true at all. Jiang *et al.* [JKS+94] write that: "Two very different looking E-patterns may generate the same language. For instance, if a terminal-free pattern α contains exactly one occurrence of some variable, then α is equivalent to the pattern x". The fact that an E-pattern may contain many superfluous variables seems to be the major problem in deciding whether two E-patterns generate the same language. We call a variable x in an E-pattern $\alpha \in (\Sigma \cup V)^+$ *superfluous*, if the E-pattern α' obtained from α by deleting all occurrences of x still generates $L_{E,\Sigma}(\alpha)$. We next tackle the equivalence problem of two E-patterns α and β by eliminating superfluous variables in α and β. The first difficulty one encounters in this approach is of course the decidability of whether a variable is superfluous or not. Since we do not know how to decide this in general, we by-pass the problem by defining a reduction relation $\rightarrow \subset (\Sigma \cup V)^+ \times (\Sigma \cup V)^+$ such that if pattern α reduces to pattern α', then the erased variables are superfluous.

Definition 9. The reduction relation $\rightarrow \subset (\Sigma \cup V)^+ \times (\Sigma \cup V)^+$ is defined by: $\alpha \rightarrow \alpha'$ if and only if

1. there is a non-empty proper subset V_d of $var(\alpha)$ such that erasing all occurrences of variables of V_d in α yields α' (that is to say, $\alpha' = d(\alpha)$, where $d = \{x \mapsto \lambda \mid x \in V_d\}$), and
2. there is a linear substitution $\sigma : var(\alpha') \rightarrow var(\alpha)$ such that $\sigma(\alpha') = \alpha$. Hereby σ is called *linear* if it has the form

$$\sigma = \{x_i \mapsto \gamma_i x_i \gamma_i' \mid x_i \in var(\alpha'), \ \gamma_i, \gamma_i' \in V_d^*, \ 1 \leq i \leq n\},$$

   where $var(\alpha') = \{x_1, \ldots, x_n\}$.

If $\alpha \rightarrow \alpha'$, then we say that α *reduces* to α' (by deleting the variables $V_d = var(\alpha) \setminus var(\alpha')$). Pattern α is said to be *reducible*, if there exists a pattern α' such that $\alpha \rightarrow \alpha'$.

We illustrate the definition by a small example which also shows that sometimes it is necessary to eliminate more than one variable in one reduction step.

Example 3. Let $\alpha = xzyzyzz$. Then $\alpha \to \alpha' = x$ because the linear substitution $\sigma = \{x \mapsto xzyzyzz\}$ satisfies $\sigma(\alpha') = \alpha$. Note that neither $\alpha \to xyy$ nor $\alpha \to xzzzz$, although xyy and $xzzzz$ generate $L_{E,\Sigma}(\alpha)$.

Given α and α', it is decidable whether $\alpha \to \alpha'$: Condition (1) can be checked in linear time and condition (2) is decidable because it suffices to check the equality $\sigma(\alpha') = \alpha$ merely for the finitely many linear substitutions σ which satisfy $\sum_{i=1}^{n} |\gamma_i \gamma_i'| = |\alpha| - |\alpha'|$ (since $\sum_{i=1}^{n} |\gamma_i \gamma_i'| \neq |\alpha| - |\alpha'|$ implies $\sigma(\alpha') \neq \alpha$ already). Hence it is also decidable whether α is reducible since there are only finitely many non-empty proper subsets V_d of $var(\alpha)$.

Lemma 10. *If $\alpha \to \alpha'$, then $L_{E,\Sigma}(\alpha) = L_{E,\Sigma}(\alpha')$. (Hence the variables $V_d = var(\alpha) \setminus var(\alpha')$ are superfluous in α.)*

Lemma 11. *The reduction relation $\to \subset (\Sigma \cup V)^+ \times (\Sigma \cup V)^+$ is a length-decreasing (i.e., $\alpha \to \alpha'$ implies $|\alpha| > |\alpha'|$) partial ordering.*

Definition 12. A pattern α is said to be in *normal form* (w.r.t. $\to$), if there is no pattern α' such that $\alpha \to \alpha'$.

Since $\to$ is length-decreasing, the successive reduction of a pattern must end in a normal form (an irreducible pattern). Clearly, a pattern α may have two distinct normal forms. Just consider $\alpha = xy$: $\alpha \to x$ and $\alpha \to y$. Our next goal is to show that the normal forms of a pattern are identical up to variable renaming. In this context, the following technical lemma plays a key rôle. Its proof can be found in [OU95].

Lemma 13. *Let $\alpha \in (\Sigma \cup V)^+$ be a pattern and let $h : var(\alpha) \to var(\alpha)^*$ be a substitution such that $h(\alpha) = \alpha$. Then the following statements hold.*

1. *α has a partition $\alpha = \beta_1 \beta_2 \ldots \beta_k$ into minimal fixpoints of h, i.e., for each $i \in \{1, \ldots, k\}$, $\beta_i \in (\Sigma \cup V)^+$, the equality $h(\beta_i) = \beta_i$ holds, and for every proper subword γ of β_i we have $h(\gamma) \neq \gamma$.*
2. *The partition is unique.*
3. *For every $i \in \{1, \ldots, k\}$, either $\beta_i \in \Sigma$ or $\beta_i \in V^+$. Moreover, if $\beta_i \in V^+$, then it contains exactly one variable x such that*

   $(*) \qquad h(x) = \gamma_1 x \gamma_2$, *where* $\gamma_1, \gamma_2 \in (var(\beta_i) \setminus \{x\})^*$ *and*
   $\qquad \qquad \beta_i = \delta_1 \gamma_1 x \gamma_2 \delta_2$ *for some* $\delta_1, \delta_2 \in (var(\beta_i) \setminus \{x\})^*$.

4. *If x is a variable such that $h(x) = \gamma_1 x \gamma_2$, where $\gamma_1, \gamma_2 \in (var(\beta_i) \setminus \{x\})^*$, and x occurs in some β_i and β_j, $i \neq j$, then $\beta_i = \beta_j$.*

Theorem 14. *Let $\alpha \in (\Sigma \cup V)^+$ be a pattern in normal form and let $h : var(\alpha) \to var(\alpha)^*$ be a substitution such that $h(\alpha) = \alpha$. Then h is the identity function on $var(\alpha)$.*

Proof. According to Lemma 13, α has a unique partition $\alpha = \beta_1\beta_2\ldots\beta_k$ into minimal fixpoints of h such that either $\beta_i \in \Sigma$ or $\beta_i \in V^+$ where β_i contains exactly one variable x_i which occurs in $h(x_i)$. Let $i_1,\ldots,i_p$ be exactly those indices for which $\beta_{i_j} \in V^+$ and let x_{i_j} be the corresponding variables satisfying (*). Let $V_{nd} = \{x_{i_1},\ldots,x_{i_p}\}$ (note that $x_{i_j} = x_{i_q}$, $i_j \neq i_q$, is possible) and $V_d = var(\alpha) \setminus V_{nd}$. Suppose that $V_d \neq \emptyset$ and let α' be the pattern obtained from α by deleting all the variables from V_d, i.e., $\alpha' = d(\alpha)$ where $d = \{x \mapsto \lambda \mid x \in V_d\}$. Note that $var(\alpha') = V_{nd}$. Define $g : var(\alpha') \to var(\alpha)^*$ by $g(x_{i_j}) = \beta_{i_j}$. According to Lemma 13, g is well-defined because $x_{i_j} = x_{i_q}$ implies $\beta_{i_j} = \beta_{i_q}$. Furthermore, we claim $g(\alpha') = \alpha$. In order to prove this claim, it suffices to show $g(d(\beta_{i_j})) = \beta_{i_j}$ for all $1 \leq j \leq p$. We have $d(\beta_{i_j}) = x_{i_j}$ because x_{i_j} is the only variable from V_{nd} in β_{i_j}. Hence $g(d(\beta_{i_j})) = g(x_{i_j}) = \beta_{i_j}$ and the claim is proved. Since β_{i_j} has a representation $\delta_1\gamma_1 x_{i_j}\gamma_2\delta_2$, where $\gamma_1, \gamma_2, \delta_1, \delta_2 \in V_d^*$, g is a linear substitution. Thus it follows $\alpha\to\alpha'$. This, however, contradicts the fact that α is in normal form. So $V_d = \emptyset$. It further follows that $\beta_{i_j} = x_{i_j}$ and therefore $h(x_{i_j}) = x_{i_j}$ for $1 \leq j \leq p$. Hence h is the identity on $var(\alpha) = \{x_{i_1},\ldots,x_{i_p}\}$.

The simple example $\alpha = xy$, $h(x) = xy$ and $h(y) = \lambda$ shows that in Theorem 14 the normal form requirement on α cannot be dropped.

Corollary 15. *If $\alpha \in (\Sigma \cup V)^+$ reduces to distinct normal forms α' and α'', then there is a variable renaming $h : var(\alpha') \to var(\alpha'')^*$ such that $h(\alpha') = \alpha''$.*

Proof. Consider two reductions of α to normal forms α' and α'':

$$\alpha' = \alpha'_n \leftarrow \alpha'_{n-1} \leftarrow \ldots \leftarrow \alpha'_1 \leftarrow \alpha'_0 = \alpha = \alpha''_0 \to \alpha''_1 \to \ldots \to \alpha''_{m-1} \to \alpha''_m = \alpha''.$$

Since $\to$ is transitive, we obtain $\alpha' \leftarrow \alpha \to \alpha''$. By definition of reduction, there are substitutions d', d'' and σ', σ'' such that $d'(\alpha) = \alpha'$, $d''(\alpha) = \alpha''$, $\sigma'(\alpha') = \alpha$ and $\sigma''(\alpha'') = \alpha$. Define $f : var(\alpha') \to var(\alpha'')^*$ by $f = d'' \circ \sigma'$. Analogously, define $g : var(\alpha'') \to var(\alpha')^*$ by $g = d' \circ \sigma''$. Then $f(\alpha') = \alpha''$ and $g(\alpha'') = \alpha'$. Hence $f(g(\alpha'')) = \alpha''$ and $g(f(\alpha')) = \alpha'$. By Theorem 14, $f \circ g$ is the identity on $var(\alpha'')$ and $g \circ f$ is the identity on $var(\alpha')$. Thus f and g are bijections, or in other words, variable renamings.

Thus the normal form of pattern α is unique (up to variable renamings); from now on it will be denoted by $\alpha\!\downarrow$. Next, we derive equivalent formulations of Conjecture 1 of Section 1.

Proposition 16. *Let $\alpha, \beta \in (\Sigma \cup V)^+$ be patterns and let $card(\Sigma) \geq 3$. The following statements are equivalent.*

1. *$\alpha\!\downarrow$ and $\beta\!\downarrow$ are identical up to variable renaming.*
2. *There are substitutions $f' : var(\alpha\!\downarrow) \to var(\beta\!\downarrow)^*$ and $g' : var(\beta\!\downarrow) \to var(\alpha\!\downarrow)^*$ such that $f'(\alpha\!\downarrow) = \beta\!\downarrow$ and $g'(\beta\!\downarrow) = \alpha\!\downarrow$.*
3. *There are substitutions $f : var(\alpha) \to var(\beta)^*$ and $g : var(\beta) \to var(\alpha)^*$ such that $f(\alpha) = \beta$ and $g(\beta) = \alpha$.*

Proof. The implication (1) $\Rightarrow$ (2) holds trivially. The converse implication (2) $\Rightarrow$ (1) is a consequence of Theorem 14 (cf. also proof of Corollary 15). The equivalence (2) $\Leftrightarrow$ (3) follows from the equalities $d(\alpha) = \alpha\downarrow$, $\sigma(\alpha\downarrow) = \alpha$, $d'(\beta) = \beta\downarrow$ and $\sigma'(\beta\downarrow) = \beta$. The existence of a substitution $f' : var(\alpha\downarrow) \to var(\beta\downarrow)^*$ with $f'(\alpha\downarrow) = \beta\downarrow$ implies for instance the existence of a substitution $f : var(\alpha) \to var(\beta)^*$ such that $f(\alpha) = \beta$; just define $f = \sigma' \circ f' \circ d$.

Consequently, in the situation of Corollary 6 (resp. of Propositions 7 and 8) language equivalence can also be decided by comparing normal forms instead of testing membership of certain words: $L_{E,\Sigma}(\alpha) = L_{E,\Sigma}(\beta)$ if and only if $\alpha\downarrow$ and $\beta\downarrow$ are identical up to variable renaming. We have not investigated yet whether reduction to normal form is less complex than solving the membership problem which is NP-complete (see [Ang80a, JKS+94]).

5 Equivalence Tests Based on Normal Forms

We next show the benefits of our normal form approach.

Proposition 17. *Let $\alpha, \beta \in (\Sigma \cup V)^+$ and let $card(\Sigma) \geq 4$. Suppose (i) $\alpha\downarrow$ has independent variable segments or (ii) $\alpha\downarrow$ is linear. Then $L_{E,\Sigma}(\alpha) = L_{E,\Sigma}(\beta)$ if and only if $\alpha\downarrow$ and $\beta\downarrow$ are identical up to variable renaming.*

Proof. (i) We show $L_{E,\Sigma}(\alpha) = L_{E,\Sigma}(\beta)$ if and only if there are substitutions $f : var(\alpha\downarrow) \to var(\beta\downarrow)^*$ and $g : var(\beta\downarrow) \to var(\alpha\downarrow)^*$ such that $f(\alpha\downarrow) = \beta\downarrow$ and $g(\beta\downarrow) = \alpha\downarrow$. The *if* direction is trivially true, so we prove the *only if* direction. Let $L_{E,\Sigma}(\alpha) = L_{E,\Sigma}(\beta)$, hence $L_{E,\Sigma}(\alpha\downarrow) = L_{E,\Sigma}(\beta\downarrow)$. According to Theorem 1, $\alpha\downarrow$ and $\beta\downarrow$ are similar, so let $\alpha\downarrow = \alpha_0 u_1 \alpha_1 u_2 \ldots \alpha_{m-1} u_m \alpha_m$ and $\beta\downarrow = \beta_0 u_1 \beta_1 u_2 \ldots \beta_{m-1} u_m \beta_m$. We show:

1. For any α_i, there is an $f_i : var(\alpha_i) \to var(\beta_i)^*$ such that $f_i(\alpha_i) = \beta_i$.
2. For any β_i, there is a $g_i : var(\beta_i) \to var(\alpha_i)^*$ such that $g_i(\beta_i) = \alpha_i$ and moreover $g_i(x) = \lambda$ whenever x also occurs in β_j, $j \neq i$.
3. The local substitutions f_i can be combined into a global substitution f such that $f(\alpha\downarrow) = \beta\downarrow$.
4. The local substitutions g_i can be combined into a global substitution g such that $g(\beta\downarrow) = \alpha\downarrow$.

In order to prove (1) and (2), we consider only the case $0 < i < m$ (the cases $i = 0$ and $i = m$ are similar and simpler). Let c be the last letter of u_i and d be the first letter of u_{i+1}. We choose $a, b \in \Sigma \setminus \{c, d\}$ such that $a \neq b$.

1. Consider $\eta_{|\alpha\downarrow|,a,b}(\beta\downarrow)$. $\eta_{|\alpha\downarrow|,a,b}(\beta\downarrow)$ can be written as $w_0 u_1 w_1 \ldots w_{m-1} u_m w_m$, where $w_i = \eta_{|\alpha\downarrow|,a,b}(\beta_i)$. Since $L_{E,\Sigma}(\alpha\downarrow) = L_{E,\Sigma}(\beta\downarrow)$, there is a σ such that $\sigma(\alpha\downarrow) = w_0 u_1 w_1 \ldots w_{m-1} u_m w_m$. Note that if α_i is between the kth appearance of c and the lth appearance of d in $\alpha\downarrow$, then w_i is between the kth appearance of c and the lth appearance of d in $\eta_{|\alpha\downarrow|,a,b}(\beta\downarrow)$. Since σ cannot introduce the constants c and d, it follows $\sigma(\alpha_i) = w_i$. As in the proof of Lemma 4, we obtain a substitution $f_i : var(\alpha_i) \to var(\beta_i)^*$ such that $f_i(\alpha_i) = \beta_i$.

2. Let τ be the substitution which encodes only variables from $var(\alpha_i)$. That is, $\tau(x) = \tau_{|\beta\downarrow|,a,b}(x)$ if $x \in var(\alpha_i)$ and $\tau(x) = \lambda$ otherwise. Since $\alpha\downarrow$ has independent variable segments, it follows

$$\tau(\alpha\downarrow) = u_0 u_1 \ldots u_i \tau_{|\beta\downarrow|,a,b}(\alpha_i) u_{i+1} \ldots u_m.$$

Since $L_{E,\Sigma}(\alpha\downarrow) = L_{E,\Sigma}(\beta\downarrow)$, there is a substitution σ such that $\sigma(\beta\downarrow) = u_0 u_1 \ldots u_i \tau_{|\beta\downarrow|,a,b}(\alpha_i) u_{i+1} \ldots u_m$. Since σ cannot introduce the constants c and d, it follows $\sigma(\beta_i) = \tau_{|\beta\downarrow|,a,b}(\alpha_i)$ and $\sigma(\beta_j) = \lambda$ for every $j \neq i$. As in the proof of Lemma 4, we obtain a substitution $g_i : var(\beta_i) \to var(\alpha_i)^*$ such that $g_i(\beta_i) = \alpha_i$. Moreover $g_i(x) = \lambda$ whenever $x \in var(\beta_i) \cap var(\beta_j)$ for some $j \neq i$.

3. Let $x \in var(\alpha\downarrow)$. Since $\alpha\downarrow$ has independent variable segments, there is exactly one index i such that $x \in var(\alpha_i)$. Define $f(x) = f_i(x)$. The substitution f obtained in this manner is well-defined and $f(\alpha\downarrow) = \beta\downarrow$.

4. Define

$$g(x) = \begin{cases} g_i(x) & \text{if } x \in var(\beta_i) \setminus \bigcup_{j \neq i} var(\beta_j) \\ \lambda & \text{otherwise} \end{cases}$$

The substitution g is well-defined and furthermore $g(\beta\downarrow) = \alpha\downarrow$.

(ii) If $\alpha\downarrow$ is linear, then it has independent variable segments.

It has already been shown by Jiang *et al.* [JKS+94] that $L_{E,\Sigma}(\alpha) \subseteq L_{E,\Sigma}(\beta)$ is decidable if β is a one-variable pattern and $card(\Sigma) \geq 2$. Thus our next result is not surprising. Its proof, however, is much more complicated than expected because only one normal form is assumed to be a one-variable pattern.

Proposition 18. *Let $\alpha, \beta \in (\Sigma \cup V)^+$ be patterns and let $card(\Sigma) \geq 3$. If $\alpha\downarrow$ is a one-variable pattern, then $L_{E,\Sigma}(\alpha) = L_{E,\Sigma}(\beta)$ if and only if $\alpha\downarrow$ and $\beta\downarrow$ are identical up to variable renaming.*

We conclude with a result similar to Proposition 18; again we have to omit the proof due to space limitations.

Proposition 19. *Let $\alpha, \beta \in (\Sigma \cup V)^+$ be patterns and let $card(\Sigma) \geq 4$. Suppose $\alpha\downarrow = \alpha_0 u_1 \alpha_1 u_2 \ldots \alpha_{m-1} u_m \alpha_m$, where $|\alpha_0| \leq 1$, $|\alpha_m| \leq 1$, and $|\alpha_i| = 1$, $0 < i < m$. In other words, $\alpha\downarrow$ does not contain consecutive variables. Then $L_{E,\Sigma}(\alpha) = L_{E,\Sigma}(\beta)$ if and only if $\alpha\downarrow$ and $\beta\downarrow$ are identical up to variable renaming.*

Note that if a pattern in normal form is linear, then its canonical form is $x_0 u_1 x_1 u_2 \ldots x_{m-1} u_m x_m$. Therefore, Proposition 19 also generalizes Proposition 17 (ii).

6 Related Work

In this paper, we did not investigate the cases $card(\Sigma) \leq 2$. In [DF96], it is shown that the equivalence of E-patterns is decidable for $card(\Sigma) = 1$ and a necessary condition for the equivalence of E-patterns in case $card(\Sigma) = 2$

is given. Moreover, Dányi and Fülöp [DF96] claim that for $card(\Sigma) \geq 2$ the inclusion and hence the equivalence problem for similar E-patterns is decidable by testing membership of a certain word; see below for the exact claim. However, Example 2 refutes the claim. To see this, let us first recall the definition of the substitution $\tau_{k,a,b}$ in [DF96] (which is a modification of Definition 3).

Definition 20. Let $\alpha = \alpha_0 u_1 \alpha_1 u_2 \ldots \alpha_{m-1} u_m \alpha_m \in (\Sigma \cup V)^+$, $V = \{x_1, \ldots, x_n\}$, and $card(\Sigma) \geq 2$. Moreover, let $max = \max\{|u_i| \mid 1 \leq i \leq m\}$. For every a, b in Σ, $a \neq b$, and $k > 0$, define $\tau_{k,a,b} : V \to \{a, b\}^*$ by

$$\tau_{k,a,b}(x_i) = ab^{max}b^{k*i+1}aab^{max}b^{k*i+2}a\ldots ab^{max}b^{k*(i+1)}a, \quad 1 \leq i \leq n.$$

Claim: Let $card(\Sigma) \geq 2$ and let $\alpha, \beta \in (\Sigma \cup V)^+$ be similar patterns, where $\alpha = \alpha_0 u_1 \alpha_1 u_2 \ldots \alpha_{m-1} u_m \alpha_m$. Let k be the number of variable occurrences in β and fix two distinct letters $a, b \in \Sigma$. Then $\tau_{k+m,a,b}(\alpha) \in L_{E,\Sigma}(\beta)$ if and only if there is a substitution $h : var(\beta) \to var(\alpha)^*$ such that $h(\beta) = \alpha$.

Let α and β be as in Example 2 and note that $\tau_{6,a,b}(y_i) = abw_{y_i}ba$ for some subword w_{y_i} of $\tau_{6,a,b}(y_i)$. It follows as in Example 2 that $\tau_{6,a,b}(\alpha) \in L_{E,\Sigma}(\beta)$ but there is no substitution h with $h(\beta) = \alpha$.

References

[Ang80a] D. Angluin. Finding patterns common to a set of strings. *Journal of Computer and System Sciences* **21**, pages 46–62, 1980.

[Ang80b] D. Angluin. Inductive inference of formal languages from positive data. *Information and Control* **45**, pages 117–135, 1980.

[DF96] G. Dányi and Z. Fülöp. A note on the equivalence problem of E-patterns. *Information Processing Letters* **57**, pages 125–128, 1996.

[IJ91] O. Ibarra and T. Jiang. Learning regular languages from counterexamples. *Journal of Computer and System Sciences* **43**, pages 299–316, 1991.

[JKS+94] T. Jiang, E. Kinber, A. Salomaa, K. Salomaa, and S. Yu. Pattern languages with and without erasing. *Intern. J. Computer Math.* **50**, pages 147–163, 1994.

[JSSY95] T. Jiang, A. Salomaa, K. Salomaa, and S. Yu. Decision problems for patterns. *Journal of Computer and System Sciences* **50**, pages 53–63, 1995.

[KMU95] P. Kilpeläinen, H. Mannila, and E. Ukkonen. MDL learning of unions of simple pattern languages from positive examples. In *Proceedings of the 2nd European Conference on Computational Learning Theory*, pages 252–260. Lecture Notes in Computer Science **904**, Berlin: Springer Verlag, 1995.

[OU95] E. Ohlebusch and E. Ukkonen. On the the equivalence problem for E-pattern languages. Report 95-04, Forschungsberichte der Technischen Fakultät, Abteilung Informationstechnik, Universität Bielefeld, 1995.

[Sal94] A. Salomaa. Patterns. *Bulletin of the European Association for Theoretical Computer Science* **54**, pages 194–206, 1994.

[Sal95] A. Salomaa. Return to patterns. *Bulletin of the European Association for Theoretical Computer Science* **55**, pages 144–155, 1995.

Specifying and Verifying Parametric Processes[*]

Wiesław Pawłowski[1]   Paweł Pączkowski[2]   Stefan Sokołowski[1]

[1] Institute of Computer Science, Polish Academy of Sciences,
ul. Abrahama 18, 81-825 Sopot, Poland.
E-mail: {w.pawlowski,s.sokolowski}@ipipan.gda.pl
[2] Department of Computing Science, Chalmers University of Technology
and Göteborgs University, 412-96 Göteborg, Sweden, and
Institute of Mathematics, University of Gdańsk,
ul. Wita Stwosza 57, 80-952 Gdańsk, Poland,
E-mail: matpmp@univ.gda.pl

Abstract. A process algebra and readiness specifications introduced by
Olderog are extended to a framework in which processes parameterized
with other processes can be specified, defined and verified. Higher order
process-parameters are allowed. The formalism resembles typed lambda
calculus built on top of a process algebra, where specifications play the
role of types. A proof system for deriving judgements "parametric process
meets a specification" is given.

1 Introduction

A typical approach to software development is that of decomposition of large
tasks into smaller subtasks. To formalize the decomposition principle, let us
consider a set of constructions *Constr* (programs, processes, etc.), a set of specifications *Spec*, and satisfaction relation **sat** between constructions and specifications (**sat** can be given, for example, by a proof system). The triple

$$(Constr, Spec, \textbf{sat}) \tag{1}$$

will be referred to in the sequel as *specification system* (see [8]). Assume we are
faced with a task of providing a construction c that satisfies a given specification
S (denoted c **sat** S). Decomposing this task amounts to dividing it into the
following subtasks:

1. provide sub-constructions $c_1, \ldots, c_n$ that satisfy some sub-specifications
   $S_1, \ldots, S_n$ into which the original specification S is decomposed, and
2. provide a method of combining $c_1, \ldots, c_n$ so as to obtain such a c that
   c **sat** S.

After [8] we adopt a point of view that the method mentioned in point 2
above should be a *construction* itself, i.e. belong to *Constr* and we rephrase 2 as

---

[*] This work was partially supported by the following grants: KBN Grant No. PB-
1312/P3/92/02, CRIT IC 1010/II, and (in case of the second author) ESPRIT BRA
CONCUR2.

2′. provide a *parametric construction* f such that

$$f \text{ sat } S_1 \to S_2 \to \cdots \to S_n \to S$$

The "functional" specification $S_1 \to S_2 \to \cdots \to S_n \to S$, which is assumed to belong to *Spec*, is intended to specify precisely those parametric constructions that yield objects satisfying S when applied to sub-constructions satisfying $S_1, \ldots, S_n$. Thus, if the decomposition steps 1 and 2′ are realized then $f c_1 \ldots c_n$ defines a c we were looking for.

The specifications $S_1, \ldots, S_n$ and $S_1 \to S_2 \to \cdots \to S_n \to S$, may be further decomposed to ease the realization; the decomposition of the functional specification $S_1 \to S_2 \to \cdots \to S_n \to S$ may lead to higher order parameterization.

The purpose of this work is to apply the methodology sketched above in the context of concurrent process specification and verification. As a starting point we adopt a process algebra and readiness logic for specifying processes proposed in [5]. Olderog extends the process algebra by introducing mixed terms, where a specification can become a part of a process term. This enables transformation based development of processes. We consider a different option and introduce parametric processes that contain process variables. We allow higher order parameters: process variables can represent also parametric processes. Technically speaking, Olderog's process algebra is extended with abstraction and application primitives. This leads to a formalism resembling a typed lambda calculus built on top of a process algebra, where specifications play the role of types. Mixed terms can be viewed as a special case of parametric processes.

We provide a proof system for deriving judgements "(parametric) process meets a specification". To this end, we needed to reformulate the proof system for nonparametric processes that was given in the final chapter of [5]. Derivations in that proof system use mixed terms. We provide a direct proof system for nonparametric processes, which does not appeal to mixed terms and which is shown to be equivalent to that of [5]. The direct proof system is extended to the parametric case. These are the main technical contributions of this paper.

The parametric framework we introduce allows us to express intermediate stages of process development in a more explicit way than in Olderog's mixed-term approach. Although we do not use mixed terms, we retain the ability to construct processes in a similar manner as with mixed-term transformations, adding the capacity for higher order parameters. To illustrate these points let us sketch an example inspired by [5].

Consider a sequence of transformations that derives cnt_2, a counter of capacity 2, as a parallel composition of two counters of capacity 1.

$$cnt_2 \equiv> \cdots \equiv> m \equiv> \cdots \equiv> CNT_2$$

This derivation starts from a specification CNT_2 of a counter of capacity 2 and progresses through a number of transformation steps. The crucial intermediate stage is a mixed term m such that $cnt_2 \equiv> \cdots \equiv> m$ and the only specification appearing in m is CNT_1, a specification of a counter of capacity 1.

In our approach, an analogous process construction is rendered as follows. We formulate our task as: provide cnt_2 such that

$$cnt_2 \text{ sat } CNT_2$$

This task is decomposed into two subtasks: constructing a counter of capacity 1, cnt_1, and a parametric process $double_1$ that constructs a counter of capacity 2 given a counter of capacity 1, i.e. cnt_1 and $double_1$ such that

$$cnt_1 \text{ sat } CNT_1$$

$$double_1 \text{ sat } CNT_1 \to CNT_2 \tag{2}$$

Parametric process $double_1$ is a counterpart of m. The assertion (2), which expresses the required property of $double_1$, is equivalent to the more verbose description of m given above.

We can easily specify more elaborate tasks such as

$$(CNT_1 \to CNT_2) \to (CNT_1 \to CNT_4) \tag{3}$$

which means "combine counters of capacity 1 to get a counter of capacity 4, assuming that we can construct counter of capacity 2 using counters of capacity 1", The ability to specify such more involved tasks as well as intermediate construction stages such as $double_1$ clarifies the decomposition steps and allows for the reuse of intermediate construction stages.

The concept of processes with process parameters (but only first order ones) has already appeared in [4] (Chapter 9). However, this method of constructing CCS processes is not reflected in their verification because the bisimulation proof technique used in this reference requires a considerable effort in decomposing large verification tasks into smaller ones (see e.g. [3]).

The formalism we develop allows decomposition steps of the form 1 and 2′ that may involve processes with higher-order parameters and helps one to do a systematic book-keeping of process dependencies, which is useful in higher-order constructions. We view our approach as an alternative to process development methods that use transformation or refinement techniques (see e.g. [2], [5], [6]).

The paper is organized as follows. In Section 2 we give an overview of a process algebra of nonparametric processes and a logic for specifying them that appear in [5]. A new, direct proof system for nonparametric processes is given in Section 3. In Section 4 we define parametric processes and extend the proof system to the parametric case. Section 5 contains examples.

Due to space limitations most proofs were omitted. They can be found in [7].

2 Nonparametric processes and specifications

The class of processes that we consider and a logic for specifying their properties are taken from [5]. We give a brief overview of the main notions, referring to [5] for a comprehensive exposition. [7] contains an extended version of this section.

Processes. Let *Comm* be an infinite set of *communications*. Together with a special symbol τ representing a hidden or internal activity of a process it will constitute the set of process *actions*.

Processes will be defined as a subset of the set of *recursive terms*. Recursive terms have the following abstract syntax

$$P ::= stop_A \mid a.P \mid P + Q \mid P \parallel Q \mid P[b/a] \mid P\backslash a \mid X \mid \mu X.P$$

where P and Q range over recursive terms, X ranges over process variables, A over finite sets of communications, a and b over communications. Essentially, the process operators in the syntax above are those of CSP apart from plus, which is given semantics as in CCS.

Iterated applications of renaming and hiding will be abbreviated as follows: $P[\bar{b}/\bar{a}]$ will stand for $P[b_1/a_1]\ldots[b_n/a_n]$, where $\bar{b} = b_1,\ldots,b_n, \bar{a} = a_1,\ldots,a_n$; $P\backslash A$ will denote $P\backslash a_1 \ldots \backslash a_n$, where $A = \{a_1,\ldots a_n\}$. Notation $\{\ldots/\ldots\}$ is used for syntactic substitutions.

With each recursive term P we associate its *(communication) alphabet* $\alpha(P)$. It is assumed that each process variable X has an associated alphabet $\alpha(X)$. The communication alphabets of remaining terms are defined inductively as follows:

$$\begin{aligned}
\alpha(stop_A) &= A & \alpha(P[b/a]) &= (\alpha(P) - \{a\}) \cup \{b\} \\
\alpha(a.P) &= \{a\} \cup \alpha(P) & \alpha(P\backslash a) &= (\alpha(P) - \{a\}) \\
\alpha(P + Q) &= \alpha(P) \cup \alpha(Q) & \alpha(\mu X.P) &= \alpha(X) \cup \alpha(P) \\
\alpha(P \parallel Q) &= \alpha(P) \cup \alpha(Q)
\end{aligned}$$

Definition 1. We say that a recursive term P is a *process term* (or simply, a process) if it satisfies the following conditions: *(1)* P is action guarded, i.e. appearances of process variables in recursive subprocesses of P are guarded, *(2)* every subterm $a.Q$ of P satisfies $a \in \alpha(Q)$, *(3)* every subterm $Q + R$ of P satisfies $\alpha(Q) = \alpha(R)$, *(4)* every subterm $\mu X.Q$ of P satisfies $\alpha(X) = \alpha(Q)$.

The set of all process terms will be denoted by *Proc*. A process term is *closed* if it does not contain free process identifiers. The set of all closed process terms will be denoted by *CProc*.

Semantics. Processes are given *readiness semantics*. It describes process behaviour in terms of *ready pairs* $(tr, \mathcal{F})$, where tr is finite sequence of communications called *trace*, and $\mathcal{F}$, called *ready set*, is a set of communications in which a process is ready to engage after "performing" the trace tr. Besides the ready pairs the semantic domain $\text{DOM}_{\mathcal{R}}$ contains elements representing undesirable behaviours of processes: initial instability and divergence.

The readiness semantics for closed process terms is a function:

$$\mathcal{R}^*[\![\cdot]\!] : CProc \to \text{DOM}_{\mathcal{R}}$$

which can be given both both operational and denotational definitions. There is a natural partial ordering on the semantic domain $\text{DOM}_{\mathcal{R}}$ denoted by $\sqsubseteq$.

Processes. Let *Comm* be an infinite set of *communications*. Together with a special symbol τ representing a hidden or internal activity of a process it will constitute the set of process *actions*.

Processes will be defined as a subset of the set of *recursive terms*. Recursive terms have the following abstract syntax

$$P ::= stop_A \mid a.P \mid P + Q \mid P \parallel Q \mid P[b/a] \mid P\backslash a \mid X \mid \mu X.P$$

where P and Q range over recursive terms, X ranges over process variables, A over finite sets of communications, a and b over communications. Essentially, the process operators in the syntax above are those of CSP apart from plus, which is given semantics as in CCS.

Iterated applications of renaming and hiding will be abbreviated as follows: $P[\bar{b}/\bar{a}]$ will stand for $P[b_1/a_1]\dots[b_n/a_n]$, where $\bar{b} = b_1,\dots,b_n, \bar{a} = a_1,\dots,a_n$; $P\backslash A$ will denote $P\backslash a_1 \dots \backslash a_n$, where $A = \{a_1,\dots a_n\}$. Notation $\{\dots/\dots\}$ is used for syntactic substitutions.

With each recursive term P we associate its *(communication) alphabet* $\alpha(P)$. It is assumed that each process variable X has an associated alphabet $\alpha(X)$. The communication alphabets of remaining terms are defined inductively as follows:

$$\begin{aligned}
\alpha(stop_A) &= A & \alpha(P[b/a]) &= (\alpha(P) - \{a\}) \cup \{b\} \\
\alpha(a.P) &= \{a\} \cup \alpha(P) & \alpha(P\backslash a) &= (\alpha(P) - \{a\}) \\
\alpha(P + Q) &= \alpha(P) \cup \alpha(Q) & \alpha(\mu X.P) &= \alpha(X) \cup \alpha(P) \\
\alpha(P \parallel Q) &= \alpha(P) \cup \alpha(Q)
\end{aligned}$$

Definition 1. We say that a recursive term P is a *process term* (or simply, a process) if it satisfies the following conditions: *(1)* P is action guarded, i.e. appearances of process variables in recursive subprocesses of P are guarded, *(2)* every subterm $a.Q$ of P satisfies $a \in \alpha(Q)$, *(3)* every subterm $Q + R$ of P satisfies $\alpha(Q) = \alpha(R)$, *(4)* every subterm $\mu X.Q$ of P satisfies $\alpha(X) = \alpha(Q)$.

The set of all process terms will be denoted by *Proc*. A process term is *closed* if it does not contain free process identifiers. The set of all closed process terms will be denoted by *CProc*.

Semantics. Processes are given *readiness semantics*. It describes process behaviour in terms of *ready pairs* $(tr, \mathcal{F})$, where tr is finite sequence of communications called *trace*, and $\mathcal{F}$, called *ready set*, is a set of communications in which a process is ready to engage after "performing" the trace tr. Besides the ready pairs the semantic domain $\text{DOM}_{\mathcal{R}}$ contains elements representing undesirable behaviours of processes: initial instability and divergence.

The readiness semantics for closed process terms is a function:

$$\mathcal{R}^*[\![\cdot]\!] : CProc \to \text{DOM}_{\mathcal{R}}$$

which can be given both both operational and denotational definitions. There is a natural partial ordering on the semantic domain $\text{DOM}_{\mathcal{R}}$ denoted by $\sqsubseteq$.

Example 1. Below we define a counter of capacity 1. Upon communication *up* (*dn*) the value stored in the counter is increased (decreased) by 1. Symbol $\equiv$ stands for the definitional equality.

$$cnt_1 \equiv \mu X \cdot up.dn.X$$

$\alpha(cnt_1) = \{up, dn\}$, $\mathcal{R}^*[cnt_1] = (\alpha(cnt_1), R)$, where R, is the set of ready pairs of cnt_1:

$$R = \{(tr, \mathcal{F}) \mid \exists k \geq 0 \ (tr = (up \, dn)^k \wedge \mathcal{F} = \{up\}) \vee (tr = (up \, dn)^k up \wedge \mathcal{F} = \{dn\})\}$$

Readiness specifications. Processes are specified by formulas of a many-sorted predicate logic, called *readiness logic*, which has sorts of traces and ready sets, among others, and allows one to define sets of ready pairs in a direct manner. Readiness logic is fairly expressive, for example, regular sets over *Comm* can be expressed in it.

A *readiness specification* is a formula of readiness logic, in which at most the distinguished trace variable h and the distinguished ready set variable F are free, and which satisfies some natural syntactic restrictions. Since a specification has at most two free variables h and F, it can be interpreted semantically with respect to an evaluation $(tr, \mathcal{F})$ of h and F. We will write

$$(tr, \mathcal{F}) \models S \quad \text{and} \quad \models S$$

if S is satisfied by, respectively, a ready pair $(tr, \mathcal{F})$ and all ready pairs.

A specification determines a set of ready pairs which satisfy it. Moreover, with every specification S its alphabet, denoted by $\alpha(S)$, can be associated. These two facts allow one to define *readiness semantics of readiness specifications*

$$\mathcal{R}^*[\cdot] : Spec \to \mathrm{DOM}_\mathcal{R}$$

where *Spec* denotes the set of specifications.

Example 2. CNT_n below is a formula of readiness logic that instantiated with any natural number n gives a specification of a counter of capacity n. The alphabet of CNT_n consists of communications *up* and *dn*, which have the same meaning as in Example 1. The symbol $up\#h$ ($dn\#h$) denotes the number of *up* (*dn*) communications in trace h.

$$CNT_n \equiv (0 \leq up\#h - dn\#h \leq n) \wedge (up\#h - dn\#h < n \ \Rightarrow \ up \in F) \wedge$$
$$(up\#h - dn\#h > 0 \ \Rightarrow \ dn \in F)$$

By instantiating n to 1 or 2 in CNT_n we get CNT_1 and CNT_2 that appeared in Introduction.

Satisfaction relation. Since processes as well as specifications are interpreted semantically over readiness domain, the satisfaction relation has a simple semantic definition:

$$P \text{ sat } S \text{ iff } \mathcal{R}^*[\![P]\!] \sqsupseteq \mathcal{R}^*[\![S]\!],$$

which means that

- process P and specification S have the same alphabets, i.e. $\alpha(P) = \alpha(S)$,
- $\mathcal{R}^*[\![P]\!]$ contains only ready pairs (no divergence or initial instability) and for each ready pair $(tr, \mathcal{F})$ in $\mathcal{R}^*[\![P]\!]$ we have $(tr, \mathcal{F}) \models S$.

Proof system for sat via mixed terms. *Mixed terms* are a mixture of readiness specifications and process terms and are formally defined by allowing specifications in recursive terms

$$P ::= S \mid stop_A \mid a.P \mid \ldots$$

and generalizing the definitions of alphabets and processes in a straightforward way.

Since processes and specifications are interpreted in the same semantic domain, there is no difficulty in extending readiness semantics to mixed terms. The satisfaction relation on mixed terms, denoted by $\equiv>$, is defined as

$$M_1 \equiv> M_2 \text{ iff } \mathcal{R}^*[\![M_1]\!] \sqsupseteq \mathcal{R}^*[\![M_2]\!],$$

where M_1, M_2 are mixed terms. The relation **sat** can be viewed as a special case of $\equiv>$ restricted to the pairs $P \equiv> S$ where P is a process and S is a specification.

In the final chapter of [5] a proof system for $\equiv>$ is given, which obviously allows us to derive assertions of the form P **sat** S, but mixed terms will be used in the derivations.

3 Proof system for sat

The first step towards introducing parametricity into the formalism proposed by Olderog is to provide a proof system for **sat**, which does not appeal to mixed terms. In Table 1 we propose such a proof system, whose rules bear a clear resemblance to Olderog's proof rules for $\equiv>$.

The judgements of the proof system for **sat** have the form

$$\Gamma \vdash P \text{ sat } S, \tag{4}$$

where $P \in Proc$, $S \in Spec$ and Γ is a *set of assumptions*,

$$\Gamma = \{X_1 \text{ sat } S_1, \ldots, X_n \text{ sat } S_n\}.$$

The sets of assumptions are needed to handle recursion.

(Consequence) $\dfrac{\models S \Rightarrow T \quad \Gamma \vdash P \textbf{ sat } S}{\Gamma \vdash P \textbf{ sat } T}$, where $\alpha(S) = \alpha(T)$

(Deadlock) $\Gamma \vdash stop_A \textbf{ sat } h\!\downarrow_A \leq \varepsilon$

(Prefix) $\dfrac{\models S\{\varepsilon/h, \{a\}/F\} \quad \Gamma \vdash P \textbf{ sat } S\{ah/h\}}{\Gamma \vdash a.P \textbf{ sat } S}$, where $a \in \alpha(S)$

(Choice) $\dfrac{\Gamma \vdash P \textbf{ sat } S \quad \Gamma \vdash Q \textbf{ sat } T}{\Gamma \vdash P + Q \textbf{ sat } \varphi_+(S,T)}$, where $\alpha(S) = \alpha(T)$

(Parallel) $\dfrac{\Gamma \vdash P \textbf{ sat } S \quad \Gamma \vdash Q \textbf{ sat } T}{\Gamma \vdash P \parallel Q \textbf{ sat } \varphi_\parallel(S,T)}$

(Renaming) $\dfrac{\Gamma \vdash P \textbf{ sat } S}{\Gamma \vdash P[\bar{b}/\bar{a}] \textbf{ sat } \varphi_{[]}(S,\bar{b},\bar{a})}$

(Hiding) $\dfrac{\begin{array}{l} \models S\{F - B/F\} \Rightarrow T \\ \models S \Rightarrow \forall b \in B \ h\!\downarrow_A \neq b \\ \models S \Rightarrow \exists n \forall b_1 \ldots b_n \in B^* \ \neg\exists F \ S\{h\,b_1 \ldots b_n/h\} \\ \Gamma \vdash P \textbf{ sat } S \end{array}}{\Gamma \vdash P\backslash B \textbf{ sat } T}$,

where $\alpha(T) = \alpha(S) - B$ and $\alpha(S) = A$

(Recursion) $\dfrac{\Gamma, \ X \textbf{ sat } S \vdash P \textbf{ sat } S}{\Gamma \vdash \mu X.P \textbf{ sat } S}$,

where $\mu X.P$ is communication guarded and $\alpha(S) = \alpha(X)$

(Assumption) $\Gamma, \ X \textbf{ sat } S \vdash X \textbf{ sat } S$, where $\alpha(S) = \alpha(X)$

$\varphi_+(S,T)$, $\varphi_\parallel(S,T)$ and $\varphi_{[]}(S,\bar{b},\bar{a})$ abbreviate formulas that appear also in Olderog's proof system for $\equiv\!\!>$ and are defined below

$\varphi_+(S,T) \equiv (h\!\downarrow_A = \varepsilon \Rightarrow S \wedge T) \wedge (h\!\downarrow_A \neq \varepsilon \Rightarrow S \vee T)$, where $A = \alpha(S) = \alpha(T)$
$\varphi_\parallel(S,T) \equiv \exists G \exists H \, (S\{G/F\} \wedge T\{H/F\} \wedge G \cap H \subseteq F \wedge (G \cup H) - (\alpha(S) \cap \alpha(T)) \subseteq F)$
$\varphi_{[]}(S,\bar{b},\bar{a}) \equiv \exists t \, \exists G \, (S\{t/h, G/F\} \wedge t[\bar{b}/\bar{a}] = h\!\downarrow_A \wedge G[\bar{b}/\bar{a}] \subseteq F)$, where $A = \alpha(S)\{\bar{b}/\bar{a}\}$

Projection operator $\downarrow_A$ and communication guardedness are defined as in [5], ε is an empty trace.

Table 1.

The theorems below relate the proposed proof system for **sat** to Olderog's proof system for ≡>. It will follow that the special case of (4) with empty Γ coincides with Olderog's "$P \Longrightarrow S$ is derivable".

To state the theorems we need the following auxiliary notation: for a process term P and $\Gamma = \{X_1 \text{ sat } S_1, \ldots, X_n \text{ sat } S_n\}$, by $P\{\Gamma\}$ we shall denote the mixed term $P\{S_1/X_1, \ldots, S_n/X_n\}$, obtained from P by replacing free occurrences of variables $X_1, \ldots, X_n$ by specifications $S_1, \ldots, S_n$. We will also write $\vdash P \Longrightarrow S$ to denote that $P \Longrightarrow S$ is derivable in Olderog's proof system.

Theorem 2. *If* $\Gamma \vdash P$ **sat** S *then* $\vdash P\{\Gamma\} \Longrightarrow S$.

Proof. Follows by by induction on the derivation of $\Gamma \vdash P$ **sat** S.

Corollary 3. *Proof system for* **sat** *is sound, that is,*

$$\text{if } \vdash P \text{ \textbf{sat} } S \text{ then } \mathcal{R}^*[P] \sqsupseteq \mathcal{R}^*[S]$$

Proof. Follows from Theorem 2 and soundness of Olderog's proof system for ≡>.

Theorem 4. *Let* P *be a process and* Γ *a set of assumptions such that* $P\{\Gamma\}$ *is a closed term. If* $\vdash P\{\Gamma\} \Longrightarrow S$ *for a specification* S *then* $\Gamma \vdash P$ **sat** S.

Proof. The proof is long and omitted (see [7]).

We end this section with the following useful observation, which can be proved by straightforward induction on the derivation of $\Gamma \vdash P$ **sat** S

Proposition 5. *If for a recursive term* P *a judgement* $\Gamma \vdash P$ **sat** S *can be derived for some* S *and* Γ, *then* $P \in Proc$, *that is,* P *satisfies the requirements of Definition 1.*

4 Parametric Processes and Specifications

The triple

$$(CProc, Spec, \text{sat}) \tag{5}$$

can be seen as a specification system that does not support parametric constructions, in the sense of this notions explained in Introduction. We lift (5) to the parametric case by extending the process algebra with abstraction and application primitives, in a manner resembling the typed lambda calculus, where types are replaced with specifications.

First we introduce a syntactic class *PSpec* of *parametric specifications* ranged over by σ and defined by the following abstract syntax:

$$\sigma ::= S \mid \sigma \to \sigma$$

Intuitively, specification $\sigma_1 \to \sigma_2$ describes a set of parametric processes, each of which when supplied with an argument, a process that satisfies (possibly parametric) specification σ_1, yields a process that satisfies specification σ_2.

The syntax of recursive terms is extended with two new clauses

$$P ::= \ldots \mid \lambda X : \sigma . P \mid PP$$

The set of *parametric processes* will be defined as a subset of so extended recursive terms. In the case of nonparametric processes, syntactic restrictions enumerated in Definition 1 were used to distinguish the set *Proc* of processes in the set of (nonparametric) recursive terms. However, as it has been noted in Proposition 5, these restrictions are encoded also in proof system for **sat**. In the parametric case, it is more convenient to adopt a counterpart of Proposition 5 as a definition rather than separately formulate syntactic restrictions that define the desired set of parametric processes. This is the approach we are going to take, but first we need to refine some notions.

So far the alphabets of processes and specifications were just finite subsets of *Comm*, the set of communications. Alphabets of parametric processes (and specifications), we shall call them *hyper-alphabets* in the sequel, have to be more complex objects, indicating the alphabets of possible arguments of a process. We define the set $\mathcal{H}$ of all hyper-alphabets as follows:

$$
\begin{aligned}
\mathcal{H}_0 &= \text{set of finite subsets of } Comm \\
\mathcal{H}_{n+1} &= \{(A, B) \mid A, B \in \mathcal{H}_n\} \cup \mathcal{H}_n \\
\mathcal{H} &= \bigcup_{n=0}^{\infty} \mathcal{H}_n
\end{aligned}
$$

The hyper-alphabet $\alpha^*(\sigma)$ of a specification σ is defined recursively by the following clauses:

$$
\begin{aligned}
\alpha^*(\sigma) &= \alpha(S), & \text{if } \sigma = S \\
\alpha^*(\sigma) &= (\alpha^*(\sigma_1), \alpha^*(\sigma_2)), & \text{if } \sigma = \sigma_1 \to \sigma_2
\end{aligned}
$$

The hyper-alphabet $\alpha^*(P)$ of a recursive term P is defined by the clauses given below. If it is not possible to associate an alphabet with a term using these defining clauses, $\alpha^*(P)$ is considered to be undefined. Just as in the nonparametric case we assume that every process variable X has an associated hyper-alphabet $\alpha^*(X)$.

$$
\begin{aligned}
\alpha^*(stop_A) &= A \\
\alpha^*(a . P) &= \alpha^*(P) \cup \{a\} & \text{if } \alpha^*(P) \in \mathcal{H}_0 \\
\alpha^*(P + Q) &= \alpha^*(P) \cup \alpha^*(Q) & \text{if } \alpha^*(P), \alpha^*(Q) \in \mathcal{H}_0 \\
\alpha^*(P \| Q) &= \alpha^*(P) \cup \alpha^*(Q) & \text{if } \alpha^*(P), \alpha^*(Q) \in \mathcal{H}_0 \\
\alpha^*(P[b/a]) &= (\alpha^*(P) - \{a\}) \cup \{b\} & \text{if } \alpha^*(P) \in \mathcal{H}_0 \\
\alpha^*(P \backslash a) &= (\alpha^*(P) - \{a\}) & \text{if } \alpha^*(P) \in \mathcal{H}_0 \\
\alpha^*(\mu X . P) &= \alpha^*(P) \cup \alpha^*(X) & \text{if } \alpha^*(P) \in \mathcal{H}_0 \\
\alpha^*(\lambda X : \sigma . P) &= (\alpha^*(X), \alpha^*(P)) & \text{if } \alpha^*(X) = \alpha^*(\sigma) \\
\alpha^*(PQ) &= B & \text{if } \alpha^*(P) = (A, B) \text{ and } \alpha^*(Q) = A
\end{aligned}
$$

Note that $\alpha^*(P) = \alpha(P)$ for $P \in Proc$ and that alphabets of terms such as $(\lambda X : \sigma . P) + Q$, where a process algebra operator $+$ is applied to a higher order term, are not defined.

In the sequel we will consider only those recursive terms, whose hyper-alphabets are defined, and the notion of recursive terms will refer to such terms from now on. The set of so understood recursive terms can be stratified into two subsets: the subset of terms whose alphabet is in $\mathcal{H}_0$ and to which process constructors $+$, $a.$, $\|$, $\backslash$, $[]$ and μ can be applied, and the subset of higher order terms, on which only abstractions and applications can be performed. Note however, that the hyper-alphabet of an application PQ, where P is a higher order term, can belong to $\mathcal{H}_0$ so process algebra operators can be applied to such PQ.

The syntactic notion of communication guardedness that appears as a side condition for Recursion rule of Table 1 also needs to be redefined. We extend Olderog's definition of communication guardedness by requiring that for every recursive subterm $\mu X.P$ of a communication guarded term no free occurrence of X in P lies within an application RQ, thus restricting the interference between process recursion and parametricity.

Now we can formulate the proof system for the satisfaction relation **sat** extended to the parametric case. The judgements will have the form

$$\Gamma \vdash P \text{ \bf sat } \sigma$$

where P is a (parametric) recursive term and the set of assumptions Γ is extended in the obvious way to contain assumptions of the form X **sat** σ.

The proof system contains all the rules collected in Table 1, which were used for nonparametric processes, but this time we assume that S and T range over nonparametric specifications, P and Q range over parametric recursive terms, Γ is extended as explained above, and the side conditions concerning alphabets refer to hyper-alphabets, i.e. α is replaced with α^*. The side condition of communication guardedness in Recursion rule refers now to the modified version of this notion.

Moreover, the proof system of Table 1 is extended with the following rules, which are essentially λ-introduction and λ-elimination rules of the proof system for Church typed λ calculus (see for example [1]), plus a modified Assumption rule, which replaces the Assumption rule from Table 1.

(Assumption) $\qquad \Gamma, X \text{ \bf sat } \sigma \vdash X \text{ \bf sat } \sigma, \qquad$ where $\alpha^*(X) = \alpha^*(\sigma)$

(Abstraction) $\qquad \dfrac{\Gamma, X \text{ \bf sat } \sigma_1 \vdash P \text{ \bf sat } \sigma_2}{\Gamma \vdash \lambda X : \sigma_1 . P \text{ \bf sat } \sigma_1 \to \sigma_2}, \qquad$ where $\alpha^*(X) = \alpha^*(\sigma_1)$

(Application) $\qquad \dfrac{\Gamma \vdash P \text{ \bf sat } \sigma_1 \to \sigma_2 \qquad \Gamma \vdash Q \text{ \bf sat } \sigma_1}{\Gamma \vdash PQ \text{ \bf sat } \sigma_2}$

Definition 6. The set of *parametric processes* PProc consists of those recursive terms P, for which $\Gamma \vdash P$ **sat** σ can be derived for some Γ and σ. *CPProc*, the set of *closed parametric processes*, is defined as the set of those parametric processes that have no free variables.

Note that the above definitions and Proposition 5 imply that $Proc \subseteq PProc$ and $CProc \subseteq CPProc$.

We have just defined all components of a parametric specification system

$$(CPProc, PSpec, \mathbf{sat})$$

In this report we do not provide semantical account of parametric processes. This can be done by extending the readiness denotational semantics so that parametric processes are interpreted as process-valued functions on processes. Such a construction, which is an adaptation of a general procedure described in [8], will be reported separately together with soundness proof of the proof system for **sat** . Here, Theorem 7 below, provides syntactic arguments indicating that the proposed formalism is sound. In Theorem 7, we use the arrow $\xrightarrow{\beta^*}$ to denote β-reducibility in a context and in possibly many steps, where the notion of β-reduction is defined in a standard manner: $(\lambda X : \sigma \,.\, P)\, Q \xrightarrow{\beta} P\{Q/X\}$.

Theorem 7.

(a) If $\vdash P$ **sat** S then there exists $P' \in Proc$ such that $P \xrightarrow{\beta^*} P'$. For every such P' $\mathcal{R}^*[P'] \sqsupseteq \mathcal{R}^*[S]$, i.e. $\vdash P'$ **sat** S is justified semantically.

(b) If $\vdash P$ **sat** $\sigma_1 \to \sigma_2$ then for every Q , whenever $\vdash Q$ **sat** σ_1 then $\vdash PQ$ **sat** σ_2.

Point (a) of Theorem 7 ensures that if a parametric process P can be proved to satisfy a nonparametric specification S then, up to β equivalence, this assertion is semantically justified. Point (b) relates satisfaction of higher order terms to lower order ones, which is a natural requirement if parametric processes are interpreted as functions.

The next proposition shows that there is an exact match between the proof systems for $\equiv>$ on mixed terms and the proof system for **sat** on a restricted class of processes of the form

$$\lambda X_1 : S_1 \ldots \lambda X_n : S_n \,.\, P \tag{6}$$

This class of parametric processes with no higher-order parameters corresponds to mixed terms of [5] as (6) can be syntactically translated into a mixed term $P\{S_1/X_1, \ldots, S_n/X_n\}$ and, conversely, a mixed term P can be translated into a process having the form (6) by replacing every occurrence of any specification S_i appearing in P with a fresh variable X_i.

Proposition 8. $\vdash \lambda X_1 : S_1 \ldots \lambda X_n : S_n \,.\, P$ **sat** $S_1 \to \cdots \to S_n \to S$ if and only if $\vdash P\{S_1/X_1, \ldots, S_n/X_n\} \equiv> S$.

Proof. It can be shown by induction that

$$\vdash \lambda X_1 : S_1 \ldots \lambda X_n : S_n \,.\, P \text{ **sat** } S_1 \to \cdots \to S_n \to S$$

if and only if

$$S_1 \text{ **sat** } X_1, \ldots, S_n \text{ **sat** } X_n \vdash P \text{ **sat** } S.$$

By Theorems 2 and 4 this is equivalent to $\vdash P\{S_1/X_1, \ldots, S_n/X_n\} \equiv> S$.

5 Examples

Now we are in a position to fill in the details of the example that was sketched in Introduction. Part of the work has been already done in Examples 1 and 2. The processes $double_1$ and cnt_2 can be defined as follows:

$$double_1 \equiv \lambda X : CNT_1 . (X[link/dn] \parallel X[link/up]) \backslash link$$
$$cnt_2 \equiv double_1 \, cnt_1$$

Using the proof system for parametric processes we may:

(i) derive the satisfaction assertion $\vdash cnt_1$ **sat** CNT_1,

(ii) independently derive the assertion $\vdash double_1$ **sat** $CNT_1 \rightarrow CNT_2$.

By virtue of Application rule, these imply $\vdash cnt_2$ **sat** CNT_2.

As another example we give a process qdr_1 which can be shown to satisfy (3).

$$qdr_1 \equiv \lambda Y : CNT_1 \rightarrow CNT_2 . \lambda X : CNT_1 . ((Y\,X)[link/up] \parallel (Y\,X)[link/dn]) \backslash link$$

A larger example of digital gates modelling, is developed in [7].

Acknowledgements. We would like to thank Thorsten Altenkirch, Kārlis Černās and Uno Holmer for helpful comments on this work.

References

1. H.P. Barendregt, *Lambda Calculi with Types*, in: Handbook of Logic in Computer Science, vol. 2 (S. Abramsky, Dov M. Gabbay, T. S. E. Maibaum, Eds.) pp. 117–309, Calderon Press, Oxford 1992.
2. S. Holmström, A refinement calculus for specification in Hennessy-Milner logic with recursion, *Formal Aspects of Computing*, vol. 1 (3). pp. 242–272, 1989.
3. K. Larsen and R. Milner, A Complete Protocol Verification using Relativised Bisimulations, R 86-12, Institute of Electrical Systems, Aalborg University Centre, 1986.
4. R. Milner, *Communication and Concurency*, Prentice Hall, 1989.
5. E.-R. Olderog, *Nets, Terms and Formulas*, Cambridge University Press, Cambridge 1991.
6. E.-R. Olderog, Towards a design calculus for communicating programs, in: *Proc. CONCUR '91*, LNCS, Springer, 1991.
7. W. Pawłowski, P. Pączkowski, S. Sokołowski, A specification system for parametric concurrent processes, electronically available as `ftp://ipipan.gda.pl/pub/local_users/Pawlowski-W./parametric.ps.gz`.
8. S. Sokołowski, Requirement specifications and their realizations: toward a unified framework. Submitted to *Theoretical Computer Science*. A preliminary version has appeared as: The GDM approach to specifications and their realizations. Part I: Specification systems. ICS PAS Report 797, Warszawa 1995.

On Saturation with Flexible Function Symbols

Regimantas Pliuškevičius

Institute of Mathematics and Informatics
Akademijos 4, Vilnius 2600, LITHUANIA
email: regis@ktl.mii.lt

Abstract. The paper extends the similarity saturation calculus [8] by including flexible function symbols. The most attractive property of the similarity saturation calculus consists in that it allows to build derivations constructively both for a finitary complete and finitary incomplete first order linear temporal logic. The saturation calculus contains the so-called alternating rule and this rule splits off the logical part from the temporal one in the quasi-primary sequents. If the saturation process involves unification and flexible function symbols, then the alternating rule includes (in the temporal part of the rule) renaming of variables and a replacement of flexible function symbols by rigid ones. The employment of flexible symbols means the introduction of some kind of constraints in the saturation process.

1 Introduction

It is well known that temporal logics became a very important tool for solving various problems of computer science. It is also known (see, e.g., [1, 11]) that the first order linear temporal logic is incomplete, in general. In [5] it is proved that it becomes complete after adding the following ω-type rule of inference;

$$\frac{\Gamma \to \Delta, A; \ldots; \Gamma \to \Delta, \bigcirc^k A, \ldots}{\Gamma \to \Delta, \Box A} \ (\to \Box_\omega),$$

where $\bigcirc$ is a temporal operator "next", $\Box$ is a temporal operator "always". In some cases the first order linear temporal logic is finitary complete (see e.g. [6, 7]). The main rule of inference in this case is the following:

$$\frac{\Gamma \to \Delta, R; R \to \bigcirc R; R \to A}{\Gamma \to \Delta, \Box A} \ (\to \Box).$$

This rule corresponds to the induction axiom: $A \wedge \Box (A \supset \bigcirc A) \supset \Box A$. The formula R is called an *invariant formula* and the main problem (in the finitary case) is to find a way to construct this formula. The process of derivability in the finitary case (i.e., with the rule of inference $(\to \Box)$) differs in principle from the one in the infinitary case (i.e., with the rule of inference $(\to \Box_\omega)$). Moreover, it is not clear in which case we can apply $(\to \Box)$ and when we have to apply $(\to \Box_\omega)$.

In [6, 7] simple (i.e. without unification and similarity) saturation calculi for a restricted first order linear temporal logic, containing $\bigcirc$ "next" and $\Box$ ("always")) were described. The saturation calculi were developed specially for the temporal logics containing induction-like postulates. Instead of induction-type postulates the saturation calculi contain a deductive saturation procedure, indicating some form of regularity in the derivations of the logic.

The saturation calculi involve the process of calculation of some sequents called k-th resolvents analogously as in the resolution-like calculus. Therefore, induction-like rules of inference are replaced by a constructively defined deductive process of finding k-th resolvents. The main rule of inference for calculating k-th resolvents in the saturation calculus is the so-called *subsumption rule* (based on the structural rule *"weakening"*) which allows to remove not the atomic formula (as in the traditional "clausal" resolution rule) but the whole sequent from the obtained set of so-called *primary sequents* (analogous to the clauses of resolution-type calculi). Namely, we can delete from the set of primary sequents $S_1, \ldots, S_n$ a sequent S_i such that is "subsumed" by S_k $(1 \leqslant k \leqslant n, \ k \neq i)$ (in symbols $S_k \succcurlyeq u S_k$), i.e., S_i can be obtained from S_k by means of "unified" structural rule "weakening" (coinciding with the ordinary one up to unification). It should be noted that any application of the subsumption rule is deterministic, opposite to the traditional resolution-like calculus where the application of the resolution rule is highly non-deterministic because of the essential non-determinism in the choice of literals and clauses. In the traditional clausal resolution-type calculus preparation steps are taken only once, i.e., reducing a formula to the clause form that is not destroyed in the derivation process. In the saturation calculi the form of primary sequents (analogous to the clauses) is destroyed in the derivation process and we again must reduce sequents to the primary ones. Derivations in the traditional resolution-like calculi are organized as forward reasoning and therefore they are not goal-oriented. Derivations realized by auxiliary rules of inference of saturation calculi with a view to reduce sequents to the primary ones are organized as backward reasoning and therefore are goal-oriented. These auxiliary rules of inference contain the so-called *alternating rule* (similar to the seperation principle [4]) which is bottom-up applied to another canonical form of sequents, i.e., so-called *quasi-primary* sequents, and this rule splits off the logical part from the temporal one in the quasi-primary sequents. The application of the alternating rule is non-deterministic, in general, i.e., we can continue the derivation of the logical part or the temporal one. If the saturation process involves unification and flexible function symbols, then the alternating rule includes (in the temporal part of the rule) renaming of variables and a replacement of flexible function symbols by rigid ones. The employment of flexible symbols means the introduction of some kind of constraints in the saturation process.

The results of [6, 7] were extended in [8], where the similarity saturation calculus for the unrestricted first order temporal logic was described. The similarity saturation is very powerful because it allows to get derivations of valid sequents both in a finitary and infinitary case. In the finitary case we obtained the set of saturated sequents, showing that "almost nothing new" can be obtained continuing the derivation process. In this case using only simple or unified saturation (i.e., based only on unification) we get a derivation of the empty k-th resolvent (in symbols: $Re^k(S) = \varnothing$). In the infinitary case, the similarity saturation calculus generates an infinite set of saturated sequents, showing that only "similar" sequents can be obtained continuing the derivation process. In this case we say the empty ω-resolvent of the given sequent S to be obtained (in symbols: $Re^\omega(S) = \varnothing$). It must be stressed that the way of obtaining saturation in the infinitary case is, in principle, the same as in the finitary case. *The most attractive property of the similarity saturation calculus consists in that it allows to build derivations constructively both for a finitary complete and finitary incomplete first order linear temporal logic.* In the finitary incomplete case the similarity saturation has some resemblance with a ω-type rule of inference. But it is only an external analogy. *The main difference between the similarity saturation and ω-like rules of inference is the following: the generation of inductive hypothesis using the similarity saturation are obtained automatically, while using ω-like rule of inference the generation of inductive hypothesis is not automatical, in general.*

In the unrestricted first order case, the initial sequent must be reduced (or represented a priori) to Fisher's normal form [3]. The notions related to the similarity saturation are very simple and natural ones. Namely, we say that the sequent S_1 is similar to S_2, if $S_1\sigma \succcurlyeq uS_2$; σ is called a similarity substitution. On the left-hand side of similarity substitutions some constants will be allowed apart from variables. Let $t_1, \ldots, t_m$ be the right-hand side of similarity substitution, then σ^n is the similarity substitution the left-hand side of which is the same as in σ, and the right-hand side is the terms $t_1^n, \ldots, t_m^n$. The foundation of the fact that $Re^\omega(S_0) = \varnothing$, in principle is nothing more than a very effective application (using a similarity principle) of the usual induction principle. Using the similarity principle we get (if possible) two sets $\{S_1, \ldots, S_n\}$ $(n \geqslant 1)$ and $\{S_1', \ldots S_m'\}$ of different primary sequents such that $S_i\sigma \succeq uS_j'$ $(1 \leqslant i \leqslant n; \ 1 \leqslant j \leqslant m)$, where σ is a similarity substitution (basis of induction). Using the substitution σ^n (where n is a meta-variable for a natural number) we automatically get the set $\{S_1^*, \ldots, S_n^*\}$ $(n \geqslant 1)$ of primary sequents where $S_i^* = S_i\sigma^n$ (induction hypothesis). Repeating the similarity saturation process constructed in the basis we get (if possible) the set $\{S_1'', \ldots, S_m''\}$ of primary sequents such that $\forall i (1 \leqslant i \leqslant n) \ S_i^*\sigma \succcurlyeq uS_j''$ $(1 \leqslant i \leqslant n, 1 \leqslant j \leqslant m)$ (step of induction). It means that $Re^\omega(S) = \varnothing$.

The purpose of this paper is to extend [8] by including flexible function symbols. The completeness issues of the first order temporal logic with flexible function symbols were considered also in [9, 10]. The tableaux deductive systems not containing an induction-like rule of inference for a propositional linear temporal logic were proposed in [2, 12], replacing the rule ($\rightarrow \Box$) by some graph pruning procedure, that does not involve an explicit method for generating invariant formulas.

2 Description of the infinitary sequent calculus $G^*_{L\omega}$ with implicit quantifiers

Saturation calculi Sat_λ ($\lambda \in \{\emptyset, \omega\}$ are founded with the help of the infinitary calculus $G^*_{L\omega}$ containing the ω-type rule of inference and not containing rules of inference for quantifiers in explicit form.

Definition 2.1 (elementary and atomic formulas, formulas, language, flexible, rigid symbols, associated rigid symbols). Elementary formulas are expressions of the type $P(t_1, \ldots, t_n)$, where P is an n-place predicate symbol, $t_1, \ldots, t_n$ are arbitrary terms. Formulas are defined starting from the elementary formulas by means of logical symbols and temporal operators $\bigcirc$ and $\Box$ as usual. The formula $\bigcirc \ldots \bigcirc A$ (n-times "next") will be abbreviated as $\bigcirc^n A$. A language is a countable collection of predicate and function symbols that may occur in formulas. The language includes flexible predicate and function symbols (i.e., symbols that may change the value with time) and rigid predicate and function ones (with time-independent meanings). It is assumed that for each flexible n-place ($n \geqslant 0$) function symbol f^n there exists an infinite sequence of rigid function symbols $f^n_0, f^n_1, \ldots, f^n_k, \ldots$ such that values of f^n and f^n_i concide at the i-th state $i \in \omega$ ($\omega := \{0, 1, \ldots\}$); f^n_i is called rigid function symbol associated with f^n [10]. It is assumed that all bounded variables are rigid ones.

Definition 2.2 (sequent). A *sequent* is an expression of the form $\Gamma \rightarrow \Delta$, where Γ, Δ are arbitrary finite sets (i.e., not sequences or multisets) of formulas.

Definition 2.3 (normal form, N-sequents). A sequent S will be in the *normal form*, (in brief: *N-sequent*), if
1) S contains neither positive occurrences of $\forall$ nor negative occurrences of $\exists$ (skolemization condition);
2) there are no occurrences of $\Box$ in the scope of quantifiers (weak miniscopization condition);
3) all negative (positive) occurrences of $\Box A$ in S have the form $\Box \forall \bar{x} A(\bar{x})$ ($\Box \exists \bar{x}$ $A(\bar{x})$, respectively), where $\bar{x} = x_1, \ldots, x_n$ ($n \geqslant 0$) and $A(\bar{x})$ does not contain quantifiers (temporal prenex-form condition);
4) all formulas not containing $\Box$ have the form $Q\bar{x}A(\bar{x})$ ($Q \in \{\forall, \exists\}$, $\bar{x} = x_1, \ldots, x_n$ ($n \geqslant 0$)), where $A(\bar{x})$ does not contain quantifiers.

Theorem 2.1. Let S be a sequent, then there exists S^+ such that S^+ is an N-sequent and S is satisfiable iff S^+ is satisfiable.
Proof: follows from [3].

Definition 2.4 (quantifier-free form of a N-sequent). Let S be a N-sequent (we can assume, without loss of generality, that in S different occurrences of quantifiers bind different variables), then a sequent S^* obtained from S by dropping all occurrences of $Q\bar{x}$ ($Q \in \{\forall, \exists\}$) will be called a *quantifier-free form* of N-sequent.

Remark 2.1: It is easy to verify that the unique (up to the renaming of variables) quantifier-free form S^* corresponds to a N-sequent S and vice versa.

In this paper we consider only N-sequents in quantifier-free form.

Definition 2.5 (substitution, formula and sequent with substitution, LHS and RHS of substitution). A *substitution* is an expression of the form $\sigma = (x_1 \leftarrow t_1; \ldots; x_n \leftarrow t_n)$, where x_i is a variable, t_i is a term ($1 \leqslant i \leqslant n$). If $\forall i$ ($1 \leqslant i \leqslant n$) t_i is a variable, then such a substitution is called a simple one. Explicitly indicated variables (terms) $x_1, \ldots, x_n$ ($t_1, \ldots, t_n$) in σ will be called *left hand side (right hand side*, correspondingly) (in short: *LHS and RHS*) of substitution σ. If A is a formula, $\sigma = (x_1 \leftarrow t_1; \ldots; x_n \leftarrow t_n)$ is a substitution, then the expression $A\sigma$ will be called a *formula with a substitution* what means that all occurrences of x_i are replaced by t_i ($1 \leqslant i \leqslant n$). The sequent, each member of which is a formula or a formula with a substitution σ, will be called a *sequent with a substitution* and denoted by $S\sigma$.

Definition 2.6 (axiomatic substitutions). Let A be an atomic formula, $\bar{p} = p_1, \ldots, p_n$; $\bar{q} = q_1, \ldots, q_n$; p_i, q_i be some terms, let $S = \Gamma, A(\bar{p})\sigma_1 \to A(\bar{q})\sigma_2, \Delta = \Gamma, A(\bar{t}) \to A(\bar{t}), \Delta$ (where $t = t_1, \ldots, t_n$; and t_i be some terms ($1 \leqslant i \leqslant n$), then the substitution $\sigma_1 \circ \sigma_2$ will be called *axiomatic one*; the formula A is called the main formula of axiomatic substitution.

Definition 2.7 (rule (*r.v*): rule of renaming of rigid variables). Let us introduce the following rule:

$$\frac{\Pi, \Gamma(\bar{x}), \Gamma'(\bar{y}) \to \Delta(\bar{x}), \Delta'(\bar{y}), \Theta}{\Pi, \Gamma(\bar{x}) \to \Delta(\bar{x}), \Theta} \; (r.v)$$

where $\Gamma' \subseteq \Gamma$; $\Delta' \subseteq \Delta$; $\bar{x} = x_1, \ldots x_n$; $\bar{y} = y_1, \ldots, y_n$; $\bar{x}, \bar{y} \notin \Pi, \theta$; $\bar{y} \notin \Gamma, \Delta$; variables $\bar{x}$ are called renaming ones.

Remark 2.2: The rule (*r.v*) corresponds to the factorization rule in a traditional resolution-like calculus, and a duplicate of the main formulas in the rules of inference $(\forall \to), (\to \exists)$ in the sequent calculus.

Definition 2.8 (rule (*r.f.s*): rule of replacing of flexible function symbols). Let us introduce the following rule of inference [10]:

$$\frac{\Gamma_1(f_k/f), \Gamma \to \Delta, \Delta_1(f_k/f)}{\Gamma_1, \Gamma \to \Delta, \Delta_1} \; (r.f.s),$$

where Γ_1, Δ_1 are sets of atomic formulas f is a n-place ($n \geqslant 0$) flexible functional symbol; f_k is a rigid function symbol, associated with f.

Definition 2.9 (rule (f): frame rule of inference). Let us introduce the following rule of inference:

$$\frac{\Pi_1, \Gamma \to \Delta, \Pi_2}{\bigcirc^k \Pi_1, \Gamma \to \Delta, \bigcirc^l \Pi_2} \ (f),$$

where Π_i $(i = 1, 2)$ consist of atomic formulas not containing flexible symbols.

Definition 2.10 (unifier of formulas). Let E_1, E_2 be terms or formulas, σ_1, σ_2 be substitutions. Then, the pair $\langle \sigma_1, \sigma_2 \rangle$ is called a *unifier* of E_1, E_2, if $E_1 \sigma_1 = E_2 \sigma_2$. Such a unifier is the *most general unifier* (in short: *mgu*), if for any unifier (σ_1', σ_2') of E_1, E_2 there is a substitution σ such that $\sigma_i' \circ \sigma = \sigma_i$ $(i = 1, 2)$.

Definition 2.11 (rule (*subs*): substitution rule). Let A, $\bar{p}$, $\bar{q}$, σ_1, σ_2 be the same as in Definition 2.6. Then let us introduce the following rule:

$$\frac{\Gamma, A(\bar{p})\sigma_1 \to \Delta, A(\bar{q})\sigma_2}{\Gamma, A(\bar{p}) \to \Delta, A(\bar{q})} \ (subs),$$

where $(\sigma_1 \circ \sigma_2)$ is the unifier of $A(\bar{p})$, $A(\bar{q})$.

Remark 2.3: The rule (*subs*) corresponds to the finding of the contrary pair in the resolution rule in a traditional resolution-like calculus.

Definition 2.12 (calculus $G_{L\omega}^*$). The *calculus* $G_{L\omega}^*$ is defined by the following postulates.
Axiom: $\Gamma, A \to A, \Delta$ or the premise of (*subs*).
Rules of inference:
1) temporal rules:

$$\frac{\bigcirc^l A, \bigcirc^{l+1} \square A, \Gamma \to \Delta}{\bigcirc^l \square A, \Gamma \to \Delta} \ (\square \to)$$

$$\frac{\Gamma \to \Delta, \bigcirc^l A; \dots; \Gamma \to \Delta, \bigcirc^{l+k} A; \dots}{\Gamma \to \Delta, \bigcirc^l \square A} \ (\to \square_\omega), \quad k, l \in \omega;$$

2) logical rules of inference consist of traditional invertible rules of inference (for $\supset, \wedge, \vee, \neg$);
3) rules of inference $(r.f.s)$ and (f);
4) implicit rules of inference for quantifiers: $(r.v)$ and (*subs*);
5) structural rules: it follows from the definition of a sequent that $G_{L\omega}^*$ implicitly contains the structural rules "exchange" and "contraction".

Definition 2.13 (quasi-different axiomatic substitution, correct application of $(r.v.)$, correct derivation in $G_{L\omega}^*$). Let $\sigma_1 = \bar{x}_1 \leftarrow \bar{t}_1$ and $\sigma_2 = \bar{x}_2 \leftarrow \bar{t}_2$ $(\bar{x}_i = x_{i1}, \dots, x_{in}; \bar{t}_i = t_{i1}, \dots, t_{in}; i \in \{1, 2\})$ be two axiomatic substitutions with different main formulas, then σ_1, σ_2 will be called *quasi-different*, if $\bar{t}_1 \neq \bar{t}_2$ and $\bar{x}_1 = \bar{x}_2$. Let $\bar{x}$ be renaming variables of *(r.v.)*, then the application of *(r.v.)* in some branch I of derivation D is correct if $\bar{x} \notin I_1$ and $\bar{x} \notin B$, where I_1 is a branch of D $(I_1 \neq I)$ and B is the main formula of an application of splitting rule of inference in D. The derivation D in $G^*_{L\omega}$ will be called *correct* if D (1) does not contain any two quasi-different axiomatic substitutions; (2) contains only correct applications of *(r.v.)*. By derivation in $G^*_{L\omega}$ will be meant the correct derivation in $G^*_{L\omega}$.

Theorem 2.2 (soundeness and completeness of $G^*_{L\omega}$, admisibility of *(cut)* in $G^*_{L\omega}$). (a) *Let S be a quantifier-free variant of N-sequent, then the sequent S is universally valid iff $G^*_{L\omega} \vdash S$*; (b) $G^*_{L\omega} + (cut) \vdash S \Rightarrow G^*_{L\omega} \vdash S$.

Proof. The point (a) is carried out in two steps. In first step the soundness and completeness of $G_{L\omega}$ with explicit rules for quantifiers is proved (see, [6, 7]). In second step the deductive equivalence between calculi $G_{L\omega}$ and $G^*_{L\omega}$ is proved. The point (b) follows from (a).

3 Description and investigation of saturated calculi Sat and Sat_ω

In this section finitary and infinitary saturation calculi Sat and Sat_ω will be introduced and investigated.

Definition 3.1 (marked occurrences of $\square$, marked formulas). The positive occurrence of $\square$ entering the negative occurrence of $\square$ will be called *marked* one and will be denoted by $\square^+$. Let a formula $\square A$ enters the scope of action of $\square$, then elementary subformulas $E_i(\bar{x})$ (containing the variables) of $\square A$ will be called *marked* one and will be denoted by $E_i^+(\bar{x})$. The operator $+$ is defined as follows. 1. $A^+ := A$, if A does not contain variables; 2. $A^+ = A^+$, if A is atomic formula with variables; 3. $(A \odot B)^+ := A^+ \odot B^+$ $(\odot \in \{\supset, \wedge, \vee\})$; 4. $(\sigma A)^+ := \sigma A^+$ $(\sigma \in \{\daleth; \square\})$.

Remark 3.1: The marked formulas prevents from incorrect bottom-up application of rule of inference (A_i^*) (see Definition 3.8, below). The marked occurrences of $\square^+$ do not allow to derive invalid sequents in Sat_λ $(\lambda \in \{\varnothing, \omega\})$.

Definition 3.2 (calculus I^*). The calculus I^* is obtained from the calculus $G^*_{L\omega}$ by (1) dropping $(\rightarrow \square_\omega)$ and changing the rule of inference $(\square \rightarrow)$ by the rule of inference $(\square^* \rightarrow)$ obtaining from $(\square \rightarrow)$ by adding the following restriction: the main formula is such that $\square A = \square^* A$ $(* \in \{\varnothing, +\})$; (2) adding the following rule of inference:

$$\frac{\Gamma \rightarrow \Delta, A; \Gamma \rightarrow \Delta, \bigcirc \square A}{\Gamma \rightarrow \Delta, \square^* A} (\rightarrow \bigcirc \square) \ (* \in \{\varnothing, +\}).$$

The correct derivation in I^* is defined as in Definition 2.13.

Remark 3.2: The right premise of $(\rightarrow \bigcirc\square)$ dropp the mark sign from $\square$ (if the main formula was $\square^+ A$) but preserves marked formulas (if they enter the main formula of $(\rightarrow \bigcirc\square)$).

Definition 3.3 (canonical forms: primary and quasi-primary sequent, degenerate primary sequent). Let S be a sequent, then S will be called *primary*, if $S = \Sigma_1, \bigcirc^n \Pi_1, \square\Omega_1 \rightarrow \Sigma_2, \bigcirc^n \Pi_2, \square\Omega_2$, where $n \geqslant 1$, $\Sigma_i = \varnothing$ ($i \in \{1,2\}$) or consist of logical formulas; $\bigcirc^n \Pi_i = \varnothing$ ($i = \{1,2\}$) or consist of atomic formulas; $\square\Omega_i = \varnothing$ ($i = \{1,2\}$) or consist of formulas of the shape $\bigcirc^k \square B$ ($k \geqslant 0$). If the set $\square\Omega_1, \square\Omega_2$ in the sequent S consists of the formulas of the shape $\bigcirc^n \square B$ ($n \geqslant 1$), then such a sequent S will be called *quasi-primary*. Let S be a primary sequent, then S is *degenerate* one, if all positive occurrences of $\square$ are marked.

Definition 3.4 (reduction of a sequent S to sequents $S_1, \ldots, S_n$). Let $\{i\}$ denote the set of rules of inference of a calculus I^*. The $\{i\}$-*reduction* (or, briefly, *reduction*) of S to $S_1, \ldots, S_n$ denoted by $R(S)\{i\} \Rightarrow \{S_1, \ldots S_n\}$ or briefly by $R(S)$, is defined to be a tree of sequents with the root S and leaves $S_1, \ldots, S_n$, and possibly some logical axioms such that each sequent in $R(S)$, different from S, is the "upper sequent" of the rule of inference in $\{i\}$ whose "lower sequent" also belongs to $R(S)$.

Lemma 3.1. *Let S be a sequent, then there exists $R(S)\{i\} \Rightarrow \{S_1, \ldots, S_n\}$ such that (1) $\forall j$ ($1 \leqslant j \leqslant n$) S_j is a quasi-primary (primary) sequent; $\{i\}$ is the set of rules of the calculus I^* (the set of logical rules of inference of I^*, respectively); (2) $G^*_{L\omega} \vdash S \Rightarrow G^*_{L\omega} \vdash S_j$ ($j = 1, \ldots, n$).*
Proof: using given derivation and bottom-up applying the rules of inference from $\{i\}$.

Definition 3.5 (primary and quasi-primary reduction, primary (quasi-primary) reduction-tree of a primary sequent S). Let $P(S)(QP(S)$, respectively) be the set of primary (quasi-primary) sequents from Lemma 3.1. Then the set $P^*(S)(Q^*P(S))$, obtaining from $P(S)$ ($QP(S)$, respectively) by dropping logical axioms will be called a proper primary (quasi-primary, respectively) reduction of S. The derivation consisting of bottom-up applications of rules of inference of the calculus I^* and having the sequent S as the rooth and the set $P^*(S)$ ($Q^*P(S)$) as the leaves will be called primary (quasi-primary) reduction tree of S and will be denoted by $P^*T(S)$ ($Q^*PT(S)$, respectively).

Definition 3.6 (unifiable formulas and sequents). We say that formulas A, B are *unifiable*, if there exists a unifier of *formulas* A, B (in symbols: $A \approx_u B$). We say that *sequents* $S_1 = A_1, \ldots, A_n \rightarrow A_{n+1}, \ldots, A_{n+m}$ and $S_2 = B_1, \ldots, B_n \rightarrow B_{n+1}, \ldots, B_{n+m}$ are *unifiable* (in symbols: $S_1 \approx_u S_2$), if $\forall i$ ($1 \leqslant i \leqslant n+m$) $A_i \approx_u B_i$.

Definition 3.7. (structural rule (W^*), u-subsumed sequent). Let us introduce the following rule of inference:

$$\frac{\Gamma \rightarrow \Delta}{\Pi, \Gamma' \rightarrow \Delta', \Theta} \ (W^*),$$

where $\Gamma \to \Delta \approx_u \Gamma' \to \Delta'$.

Let S_1 and S_2 be non-degenerate primary sequents. We say that a sequent S_1 u-subsumes a sequent S_2, if $\frac{S_1}{S_2}$ (W^*) (in symbols $S_1 \succcurlyeq uS_2$), or S_2 is u-subsumed by S_1. If S_1, S_2 do not contain the variables, then we say that S_1 subsumes S_2 and denote it by $S_1 \succcurlyeq S_2$.

Definition 3.8 (alternating rule: rule of inference (A_i^*)). Let us introduce the following rule of inference

$$\frac{S_i^*}{S^*} (A_i^*), \quad (i = 1 \text{ or } i = 2)$$

where S^* is a quasi-primary sequent, with explicitly indicated marked formulas i.e., $S^* = \Sigma_1, \bigcirc^p \nabla_1, \bigcirc^n \Pi_1, \bigcirc^n \square \Omega_1(\overline{x}), \bigcirc^n \square^* \Omega_2^+(\overline{v}) \to \Sigma_2, \bigcirc^q \nabla_2, \bigcirc^n \Pi_2, \bigcirc^n \square \Delta_1(\overline{y}), \bigcirc^n \square^* \Delta_2^+(\overline{w}); S_1^* = \Sigma_1, \nabla \to \Sigma_2, \nabla_2; S_2^* = \Pi_1^\circ, \square \Omega_1(\overline{z}), \square^* \Omega_2^+(\overline{v}) \to \Pi_2^\circ, \square \Delta_1(\overline{u}), \square^* \Delta_2^+(\overline{w})$, where Π_i° is obtained from Π_i $(i \in \{1,2\})$ replacing all flexible function symbols f^n $(n = 1, 2, \ldots, l)$ by rigid function symbols f_k^n, respectively, associated with f^n; ∇_i $(i = 1, 2) = \varnothing$ or contain atomic formulas with rigid symbols only; $\overline{x} = x_1, \ldots, x_n; \overline{y} = y_1, \ldots, y_m; \overline{z} = z_1, \ldots, z_n; \overline{u} = u_1, \ldots, u_m, \overline{v} = v_1, \ldots, v_k, \overline{w} = w_1, \ldots, w_l; n, m \geqslant 0; k, l \geqslant 1 \ p, q \geqslant 0, (* \in \{\varnothing, +\})$; variables $\overline{x}, \overline{y}$ are called renaming ones.

Lemma 3.2 (disjunctional invertibility of (A_i^*)). *Let S^* be a conclusion of (A_i^*), S_i^* $(i = 1 \text{ or } i = 2)$ be a premise of (A_i^*), then (1) $G_{L\omega}^* \vdash S^* \Rightarrow Log \vdash S_1^*$, or (2) $G_{L\omega}^* \vdash S^* \Rightarrow G_{L\omega}^* \vdash S_2^*$.*

Proof: is proved by induction on $O(D)$, where D is the given derivation.

Definition 3.9 (similar sequents, similarity substitution, similar set of sequents, similar terms). We say that the sequent S_1 is similar to S_2, if $S_1\sigma \succcurlyeq uS_2$; σ is called a similarity substitution. In LHS of similarity substitution σ besides variables some constants will be allowed. Let $t_1, \ldots, t_m$ be RHS of similarity substitution σ, then σ^n is similarity substitution which LHS is the same as LHS of σ and RHS $= t_1^n, \ldots, t_m^n$ $(n \in \omega)$. If $S_i\sigma \succcurlyeq uS_j$ then we say that S_i σ-similar u-subsumes S_2 or S_2 is σ-similar u-subsumed by S_1. Let $M_1 = \{S_1, \ldots, S_n\}$ and $M_2 = \{S_1^*, \ldots, S_n^*\}$ be sets of sequents. We say that the set M_1 is σ-similar to M_2 (in symbols $M_1 \approx \sigma M_2$) if $\forall i \ (1 \leqslant i \leqslant n)$ $S_i\sigma \succcurlyeq uS_i^*$. We say that the term t_1 is σ-similar to term t_2 if $t_1\sigma = t_2$.

Definition 3.10. (subsumption rule). Let us defines the following rule, which will be applied in the bottom-up manner:

$$\frac{\{S_1, \ldots, S_j, \ldots, S_n\}}{\{S_1, \ldots, S_{j-1}, S_{j+1}, \ldots, S_n\}} (S),$$

where $S_i \succcurlyeq uS_j$ $(i = 1, \ldots, j-1, j+1, \ldots, n)$; $S_1, \ldots, S_n$ are non-degenerate primary sequents. The bottom-up applications of S will be denoted by S^-.

Definition 3.11 (subsumption-tree (ST)). The *subsumption-tree* of a primary sequent is defined as deduction-tree (and denoted by (ST)), consisting of all possible applications of (S^-). The sequents of (ST) which subsumes some sequents from (ST) will be called *active part* of (ST), and the sequents of (ST) which are subsumed will be called *passive part* of (ST).

Definition 3.12 (resolvent-tree $ReT(S)$ and resolvent $Re(S)$ of a primary sequent S). The *resolvent-tree* of a *primary sequent* S is defined by the following bottom-up deduction (denoted by $ReT(S)$) :

$$
\cfrac{
\cfrac{
\cfrac{\cfrac{S_{11},\ldots,S_{1n_1}}{S_1^+}\,P^*T(S)}{S_1^*}\,(A_i^*)
\qquad\quad
\cfrac{\cfrac{S_{m1},\ldots,S_{mn_m}}{S_m^+}\,P^*T(S)}{}\,(A_i^*)
}{
S_1^*,\ \ldots\ldots\ldots\ldots\ldots\ldots,\ S_n^*
}\,Q^*PT(S)
}{S}
$$

where $S_1^*,\ldots,S_n^*$ are quasi-primary sequents; moreover $S_j^* = \varnothing$ $(1 \leqslant j \leqslant n)$ if $Q^*P(S) = \varnothing$; $S_1^+,\ldots,S_m^+$ are some sequents; moreover S_l^+ $(1 \leqslant l \leqslant m)$ is empty if $(A_i^*) = (A_1^*)$ and non-empty if $(A_i^*) = (A_2^*)$; S_{hl} $(1 \leqslant h \leqslant m, \ n_1 \leqslant l \leqslant n_m)$ are primary sequents. If $\forall j, h, l$ $(1 \leqslant j \leqslant n; \ 1 \leqslant h \leqslant m; \ n_1 \leqslant l \leqslant n_m)$ $S_j^* = \varnothing$ or $S_{hl} = \varnothing$, then we say that *resolvent* of the sequent S is *empty* (in symbols: $Re(S) = \varnothing$). The set of primary sequents $S_1,\ldots,S_k$ $(k \geqslant 0)$ will be called the *resolvent of S* and will be denoted by $Re(S)$. If $\exists S_i \in Re(S)$ such that S_i is degenerate then we say that resolvent of the sequent S does not exist (in symbols $\rceil\exists Re(S)$).

Definition 3.13 (k-th resolvent tree $Re^kT(S)$ and k-th resolvent $Re^k(S)$ of a primary sequent S). Let $k = 0$, then $Re^0(S) = Re^0T(S) = S$; let $Re^k(S) = \{S_1,\ldots,S_n\}$. The $Re^{k+1}T(S)$ is defined by the following bottom-up deduction:

$$
\cfrac{
\cfrac{
Re^{k+1}(S)
}{
\cfrac{R^*e^{k+1}(S)}{\underbrace{\cfrac{Re(S_1)}{S_1}\,ReT(S_1) \quad\cdots\quad \cfrac{Re(S_n)}{S_n}\,ReT(S_n)}_{Re^k(S)}}\,(ST)
\qquad
\bigcup_{i=0}^{k} Re^i(S)
}\,(ST)
}{}
$$

The lower bottom-up application of (ST) reduces the set $\bigcup_{i=1}^{n} Re(S_n)$ to the set $R^* e^{k+1}(S)$. The upper bottom-up application of (ST) reduces the set $R^* e^{k+1}(S) \cup \bigcup_{i=0}^{k} Re^i(S)$ to the set of primary sequents which will be called $k+1$-*th resolvent* of S and denoted by $Re^{k+1}(S)$. Moreover, the members of $\bigcup_{i=0}^{k} Re^i(S)$ are active part of the upper application of (ST) and members of $R^* e^{k+1}(S)$ are passive one of the upper application of (ST). If $\daleth\exists Re(S_i)(0 \leqslant i \leqslant n)$, then $\daleth\exists Re^{k+1}(S)$. The set $Re^{k+1}(S)$ is empty in two following cases: (1) $\forall i \, (1 \leqslant i \leqslant n) \, Re(S_i) = \varnothing$ or (2) the upper application of (ST) in $Re^{k+1}T(S)$ yields the empty set.

Hypothetical k-th resolvent tree $HRe^k T(S)$ and hypothetical k-th resolvent $HRe^k(S)$ are defined quite analogously as $Re^k T(S)$ and $Re^k(S)$.

Definition 3.14 (Calculi Sat and Sat_ω). The postulates of *calculi Sat and Sat_ω* consist of two parts:

– an auxiliary part, consisting of the postulates of the calculus I^* (see Definition 3.2) removing rule of inference (f) and adding the rule of inference (A_i^*) (see Definition 3.8); the auxiliary part is common for both calculi Sat and Sat_ω;

– the main part, consisting of the two following postulates

1) The rule of *subsumption* (which is common for both calculi Sat and Sat_ω, see Definition 3.10);

2) the axiom $Re^k(S) = \varnothing$ for the finitary calculus Sat the axiom $Re^\omega(S) = \varnothing$ for infinitary calculus Sat_ω.

The correct derivation in Sat_λ ($\lambda \in \{\varnothing, \omega\}$) is defined analogously as in Definition 2.13.

Example 3.1. (a) Let $S = P(c), \Box\Omega(x) \to \Box\daleth P(y)$, where $\Omega(x) = (P(x) \supset \daleth\bigcirc\Box^+ \daleth P^+(f(x)))$, where P is a flexible predicate symbol; f is a rigid function symbol, then $Re^1(S) = S_1 = \Box\Omega(x_1) \to \Box^+ \daleth P^+(f(x))$, $\Box\daleth P(y_1)$. Since $\Box\Omega(x_2) \to \Box^+ \daleth P^+(f(x)) \in Re(S_1)$, $Sat_\lambda \nvdash S$.

(b) Let $S_1 = P(c), \Box\Omega(x) \to \Box P(y)$, where $\Omega(x) = P(x) \supset \bigcirc P(f(x))$; where P is a flexible predicate symbol, f is a flexible function symbol, c is a rigid constant. It is easy to verify that $Re^1(S_1) = S_2 = \Box\Omega(x_1), P(f(x)) \to \Box P(y_1)$, where $x \leftarrow c$; f_1 is the rigid function symbol, associated with f; $S_1\sigma \succcurlyeq uS_2$, where $\sigma = c \leftarrow f_1(c)$ (basis of induction). Therefore $HRe^m(S) = S_m = \Box\Omega(x_m), P(f_1^m(c)) \to \Box P(y_m)$ (induction hypothesis). Analogously we get $HRe^{m+1}(S) = S_{m+1} = \Box\Omega(x_{m+1}), P(f(x_m)) \to \Box P(y_{m+1})$, where $x_m \leftarrow f_1^m(c)$; $S_m\sigma \succcurlyeq uS_{m+1}$ (step of induction). Therefore $Re^\omega(S) = \varnothing$ and $Sat_\omega \vdash S$. If c would be a flexible constant then it would be necessary to replace c by the rigid constant c_0 associated with c.

(c) Let $S = \Box\Omega \to A, \Box\daleth\Box A$, where $\Omega = \daleth(\daleth A \wedge \Box^+(A \supset \Box^+ A))$, where A is a flexible propositional symbol (the sequent S express discreteness property), then $Re^1(S) = S_1 = \Box\Omega \rightarrow \Box^+(A \supset \Box^+A), \Box\daleth\Box A$; $Re^2(S) = \{S_2, S_3\}$, where $S_2 = \Box\Omega \rightarrow \Box^+A, \Box\daleth\Box A$; $S_3 = \Box\Omega, \Box A \rightarrow \Box(A \supset \Box A)$; $Re^3(S) = \varnothing$, therefore $Sat \vdash S$.

Theorem 3.1. $Sat_\lambda \vdash^D \Rightarrow G^*_{L\omega} \vdash^{D^*} S$.

Proof. The proof is carried out in two steps. In the first step the derivation D is transformed into derivation D_1 in "invariant" calculus IN_ρ ($\rho \in \{\varnothing, \infty\}$), obtained from $G^*_{L\omega}$ replacing $(\rightarrow \Box_\omega)$ by "invariant" rule $(\rightarrow \Box)$ (see section 1). Moreover the invariant formula has the shape of infinite disjunctions when $\lambda = \omega$. In the second step the derivation D_1 is transformed into the derivation D^* in $G^*_{L\omega}$ (see [6, 7, 8]).

Theorem 3.2. Let S be a primary sequent then $G^*_{L\omega} \vdash S \Rightarrow Sat_\lambda \vdash S$.

Proof. Applying to S the process of calculating k-th resolvents we get either $Re^k(S) = \varnothing$ or $Re^\omega(S) = \varnothing$. Therefore $Sat_\lambda \vdash S$ ($\lambda = \varnothing$ or $\lambda = \omega$).

Theorem 3.3 (soundness and completness of Sat_λ). Let S be a N-sequent then $Sat_\lambda \vdash S \Leftrightarrow \forall M \vDash S$.

Proof: follows from soundness and completness of $G^*_{L\omega}$ and Theorem 3.1, 3.2.

References

1. H. Andreka, J. Nemeti, J. Sain: On the strenghth of temporal proof, LNCS, 379, 135–144 (1989).
2. Ben-Ari: Mathematical Logic for Computer Science, Prentice Hall, 1993.
3. M. Fisher: A normal form for first order temporal formulae, LNCS, 607, 370–384 (1992).
4. D.M. Gabbay: The declarative past and the imperative future, LNCS, 398, 409–448 (1987).
5. H. Kawai: Sequential calculus for a first order infinitary temporal logic, Zeitchr. fur Math. Logic und Grundlagen der Math., 33, 423–452 (1987).
6. R. Pliuškevičius: On saturated calculi for a linear temporal logic, LNCS 711, 640–649 (1993).
7. R. Pliuškevičius: The saturated tableaux for linear miniscoped Horn-like temporal logic, Journal of Automated Reasoning, 13, 51–67 (1994).
8. R. Pliuškevičius: On the replacement of induction for a first order linear temporal logic, Proc. of Second World Conf. on the Fund. of Artif. Intell., 331–342 (1995).
9. I. Sain: Temporal logics need their clocks. Theoret. Comput. Sci., 95, 75–95 (1992).
10. J. Sakalauskaitė: A sequent calculus for a first order linear temporal logic with equality. LNCS, 620, 430–440 (1992).
11. A. Szalas: Concerning the semantic consequence relation in first-order temporal logic, Theoret. Comput. Sci. 47, 329–334 (1986).
12. P. Wolper: The tableaux method for temporal logic: an overview, Log. et anal. 28, 119–136 (1985).

Approximating Good Simultaneous
Diophantine Approximations Is Almost NP-Hard

Carsten Rössner and Jean-Pierre Seifert*

Dept. of Math. Comp. Science,
University of Frankfurt, P. O. Box 111932,
60054 Frankfurt/Main, Germany
{roessner,seifert}@cs.uni-frankfurt.de

Abstract. Given a real vector $\alpha = (\alpha_1, \ldots, \alpha_d)$ and a real number $\varepsilon > 0$ a good Diophantine approximation to α is a number Q such that $\|Q\alpha \bmod \mathbf{z}\|_\infty \leq \varepsilon$, where $\| \cdot \|_\infty$ denotes the ℓ_∞-norm $\|\mathbf{x}\|_\infty := \max_{1 \leq i \leq d} |x_i|$ for $\mathbf{x} = (x_1, \ldots, x_d)$.

Lagarias [12] proved the NP-completeness of the corresponding decision problem, i.e., given a vector $\alpha \in \mathbb{Q}^d$, a rational number $\varepsilon > 0$ and a number $N \in \mathbb{N}_+$, decide whether there exists a number Q with $1 \leq Q \leq N$ and $\|Q\alpha \bmod \mathbf{z}\|_\infty \leq \varepsilon$.

We prove that, unless $\mathbf{NP} \subseteq \mathbf{DTIME}(n^{\mathrm{poly}(\log n)})$, there exists no polynomial-time algorithm which computes on inputs $\alpha \in \mathbb{Q}^d$ and $N \in \mathbb{N}_+$ a number Q^* with $1 \leq Q^* \leq 2^{\log^{0.5-\gamma} d} N$ and

$$\|Q^*\alpha \bmod \mathbf{z}\|_\infty \leq 2^{\log^{0.5-\gamma} d} \min_{1 \leq q \leq N} \|q\alpha \bmod \mathbf{z}\|_\infty,$$

where γ is an arbitrary small positive constant. To put it in other words, it is almost NP–hard to approximate a minimum good Diophantine approximation to α in polynomial-time within a factor $2^{\log^{0.5-\gamma} d}$ for an arbitrary small positive constant γ.

We also investigate the nonhomogeneous variant of the good Diophantine approximation problem, i.e., given vectors $\alpha, \beta \in \mathbb{Q}^d$, a rational number $\varepsilon > 0$ and a number $N \in \mathbb{N}_+$, decide whether there exists a number Q with $1 \leq Q \leq N$ and $\|Q\alpha - \beta \bmod \mathbf{z}\|_\infty \leq \varepsilon$.

This problem is particularly interesting since finding good nonhomogeneous Diophantine approximations enables us to factor integers and compute discrete logarithms (see Schnorr [17]).

We prove that the problem Good Nonhomogeneous Diophantine Approximation is NP-complete and even approximating it in polynomial-time within a factor $2^{\log^{1-\gamma} d}$ for an arbitrary small positive constant γ is almost NP-hard.

Our results follow from recent work in the theory of probabilistically checkable proofs [4] and 2-prover 1-round interactive proof-systems [7, 14].

Key Words. approximation algorithm, computational complexity, NP-hard, probabilistically checkable proofs, Diophantine approximation, 2-prover 1-round interactive proof-systems

* Supported by DFG under grant DFG-Leibniz-Programm Schn 143/5-1

1 Introduction

Since **NP** optimization problems are unlikely to be solved in polynomial-time, unless **P = NP**, a lot of work has been done to find polynomial-time approximation algorithms for these problems. An algorithm is said to *approximate* a positive real-valued function $opt(\cdot)$ *within a factor* f if on every input I its output is within a factor f of $opt(I)$.

Unfortunately, for many **NP**-hard optimization problems it is even **NP**-hard or almost **NP**-hard to compute such approximate solutions, see, e.g., Crescenzi and Kann [6] or Arora and Lund [3]. Therefore, it is quite important, both from the practical point of view and from the point of view of complexity theory, to find conditions which enable or disable us to design polynomial-time approximation algorithms for **NP**-hard optimization problems

In this paper we investigate the approximability of the following **NP** optimization problems:

MINIMUM GOOD DIOPHANTINE APPROXIMATION in ℓ_∞-norm (MINGDA$_\infty$)
INSTANCE: A rational vector $\alpha = (\alpha_1, \ldots, \alpha_d) \in \mathbb{Q}^d$ and a number $N \in \mathbb{N}$
SOLUTION: A number $Q \in [1, N] \cap \mathbb{Z}$
MEASURE: The ℓ_∞-norm $\|Q\alpha \bmod \mathbb{Z}\|_\infty := \max_{1 \le i \le d} \min_{n \in \mathbb{Z}} |Q\alpha_i - n|$.

MINIMUM GOOD NONHOMOGENEOUS DIOPHANTINE APPROXIMATION in ℓ_∞-norm (MINGNDA$_\infty$)
INSTANCE: Rational vectors $\alpha = (\alpha_1, \ldots, \alpha_d), \beta = (\beta_1, \ldots, \beta_d) \in \mathbb{Q}^d$ and a number $N \in \mathbb{N}$
SOLUTION: A number $Q \in [1, N] \cap \mathbb{Z}$
MEASURE: The ℓ_∞-norm $\|Q\alpha - \beta \bmod \mathbb{Z}\|_\infty := \max_{1 \le i \le d} \min_{n \in \mathbb{Z}} |Q\alpha_i - \beta_i - n|$.

We refer to MINGDA$_\infty$ and MINGNDA$_\infty$ also as the problem Minimum Good Simultaneous Diophantine Approximation and Minimum Good Nonhomogeneous Simultaneous Diophantine Approximation, respectively, and to the solution $Q \in [1, N] \cap \mathbb{Z}$ as the common denominator of the good (nonhomogeneous) simultaneous diophantine approximation.

In fact, good simultaneous diophantine approximations have wide practical impact. Algorithms for finding such approximations may be used to find strongly polynomial-time algorithms in combinatorial optimization [8], to factor univariate integer polynomials [18] and to compute minimal polynomials of an algebraic number [11].

The motivation for our first result comes from the following conjecture raised by Lagarias [12]: If there is a polynomial-time algorithm which computes on inputs $\alpha \in \mathbb{Q}^d$ and $N \in \mathbb{N}_+$ a denominator $Q^* \in [1, f(d)N]$ satisfying

$$\|Q^*\alpha \bmod \mathbb{Z}\|_\infty \le f(d) \min_{1 \le q \le N} \|q\alpha \bmod \mathbb{Z}\|_\infty,$$

where $f(d)$ is some polynomial in d, then **P = NP**. Conversely, Lagarias gave an algorithm which computes for inputs $\alpha \in \mathbb{Q}^d$ and $N \in \mathbb{N}_+$ a denominator

$Q^* \in [1, 2^{d/2}N]$ satisfying

$$\|Q^*\alpha \bmod \mathbf{Z}\|_\infty \leq \sqrt{5d}\, 2^{(d-1)/2} \min_{1 \leq q \leq N} \|q\alpha \bmod \mathbf{Z}\|_\infty.$$

We prove, that approximating MINGDA_∞ in polynomial-time within a factor $2^{\log^{0.5-\gamma} d}$ for an arbitrary small positive constant γ implies $\mathbf{NP} \subseteq \mathbf{DTIME}(n^{\text{poly}(\log n)})$. Thus, in the sense of Lagarias' conjecture, our result may be regarded as a step towards narrowing the gap between approximability and inapproximability of MINGDA_∞ in polynomial-time.

Our results follow by a chain of gap-preserving reductions from two well-known lattice problems: SHORTEST VECTOR in ℓ_∞-norm and NEAREST VECTOR in ℓ_∞-norm. Using previous results [7, 4] on interactive proof-systems, Arora et al. [2] proved that, unless $\mathbf{NP} \subseteq \mathbf{DTIME}(n^{\text{poly}(\log n)})$, no polynomial-time algorithm can approximate the shortest non-trivial vector in the ℓ_∞-norm in a lattice within a factor $2^{\log^{0.5-\gamma} n}$ for an arbitrary small positive constant γ. They also showed the same inapproximability result for the nearest vector problem in the ℓ_∞-norm. By a recent result of Raz [14] the inapproximability factor in case of approximating the nearest vector in the ℓ_∞-norm in a lattice can be amplified to $2^{\log^{1-\gamma} n}$ for an arbitrary small positive constant γ.

We transfer these inapproximability gaps to MINGDA_∞ and MINGNDA_∞, respectively, via two intermediate problems.

Roadmap In section 2 we introduce some notation and the problem SHORTEST INTEGER RELATION in ℓ_∞-norm (SIR_∞) which is known to be almost **NP**-hard to approximate within a factor $2^{\log^{0.5-\gamma} n}$, for γ an arbitrary small positive constant and n the input size, see Rössner and Seifert [16]. In section 3 we give a gap-preserving reduction from SIR_∞ to MINGDA_∞ proving the first result. In section 4 we define the problem MINIMUM DIOPHANTINE EQUATION SOLUTION in ℓ_∞-norm (MINDES_∞) and sketch a gap-preserving reduction from MINDES_∞ to MINGNDA_∞. This implies our second result.

2 Preliminaries

2.1 Definitions

We briefly introduce some notation, see [5].

Definition 1. An *optimization problem* Π is a set $\mathcal{I} \subseteq \{0,1\}^*$ of instances, a set $\mathcal{S} \subseteq \{0,1\}^*$ of feasible solutions and a polynomial-time computable positive measure function $m : \mathcal{I} \times \mathcal{S} \to \mathbb{R}_+$, that assigns each tuple of an instance I and a solution S, a positive real number $m(I, S)$, called the *value* of the solution S. The optimization problem is to find, for a given input $I \in \mathcal{I}$ a solution $S \in \mathcal{S}$ such that $m(I, S)$ is optimum over all possible $S \in \mathcal{S}$.

If the optimum is $\min_{S \in \mathcal{S}}\{m(I, S)\}$ (resp. $\max_{S \in \mathcal{S}}\{m(I, S)\}$) we refer to Π as a *minimization* (resp. *maximization*) problem.

Definition 2. For an input I of a minimization (resp. maximization) problem Π whose optimal solution has value $opt(I)$, an algorithm A is said to *approximate* $opt(I)$ *within a factor* $f(I)$ iff

$$opt(I) \leq A(I) \leq opt(I)f(I) \qquad (\text{resp. } opt(I)/f(I) \leq A(I) \leq opt(I)),$$

where $f(I) \geq 1$ and $A(I) > 0$.

For exhibiting the hardness of approximation problems we introduce the following reduction due to Arora [1].

Definition 3. Let Π and Π' be two minimization problems and ρ, $\rho' \geq 1$. A *gap-preserving reduction from* Π *to* Π' *with parameters* $((c, \rho), (c', \rho'))$ is a polynomial-time transformation τ mapping every instance I of Π to an instance $I' = \tau(I)$ of Π' such that for the optima $opt_\Pi(I)$ and $opt_{\Pi'}(I')$ of I and I', respectively, the following holds:

$$opt_\Pi(I) \leq c \Longrightarrow opt_{\Pi'}(I') \leq c'$$
$$opt_\Pi(I) > c \cdot \rho \Longrightarrow opt_{\Pi'}(I') > c' \cdot \rho',$$

where c, ρ and c', ρ' depend on the instance sizes $|I|$ and $|I'|$, respectively.

2.2 Previous Results

The proof of our first result will mainly rely on a gap-preserving reduction to MINGDA_∞ from the problem SHORTEST INTEGER RELATION in ℓ_∞-norm stated as follows:

SHORTEST INTEGER RELATION in ℓ_∞-norm (SIR$_\infty$)
INSTANCE: A rational vector $\mathbf{a} \in \mathbb{Q}^d$
SOLUTION: A nonzero vector $\mathbf{x} \in \mathbb{Z}^d$ such that $\langle \mathbf{a}, \mathbf{x} \rangle = 0$
MEASURE: The ℓ_∞-norm $\|\mathbf{x}\|_\infty := \max_{1 \leq i \leq n} |x_i|$ of the vector $\mathbf{x}$

The SHORTEST INTEGER RELATION problem in ℓ_∞-norm was proven to be NP-complete by van Emde Boas [19]. Very recently, Rössner and Seifert [16] showed the following Theorem, stating that it is even almost NP-hard to approximate SIR$_\infty$ in polynomial-time within a factor $2^{\log^{0.5-\gamma} n}$, where γ is an arbitrary small positive constant and n the size of the SIR$_\infty$ instance I.

Theorem 4. *There exists an almost polynomial-time, i.e.,* $\mathbf{DTIME}(n^{\text{poly}(\log n)})$ *transformation τ from* 3-SAT *to* SHORTEST INTEGER RELATION *in ℓ_∞-norm such that, for all instances I,*

$$I \in \text{3-SAT} \Longrightarrow opt_{\text{SIR}_\infty}(\tau(I)) = 1$$
$$I \notin \text{3-SAT} \Longrightarrow opt_{\text{SIR}_\infty}(\tau(I)) > 2^{\log^{0.5-\gamma} |\tau(I)|},$$

where γ is an arbitrary small positive constant.

The above Theorem, in turn, was proven by a reduction from the SHORTEST VECTOR problem in the ℓ_∞-norm, involving techniques from the Feige and Lovász [7] 2-prover 1-round interactive proof-system, see [2, 16] for more details.

3 The Reduction

3.1 Reducing SIR_∞ to MinGDA_∞

Theorem 5. *There exists a polynomial-time transformation τ from* SHORTEST INTEGER RELATION *in ℓ_∞-norm to* MINIMUM GOOD DIOPHANTINE APPOXIMATION *in ℓ_∞-norm, $\tau : I \mapsto \langle (a_0/b_0, \ldots, a_d/b_d), N \rangle$, such that, for all instances I and for all $\rho \geq 1$,*

$$opt_{\text{SIR}_\infty}(I) = 1 \implies \min_{1 \leq q \leq N} \| q\alpha \bmod \mathbf{Z} \|_\infty \leq \tfrac{1}{b_1}$$

$$opt_{\text{SIR}_\infty}(I) > \rho \implies \min_{1 \leq Q^* \leq \rho N} \| Q^*\alpha \bmod \mathbf{Z} \|_\infty > \rho \tfrac{1}{b_1}.$$

Proof. Our proof follows closely [12]. Due to a few changes specific to our claim, we include the complete proof here. Let $\mathbf{a} = (a_1, \ldots, a_d) \in \mathbf{Z}^d$ be the vector of a given SIR_∞ instance I. First, we encode the task to find a non-trivial $\mathbf{x} \in \mathbf{Z}^d$ with $\|\mathbf{x}\|_\infty \leq \rho$ and

$$\sum_{j=1}^{d} x_j a_j = 0 \tag{1}$$

as a congruence. Let $A := \rho \sum_{j=1}^{d} |a_j|$ and let p_0 be the smallest prime with $p_0 \nmid \prod_{j=1}^{d} a_j$. We set $R := \lfloor \log_{p_0} A \rfloor + 1$. The following steps will crucially use the following Lemma whose proof is deferred to the Appendix.

Lemma A. *There exists a polynomial-time (polynomial in $|I|$) computable set of primes $\{Q_1, \ldots, Q_d\}$ and an integer $T \in \mathbf{N}_+$ such that*

(a) $Q_i < Q_{i+1}, i = 1, \ldots, d-1$,
(b) $\gcd(Q_i, p_0 \prod_{j=1}^{d} a_j) = 1, i = 1, \ldots, d$,
(c) $Q_1^T \geq 4\rho(d+1)p_0^R$ *and*
(d) $\rho^{1/T} Q_d < (\rho+1)^{1/T} Q_1$.

By the Chinese Remainder Theorem we find for every $j = 1, \ldots, d$ a smallest positive integer r_j satisfying

$$r_j \equiv 0 \quad \left(\bmod \prod_{\substack{i=1 \\ i \neq j}}^{d} Q_i^T\right) \tag{2a}$$

$$r_j \equiv a_j \quad (\bmod\ p_0^R) \tag{2b}$$

$$r_j \not\equiv 0 \quad (\bmod\ Q_j), \tag{2c}$$

where $Q_1, \ldots, Q_d$ are given as above. (2c) is a consequence of (2a) and (2b), for if r_j^0 is the smallest positive solution satisfying (2a) and (2b), we set

$$r_j := \begin{cases} r_j^0, & \text{if } r_j^0 \not\equiv 0 \pmod{Q_j}; \\ r_j^0 + p_0^R \left(\prod_{\substack{i=1 \\ i \neq j}}^{d} Q_i^T\right), & \text{otherwise.} \end{cases}$$

As $\gcd(p_0^R \prod_{i=1}^d Q_i^T / Q_j^T, Q_j) = 1$ by (b) of Lemma A, we infer that $r_j \not\equiv 0 \ (\mathrm{mod}\ Q_j)$, $j = 1, \ldots, d$, i.e., (2c) holds for either choice of r_j.

By (2b) and $A < p_0^R$, we see that the systems

$$\sum_{j=1}^d x_j a_j = 0, \qquad (1a) \quad \text{and} \quad \sum_{j=1}^d x_j r_j \equiv 0 \ (\mathrm{mod}\ p_0^R), \quad (3a)$$

$$1 \le \|\mathbf{x}\|_\infty \le \rho \qquad (1b) \qquad\qquad 1 \le \|\mathbf{x}\|_\infty \le \rho. \qquad (3b)$$

have identical integral solutions sets.

For an integral vector $\mathbf{x}$ with $1 \le \|\mathbf{x}\|_\infty \le \rho$ we define

$$Z := \sum_{j=1}^d x_j r_j, \qquad H := \sum_{j=1}^d r_j \quad \text{and} \quad B := \prod_{j=1}^d Q_j^T.$$

We clearly have $|Z| \le \rho H$. Moreover, (c) of Lemma A implies

$$r_j \le r_j^0 + p_0^R \left(\prod_{\substack{i=1 \\ i \ne j}}^d Q_i^T \right) \le 2 p_0^R \frac{B}{Q_j^T} \le \frac{1}{2\rho(d+1)} B, \qquad \text{thus} \qquad \rho H < 1/2B.$$

Lemma 3.6. Let $opt_{\mathrm{modSIR}_\infty}(3a)$ denote the ℓ_∞-norm of the ℓ_∞-shortest nontrivial integral solution of (3a). Then, we have

$$opt_{\mathrm{modSIR}_\infty}(3a) = 1$$
$$\Longrightarrow \exists Z: Z \ne 0 \wedge |Z| \le H \wedge Z \equiv 0 \ (\mathrm{mod}\ p_0^R) \wedge \bigwedge_{1 \le j \le d} Z \equiv \hat{x}_j r_j \ (\mathrm{mod}\ Q_j^T)$$
$$\wedge \bigwedge_{1 \le j \le d} \hat{x}_j \in \{0, \pm 1\}$$

$$opt_{\mathrm{modSIR}_\infty}(3a) > \rho$$
$$\Longrightarrow \forall Z: Z \ne 0 \wedge |Z| \le \rho H \wedge Z \equiv 0 \ (\mathrm{mod}\ p_0^R) \wedge \bigwedge_{1 \le j \le d} Z \equiv \hat{x}_j r_j \ (\mathrm{mod}\ Q_j^T)$$
$$\Rightarrow \mathop{\exists}_{1 \le j \le d} \hat{x}_j \notin [-\rho, \rho] \cap \mathbf{Z}$$

Proof. First, assume that $opt_{\mathrm{modSIR}_\infty}(3a) = 1$ and let $\mathbf{x}$ be the corresponding solution of (3a). For $Z := \sum_{j=1}^d x_j r_j$ we have

- $Z \ne 0$ by (2a), (2c) and since there exists an index j with $x_j \ne 0$,
- $|Z| \le H$ as $\|\mathbf{x}\|_\infty \le 1$,
- $Z \equiv \sum_{j=1}^d x_j r_j \equiv 0 \ (\mathrm{mod}\ p_0^R)$ by definition and,
- $\bigwedge_{1 \le j \le d} Z \equiv \hat{x}_j r_j \ (\mathrm{mod}\ Q_j^T) \wedge \bigwedge_{1 \le j \le d} \hat{x}_j \in \{0, \pm 1\}$ by (2a) and $\|\mathbf{x}\|_\infty \le 1$.

In order to show the second implication let us assume it exists $Z \ne 0$ with

$$|Z| \le \rho H \wedge Z \equiv 0 \ (\mathrm{mod}\ p_0^R) \wedge \bigwedge_{1 \le j \le d} Z \equiv \hat{x}_j r_j \ (\mathrm{mod}\ Q_j^T) \wedge \bigwedge_{1 \le j \le d} \hat{x}_j \in [-\rho, \rho] \cap \mathbf{Z}.$$

To prove the claim we will show the existence of a solution $\mathbf{x} \in \mathbf{Z}^d$ for $\sum_{j=1}^d x_j r_j \equiv 0 \ (\mathrm{mod} p_0^R)$ satisfying $1 \leq \|\mathbf{x}\|_\infty \leq \rho$. For that we consider a candidate solution $\mathbf{x} = (x_1, \ldots, x_d) \in \mathbf{Z}^d$ by setting $\sum_{j=1}^d x_j r_j := Z$. Then, by (2a) we have $x_j r_j \equiv Z \ (\mathrm{mod} Q_j^T)$, $1 \leq j \leq d$.

We show how to uniquely recover $x_j \ (\mathrm{mod} Q_j^T)$ from the given Z. By (2c) and $\gcd(r_j, Q_j^T) = 1$ we can find the unique r_j^* with $1 \leq r_j^* < Q_j^T$ satisfying $r_j r_j^* \equiv 1 \ (\mathrm{mod} Q_j^T)$, $1 \leq j \leq d$, using, e.g., the Extended Euclidean Algorithm. Consequently, we have

$$\bigvee_{1 \leq j \leq d} x_j \equiv x_j r_j r_j^* \equiv Z r_j^* \equiv \hat{x}_j r_j r_j^* \equiv \hat{x}_j \quad (\mathrm{mod} \ Q_j^T) \text{ with } \bigvee_{1 \leq j \leq d} \hat{x}_j \in [-\rho, \rho] \cap \mathbf{Z}.$$

We now prove that even $x_j \in [-\rho, \rho] \cap \mathbf{Z}$. From the Chinese Remainder Theorem we infer that the system of congruences

$$Z \equiv \hat{x}_j r_j \quad (\mathrm{mod} \ Q_j^T), \quad \hat{x}_j \in [-\rho, \rho] \cap \mathbf{Z}, \quad 1 \leq j \leq d$$

has exactly $(2\rho + 1)^d$ solutions in the interval

$$-1/2B < Z < 1/2B$$

since $B := \prod_{j=1}^d Q_j^T$. From the inequality $\rho H < 1/2B$ we see that we have at most $(2\rho + 1)^d$ solutions for the system

$$|Z| \leq \rho H,$$
$$Z \equiv \hat{x}_j r_j \quad (\mathrm{mod} \ Q_j^T), \quad \hat{x}_j \in [-\rho, \rho] \cap \mathbf{Z}, \quad 1 \leq j \leq d.$$

But it is an easy task to come up with $(2\rho + 1)^d$ distinct solutions, namely those with all

$$x_j \in [-\rho, \rho] \cap \mathbf{Z}.$$

These solutions are all distinct by $x_j r_j \equiv Z \ (\mathrm{mod} Q_j^T)$, for if

$$x_j' \neq x_j'' \quad \text{then} \quad Z' \equiv x_j' r_j \ (\mathrm{mod} \ Q_j^T) \neq Z'' \equiv x_j'' r_j \ (\mathrm{mod} \ Q_j^T).$$

This means that we have found all $(2\rho + 1)^d$ solutions which, in fact, satisfy $x_j \in [-\rho, \rho] \cap \mathbf{Z}$. Also note that $Z \neq 0$ if and only if $\mathbf{x}$ is not the all-zero vector. Since $Z = \sum_{j=1}^d x_j r_j$ and $Z \equiv 0 \ (\mathrm{mod} p_0^R)$ we have shown that $opt_{\mathrm{modSIR}_\infty}(3a) \leq \rho$.
□

Lemma 3.7. *Let I be the* MINIMUM GOOD DIOPHANTINE APPOXIMATION *instance defined by*

$$\alpha_0 := \frac{1}{p_0^R},$$

$$\alpha_j := \frac{r_j}{Q_j^T}, \quad 1 \leq j \leq d,$$

where r_j^*, $1 \leq r_j^* < Q_j^T$, is the unique inverse of r_j $(\mathrm{mod}\, Q_j^T)$. Then, we have

$$\exists Z: Z \neq 0 \wedge |Z| \leq H \wedge Z \equiv 0 \,(\mathrm{mod}\, p_0^R) \wedge \bigvee_{1 \leq j \leq d} Z \equiv \hat{x}_j r_j \,(\mathrm{mod}\, Q_j^T)$$

$$\wedge \bigvee_{1 \leq j \leq d} \hat{x}_j \in \{0, \pm 1\}$$

$$\Longrightarrow \exists Z: Z \neq 0 \wedge |Z| \leq H \wedge \bigvee_{0 \leq j \leq d} \min_{n \in \mathbf{Z}} |Z\alpha_j - n| \leq \tfrac{1}{Q_1^T}$$

$$\forall Z: Z \neq 0 \wedge |Z| \leq \rho H \wedge Z \equiv 0 \,(\mathrm{mod}\, p_0^R) \wedge \bigvee_{1 \leq j \leq d} Z \equiv \hat{x}_j r_j \,(\mathrm{mod}\, Q_j^T)$$

$$\Rightarrow \mathop{\exists}_{1 \leq j \leq d} \hat{x}_j \notin [-\rho, \rho] \cap \mathbf{Z}$$

$$\Longrightarrow \forall Z: Z \neq 0 \wedge |Z| \leq \rho H \Rightarrow \mathop{\exists}_{0 \leq j \leq d} \min_{n \in \mathbf{Z}} |Z\alpha_j - n| > \tfrac{\rho}{Q_1^T}$$

Proof. First, assume there exists a $Z \neq 0$, such that:

$$|Z| \leq H \wedge Z \equiv 0 \,(\mathrm{mod}\, p_0^R) \wedge \bigvee_{1 \leq j \leq d} Z \equiv \hat{x}_j r_j \,(\mathrm{mod}\, Q_j^T) \wedge \bigvee_{1 \leq j \leq d} \hat{x}_j \in \{0, \pm 1\}.$$

Obviously, we have $Z \neq 0 \wedge |Z| \leq H$ and also by $Z \equiv 0 \,(\mathrm{mod}\, p_0^R)$

$$\min_{n \in \mathbf{Z}} \left| Z \tfrac{1}{p_0^R} - n \right| = 0.$$

Moreover, by (2c) and (a) of Lemma A we infer for $1 \leq j \leq d$

$$\min_{n \in \mathbf{Z}} \left| Z \tfrac{r_j^*}{Q_j^T} - n \right| = \min_{n \in \mathbf{Z}} \left| \tfrac{\hat{x}_j r_j r_j^*}{Q_j^T} - n \right| = \min_{n \in \mathbf{Z}} \left| \tfrac{\hat{x}_j}{Q_j^T} - n \right| \leq \tfrac{1}{Q_j^T} \leq \tfrac{1}{Q_1^T}.$$

Thus, there exists a denominator Z with the required properties.

In order to prove the second implication let us now assume

$$\exists Z: Z \neq 0 \wedge |Z| \leq \rho H \wedge \bigvee_{0 \leq j \leq d} \min_{n \in \mathbf{Z}} |Z\alpha_j - n| \leq \tfrac{\rho}{Q_1^T}.$$

Obviously, again we have $Z \neq 0 \wedge |Z| \leq \rho H$ and by (c) of Lemma A we have

$$\frac{1}{p_0^R} > \frac{\rho}{Q_1^T},$$

which together with $\min_{n \in \mathbf{Z}} |Z \tfrac{1}{p_0^R} - n| \leq \tfrac{\rho}{Q_1^T}$ forces $\min_{n \in \mathbf{Z}} |Z \tfrac{1}{p_0^R} - n| = 0$. Thus, $Z \equiv 0 \,(\mathrm{mod}\, p_0^R)$. By (a) and (d) of Lemma A it follows that

$$\frac{\rho + 1}{Q_j^T} > \frac{\rho}{Q_1^T},$$

which together with $\min_{n \in \mathbf{Z}} |Z \tfrac{r_j^*}{Q_j^T} - n| \leq \tfrac{\rho}{Q_1^T}$ enforces $\min_{n \in \mathbf{Z}} |Z \tfrac{r_j^*}{Q_j^T} - n| \leq \tfrac{\rho}{Q_j^T}$. But this is only possible if

$$Z \equiv \hat{x}_j r_j \pmod{Q_j^T} \wedge \hat{x}_j \in [-\rho, \rho] \cap \mathbf{Z}, \quad 1 \leq j \leq d.$$

This of course proves the lemma. $\qquad\qquad\square$

Combining the solution equivalence of the systems $(1a, 1b)$ and $(3a, 3b)$ with Lemma 3.6 and Lemma 3.7 yields the desired polynomial-time transformation τ, since all operations of our reduction can clearly be carried out in time polynomial in $|I|$. □

3.2 Hardness of Approximating Diophantine Approximations

By piecing together the results of Theorem 4 and Theorem 5, we obtain the following:

Main Theorem 8 *Unless* $\mathbf{NP} \subseteq \mathbf{DTIME}(n^{\mathrm{poly}(\log n)})$, *there exists no polynomial-time algorithm which on input* $\alpha \in \mathbb{Q}^d$ *and* $N \in \mathbb{N}_+$ *computes a denominator* Q^* *with* $1 \leq Q^* \leq 2^{\log^{0.5-\gamma} d} N$ *such that*

$$\|Q^*\alpha \bmod \mathbb{Z}\|_\infty \leq 2^{\log^{0.5-\gamma} d} \min_{1 \leq q \leq N} \|q\alpha \bmod \mathbb{Z}\|_\infty,$$

where γ *is an arbitrary small positive constant.*

Corollary 9. *Approximating* MINGDA_∞ *in polynomial-time within a factor* $2^{\log^{0.5-\gamma} d}$ *for an arbitrary small positive constant* γ *is almost* **NP**-*hard.*

4 The Nonhomogeneous Case

To capture the nonhomogeneous case, i.e., the problem $\mathrm{MINGNDA}_\infty$, we will reduce from a well-suited problem, namely:

MINIMUM DIOPHANTINE EQUATION SOLUTION in ℓ_∞-norm (MINDES_∞)
INSTANCE: An equation $x_1 a_1 + \cdots + x_n a_n = b$ with $a_1, \ldots, a_n, b \in \mathbb{Z}$
SOLUTION: A vector $\mathbf{x} \in \mathbb{Z}^n$ such that $\sum_{i=1}^n x_i a_i = b$
MEASURE: The ℓ_∞-norm $\|\mathbf{x}\|_\infty := \max_{1 \leq i \leq n} |x_i|$ of the vector $\mathbf{x}$

Majewski and Havas [13] proved the **NP**-completeness of MINDES_∞ in its feasibility recognition form. Using the Parallel Repetition Theorem of Raz [14] and the techniques of Arora et al. [2] it is not difficult to modify the proof of Theorem 4 from [16] such that even the following holds, see [15] for a detailed proof.

Theorem 10. *There exists an almost polynomial-time, i.e.,* $\mathbf{DTIME}(n^{\mathrm{poly}(\log n)})$ *transformation* τ *from* 3-SAT *to* MINIMUM DIOPHANTINE EQUATION SOLUTION *in* ℓ_∞-*norm such that, for all instances* I,

$$I \in 3\text{-SAT} \implies opt_{\mathrm{MinDES}_\infty}(\tau(I)) = 1$$
$$I \notin 3\text{-SAT} \implies opt_{\mathrm{MinDES}_\infty}(\tau(I)) > 2^{\log^{1-\gamma} |\tau(I)|},$$

where γ *is an arbitrary small positive constant.*

Adapting the reduction in the proof of Theorem 5 to the nonhomogeneous case, the following can be shown:

Theorem 11. *There exists a polynomial-time transformation τ from* SHORTEST INTEGER RELATION *in ℓ_∞-norm to* MINIMUM GOOD NONHOMOGENEOUS DIOPHANTINE APPOXIMATION *in ℓ_∞-norm, $\tau : I \mapsto \langle (a_0/b_0,\ldots,a_d/b_d),\beta,N\rangle$, such that, for all instances I and for all $\rho \geq 1$,*

$$opt_{\mathrm{MinDES}_\infty}(I) = 1 \implies \min_{1 \leq q \leq N} \|q\alpha - \beta \bmod \mathbb{Z}\|_\infty \leq \tfrac{1}{b_1}$$

$$opt_{\mathrm{MinDES}_\infty}(I) > \rho \implies \min_{1 \leq Q^* \leq \rho N} \|Q^*\alpha - \beta \bmod \mathbb{Z}\|_\infty > \rho\tfrac{1}{b_1}.$$

(For the reduction from the MinDES_∞–instance $\langle(a_1,\ldots,a_n,b)\rangle$ we have to ensure that $p_0 \nmid b$. Then, defining the vector β of the instance of MINIMUM GOOD NONHOMOGENEOUS DIOPHANTINE APPROXIMATION in ℓ_∞-norm by $\beta_0 = b/p_0^R$, $\beta_i = 0$, $i = 1,\ldots,d$, admits a straightforward adaption of the proof of Theorem 5.)

By the last Theorem and the NP-completeness of MinDES_∞ we infer:

Main Theorem 12 $\mathrm{MinGNDA}_\infty$ *is NP-complete (in its feasibility recognition form).*

Moreover, Theorem 10 and Theorem 11 imply:

Main Theorem 13 *Unless* $\mathrm{NP} \subseteq \mathrm{DTIME}(n^{\mathrm{poly}(\log n)})$, *there exists no polynomial-time algorithm which on input $\alpha,\beta \in \mathbb{Q}^d$ and $N \in \mathbb{N}_+$ computes a denominator Q^* with $1 \leq Q^* \leq 2^{\log^{1-\gamma} d} N$ such that*

$$\|Q^*\alpha - \beta \bmod \mathbb{Z}\|_\infty \leq 2^{\log^{1-\gamma} d} \min_{1 \leq q \leq N} \|q\alpha - \beta \bmod \mathbb{Z}\|_\infty,$$

where γ is an arbitrary small positive constant.

Corollary 14. *Approximating* $\mathrm{MinGNDA}_\infty$ *in polynomial-time within a factor $2^{\log^{1-\gamma} d}$ for an arbitrary small positive constant γ is almost NP-hard.*

Acknowledgment

We would like to thank Jeff Lagarias for several helpful comments on possible improvements of this paper.

References

1. S. Arora. *Probabilistic Checking of Proofs and Hardness of Approximation Problems.* Ph.D. thesis, University of California at Berkeley, 1994.
2. S. Arora, L. Babai, J. Stern and Z Sweedyk. The hardness of approximate optima in lattices, codes and systems of linear equations. In *Proc. 34th IEEE Symp. on Foundations of Computer Science*, pages 724–730, 1993.

3. S. Arora and C. Lund. Hardness of approximation. In D. Hochbaum (editor), *Approximation Algorithms for NP-hard problems*, Chapter 11. PWS Publ., 1996.
4. S. Arora, C. Lund, R. Motwani, M. Sudan and M. Szegedy. Proof verification and hardness of approximation problems. In *Proc. 33rd IEEE Symp. on Foundations of Computer Science*, pages 14–23, 1992.
5. G. Ausiello, P. Crescenzi and M. Protasi. Approximate solutions of NP optimization problems. *Theoretical Computer Science*, Volume 150, pages 1–55, 1995.
6. P. Crescenzi and V. Kann. A list of NP-complete optimization problems. Surveys on complexity, Electronic Colloqium on Computational Complexity, http://www.informatik.uni-trier.de/eccc/, 1996.
7. U. Feige and L. Lovász. Two-prover one-round proof systems: Their power and their problems. In *Proc. 24th ACM Symp. Theory of Computing*, pages 643–654, 1992.
8. A. Frank and É. Tardos. An application of simultaneous diophantine approximation in combinatorial optimization. *Combinatorica*, Volume 7, pages 49–65, 1987.
9. D. R. Heath-Brown. The number of primes in a short interval. *J. reine angew. Math.*, Volume 389, pages 22–63, 1988.
10. D. R. Heath-Brown and H. Iwaniec. On the difference between consecutive primes. *Inventiones math.*, Volume 55, pages 49–69, 1979.
11. R. Kannan, A. K. Lenstra and L. Lovász. Polynomial factorization and nonrandomness of bits of algebraic and some transcendental numbers. *Math. Comp.*, Volume 50, pages 235–250, 1988.
12. J. C. Lagarias. The computational complexity of simultaneous diophantine approximation problems. *SIAM J. Comput.*, Volume 14, pages 196–209, 1985.
13. B. S. Majewski and G. Havas. The complexity of greatest common divisor computations. In *Proc. 1st International Symposium on Algorithmic Number Theory*, pages 184–193. Springer, 1994. LNCS 877.
14. R. Raz. A parallel repetition theorem. In *Proc. 27th ACM Symp. Theory of Computing*, pages 447–456, 1995.
15. C. Rössner and J.-P. Seifert. The complexity of approximate optima for greatest common divisor computations. In *Proc. 2nd Algorithmic Number Theory Symposium*, pages ?–? Springer, 1996. LNCS.
16. C. Rössner and J.-P. Seifert. On the hardness of approximating shortest integer relations among rational numbers. In *Proc. CATS'96 (Computing: The Australasian Theory Symposium)*, pages 180–186, 1996.
17. C. P. Schnorr. Factoring integers and computing discrete logarithms via diophantine approximations. *AMS DIMACS Series in Disc. Math. and Theoretical Comp. Science*, Volume 13, pages 171–181, 1993.
18. A. Schoenhage. Factorization of univariate integer polynomials by diophantine approximation and an improved basis reduction algorithm. In *11th ICALP*, pages 436–447. Springer, 1987. LNCS 172.
19. P. van Emde Boas. Another NP-complete partition problem and the complexity of computing short vectors in a lattice. Technical Report 81-04, Math. Inst., University of Amsterdam, 1981.

Appendix

We will prove that for suitable choices of T we can find in $O(n^{50})$ bit operations an interval containing d prime numbers $Q_1, \ldots, Q_d$ satisfying the conditions (a)-

(d) of Lemma A.

Proof. (of Lemma A) Let $n := |I|$ denote the length of the given SIR_∞ instance I, i.e., the vector $\mathbf{a} = (a_1, \ldots, a_d) \in \mathbf{Z}^d$. Obviously, $n \geq d$.

As the binary length of $\prod_{i=1}^{d} a_i$ is bounded by dn, this product has at most $dn \leq n^2$ distinct prime factors. Therefore, p_0 will be one of the first $(n^2 + 1)$ primes which can be found by a brute force trial division in $O(n^4)$ bit operations. Using $\rho \leq n$ and the specific choice of p_0 and R we have

$$2^{3n^2} \geq 2^{n^2+1} 2^{\log \rho} 2^{n \log d} \geq 2^{n^2+1} \rho \sum_{j=1}^{d} |a_j| \geq p_0^R.$$

Hence, setting $T := 4n^2$, guarantees $Q_1^T \geq 4\rho(d+1)p_0^R$, i.e., condition (c) holds.

In order to find a set of primes $\{Q_1, \ldots, Q_d\}$ satisfying the remaining conditions of Lemma A we invoke the following primitive search routine:

for every $x = 1, 2, \cdots$
  if $[\rho^{1/T} x, (\rho + 1)^{1/T} x] =: I_x$ contains $\geq d + n^2 + 1$ distinct primes **then** stop;

If this search stops with x, we are guaranteed that for this choice of x at least d primes in I_x satisfy the condition (b) since $p_0 \prod_{i=1}^{d} a_i$ has at most n^2+1 distinct prime factors. Moreover, the conditions (a) and (d) are satisfied by selecting the suited primes in the interval I_x.

The main difficulty is now to prove that the above search routine performs at most n^k bit operations for some $k \in \mathbf{N}$. Thus, we must give an upper bound for the value of x for which the search algorithm stops. We use the following number-theoretic result on the number of primes in a short interval.

Theorem [10, 9]. *For each $\delta > \frac{11}{20}$ there exists a constant x_δ such that the interval $[x, x + x^\delta]$ contains for all $x > x_\delta$ a prime.*

From $\rho \leq n$, we derive

$$\left(\frac{\rho+1}{\rho}\right)^{1/T} \geq 1 + \ln\left(\frac{\rho+1}{\rho}\right)\frac{1}{T} \geq 1 + \frac{1}{2\rho}\frac{1}{4n^2} \geq 1 + \frac{1}{8n^3}.$$

Setting $x := \frac{n^{20}}{\rho^{1/T}}$, we infer

$$I_x = [\rho^{1/T} x, (\rho + 1)^{1/T} x] = [n^{20}, (\frac{\rho+1}{\rho})^{1/T} n^{20}] \supseteq [n^{20}, n^{20} + \frac{1}{8}n^{17}].$$

For the choice of $\delta := \frac{3}{5}$ the above Theorem guarantees that we can find in the interval $[n^{20}, n^{20} + n^{12}]$ a prime, if n is sufficiently large. Since we can locate in the interval $[n^{20}, n^{20} + n^{17}]$ all the intervals of the form

$$[n^{20} + i\, n^{12}, n^{20} + (i + 1)\, n^{12}], \qquad i = 0, \ldots, n^5 - 1,$$

we will be able to find at least n^5 distinct primes.

As the primality of each number x in I_x can be tested in $O((x^{3/2}+xn^2)(\log x)^2)$ bit operations, the above search routine uses at most $O(n^{50})$ bit operations. $\square$

On the Conjugation of Standard Morphisms*

Patrice Séébold
LaRIA
Université de Picardie
CURI
5 rue du Moulin Neuf
F-80000 Amiens

May 7, 1996

Abstract

Let $A = \{a, b\}$ be an alphabet. An infinite word on A is *Sturmian* if it contains exactly $n + 1$ distinct factors of length n for every integer n. A morphism f on A is *Sturmian* if $f(\mathbf{x})$ is Sturmian whenever $\mathbf{x}$ is. A morphism on A is *Standard* if it is an element of the monoid generated by the two elementary morphisms E, which exchanges a and b, and ϕ, the Fibonacci morphism defined by $\phi(a) = ab$ and $\phi(b) = a$. The set of Standard morphisms is a proper subset of the set of Sturmian morphisms. In the present paper, we give a characterization of Sturmian morphisms as conjugates of Standard ones. Sturmian words generated by Standard morphisms are characteristic words. The previous result allows to prove that a morphism f generates an infinite word having the same set of factors as a characteristic word generated by a Standard morphism g if and only if f is a conjugate of g.

1 Introduction

A Sturmian word is an infinite word which contains exactly $n+1$ distinct factors of length n for every integer n. So a sturmian word is necessarily a binary one. Many other characterizations of the sturmian words can be found in the literature (see, for example, [14, 15, 20, 25, 31, 34]). These words have numerous properties (see, for example, [4, 5]) and they are used in various fields of Mathematics such as the symbolic dynamics (see [15, 20, 19, 30]), the study of continued fraction expansion (see [3, 26, 35] and also a lot of recent works [1, 6, 7, 10, 11, 13, 14, 16, 21, 22, 28]) or others ([11, 33, 36]), but also in Physics ([2, 9]) and, of course, in numerous domains of Computer Science as infography ([12]), formal

*Partially supported by PRC "Mathématiques et Informatique" and by ESPRIT BRA working group 6317 - ASMICS 2.

languages theory and algorithms on words ([18, 23, 27]) or combinatorics on words ([17, 24]).

Here, we are interested in Sturmian morphisms, i.e. morphisms such that the image of all Sturmian words are Sturmian words. These morphisms have been recently widely studied ([1, 6, 7, 8, 16, 29, 32]). In particular, some of such morphisms have been obtained, in [29], from the rules of Rauzy [31]. Here, we greatly complete and generalize the results of [29]. More precisely, we use the characterization of Standard morphisms [8] to prove that the infinite words generated by these morphisms are characteristic words. Moreover, we prove that all the Sturmian morphisms are conjugates of the Standard ones, which gives a new characterization of Sturmian morphisms and allows to establish that if two morphisms generate Sturmian words having the same set of factors then one of them is a conjugate of a power of the other.

After the main definitions (section 2), we give some details about the notions of Sturmian morphisms (section 3) and Standard morphisms (section 4). Section 5 is dedicated to the results and some elements of the proofs are given in section 6.

2 Definitions and notations

Let $A = \{a, b\}$ be a binary alphabet. A^*, set of all the words on A, is the free monoid generated by A, ε the empty word and $A^+ = A^* \setminus \{\varepsilon\}$. For every $u \in A^*$, $|u|$ is the number of letters in the word u ($|\varepsilon| = 0$) and if $c \in A$, $|u|_c$ is the number of occurrences of the letter c in the word u.

An *infinite word* on A is an application $\mathbf{x} : \mathbb{N} \to A$. We denote by A^ω the set of infinite words on A and $A^\infty = A^\omega \cup A^*$.

Let $\mathbf{x}$ be an infinite word on A. $\mathbf{x}$ is *ultimately periodic* if there exist two words $u \in A^*$ and $v \in A^+$ such that $\mathbf{x} = uv^\omega$.

Let $u \in A^\infty$ and $v \in A^*$. v is a *factor* of u if there exist $u_1 \in A^*$ and $u_2 \in A^\infty$ such that $u = u_1vu_2$; if $u_1 = \varepsilon$, v is a *left factor* of u and if $u_2 = \varepsilon$, v is a *right factor* of u.

A *morphism* f is an application from A^* onto itself such that $f(uv) = f(u)f(v)$ for all words $u, v \in A^*$. The morphism f is *non-erasing* if neither $f(a)$ nor $f(b)$ is the empty word.

If there exist a letter $c \in A$ and a word $u \in A^+$ such that $f(c) = cu$ and, for all $n \in \mathbb{N}$, $|f^{n+1}(c)| > |f^n(c)|$ then f is *prolongeable on c* (this is in particular the case if f is non-erasing and $|f(c)| \geq 2$). In this case, f *generates* an infinite word $\mathbf{x} = \lim_{n \to \infty} f^n(c)$. Remark that on A, binary alphabet, every morphism f is such that, for all letters $x \in A$, either f or f^2 is prolongeable on x, or not any power of f is prolongeable on x (thus either f or f^2 generates an infinite word, or not any power of f generates an infinite word). Remark also that if f generates an infinite word $\mathbf{x}$ then $\mathbf{x}$ is a fixed point of f, i.e. $\mathbf{x} = f(\mathbf{x})$.

3   Sturmian Morphisms

Let u and v be two words with the same length. The *difference* between u and v is the number

$$\delta(u, v) = \Big||u|_a - |v|_a\Big| = \Big||u|_b - |v|_b\Big|$$

A set of words X is *balanced* if

$$u, v \in X, |u| = |v| \Rightarrow \delta(u, v) \leq 1$$

An infinite word $\mathbf{x}$ is balanced if the set $Fact(\mathbf{x})$ of its factors is balanced.

The following property is a well known characterization of the Sturmian words.

Property 1 *Let* $\mathbf{x} \in A^\omega$. $\mathbf{x}$ *is Sturmian if and only if* $\mathbf{x}$ *is a non ultimately periodic balanced word.*

Let us now consider the two morphisms

$$\phi : \begin{array}{l} a \mapsto ab \\ b \mapsto a \end{array} \qquad \tilde{\phi} : \begin{array}{l} a \mapsto ba \\ b \mapsto a \end{array}$$

A morphism $f : A^* \to A^*$ is *Sturmian* if $f(\mathbf{x})$ is Sturmian for every sturmian word $\mathbf{x}$. f is *locally Sturmian* if there exists at least one Sturmian word $\mathbf{x}$ such that $f(\mathbf{x})$ is Sturmian. The identity morphism Id_A and the morphism E which exchanges the letters a and b are evidently Sturmian (and, thus, also locally Sturmian). Moreover, one has

Property 2 ([32]) *The morphisms* ϕ *and* $\tilde{\phi}$ *are Sturmian.*

Let us denote $St = \{E, \phi, \tilde{\phi}\}^*$ the set of morphisms obtained by composition of E, ϕ and $\tilde{\phi}$ in any number and order. St, which is called the *monoid of Sturm*, is the set of all the Sturmian morphisms and it has the following characterization:

Theorem 1 ([6, 29]) *Let* $f : A^* \to A^*$ *be a morphism. The following four conditions are equivalent:*

(i) $f \in St$;

(ii) f *is Sturmian;*

(iii) f *is locally Sturmian;*

(iv) $f(ab) \neq f(ba)$ *and the word* $f(ba^2ba^2baba^2bab)$ *is balanced.*

We just add here a little new result which will be useful:

Lemma 1 *Let* $g : A^* \to A^*$ *be a morphism. If* $\mathbf{s} = (g^2)^\omega(x)$ *is a Sturmian word for a letter* $x \in A$ *then* $g \in St$.

4   Standard morphisms

An interesting particular case of Sturmian morphisms is that of *Standard morphisms*.

Let us consider the family R of (unordered) pairs of words of A^* defined as the smallest set of pairs of words such that:

1. $\{a, b\} \in R$;

2. $\{u, v\} \in R \Rightarrow \{uv, u\} \in R$.

The elements of R are called *Standard pairs* and the relation between Standard pairs and Sturmian morphisms is given by the following theorem:

Theorem 2 ([8]) *Let f be a morphism on A. $f \in \{E, \phi\}^*$ if and only if the set $\{f(a), f(b)\}$ is a Standard pair.*

The elements of $\{E, \phi\}^*$ are called *Standard morphisms*. What is new is that the morphism $\tilde{\phi}$ is not used in the decomposition of a Standard morphism. These morphisms will have a central part in what follows and we just give here a property which is very easy to prove

Property 3 *Let f be a Standard morphism. One and only one of the two following assertions is true:*

- $f(a) = a$ *or* $f(b) = b$, *and in this case not any power of f is prolongeable on a nor on b;*

- f *is prolongeable on one and only one of the two letters a or b.*

Standard pairs are also closely related to the construction of characteristic Sturmian words by using the *rules of Rauzy* [31]: a Sturmian word **x** is *characteristic* if both $a\mathbf{x}$ and $b\mathbf{x}$ are Sturmian (see, for example, [8]); two sequences of words $(A_n)_{n\in\mathbb{N}}$ and $(B_n)_{n\in\mathbb{N}}$ are constructed as follows

$$A_0 = a, \quad B_0 = b$$

and

$$
\begin{array}{ccc}
A_{n+1} = A_n & & A_{n+1} = B_n A_n \\
B_{n+1} = A_n B_n & \text{or} & B_{n+1} = B_n
\end{array}
$$

the two rules being both used infinitely often.

Clearly, for all $n \in \mathbb{N}$ and for every sequence of rules of Rauzy, $\{A_n, B_n\}$ is a Standard pair. Moreover

Proposition 1 ([31]) *The two sequences $(A_n)_{n\in\mathbb{N}}$ and $(B_n)_{n\in\mathbb{N}}$ have the same limit which is a characteristic word; conversely, any characteristic word is the limit of two such sequences.*

The relation between characteristic words and Standard morphisms is given by the following

Proposition 2 ([6]) *Let s, t be two characteristic words and $f : A^* \to A^*$ a morphism. If $\mathbf{t} = f(\mathbf{s})$ then f is Standard.*

5 Results

Our aim here is to characterize Sturmian morphisms from Standard ones and to cleverly study the relations between Sturmian morphisms which generate words having the same set of factors. The central notion is that of *conjugates* of a morphism.

Let f and g be two morphisms on A. g is a *conjugate* of f if there exists $s \in A^*$ such that $sg(ab) = f(ab)s$ and $|g(a)| = |f(a)|$ (which of course implies that $|g(b)| = |f(b)|$).

A morphism f has $|f(ab)|$ conjugates (including itself when $s = \varepsilon$). With each conjugate is associated a unique word s such that $|s| < |f(ab)|$ and this conjugate is denoted $f_{|s|}$. In particular, in what follows, we call *first conjugate* of f the conjugate f_1 and *last conjugate* of f the conjugate $f_{|f(ab)|-1}$. All the conjugates except the last one are called *good conjugates*.

Example Let $f : A^* \to A^*$ be defined by $f(a) = ababa$, $f(b) = ab$.

f has 7 conjugates :

$f_0 : a \mapsto ababa$, $b \mapsto ab$
$f_1 : a \mapsto babaa$, $b \mapsto ba$
$f_2 : a \mapsto abaab$, $b \mapsto ab$
$f_3 : a \mapsto baaba$, $b \mapsto ba$
$f_4 : a \mapsto aabab$, $b \mapsto ab$
$f_5 : a \mapsto ababa$, $b \mapsto ba$
$f_6 : a \mapsto babab$, $b \mapsto aa$

$f_0 = f$, the first conjugate is f_1, the last conjugate is f_6.

In what follows, we call *non trivial* any morphism different from E and Id_A. Let us start with a result which indicates that it is necessary to distinguish the last conjugate of a Standard morphism.

Proposition 3 *Let f be a non trivial Standard morphism. The last conjugate of f is not Sturmian.*

The first fundamental result is the following

Theorem 3 *A non trivial morphism g is Sturmian if and only if there exists a Standard morphism f such that g is a good conjugate of f (in other words, $g = f_k$ with $0 \le k \le |g(ab)| - 2$).*

Moreover

Proposition 4 *Let $g \in St$. g is a good conjugate of a power of a Standard morphism f if and only if g is a composition of good conjugates of f.*

And

Proposition 5 *Let f and g be two Sturmian morphisms, good conjugates of the same Standard morphism. Then, for all $x, y \in A$, $|g(x)|_y = |f(x)|_y$.*

The problem is now to find which of these morphisms generate Sturmian words and what are the relations between these. The first results concern Standard morphisms which have here a central part.

Let us call G, $\bar{\phi}$ and $\bar{G}$ the morphisms $\phi \circ E$, $E \circ \phi$ and $E \circ G = E \circ \phi \circ E$.

The set of non trivial Standard morphisms is clearly equal to the set of all the compositions of ϕ, G, $\bar{\phi}$ and $\bar{G}$:

$$\{E, \phi\}^* = \{\phi, G, \bar{\phi}, \bar{G}\}^+ \cup \{E\}^*.$$

Some of these morphisms do not generate any Sturmian word: this is the case, for example, with G and $\bar{\phi}$ since no power of these two morphisms is prolongeable on a, nor on b. More precisely,

Proposition 6 *Let f be a Standard morphism. f generates a Sturmian word if and only if $f \in \{\phi, G, \bar{\phi}, \bar{G}\}^+ \setminus (\{G\}^+ \cup \{\bar{\phi}\}^+)$.*

This allows to establish what are the Sturmian words generated by Standard morphisms:

Theorem 4 *Let f be a Standard morphism. If f generates a Sturmian word then this is a characteristic word.*

Such Standard morphisms are called *characteristic morphisms*. Then we have a second fundamental result (a morphism is *primitive* if it is not a power of another morphism):

Theorem 5 *Let f be a characteristic morphism and $\mathbf{x}$ the characteristic word generated by f. Then there exists a primitive characteristic morphism h such that*

1. *$f = h^n$ for an integer n ;*

2. *a morphism $g : A^* \to A^*$ generates an infinite word having the same set of factors as $\mathbf{x}$ if and only if g is a good conjugate of a power of h.*

In particular, every Sturmian morphism which generates an infinite word having the same set of factors as the word generated by a primitive characteristic morphism is a conjugate of a power this morphism.

An infinite word generated by a morphism is *rigid* if all the morphisms which generate this word are powers of the same unique morphism. A direct consequence of the previous result is then

Theorem 6 *Sturmian words generated by morphisms are all rigid.*

6 Proofs

We just give here the main points of the proofs, details will be given in the final version.

We start with the proof of theorem 3 which is an immediate corollary of the two following propositions.

Proposition 7 *For every Sturmian morphism g there exists a Standard morphism f such that g is a conjugate of f.*

Proposition 8 *Let f be a non trivial Standard morphism. All the conjugates of f except the last one are Sturmian morphisms.*

Proposition 3 is of course a direct consequence of proposition 8.

Proposition 7 is based on the following lemma:

Lemma 2 *Let $g : A^* \to A^*$ be a morphism. If g is a conjugate of a Standard morphism f, then $\phi \circ g$ and $\tilde{\phi} \circ g$ are conjugates of $\phi \circ f$ and $E \circ g$ is a conjugate of $E \circ f$.*

The proof of proposition 8 is more difficult and requires a careful study of the conjugates of a Standard morphism.

We denote $\Sigma = \Sigma_1 \cup \Sigma_2$ with $\Sigma_1 = \{\phi, G, \tilde{\phi}, \tilde{G}\}$ and Σ_2 is such that, for all $\sigma \in \Sigma_1$, $\tilde{\sigma} \in \Sigma_2$ (if f is a morphism on A, $\tilde{f}$ is defined by: $\tilde{f}(a)$ is the mirror image of $f(a)$ and $\tilde{f}(b)$ is the mirror image of $f(b)$). It is known [29] that Σ^+ is the set of non trivial Sturmian morphisms: $St = \Sigma^+ \cup \{E\}^*$.

The proof of proposition 8 is decomposed in three lemmas.

Lemma 3 *Let f be a non trivial Sturmian morphism, then one of the two words $f(a)$ or $f(b)$ is a left or a right factor of the other one.*

Lemma 4 *Let f be a non trivial Sturmian morphism and f_1 its first conjugate. f_1 is Sturmian if and only if $f(a)$ and $f(b)$ start with the same letter.*

Lemma 5 *Let f be a non trivial Standard morphism and g a conjugate of f. $g(a)$ and $g(b)$ do not start with the same letter if and only if g is one of the last two conjugates of f.*

The proof of proposition 6 is based on the following lemma:

Lemma 6 *Let f be a Standard morphism :*

- *$f(a) = a$ if and only if there exists $n \in \mathbb{N}$ such that $f = G^n$;*
- *$f(b) = b$ if and only if there exists $n \in \mathbb{N}$ such that $f = \tilde{\phi}^n$.*

We now turn to the proof of theorems 4 and 5.

The proof of theorem 4 is just an application of propositions 1 and 6.

The proof of theorem 5 is more complicated and needs some intermediate results.

Lemma 7 *Let f be a characteristic morphism. Then any primitive morphism g on A, such that f is a power of g, is Standard.*

The two following results are rather technical and constitute the central part of the proof.

Proposition 9 *Two primitive characteristic morphisms generate the same word if and only if they are equal.*

Lemma 8 *Let $g \in St$ be a morphism which generates a Sturmian word $\mathbf{x}$. Then g is a conjugate of a characteristic morphism f which generates a word $\mathbf{y}$ having the same set of factors as $\mathbf{x}$.*

The following result is an analogous, for the Sturmian morphisms, of proposition 6:

Proposition 10 ([29]) *Let $f : A^* \to A^*$ be a Sturmian morphism. f generates a Sturmian word if and only if $f \in \Sigma^+ \setminus (\{G, \tilde{G}\}^+ \cup \{\bar{\phi}, \tilde{\phi}\}^+)$.*

Moreover

Lemma 9 *Let $f : A^* \to A^*$ be a morphism.*
If $f \in \{G, \tilde{G}\}^+$ then there exists $n \in \mathbb{N}$ such that f is a conjugate of G^n.
If $f \in \{\bar{\phi}, \tilde{\phi}\}^+$ then there exists $n \in \mathbb{N}$ such that f is a conjugate of $\bar{\phi}^n$.

A last result gives the relation between the morphisms that are used in the decomposition of a Sturmian morphism and these that are used in the decomposition of the Standard morphism of which it is a conjugate.

Lemma 10 *Let g be a Sturmian morphism. g is a good conjugate of a Standard morphism f if and only if there exists $n \in \mathbb{N}$ such that $f = f_1 \circ \ldots \circ f_n$ and $g = g_1 \circ \ldots \circ g_n$ where, for all $i \in \mathbb{N}$, $1 \le i \le n$, $f_i \in \{E, \phi\}$, $g_i \in \{E, \phi, \bar{\phi}\}$ and, if $g_i = \phi$ or $g_i = \tilde{\phi}$ then $f_i = \phi$, if $g_i = E$ then $f_i = E$.*

This ends the proof of theorem 5 and propositions 4 and 5 are also direct consequences of this lemma.

Acknowledgements

The author is greatly indebted to Jean Berstel for very helpful discussions during the preparation of this work and thanks an anonymous referee for a careful reading of the submitted version.

References

[1] J.-P. ALLOUCHE, Sur la complexité des suites infinies, *Bull. Belg. Math. Soc.* **1** (1994), 133–143.

[2] J.-P. ALLOUCHE, M. MENDÈS FRANCE, Quasicrystal Ising Chain and Automata Theory, *J. Stat. Phys.* **42** (1986), 809–821.

[3] J. BERNOULLI III, Sur une nouvelle espèce de calcul, in *Recueil pour les astronomes, vol. 1*, Berlin (1772), 255–284.

[4] J. BERSTEL, Mots de Fibonacci, in *Séminaire d'Informatique Théorique, L.I.T.P. Universités Paris 6 et 7*, Année 1980-81, 57–78.

[5] J. BERSTEL, Recent results on Sturmian words, in *Proceedings DLT'95* (J. Dassow Ed.), (1995), to appear.

[6] J. BERSTEL, P. SÉÉBOLD, A Characterization of Sturmian Morphisms, in *MFCS'93* (A. Borzyskowski, S. Sokolowski eds.), *Lect. Notes Comp. Sci.* **711** (1993), 281–290.

[7] J. BERSTEL, P. SÉÉBOLD, Morphismes de Sturm, *Bull. Belg. Math. Soc.* **1** (1994), 175–189.

[8] J. BERSTEL, P. SÉÉBOLD, A Remark on Morphic Sturmian Words, *Informatique théorique et applications* **28** (1994), 255–263.

[9] E. BOMBIERI, J. E. TAYLOR, Which distributions of matter diffract ? An initial investigation, *J. Phys.* **47** (1986), Colloque C3, 19–28.

[10] J.-P. BOREL, F. LAUBIE, Construction de mots de Christoffel, *C. R. Acad. Sci. Paris* **313** (1991), 483–485.

[11] J.-P. BOREL, F. LAUBIE, Quelques mots sur la droite projective réelle, *J. Théorie des Nombres de Bordeaux* **5** (1993), 23–51.

[12] J. E. BRESENHAM, Algorithm for computer control of a digital plotter, *IBM Systems J.* **4** (1965), 25–30.

[13] T. C. BROWN, A characterization of the quadratic irrationals, *Canad. Math. Bull.* **34** (1991), 36–41.

[14] T. C. BROWN, Descriptions of the Characteristic Sequence of an Irrational, *Canad. Math. Bull.* **36** (1993), 15–21.

[15] E. COVEN, G. HEDLUND, Sequences with minimal block growth, *Math. Systems Theory* **7** (1973), 138–153.

[16] D. CRISP, W. MORAN, A. POLLINGTON, P. SHIUE, Substitution Invariant Cutting Sequences, *J. Théorie des Nombres de Bordeaux* **5** (1993), 123–138.

[17] X. DROUBAY, Palindromes in the Fibonacci word, *Inform. Proc. Letters* **55** (1995), 217–221.

[18] S. DULUCQ, D. GOUYOU-BEAUCHAMPS, Sur les facteurs des suites de Sturm, *Theoret. Comput. Sci.* **71** (1990), 381–400.

[19] G.A. HEDLUND, Sturmian minimal sets, *Amer. J. Math* **66** (1944), 605–620.

[20] G.A. HEDLUND, M. MORSE, Symbolic dynamics II - Sturmian trajectories, *Amer. J. Math* **62** (1940), 1–42.

[21] S. ITO, S. YASUTOMI, On continued fractions, substitutions and characteristic sequences, *Japan. J. Math.* **16** (1990), 287–306.

[22] F. LAUBIE, É. LAURIER, Calcul de multiples de mots de Christoffel, *C. R. Acad. Sci. Paris* **320** (1995), 765–768.

[23] M. LOTHAIRE, *Combinatorics on Words*, Encyclopedia of Mathematics and Applications vol. 17, Addison-Wesley, Reading, Mass., 1983.

[24] A. DE LUCA, F. MIGNOSI, Some combinatorial properties of Sturmian words, *Theoret. Comput. Sci.* **136** (1994), 361–385.

[25] W. F. LUNNON, P. A. B. PLEASANTS, Characterization of two-distance sequences, *J. Australian Math. Soc.* **53** (1992), 198–218.

[26] A. A. MARKOFF, Sur une question de Jean Bernoulli, *Math. Ann.* **19** (1882), 27–36.

[27] F. MIGNOSI, Infinite words with linear subwords complexity, *Theoret. Comput. Sci.* **65** (1989), 221–242.

[28] F. MIGNOSI, On the number of factors of Sturmian words, *Theoret. Comput. Sci.* **82** (1991), 71–84.

[29] F. MIGNOSI, P. SÉÉBOLD, Morphismes sturmiens et règles de Rauzy, *J. Théorie des Nombres de Bordeaux* **5** (1993), 221–233.

[30] M. QUEFFÉLEC, *Substitution Dynamical Systems – Spectral Analysis* Lecture Notes Math.,vol.1294, Springer-Verlag, 1987.

[31] G. RAUZY, Mots infinis en arithmétique, in :*Automata on infinite words* (D. Perrin ed.), *Lect. Notes Comp. Sci.* **192** (1985), 165–171.

[32] P. SÉÉBOLD, Fibonacci morphisms and Sturmian words, *Theoret. Comput. Sci.* **88** 1991, 365–384.

[33] C. SERIES, The geometry of Markoff numbers, *The Mathematical Intelligencer* **7** (1985), 20–29.

[34] K. B. STOLARSKY, Beatty sequences, continued fractions, and certain shift operators, *Canad. Math. Bull.* **19** (1976), 473–482.

[35] B. A. VENKOV, *Elementary Number Theory*, Wolters-Noordhoff, Groningen, 1970.

[36] Z.-X. WEN, Z.-Y. WEN, Local isomorphisms of invertible substitutions, *C. R. Acad. Sci. Paris* **318** (1994), 299–304.

A Semantic Matching Algorithm:
Analysis and Implementation

Hui Shi*

Universität Bremen, FB3 Informatik, Bibliothekstr. 4, 28359 Bremen, Germany
E-mail: shi@informatik.uni-bremen.de

Abstract. A decision procedure for a class of semantic matching problems
was proposed in [3], but it yielded efficiency problems, principally because of
redundancies. We present in this paper a new semantic matching algorithm
for a restricted class of convergent rewrite systems, its theoretical properties,
and an efficient implementation. This class of rewrite systems is particularly interesting for functional programming and for program manipulations.
To make the algorithm efficient, two techniques are introduced: *dependency
analysis* and *sharing analysis*. The possibility to divide the semantic matching problems into three complexity classes *linear*, *polynomial*, and *decidable* is
also an advantage of our approach. A number of test results shed some light
on the efficiency of an ML implementation in the theorem prover Isabelle.

1   Introduction

Semantic matching is a useful mechanism. It is well-known that any strategy for
finding a complete set of matches with respect to a first-order theory may not terminate, even when the theory is presented as a convergent rewrite system. A decision
procedure has been proposed in [3] for a class of convergent rewrite systems. However, it is not always decidable, whether a rewrite system belongs to this class or
not, nor efficiency problems have been considered there.

In this paper we modify the class of rewrite systems presented in [3] in three
aspects. First, we study a class of *constructor-based* rewrite systems in consideration of the efficience of its semantic matching procedure and the application in
functional programming and program manipulations [8]. Second, we treat variable-
dropping and variable-preserving rewrite systems uniformly. Third, it is always decidable whether a rewrite system belongs to this class or not. In order to get a
decidable semantic matching procedure, we restrict constructor-based rewrite systems to *acyclic* ones.

The following rewrite system defines the maximum function '*max*' over the natural numbers with the constructors '0' and '*s*' and the height function '*ht*' over binary
trees with the constructors '*empty*' and '*node*' forming a tree from two trees.

$$
\begin{array}{ll}
max(0, y) \rightarrow y & ht(empty) \rightarrow 0 \\
max(s(x), 0) \rightarrow s(x) & ht(node(x, y)) \rightarrow s(max(ht(x), ht(y))) \\
max(s(x), s(y)) \rightarrow s(max(x, y)) &
\end{array}
$$

* Research partially supported by BMBF Project *UniForM*.

Although the algorithm described in Section 3 has some nice properties, such as completeness and termination, after implementing the algorithm in SML/NJ [5] directly, we were very surprised and also disappointed at how inefficient it could be even for some simple matching problems. For instance, it takes more than half an hour to compute all the possible trees of height 4. A combinatorial investigation shows that there are 651 such trees! For practical use, additional techniques to improve the algorithm are indispensable. The techniques we applied are:

- dependency analysis – independent subgoals will be solved independently; and
- sharing analysis – a subgoal will never be solved more than once.

After applying these techniques, the algorithm has been obviously improved: all 651 trees of height 4 are produced within 2 minutes. A number of test results presented in Section 5 give an intuitive view on the performance gain.

An additional benefit of theses techniques is that they make it possible to classify matching problems according to their complexities: linear, polynomial, and decidable. Such a refinement classification has rarely been done in the literature about semantic unification or semantic matching, though it is important for promoting the application of them.

The rest of this paper is organized as follows. Section 2 introduces the basic definitions and notations that will be used in this paper. Section 3 presents a class of convergent rewrite systems, the matching algorithm, and its properties. The improvement techniques are discussed in Section 4. In Section 5 the algorithm will be analysed and some test results will be presented. We conclude in Section 6.

2 Preliminaries

In this section we briefly review the relevant basic notations, terminology, and results. For surveys of this area, see [2, 7].

Terms are constructed from a given set of function symbols and variables. We normally use s, t, l, and r for terms, and x, y, z for variables. A *ground* term is one containing no variables. A term is said to be *linear* if no variables occur more than once. The *set of variables* occurring in a term t is denoted by $\mathcal{V}(t)$.

A precise formalism for describing a subterm is obtained through the notion of *occurrence*, which describes the path from the root to the subterm. We just briefly review the notation, details can be found in [2]. The set of all occurrences in a term t is denoted by $\mathcal{O}(t)$. The letters p and q stand for occurrences. The root occurrence is ε. The *subterm* of t at occurrence p will be denoted as $t|_p$. The replacement of a subterm of t at occurrence p with s will be written as $t[s]_p$. The *depth* of a term t is the length of the longest occurrence of t, denoted as $|t|$.

A *Substitution* σ is a function denoted by $\{(x_1, s_1), \cdots, (x_n, s_n)\}$. The *domain* and the *range* of σ are defined as $dom(\sigma) = \{x_1, \cdots, x_n\}$ and $ran(\sigma) = \mathcal{V}(s_1) \cup \cdots \cup \mathcal{V}(s_n)$ respectively. Renamings are special substitutions, which are functions from variables to variables. The composition of two substitutions σ and τ, denoted as $\sigma \circ \tau$, and the restriction of a substitution σ to a set of variables W, written as $\sigma[W]$, are defined as usual: $\sigma \circ \tau\ (x) = \sigma(\tau(x))$ and $\sigma[W]\ (x) = \sigma(x)$ if $x \in W$, x

otherwise. We say σ is more general than τ, denoted as $\sigma \leq \tau$, if there exists some ρ such that $\rho \circ \sigma = \tau$.

A *rewrite rule* is an oriented pair of terms, written as $l \to r$ such that $\mathcal{V}(r) \subseteq \mathcal{V}(l)$. A *rewrite system* is a set of such rules. A rewrite rule is *left linear* if its left hand side is linear; a rewrite system is called *left linear* if all its rules are left linear. $l \to r$ is a *variable-preserving* rule if $\mathcal{V}(l) = \mathcal{V}(r)$, otherwise it is *variable-dropping*. Two rules $l_1 \to r_1$ and $l_2 \to r_2$ are *overlapping*, if there exists a substitution σ such that $\sigma(\tau(l_1|_p)) = \sigma l_2$ for some occurrence p and renaming τ. A rewrite system $\mathcal{R}$ is *non-overlapping* if any two rules of $\mathcal{R}$ are not overlapping. A function symbol f is said to be a *defined function* (short: *function*) with respect to a rewrite system $\mathcal{R}$ if there is a rule in $\mathcal{R}$ with f as the top-level symbol of its left hand side; if there is no such rule, then f is called a *constructor*. In the sequel, f and g will be used for functions, c for constructors.

Several well-known concepts about rewrite systems like *normal form, terminating, confluent,* and *convergent* will also be used in this paper, details can be found in [2, 7]. We write $s \to t$, if s rewrites to t in one step; $s \to^* t$, if s rewrites to t in zero or more steps; $s \to^! t$, if $s \to^* t$ and t is a normal form. In addition, we are only interested in convergent rewrite systems, i.e., every term can be normalized uniquely. A substitution σ is said to be *normalized* if $\sigma(x)$ is a normal form for each $x \in dom(\sigma)$. Only normalized substitutions are considered in this paper.

A *matching pair* is an ordered pair of terms (s, t). A *matching problem* Γ is a set of matching pairs: $\{(s_1, t_1), \cdots, (s_n, t_n)\}$. If all t_i $(1 \leq i \leq n)$ are ground terms, then Γ is called a *ground matching problem*. $\mathcal{V}(\Gamma) = \mathcal{V}(s_1) \cup \cdots \cup \mathcal{V}(s_n)$.

As an abbreviation, we represent lists of terms using a "vector" notation, so that $f(s_1, \cdots, s_n)$ will be written as $f(\overline{s_n})$, and we shall even use $\{\overline{(x_n, s_n)}\}$ for the substitution $\{(x_1, s_1), \cdots, (x_n, s_n)\}$, and $\{\overline{(s_n, t_n)}\}$ for the matching problem $\{(s_1, t_1), \cdots, (s_n, t_n)\}$.

The set $\mathcal{M}$ of substitutions is a *complete set of minimal R-matches for Γ* if the following holds:

- for all $\sigma \in \mathcal{M}$, $dom(\sigma) \subseteq \mathcal{V}(\Gamma)$;
- for all $\sigma \in \mathcal{M}$ and $(s, t) \in \Gamma$, $\sigma(s) \to^*_R t$;
- if τ is a substitution and $\tau(s) \to^*_R t$ for every $(s, t) \in \Gamma$, then there is a $\sigma \in \mathcal{M}$ such that $\sigma \leq \tau$; and
- for any two substitutions $\sigma, \sigma' \in \mathcal{M}$, $\sigma \not\leq \sigma'$.

3 Definitions and algorithm

A rewrite system is *constructor-based* if it is left linear, non-overlapping, and has no defined function symbols appearing in the inner part of the left hand side of any rule. Some of the functional languages primarily based on recursive equations, such as HOPE, ML, and Haskell, fall into this category when restricted to defining first-order non-overlapping functions. Well known results (e.g. [4, 6]) ensure that complete set of matches can always be found for such class of semantic matching problems, but it is not necessarily finite. For example, the matching problem $\{(x - y, s(0))\}$ may have infinitely many matches: $\{(x, s(0)), (y, 0)\}$, $\{(x, s(s(0))), (y, s(0))\}$,

$\cdots$, where '$-$' is defined on the natural numbers in the usual way:

$$x - 0 \to x, 0 - y \to 0, s(x) - s(y) \to x - y$$

To solve this problem, the concept of *non-decreasing functions with respect to a suitable property* is introduced in [3]. Unfortunately, it is not possible to always decide whether a function is non-decreasing with respect to a property P, even for some simple suitable property like depth. Moreover, according to this definition all functions containing variable-dropping rules, such as the function *len* over lists, are decreasing with respect to depth.

$$len(nil) \to 0, \ len(x : x_s) \to s(len(x_s)),$$

where *nil* and : are constructors of lists.

In the following, we are going to introduce a new definition of *non-decreasing functions*, which solves the above two problems.

Definition 1 (Subterm property) A *subterm property* P is a measure of terms, along with a well-founded total ordering $\rhd$ which compares values of P, such that P is strictly larger, under $\rhd$, for terms than for its subterms.

Definition 2 (Non-decreasing function) Let f be a function defined by a constructor-based rewrite system $\mathcal{R}$, P a subterm property. f is said to be *non-decreasing* (with respect to P) (short: *non-decreasing*), if for any $f(\overline{l_n}) \to r \in \mathcal{R}$ the following holds: $P(l_i) \not\rhd P(r)$ for any $1 \leq i \leq n$, and any other function in r is non-decreasing. Otherwise, f is called *decreasing* (with respect to P) (short: *decreasing*).

According to this definition, *len* is a non-decreasing function with respect to depth along with the greater-than function on the natural numbers. The following function *half* is decreasing, since $|s(s(0))| > |s(0)|$.

$$half(s(s(0))) \to s(0), \ half(s(s(s(x)))) \to s(half(s(x)))$$

Although the definition of non-decreasing functions has been extended to cover functions with variable-dropping rules, we can still prove the following theorem.

Theorem 1 Let s be a normal form, in which all functions are non-decreasing with respect to some subterm property P. If $\sigma(s) \to^! t$ and for any other $\sigma'(s) \to^! t$, $\sigma' \not\leq \sigma$, then $P(\sigma x) \not\rhd P(t)$ for any $x \in dom(\sigma)$.

Proof We proof the theorem by an induction on the structure of s.

- Let s be a constant or a variable. Then the theorem holds trivially.
- Let s be of the form $c(\overline{s_n})$. Then t must have the form $c(\overline{t_n})$, where each t_i ($1 \leq i \leq n$) is a normal form. By induction assumption, we can prove the theorem.
- Let s be of the form $f(\overline{s_n})$, and $\sigma(s_i) \to^! s'_i$ for all $1 \leq i \leq n$. We prove that $P(s_i) \not\rhd P(t)$ by an induction on rewriting steps.
  - Let $f(\overline{s'_n}) = t$. Then the theorem holds trivially.

- Let $f(\overline{s_n'}) \to t'$ with some rule $f(\overline{l_n}) \to r$. Then there exists a substitution θ such that $\theta(l) = f(\overline{s_n'})$ and $t' = \theta(r)$. If $f(\overline{l_n}) \to r$ is variable-preserving, then $P(\theta(l_i)) \not\triangleright P(\theta(r))$, since f is a non-decreasing function. By induction assumption, the assertion holds. If $f(\overline{l_n}) \to r$ is a variable-dropping rule, the assertion holds by induction assumption and the fact that σ is a minimal substitution such that $\sigma(s) \to^! t$.

□

Now we can define a class of *acyclic rewrite systems*, for which semantic matching is always decidable.

Definition 3 (Acyclic rewrite system) Let P be a subterm property. A constructor-based rewrite system $\mathcal{R}$ is called *acyclic* (with respect to P), if each $f(\overline{l_n}) \to r \in \mathcal{R}$ satisfies the following conditions:

- the top-level symbol of r is a variable or a constructor; and
- there is no function in r nested below any decreasing function with respect to P.

According to this definition, $-$ is not defined by an acyclic rewrite system, since the right hand side of its third rule begins with $-$; nor is the following function *isEq*:

$$isEq(0,0) \to s(0), \quad isEq(s(x),s(y)) \to s(half(isEq(x,y))),$$

because *isEq* occurs below the decreasing function *half* on the right hand side of the second rule. On the contrary, the rewrite systems defining *max* and *ht*, *len*, and *half* are all acyclic, so are the following two rewrite systems:

$$
\begin{aligned}
0 + y &\to y & 0 * y &\to 0 \\
s(x) + y &\to s(x + y) & s(x) * 0 &\to 0 \\
& & s(x) * s(y) &\to s(y + (x * s(y)))
\end{aligned}
$$

$$
nil \mathbin{+\!\!+} y_s \to y_s, \qquad (x : x_s) \mathbin{+\!\!+} y_s \to x : (x_s \mathbin{+\!\!+} y_s)
$$

Instead of using an existing narrowing strategy, such as basic narrowing or innermost narrowing, we introduce a new narrowing strategy — *structural narrowing*. We use $mgu(t_1, t_2)$ to denote the *most general (syntactical) unifier* of t_1 and t_2.

Definition 4 (Structural narrowing) A non-variable term t is *structurally narrowable* into t' with respect to $\mathcal{R}$, if there is a renaming $l \to r$ of a rule in $\mathcal{R}$ such that, t and l have no common variables and are unifiable, and $t' = mgu(t, l)(r)$. It is denoted as $t \rightsquigarrow_{l \to r} t'$. A term that is not structurally narrowable is called *structurally normalized*.

In fact, structural narrowing is a lazy narrowing strategy, which always occurs at the top-level of a term, i.e., considers the term as a whole. We give some examples to show the idea.

$$x - y \rightsquigarrow_{x_1 - 0 \to x_1} x_1$$
$$x - y \rightsquigarrow_{s(x_1) - s(y_1) \to x_1 - y_1} x_1 - y_1 \rightsquigarrow_{s(x_2) - s(y_2) \to x_2 - y_2} x_2 - y_2 \rightsquigarrow_{s(x_3) - s(y_3) \to x_3 - y_3} \cdots$$

The first narrowing sequence terminates with x_1 which is structurally normalized, while the second narrowing sequence does not terminate. Structural narrowing is particularly effective for solving matching problems with respect to constructor-based systems, since if t is structurally narrowable with $l \to r$ and $\sigma = mgu(t, l)$, $\sigma(x)$ contains no function symbols for any $x \in dom(\sigma)$.

We present our matching algorithm as a set of transformations for transforming a ground matching problem into an explicit representation of a solution. A pair (x, s) in a matching problem Γ is in *solved form* and x is called a *solved variable*, if x is a variable that occurs neither in s nor anywhere else in Γ. A matching problem is in *solved form* if all its pairs are in solved form. For a given matching problem Γ, the algorithm *succeeds* when there is a sequence of transformations terminating with a matching problem in solved form, say σ, in which case σ is a match of the initial problem Γ, fails when there exist no sequences of transformations that terminate with a matching problem in solved form. In the case of failure, Γ is not matchable. In the sequel, we write $l \to r \in \mathcal{R}$ to denote a renaming of a rule in $\mathcal{R}$, whose variables do not occur in other related terms. It is assumed that terms in a matching problem are always automatically normalized.

(Elimination) finds a partial solution for a variable. If x occurs in Γ,

$$\{(x, t)\} \cup \Gamma \Longrightarrow \{(x, t)\} \cup \sigma\Gamma, \quad \text{where } \sigma = \{(x, t)\}.$$

(Decomposition) breaks a matching pair down into simpler ones. If $c \in \mathcal{C} \cup \mathcal{F}$,

$$\{(c(\overline{t_n}), c(\overline{s_n}))\} \cup \Gamma \Longrightarrow \{\overline{(t_n, s_n)}\} \cup \Gamma.$$

(Narrowing) realizes structural narrowing of a pair. If each t_i $(1 \le i \le n)$ contains no function symbols,

$$\{(f(\overline{t_n}), s)\} \cup \Gamma \Longrightarrow \sigma \cup \{(\sigma r, s)\} \cup \sigma\Gamma, \quad \text{where } l \to r \in \mathcal{R}, \sigma = mgu(f(\overline{t_n}), l).$$

(Deletion) removes a matching pair playing no role in solutions. If $x \notin \mathcal{V}(\Gamma)$,

$$\{(t, x)\} \cup \Gamma \Longrightarrow \Gamma.$$

(Abstraction) abstracts a matching pair. If the head of $f(\overline{t_n})|_p$ is a function symbol and $p \ne \epsilon$,

$$\{(f(\overline{t_n}), s)\} \cup \Gamma \Longrightarrow \{(f(\overline{t_n})[x]_p, s), (f(\overline{t_n})|_p, x)\} \cup \Gamma, \quad \text{where } x \text{ is a new variable.}$$

As usual, $\Longrightarrow^*$ and $\Longrightarrow^+$ denotes the reflexive-transitive and transitive closure of $\Longrightarrow$ respectively. The first three transformations are very similar to a standard semantic matching algorithm. The use of structural narrowing requires two additional transformations (Deletion) and (Abstraction). The rule (Abstraction) ensures the completeness of the algorithm. (Deletion) removes the matching pairs which are introduced by applying variable-dropping rules.

The rule (Narrowing) can not be used to the matching problem $\{((x + y) + z), u)\}$ directly. Applying (Abstraction) to it, we get the new problem $\{(x_1 + z, u), (x + y, x_1)\}$. Now, (Narrowing) can be applied to both pairs. The next example shows the application of the transformation (Deletion). Let $\{(len((x + y) : zs), s(0))\}$ be a matching problem. After applying (Abstraction) and (Narrowing) with the rewrite rule $len(x : xs) \rightarrow s(len(xs))$, we get the new problem $\{(s(len(xs)), s(0)), (x+y, z)\}$. Because the rewrite rule is variable-dropping, so the second pair can be removed by (Deletion) without loss of the completeness.

Theorem 2 For any matching problem Γ over an acyclic constructor-based rewrite system, the algorithm terminates with a complete set of minimal matches of Γ.

Proof The proofs of the termination and the completeness are similar to the proofs in [3]. A complete proof of this theorem can be found in [12]. We present here a proof of minimality. The only two rules to be considered are (Elimination) and (Narrowing). The application of any other rule does not influence the matches.

- Let (x, s) be a matching pair in a matching problem. Then the minimal match of the pair is $\{(x, s)\}$. That is to say, the (Elimination) transformation produces a minimal partial solution.
- Let $(f(\overline{t_n}), s)$ be a matching pair in a matching problem, $l_1 \rightarrow r_1$ and $l_2 \rightarrow r_2$ be two rules that can be applied to narrow it. If $\sigma_1 = mgu(f(\overline{t_n}), l_1)$ and $\sigma_2 = mgu(f(\overline{t_n}), l_2)$, then $\sigma_1 \not\leq \sigma_2$, since $\mathcal{R}$ is a non-overlopping rewrite system.

$\square$

4 Improvement

We present in this section two techniques *dependency analysis* and *sharing analysis* to improve our algorithm.

Dependency analysis is motivated by the partial ordering for first-order unification problems from Martelli and Montanari [9]. One reason to cause our algorithm inefficient is the random selection of matching pairs to apply structural narrowing. Especially, the use of the transformations (Narrowing) and (Abstraction) may transform ground matching problems into non-ground ones. See the following sequence of the (Abstraction) transformations:

$$\{(max(ht(t_1), ht(t_2)), s(s(0)))\} \Longrightarrow^+ \{(max(x, y), s(s(0))), (ht(t_1), x), (ht(t_2), y)\}$$

The right hand side of the transformation is not a ground matching problem, which contains new variables x and y. Solving $(ht(t_1), x)$ at first without knowing x could be very inefficient through many useless narrowing steps. Instead, the pair $(max(x, y), s(s(0)))$, whose right term contains no variables that occur in the left terms of other pairs, should be considered first.

Definition 5 (Solvable matching pair) A matching pair (s, t) is called *dependent on* (s', t') in Γ, denoted as $(s, t) \prec (s', t')$, if $\mathcal{V}(t) \cap \mathcal{V}(s') \neq \emptyset$. A matching pair $(s, t) \in \Gamma$ is *solvable*, if it is not in solved form and there is no $(s', t') \in \Gamma$ such that $(s, t) \prec (s', t')$.

A solvable matching pair can be solved independently. We can treat a solvable pair as a new matching problem, solve it using the matching algorithm, and then substitute its matches into the original problem. For instance, to solve $\{(max(x, y),$

$s(s(0))), (ht(t_1), x), (ht(t_2), y)\}$, we shall first solve $\{(max(x, y),\ s(s(0)))\}$. One of the matches is $\{(x, s(s(0))), (y, s(s(0)))\}$. Substituting it into the original problem we get the new problem:

$$\{(x, s(s(0))), (y, s(s(0))), (ht(t_1), s(s(0))), (ht(t_2), s(s(0)))\}$$

We add an additional condition to the transformation (Narrowing) such that it can only be applied to solvable matching pairs.

For proving Theorem 2 the order of applying transformations is unimportant. If we can show that there always exists a solvable matching pair during solving a ground matching problem, unless it is not matchable, we can conclude that the theorem still holds with the additional condition on the transformation (Narrowing). At first, we present some lemmas.

Lemma 1 Let t be any term and l a linear term such that $\mathcal{V}(t) \cap \mathcal{V}(l) = \emptyset$. If $\sigma = mgu(t, l)$, then

- σx is linear for any $x \in \mathcal{V}(t)$;
- $\mathcal{V}(\sigma x_1) \cap \mathcal{V}(\sigma x_2) = \emptyset$ for any x_1 and $x_2 \in \mathcal{V}(t)$.

Proof By an induction on the structure of l.

- If l is a variable, then $\sigma = \{(x, t)\}$, the lemma holds obviously.
- If l is a constant, then $ran(\sigma) = \emptyset$.
- If l has the form $f(l_1, l_2)$ (for simplicity), f is a constructor or a function symbol. t must have the form $f(t_1, t_2)$. Otherwise, they are not unifiable. Let $\sigma_1 = mgu(t_1, l_1)$ and $\sigma_2 = mgu(\sigma_1 t_2, \sigma_1 l_2)$, then $\sigma = \sigma_2 \circ \sigma_1$. Because $\mathcal{V}(l_1) \cap \mathcal{V}(l_2) = \emptyset$, $\mathcal{V}(t) \cap \mathcal{V}(l) = \emptyset$, and $dom(\sigma_1) \cup ran(\sigma_1) = \mathcal{V}(t_1) \cup \mathcal{V}(l_1)$ (see [10]), we have $dom(\sigma_1) \cap \mathcal{V}(l_2) = \emptyset$ and $\sigma_1 l_2 = l_2$. By induction assumption, it is easy to show that σx is linear for any $x \in \mathcal{V}(t)$. Now, we prove $\mathcal{V}(\sigma x_1) \cap \mathcal{V}(\sigma x_2) = \emptyset$ for any x_1 and $x_2 \in \mathcal{V}(t)$. The case of x_1 and $x_2 \in \mathcal{V}(t_1)$ or x_1 and $x_2 \in \mathcal{V}(t_2)$ can be proved directly by induction assumption. If $x_1 \in \mathcal{V}(t_1)$ and $x_2 \in \mathcal{V}(t_2)$, we have $\sigma x_2 = \sigma_2 x_2$ and $\sigma x_1 = \sigma_2(\sigma_1 x_1)$. Since $x_2 \notin \mathcal{V}(\sigma_1 x_1)$, so $\mathcal{V}(\sigma x_1) \cap \mathcal{V}(\sigma x_2) = \emptyset$ by induction assumption.

□

Following from this lemma and using the fact that each constructor-based rewrite system is left linear, the following lemma can be proved by an induction on the steps of transformations.

Lemma 2 Let Γ be a ground matching problem and $\Gamma \Longrightarrow^* \Gamma'$.

- If $(t, s) \in \Gamma'$, then s is linear.
- If $(t_1, s_1), (t_2, s_2) \in \Gamma'$, then $\mathcal{V}(s_1) \cap \mathcal{V}(s_2) = \emptyset$.

Lemma 3 Let Γ be a ground matching problem and $\Gamma \Longrightarrow^* \Gamma'$, then the dependent relation $\prec$ is well-founded on the matching pairs in Γ'.

Proof By an induction on transformations. The theorem holds for the case $\Gamma' = \Gamma$, since Γ is a ground matching problem. Let $\Gamma_1 \Longrightarrow \Gamma_2$ with the transformation (Elimination), (Abstraction), or (Narrowing), since any other rule does not change the dependent relation between matching pairs.

- (Elimination). Let $\sigma = \{(x,t)\}$, (s_1, s_2) and (s_1', s_2') be two pairs in Γ_1. We show that if $(s_1, s_2) \not\prec (s_1', s_2')$, i.e., $\mathcal{V}(s_2) \cap \mathcal{V}(s_1') = \emptyset$, then $(\sigma s_1, \sigma s_2) \not\prec (\sigma s_1', \sigma s_2')$. There are three cases to be considered:
  - If $x \notin \mathcal{V}(s_2)$ and $x \notin \mathcal{V}(s_1')$, then $\mathcal{V}(\sigma s_2) \cap \mathcal{V}(\sigma s_1') = \emptyset$.
  - If $x \in \mathcal{V}(s_2)$, but $x \notin \mathcal{V}(s_1')$, because $\mathcal{V}(t) \cap \mathcal{V}(s_1') = \emptyset$ ((x,t) is solvable), so $\mathcal{V}(\sigma s_2) \cap \mathcal{V}(\sigma s_1') = \mathcal{V}(\sigma s_2) \cap \mathcal{V}(s_1') = \emptyset$.
  - If $x \notin \mathcal{V}(s_2)$, but $x \in \mathcal{V}(s_1')$, because $\mathcal{V}(t) \cap \mathcal{V}(s_2) = \emptyset$ (lemma 2), so $\mathcal{V}(\sigma s_2) \cap \mathcal{V}(\sigma s_1') = \mathcal{V}(s_2) \cap \mathcal{V}(\sigma s_1') = \emptyset$.

  That is, $\mathcal{V}(\sigma s_2) \cap \mathcal{V}(\sigma s_1') = \emptyset$.
- (Abstraction). Let $\{(f(\overline{t_n}), s)\} \cup \Gamma \implies \{(f(\overline{t_n})[x]_p, s), (f(\overline{t_n})|_p, x)\} \cup \Gamma$ be a transformation step using (Abstraction). For any $(s_1, s_2) \in \Gamma$, $(f(\overline{t_n})|_p, x) \not\prec (s_1, s_2)$, since x is a new variable; if $(s_1, s_2) \not\prec (f(\overline{t_n}), s)$, then $(s_1, s_2) \not\prec (f(\overline{t_n})[x]_p, s)$ and $(s_1, s_2) \not\prec (f(\overline{t_n})|_p, x)$. So, if $\prec$ is well-founded on $\{(f(\overline{t_n}), s)\} \cup \Gamma$, it is also well-founded on $\{(f(\overline{t_n})[x]_p, s), (f(\overline{t_n})|_p, x)\} \cup \Gamma$.
- (Narrowing). The proof is similar to the first case.

$\square$

Theorem 3 Let Γ be a ground matching problem and $\Gamma \implies^* \Gamma'$, then Γ' contains at least one solvable matching pair as long as it is not in solved form.

Proof Follows from the above lemma. $\square$

During solving a matching problem, there are often some *similar matching pairs* that are identical under renaming. The matching pairs $(ht(t_1), s(s(0)))$ and $(ht(t_2), s(s(0)))$ are similar in the above new matching problem. Usually, all these matching pairs will be solved independently, because our matching algorithm does not "remember" the matching pairs which have been solved already. The introduction of *sharing analysis* can avoid such redundant computation of similar matching pairs. We simply define a *solution list* for each given matching problem, in which solved pairs with their matches will be recorded. In the above example, the solution list contains matches of $(max(x,y), s(s(0)))$, $(max(x,y), s(0))$, $(max(x,y), 0)$, $(ht(t_1), 0)$, $(ht(t_1), s(0))$ and $(ht(t_1), s(s(0)))$. The number of matching pairs to be solved is now 6 instead of more than 60. Without solution list, solving $(ht(t_1), s(s(0)))$, for instance, contains again to solve $(max(x,y), s(0))$, $(ht(t_1), s(0))$, and $(ht(t_1), 0)$.

Sharing analysis is different from matching modulo commutativity, e.g., $(x + 0, s(0))$ and $(0 + x, s(0))$ are not similar matching pairs. In addition, sharing analysis can sometimes be expensive, if the number of matching pairs is quite large and there are rarely similar pairs. We believe that sharing analysis is useful for solving matching problems containing functions that are defined through commutative functions. ht and $*$ are examples of such functions. They are defined through the commutative functions max and $+$ respectively. We take the matching problem $\{(max(x,y), s(s(0))), (ht(t_1), x), (ht(t_2), y)\}$ as an example. If $\{(x, s_1), (y, s_2)\}$ is a match of the matching pair $max(x,y)$, so is $\{(x, s_2), (y, s_1)\}$, since max is commutative. We should solve the matching problems $\{(ht(t_1), s_1), (ht(t_2), s_2)\}$ and $\{(ht(t_1), s_2), (ht(t_2), s_1)\}$. Applying sharing analysis only one of them will be solved, since their matching pairs are pairwise similar.

5 Algorithm analysis and test results

Dependency analysis also offers a direct way to analyse the algorithm. For simplicity, we only consider narrowing steps. Let $f(\overline{t_n})$ and t be two normal forms, where f is a function symbol and $t_1, \cdots, t_n$ contain no function symbols. To solve the matching pair $(f(\overline{t_n}), t)$, we should consider the following factors:

- The number M_f of rules defining f. There are at most M_f equations that can be used to narrow the matching pair. For instance, $M_{max} = 3$, $M_{ht} = 2$.
- The multiset S_r of the occurrences of function symbols in the right hand side of $l \rightarrow r$. It decides matching pairs in the new problem after applying (Narrowing) with the rule and (Abstraction). For instance,

$$S_y = S_{s(x)} = S_0 = \varnothing, S_{s(max(x,y))} = \{max\}, S_{s(max(ht(x), ht(y)))} = \{max, ht, ht\}$$

- The depth of t, which provides the upper bound on times that the transformation (Narrowing) can be applied. Because after each (Narrowing) transformation, it may follow some (Decomposition) transformation, which reduces $|t|$ at least one. On the other hand, the application of (Abstraction) introduces dependent pairs. Since we only consider functions defined by acyclic rewrite systems, if (t_1, s_1) is dependent on (t_2, s_2), then $|\sigma s_1| \leq s_2$ (see theorem 1), where σ is a minimal match of (t_2, s_2).

Coarsely, any matching problem associated with an acyclic rewrite system is decidable. According to the above analysis, it is possible to make a refinement classification. Let f be a function defined by the set of rules $\{l_1 \rightarrow r_1, \cdots, l_k \rightarrow r_k\}$ in an acyclic rewrite system $\mathcal{R}$, i.e., $M_f = k$. We use $T_f(m)$ to denote the upper bound of the narrowing steps to solve the matching pair $(f(\overline{t_n}), t)$ with $|t| = m$, which can be computed by the following two equations:

$$\begin{align}
T_f(0) &= k \tag{1}\\
T_f(m+1) &= k + \Sigma_{1 \leq i \leq k}(\Pi_{g \in S_{r_i}}(T_g(m))) \tag{2}
\end{align}$$

where Σ and Π are the finite addition and multiplication of the natural numbers. If the length of $|t| = 0$, we only need k narrowing steps, since there are at most k rules can be applied, see equation (1). If $|t| = m+1$, in addition to k narrowing steps, we still need to solve all new problems associated with functions in S_{r_i} for each $l_i \rightarrow r_i$ $(1 \leq i \leq k)$, see equation (2).

Definition 6 (Classification) Let f be a function symbol defined by an acyclic rewrite system. If $T_f(m)$ can be presented as a linear expression, we call f a *linear function*. If $T_f(m)$ can be presented as a polynomial expression, then f is called a *polynomial function*. Otherwise, f is a *decidable function*.

Surprisingly, there are a lot of useful functions, such as $+$, max, len, and $+\!\!+$, are linear functions. The use of linear functions is unproblematic in practice. Unfortunately, there are still a lot of functions that are neither linear nor polynomial, such as $*$ and ht. We doubt the existence of a matching algorithm with polynomial complexity. For instance, the number of trees increases exponentially with respect to the height. In practice, solving matching problems with such kind of functions can be very inefficient.

The matching algorithm and its improved versions have been implemented in SML/NL [5]. We use the data structures for terms in the theorem prover Isabelle [11].

A number of tests have been done on a SUN SPARC-Station-20 with 64 megabytes of memory. In order to find out how much performance gain we can get, we have measured the CPU-time-costs in seconds on more than 30 examples. Figure 1 presents some test results. Column T_0 is for the original algorithm, column T_1 for the algorithm with dependency analysis, and column T_2 for the algorithm with both dependency analysis and sharing analysis. '–' stands for tests consuming too much time to finish within 4 hours. '[' and ']', and '0', '1', '2', $\cdots$ are used to present lists and the natural numbers shortly. We can draw the following conclusions, which are in accordance with the analysis at the beginning of this section.

	Matching pairs	T_0	T_1	T_2
1	$(x +\!\!+ [0] +\!\!+ z, [])$	0.22	0.22	0.23
2	$(x +\!\!+ [0] +\!\!+ z, [1, 0, 2])$	0.35	0.34	0.33
3	$(x +\!\!+ y +\!\!+ z, [0, 1, 2, 3])$	0.57	0.58	0.54
4	$(ht(node(x, y)), 1)$	0.26	0.20	0.23
5	$(x * y, 1)$	0.26	0.28	0.26
6	$(x +\!\!+ y +\!\!+ z +\!\!+ w, [0, 1, 2, 3, 4, 5])$	61.10	3.39	3.54
7	$(x +\!\!+ y +\!\!+ z +\!\!+ w, [0, 1, 2, 2, 1, 0])$	53.62	2.97	3.60
8	$(ht(node(x, y)), 3)$	30.24	2.70	2.62
9	$(ht(node(x, y)), 4)$	–	94.57	92.50
10	$(x * y, 12)$	7.64	3.99	1.91
11	$(x * y, 36)$	493.35	91.22	51.11
12	$(x * y, 48)$	–	247.48	139.69

Fig. 1. Test results

- For simple matching problems, all these algorithms need less than one second. This can been see from the first 5 examples.
- In general, the more complicated matching problems are, the better the algorithms with dependency analysis are. The second group contains examples, for which algorithms with dependency analysis get about 93% speedup in execution time.
- For some matching problems the performance gain through both dependency analysis and sharing analysis is very significant. This can be seen from the examples in the third test group. The algorithm with only dependency analysis gets from 50% to 85% speedup in execution time. The algorithm with both dependency analysis and sharing analysis gets about 46% speedup over the algorithm with only dependency analysis.

6 Conclusion

We have presented a new semantic matching algorithm. Our implementation shows that the new algorithm can solve many matching problems efficiently. The implementation will be used in a prototypical program transformation system [8, 12].

Now, we are considering an efficient implementation of a second-order semantic matching algorithm. Because of the complexities of the second-order matching and the semantic matching, efficiency is particularly important to make it practically useful. In addition, the number of matches of a given matching problem may be very large. Usually only a few of them, or even only one of them, are interesting in practice. How to make use of context information or additional restrictions on matching problems during matching process will be considered.

Acknowledgement. We thank the anonymous referees for their corrections and comments.

References

1. Bockmayr, B.: A Note on a Canonical Theory with Undecidable Unification and Matching Problem. In *Journal of Automated Reasoning*, Vol. 3, 379-381, 1987.
2. Dershowitz, N., Jouannaud, J.-P.: Rewrite System. In I. van Leeuwen, editor, *Formal Models and Semantics, Handbook of Theoretical Computer Science, Vol. B*, 243-320. The MIT press, 1990.
3. Dershowitz, N., Mitra, S. and Sivakumar, G.: Decidable Matching for Convergent Systems. In *Proc. of 11th International Conference on Automated Deduction*, LNCS 607, 589-602, 1992.
4. Fay, M.: First Order Unification in an Equational Theory. In *Proc. 4th Workshop on Automated Deduction* (1979), 161-167.
5. Harper, R., MacQueen, D., Miller, R.: Standard ML. LFCS Report Series, ECS-LFCS-86-2. Dept. of Computer Science, University of Edinburgh, 1986.
6. Hullot, J.-M.: Canonical Forms and Unification. In *Proc. 5th Workshop on Automated Deduction* (1980), 318-334.
7. Klop, J. W.: Term Rewriting Systems. In Abramski, S., Gabbay, Dov. M. and Maibaum, T.S.E. (eds): *Handbook of Logic in Computer Science, Chapter 1*. Oxford Science Publications, (1992).
8. Krieg-Brückner, B., Liu, J., Shi, H. and Wolff, B.: Towards Correct, Efficient and Reusable Transformational Developments. In Broy, M. and Jähnichen, S. (ed.): *KORSO, Correct Software by Formal Methods*, LNCS 1009, 1995.
9. Martelli, A. and Montanari, U.: An Efficient Unification Algorithm. *ACM Transactions on Programming Languages and Systems*, Vol 4, No. 2, (1982), 258-282.
10. Middeldorp, A. and Hamoen, E.: Completeness Results for Basic Narrowing. In *Proc. 3rd International Conference on Algebraic and Logic Programming*, LNCS 632, 244-258, (1992).
11. Paulson, Lawrence C.: *Isabelle - A Generic Theorem Prover*. LNCS 828, Springer-Verlag, 1994.
12. Shi, H.: *Extended Matching with Applications to Program Transformation*. Ph.D. thesis, Universität Bremen, 1994.

Routing on Triangles, Tori and Honeycombs

Jop F. Sibeyn*

Abstract

The standard $n \times n$ torus consists of two sets of axes: horizontal and vertical ones. For routing h-relations, the bisection bound gives a lower bound of $h \cdot n/4$. Several algorithms nearly matching this bound have been given.

In this paper we analyze the routing capacity of modified tori: tessellations of the plane with triangles or hexagons and tori with added diagonals. On some of these networks the ratio of routing capacity and degree is higher than for ordinary tori, even though they are as easily constructed. Hence, they may constitute more cost effective alternatives.

For networks with n^2 PUs, we get the following results: on a torus of hexagons, node degree 3, h-relations are performed in $0.37 \cdot h \cdot n$ steps; on a torus of triangles, node degree 6, in $0.13 \cdot h \cdot n$; and on a torus with added diagonals, node degree 8, in $h \cdot n/12$. The latter result matches the bisection bound for this network. Even faster is the routing on a torus of hexagons with diagonals, node degree 12: $0.053 \cdot h \cdot n$.

The algorithm is simple, inspired by the algorithm of Valiant and Brebner. The results can easily be extended to sorting, dynamic routing or routing for average-case inputs.

Classification: Theory of parallel and distributed computation, VLSI structures, parallel algorithms.

1 Introduction

Routing Problem. The *routing problem* is the problem of rearranging a set of information packets in a network, such that every packet ends up at the processor specified in its destination address. This problem is of fundamental importance in the design of efficient algorithms for realistic models of parallel computation, as well as in the simulation of more powerful, idealistic models. Consequently, the routing problem has received considerable attention, and a variety of algorithms have been proposed for several variants of the problem. The performance of a routing algorithm is measured by its *running time* (the maximum time a packet may need to reach its destination) and its *queue size* (the maximum number of packets any node may have to store during the routing).

Practically the most important variant of the routing problem is the routing of h-relations, in which every processor is the source and destination of at most h packets. In Valiant's BSP-model for parallel computation [11], h-relations constitute the basic communication primitive. In this model the algorithms designer only knows three system parameters, one of them expressing the time required to perform arbitrary h-relations. Clearly it is of utmost importance that this primitive operation is performed optimally.

---

*Max-Planck-Institut für Informatik, Im Stadtwald, 66123 Saarbrücken, Germany. Email: jopsi@mpi-sb.mpg.de.

Typically there are three ranges of h values. Of great theoretical interest is the case $h = 1$: the *permutation routing problem*. It has been considered extensively for all types of networks. The task is to route the packets in a time given by the diameter of the network. Less interesting, and involving many specific details are small values of h other than $h = 1$. For sufficiently large h, for meshes and tori for $h = \mathcal{O}(\text{dimension of the network})$, the routing time is bounded by the bisection width of the network. In practice, a parallel system is always sufficiently loaded, so that the h values lie in this range. In this paper we focus on such h values.

Machine Model. We consider grid-like networks. A two-dimensional *mesh* is a parallel computer consisting of n^2 processing units, *PUs*, arranged in an $n \times n$ grid. Each PU is connected to its (at most) four neighbors in the grid. A *torus* is a mesh with *wrap-around connections*. That is: the left-most PU of each row is connected to the right-most PU of this row, and similarly in the columns. d-dimensional meshes and tori are the immediate generalizations.

In this paper we focus on generalizations of two-dimensional tori. In the standard torus the nodes are placed at the corner points of a tessellation of the plane with squares, and they are interconnected by two sets of orthogonal axes. We call this a *bi-axial* torus. There are other possibilities, however. We can increase the number of axes to four, by adding diagonals: the *quadri-axial* torus. Alternatively, the nodes can be placed at the corner points of a tessellation by hexagons or triangles. In their sparsest form they give networks of degree three and six, respectively. But 'diagonals' can also be added here.

Previous Work. For routing h-relations on normal meshes and tori of degree four, there are numerous results [6, 7, 1, 2, 4, 3]. At first it was not clear, but later it was understood that the best approach to routing h-relations on meshes and tori were suitable (derandomized) implementations of the Valiant and Brebner routing paradigm [12].

A paper by Kunde, Niedermeier and Rossmanith [5] inspired the research presented here. They analyze routing and sorting on quadri-axial meshes and tori. For h-relations on meshes they achieve $(1/9 + o(1)) \cdot h \cdot n$, on tori $(1/10 + o(1)) \cdot h \cdot n$. The great importance of this paper lies in showing that by doubling the number of connections one can reduce the routing time by more than a factor of two. In general this is not surprising (a four-dimensional mesh also has degree eight and is even much faster), but that it could be achieved so easily was not known. Even though these are good results, they do not match the bisection lower bound (the number of steps required when all packets from two halves of the network have to be exchanged), and they are fairly complicated.

Ideas and Results. For generalizations of meshes we do not know how to improve the results of [5]. But for tori, which have the advantage of node symmetry, we present easier, more general and faster algorithms. We apply suitable implementations of Valiant and Brebner: first route all packets to randomly chosen intermediate destinations; then from there to their actual destinations. We call an algorithm that works according to this paradigm a *VB-algorithm*. We strongly believe in VB-algorithms: they are simple and for general permutations on most networks they perform optimally. Both conditions are essential for a possible application in practice. We develop an elegant technique for analyzing VB-algorithms.

On the torus of triangles every PU has degree six. The square torus of triangles has a very special property that we have not encountered before: VB-algorithms provably require more than the bisection bound by a factor larger than one. We show that h-relations can be routed in $(7/48 + o(1)) \cdot h \cdot n$. The bisection bound gives $h \cdot n/8$. On rectangular $n \times 0.91 \cdot n$

tori, the routing time is optimal: $(1/8 + o(1)) \cdot h \cdot n$. As this network is almost as easy to build as an ordinary torus, it may be a cost-effective alternative.

On the quadri-axial torus every PU has degree eight. The bisection bound is $h \cdot n/12$. We show that a VB-algorithm matches this bound up to a lower-order term, improving the result of [5].

Now that we know how to route on tori of triangles, the problem of routing on tori of hexagons has also become easy. Each node has a degree of only three: from now on, we will refer to it as the *honeycomb*. The honeycomb can be enriched with diagonals. This gives a network of degree 12. For routing h-relations honeycombs are as competitive as the others.

The developed ideas for routing h-relations are general. The most important observation is that in order to achieve optimality, on the average packets should only pass once over any bisection. This implies that packets must be routed along the shortest paths, with at most one bend of minimal angle. So far we have utilized a 'randomized vocabulary'. However, applying the techniques from [3], it is trivial to turn any randomized mesh algorithm of this type into a deterministic one. Furthermore, the algorithms can be extended to sorting in essentially the same time by applying the techniques from [4, 3], or splitter-based sorting algorithms [9, 8, 1, 10]. For average-case inputs, one should simply leave out the randomization of the packets in order to halve all routing times. For dynamic routing the packets should be provided with time stamps to guarantee that packets do not continue to hang around.

The general conclusion is that a VB-algorithm in which the packets are routed in each phase along a shortest path to their destinations is suited for routing on any of the considered tori of tessellations: the routing time lies close to the respective bisection bounds. With increasing node degree the number of steps for an h-relation decreases more than proportionally

$$T_{\text{route}}(deg, n^2, h) \simeq 2 \cdot h \cdot n/deg^{3/2}, \tag{1}$$

for all of them. Here $deg \in \{3, 4, 6, 8, 12\}$, denotes the degree of the network. The higher the degree of a network, the more expensive it is to build. But maybe the increased routing power makes worth the investment. Actually, the MasPar parallel computer is a kind-of quadri-axial torus. For a particular construction-cost function, (1) allows a computer constructor to choose the best trade-off.

Contents. After some preliminaries, we consider in Section 3 lower and upper bounds for routing h relations on tori of triangles. In Section 4 we analyze quadri-axial tori. Then, we give the results for routing on honeycombs, and finally we consider further extensions.

2 Preliminaries

Routing Model. The basic unit of information is a *packet*. A packet cannot be split or recombined. It takes one *(routing) step* to transfer a packet over a connection. If several packets compete for the use of the same connection, then only one can proceed immediately. The others are stored in a queue and forwarded at a later step: the *store-and-forward model*. Throughout this paper we assume that the packet that has to go furthest in the present direction gets priority: the *farthest-first strategy*. All our connections are full-duplex: they allow data transmission in both directions at the same time. We consider routing in the *full-port model*. This means that a PU can communicate with all its neighbors in a single step. As we are interested in the analysis of routing algorithms, we do not consider internal computation, but in all cases it is limited.

Networks. A processor network is identified with the underlying graph. We apply graph terminology to describe a network. Particularly, the *degree* of a network is the maximum of the number of connections of the nodes. Mostly networks are *regular*, that is all nodes have the same degree. The diameter of a network is the maximum distance between any pair of nodes. The *bisection width* is the number of connections that must be severed in order to split a network into two subnetworks with each (approximately) half the number of nodes. A network is said to be *node-symmetric*, if for each pair of nodes, (P_1, P_2), there is a automorphism mapping P_1 on P_2. Tori are node-symmetric, but meshes are not.

Lower Bounds. Let $\mathcal{N}$ be a class of networks, and let $\mathcal{N}(N)$ be the instance with N nodes (possibly not all values of N are allowed). Denote by $D_{\max}(\mathcal{N}, N)$ the diameter of $\mathcal{N}(N)$, by $D_{av}(\mathcal{N}, N)$ the average distance between nodes, by $deg(\mathcal{N}, N)$ its degree and by $EB(\mathcal{N}, N)$ its edge bisection. The time for routing h-relations on $\mathcal{N}(N)$ is $T(\mathcal{N}, N, h)$. For routing h-relations we have the distance and the bisection bound:

$$T(\mathcal{N}, N, h) \geq \max\{D_{\max}(\mathcal{N}, N), h \cdot N/(2 \cdot EB(\mathcal{N}, N))\}. \qquad (2)$$

We consider the class of VB-algorithms, in which packets are first completely randomized/unshuffled (in the context of routing algorithms, unshuffling is a deterministic equivalent of randomization, see [3]), before they are routed to their destinations. Clearly in each phase every packet covers an expected distance D_{av}. Thus,

Lemma 1 *For a VB-algorithm*

$$T_{VB}(\mathcal{N}, N, h) \geq 2 \cdot h \cdot D_{av}(\mathcal{N}, N)/deg(\mathcal{N}, N).$$

Proof: In both phases together the sum of the distances covered by all packets is about $2 \cdot D_{av} \cdot N$. Dividing by the number of connections, $deg \cdot N$, gives the answer. $\qquad \square$

Corollary 1 *A VB-algorithm cannot match the bisection bound if*

$$4 \cdot D_{av}(\mathcal{N}, N) \cdot EB(\mathcal{N}, N) > deg(\mathcal{N}, N) \cdot N.$$

Rounding Policy. In order to get a pleasant notation which allows us to concentrate on the main points, we frequently simplify mathematical expressions. Such a simplification is expressed by the use of $\simeq$ instead of $=$. For example, we use $\sum_{i=1}^{x} i \simeq x^2/2$ and $\sum_{i=1}^{x} i^2 \simeq x^3/3$. We also normally compute only expected values instead of maximum values. In all cases this is justified by the knowledge that by Chernoff bounds, the maximum value is larger by only a lower order term, typically $\mathcal{O}((h \cdot n \cdot \log n)^{1/2})$. In the deterministic version of such an algorithm, this term must be replaced by $\mathcal{O}((h \cdot n)^{2/3})$. See [1, 3] for details. Similarly, the given lower bounds refer to worst-case inputs, but the corresponding result that holds for almost all inputs is smaller only by a 'Chernoff deviation': $\mathcal{O}((h \cdot n \cdot \log n)^{1/2})$.

Analysis of One-Dimensional Routing Phases. From earlier work [1, 2, 3] we know that to analyze a one-dimensional routing problem (under a weak regularity condition on the distribution), one only has to consider the maximal distance a packet may have to go, and the maximal number of packets that may have to traverse a connection. This goes back to the 'routing-lemma' for one-dimensional routing (see [2] for a precise formulation and a proof in the case of routing on cycles, one-dimensional tori).

3 Tori of Triangles

In this section we analyze the properties of tori of triangles. In the simplest of them, the nodes have degree six: it has three routing axes. The tri-axial $n_1 \times n_2$ torus of triangles is denoted by $Tri(3, n_1, n_2)$. $Tri(3, n_1, n_2)$ consists of n_1 rows, each with n_2 PUs. Consecutive rows are shifted with respect to each other by half a position. An example is given in

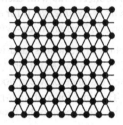

Figure 1: $Tri(3, 8, 8)$: the 8×8 tri-axial torus. The top and bottom row are to be identified, and likewise the left- and rightmost column.

Figure 1. Adding diagonals yields tori of triangles with $6, 12, \ldots$ axes (see Section 6 for a discussion). In our analyses we often concentrate on a central node. Such considerations are general because of the node symmetry.

3.1 Basic Results

Lemma 2 $Tri(3, n, n)$ has the following basic parameters:

$$
\begin{aligned}
deg(Tri, 3, n, n) &= 6, \\
EB(Tri, 3, n, n) &\leq 4 \cdot n, \\
D_{max}(Tri, 3, n, n) &= 3/4 \cdot n, \\
D_{av}(Tri, 3, n, n) &\simeq 19/48 \cdot n.
\end{aligned}
\tag{3}
$$

Proof: For the diameter we consider the path from a node drawn in the center to any corner. It goes over $n/2$ diagonal connections and $n/4$ horizontal connections. All other nodes are closer.

In order to minimally bisect $Tri(3, n, n)$, we must make two vertical or two horizontal cuts, $n/2$ apart. Each horizontal or vertical cut is crossed by $2 \cdot n$ connections: for a horizontal cut, by two edges from every node below and above the cut; for a vertical cut, by three connections from the nodes closest to the cut, and by one connection from every node half a position further away.

We compute the average distance of a node P to all other nodes. At distance i, for $0 \leq i \leq n/2$, from a PU there are $6 \cdot i$ PUs. At distance $n/2 < i \leq 3/4 \cdot n$, there are four corner sections, containing $2 \cdot (3/4 \cdot n - i)$ PUs each. For the sum of the distances from P to all other PUs this gives

$$
\sum_{i=1}^{n/2} 6 \cdot i^2 + \sum_{i=1}^{n/4-1} 8 \cdot i \cdot (3/4 \cdot n - i) \simeq n^3/4 + 3/16 \cdot n^3 - n^3/24.
$$

$\square$

Substituting in (2) gives

Corollary 2 *Routing h-relations on Tri(3, n, n) takes*

$$T(Tri, 3, n, n, h) \geq \max\{3/4 \cdot n, h \cdot n/8\}.$$

Lemma 1 and Corollary 1 give

Theorem 1 *On Tri(3, n, n) a VB-algorithm takes at least $19/144 \cdot h \cdot n$ steps. If (3) is an equality, then a VB-algorithm cannot route h-relations in the time given by the bisection bound.*

3.2 Algorithm

We sketch and analyze a VB-algorithm for routing on *Tri(3, n, n)*:

Algorithm VB-ROUTE
1. For every packet select uniformly a random intermediate destination. Route all packets to their intermediate destination along a shortest path with at most one bend.

2. Route all packets to their destinations along a shortest path

The packets are routed along a shortest path with at most one change of direction. From the two such paths, either one is selected with probability 1/2. Thus, the routing in each step consists of two one-dimensional phases. This choice of paths is motivated by its extreme simplicity. We will see that VB-ROUTE nevertheless performs well.

Edge contentions are resolved with the farthest-first strategy. The four one-dimensional phases can be separated by not starting Phase $f + 1$ before we can be (almost) sure that Phase f, $1 \leq f \leq 4$, has terminated (after a fraction $f/4$ of the total routing time). However, it is more practical to coalesce the phases. In that case, the packets must be equipped with the number of the phase they are performing, and packets still performing Phase f should get absolute priority over packets that are performing Phase f', for some $f' > f$.

3.3 Analysis

We have to determine the maximum number of packets that may have to go over a connection in any of the four one-dimensional routing phases. As the network is node-symmetric, we can consider the connections of the PU in the middle. Both diagonal axes play a similar role, but they have to be distinguished from the horizontal axis. We only compute expected values. Applying Chernoff bounds, or using a deterministic algorithm, the actual maximum number of packets is only larger by a lower-order term. We analyze the case that $h = 2 \cdot n^2$, that means that each PU sends one packet to each PU along either of the two shortest paths. Since all involved operations are linear, we can multiply the obtained result by $h/(2 \cdot n^2)$ to get the desired estimate.

Consider the subdivision of the torus given in Figure 2. The central PU, P sends one packet over the indicated axes to each of the PUs in each region. As not all regions are equally large, we cannot expect that the routing is perfectly balanced in the sense that all axes have to transfer the same number of packets.

We analyze the number of packets passing through the leftwards connection lc of P during the first routing phase. P itself sends one packet to each PU in $L = UL \cup DL$ over lc. A PU P' i positions to the right of P, $0 \leq i \leq n/4$, sends $i \cdot n$ packets fewer. More generally, the number of packets sent over lc by a PU P' i positions to the right of P can be determined as the area of the intersection of L, and a slab of the same size shifted i positions rightwards. This is illustrated in Figure 2.

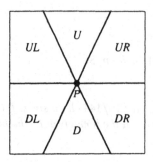 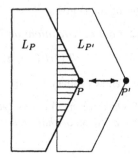

Figure 2: **Right:** The routing regions for $Tri(3, n, n)$. **Left:** The number of packets that P' sends over the left connection lc of P, equals the area of the intersection of L_P and $L_{P'}$.

Lemma 3 *In the first routing phase for $h = 2 \cdot n^2$, lc has to transfer an expected number of $7/96 \cdot n^3$ packets.*

Proof: The expected number is approximately given by

$$\sum_{i=1}^{n/4} \sum_{j=1}^{i} 4 \cdot j + \sum_{i=1}^{n/4} (i \cdot n + n^2/8) \simeq n^3/96 + n^3/32 + n^3/32.$$

$\square$

In a similar way we can compute the expected number of packets traversing the connection ulc to the upper-left out of P:

Lemma 4 *In the first routing phase for $h = 2 \cdot n^2$, ulc has to transfer an expected number of $n^3/16$ packets.*

Proof: We now have to analyze how many packets each of the PUs to the lower-right of P want to transfer over ulc. For a PU P' at distance i from P, this number equals the area of the intersection of $ULU = LU \cup U$ and a slab of the same size shifted by i positions to the lower-right. The situation is illustrated in Figure 3. Now we easily see that the expected

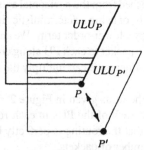

Figure 3: The number of packets that P' sends over the upper-left connection ulc of P, equals the area of the intersection of ULU_P and $ULU_{P'}$.

number is approximately given by

$$\sum_{i=1}^{n/2} ((n/4 + i/2) \cdot i + (i/2)^2) = \sum_{i=1}^{n/2} (i \cdot n/4 + 3/4 \cdot i^2) \simeq n^3/32 + n^3/32.$$

$\square$

Lemma 5 *The first phase of routing an h-relation with VB-route takes*

$$T_1(Tri, 3, n, n, h) = (1 + o(1)) \cdot \max\{7/192 \cdot h \cdot n, n/2\}.$$

For the second phase, in which the routing of the packets to their intermediate destination is completed, we could give a similar analysis. However, it is not hard to see that the situation is symmetrical with the first phase: in Phase 2, a PU P', at i positions to the *left* of P is going to receive as many packets that were routed through lc, as the PU P'' at i positions to the *right* has sent through lc during Phase 1. Thus, Phase 2 takes approximately the same number of steps as Phase 1. Routing packets from random positions to given destinations is as expensive as routing them from given positions to random destinations. So, Step 2 of VB-ROUTE is as expensive as Step 1. So, adding the times for the in total four routing phases,

Theorem 2 *Applying* VB-ROUTE *for routing h-relations on* $Tri(3, n, n)$ *takes*

$$T(Tri, 3, n, n, h) = (4 + o(1)) \cdot \max\{7/192 \cdot h \cdot n, n/2\}.$$

For $h \geq 14$, the distance component in the routing time no longer plays a role:

Corollary 3 *For* $h \geq 14$,

$$T(Tri, 3, n, n, h) = (7/48 + o(1)) \cdot h \cdot n.$$

If we compare with the derived lower bounds, then we see that VB-ROUTE is not optimal: the bisection bound gives us a leading factor $1/8 = 0.125$; the connection-availability bound for VB-algorithms gives $19/144 \simeq 0.132$; and we have found $7/48 \simeq 0.146$. But

Corollary 4 VB-ROUTE *is an optimal VB-algorithm routing all packets along the shortest paths to their intermediate and final destinations.*

Proof: If a packet p is routed to a randomly chosen destination, then $7/96 \cdot n$ is just the expected distance p has to travel over a leftward connection if it takes a shortest path. □

3.4 Refinements

In order to reduce the routing time, we somehow have to reduce the number of packets that are routed over the horizontal connections. There are the following obvious possibilities:

1. Route some packets over two diagonal connections.

2. Take a torus with reduced width.

3. Do not randomize the packets completely.

We only describe the second approach. The optimal width w of the torus is such that the number of packets that have to be routed along the leftward connection lc of P equals the number of packets that have to be routed along its connection ulc to the upper-left. Substituting the width as a variable in the analysis of Lemma 3 and Lemma 4, we find that this happens for $w = (1/2 + 1/\sqrt{6}) \cdot n \simeq 0.91 \cdot n$:

Theorem 3 *For routing h-relations on an* $n \times (1/2 + 1/\sqrt{6}) \cdot n$ *torus of triangles* VB-ROUTE *takes*

$$T(Tri, 3, n, (1/2 + 1/\sqrt{6}) \cdot n, h) = (4 + o(1)) \cdot \max\{h \cdot n/32, n/2\}.$$

Proof: Taking $w = (1/2 + 2/\sqrt{24}) \cdot n$, we can recompute the results of Lemma 3 and Lemma 4. It turns out that both lc and ulc have to transfer

$$(3 + \sqrt{6})/96 \cdot n^3$$

packets if each PU routes two packets to every PU. But now the number of PUs is only $n \cdot w$, so we must multiply by $4 \cdot h/(2 \cdot n \cdot w)$ to compute the routing time of the whole algorithm from this. □

Corollary 5 *For* $h \geq 16$, VB-ROUTE *matches the bisection bound for routing h-relations on* $Tri(3, n, (1/2 + 1/\sqrt{6}) \cdot n)$:

$$T(Tri, 3, n, (1/2 + 1/\sqrt{6}) \cdot n, h) = (1/8 + o(1)) \cdot h \cdot n.$$

4 Quadri-Axial Tori

Now we analyze the properties of the quadri-axial $n \times n$ torus. We only consider the $n \times n$ version, denoted by $Tor(4, n)$. An example is given in Figure 4. After the examplary exposition in the previous section, we can present in the remainder only a few key results.

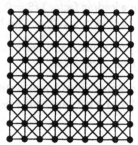

Figure 4: $Tor(4, 8)$: the 8×8 quadri-axial torus. The top and bottom row are to be identified, and likewise the left- and rightmost column.

Lemma 6 $Tor(4, n)$ *has the following basic parameters:*

$$
\begin{aligned}
deg(Tor, 4, n) &= 8, \\
EB(Tor, 4, n) &\leq 6 \cdot n, \\
D_{max}(Tor, 4, n) &= n/2, \\
D_{av}(Tor, 4, n) &\simeq n/3.
\end{aligned}
\tag{4}
$$

Substituting in (2) gives the distance and bisection bound:

Corollary 6 *Routing h-relations on* $Tor(4, n)$ *takes*

$$T(Tor, 4, n, h) \geq \max\{n/2, h \cdot n/12\}.$$

We apply the same algorithm VB-ROUTE as for routing on $Tri(3, n, n)$, with the additional specification that the bend in a path has the minimal angle of $45°$. This guarantees that again there are at most two shortest paths between any pair of PUs. Each of them is selected with probability $1/2$.

We have to determine the maximum number of packets that may have to go over a connection in any of the four one-dimensional routing phases. Though intuitively it appears clear that all axes play a similar role, we cannot demonstrate this: $Tor(4, n)$ is *not* edge-symmetric. So, we must distinguish the 'main' axes from the diagonal axes. Performing an analysis similar to the one in Section 3.3, we find

Lemma 7 *The first phase of routing an h-relation with VB-route takes*

$$T_1(Tor, 4, n, h) = (1 + o(1)) \cdot \max\{h \cdot n/48, n/2\}.$$

The other three one-dimensional routing phases take the same number of steps. Thus,

Theorem 4 *Applying* VB-ROUTE *for routing h-relations on $Tor(n, 4)$ takes*

$$T(Tor, 4, n, h) = (4 + o(1)) \cdot \max\{h \cdot n/48, n/2\}.$$

Corollary 7 *For $h \geq 24$,*

$$T(Tor, 4, n, h) = (1/12 + o(1)) \cdot h \cdot n.$$

As this matches the bisection bound, we have achieved an important result: h-relations can be routed optimally on quadri-axial tori, by a simple VB-algorithm.

5 Honeycombs

In a honeycomb, the nodes are placed on the corners of a regular tessellation of the plane by hexagons. In the simplest version the nodes have degree three. Three axes can be distinguished, each consisting of an alternation of two types of connections. A square honeycomb has too small a bisection width. Therefore, we consider $n_1 \times n_2$ versions, denoted $Hon(3, n_1, n_2)$, with $n_2 > n_1$. An example is given in Figure 5. Notice the similarity to $Tri(3, n, n)$. Honeycombs with more axes can be obtained by adding 'diagonals' (see Section 6). The analysis of honeycombs completes our picture of routing on tori of planar tessellations. The routing algorithm for honeycombs is based on the following lemma, which is illustrated in Figure 5.

Lemma 8 *$Tri(3, n_1, n_2)$ can be embedded in $Hon(3, n_1, 2 \cdot n_2)$ with dilation, congestion and expansion two.*

Now, to route an h-relation on $Hon(3, n_1, 2 \cdot n_2)$, we can apply the algorithm for routing a $2 \cdot h$-relation on the embedded $Tri(3, n_1, n_2)$. Because every connection is doubly used, this routing is delayed by a factor of two.

Theorem 5 *For routing h-relations on an $n \times (1 + 2/\sqrt{6}) \cdot n$ honeycomb* VB-ROUTE *takes*

$$T(Hon, 3, n, (1 + 2/\sqrt{6}) \cdot n, h) = (4 + o(1)) \cdot \max\{h \cdot n/8, n\}.$$

So, for $h \geq 8$, $2 \cdot h \cdot n^2$ packets are routed in $(1/2 + o(1)) \cdot h \cdot n$ time. This is not very competitive in comparison with the other networks (see Section 7 for a comparison).

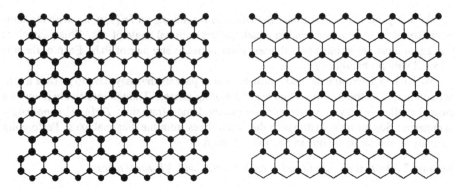

Figure 5: **Left:** $Hon(3, 8, 16)$: the 8×16 honeycomb. The top and bottom row are to be identified, and likewise the left- and rightmost column. The three main axes are highlighted: horizontal, upper-left to lower-right and lower-left to upper-right. **Right:** $Tri(3, 8, 8)$ embedded in $Hon(3, 8, 16)$: dilation, congestion and expansion are two.

6 Extensions?

The considered networks can be further 'enriched' with diagonals. At the expense of a more complicated structure, this improves their routing capacity. The larger the number of axes, the better the ratio of degree and bisection bound becomes. This is the underlying reason that going from $Tor(2, n)$ (the bi-axial $n \times n$ torus) to $Tor(4, n)$ gives more than linear speed-up.

When generalizing, we must be careful: not all generalizations make sense. So far we have that at all times the shortest path to a node is also a shortest path if we have measured the Euclidian distance. Such a network we call *natural*. Naturalness is a valuable property which we want a network to have. It allows for straightforward, easily analyzable algorithms. On a non-natural network some connections must be used as 'expressways' while others are more for local communication. This is one step too many in the direction of a three-dimensional torus. In Figure 6 we give an example that demonstrates that neither

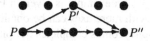

Figure 6: The indicated 'detour' from P to P'' via P' is shorter than the direct path.

$Tor(8, n)$ nor $Tri(6, n, n)$ are natural.

The only extension that makes sense is to enrich the honeycombs with diagonals, replacing them by complete graphs of six nodes. This is similar to the enrichment of $Tor(2, n)$ with diagonals in order to obtain $Tor(4, n)$. It is called the *enriched honeycomb*. It has six routing axes. The $n_1 \times n_2$ version is denoted by $Hon(6, n_1, n_2)$. An example is given in Figure 7.

For routing h-relations we again apply the algorithm VB-ROUTE. In the analysis we must distinguish four types of axes: horizontal, 30° vertical, 60° vertical and vertical. The width w of the honeycomb must be chosen so that, for given height n, the number of packets moving over each of these axes is as equal as possible.

Theorem 6 *For routing h-relations, $h \geq 40$, on an $n \times 1.61 \cdot n$ enriched honeycomb,*

$$T(Hon, 6, n, 1.61 \cdot n, h) = (0.066 + o(1)) \cdot h \cdot n.$$

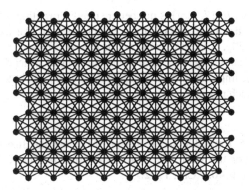

Figure 7: $Hon(6, 8, 16)$: the enriched 8×16 honeycomb. The top and bottom row are to be identified, and likewise the left- and rightmost column.

Is this a good result? Routing on enriched honeycombs is considerably faster than any of the other networks. But with degree 12 this was to be expected. The bisection bound for this value of w gives $h \cdot n/16$. Our algorithm requires about 6% more.

7 Conclusion

We considered routing h-relations on tori of degree three, six, eight and twelve. We have added considerably to our understanding, but there are also possible practical aspects.

Typically, a PU consists of a CPU, a local memory, a bus, and a dedicated link processor for each connection. So, the amount of hardware increases linearly with the degree. For reasons of scalability and constructibility simple planar (two-dimensional meshes) and quasi-planar (two-dimensional tori) architectures are often preferred over cleverer architectures with complicated wiring (such as cube-connected cycles). The *cost-effectiveness* of a network can be defined as $ce = \#\{\text{delivered packets per step}\}/\#\{\text{connections}\}$. Let g be the parameter such that h-relations can be performed in $g \cdot h$ time. So, $ce = 1/(g \cdot deg)$. The cost-effectiveness has a second meaning: if the actual routing does not allow full-port communication, but only single-port, then a routing algorithm can be emulated with slow-down equal to the degree of the network. That is, h-relations will be performed with parameter $g' = deg \cdot g = 1/ce$.

	deg	g/n	$n \cdot ce$
$Hon(3, 0.74 \cdot n, 1.35 \cdot n)$	3	0.37	0.90
$Tor(2, n, n)$	4	0.25	1.00
$Tri(3, 1.05 \cdot n, 0.95 \cdot n)$	6	0.13	1.27
$Tor(4, n, n)$	8	0.083	1.50
$Hon(6, 0.79 \cdot n, 1.27 \cdot n)$	12	0.052	1.60

Table 1: Overview for the considered networks. All networks are the optimal choices with n^2 PUs.

In Table 1 we list the results for the considered networks. The cost-effectiveness increases with the degree of the network: $Hon(6, 0.79 \cdot n, 1.27 \cdot n)$ is clearly the most cost-effective, but it might be easier to construct a network with fewer crossings, such as $Tri(3, 1.05 \cdot n, 0.95 \cdot n)$.

The standard torus $Tor(2, n, n)$ is not better suited for routing h-relations than any of the other networks. It has the advantage that it allows problems like matrix multiplication to be solved more naturally, but these should not be hard on $Tri(3, n, n)$ either. Furthermore, if parallel computers are going to be used more and more as general purpose machines, it becomes less and less important that some special problems can be solved in a pretty way. h-relations matter much more!

References

[1] Kaufmann, M., S. Rajasekaran, J.F. Sibeyn, 'Matching the Bisection Bound for Routing and Sorting on the Mesh,' *Proc. 4th Symposium on Parallel Algorithms and Architectures*, pp. 31–40, ACM, 1992.

[2] Kaufmann, M., J.F. Sibeyn, 'Optimal Multi-Packet Routing on the Torus', *Proc. 3rd Scandinavian Workshop on Algorithm Theory*, LNCS 621, pp. 118–129, Springer-Verlag, 1992.

[3] Kaufmann, M., J.F. Sibeyn, T. Suel, 'Derandomizing Routing and Sorting Algorithms for Meshes,' *Proc. 5th Symposium on Discrete Algorithms*, pp. 669–679, ACM-SIAM, 1994.

[4] Kunde, M., 'Block Gossiping on Grids and Tori: Deterministic Sorting and Routing Match the Bisection Bound,' *Proc. European Symp. on Algorithms*, LNCS 726, pp. 272–283, 1993.

[5] Kunde, M., R. Niedermeier, P. Rossmanith, 'Faster Sorting and Routing on Grids with Diagonals,' *Proc. 11th Symposium on Theoretical Aspects of Computer Science*, LNCS 775, pp. 225–236, Springer Verlag, 1994.

[6] Kunde, M., T. Tensi, 'Multi-Packet Routing on Mesh Connected Processor Arrays,' *Proc. Symposium on Parallel Algorithms and Architectures*, pp. 336–343, ACM, 1989.

[7] Rajasekaran, S., M. Raghavachari, 'Optimal Randomized Algorithms for Multipacket and Cut Through Routing on the Mesh', *Journal of Parallel and Distributed Computing*, 26(2), pp. 257–260, 1995.

[8] Reif, J., L.G. Valiant, 'A logarithmic time sort for linear size networks,' *Journal of the ACM*, 34(1), pp. 68–76, 1987.

[9] Reischuk, R., 'Probabilistic Parallel Algorithms for Sorting and Selection,' *SIAM Journal of Computing*, 14, pp. 396–411, 1985.

[10] Sibeyn, J.F., 'Sample Sort on Meshes,' *Proc. Computing Science in the Netherlands*, SION, Amsterdam, 1995. Full version in *Techn. Rep. MPI-I-95-112*, Max-Planck Institut für Informatik, Saarbrücken, Germany, 1995.

[11] Valiant, L.G., 'A Bridging Model for Parallel Computation,' *Communications of the ACM*, 33(8), pp. 103–111, 1990.

[12] Valiant, L.G., G.J. Brebner, 'Universal Schemes for Parallel Communication,' *Proc. 13th Symposium on Theory of Computing*, pp. 263–277, ACM, 1981.

A Uniform Analysis of Trie Structures that Store Prefixing-Keys with Application to Doubly-Chained Prefixing-Tries

Pilar de la Torre[1] and David T. Kao[2]

Department of Computer Science, University of New Hampshire, Durham, NH 03824, U.S.A.

Abstract

Tries are data structures for storing sets where each element is represented by a key that can be viewed as a string of characters over a finite alphabet. These structures have been extensively studied and analyzed under several probability models. All of these models, however, preclude the occurrence of sets in which the key of one element is a prefix of that of another – such a key is called a *prefixing-key*.

This paper presents an average case analysis of several trie varieties, which we generically called *doubly-chained prefixing-tries*, for representing sets with possible prefixing-keys. The underlying probability model, which we call the *prefix model*, $\mathcal{P}_{h,n,m}$ assumes as equally likely all n-element sets whose keys are composed of at most h characters from a fixed alphabet of size m. For each of the trie varieties analyzed, we derive exact formulas for the expected space required to store such a set, and the average time required to retrieve an element given its key, as functions of h, n, and m.

Our approach to the analysis is of interest in its own right. It provides a unifying framework for computing the expectations of a wide class of random variables with respect to the prefix model. This class includes the cost functions of the trie varieties analyzed here.

1 Introduction

Tries are classical data structures (see, for example, [dlB59, Fre60, Sus63, Gwe68, Mor68, Knu73b, GBY91]) for representing sets of string keys, and their analysis (see, for example, [Knu73b], and [GBY91] for an extensive bibliography) has been carried out with respect to several probability models. These include Knuth's *infinite key-length model* [Knu73b], the *finite key-length model* and the *set-intersection model* introduced by Trabb Pardo [Tra78], and the *biased bit model* of Flajolet, Regnier and Sotteau [FRS85]. All these models, however, preclude the occurrence of sets in which the key of one element is a prefix of that of another – such a key is called a *prefixing-key*.

Knott has proposed [Kno86] three trie structures for representing sets of string keys that may include prefixing-keys: *full doubly-chained prefixing-tries*, *compact doubly-chained prefixing-tries*, and *patrician doubly-chained prefixing-tries*. Although a way of getting around the no-prefixing-key restriction has been part of the programming folklore (namely by ending each key with a special symbol, the *endmarker*, to mark the key's end), the analysis of tries built from sets that include prefixing-keys seems to have received no attention so far. (For the basic background material on tries and doubly-chained tries, see, for example, [Knu73b].).

---

[1] The research of this author was supported in part by the National Science Foundation under Grant CCR-9010445, and CCR-9410592.

[2] The research of this author was supported in part by the National Science Foundation under Grant IRI-9117153.

This research makes two main contributions. One is the analysis of three variants of doubly-chained tries for storing prefixing-keys: *full doubly-chained prefixing-tries*, *compact doubly-chained prefixing-tries*, and *patrician doubly-chained prefixing-tries*. We analyze the time and space requirement for the retrieval algorithm of each of these data structures under the *prefix model*, a probability model which takes into account sets of keys that include prefixing-keys. The sample space of this model consists of all the n-element sets of string keys of length no greater than h that can be composed from a fixed finite alphabet; all such sets are assumed to be equally probable. We derive exact expressions for the average space complexity, $E_{h,n}[S]$, and average time complexity, $E_{h,n}[T]$, of the retrieval algorithms for each of the above trie varieties.

The other contribution is our approach to the analysis, the *root-function method*, which is of interest in its own right. It provides a unifying framework for computing the expectations of a wide class of random variables with respect to the prefix model. In this paper, we apply this method to analyze doubly-chained prefixing-tries. We have also applied it in [dlT87a, dlTK95a] to the analysis of *full prefixing-tries*, *compact prefixing-tries*, *patrician prefixing-tries*, and *bucket tries*.

A brief historical perspective is in order. The root-function method presented in this paper, which was developed in de la Torre's doctoral thesis [dlT87a, dlT87b], is a recurrence-based approach close in spirit to Knuth's and Trabb Pardo's trie analyses. Independent and separate work of Flajolet-Regnier-Sotteau [FRS85] proposed a systematic generating-function approach to the analysis of tries and set-intersection algorithms, and developed it for Knuth's model [Knu73b], the two Trabb Pardo's models [Tra78], and their own biased-bit model; all of these models preclude the occurrence of sets with prefixing-keys. Post-doctoral work of de la Torre [dlT87c, dlTK94, dlTK95b] extended their approach to the prefix model, the *set-intersection prefix model*, and applied the extended framework to the analysis of set-intersection algorithms based on *full prefixing-tries* and *compact prefixing-tries*.

This paper is organized as follows. Section §2 introduces the root-function method for calculating expectations of a wide class of random variables under the prefix model. Section §3 introduces the *prefixing-tree*, which is the logical structure underlying the construction of the trie data structures, and gives some terminology concerning tries. The root-function method is applied to analyze doubly-chained prefixing-tries in Section §4. Section §5 contains concluding remarks. Due to the space constraints we have omitted the proofs which can be found in [dlTK95a].

2 Root-function method

In this section we develop the main tool, which we called *the root-function method*. This method provides a unified approach for computing the expectations of a wide class of random variables with respect to the prefix model. This class includes the cost functions emerging in our analysis of tries.

2.1 Prefix model

The sample space of the *prefix model* consists of a class of sets of keys described by three parameters: the common size n of the key sets, the maximum length h of the keys that are allowed in these sets, and the size m of the underlying alphabet $\mathcal{A} = \{0, \ldots, m-1\}$. For a nonnegative integer h, the set of all strings of length less than or equal to h composed from the alphabet $\mathcal{A} = \{0, \ldots, m-1\}$ will be denoted by $\mathcal{A}^{[h]}$, i.e., $\mathcal{A}^{[h]} \equiv \mathcal{A}^0 \cup \mathcal{A}^1 \cup \ldots \cup \mathcal{A}^h$.

(Note that $\mathcal{A}^{[h]}$ equals the set of prefixes of the strings in $\mathcal{A}^h$.) The set of finite length strings over $\mathcal{A}$ will be denoted by $\mathcal{A}^*$.

Given the integers h, n, m, with $h, n \geq 0$ and $m \geq 2$, the probability space for the prefix model consists of the n-element subsets of $\mathcal{A}^{[h]}$, all of which are assumed to be equally probable. If B is a finite set, $|B|$ will denote the number of elements in B, and $\mathcal{R}_n(B)$ the set of its n-element subsets. It can be readily seen that

$$m^{[h]} \equiv |\mathcal{A}^{[h]}| = \frac{m^{h+1} - 1}{m - 1}, \qquad \text{and} \qquad |\mathcal{R}_n(\mathcal{A}^{[h]})| = \binom{m^{[h]}}{n}.$$

The *expectation* of a mapping X defined on the n-element subsets of $\mathcal{A}^{[h]}$ is denoted by $E_{h,n}[X]$. Whenever it is clear from the context, we will use $E[X]$ instead of $E_{h,n}[X]$. The *normalized expectation* of X is defined as $N_{h,n}[X] = \sum_{s \in \mathcal{R}_n(\mathcal{A}^{[h]})} X(s) = \binom{m^{[h]}}{n} E_{h,n}[X]$.

2.2 Root-Functions

Definition 2.1 THE s_i NOTATION. *For each integer $0 \leq i \leq m - 1$, and $s \in P(\mathcal{A}^*)^3$, where $\mathcal{A} = \{0, \ldots, m-1\}$, we define*

$$s_i = \{\sigma \in \mathcal{A}^* | i\sigma \in s\}$$

Note that $\forall i \in \mathcal{A}$, is$_i = s \cap i\mathcal{A}^*$ (i.e., s_i is the set of tails, or suffixes, of the strings in s which have prefix i). In general, for $x \in \mathcal{A}^*$, we define $s_x = \{\sigma \in \mathcal{A}^* | x\sigma \in s\}$. (Note that $s_\varepsilon = s$.)

The decomposition $s = ((s \cap \{\varepsilon\}) \cup 0s_0 \cup \ldots \cup (m-1)s_{m-1})$ yields a natural bijection:

$$\beta : P(\mathcal{A}^*) \to P(\{\varepsilon\}) \times P(\mathcal{A}^*) \times \ldots \times P(\mathcal{A}^*),$$

where $\beta(s) = (s \cap \{\varepsilon\}, s_0, \ldots, s_{m-1})$. In particular, note that $\beta(P(\mathcal{A}^{[h]})) \subseteq P(\{\varepsilon\}) \times P(\mathcal{A}^{[h-1]}) \times \ldots \times P(\mathcal{A}^{[h-1]})$. This mapping relates to the partition induced on a set s by its representation as a trie $t(s)$, namely each s_i can be viewed as the set of strings represented by the $i + 2$-th subtree of $t(s)$. Throughout the remainder of this section, X will denote a function $X : P(\mathcal{A}^*) \to R$, where R are the real numbers.

Definition 2.2 ROOT-FUNCTION. *The root-function of X is the function $\bar{X} : P(\mathcal{A}^*) \to R$ defined by*

$$\bar{X}(s) \equiv X(s) - \sum_{i \in \mathcal{A}} X(s_i), \tag{1}$$

The intuition behind the notion, root-functions, is given in Figure 1. The following theorem gives an expression of value of X in terms of $\bar{X}$.

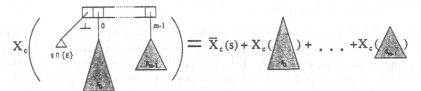

Figure 1. Let X_C denote a cost function defined on the set of strings in trie $t(s)$ – such as the number of strings (i.e., leaf nodes) of $t(s)$. Then $X_C(s)$ equals the value of the root-function $\bar{X}_C(s) = |s \cap \{\varepsilon\}|$ plus the sum of costs $X_C(s_0) + \ldots + X_C(s_{m-1})$ on the strings in the subtrees.

[3]Given an arbitrary set B, $P(B)$ denotes its power set.

Theorem 2.3 *Let $\bar{X}$ be the root-function of X.*

1. *If $X(\emptyset) = 0$, then $X(s) = \sum_{z \in A^*} \bar{X}(s_z)$.*

2. *In general we have $X(s) = \frac{1}{1-m}\bar{X}(\emptyset) + \sum_{z \in A^*}[\bar{X}(s_z) - \bar{X}(\emptyset)]$, and $\bar{X}$ is the unique function with this property.*

2.3 Reduction to root-functions

Our intention is to express $E[X]$ in terms of the expectation $E[\bar{X}]$ of the root-function $\bar{X}$ of X. The desired expression for $E[X]$ will be obtained by solving the recurrence relation for $E[X]$ provided by the following theorem.

Theorem 2.4 *Let $\bar{X}$ be the root-function of X. Then $N_{0,n}[X] = \delta_{0,n}X(\emptyset) + \delta_{1,n}X(\{\varepsilon\})$ and*

$$N_{h,n}[X] = N_{h,n}[\bar{X}] + m\sum_{k \geq 0}\binom{m^h}{n-k}N_{h-1,k}[X] \qquad (h \geq 1). \tag{2}$$

Let $f(x, y)$ and $g(x)$ be real-valued functions. The unique solution to the recurrence

$$
\begin{aligned}
y(0, n) &= f(0, n) \qquad (n \geq 0), \\
y(h, n) &= f(h, n) + m\sum_{k \geq 0}\binom{g(h)}{n-k}y(h-1, k) \qquad (h \geq 1, n \geq 0),
\end{aligned}
$$

can be calculated from the *independent term* $f(h, n)$ using iteration [dlT87a]

$$y(h, n) = \sum_{0 \leq i \leq h} m^i \sum_{k \geq 0}\binom{\sum_{j=0}^{i-1} g(h-j)}{n-k}f(h-i, k). \tag{3}$$

Solving recurrence (2) with the help of equation (3) we arrive at the following expression of $N_{h,n}[X]$ in terms of $N_{h,n}[\bar{X}]$.

Theorem 2.5 *If $\bar{X}$ is the root-function of X then*

$$N_{h,n}[X] = \sum_{\substack{0 \leq j \leq h \\ 0 \leq k}} m^j\binom{m^{[h]} - m^{[h-j]}}{n-k}N_{h-j,k}[\bar{X}], \tag{4}$$

where $N_{0,k}[\bar{X}] = \delta_{0,k}(1 - m)X(\emptyset) + \delta_{1,k}[X(\{\varepsilon\}) - mX(\emptyset)]$.

Consider the function $Z(s) = |s|$, for instance. We have $Z(s) = \bar{Z}(s) + Z(s_0) + \ldots + Z(s_{m-1})$ with $\bar{Z}(s) = 1$, if $\varepsilon \in s$, and $\bar{Z}(s) = 0$, otherwise. Substitution of $N_{h,n}[\bar{Z}] = \binom{m^{[h]}-1}{n-1}$ in (4) followed by the application of *Vandermonde* convolution yields $N_{h,n}[Z] = m^{[h]}\binom{m^{[h]}-1}{n-1}$, which is as expected.

Example 2.6 A nontrivial application of Theorem 2.5 may be found in the following computation of the average value of $Tl(s) \equiv \sum_{x \in s} l(x)$, the total sum of the lengths of the strings in an n-element set $s \subseteq A^{[h]}$. This random variable Tl will be used later to compute the expectations of trie cost functions (see Table III in Figure 6 and Table V in Figure 10). ∎

Theorem 2.7 *The average value of $Tl(s)$ over the n-element subsets $s \subseteq A^{[h]}$, $0 \leq n \leq m^{[h]}$, equals*

$$E[Tl] = n\left[h - \frac{1}{m-1}\left(1 - \frac{h+1}{m^{[h]}}\right)\right].$$

2.4   Computing expectation of root-functions

Theorem 2.5 reduces the expression of expectation $E[X]$ of X in terms of the expectation of its root-function $\bar{X}$. For some mappings (as seen earlier for Tl, for instance), the root-function is simple enough so that its expectation can be easily determined by direct counting. As we shall see below, a more systematic approach to calculating expectations of root-functions is often possible. The following property is enjoyed by many functions related to trie analysis by trie-related cost functions.

Definition 2.8 CARDINALITY-DEPENDENT FUNCTION AND ITS COUNTING-FUNCTION. *Let the function* $C : P(\mathcal{A}^*) \to N^{m+1}$, *defined by* $C(s) = (|s \cap \{\varepsilon\}|, |s_0|, \ldots, |s_{m-1}|)$. *Function* $Y : P(\mathcal{A}^*) \to R$ *is cardinality-dependent iff there exists* $\check{Y} : N^{m+1} \to R$ *such that* $Y = \check{Y} \circ C$, *i.e.*,

$$Y(s) = \check{Y}(|s \cap \{\varepsilon\}|, |s_0|, \ldots, |s_{m-1}|).$$

$\check{Y}$ *is called the counting-function of* Y.

It is easy to verify the following lemma.

Lemma 2.9 *Function* $Y : P(\mathcal{A}^*) \to R$ *is cardinality-dependent iff for every pair of sets* s *and* s' *in* $P(\mathcal{A}^*) : Y(s) = Y(s') \Leftrightarrow C(s) = C(s')$.

Example 2.10 The mapping $Tl(s) \equiv \sum_{x \in s} l(x)$ is not cardinality-dependent, for instance. However, $\bar{Tl}$ is, and $\check{\bar{Tl}}(\nu, n_0, \ldots, n_{m-1}) = n_0 + \cdots + n_{m-1}$ is its counting-function. ∎

As it turns out, the functions X emerging from the analysis of tries have root-functions $Y = \bar{X}$ which are cardinality-dependent. Moreover, their counting-functions $\check{Y} = \check{\bar{X}}$ can be easily derived from the recursive definition of the data structures. The following lemma expresses $N_{h,n}[Y]$ in terms of $\check{Y}$.

Lemma 2.11 *Let* $h \geq 1$ *and* $1 \leq n \leq m^{[h]}$. *If* $Y : P(\mathcal{A}^*) \to R$ *is a cardinality-dependent function and* $\check{Y}$ *is its counting-function, then*

$$N_{h,n}[Y] = \sum_{\substack{\nu + n_0 + \cdots + n_{m-1} = n \\ \nu = 0,1}} \prod_{0 \leq i \leq m-1} \binom{m^{[h-1]}}{n_i} \check{Y}(\nu, n_0, \ldots, n_{m-1}) \qquad (h \geq 1)$$

This suggests the introduction of the following operator $\check{N}_{h,n}$ which maps each counting-function $\check{Y}$ to $N_{h,n}[Y]$.

Definition 2.12 THE $\check{N}_{h,n}$ OPERATOR. *For every function* $f : N^{m+1} \to R$,

$$\check{N}_{h,n}[f] = \sum_{\substack{\nu + n_0 + \cdots + n_{m-1} = n \\ \nu = 0,1}} \prod_{0 \leq i \leq m-1} \binom{m^{[h-1]}}{n_i} f(\nu, n_0, \ldots, n_{m-1}) \qquad (h \geq 1),$$

which equals $N_{h,n}[f \circ C]$ *(Lemma 2.11) with* C *is defined in Definition 2.8.*

It follows directly that $N_{h,n}[Y] = \check{N}_{h,n}[\check{Y}]$, if Y is cardinality-dependent. Operator $\check{N}_{h,n}$ inherits the linearity of $N_{h,n}$, a property which can be easily verified and will be useful in our calculations later.

   Finally we obtain the desired expression of the expectation $E[Y]$ of a cardinality-dependent function Y in terms of values $\check{N}_{i,j}[\check{Y}]$ which are often easier to compute for random variables associated to our analysis of tries. For the purpose of future reference, the theorem below follows directly from Theorem 2.5.

Theorem 2.13 *Let* $\tau(a,b,c,d) = \binom{a-b}{c-d}/\binom{a}{c}$ *and* $0 \le n \le m^{[h]}$. *If* $\check{Y}$ *is the counting-function of* Y *then*

$$E[Y] = m^h\left[\left(1 - \frac{n}{m^{[h]}}\right)Y(\emptyset) + \frac{n}{m^{[h]}}Y(\{\varepsilon\})\right] + \sum_{\substack{0 \le j < h \\ 0 \le k}} m^j \tau(m^{[h]}, m^{[h-j]}, n, k)\check{N}_{h-j,k}[\check{Y}].$$

Theorem 2.13 enables us to compute $E[Y]$ as a function of h and n provided we know such an expression for $\check{N}_{h,n}[\check{Y}]$. For the functions Y emerging from our trie analysis, the desired expression for $\check{N}_{h,n}[\check{Y}]$ as function of h and n can be easily derived from Table I in Figure 2 and the linearity of $\check{N}_{h,n}$.

$\check{Y}(\nu, n_0, \ldots, n_{m-1})$	$\check{N}_{h,n}[\check{Y}], \quad h \ge 1$
1	$\binom{m^{[h]}}{n}$
δ_{0,n_i}	$\binom{m^h}{n}$
δ_{n,n_i}	$\binom{m^{[h-1]}}{n}$
n_i	$\binom{m^{[h]}-1}{n-1}m^{[h-1]}$
$n_j\delta_{0,n_i}, \; i \ne j$	$\binom{m^h-1}{n-1}m^{[h-1]}$
ν	$\binom{m^{[h]}-1}{n-1}$
$\nu\delta_{n,n_i}$	0
$\nu\delta_{0,n_i}$	$\binom{m^h-1}{n-1}$
$f(n)\check{Y}_1 + g(n)\check{Y}_2$	$f(n)\check{N}_{h,n}[\check{Y}_1] + g(n)\check{N}_{h,n}[\check{Y}_2]$

Figure 2. TABLE I: ELEMENTARY VALUES OF $\check{N}_{h,n}$. Note that $f(n)$ and $g(n)$ are any two real-valued functions, and $\check{Y}_1$, $\check{Y}_2$ are any two cardinality-dependent functions.

Example 2.14 We shall illustrate our approach by calculating the average value of $K(s) = |ppref(s)|$, the number of keys in s which are prefixes of other keys in that set. Also, the functions $K(s)$ and $Tl(s)$ will be helpful for expressing the relationships among the different kinds of tries, and will add intuitive meaning to the expectations of the trie cost functions to be computed later. ∎

Let $ppref(s) = \{x \in s | |s_x| \ge 2\}$ denotes the set of proper prefixes of s, which are in s themselves, and $K(s) = |ppref(s)|$. Thus,

$$ppref(s) = \begin{cases} \emptyset & if |s| \le 1 \\ (\{\varepsilon\} \cap s) \cup \bigcup_{i \in \mathcal{A}} i \, ppref(s_i) & otherwise \end{cases}$$

which implies $|ppref(\emptyset)| = 0$ and $|ppref(s)| = |\{\varepsilon\} \cap s|(1 - \delta_{|s|,1}) + \sum_{i \in \mathcal{A}} |ppref(s_i)|$ for $s \ne \emptyset$. Therefore $\bar{K} = |\{\varepsilon\} \cap s|(1 - \delta_{|s|,1})$ is the root-function, and the counting-function of $\bar{K}(s)$ is $\check{\bar{K}}(\nu, n_0, \ldots, n_{m-1}) = \nu(1 - \delta_{n,1})$, with $n = \nu + \sum_{0 \le j \le m-1} n_j$.

Theorem 2.15 *Let* $\tau(a,b,c,d) = \binom{a-b}{c-d}/\binom{a}{c}$ *and* $0 \le n \le m^{[h]}$. *The average value of* $K(s)$ *over the n-element subsets* $s \subseteq \mathcal{A}^{[h]}$ *equals*

$$E[K] = n\frac{m^{[h-1]}}{m^{[h]}} - \sum_{0 \le j < h} m^j \tau(m^{[h]}, m^{[h-j]}, n, 1).$$

3 Prefixing-trees: Implementation and Analysis

We begin by discussing the logical tree that underlies the trie data structures. Let s be a finite set of strings composed from a finite ordered alphabet $\mathcal{A}$, which, without loss of generality, we assume $\mathcal{A} = \{0, \ldots, m-1\}$, where the characters are ranked in increasing order. The set $pref(s) \equiv \{x \in \mathcal{A}^* | xz \in s\}$ of prefixes of the elements of s supports a natural m-ary tree structure. This tree will be called the *prefix tree* of s and denoted by $t(s)$. The set of nodes of $t(s)$ is $pref(s)$, and the $i+1$-th subtree of the node $x \in pref(s)$, $0 \le i \le m - 1$, consists of those strings of $pref(s)$ that begin with xi. The root of $t(s)$ is the length zero string which will be denoted by ε.

There is a correspondence between the paths on $t(s)$ and the strings with symbols in $\mathcal{A}$. To a length zero path corresponds the length zero string ε; to the (possibly infinitely long) path $p = v_1, v_2, \ldots$, where v_{i+1} is the (l_i+1)-th son of v_i, corresponds $x = l_1 l_2 \ldots \in s$. This correspondence defines an injective mapping between the maximal paths of $t(s)$ and the keys of s. This mapping is bijective precisely when s satisfies the no-prefixing-key restriction. A *trie*, or *digital tree*, is the generic name used for data structures that implement the prefix tree.

An unrestricted finite set s of strings composed from $\mathcal{A}$ can be easily modified enabling the prefix tree to yield a representation of s as the set of maximal paths of an $(m + 1)$-ary tree, which is constructed as follows. Let $ppref(s) = \{x \in s \mid x \in pref(s - \{x\})\}$, and let $\perp$ be a symbol not belonging to $\mathcal{A}$. In the set $s[\perp] \equiv (s - ppref(s)) \cup \{x \perp | x \in ppref(s)\}$, no key is a prefix of another. The $(m + 1)$-ary prefix tree $t(s[\perp])$ with respect to the extended alphabet $\mathcal{A}^{\perp} \equiv \{\perp\} \cup \mathcal{A}$, where the endmarker symbol $\perp$ is ranked according to $rank(\perp) = 0$, is called the *prefixing-tree* of s (Compare Figure 3(a)).

The trie structures are implementations of the prefixing-tree $t(s[\perp])$ built with the set s of the keys belonging to the items. Each of the tries defined in subsequent sections is a tree, generically denoted by $g(s)$, with two kinds of nodes: *internal* nodes and *data* nodes. The internal nodes are the nonterminal nodes of $g(s)$ and hold the string matching information that drives the search algorithm. The data nodes are the terminal nodes of $g(s)$; each terminal node corresponds to a key of s and holds information about the location of the data of the item identified by this key. The retrieval algorithm traverses the unique path connecting the root and the data node corresponding to the search key. Thus, for a space and time cost analysis, the underlying structure of interest is that of a *t-ary tree with data nodes* [Knu73a, 2.3.4.5].

For the three variants: *full*, *compact* and *patrician* of *doubly-chained prefixing-tries* analyzed in this paper, an internal node v of $t(s[\perp])$ is represented by the linked list of pointers to v's non-empty subtrees. For each one of these six cost functions of the three trie variants, we shall compute the exact expected value, over the n-element subsets $s \subseteq \mathcal{A}^{[h]}$, as a function of h, n and the alphabet size m (see Table III in Figure 6). Each of these computations will proceed as follows. We first formulate a recursive structural definition for the particular trie variety. This definition implies recursive expressions for the trie cost functions from which the counting root-functions are readily apparent. The desired expectations of the cost functions are then derived from the general formula provided by Theorem 2.13.

4 Analysis of doubly-chained prefixing-tries

We shall now consider full, compact and patrician doubly-chained prefixing-tries, which were presented in [Kno86] for storing a finite collection of items whose set of keys s may include prefixing keys. If no key of s is a prefix of another, the full and compact doubly-chained prefixing-tries built from s revert to the doubly-chained tries of de la Brandais [dlB59] and Sussenguth [Sus63]. The patrician doubly-chained prefixing-trie is Knott's doubly-chained version of PATRICIA tries [Gwe68, Mor68].

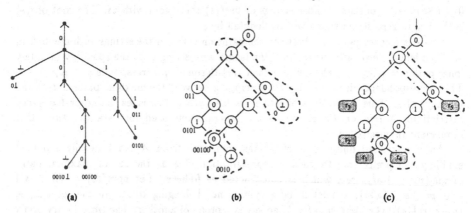

Figure 3. (a) Prefixing-tree built from $s = \{00100, 0101, 011, 0010, 0\}$. Note that the string corresponding to each terminal node is written below it. (b) $t^{bin}(s[\perp])$. (c) Full doubly-chained prefixing-trie for the items $r_1, \ldots, r_5$ with keys in s. Note that $k_1 = 00100$, $k_2 = 0101$, $k_3 = 011$, $k_4 = 0010$, and $k_5 = 0$.

Doubly-chained prefixing-tries can be thought of as implementations of a modified version of the *canonical binary representation* [Knu73b, 2.3.2] of the forest of subtrees of the prefixing-tree $t(s[\perp])$. This binary tree, denoted by $t^{bin}(s[\perp])$, is constructed as follows. A node v of $t(s[\perp])$ with non-empty subtrees $T_{l_1}, \ldots, T_{l_q}$ in order, $l_1 < \ldots < l_q$, is represented by a sequence $(v_q, l_q), \ldots, (v_1, l_1)$ of labeled nodes of $t^{bin}(s[\perp])$, where l_i is the label of node v_i. The right son of v_i is v_{i-1}, $2 \leq i \leq q$, and the right subtree of v_1 is empty. The left subtree of v_i is the binary tree representation of the forest of subtrees of T_{l_i}, $1 \leq i \leq q$. (Compare Figure 3(b)).

Full doubly-chained prefixing-tries are based on modifications of a pruned version of $t^{bin}(s[\perp])$ that will be denoted by $\bar{t}(s[\perp])$ and is constructed as follows. The tree $\bar{t}(s[\perp])$ results from $t^{bin}(s[\perp])$ by removing each node v with label $\perp$, while inserting v's only son (which is a terminal node) in its place. (Compare Figure 3(c)).

Note that the keys of s bijectively correspond to the maximal paths of $\bar{t}(s[\perp])$, where keys in $ppref(s)$ correspond to paths ending at right-son terminal nodes, and keys in $s - ppref(s)$ correspond to the remaining maximal paths.

We shall now compute the expectations of the cost functions of the doubly-chained prefixing-tries $t^{fd}(s)$, $t^{cd}(s)$, and $t^{pd}(s)$. These cost functions are $S_{fd}(s)$ and $T_{fd}(s)$, $S_{cd}(s)$ and $T_{cd}(s)$, $S_{pd}(s)$ and $T_{pd}(s)$. To make explicit the relationship between the cost functions of prefixing-tries and doubly-chained prefixing-tries, and following Knuth [Knu73b, 6.3, Exercise 24], it is valuable to break up the depth $depth(d)$ of a data node d into two components: $ldepth(d)$ and $rdepth(d)$. The left-link depth $ldepth(d)$ is the number of edges ending at left sons on the path that connects the root and d; the right-link depth $rdepth(d)$ is defined analogously, but counting path edges that end at right sons. The left-link and right-link total data node path lengths of a binary tree, g, with data nodes, d, are defined by

$$ltpl(g) \equiv \sum_{\text{all data nodes } d \text{ of } g} ldepth(d), \quad rtpl(g) \equiv \sum_{\text{all data nodes } d \text{ of } g} rdepth(d).$$

Note that $ltpl(g) + rtpl(g)$ gives the total data node path length (i.e., external path length). For the doubly-chained prefixing-tries $t^{fd}(s)$, $t^{cd}(s)$, and $t^{pd}(s)$, the respective left-link total data node path lengths will be respectively denoted by $L_{fd}(s)$, $L_{cd}(s)$, and $L_{pd}(s)$; the right-link total data node path lengths by $R_{fd}(s)$, $R_{cd}(s)$, and $R_{pd}(s)$. The corresponding total data node path lengths are given by $T_{fd}(s)$, $T_{cd}(s)$, and $T_{pd}(s)$.

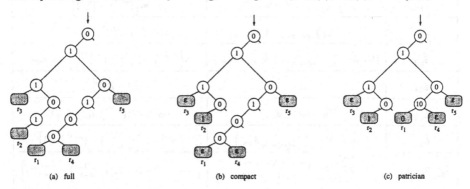

Figure 4. (a), (b), and (c) Doubly-chained prefixing-tries for the items $r_1, \ldots, r_5$ with respective keys $k_1 = 00100$, $k_2 = 0101$, $k_3 = 011$, $k_4 = 0010$, $k_5 = 0$.

Full

Definition 4.1 FULL DOUBLY-CHAINED PREFIXING-TRIE. *The full doubly-chained prefixing-trie built from a finite set of $s \subset A^*$ is the labeled binary tree with data nodes, denoted by $t^{fd}(s)$, which is recursively defined as follows:*

(i) *If $s = \emptyset$, $t^{fd}(s)$ is equal to the empty tree Λ.*

(ii) *If $s = \{\varepsilon\}$, $t^{fd}(s)$ is equal to the 'data' node $\mathcal{D}$.*

(iii) *Otherwise, let a be the highest ranked character which begins a string of s, i.e., $a \equiv max(\text{first}(s))$ where $\text{first}(s) \equiv \{c \in A | s_c \neq \emptyset\}$. Then, $t^{fd}(s)$ is the labeled binary tree having an 'internal' root node with label a, left subtree $t^{fd}(s_a)$ and right subtree $t^{fd}(s - as_a)$.*

In order to search for k in $t^{fd}(s)$, we start at the root node and proceed recursively. If the root is a nonterminal node, which we denote by r and its label l, we proceed as follows: if l is not a prefix of k, we search for k in the right subtree; if $k = lz$, we search for z in the left subtree. Otherwise, the search ends; it is successful precisely when the root is a terminal node and $k = \varepsilon$ (compare Figure 4(a)).

The counting root-functions of the full doubly-chained prefixing-tries cost functions, S_{fd}, L_{fd} and R_{fd}, can be deduced from Definition 4.1 by the following observation. Let the triple (t, a, t') denote the labeled binary tree having a root node with label a, left subtree t, and right subtree t'. If $\text{first}(s) \equiv \{c \in A | s_c \neq \emptyset\} = \{c_1, \ldots c_r\}$ with $c_1 < \ldots < c_r$, then

$$t^{fd}(s) = \left(t^{fd}(s_{c_r}), c_r, \left(t^{fd}(s_{c_{r-1}}), c_{r-1}, \ldots \left(t^{fd}(s_{c_1}), c_1, t^{fd}(s \cap \{\varepsilon\})\right) \ldots\right)\right).$$

This relation, together with Definition 4.1, yields recursive expressions for $S_{fd}(s)$, $L_{fd}(s)$ and $R_{fd}(s)$, The implied counting root-functions and their corresponding values of $\check{N}_{h,n}$ are recorded in Table II in Figure 5. The expectations $E[S_{fd}]$, $E[L_{fd}]$ and $E[R_{fd}]$ computed by means of Theorem 2.13 appear in Table III in Figure 6, which is given at the end of this section.

X	$\overset{\vee}{X}(\nu, n_0, \ldots, n_{m-1})$	$\check{N}_{h,n}[\overset{\vee}{X}]$
S_{fd}	$\sum_{0 \le i \le m-1}(1 - \delta_{0,n_i})$	$m\left[\binom{m^{[h]}}{n} - \binom{m^h}{n}\right]$
S_{cd}	$(1 - \delta_{1,n})\sum_{0 \le i \le m-1}(1 - \delta_{0,n_i})$	$m\left[\binom{m^{[h]}}{n} - \binom{m^h}{n}\right](1 - \delta_{1,n})$
$\widetilde{S_{pd}}$	$(1 - \delta_{n,0})\sum_{0 \le i \le m-1}(1 - \delta_{0,n_i} - \delta_{n,n_i})$	$m\left[\binom{m^{[h]}}{n} - \binom{m^h}{n} - \binom{m^{[h-1]}}{n}\right](1 - \delta_{n,0})$
L_{fd}	$n - \nu$	$(m^{[h]} - 1)\binom{m^{[h]}-1}{n-1}$
L_{cd}	$(n - \nu)(1 - \delta_{1,n})$	$(m^{[h]} - 1)\binom{m^{[h]}-1}{n-1}(1 - \delta_{1,n})$
$\widetilde{L_{pd}}$	$(n - \nu)\left[1 - \sum_{0 \le i \le m-1}\delta_{n,n_i}\right]$	$(m^{[h]} - 1)\left[\binom{m^{[h]}-1}{n-1} - \binom{m^{[h-1]}-1}{n-1}\right]$
R_{fd}, R_{cd}, R_{pd}	$\sum_{0 < i \le m-1}\nu(1 - \delta_{0,n_i}) +$ $\sum_{0 \le j < i \le m-1}n_j(1 - \delta_{0,n_i})$	$\left[\frac{m-1}{2}(m^{[h]} - 1) + m\right]\left[\binom{m^{[h]}-1}{n-1} - \binom{m^h-1}{n-1}\right]$

Figure 5. TABLE II: DOUBLY-CHAINED PREFIXING-TRIES. Note that $n = \nu + \sum_{0 \le j \le m-1} n_j$ and $X(\{\varepsilon\}) = X(\emptyset) = 0$.

Compact

Definition 4.2 COMPACT DOUBLY-CHAINED PREFIXING-TRIE. *A recursive definition of the compact doubly-chained prefixing-trie $t^{cd}(s)$ can be obtained from Definition 4.1 by replacing condition (ii) with the following 'compaction' condition:*

(ii*) If $s = \{x\}$, $t^{cd}(s)$ is equal to a single 'data' node $\mathcal{D}$ with a label equal to x.

To search for a key k in $t^{cd}(s)$, we proceed as we did for the full doubly-chained prefixing-tries except that, upon reaching a terminal node w, the label of w must be compared to the current search key; precisely when they are equal is the search successful (compare Figure 4(b)).

The counting root-functions of S_{cd}, L_{cd} and R_{cd}, and their corresponding values of $\check{N}_{h,n}$ are recorded in Table II in Figure 5. The expectations $E[S_{cd}]$, $E[L_{cd}]$ and $E[R_{cd}]$, deduced with the help of Theorem 2.13, appear in the Table III in Figure 6. Further observations can be made [dIT87a] on the relationship among the cost functions of tries analyzed so far.

Patrician

Definition 4.3 PATRICIAN DOUBLY-CHAINED PREFIXING-TRIE. *The patrician doubly-chained prefixing-trie built from a finite set of keys $s \subset A^*$ is the labeled binary tree with data nodes, denoted by $t^{pd}(s)$, which is recursively defined as follows:*

(i) *If $s = \emptyset$, $t^{pd}(s)$ is the empty tree Λ.*

(ii) *If $s = \{x\}$, $t^{pd}(s)$ is equal to a single 'data' node $\mathcal{D}$ with a label equal to x.*

(iii) *Otherwise, let a be the highest ranked character that begins a key of s (i.e., $a \equiv max(\text{first}(s))$), and let x be the longest prefix shared by all the keys of s that begin with a. Then, $t^{pd}(s)$ is the binary tree with data nodes having an 'internal' root node with label x, left subtree $t^{pd}(s_x)$, and right subtree $t^{pd}(s - as_a)$ (note that $t^{pd}(s - as_a) = t^{pd}(s - xs_x)$).*

To search for a key k in $t^{pd}(s)$, we start at the root node and proceed recursively. If the root is a nonterminal node, which we denote by r and its label by l, we proceed as follows: if l is not a prefix of k, the search ends unsuccessfully; if $k = l$, we search for ε in the right subtree; if $k = lz$, we search for z in the left subtree. Otherwise, the search ends: it is successful precisely when the root is a terminal node and $k = \varepsilon$ (compare Figure 4(c)).

X	$E[X]$
Tl	$n\left[h - \frac{1}{m-1}\left(1 - \frac{h+1}{m^{[h]}}\right)\right]$
K	$\frac{n}{m}\left[1 - \frac{1}{m^{[h]}}\right] - \sigma[0,1]$
F	$m\left[\tau(m^{[h]}, m^h, n, 0) - \delta_{n,0}\right]$
S_{fd}	$n\frac{m^h}{m^{[h]}} - 1 + \delta_{0,n} + m^{[h-1]} - \sigma[0,0]$
S_{cd}	$E[S_{\text{fd}}] - \frac{m}{m-1}\left(\kappa[0,1] - \sigma[0,1]\right)$
S_{pd}	$n\frac{m^h}{m^{[h]}} + m^{[h-1]} - 1 + \delta_{0,n} - m\,\theta[0,0] + (m-1)\,\sigma[0,0] + E[F]$
L_{fd}	$E[Tl]$
L_{cd}	$E[Tl] - \frac{m}{m-1}\left(\kappa[0,1] - \sigma[0,1]\right)$
L_{pd}	$E[Tl] + \frac{m}{m-1}\left(\theta[1,1] - \eta[1,1]\right) + n\,E[F]$
$R_{\text{fd}} = R_{\text{cd}} = R_{\text{pd}}$	$\frac{m-1}{2}E[Tl] + n\left(1 - \frac{1}{m^{[h]}}\right) - \frac{m}{2}\kappa[1,1] - \frac{1}{2}\sigma[1,1]$

Figure 6. TABLE III: DOUBLY-CHAINED PREFIXING-TRIES. The basic trie sums $\sigma[a,b]$, $\kappa[a,b]$, $\theta[a,b]$ and $\eta[a,b]$ are as defined in (5), (6), (7), and (8). Note that $T_{\text{fd}}(s) = L_{\text{fd}}(s) + R_{\text{fd}}(s)$, $T_{\text{cd}}(s) = L_{\text{cd}}(s) + R_{\text{cd}}(s)$, and $T_{\text{pd}}(s) = L_{\text{pd}}(s) + R_{\text{pd}}(s)$.

We close the analysis of doubly-chained prefixing-tries with a summary of the expectations computed so far. Taking $\tau(a,b,c,d) = \binom{a-b}{c-d}/\binom{a}{c}$, the expectations emerging from our analyses can be expressed in terms of a few primitives, the *basic trie sums for the prefix model*, which are

$$\sigma[a,b] \equiv \sigma[a,b;h,n] \equiv \sum_{a \leq j < h+a} m^j \tau(m^{[h]} - a, m^{[h-j]}, n, b), \tag{5}$$

$$\kappa[a,b] \equiv \kappa[a,b;h,n] \equiv m^h \sum_{a \leq j < h+a} \tau(m^{[h]} - a, m^{[h-j]}, n, b), \tag{6}$$

$$\theta[a,b] \equiv \theta[a,b;h,n] \equiv \sum_{0 \leq j < h} m^j \tau(m^{[h]} - a, m^{h-j}, n, b), \tag{7}$$

$$\eta[a,b] \equiv \eta[a,b;h,n] \equiv m^h \sum_{0 \leq j < h} \tau(m^{[h]} - a, m^{h-j}, n, b). \tag{8}$$

5 Concluding remarks and future research

We have computed exact average space complexity $E[S]$ and average time complexities $E[T]$ of the retrieval algorithms for several trie varieties that store unrestricted sets of keys. The root-function method presented in Section §2 is of interests in its own right. It provides a general framework for computing expectations of random variables belonging to a certain wide class with respect to the prefix model. For example, we have applied this method to the analysis of full prefixing-tries, compact prefixing-tries, patrician prefixing-tries, and bucket tries [dlT87a, dlTK95a].

References

[dlB59]   R. de la Brandais. File Searching Using Variable Length Keys. In *Proceedings of Western Joint Computer Conference*, volume 15, pages 295–298, 1959.

[dlT87a]   P. de la Torre. Analysis of Tries. Ph.D. Thesis CS–TR–1890, Department of Computer Science, University of Maryland, 1987.

[dlT87b]   P. de la Torre. Analysis of Tries that Store Prefixing Keys. Technical Report UMIACS–TR–87–59, Institute of Advanced Computer Studies, University of Maryland, December 1987. A revised version of this paper has appeared as Technical Report 95–18, Department of Computer Science, University of New Hampshire.

[dlT87c]   P. de la Torre. Extending the Flajolet-Regnier-Sotteau Method of Trie Analysis to the Prefix Model. Technical Report UMIACS–TR–87–60, Institute of Advanced Computer Studies, University of Maryland, December 1987. A revised version of this paper has appeared as Technical Report 94–18, Department of Computer Science, University of New Hampshire.

[dlTK94]   P. de la Torre and D. T. Kao. An Algebraic Approach to the Prefix Model Analysis of Binary Trie Structures and Set Intersection Algorithms. Technical Report 94–18, Department of Computer Science, University of New Hampshire, 1994. Submitted to *Discrete Mathematics*. This is a revised version of the University of Maryland technical report UMIACS–TR–87–60.

[dlTK95a]   P. de la Torre and D. T. Kao. A Uniform Approach to the Analysis of Trie Structures that Store Prefixing Keys. Technical Report 95–18, Department of Computer Science University of New Hampshire, 1995. To appear in *Journal of Algorithms*. This is a revised version of the University of Maryland technical report UMIACS–TR–87–59.

[dlTK95b]   P. de la Torre and D. T. Kao. Extended Abstract: An Algebraic Approach to the Prefix Model Analysis of Binary Trie Structures and Set Intersection Algorithms. In B. Leclerc and J. Y. Thibon, editors, *Formal Power Series and Algebraic Combinatorics: 7th Conference*, pages 127–138, Paris, France, 1995.

[Fre60]   E. Fredkin. Trie Memory. *CACM*, 3(9):490–499, 1960.

[FRS85]   Ph. Flajolet, M. Regnier, and D. Sotteau. Algebraic Methods for Trie Statistics. *Annals of Discrete Mathematics*, 25:145–188, 1985.

[GBY91]   G. H. Gonnet and R. Baeza-Yates. *Handbook of Algorithms and Data Structures: In Pascal and C.* Addison-Wesley, Reading, Massachusetts, 2 edition, 1991.

[Gwe68]   G. Gwegenberger. Anwendung einer binären Verweiskettenmethode beim Aufbau von Listen. *Elektronische Rechenanlagen*, 10:223–226, 1968.

[Kno86]   G. D. Knott. Including Prefixes in Doubly-Chained Tries. Technical Report CAR–TR–236, Computer Science Department, University of Maryland, 1986.

[Knu73a]   D. E. Knuth. *The Art of Computer Programming, Volume 1: Fundamental Algorithms.* Addison-Wesley, Reading, Massachusetts, 1973.

[Knu73b]   D. E. Knuth. *The Art of Computer Programming, Volume 3: Sorting and Searching.* Addison-Wesley, Reading, Massachusetts, 1973.

[Mor68]   D. R. Morrison. PATRICIA – Practical Algorithm to Retrieve Information Coded in Alphanumeric. *JACM*, 15:514–534, 1968.

[Sus63]   E. H. Sussenguth. Use of Tree Structures for Processing Files. *CACM*, 6:272–279, May 1963.

[Tra78]   L. I. Trabb Pardo. Set Representation and Set Intersection. Ph.D. Thesis STAN–CS–78–681, Department of Computer Science, Stanford University, 1978.

On Fairness in Terminating and Reactive Programs

Axel Wabenhorst

Programming Research Group
Oxford University Computing Laboratory, Oxford, UK

Abstract. In the setting of boundedly nondeterministic action systems, we give a general definition of the notion of fairness. We provide results that support the top-down development of fair designs and extend them from terminating action systems to non-terminating action systems. One strength of our approach is the similarity of the results to those used for the standard developments. Another is the generality of the notion of fairness which it captures.

1 Introduction

The aim of this paper is to provide support for the formal development of information systems which exhibit some form of fairness. Its contribution consists of a general definition of fairness and results which support development. We choose the notation of action systems [Bac89] for their expressive power and developmental clarity and because their predicate transformer semantics are concise and calculational.

By addressing fairness we are able to develop, entirely within the formalism, designs whose infinitary behaviour previously required *ad hoc* reasoning (for example, realistic forms of the readers-writers problem). Intuitively, fairness guarantees that in the repeated nondeterministic choice between statements in a program, none of them is neglected indefinitely. There are many different kinds of fairness, depending on the decision one makes about what constitutes the "neglect" of a statement. Our definition extends those formalised in [Fra86] including, in particular, the popular strong and weak fairness. The results about these are anyway restricted to verification of sequential programs, largely termination (see also [MP92]). A worked development of a readers-writers implementation, using the results of this paper, appears in [Wab96].

Related work, with similar aims, is due to [BX95] and [Mor90]. In [BX95], forward and backward simulation are used to show refinement between fair action systems. The conditions under which refinement take place are easy to check, and the proof rules are similar to our combination of monotonicity, adding skips and data refinement. However, the applicability of the results in [BX95] are limited to strong and weak fairness. In addition, it is assumed that the actions of the refining action system either refine *skip* or correspond to actions in the action system to be refined.

Morris [Mor90] gives a treatment of fairness using fixed-point theorems. The calculation of weakest preconditions sometimes involves two nested recursions, but the proof rules for termination are derived succinctly using fixed points. In addition, the results hold in a more general framework than action systems, as the syntactic shape of action systems is not exploited. However, the type of fairness catered for is restricted to strong or weak fair choice between statements.

Probabilistic semantics of sequential programs can be thought of as providing a strong form of fairness on the nondeterministic choice between two program statements. A framework for developing probabilistic programs (rather than systems) can be found, for example, in [MMSS95].

All proofs of theorems stated in this paper, together with extensions and further examples are contained in [Wab96]. In particular, there data refinement for both fair terminating and non-terminating action systems is defined and shown to distribute through the components of an action system.

1.1 Outline

We begin in Section 2 by recalling the definitions of predicate transformers [DS90] and action systems [Bac89], together with the resulting definition of refinement. We define action sequences, which are used heavily in the rest of the paper, and give a characterisation of the semantics of iteration using them.

In Section 3, we motivate and illustrate the definition of fairness by simple examples. From the definition of fairness, we prove some theorems useful in the development of terminating programs from simple specifications. In Section 4, we define refinement of non-terminating action systems in terms of refinement of terminating action systems. The benefit of that approach is that the same definitions of fairness and the same theorems for program development can be used as for terminating action systems.

2 Background

We use the following notation: Ord is the class of ordinals; X^κ denotes the type of all functions from $\kappa : Ord$ (seen as the set of ordinals less than κ) to X. Other notation will be defined as needed.

2.1 Predicate Transformers

Let Σ_A denote the state space of a program A expressed in the language of guarded commands [DS90]. A *predicate* is a function $f : \Sigma_A \to \mathbb{B}$, the set of all of which will be denoted by $Pred.A$. If there is no ambiguity, we write $Pred$ instead of $Pred.A$.

For each statement (i.e. program command) S, the weakest precondition semantics [DS90] of S is a predicate transformer, denoted by the same symbol $S : Pred \to Pred$, satisfying, for all $P, Q : Pred$, the axioms

- If $P \Rightarrow Q$, then $S(P) \Rightarrow S(Q)$ (monotonicity);
- If $I \neq \emptyset$, then $\forall \alpha \in I\ S(P_\alpha) \Rightarrow S(\forall \alpha \in I\ P_\alpha)$ (conjunctivity).

Informally, we can regard the weakest precondition $S(P)$ for $P : Pred$ to be the weakest predicate on the variable states from which execution of the statement S guarantees termination in a state satisfying P. In particular, $S(true)$ is the weakest predicate which guarantees that S terminates. Notice that we allow statements which do not satisfy strictness $S(false) \Leftrightarrow false$.

Examples of statements and their associated weakest precondition transformers are given in [DS90, Boo82]. For example,

$$(\textbf{do}\ []_{i=1}^{n}\ g_i \rightarrow S_i\ \textbf{od})(P) \mathrel{\widehat{=}} \exists \kappa : Ord\ H_\kappa(P),$$

where

$$H_0(P) \mathrel{\widehat{=}} P \wedge \neg \bigvee_{i=1}^{n} g_i,$$

$$H_{\kappa+1}(P) \mathrel{\widehat{=}} H_0(P) \vee (\textbf{if}\ []_{i=1}^{n}\ g_i \rightarrow S_i\ \textbf{fi})(H_\kappa(P))\ \text{and}$$

$$H_\delta(P) \mathrel{\widehat{=}} \exists \kappa < \delta\ H_\kappa(P),\ \text{where } \delta \text{ is a limit ordinal.}$$

An *action* is a guarded command $g \rightarrow S$, where g is a predicate and S a statement. An *action system* [Bac89] has the form

$$\mathcal{A} \mathrel{\widehat{=}} \begin{array}{ll} \textbf{var} & x_1, \ldots, x_m \\ \textbf{init} & x_1, \ldots, x_m :\in X_1, \ldots, X_m \\ \textbf{do} & []_{i=1}^{n}\ A_i\ \textbf{od,} \end{array}$$

where $x_1, \ldots, x_m$ are typed variables, $m : \mathbb{N} \setminus \{0\}$ and the A_i are actions. Sometimes we will omit the variable declaration when it can be deduced from the body of the action system. When a variable is initialised arbitrarily within its type, we will omit an explicit initialisation.

Definition 1 Refinement. A statement S *is refined by* a statement S' (notation $S \leq S'$) if for all $P : Pred\ S(P) \Rightarrow S'(P)$. □

2.2 Action Sequences

Informally, an action sequence is a sequence (either finite or infinite) of actions in an action system. In later sections, we will see that action sequences are useful for expressing fairness properties: while all finite action sequences are fair, certain infinite ones may be unfair.

Let $\mathcal{A}$ be an action system. For $\kappa \leq \omega$, $seq_\kappa \mathcal{A} \mathrel{\widehat{=}} \{A_i \mid 1 \leq i \leq n\}^\kappa$. For $s : seq_\kappa \mathcal{A}$, if $\kappa' < \kappa$, then $s|_{\kappa'}$ is defined by $s|_{\kappa'}(\alpha) = s(\alpha)$ for $\alpha < \kappa'$. We say that $s' : seq_{\kappa'}\mathcal{A}$ is a *prefix* of $s : seq_\kappa \mathcal{A}$ (notation: $s' \sqsubseteq s$) if $s' = s|_{\kappa'}$. We also define $seq\mathcal{A} \mathrel{\widehat{=}} \bigcup_{\kappa < \omega} seq_\kappa$ and $\overline{seq}\mathcal{A} \mathrel{\widehat{=}} seq\mathcal{A} \cup seq_\omega \mathcal{A}$, the latter are the *action sequences* of $\mathcal{A}$.

We define the predicate transformer semantics of action sequences as follows. Let ; denote the standard sequential composition, and let $P : Pred.\mathcal{A}$. Then

- if $s : seq_0 A$, $s(P) \; \hat{=} \; P$,
- if $\kappa < \omega$ and $s : seq_{\kappa+1} A$, $s(P) \; \hat{=} \; s|_\kappa ; s(\kappa)(P)$,
- if $s : seq_\omega A$, $s(P) \; \hat{=} \; \exists \kappa < \omega \; s|_\kappa(P)$.

We will sometimes abuse notation, so that when action sequences s and t satisfy $s(P) \Leftrightarrow t(P)$ for all $P : PredA$, we will say that $s = t$. Thus, we will identify $seq_0 A$ with $\{skip\}$. (Note that $seq_0 A$ is a singleton.) Also, we will identify $seq_\kappa A$ with the sequential composition of actions of length κ, $\kappa < \omega$. We will also use ; to denote catenation within infinite action sequences. It should be emphasised that in this context, ; is not sequential composition.

For sets S and T of finite action sequences, $S; T$ has the usual pointwise definition. Let $u[A_i/A_j]$ be the action sequence resulting from the replacement of A_j throughout u by A_i. Similarly, $u[/A_j]$ is the action sequence resulting from the removal of all occurrences of A_j in u.

With our definition of action sequences, we permit impossible juxtapositions of actions. The result is a type more general than that which contains just those sequences of actions of A which actually occur, which we exploit in the results to follow.

2.3   A Characterisation of Iteration

The following characterisation [BM95] of the weakest precondition of an action system can be thought of as a decomposition. The first conjunct is a condition on the finite action sequences which states that if A terminates at all, then it terminates in state P, while the second conjunct is a restriction on the infinite action sequences which states that A terminates eventually. In fact, the second conjunct is equivalent to $A(true)$. It is useful to isolate the conditions on the finite and infinite action sequences when we consider fairness, because we shall see that fairness is a condition on the infinite action sequences rather than the finite ones.

Theorem 2. *Let*

$$A \; \hat{=} \; \textbf{do} \; []_{i=1}^n \; g_i \to S_i \; \textbf{od}.$$

Then for all $P : PredA$,

$$A(P) \; \Leftrightarrow \; (\forall s : seqA \; s(P \vee \bigvee_{i=1}^n g_i)) \wedge A(\neg \bigvee_{i=1}^n g_i),$$

where

$$A(\neg \bigvee_{i=1}^n g_i) \; \Leftrightarrow \; (\forall s : seq_\omega A \; s(\neg \bigvee_{i=1}^n g_i))$$

if A exhibits bounded nondeterminism. □

The first conjunct $(\forall s : seq\mathcal{A}\ s(P \vee \bigvee_{i=1}^{n} g_i))$ in Theorem 2 is the *weakest errorfree precondition* of $\mathcal{A}$ with respect to $P : Pred\mathcal{A}$ (see also [Hes95]). Informally, the weakest errorfree precondition is the weakest predicate on states at which execution of $\mathcal{A}$ guarantees either termination in a state satisfying P or infinite execution of the loop of $\mathcal{A}$. The only difference between it and the weakest liberal precondition is that while the weakest liberal precondition allows non-termination of all kinds, including aborts of statements within the action system, the only non-termination allowed in the weakest errorfree precondition is non-termination of the iteration loop.

3 Fairness

In this section, we restrict our attention to action systems exhibiting bounded nondeterminism.

Before motivating our definition of fairness, we consider an example which encapsulates much of our intuition about it.

Example 1. If the action system

$$
\begin{aligned}
\mathcal{B} \;\widehat{=}\; \textbf{var} \quad & n : \mathbb{N} \\
\textbf{do} \quad & n > 0 \rightarrow n := n - 1\ (B0) \\
[\!] \quad & n > 0 \rightarrow skip \qquad (B1) \\
\textbf{od}. \quad &
\end{aligned}
$$

executes in accordance with the definition of iteration in Section 2.1, it is not guaranteed to terminate. This is because action $(B0)$ may not execute sufficiently often, the necessary number of executions being dependent on the initial value of n. In fact, no fixed finite number of executions of $(B0)$ can guarantee termination for all initial values of n. Termination in $\mathcal{B}$ is guaranteed only if $\mathcal{B}$ exhibits what is called *weak fairness*, which in this case is the same as *strong fairness*. Our definition of fair action systems should allow us to specify that $\mathcal{B}$ executes in accordance with weak fairness or, for that matter, any other kind of fairness. □

3.1 Motivation

We now motivate some of the details of our definition by considering various simple fair systems.

We shall constrain an action system to behave "fairly" by specifying those (infinite) unfair sequences which do not occur. We choose to specify the unfair sequences rather than their complement the fair ones, because we find it more convenient to specify the former. The only restriction we place on the unfair sequences is that they be infinite.

Thus, when considering the termination of a fair action system, we insist that fair infinite action sequences must terminate; in other words, a finite prefix of each such action sequence must terminate. We make no such demand on unfair infinite action sequences, because they do not occur. We then calculate the weakest precondition incorporating this consideration.

Example 2. For the action system $\mathcal{B}$ in Example 1, we take the unfair action sequences to be $\{s; B1^\omega \mid s : seq\mathcal{B}\}$. These are exactly the infinite sequences which execute statement $(B0)$ only finitely often, despite the guard of $(B0)$ being true afterwards. □

The following example demonstrates that we need the flexibility to be able to specify different unfair action sequences for different initial conditions.

Example 3. Consider the action system

$$
\begin{aligned}
\mathcal{C} \;\widehat{=}\; \mathbf{var} \quad & m : \mathbb{Z} \\
\mathbf{do} \quad & m > 0 \rightarrow m := 0 \; (C0) \\
[] \; & m \neq 0 \rightarrow skip \quad (C1) \\
\mathbf{od}. \quad &
\end{aligned}
$$

Under the initial condition $m > 0$, the guards of both $(C0)$ and $(C1)$ are true. If $\mathcal{C}$ is to execute fairly, we need to specify $C1^\omega$ as an unfair action sequence under the initial condition $m > 0$, otherwise $(C0)$ may never be executed despite its guard being true always.

Under the initial condition $m < 0$ on the other hand, only the guard of $(C1)$ is true, so that the action sequence $C1^\omega$ is the only possible execution. We are not neglecting $(C0)$ because its guard is never enabled, so there are no unfair action sequences for initial condition $m < 0$.

Under the initial condition $m = 0$, both guards are *false* and so $\mathcal{C}$ terminates immediately. Thus, we do not need to specify any unfair action sequences although it does no harm to do so. □

To allow different specifications of unfair action sequences for different initial conditions, we use the notation $P_\alpha : uf_\alpha$ corresponding to the unfair action sequences uf_α for initial condition P_α. Here, the indices α are taken from any index set I. In particular, I could be infinite. Note that once we have specified the unfair sequences for certain initial conditions, we do not have further control over the execution of the action system inside the loop — the demon choosing the execution sequences has full control subject to the constraint of the unfair sequences specified upon initialisation.

3.2 Definition

We now define fair action systems $\mathcal{A}$ which do not engage in any of the "unfair" action sequences uf_α if P_α holds initially, but otherwise behaves like ordinary action systems. To formalise this description of the behaviour of $\mathcal{A}$, we define $\mathcal{A}$ as a predicate transformer.

Definition 3. Let I be an index set. For $\alpha \in I$, let $P_\alpha : Pred$ such that $\exists \alpha P_\alpha \Leftrightarrow true$. Let

$$
\begin{aligned}
\mathcal{A} \;\widehat{=}\; \mathbf{do} \quad & []_{i=1}^n \; g_i \rightarrow S_i \\
\mathbf{uf} \quad & P_\alpha : uf_\alpha \\
\mathbf{od}, \quad &
\end{aligned}
$$

where $uf_\alpha \subseteq seq_\omega A$ for all $\alpha \in I$. We define

$$\mathcal{A}(P) \;\widehat{=}\; \forall s : seqA \; s(P \vee \bigvee_{i=1}^{n} g_i)$$

$$\wedge$$

$$\exists \alpha \in I \; P_\alpha \wedge X_\alpha,$$

where

$$X_\alpha \;\widehat{=}\; \forall s : seq_\omega A \setminus uf_\alpha \; s(\neg \bigvee_{i=1}^{n} g_i).$$

□

$\mathcal{A}(P)$ is well-defined in the sense that it is independent of the representation of the $P_\alpha : uf_\alpha$. In the definition, we think of uf_α as being the unfair infinite sequences of the action system under initial condition P_α.

It can be seen that the definition is a generalisation of the usual definition of iteration in the case of bounded nondeterminism: when $uf = \emptyset$ for all initial conditions, corresponding to the absence of unfairness, we obtain the characterisation of the standard definition given in Theorem 2. Also, when $uf_\alpha = seq_\omega A$ for all initial conditions, $X_\alpha \Leftrightarrow true$ and so the second conjunct in the definition is satisfied trivially. Thus, we obtain the weakest errorfree precondition of the action system (Section 2.3).

Since we have defined fair action systems using weakest preconditions, we can use existing definitions of refinement and data refinement. Fair action systems as defined in Definition 3 satisfy conjunctivity and monotonicity [Wab96], so they can be used as statements within other programs.

We now exemplify the calculations involved in the definition.

Example 4. Consider the action system

$$
\begin{aligned}
\mathcal{B} \;\widehat{=}\; \textbf{var} \quad & n : \mathbb{N} \\
\textbf{do} \quad & n > 0 \rightarrow n := n - 1 \qquad (B0) \\
[\![\quad & n > 0 \rightarrow skip \qquad\qquad (B1) \\
\textbf{uf} \quad & true : \{s; B1^\omega \mid s : seqB\} \\
\textbf{od.} \quad &
\end{aligned}
$$

Let $s : seq_\omega B \setminus \{s; B1^\omega \mid s : seqB\}$, note that s contains infinitely many occurrences of $(B0)$. Take any $t \sqsubseteq s$, and let $k : \mathbb{N}$ be the number of occurrences of $(B0)$ in t. Then $t(n \leq 0) \Leftrightarrow n \leq k$. Since we can choose t to contain arbitrarily many k, $s(n \leq 0) \Leftrightarrow true$. Thus the second conjunct in the definition evaluates to $true$, and the action system is guaranteed to terminate.

For $P : PredB$, it can be shown that the first conjunct, and hence $\mathcal{B}(P)$, evaluates to

$$(n \leq 0 \Rightarrow P) \wedge (n > 0 \Rightarrow P[0/n]).$$

□

3.3 Program Development

In this section, we give some simple rules by which programs may be developed.

Strengthening the fairness requirements of an action system provides a refinement:

Theorem 4. *Let I be an index set and for $\alpha \in I$, let $P_\alpha : Pred$ such that $\exists \alpha\, P_\alpha \Leftrightarrow true$. Let*

$$
\mathcal{A} \; \widehat{=} \; \text{do} \; \big[\!\big]_{i=1}^{n} \quad g_i \rightarrow S_i \\
\qquad\quad \textbf{uf} \qquad P_\alpha : uf_\alpha \\
\qquad \textbf{od}
$$

and

$$
\mathcal{B} \; \widehat{=} \; \text{do} \; \big[\!\big]_{i=1}^{n} \quad g_i \rightarrow S_i \\
\qquad\quad \textbf{uf} \qquad P_\alpha : uf'_\alpha \\
\qquad \textbf{od,}
$$

where for all $\alpha \in I$, $uf_\alpha \subseteq uf'_\alpha$. Then $\mathcal{A} \leq \mathcal{B}$. □

Thus, fair action systems with fixed initial condition and fixed body form a complete lattice under the partial order $\leq$. For any subset of points in this lattice, the least upper bound (greatest lower bound) equals the union (intersection) of corresponding uf_α's. The minimum element of the lattice is the weakest precondition of the action system given in Section 2.1; the maximum element is the weakest errorfree precondition of the action system.

Monotonicity allows us to refine actions one-by-one with the result that the action system as a whole is refined also.

Theorem 5 Monotonicity for Fair Action Systems. *Let*

$$
\mathcal{A} \; \widehat{=} \; \text{do} \; \big[\!\big]_{i=1}^{n} \quad g_i \rightarrow S_i \\
\qquad\quad \textbf{uf} \qquad P_\alpha : uf_\alpha \\
\qquad \textbf{od}
$$

and

$$
\mathcal{B} \; \widehat{=} \; \text{do} \; \big[\!\big]_{i=1}^{n}\big[\!\big]_{j=1}^{m_i} \quad h_{i_j} \rightarrow T_{i_j} \\
\qquad\quad \textbf{uf} \qquad\qquad P_\alpha : uf'_\alpha \\
\qquad \textbf{od,}
$$

where

- $\bigvee_{i=1}^{n} g_i \Leftrightarrow \bigvee_{i=1}^{n} \bigvee_{j=1}^{m_i} h_{i_j}$,
- *for all i such that $1 \leq i \leq n$, $g_i \rightarrow S_i \leq \big[\!\big]_{j=1}^{m_i} h_{i_j} \rightarrow T_{i_j}$ and*
- *for each $\alpha \in I$,*

$$
uf'_\alpha = \{u[h_{i_j} \rightarrow T_{i_j}/g_i \rightarrow S_i] \mid u : uf_\alpha \,\wedge\, 1 \leq i \leq n \,\wedge 1 \leq j \leq m_i\}.
$$

Then $\mathcal{A} \leq \mathcal{B}$. □

Note that

- uf'_α consists of all members of uf_α with all different possible substitutions of $h_{i_j} \to T_{i_j}$ for $g_i \to S_i$. In particular, different substitutions of $h_{i_j} \to T_{i_j}$ may occur for the same action $g_i \to S_i$ at different places within an unfair sequence;
- $uf_\alpha = \emptyset \Rightarrow uf'_\alpha = \emptyset$, which gives us monotonicity for action systems executing according to the standard definition of weakest preconditions;
- $uf_\alpha = seq_\omega A \Rightarrow uf'_\alpha = seq_\omega B$, which gives us monotonicity for action systems executing according to the weakest errorfree precondition.

As an application of our monotonicity theorem, we can strengthen the guards of actions in a fair action system, provided their disjunction is not changed.

Occasionally, we would like to *add* an action to an action system, rather than merely to refine existing actions. The first step in this process, that of adding *skips*, is justified by the following result:

Theorem 6. *Let*

$$A \mathrel{\hat{=}} \textbf{do} \quad []_{i=1}^n \quad g_i \to S_i$$
$$\textbf{uf} \quad P_\alpha : uf_\alpha$$
$$\textbf{od}$$

and

$$B \mathrel{\hat{=}} \textbf{do} \quad []_{i=1}^n \quad g_i \to S_i$$
$$[] \qquad \bigvee_{i=1}^n g_i \to skip \qquad (A)$$
$$\textbf{uf} \quad P_\alpha : uf'_\alpha$$
$$\textbf{od},$$

where for each $\alpha \in I$, $uf'_\alpha = \{u : seq_\omega B \mid u[/A] : uf_\alpha\} \cup \{s; A^\omega \mid s : seqB\}$. Then $A \le B$ and $B \le A$. $\qquad\square$

4 Non-Terminating Action Systems

4.1 Introduction

With some programs, such as operating systems, communications protocols and electronic banking databases, non-termination does not reflect an error (as has been convenient to assume until now), but represents normal behaviour. These programs typically involve interaction with the environment. We find that Definition 1 is inadequate as a definition of refinement of such non-terminating action systems, because

$$\forall P : PredA \ (A(P) \Leftrightarrow false)$$

for any action system A which never terminates. As a result, A is refined by any other action system. So a new definition of refinement is required.

Since there may be no final state in a non-terminating action system, the new definition needs to take into account the *intermediate* states of the action system. Informally, we propose that a concrete action system refines an abstract one if those initial conditions which guarantee that the abstract action system will eventually reach a state satisfying $P : Pred$ will also guarantee that a state satisfying P will be reached in the concrete action system.

This condition on the states of the two action systems is weaker than the condition on traces imposed by trace-refinement. The trace-refinement definition requires the concrete action system to have almost the same actions as the abstract action system, with stuttering allowed. We find this too restrictive a definition, although it is a condition which is simple to verify in practice.

According to our informal definition, the concrete action system may contain intermediate states which are not present in the abstract one. For example, a transaction in an electronic banking database which transfers money from one account to another will be refined by one which at different times withdraws the money from one account and deposits the same amount in the other. Also, a prime number generator which generates prime numbers two at a time will be refined by one which generates them one at a time, but not vice-versa.

This informal definition of refinement of non-terminating action systems places no constraint on the *order* in which the intermediate states are reached. This may or may not be undesirable. For a prime-number generator, the order of generation may not be important so that a generator which generates primes in increasing order may refine (and be refined by) a generator which generates primes in some other order. On the other hand, an electronic banking database action system which transfers money between accounts would be expected to have the same order of variable transitions in the concrete as in the abstract: this reflects our desire that the balance of the account should always be up-to-date. The constraint of preserving the order of states of variables can be incorporated into the action system by making the variables *sequences* which hold all of the different states of the variables.

4.2 Definition

We now give a formal definition of refinement of non-terminating action systems. We could proceed by defining a new predicate transformer on an unfair non-terminating action system $\mathcal{A}$ (see Section 2.1) to capture the weakest precondition $wrp.\mathcal{A}.P$ which guarantees that $\mathcal{A}$ eventually reaches (and maybe passes through) a state satisfying $P : Pred\mathcal{A}$. Thus we would have $wrp.\mathcal{A}.P \ \hat{=} \ \exists \kappa : Ord \ G_\kappa(P)$ (cf. the definition of iteration in Section 2.1), where

$$G_0(P) \ \hat{=} \ P,$$
$$G_{\kappa+1}(P) \ \hat{=} \ G_0(P) \vee (\textbf{if } []_{i=1}^n \ g_i \rightarrow S_i \ \textbf{fi})(G_\kappa(P)) \ and$$
$$G_\delta(P) \ \hat{=} \ \exists \kappa < \delta \ G_\kappa(P).$$

However, that would require us to develop a set of properties of this new predicate transformer. Instead, for $P : Pred\mathcal{A}$, we consider the *terminating* action system

$$\mathcal{A}_P \ \hat{=} \ \textbf{do } []_{i=1}^n \ \neg P \wedge g_i \rightarrow S_i \ \textbf{od},$$

and notice that $\mathcal{A}_P(P) \Leftrightarrow wrp.\mathcal{A}.P$. Since we have already developed results for the weakest precondition, we save a great deal of work if we use $\mathcal{A}_P$ instead of wrp. Thus we have the following definition:

Definition 7 Refinement for Non-Terminating Action Systems. A non-terminating action system $\mathcal{A}$ *is refined by* a non-terminating action system $\mathcal{B}$ (notation $\mathcal{A} \leq \mathcal{B}$) if for all $P : Pred\mathcal{A}$,

$$\mathcal{A}(P) \Rightarrow \mathcal{B}(P) \ \wedge \ \mathcal{A}_P(P) \Rightarrow \mathcal{B}_P(P).$$

□

Note that this definition is valid only for action systems, and not for arbitrary statements.

The first conjunct is a condition on the *final* state of the program, while the second conjunct is a condition on the *intermediate* state of the program. Definition 7 imposes a partial order on action systems, satisfies monotonicity and is stronger than Definition 1 since the first conjunct in the new definition is exactly Definition 1. This first conjunct is necessary in the case of action systems which terminate, otherwise we could have a refinement of a terminating system by another system terminating in a different state, but whose intermediate states contain those of the first system.

We now define refinement of fair non-terminating action systems.

Definition 8 Refinement for Fair Non-Terminating Action Systems.
Let
$$\mathcal{A} \ \hat{=} \ \textbf{do} \ \ []_{i=1}^n \ \ g_i \rightarrow S_i$$
$$\textbf{uf} \ \ \ \ P_\alpha : uf_\alpha$$
$$\textbf{od}.$$

We define
$$\mathcal{A}_P \ \hat{=} \ \textbf{do} \ \ []_{i=1}^n \ \ \neg P \wedge g_i \rightarrow S_i$$
$$\textbf{uf} \ \ \ \ \neg P \wedge P_\alpha : uf'_\alpha$$
$$P : \emptyset$$
$$\textbf{od},$$

where $uf'_\alpha \ \hat{=} \ \{u[\neg P \wedge g_i \rightarrow S_i/g_i \rightarrow S_i] \ | \ u : uf_\alpha \wedge 1 \leq i \leq n\}$. With this definition of $\mathcal{A}_P$, we can use Definition 7 for refinement. □

Thus, by applying the relevant theorems to each conjunct in Definition 7, we conclude that our theorems for program development, Theorems 4, 5 and 6, hold also for non-terminating action systems.

5 Application

The task of the readers-writers problem is to control access to a file which is desired by many processes. Of these processes, we distinguish between those that wish to make changes to the file, called writers, and those that merely wish to access it without making changes, called readers. If a process is writing, no other process may have access to the file, whereas if there are no writers, arbitrarily many processes may read the file.

The standard approach, extended by techniques in this paper, is used in [Wab96] to develop various designs giving different priorities to different processes. Fairness, rather than being intrinsic to the problem, is used as a technique in the development. In particular, use is made of monotonicity, adding skips and data refinement. The result is a development, entirely within the formalism, of designs whose infinitary behaviour previously required *ad hoc* reasoning.

References

[Bac89]   R.J.R. Back, *A Method of Refining Atomicity in Parallel Algorithms*, LNCS 366, Springer Verlag, 1989

[BM95]   M.J. Butler, C.C. Morgan, *Action Systems, Unbounded Nondeterminism, and Infinite Traces*, Formal Aspects of Computing, Vol. 7, 1995

[Boo82]   H.J. Boom, *A Weaker Precondition for Loops*, ACM Transactions on Programming Languages and Systems, Vol.4, 1982

[BX95]   R.J.R. Back, Q.W. Xu, *Fairness in Action Systems*, unpublished, 1995

[CM88]   K.M. Chandy, J. Misra, *Parallel Program Design: A Foundation*, Addison-Wesley, 1988

[DS90]   E.W. Dijkstra, C.S. Scholten, *Predicate Calculus and Program Semantics*, Springer-Verlag, 1990

[Fra86]   N. Francez, *Fairness*, Springer Verlag, 1986

[Hes95]   W.H. Hesselink, *Safety and Progress of Recursive Procedures*, Formal Aspects of Computing, Vol. 7, 1995

[MMSS95] C.C. Morgan, A.K. McIver, K. Seidel, J.W. Sanders, *Probabilistic Predicate Transformers*, Technical Report TR-4-95, Programming Research Group, Oxford University Computing Laboratory, 1995

[Mor90]   J.M. Morris, *Temporal Predicate Transformers and Fair Termination*, Acta Informatica, Vol. 27, 1990

[MP92]   Z. Manna, A. Pnueli, *The Temporal Logic of Reactive and Concurrent Systems*, Springer-Verlag, 1992

[Wab96]   A.K. Wabenhorst, *Developing Fairness in Terminating and Reactive Programs*, Technical Report TR-1-96, Programming Research Group, Oxford University, 1996,
          http://www.comlab.ox.ac.uk/oucl/publications/tr/TR-1-96.html

Polynomial Time Samplable Distributions

Tomoyuki Yamakami

Department of Computer Science, University of Toronto
Toronto, Canada M5S 3G4

Abstract. This paper studies distributions which can be sampled by randomized algorithms in time polynomial in the length of their outputs. Those distributions are called "polynomial-time samplable" and important to average-case complexity theory, cryptography, and statistical physics. This paper shows that those distributions are exactly as hard as #P-functions to compute deterministically and at least as hard as NP-sets to approximate by deterministic protocols.

§1. Introduction. Let us see a simple example of how to generate an "occupied territory" on a finite square board (e.g., [11]). First we randomly choose a non-negative integer n, and define the "occupied territory" at stage 0 to be the center square of the $n \times n$ board. At stage i, a walker randomly chooses a starting point which is on the boundary of the board and walks neighbor points at random. If the walker successfully reaches an adjacent points s_i of the occupied territory at stage $i-1$, then the territory is expanded to include the point s_i. We continue to the next stage. After infinitely many stages, we are able to consider the probability that a certain region becomes the occupied territory at some stage.

This type of (probability) distributions is called "samplable" and the algorithm which produces instances under this distribution is called a "sampling algorithm" [2] or a "generator" [14]. Instances occurring under samplable distributions have often been observed in statistical physics. Samplable distributions are also of importance in cryptography. It is known that the existence of complex samplable distributions leads to the existence of pseudo-random number generators (see, [2, 6]). The theory of average-case complexity is also sensitive to the choice of distributions, but few studies on samplable distributions have been done.

Especially, distributions which are sampled by a randomized Turing machines in time polynomials in the lengths of outputs are called "P-samplable" by Ben-David et al. [2]. Polynomial-time samplable distributions also play an important role in average-case complexity, cryptography, and statistical physics.

In this paper, we focus on these polynomial-time samplable distributions and study their complexity and related topics, such as the existence of universal distributions and the properties of domination relations, which are essential to Levin's average-case NP-completeness theory.

Note that the definition of samplable distributions, given by Ben-David et al. [2], deals with only dyadic rational numbers. To handle classes of distributions whose values include more than the dyadic rational numbers, Gurevich [4] applied an "approximation scheme" [7] to the polynomial-time computability of distributions. In [4], a distribution μ is called *polynomial-time computable* if we have a polynomial-time algorithm M which, on a pair of inputs x and 0^i, outputs an approximation of the value $\mu(x)$ within a factor of 2^{-i}, i.e., $|\mu(x) - M(x, 0^i)| < 2^{-i}$. In this paper, we adapt Gurevich's scheme and extend it to capture the notion of polynomial-time samplability in a more general setting. Precisely speaking, a distribution μ is *polynomial-time samplable* if there exist a polynomial p and a randomized Turing machine M such that $|\hat{\mu}(x) - \rho_{i,x}| < 2^{-i}$ for all x and i, where $\hat{\mu}$ is the associated density function of μ, and $\rho_{i,x}$ denotes the probability that M on input 0^i halts and outputs x within $p(|x|, i)$ steps.

All polynomial-time computable distributions are naturally polynomial-time samplable. However, Ben-David et al. show that, in their setting, polynomial-time samplable distributions are more complex than polynomial-time computable distributions unless P = NP [2]. This paper extends their result and shows by a simple counting argument that all polynomial-time samplable distributions are polynomial-time computable if and only if #P equals FP, where FP is the set of polynomial-time computable functions. In other words, polynomial-time samplable distributions are as hard as #P functions to compute deterministically. As a corollary, we can show that there is no "meaningful" universal, polynomial-time samplable distribution if FP = #P.

When distributions μ and ν satisfy the inequation $p(|x|) \cdot \hat{\mu}(x) \geq \hat{\nu}(x)$ for some fixed polynomial p, we simply say that μ *polynomially dominates* ν. This type of domination relation between distributions is of great importance in average-case complexity theory. The first step along this line was done by Levin [8]. He introduced the notion of many-one reducibility between distributional (or randomized) problems by requiring an additional condition based on the polynomial-domination relations. As proven in [4], distributional problems which can be solved in polynomial time on the average are invariant to polynomial-domination relations.

In Levin's definition, domination relations play a special role of "reducibility" between distributions in measuring the complexity of these distributions. For instance, every polynomial-time computable distribution is polynomially dominated by some polynomial-time samplable distribution; but whether the converse holds is not clear. Ben-David et al. show that, in their setting, if strong one-way functions exist, then there exists a polynomial-time samplable distribution which is not polynomially-dominated by any polynomial-time computable distributions [2]. We call a set A is *nearly*-RP if A is, roughly speaking, "approximated" by a nondeterministic Turing machine which, on most instances, looks like a one-sided probabilistic machine with bounded error. This paper shows that the assumption that there exists an NP set which is not nearly-RP (weaker than the assumption that strong one-way functions exist) yields the same conclusion. In this proof, we use a family of hash function to randomize nondeterminism and an amplification technique for probabilistic algorithms to reduce their error probability (as has been used in, e.g., [6]).

When two distributions polynomially dominate each other, we simply call them *polynomially equivalent*. It seems unlikely that every polynomial-time samplable distribution is polynomially equivalent to some polynomial-time computable distribution, but it is still an open question. We note that the assumption FP $\neq$ #P may not be sufficient for a negative answer. This paper shows that if NP differs from P, as is widely believed, then there is a polynomial-time samplable distribution which is not polynomially equivalent to any polynomial-time computable distribution.

§2. Foundations of Notions and Notations. We assume the reader's familiarity with computational complexity theory and average-case complexity theory.

Denote by $\mathbb{N}$ and $\mathbb{R}^+$ the set of *nonnegative integers* and the set of *nonnegative real numbers*, respectively. The cardinality of a set S is denoted by $\|S\|$. For a set S, we say that a property $\mathcal{P}(x)$ holds for *almost all* x in S if the set $\{x \in S \mid \mathcal{P}(x)$ does not hold$\}$ is finite. Let $\text{ilog}(m) = \lceil \log_2 m \rceil$ and $\ell(m) = \lfloor \log_2(m+1) \rfloor$. The notation $\log^k n$ stands for $(\log_2 n)^k$.

Fix $\Sigma = \{0,1\}$ and denote by λ the *empty string*. Let $\Sigma^+ = \Sigma^* - \{\lambda\}$. For each $n \in \mathbb{N}$, let $\Sigma^n = \{s \in \Sigma^* \mid \|s\| = n\}$ and $A^n = A \cap \Sigma^n$. The standard order $\leq$ on Σ^* (i.e., to order strings first by length and then lexicographically) is assumed, and $\mathbb{N}$ is often identified with Σ^* based on this order. Let $x_{\leftarrow i}$ be the first i bits of string x and denote by s_k^n the $(k+1)$-th string of the set $\Sigma^{\text{ilog}(n)}$ with respect to the standard order. In particular, $s_0^n = 0^{\text{ilog}(n)}$. We assume that a pairing function $\langle , \rangle$ is computable and invertible in polynomial time such that $|\langle x, y \rangle| \leq 2(|x| + |y| + 1)$ for all strings x, y. By $x \sqsubseteq y$, we mean that x is an initial segment of y, i.e., $xs = y$ for some s.

Let $\mathbb{D}$ be the set of all *dyadic rational numbers* on the unit real interval $[0,1]$, i.e., $\{\frac{m}{2^n} \mid m \leq 2^n, m, n \in \mathbb{N}\}$. We always identify a string $s_1 s_2 \cdots s_k$, where $s_i \in \{0,1\}$, with $\sum_{i=1}^{k} s_i \cdot 2^{-i}$ in $\mathbb{D}$.

For a function f, $\text{dom}(f)$ ($\text{ran}(f)$, resp.) denotes the set of instances to (from, resp.) which f maps. Denote by FP be the class of functions on Σ^* which are polynomial-time computable. A function f from Σ^* to $\mathbb{N}$ is in #P [17] if there is a set $A \in P$ and a polynomial p such that $f(x) = \|\{y \in \Sigma^{p(|x|)} \mid \langle x, y \rangle \in A\}\|$ for all x. By identifying Σ^* with $\mathbb{N}$, we have $\text{FP} \subseteq \text{\#P}$.

A function f on Σ^* is *p-honest* if there exists a polynomial p such that $|x| \leq p(|f(x)|)$ for all x, and f is *exp-honest* if $|x| \leq 2^{c|f(x)|+c}$ for some constant $c \geq 0$. A function f from Σ^* to $\mathbb{R}^+$ is called *positive* if $f(x) > 0$ for all x, and f is *polynomially bounded* (p-bounded, for short) if there exists a polynomial p such that $f(x) \leq p(|x|)$ for all x. A set A is *polynomial-time many-one reducible* (p-m-reducible, for short) to a set B if $A = \{x \mid f(x) \in B\}$ for some function f in FP. If f is further p-honest, then we say that A is *p-honest polynomial-time many-one reducible* (hp-m-reducible, for short) to B.

Let $\text{Prob}_x[E(x)]$ denote the probability that event $[E(x)]$ occurs when x is randomly chosen. The notation $\text{Prob}_x[E(x) \mid G(x)]$ denotes the conditional probability that $E(x)$ occurs when x is chosen at random from the set $\{x \mid G(x)\}$. For simplicity, we also use the notation $\text{Prob}_n[E(x)]$ for $\text{Prob}_x[E(x) \mid x \in \Sigma^n]$. For a randomized Turing machine M, $\text{Prob}_M[E(M, x)]$ expresses the probability that $E(M, x)$ holds when all possible computations of M on input x are made by flipping a fair coin.

For $n, c \in \mathbb{N}$, let $H_{n,n+c}$ denote the family of pairwise independent universal *hash functions* from Σ^n to Σ^{n+c} which is defined as follows: a hash function h in $H_{n,n+c}$ is of the form $h = (M, b)$, where M is an $n+c$ by n bit matrix and b is a bit vector, and takes its value as $h(x) = Mx \oplus b$ [6]. Hence $H_{n,n+c}$ can be identified with the set of all $n+c$ by $n+1$ matrices over $\{0,1\}$, and h is encoded to a string of length $(n+1)(n+c)$. We say that a hash function h in $H_{n,n+c}$ *i-distinguishes* x on X if $h(x)_{\leftarrow i+c} \neq h(w)_{\leftarrow i+c}$ for all $w \in X - \{x\}$.

A *distribution* μ is a nondecreasing function from Σ^* to $[0,1]$ such that $\lim_{x \to \infty} \mu(x) = 1$, and its associated *(probability) density function* $\hat{\mu}$ is defined by $\hat{\mu}(\lambda) = \mu(\lambda)$ and $\hat{\mu}(x) = \mu(x) - \mu(x^-)$ for all nonempty string x, where x^- is the predecessor of x. In other words, $\mu(x) = \sum_{z \leq x} \hat{\mu}(z)$. For practical reasons, we sometimes use semi-distributions instead of distributions: a *semi-distribution* is obtained by replacing $\lim_{x \to \infty} \mu(x) = 1$ by $\lim_{x \to \infty} \mu(x) \leq 1$. For a set S, let $\hat{\mu}(S) = \sum_{x \in S} \hat{\mu}(x)$. For a function f on Σ^*, we simply write $\mu_{f^{-1}}$ to denote the distribution defined by $\mu_{f^{-1}}(x) = \sum_{z \leq x} \hat{\mu}_{f^{-1}}(z)$, where $\hat{\mu}_{f^{-1}}(x) = \hat{\mu}(\{z \mid f(z) = x\})$.

For convenience sake, let $Tr(\lambda) = \lambda$, $Tr(00s) = 0Tr(s)$, $Tr(11s) = 1Tr(s)$, and $Tr(01s) = Tr(10s) = \#$ for any string s of even length, where $\#$ is a symbol different from 0 and 1. In this paper, we use the *standard* distribution ν_{st}, whose values $\nu_{st}(x)$ are dyadic rational numbers, which can be easily sampled by the following randomized algorithm: pick a nonnegative integer n randomly (i.e., generate a string of the form $s01$ or $s10$ such that $Tr(s)$ is the string identified with n) and then pick a string x of length n randomly. In other words, $\hat{\nu}_{st}(x) = 2^{-|x| - 2\ell(|x|) - 1}$. Notice that $\frac{2^{-|x|}}{8(|x|+1)^2} \leq \hat{\nu}_{st}(x) \leq \frac{2^{-|x|}}{2(|x|+1)^2}$.

A distribution μ *polynomially dominates* a distribution ν (denoted by $\nu \precsim^P \mu$) if there exists a p-bounded function p from Σ^* to $\mathbb{R}^+$ such that $p(x) \cdot \hat{\mu}(x) \geq \hat{\nu}(x)$ for all x [4]. Similarly, μ *average-polynomially dominates* ν (denoted by $\nu \precsim^{P,av} \mu$) if $p(x) \cdot \hat{\mu}(x) \geq \hat{\nu}(x)$ for some function p which is polynomial on ν-average [4].

Definition 1 [7, 4] A (semi-)distribution μ is *polynomial-time computable* (P-computable, for short) if there exists a deterministic Turing machine, with two input tapes, one output tape, and one work tape, which works in polynomial time (i.e., on input (x, y), the running time of M is at most $p(|x|, |y|)$ for some polynomial p) such that $|\mu(x) - M(x, 0^i)| < 2^{-i}$ for all x and $i \in \mathbb{N}$. Denote by P-comp the set of all distributions which are P-computable.

§3. Polynomial-Time Samplable Distributions.

This section formally defines the notion of polynomial-time samplability of distributions and shows that polynomial-time samplable distributions are as hard as #P functions to compute deterministically.

§3.1. Polynomial-Time Samplable Distributions.

Ben-David et al. [2] first formulated a notion of polynomial-time samplable distributions on dyadic rational numbers by using sampling algorithm. Recent work by Håstad et al. [6] on pseudo-random number generators also used an ensemble of "polynomial samplable" probability distributions. Here we use an approximation scheme to cope with real-valued distributions and give a generalized definition of the polynomial-time samplability.

Definition 2 A distribution μ is *polynomial-time samplable* (P-samplable, for short) if there exists a polynomial p and a randomized Turing machine M (it does not necessarily halt), called a *sampling algorithm*, such that

$$|\hat{\mu}(x) - \text{Prob}_M[M \text{ on input } 0^i \text{ halts and outputs } x \text{ within } p(|x|, i) \text{ steps}]| < 2^{-i}$$

for all $x \in \Sigma^*$ and $i \in \mathbb{N}$. Denote by P-samp the set of all P-samplable distributions.

From a different point of view, Impagliazzo and Levin [5] defined "polynomial samplable" distributions to be of the form $\mu_{f^{-1}}$ for some $\mu \in$ P-comp and some $f \in$

FP. This definition is, however, so broad that we can actually construct distributions that do not belong to P-comp. Moreover, we can construct a $\mu \in$ P-comp and an exp-honest $f \in$ FP such that no distributions in P-comp polynomially dominate μ_{f-1}. The notion of "polynomial-domination" will be thoroughly studied in §4.

Proposition 3 *There exists a positive distribution $\mu \in$ P-comp and a nondecreasing, exp-honest function $f \in$ FP such that μ_{f-1} is not polynomially dominated by any ν in P-comp.*

Proof Sketch. Take an effective enumeration of all polynomial-time computable semi-distributions $\{\mu_i\}_{i \in \mathbb{N}}$ [13, 16] and take a P-computable distribution μ such that $\hat{\mu}(0^{n^6}) > \frac{1}{n^{1/2}}$ for all $n \geq 2$. Let $f(x)$ be the minimal string y such that $\log n \leq |y| \leq 9 \log n$ and $|y|^{k-1} \cdot \hat{\nu}_i(y) < \hat{\mu}(0^n)$ for all $i < \log n$ and for all integer k with $1 \leq k \leq \frac{\log n}{5 + 2 \log \log n}$, where $n = \min\{r \mid r^6 \leq |x| < (r+1)^6\}$ if $n \geq 2^{13}$; otherwise, let $f(x) = x$. For every k and i, we have $|y|^{k-1} \cdot \hat{\nu}_i(y) < \hat{\mu}_{f-1}(y)$ for some y since $\hat{\mu}(0^n) \leq \hat{\mu}_{f-1}(y)$. □

In this paper, we require f to be p-honest, and take the following weaker (than in [5]) definition:

Definition 4 A distribution μ is *weakly polynomial-time samplable* (weakly P-samplable, for short) if there exists a distribution $\nu \in$ P-comp and a p-honest function $f \in$ FP such that $\mu = \nu_{f-1}$. Denote by WP-samp the set of all weakly P-samplable distributions. Let WP$_1$-samp be the set of all distributions of the form ν_{f-1} for some $\nu \in$ P-comp and some one-one, p-honest function $f \in$ FP.

Take the identity function f on Σ^*. Obviously f is p-honest and one-one. Then, we have $\mu = \mu_{f-1}$ for all distributions μ. Hence, we have the following inclusions.

Proposition 5 P-comp $\subseteq$ WP$_1$-samp $\subseteq$ WP-samp.

As shown in [18], the feasible computability of μ_{f-1}, in general, does not imply that of μ since there is a distribution μ which is not in P-comp, but μ_{f-1} is in P-comp for $f(x) = 0^{|x|}$. Moreover, Wang and Belanger show that, for every $\mu \in$ P-comp and every nondecreasing, p-honest function $f \in$ FP, μ_{f-1} belongs to P-comp [1].

By modifying the proof of [2, Theorem 7], we can prove:

Proposition 6 WP-samp $\subseteq$ P-samp.

The converse of this proposition does not seem to hold; however, we can prove that every P-samplable distribution is polynomially dominated by some weakly P-samplable distribution.

Proposition 7 *For every $\mu \in$ P-samp, there exists a distribution $\nu \in$ WP-samp such that $\mu \precsim^{\mathrm{P}} \nu$.*

Proof Sketch. For a $\mu \in$ P-samp, take a sampling algorithm M with time-bound p, where p is an increasing polynomial, which witnesses μ. Let $f(z)$ be the output x of M on input $0^{2|x|}$ on path encoded with z' in time $p(3|x|)$ if $z = z'1$ and such x exists;

λ if $z = z'1$ but no such x exists; and z' if $z = z'0$. Let $\hat{\nu}(x) = \hat{\nu}_{st}(\{w \mid f(w) = x\})$. Then, we have $\mu \precsim^p \nu$. □

Notice that we do not know whether we can replace WP-samp in Proposition 7 by WP_1-samp.

By definition, WP-samp enjoys the following closure property: if $\mu \in$ WP-samp and f is p-honest and in FP, then $\mu_{f^{-1}} \in$ WP-samp. Nevertheless, we may not show a similar property for P-samp because we use the approximation scheme to define P-samplable distributions. If we take the original definition of Ben-David et al. [2], this property holds even for the set of P-samplable distributions. Instead, it is easy to see that, by Proposition 7, if $\mu \in$ P-samp and f is p-honest in FP, then $\mu_{f^{-1}} \precsim^p \nu$ holds for some $\nu \in$ P-samp.

For a distribution μ, a function f from Σ^* to $\mathbb{R}^+$ is *polynomial on μ-average* if the expectation $\sum_{x \neq \lambda} |x|^{-1} f(x)^\delta \hat{\mu}(x)$ converges for some real number $\delta > 0$ [8]. We say that a Turing machine M works in *polynomial-time on μ-average* [4], where $\text{Time}_M(x)$ is the running time of M on input x. For a set $\mathcal{F}$ of distributions, let us denote by $P_{\mathcal{F}}$ the collection of sets A such that, for all $\mu \in \mathcal{F}$, there is a deterministic Turing machine M which computes A in time polynomial on μ-average [15]. It is known that $P_{\text{P-comp}}$ is not closed downward under p-m-reductions [16]. As an immediate consequence of the remark above, we have the following.

Lemma 8 *The classes* $P_{\text{WP-samp}}$ *and* $P_{\text{P-samp}}$ *are closed downward under hp-m-reductions.*

Proof. Let $A \leq_m^p B$ via a p-honest f in FP. Note that if B is computable in time polynomial on ν-average and $\mu_{f^{-1}} \precsim^{p,\text{av}} \nu$, then A is computable in time polynomial on $\mu_{f^{-1}}$-average [4]. The lemma follows from the above remark. □

§3.2. The P-comp = P-samp Question.

This subsection shows that P-samplable distributions are computable in polynomial time if and only if FP = #P. So, it seems unlikely that P-comp = P-samp.

Toward the goal of this subsection, we will study another category of distributions, the so-called #P-computable distributions introduced by Schuler and Watanabe [14], which seems computationally more powerful than P-samplable distributions. Again we modify their definition to fit our approximation scheme.

Definition 9 A distribution μ is *#P-computable* if there exists a function $f \in$ #P and a polynomial p such that $\left| \hat{\mu}(x) - \frac{f(x,0^i)}{2^{p(|x|,i)}} \right| < 2^{-i}$ for all x and $i \in \mathbb{N}$. Denote by #P-comp the set of all #P-computable distributions.

The main theorem of this subsection is stated as follows.

Theorem 10 *The following five statements are equivalent to each other: (1) FP = #P; (2) P-comp = #P-comp; (3) P-comp = P-samp; (4) P-comp = WP-samp; and (5) P-comp = WP_1-samp.*

The theorem immediately follows from a proposition and two lemmas below.

Proposition 11 P-samp $\subseteq$ #P-comp.

Proof. Assume that μ is P-samplable and is witnessed by a sampling algorithm M and a polynomial p. Without loss of generality, we assume that every path of M on input 0^i which outputs x has of length exactly $p(|x|, i)$. Let $f(x, 0^i)$ be the number of computation paths y such that M on 0^i outputs x and halts on path y in time $p(|x|, i)$. Clearly $f \in$ #P. It is easy to see that the probability that M on 0^i outputs x and halts in time $p(|x|, i)$ equals $\frac{f(x, 0^i)}{2^{p(|x|, i)}}$. Hence, μ belongs to #P-comp. $\square$

The converse inclusion, #P-comp $\subseteq$ P-samp, is an open question. The best-known result is, due to Schuler and Watanabe [14], that every #P-computable conditional distribution can be "approximated with a constant factor" by some sampling algorithm which, on input 0^n, works in time polynomial in n with nonadaptive queries to an NP oracle. An immediate consequence of this result will be stated as Lemma 25 in §4.2.

The next lemma establishes a basic relationship between #P and #P-comp. Its proof is straightforward.

Lemma 12 FP = #P *implies* P-comp = #P-comp.

It is known that P-samp = P-comp implies P = NP [2]. We can strengthen this result using a simple counting argument.

Lemma 13 *If* P-comp = WP_1-samp, *then* FP = #P.

Proof. Assume that P-comp = WP_1-samp. For any set A in P and any polynomial p, let $g(x) = \|\{y \in \Sigma^{p(|x|)} \mid xy \in A\}\|$. Without loss of generality, we assume that p is strictly increasing. We will show that $g \in$ FP.

Now take the standard distribution ν_{st} and define a one-one, polynomial-time computable function f as follows: let $f(xy)$ be $0xy$ if $xy \in A$ and $|y| = p(|x|)$; $1xy$ if $xy \notin A$ and $|y| = p(|x|)$; and xy otherwise. We also define η as $\hat{\eta}(y) = \hat{\nu}_{st}(\{z \mid f(z) = y\})$. By our assumption, η belongs to P-comp. For the g, we have the following simple equation: $2^{-r(|x|)-2\ell(r(|x|))-1} \cdot g(x) = \eta(0x1^{p(|x|)}) - \eta(0x^{-}1^{p(|x|)})$, where $r(n) = n + p(n) + 1$. Therefore, g is polynomial-time computable. $\square$

§3.3. Universal Distributions.

We have seen in Theorem 10 that P-comp = P-samp exactly when FP = #P. This subsection applies this theorem to polynomial-time samplable universal distributions. Universal distributions are known to be *malign*, i.e., average-case complexity equals worst-case complexity [9]. Under the assumption FP = #P, we can show that there is no "meaningful" universal distributions in P-samp. Here "meaningful" means the following "p-universal."

Definition 14 Let $\mathcal{F}$ be a set of distributions and $\mathcal{T}$ a set of functions from Σ^* to $\mathbb{R}^+$. A distribution μ is called $\mathcal{T}$-*universal for* $\mathcal{F}$ if (i) $\mu \in \mathcal{F}$, and (ii) for all $\nu \in \mathcal{F}$ there exists a function $t \in \mathcal{T}$ such that $t(x) \cdot \hat{\mu}(x) \geq \hat{\nu}(x)$ for all strings x. Especially, if $\mathcal{T}$ is the set of p-bounded functions, then μ is called p-*universal*.

By a modification of the proof of [10, Lemma 4.1], we first show that P-comp has no p-universal distributions if P = NP.

Proposition 15 *Assume that* P = NP. *For every function* $f \in o(2^n)$, *P-comp has no* $O(f)$-*universal distribution. Hence, there is no p-universal distribution for* P-comp.

Proof Sketch. Assume P = NP. Let μ_0 be $O(f)$-universal for P-comp, where $f \in o(2^n)$. Fix a sufficiently large string x_{-1}, and take a semi-distribution ν, whose value in binary can be exactly computed by a polynomial-time Turing machine, such that $\hat{\nu}(x) \geq \hat{\mu}_0(x) - 2^{-3|x|} > 0$ for all $x \geq x_{-1}$. We then define a series of strings $\{x_i\}_{i \in \mathbb{N}}$ as follows: let x_i be the minimal string y such tha $y \geq 2^{|x_{i-1}|}$ and $\nu(y) - \nu(x_{i-1}) \geq 2^{|y|} \cdot \hat{\nu}(y)$. Find a distribution η so that $\hat{\eta}(x_i) = 2^{|x_i|}(\hat{\nu}(x_i) + 2^{-3|x_i|})$ for all $i \in \mathbb{N}$. Since P = NP, the set $\{x_i\}_{i \in \mathbb{N}}$ is in P, and thus, η can be found from P-comp. By definition, this η contradicts the $O(f)$-universality of μ_0 for P-comp. □

Use Theorem 10, and we get the following corollary.

Corollary 16 *If* FP = #P, *then there is no p-universal distribution for* P-samp.

§4. Domination and Equivalence Relations.

Domination relations were explicitly introduced by Levin [8] in his theory of average-case complexity. He used domination relations as a part of a certain type of auxiliary condition on "many-one reducibility" between two distributional decision problems which guarantees that the complexity of these distributions are of the same degree. More precisely, many-one reductions map instances of high probability to instances of high probability. In this sense, two distributions which dominate each other can be considered to have almost the same degree of complexity. So, we call them "equivalent." Equivalence relations capture the closeness of two distributions and also give rise to an appropriate "approximation" between them. In this section, we study the consequences of several types of conditions of these domination and equivalence relations.

§4.1. Domination Relations.

In this subsection, we first focus on polynomial-domination relations. Polynomial-domination relations are useful in average-case complexity theory since they do not change the degree of average running time: namely, provided that μ (average-) polynomially dominates ν, if an algorithm requires polynomial-time on μ-average, then this algorithm also runs in polynomial-time on ν-average [4].

Consider the following condition:

Condition I. Every distribution in P-samp is polynomially dominated by some distribution in P-comp.

Each of the following conditions (1) and (2) is shown, by [4, Lemma 3.2], to be equivalent to Condition I: (1) for every $\mu \in$ WP-samp, there exists a $\nu \in$ P-comp such that $\mu \precsim^P \nu$; and (2) for every p-honest function $f \in$ FP and every $\mu \in$ P-comp, there exists a $\nu \in$ P-comp and a p-bounded function from Σ^* to $\mathbb{R}^+$ such that $\hat{\nu}(y) \geq \sum_{x \in f^{-1}(y)} \frac{\hat{\mu}(x)}{p(x)}$ for all y.

By Theorem 10, Condition I is derived from the assumption FP = #P. Ben-David et al. further show in [2, Theorem 8] that if Condition I holds, then no strong one-way function exists.

Definition 17 A set A is *nearly*-RP if, for every polynomial p, there exists a set S and a polynomial-time probabilistic Turing machine M such that, for each x, (i) $x \in A - S$ implies $\text{Prob}_M[M(x) \neq A(x)] < \frac{1}{2}$, (ii) $x \in \overline{A} - S$ implies $\text{Prob}_M[M(x) \neq A(x)] = 0$, and (iii) $\text{Prob}_n[x \in S] < \frac{1}{p(n)}$ for almost all n.

Note that we can amplify the success probability of M on inputs in $A - S$ up to $1 - 2^{-n^k}$, $k \in \mathbb{N}$, by repeating its computations at random. By the same amplification technique, the error probability $\frac{1}{2}$ in Definition 17 can be replaced by $1 - \frac{1}{n^k}$ for any $k \in \mathbb{N}$.

In the proof of [2, Theorem 8], the following fact was actually used.

Lemma 18 cf.[2] *Assume Condition I. Let f be any function in FP, let k be any positive number, and let q be any polynomial with $q(n) \geq 1$. Assume that $|x| \leq |f(x)| + k \log |f(x)|$ for almost all x. There exists a set S and a deterministic Turing machine M such that (i) $S \subseteq \text{ran}(f)$, (ii) $\|S \cap \Sigma^n\| < \frac{2^n}{q(n)}$ for almost all $n \in \mathbb{N}$, and (iii) M on input x correctly lists all elements of $f^{-1}(x)$ (whenever $f^{-1}(x) = \emptyset$, M outputs 0) in polynomial time unless $x \in S$.*

Using the hash-function technique of [6] and the amplification technique for probabilistic algorithms from [12], we can show that Condition I leads to the consequence below, which is stronger than the result stated in [2].

Theorem 19 *Assume Condition I. Let A be any set in NP. Then, for every polynomial q, there exists a set D and a polynomial-time randomized Turing machine M such that $D \subseteq A$, and, for each x, $x \in A - D$ implies $\text{Prob}_M[M(x) = A(x)] \geq \frac{1}{2(2n+3)}$, $x \notin A$ implies $\text{Prob}_M[M(x) = A(x)] = 1$, and $\text{Prob}_n[x \in D] < \frac{1}{q(n)}$ for almost all n. Hence, A is nearly-RP.*

Proof. Assume Condition I. Take arbitrarily a set A in NP and a polynomial q, and we will show that there exist a set D and a Turing machine M which satisfy the conditions of the theorem. There exists a set $B \in$ P such that $A = \{x \mid \exists z \in \Sigma^{|x|}[xz \in B]\}$. Let $B_x = \{z \mid xz \in B\}$ for each x. We take any sufficiently large n and fix it through the following argument.

We take the set $H_{n,n+c}$ of hash functions. Define $f(x') = 1xs_k^n hh(y)_{\leftarrow k+c}0^{n-k}$ if $x' = xys_k^n h$ and $y \in B_x$; otherwise $0x'$, where $x \in \Sigma^n$, $h \in H_{n,n+c}$, and $c = \text{ilog}(n)$. Notice that $|x'| \leq |f(x')|$ for all x'. For each x, let $g(x) = \|f^{-1}(x)\|$. For brevity, write $t(n) = 1 + n + \ell(n) + (n+1)(n + \text{ilog}(n)) + n + \text{ilog}(n)$.

Now fix k ($\text{ilog}(\|B_x\|) \leq k \leq n$) and x ($|x| = n$). We first show that the probability $\rho_{k,x} = \text{Prob}_{hw}[g(1xs_k^n hw_{\leftarrow k+c}0^{n-k}) = 1 \mid h \in H_{n,n+c}, w \in \Sigma^{n+c}]$ is at least $\frac{1}{2n+3}$. If $\|B_x\| > 0$, then $\rho_{k,x}$ is larger than or equal to the probability over all hw that, for each y in B_x, $h(y)_{\leftarrow k+c} = w_{\leftarrow k+c}$, and h k-distinguishes y on B_x. Thus, we have $\rho_{k,x} \geq \|B_x\| \cdot 2^{-(k+c)} \cdot (1 - 2^{-c}) \geq \frac{n-1}{2n^2} \geq \frac{1}{2n+3}$. Obviously, if $\|B_x\| = 0$, then $g(1xs_k^n hw_{\leftarrow k+c}0^{n-k}) = 0$ for all k, h, and w.

By Lemma 18, there is a set S and a polynomial-time deterministic Turing machine N such that $S \subseteq \mathrm{ran}(f)$, $\|S \cap \Sigma^{t(n)}\| < \frac{2^{t(n)}}{2(2n+3)q(n)}$, and moreover, if $x \in \overline{S}$, then N accepts x iff $f^{-1}(x) \neq \emptyset$. We define a randomized polynomial-time algorithm M as follows: on input x (say, $n = |x|$), (i) choose w, h at random, where $w \in \Sigma^{n+c}$, $h \in H_{n,n+c}$, and $c = \mathrm{ilog}(n)$; (ii) for all k (with $1 \leq k \leq n$), run N on x_k', where $x_k' = 1xs_k^n hw_{\leftarrow k+c}0^{n-k}$; and (iii) output $\mathrm{OR}_{k=1}^n N(x_k')$. Let $\delta_{k,x} = \mathrm{Prob}_{hw}[1xs_k^n hw_{\leftarrow k+c}0^{n-k} \in S \mid h \in H_{n,n+c}, w \in \Sigma^{n+c}]$. For the desired set D, let $D = \{x \in \Sigma^+ \cap A \mid \exists k \text{ s.t. } \mathrm{ilog}(\|B_x\|) \leq k \leq n \wedge \delta_{k,x} \geq \frac{1}{2(2n+3)}, |x| = n\}$. Since $\max\{\delta_{k,x} \mid \mathrm{ilog}(\|B_x\|) \leq k \leq n, x \in \Sigma^n\} \cdot \mathrm{Prob}_n[x \in D] \leq \frac{\|S^{t(n)}\|}{2^{t(n)}}$, we have $\mathrm{Prob}_n[x \in D] < \frac{1}{q(n)}$.

Our goal is to prove that (i) $\mathrm{Prob}_M[M(x) = A(x)] \geq \frac{1}{2(2n+3)}$ for all x in $A - D$, and (ii) $\mathrm{Prob}_M[M(x) = A(x)] = 1$ for all $x \notin A$. Take any input x of length n. Let $\rho_x = \mathrm{Prob}_{hw}[A(x) = \mathrm{OR}_{k=1}^n N(x_k')) \mid h \in H_{n,n+c}, w \in \Sigma^{n+c}]$. It is easy to see that the probability $\mathrm{Prob}_M[M(x) = A(x)]$ is at least ρ_x. Assume $A(x) = 1$ for a string $x \in \overline{D}$. Note that if $x_{k'}' \in \overline{S}$ and $g(x_{k'}') = 1$ for some k', then $\mathrm{OR}_{k=1}^n N(x_k') = 1$. Hence, $\rho_x \geq \max\{\rho_{k,x} - \delta_{k,x} \mid \mathrm{ilog}(\|B_x\|) \leq k \leq n\} \geq \frac{1}{2(2n+3)}$. For the other case $A(x) = 0$, $N(xs_k^n hw_{\leftarrow k+c}0^{n-k}) = 0$ for all h, w, and k; and thus $\rho_x = 1$. This completes the proof. □

Here we show that the assumption that every NP set is nearly-RP implies that no strong one-way functions exist.

Lemma 20 *If every* NP *set is nearly-RP, then there is no strong one-way function.*

Proof Sketch. Assuming the existence of a strong one-way function, we first construct a length-preserving strong one-way function f which is one-one on at least $\frac{2^n}{p(n)}$ elements of each Σ^n, where p is an adequate increasing polynomial (see [2]). Second, we encode into an NP set A each bit of the output value of f on input x. The cardinality of A^n is at least $\frac{2^n}{n \cdot p(n)}$. Using this set A, we define an adequate randomized algorithm which computes f with high probability. This contradicts the one-wayness of f. □

Polynomial-domination relations are useful but too tight to be considered as an effective measure of "approximation" or "reducibility" between distributions in average-case complexity theory. Gurevich [4] later introduced a weaker form of domination relations – average-polynomial-domination, defined in §2. We will relax Condition I to allow average-polynomial-domination relations.

Condition I'. Every distribution in P-samp is average-polynomially dominated by some distribution in P-comp.

Note by [4, Lemma 7.3] that the following two conditions are equivalent to each other: (1) every distribution in WP$_1$-samp is average-polynomially dominated by some distribution in P-comp; and (2) for every p-honest function $f \in \mathrm{FP}$ and every $\mu \in$ P-comp, there exists a $\nu \in$ P-comp and a function p which is polynomial on

μ-average such that $\hat{\nu}(y) \geq \sum_{x \in f^{-1}(y)} \frac{\hat{\mu}(x)}{p(x)}$ for all y. Clearly Condition I' infers (1), but we do not know whether (1) infers Condition I'.

Notice that Condition I implies Condition I'.

We have seen in §3.1 that the class $P_{\text{P-samp}}$ is closed under hp-m-reductions. Under the assumption that Condition I' holds, a similar argument used in Lemma 8 shows that the class $P_{\text{P-comp}}$ is also closed under hp-m-reductions.

Proposition 21 *If Condition I' holds, then $P_{\text{P-comp}}$ is closed downward under hp-m-reductions.*

§4.2. Equivalence Relations. As seen in §4.1., domination relations can be viewed as an "approximation" or a "reducibility" between two distributions in average-case complexity theory. If two distributions dominate each other, we call them "equivalent" since they are close to each other and have almost the same degree of complexity. Equivalence relations was first discussed in [14] under the terminology "approximation within constant factor" to show the closeness of two distributions.

We say that μ is *polynomially equivalent* to ν (denoted by $\mu \simeq^P \nu$) if μ and ν polynomially dominate each other, i.e., $\mu \precsim^P \nu \precsim^P \mu$. Similarly, μ is *average-polynomially equivalent* to ν (denoted by $\mu \simeq^{P,\text{av}} \nu$) if $\mu \precsim^{P,\text{av}} \nu \precsim^{P,\text{av}} \mu$.

In this subsection, we study the following conditions of equivalence relations:

Condition II. Every distribution in P-samp is polynomially equivalent to some distribution in P-comp.

Condition II'. Every distribution in P-samp is average-polynomially equivalent to some distribution in P-comp.

Note that, similar to Condition I, Condition II is equivalent to each of the following conditions: (1) for every $\mu \in \text{WP}_1$-samp, there is a $\nu \in$ P-comp such that $\mu \simeq^P \nu$; and (2) for every p-honest $f \in \text{FP}$ and every $\mu \in$ P-comp, there exists $\nu \in$ P-comp and p-bounded functions p, q from Σ^* to $\mathbb{R}^+$ such that $\sum_{x \in f^{-1}(y)} q(x)\hat{\mu}(x) \geq \hat{\nu}(y) \geq \sum_{x \in f^{-1}(y)} \frac{\hat{\mu}(x)}{p(x)}$ for all y. A similar observation can be made for Condition II' as made for Condition II. Later we will show that the set P-samp in the definition of Condition II can be replaced by the set #P-comp.

Clearly Condition II implies Condition I, and Condition II' implies Condition I'. By Theorem 10, Condition II is true if FP = #P.

In what follows, we will show that Condition II implies P = NP and that Condition II' implies P = RP. We prove the latter claim first. In this case, the amplification lemma for one-sided bounded-error probabilistic algorithms is effectively used to make the error probability exponentially small.

Proposition 22 *Condition II' implies P = RP.*

Proof Sketch. Take an arbitrary $A \in$ RP, and we will prove that A belongs to P. By the amplification lemma [12], there is a strictly increasing polynomial p and a set $B \in$ P such that, for every $x \in \Sigma^n$, $\text{Prob}_y[\langle x,y \rangle \notin B \mid y \in \Sigma^{p(n)}] \leq 2^{-n}$ if $x \in A$; $\text{Prob}_y[\langle x,y \rangle \in B \mid y \in \Sigma^{p(n)}] = 0$ otherwise.

Take the distribution μ defined as $\hat{\mu}(xy) = \hat{\nu}_{st}(x) \cdot 2^{-p(|x|)}$ if $y \in \Sigma^{p(|x|)}$, or else $\hat{\mu}(xy) = 0$. Let $f(xy)$ be $x1^{p(|x|)}$ if $y \in \Sigma^{p(|x|)}$ and $\langle x, y \rangle \in B$; $x0^{p(|x|)}$ if $y \in \Sigma^{p(|x|)}$ and $\langle x, y \rangle \notin B$; and xy otherwise. By Condition II', we have a distribution $\nu \in$ P-comp and a function q which is polynomial on μ-average such that $\sum_{x \in f^{-1}(y)} q(x)\hat{\mu}(x) \geq \hat{\nu}(y) \geq \sum_{x \in f^{-1}(y)} \frac{\hat{\mu}(x)}{q(x)}$ for all y. Let k be the minimal positive integer such that $\sum_{x \neq \lambda} |x|^{-1} q(x)^{1/k} \hat{\mu}(x) < \infty$. Thus, for almost all x and all y of length $p(|x|)$, $q(xy) \leq 2^{r(|x|)}$, where $r(n) = (2n + p(n))^k$. Since $\nu \in$ P-comp, there exists a deterministic polynomial-time Turing machine M such that $|\hat{\nu}(x) - M(x, 0^i)| < 2^{-i}$. Let $M'(x) = M(x, 0^{r(|x|)+2|x|})$. By definition, $|\hat{\nu}(x) - M'(x)| < 2^{-r(|x|)-2|x|}$ for all x. Now we have a complete characterization of A in terms of M'; namely, $A \cap \Sigma^n = \{x \in \Sigma^n \mid M'(x) \geq 2^{-r(n)-2n}(2^n - 2)\}$ for almost all n. Since M' halts in polynomial-time, A is also computable in polynomial-time. $\square$

The above proposition may not be simply improved to conclude that P = PP since the amplification lemma may not hold for PP sets.

To close this subsection, we prove that Condition II yields the consequence that NP collapses to P. We first strengthen Lemma 18 under Condition II as follows.

Lemma 23 *Assume that Condition II holds. For every set $B \in$ P and every polynomial p, let $S_B = \{x \mid \|B_x\| \leq p(|x|)\}$, where $B_x = \{z \in \Sigma^{|x|} \mid xz \in B\}$. There exists a deterministic Turing machine M such that, for each $n \in \mathbb{N}$, M on input x in $S_B \cap \Sigma^n$ lists all elements of B_x (whenever $B_x = \emptyset$, M outputs 0) in polynomial time.*

Theorem 24 *Condition II implies P = NP.*

Proof Sketch. Assume Condition II. Take any set A in NP, and we will show that $A \in$ RP since P = RP by Proposition 22. It suffices to consider a set A of the form $A = \{x \mid \exists z \in \Sigma^{|x|}[xz \in B]\}$ for some $B \in$ P. Let $B_x = \{z \in \Sigma^{|x|} \mid xz \in B\}$.

Define $\hat{B} = \{x'z' \mid \exists xzkh \text{ s.t. } x' = xs_k^n hh(z)_{\leftarrow k + c}0^{n-k}, z' = z10^{|x'|-|z|}, xz \in B, x, z' \in \Sigma^n, h \in H_{n,n+c}, c = i\log(n)\}$. Let $S_{\hat{B}} = \{x' \mid \|\hat{B}_{x'}\| \leq 1\}$, where $\hat{B}_{x'}$ is defined similar to B_x. Fix any sufficiently large n, any string x ($|x| = n$), and any integer k ($i\log(\|B_x\|) \leq k \leq n$). Let $\rho_{k,x} = \text{Prob}_{hw}[xs_k^n hw_{\leftarrow k+c}0^{n-k} \in S_{\hat{B}} \mid h \in H_{n,n+c} \wedge w \in \Sigma^{n+c}]$. It is not difficult to see that $\rho_{k,x} \geq \frac{1}{2n+3}$ if $\|B_x\| > 0$. By Lemma 23, there is a polynomial-time deterministic Turing machine N which recognizes $S_{\hat{B}}$. A randomized algorithm, based on this N, which is similar to M in the proof of Theorem 19 accepts A with high probability. Therefore, A is in RP. $\square$

We say that μ is *constantly equivalent* to ν if there are constants $c, d \geq 0$ such that $c \cdot \hat{\mu}(x) \leq \hat{\nu}(x) \leq d \cdot \hat{\mu}(x)$ for all x. The sampling algorithm used in the proof of [14, Theorem 5.2] defines only a set of conditional distributions, but it can be easily extended into a full distribution by inductively simulating each conditional distribution. Therefore, we conclude the following.

Lemma 25 cf.[14] *Assume P = NP. Every distribution in #P-comp is constantly equivalent to some distribution in P-samp.*

Corollary 26 *Condition II holds if and only if every #P-computable distribution is polynomially equivalent to some P-computable distribution.*

Proof. (If – part) This is obvious since P-samp ⊆ #P-comp by Proposition 11. (Only if – part) Assume Condition II. By Theorem 24, we have P = NP. Then, by Lemma 25, every #P-computable distribution is polynomially equivalent to some P-samplable distribution. Combining Condition II, we have the desired consequence. □

References

1. J. Belanger and J. Wang, Isomorphisms of NP complete problems on random instances, in: *Proceedings, 8th Conference on Structure in Complexity Theory*, 1993, pp.65–74.
2. S. Ben-David, B. Chor, O. Goldreich, and M. Luby, On the theory of average case complexity, *J. Comput. System Sci.*, **44** (1992), pp.193–219.
3. J. Carter and M. Wegman, Universal classes of hash functions, *J. Comput. System Sci.*, **18** (1979), pp.143–154.
4. Y. Gurevich, Average case complexity, *J. Comput. System Sci.*, **42** (1991), pp.346–398.
5. R. Impagliazzo and L. A. Levin, No better ways to generate hard NP instances than picking uniformly at random, in: *Proceedings, 31st IEEE Conference on Foundation of Computer Science*, pp. 812–821, 1990.
6. J. Håstad, R. Impagliazzo, L. A. Levin, and M. Luby, Construction of a pseudo-random generator from any one-way function, Technical Report, TR-91-068, International Computer Science Institute, Berkeley, California, 1991.
7. K. Ko and H. Friedman, Computational complexity of real functions, *Theoretical Computer Science*, **20** (1982), pp.323–352.
8. L. Levin, Average case complete problems, *SIAM J. Comput.* **15** (1986), pp.285–286.
9. M. Li and P. M.B. Vitányi, Average case complexity under the universal distribution equals worst-case complexity, *Information Processing Letters*, **42** (1992), pp.145–149.
10. M. Li and P. Vitányi, *An introduction to Kolmogorov complexity and its applications*, Springer-Verlag, New York, 1993.
11. J. Machta and R. Greenlaw, The computational complexity of generating random fractals, Technical Report, TR 93-04, University of Massachusetts at Amherst, 1993.
12. U. Schöning, Complexity and Structure, LNCS, Vol.211, 1986.
13. R. Schuler, Some properties of sets tractable under every polynomial-time computable distribution, *Information Processing Letters*, **55** (1995), pp.179-184.
14. R. Schuler and O. Watanabe, Towards average-case complexity analysis of NP optimization problems, in: *Proceedings, 10th on Structure in Complexity Theory Conference*, pp.148–159, 1995.
15. R. Schuler and T. Yamakami, Structural average case complexity, to appear in *Journal of Computer and System Sciences*. A preliminary version appeared in: *Proceedings, 12th Foundations of Software Technology and Theoretical Computer Science*, LNCS, Vol.652, 1992, pp.128–139.
16. R. Schuler and T. Yamakami, Sets computable in polynomial time on average, in: *Proceedings, 1st Annual International Computing and Combinatorics Conference*, LNCS, Vol.959, pp.400-409, 1995.
17. L. Valiant, The complexity of computing the permanent, *Theoretical Computer Science*, **5** (1979), pp.189–201.
18. J. Wang and J. Belanger, On average P vs. average NP, in *Complexity Theory – Current Research*, editors K. Ambos-Spies, S. Homer and U. Schöning, Cambridge University Press, pp.47–67, 1993.

From Static to Dynamic Abstract Data-Types*

Elena Zucca

DISI - Università di Genova
Via Dodecaneso, 35 - 16146 Genova (Italy)
email: zucca@disi.unige.it

Abstract. We show how to extend in a canonical way a given formalism for specifying (static) data types (like usual algebraic specification frameworks) with dynamic features. We obtain in this way a corresponding formalism for specifying *dynamic data-types* based on the "state-as-algebra" approach: a dynamic data-type models a dynamically evolving system in which any state can be viewed as a static data type in the underlying formalism, and the dynamic evolution is given by operations on states. Formally, our construction is a transformation of (pre)institutions.

Introduction

This paper deals with the extension to the dynamic case of algebraic specification techniques. In particular, we consider the "state-as-algebra" approach, which is at the basis of several recent proposals (see [1, 6, 11, 3], and [7] for an informal presentation and motivation).

As suggested by the name, the basic idea is to model states in the life of a dynamic system as algebras. The signature of these algebras (*static signature*) represents the static interface of the system, intuitively the *observations* that one may perform, getting answers which are depending on the current state. Dynamic evolution is modelled consequently by transformation between algebras.

This view of a dynamic system has in our opinion many advantages. First of all, it is natural to model a state as an algebra when thinking of "real large systems", like information systems, where a single snapshot is something complicated and structured. The static and the dynamic features of a system are kept distinct in a clean way, and modelled by semantic entities of different nature. More importantly, this separation allows to combine in a modular way different specification techniques for static and dynamic requirements, in the spirit of the integration of different formalisms which is now emerging as a fundamental topic ([5]).

Static requirements are conditions that each state must satisfy, like integrity constraints in the case of databases. Since each state is an algebra, they can be expressed in one of the many well-established algebraic formalisms (i.e., an institution in the now well–known sense of [8]), choosing the most adequate for any particular situation. Thus it is important to be able to specify dynamic aspects in a way that can be actually integrated with any of these formalisms.

* This work has been partially supported by WG n.6112 COMPASS, Murst 40% - Modelli della computazione e dei linguaggi di programmazione and CNR, Italy.

The contribution of our paper in this respect can be summarized as follows.

- We give a formal definition of *static framework* intended to be an abstraction of concrete formalisms for the description of static aspects of a system. We require some additional structure w.r.t. a generic institution: models have carriers which are (sorted) sets and there is an explicit notion of *variables* in sentences with related *valuations*. These requirements correspond to a notion of "institution with variables" which we believe of independent interest (see [4] and [10] for related work).
- We give a formal definition of *dynamic framework* over a given static framework, intended to be an abstraction of possible formalisms defined adding dynamics "on top" of a given formalism for defining static aspects. Formally, a dynamic framework is again a special case of institution: models have carriers which are classes of states (which can be viewed as models in the underlying static framework). The above notion of valuation of variables is lifted in a natural way to the dynamic case.
- We present a concrete example of dynamic framework which can be defined over any static framework, i.e., a logical formalism which is parametric over the formalism chosen for specifying the static aspects. The models of this framework are based on *d-oids* presented in [1].

The paper is structured as follows. In Sect. 1 we introduce our ideas by means of a toy example of dynamic system. In Sect. 2 and 3 we define static and dynamic frameworks. In Sect. 4 we define our parameterized dynamic framework.

Only for lack of space, we assume that models over a signature form simply a set instead of a category (as in *preinstitutions*), and present our construction associating with any static a corresponding dynamic framework as a map, instead of a functor. We refer to an extended version of this paper ([12]) for a full presentation.

1 An Introductory Example

In order to present our ideas, we use throughout the paper a toy example of graphical system. At any instant of time, there exists a finite number of circles, each one having a position, a size and a colour (red or green). The user can either move or resize circles, change their colour, create copies of existing circles and delete simultaneously all the, say, green circles. In the initial state only one circle exists, say, with center in the origin, radius one and green colour.

A natural way of representing a state of the system is as an algebra A over Σ_C:

$$\begin{aligned}
&\textbf{sig } \Sigma_C = \\
&\quad \textbf{sorts } \textit{real, colour, circle} \\
&\quad \textbf{opns} \\
&\quad\quad X, Y \colon \textit{circle} \to \textit{real} \\
&\quad\quad \textit{radius} \colon \textit{circle} \to \textit{real}, \textit{ col} \colon \textit{circle} \to \textit{colour}
\end{aligned}$$

s.t. A_{real} and A_{colour} are the sets of real and colour values, respectively, and A_{circle} is a (finite) set representing the existing circles. Set C the class of such algebras.

Dynamic evolution is defined by giving a family of *dynamic operations* DOP_C, each one modelling one possible basic way of modifying the system.

$DOP_C=$
    move: circle real real $\Rightarrow$, *resize: circle real* $\Rightarrow$
    changeCol: circle $\Rightarrow$, *copy: circle* $\Rightarrow$ *circle, delGreen:* $\Rightarrow$ *real*

Dynamic operations have functionality of the form $s_1 \ldots s_n \Rightarrow [s]$, with $[-]$ denoting optionality, since they may return a final result or just have a side-effect on the state (like methods in object oriented languages). The final result of *delGreen* is the total area of the green circles which have been deleted.

The interpretation of a dynamic operation, say $dop: s_1 \ldots s_n \Rightarrow [s]$, is, accordingly with our view of a state as an algebra, a map associating with each pair $\langle A, \langle a_1, \ldots, a_n \rangle \rangle$, where A is a state (algebra) and $\langle a_1, \ldots, a_n \rangle \in A_{s_1 \ldots s_n}$, a new state modelled by an algebra B and, if $[s]$ not null, a value $b \in B$. In addition, we want to be able to recover how the elements of A are transformed when passing to B; formally, we assume a partial map $f: |A| \to |B|$, called *tracking map*. This map is significant only for entities which may change (like circles in the example); basic values (like reals and colours in the example) always exist and never change (i.e. the tracking map is the identity over them).

Consider, for example, the dynamic operation *move*. The expected interpretation of this operation associates, with each pair $\langle A, \langle c, x, y \rangle \rangle$, where A represents a state of the system, $c \in A_{circle}$ an existing circle and x, y two integer values, a new state $B \in C$ defined as follows:

$$|B| = |A|$$
$$X^B(c) = X^A(c) + x, \; Y^B(c) = Y^A(c) + y$$
$$X^B(c') = X^A(c'), \; Y^B(c') = Y^A(c'), \; \forall c' \in B_{circle}, c' \neq c$$
$$radius^B = radius^A, \; col^B = col^A.$$

In this case the tracking map from $|A|$ into $|B|$ is just the identity: no new elements are created. An equivalent definition is to take B_{circle} any set in bijection with A_{circle}, with the tracking map being this bijection, say f, and operations in B defined by homomorphism. This equivalence can be formally expressed as an isomorphism in the category of d-oids over a signature (see [1, 12]).

We consider now a dynamic operation creating new objects, like e.g. *copy*. The expected interpretation associates, with each pair $\langle A, c \rangle$, where A represents a state and $c \in A_{circle}$ an existing circle, a pair $\langle B, c' \rangle$ defined as follows:

$$B_{circle} = A_{circle} \cup \{c'\}, \text{ with } c' \notin A_{circle}$$
$$X^B(c') = X^A(c), \; Y^B(c') = Y^A(c), \; radius^B(c') = radius^A(c), \; col^B(c') = col^A(c)$$
$$X^B(c'') = X^A(c''), \; Y^B(c'') = Y^A(c''),$$
$$radius^B(c'') = radius^A(c''), \; col^B(c'') = col^A(c''), \; \forall c'' \in A_{circle}.$$

In this case the tracking map from A_{circle} into B_{circle} is the inclusion. Analogously to the case above, an equivalent definition can be obtained by replacing the inclusion by any injective function from A_{circle} into a set having one more element, and defining operations over images of old elements by homomorphism. Finally, we consider constant dynamic operation symbols, with functionality written $[s]$, to be distinguished from parameterless operations (with functionality $\Rightarrow [s]$, like *delGreen* in our example) which always have the state as implicit argument. Constant operations just define a particular starting state of the system (possibly together with a value). In our example, we add to DOP_C one constant dynamic operation symbol, *start*: *circle*, whose interpretation is the pair $\langle A, c \rangle$ where

$$A_{circle} = \{c\},$$
$$X^A(c) = 0, \ Y^A(c) = 0, \ radius^A(c) = 1, \ col^A(c) = green.$$

The structure obtained enriching C by the interpreation in C of the dynamic operation symbols is called a *d-oid* over the *d-oid signature* $D\Sigma_C = \langle \Sigma_C, DOP_C \rangle$. In the example considered until now, states in C are exactly algebras over Σ_C. Anyway, in general this is a too strong requirement; to see this, let us consider the following subsignature $\overline{D\Sigma} = \langle \overline{\Sigma}, \overline{DOP} \rangle$ of $D\Sigma_C$:

> **sig** $\overline{\Sigma}$ =
>> **sorts** *real, circle*
>> **opns**
>>> X, Y: *circle* $\rightarrow$ *real*, *radius*: *circle* $\rightarrow$ *real*
>
> $\overline{DOP}$=
>> *move*: *circle real real* $\Rightarrow$, *resize*: *circle real* $\Rightarrow$, *delGreen*: $\Rightarrow$ *real*

It seems intuitively reasonable to see C also as a system supporting this restricted interface, just "forgetting" some operations; formally, to define the *reduct* of C w.r.t. $\overline{D\Sigma}$, denoted $C_{|\overline{D\Sigma}}$. Anyway, it is not possible to take, as states of $C_{|\overline{D\Sigma}}$, algebras over the restricted static signature $\overline{\Sigma}$. Indeed, the interpretation of *delGreen* in C strictly depends on the *col* operation, hence, if we forget *col*, there is no canonical way of deriving an interpretation of *delGreen* over $\overline{\Sigma}$-algebras. This is a simple example of a general situation, i.e. the fact that the definition of dynamic operations (e.g. procedures in a software module) may depend on some state components which are hidden, i.e. not declared in the interface.

The solution we take is to "relax" the definition of d-oid over $D\Sigma = \langle \Sigma, DOP \rangle$; we do no longer require that the states are Σ-algebras, but only that there is a mapping from the class of the states into Σ-algebras, intuitively giving, for any state, its "view" as Σ-algebra (i.e. its observable view). With this new definition of d-oid, we can take as reduct of C w.r.t. $\overline{D\Sigma}$ the d-oid which keeps the same states of C, but where the view of a state, say A (which was A itself in C) is the reduct $A_{|\overline{\Sigma}}$. See Def/Prop.5 for the formalization of this idea.

We consider now which should be intuitively a *specification* of a dynamic system with interface $\langle \Sigma_C, DOP_C \rangle$. In other words, we want to specify, by means of logical sentences, some requirements over the dynamic system. For instance, these requirements may be static sentences interpreted as invariants, like (assuming standard operations over basic values)

$$c_1 \neq c_2 \supset \sqrt{(X(c_1) - X(c_2))^2 + (Y(c_1) - Y(c_2))^2} > radius(c_1) + radius(c_2)$$

requiring no overlapping of circles (not satisfied by C above).
Other requirements may specify some expected behaviour of a dynamic operation, like

$$\{X(c) = x \wedge Y(c) = y\} \, move(c, x', y') \, \{X(c) = x + x' \wedge Y(c) = y + y'\}$$
$$move(c, x, y); move(c, x', y') = move(c, x + x', y + y')$$
$$\{true\} \, c' \leftarrow copy(c) \, \{X(c') = X(c) \wedge Y(c') = Y(c)\}.$$

We present in the sequel a logical formalism for specifiying dynamic data-types based on the ideas informally introduced so far, parameterized on the underlying framework for the static aspects. Indeed, we want to be free in the choice of the models (algebras) corresponding to states and of related static sentences.

2 Static Frameworks

A static framework is a logical formalism with all the components required for being a (pre)institution (signatures, models, sentences and satisfaction relation), and some additional features:

- a signature Σ has an underlying set of *sorts* S, and a model over Σ has an underlying *carrier* which is an S-sorted set;
- instead of considering sentences over a signature, we consider sentences over a *signature with variables*, i.e. a pair $\langle \Sigma, X \rangle$ where Σ is a signature over S and X is an S-sorted set; correspondingly, a sentence is evaluated w.r.t. a *valuation*, which is a pair $\langle A, r: X \to |A| \rangle$ where A is a Σ-model and r is a map associating values with variables. Sentences as usually defined ("constant" sentences in the sense that their evaluation does not depend on a valuation but only on a Σ-model) are sentences without free variables ($X = \emptyset$).

Notation. If $\mathbf{C}$ is a category, then $|\mathbf{C}|$ is the class of its objects. The functor $SSet$ gives, for any set S, the class of S-sorted sets. If $\sigma: S \to S'$ is a map, X is an S'-sorted set, then $X_{|\sigma}$ denotes the S-sorted set whose s-component is the $\sigma(s)$-component of X, $\forall s \in S$.
We define first the model part of a static framework, then we add sentences.

Definition 1. A *static model part* is a 4-tuple $\langle \mathbf{Sig}, Sorts, Mod, |-| \rangle$ where

- **Sig** is a category whose objects are called *signatures*;
- *Sorts* is a functor, $Sorts: \mathbf{Sig} \to \mathbf{Set}$, called a *sort functor for* **Sig**; for any signature Σ, the elements of $S = Sorts(\Sigma)$ are called the *sorts* of Σ; we also say that Σ is *over* S; for any σ morphism in **Sig**, $Sorts(\sigma)$ is denoted by σ when there is no ambiguity;

- *Mod* is a functor, $Mod: \mathbf{Sig}^{op} \to \mathbf{Set}$; for any signature Σ, elements of $Mod(\Sigma)$ are called *models* over Σ or Σ-*models*; for any $\sigma: \Sigma_1 \to \Sigma_2$ in $\mathbf{Sig}$, the map $Mod(\sigma): Mod(\Sigma_2) \to Mod(\Sigma_1)$ is called the *reduct* and denoted by $-|_\sigma$;
- $|-|$ is a natural transformation, $|-|: Mod \to SSet \circ (Sorts)^{op}$; for any Σ-model A, $|A|_\Sigma$ is called the *carrier* of A, and denoted by $|A|$ or even A when there is no ambiguity.

Typical examples of static model parts are total or partial algebras over a many-sorted algebraic signature.

Since a signature has an underlying set of sorts, it is possible to extend in a canonical way signatures to *signatures with variables*, as shown below.

Def/Prop. 2. If $\mathbf{Sig}$ is a category of signatures with a sort functor, then $\mathbf{Sig}^{Var}$ denotes the category where an object is a pair $\langle \Sigma, X \rangle$ with Σ signature over S and X sorted set over S whose elements are called *variables*, and a morphism from $\langle \Sigma_1, X_1 \rangle$ into $\langle \Sigma_2, X_2 \rangle$ is a pair $\langle \sigma, h \rangle$ with $\sigma: \Sigma_1 \to \Sigma_2$ and $h: X_1 \to (X_2)_{|\sigma}$. We call an object in $\mathbf{Sig}^{Var}$ a *signature with variables*.

Since models have a carrier which is a sorted set, it is possible to extend in a canonical way models to *valuations*, as shown below.

Def/Prop. 3. Let $\langle \mathbf{Sig}, Sorts, Mod, |-| \rangle$ be a static model part. Then

$$Val: (\mathbf{Sig}^{Var})^{op} \to \mathbf{Set}$$

is the functor defined by:

- for any $\langle \Sigma, X \rangle \in |\mathbf{Sig}^{Var}|$, an element of $Val(\langle \Sigma, X \rangle)$ is a pair $\langle A, r \rangle$, with $A \in Mod(\Sigma)$ and $r: X \to |A|$, called a *valuation* of X (into A);
- for any $\langle \sigma, h \rangle: \langle \Sigma_1, X_1 \rangle \to \langle \Sigma_2, X_2 \rangle$ in $\mathbf{Sig}^{Var}$,

$$\forall \langle A, r \rangle \in Val(\langle \Sigma_2, X_2 \rangle), \langle A, r \rangle_{|\langle \sigma, h \rangle} = \langle A_{|\sigma}, r_{|\sigma} \circ h \rangle.$$

Definition 4. A *static framework* is a tuple $\mathsf{SF} = \langle \mathbf{Sig}, Sorts, Mod, |-|, Sen, \Vdash \rangle$ where

- $\langle \mathbf{Sig}, Sorts, Mod, |-| \rangle$ is a static model part;
- *Sen* is a functor, $Sen: \mathbf{Sig}^{Var} \to \mathbf{Set}$; $\forall \langle \Sigma, X \rangle \in |\mathbf{Sig}^{Var}|$, the elements of $Sen(\langle \Sigma, X \rangle)$ are called *sentences* over Σ and X; for any morphism $\langle \sigma, h \rangle$ in $\mathbf{Sig}^{Var}$, $Sen(\langle \sigma, h \rangle)$ is denoted by $\langle \sigma, h \rangle$ when there is no ambiguity;
- for any $\langle \Sigma, X \rangle \in |\mathbf{Sig}^{Var}|$, $\Vdash_{\langle \Sigma, X \rangle}$ is a relation over $Val(\langle \Sigma, X \rangle) \times Sen(\langle \Sigma, X \rangle)$ s.t., for any $\langle \sigma, h \rangle: \langle \Sigma_1, X_1 \rangle \to \langle \Sigma_2, X_2 \rangle$ in $\mathbf{Sig}^{Var}$, the satisfaction condition

$$\langle A, r \rangle \Vdash_{\langle \Sigma_2, X_2 \rangle} \langle \sigma, h \rangle(\phi) \text{ iff } \langle A, r \rangle_{|\langle \sigma, h \rangle} \Vdash_{\langle \Sigma_1, X_1 \rangle} \phi$$

holds for any $\langle A, r \rangle \in Val(\langle \Sigma_2, X_2 \rangle)$ and for any $\phi \in Sen(\langle \Sigma_1, X_1 \rangle)$.

3   Dynamic Frameworks

Let us assume a fixed static framework $SF = \langle \mathbf{Sig}, Sorts, Mod, |-|, Sen, \Vdash \rangle$.
Accordingly with our informal discussion in Sect. 1, the state set of a dynamic system with static interface Σ is a class of arbitrary structures which can be viewed as Σ-models in the underlying static framework: formally, it is a map from this class of structures into $Mod(\Sigma)$.

Def/Prop. 5. The *state set* functor over SF, $\widetilde{Mod} : \mathbf{Sig}^{op} \to \mathbf{Set}$, is defined as follows:

- $\forall \Sigma \in |\mathbf{Sig}|$, an element $\mathcal{A}$ in $\widetilde{Mod}(\Sigma)$ is a pair $\langle Dom(\mathcal{A}), View_{\mathcal{A}} \rangle$, with $Dom(\mathcal{A})$ class of structures called *states* and $View_{\mathcal{A}} : Dom(\mathcal{A}) \to Mod(\Sigma)$; we call $\mathcal{A}$ a *state set* and write $A \in \mathcal{A}$ instead of $A \in Dom(\mathcal{A})$;
- $\forall \sigma : \Sigma_1 \to \Sigma_2$ morphism in $\mathbf{Sig}$, the reduct $-_{|\sigma} : \widetilde{Mod}(\Sigma_2) \to \widetilde{Mod}(\Sigma_1)$ is defined by: $\forall \mathcal{A}$ in $\widetilde{Mod}(\Sigma_2)$, $\mathcal{A}_{|\sigma} = \langle Dom(\mathcal{A}), -_{|\sigma} \circ View_{\mathcal{A}} \rangle$.

The above definition is very general, since states over Σ are allowed to be arbitrary structures. A more concrete definition of $\widetilde{Mod}$, which we consider in the examples, is the following. For any signature Σ, a state set $\mathcal{A}$ over Σ is a pair $\langle Dom(\mathcal{A}), View_{\mathcal{A}} \rangle$, where $Dom(\mathcal{A})$ is a class of $\Sigma^{\mathcal{A}}$-models, for some signature $\Sigma^{\mathcal{A}}$, and $View_{\mathcal{A}}$ is the $\sigma^{\mathcal{A}}$-reduct, for some signature morphism $\sigma^{\mathcal{A}} : \Sigma \to \Sigma^{\mathcal{A}}$. In this case, $\mathcal{A}$ can be equivalently represented by the pair $\langle Dom(\mathcal{A}), \sigma^{\mathcal{A}} \rangle$.

Intuitively, that corresponds to assume that, in a dynamic system $\mathcal{A}$ having static interface Σ, states are models over an internal signature $\Sigma^{\mathcal{A}}$, and their view as Σ-models can be obtained via the reduct.

As an example, consider the class $\mathcal{C}$ of Σ_C-algebras of Sect. 1. This class can be seen as a state set over Σ_C taking as signature morphism the identity. If $\sigma : \Sigma_1 \to \Sigma_2$ is a signature morphism, then the reduct w.r.t. σ of a state set $\mathcal{A}$ over Σ_2 keeps the same states of $\mathcal{A}$, but viewed now as Σ_1-models. For example, if ι denotes the inclusion morphism from $\overline{\Sigma}$ into Σ_C of Sect. 1, then $\langle \mathcal{C}, \iota \rangle$ is a state set over $\overline{\Sigma}$.

Analogously to the static case, a dynamic framework is a logical formalism with all the components required for being a (pre)institution and some additional features:

- a signature $D\Sigma$ has a *static part* Σ which is a signature in the underlying static framework;
- a model over $D\Sigma$ has a carrier which is a state set over Σ;
- as in the static case, we consider sentences over a signature with variables $\langle D\Sigma, X \rangle$; in this case, accordingly with the intuition, a valuation is a triple $\langle \mathcal{A}, A, r \rangle$ where $\mathcal{A}$ is a $D\Sigma$-model, A is a state of $\mathcal{A}$ (intuitively the current state in which to evaluate the sentence) and $\langle View_{\mathcal{A}}(A), r \rangle$ is a (static) valuation of X into the Σ-model corresponding to A.

Note that, differently from the static case, constant sentences (i.e. sentences which can be evaluated just w.r.t. a $D\Sigma$-model) do not coincide with sentences without free variables. For instance,

$$\{c \leftarrow start\}X(c) = 0 \wedge Y(c) = 0 \wedge radius(c) = 1 \wedge col(c) = green$$

is a constant sentence, intuitively stating a property of the constant dynamic operation *start*, while $(/\exists c)col(c) = green$ (where we assume the existential quantifier in the underlying static sentences) has no free variables, but its truth value depends on the current state.

Hence we need to keep explicitly the two kinds of sentences in a dynamic framework.

Definition 6. A *dynamic model part* based on SF is a 4-tuple consisting of $\langle \mathbf{DSig}, St, DMod, |-| \rangle$ where

1. **DSig** and *DMod* are like **Sig** and *Mod* in Def. 1;
2. $St: \mathbf{DSig} \to \mathbf{Sig}$ is a functor, called a *static part* functor; $\forall D\Sigma \in |\mathbf{DSig}|$, $St(D\Sigma)$ is called the *static part* of $D\Sigma$, and analogously for a morphism;
3. $|-|$ is a natural transformation, $|-|: DMod \to \widetilde{Mod} \circ (St)^{op}$; for any $D\Sigma$-model $\mathcal{A}$, $|\mathcal{A}|_{D\Sigma}$ is called the *carrier* of $\mathcal{A}$, and denoted by $|\mathcal{A}|$ or even $\mathcal{A}$ when there is no ambiguity.

It is straightforward to define, analogously to what we have done for a static model part in Def/Prop. 3, a functor $DVal: (\mathbf{DSig}^{Var})^{op} \to \mathbf{Set}$. In this case, a dynamic valuation over a dynamic signature with variables $\langle D\Sigma, X \rangle$ is a triple $\rho = \langle \mathcal{A}, A, r \rangle$, with $\mathcal{A} \in DMod(D\Sigma)$, $A \in \mathcal{A}$ and $r: X \to |View_{\mathcal{A}}A|$. We refer to [12] for the details.

Definition 7. A *dynamic framework* based on SF is a 6-tuple consisting of $\langle \mathbf{DSig}, St, DMod, |-|, DSen, \Vdash \rangle$ where

- $\langle \mathbf{DSig}, St, DMod, |-| \rangle$ is a dynamic model part based on SF;
- *DSen* is a functor, $DSen: \mathbf{DSig} + \mathbf{DSig}^{Var} \to \mathbf{Set}$; $\forall \langle D\Sigma, X \rangle \in |\mathbf{DSig}^{Var}|$ the elements of $DSen(\langle D\Sigma, X \rangle)$ are called *non-constant dynamic sentences* over $\langle D\Sigma, X \rangle$ and analogously for the constant case;
- $\forall \langle D\Sigma, X \rangle \in |\mathbf{DSig}^{Var}|$, $\Vdash_{\langle D\Sigma, X \rangle} \subseteq DVal(\langle D\Sigma, X \rangle) \times DSen(\langle D\Sigma, X \rangle)$ s.t., for any $\langle d\sigma, h \rangle: \langle D\Sigma_1, X_1 \rangle \to \langle D\Sigma_2, X_2 \rangle$ in $\mathbf{DSig}^{Var}$, the satisfaction condition

$$\rho \Vdash_{\langle D\Sigma_2, X_2 \rangle} \langle d\sigma, h \rangle (d\phi) \text{ iff } \rho_{|\langle d\sigma, h \rangle} \Vdash_{\langle D\Sigma_1, X_1 \rangle} d\phi$$

holds for any $\rho \in DVal(\langle D\Sigma_2, X_2 \rangle)$ and for any $d\phi \in DSen(\langle D\Sigma_1, X_1 \rangle)$;
- $\forall D\Sigma \in |\mathbf{DSig}|$, $\Vdash_{D\Sigma}$ is a relation over $DMod(D\Sigma) \times DSen(D\Sigma)$ for which an analogous satisfaction condition holds.

Here above $\mathbf{DSig} + \mathbf{DSig}^{Var}$ denotes the sum category (coproduct in **Cat**); analogously below $DMod + DVal$ denotes the coproduct of functors.

4 A Parameterized Dynamic Framework

In this section, we outline the construction of a dynamic framework $\overline{DF} = \langle \overline{\mathbf{DSig}}, \overline{St}, \overline{DMod}, |-|, \overline{DSen}, |\vdash \rangle$ on top of a given static framework SF. More details can be found in [12]. First we define the model part. As informally presented in Sect. 1, signatures (*d-oid signatures*) are pairs consisting of a static signature and a family of so-called *dynamic operation symbols* (like usual operation symbols, but with an hidden parameter, the state); models (*d-oids*) are classes of states enriched by the intepretation of dynamic operation symbols. D-oids have been firstly introduced in [1]; we keep here the same name, since the basic idea is the same, even if the version presented in this paper is modified in order to fill in the institutional framework (states are arbitrary structures which can be viewed as static models, as defined in Def/Prop. 5).
If S is a set, then $[S]$ denotes $S \cup \{A\}$; we use $[s]$ for ranging over $[S]$, i.e. $[s]$ stands for either an element of S or for the empty string.

Definition 8. A *d-oid signature* is a pair $\langle \Sigma, DOP \rangle$ where Σ is a static signature and DOP is a $[S] \cup (S^* \times [S])$-sorted set of *dynamic operation symbols*; if $dop \in DOP_{[s]}$, then we write $dop: [s]$, and say that dop is a *constant* dynamic operation symbol; if $dop \in DOP_{s_1 \ldots s_n, [s]}$, then we write $dop: s_1 \ldots s_n \Rightarrow [s]$. We denote by $\overline{\mathbf{DSig}}$ the category of d-oid signatures, defined in the obvious way, and by $\overline{St}$ the functor giving the first component.

Definition 9. Let $D\Sigma = \langle \Sigma, DOP \rangle$ be a d-oid signature. For any $\mathcal{A}$ state set over Σ, $w \in S^*$, set $\mathcal{A}_w = \{\langle A, \overline{a} \rangle \mid A \in \mathcal{A}, \overline{a} \in (View_{\mathcal{A}}(A))_w\}$.
Then, a *d-oid* over $D\Sigma$ is a pair $\mathcal{A} = \langle |\mathcal{A}|, \{dop^{\mathcal{A}}\}_{dop \in DOP} \rangle$ where:

- $|\mathcal{A}|$ is a state set over Σ (denoted by $\mathcal{A}$ when there is no ambiguity);
- for any $dop: [s]$, $dop^{\mathcal{A}} \in \mathcal{A}_{[s]}$; $dop^{\mathcal{A}}$ is called a *constant dynamic operation*;
- for any $dop: w \Rightarrow [s]$, $dop^{\mathcal{A}}$ is a map which associates with any $\langle A, \overline{a} \rangle \in \mathcal{A}_w$ a triple $\langle B, f[, b] \rangle$ where $\langle B[, b] \rangle \in \mathcal{A}_{[s]}$ and $f: |View_{\mathcal{A}}(A)| \to |View_{\mathcal{A}}(B)|$; we write $dop^{\mathcal{A}}(\langle A, \overline{a} \rangle) = \langle f: A \Rightarrow B[, b] \rangle$; $dop^{\mathcal{A}}$ is called a *(non-constant) dynamic operation*, and f is called *tracking map*.

We denote by $\overline{DMod}(D\Sigma)$ the class of the d-oids over $D\Sigma$.

The above definition can be easily extended to a functor $\overline{DMod}: \overline{\mathbf{DSig}}^{op} \to \mathbf{Set}$. Moreover, let $|-|: \overline{DMod} \to \widetilde{Mod} \circ (St)^{op}$ be the natural transformation mapping d-oids in underlying state sets.
Analogously to what is usually done in the static case, we can define, for any d-oid signature with variables $\langle D\Sigma, X \rangle$, the sets of the *constant dynamic terms over $D\Sigma$* and *non-constant dynamic terms* over $D\Sigma$ and X, denoted $DT(D\Sigma)$, $DT(\langle D\Sigma, X \rangle)$, respectively. We omit here the detailed formal definition of dynamic terms and their interpretation, which is straightforward (see [2, 12]) and give just the intuition.
A dynamic term is of the form $dt = x_1 \leftarrow dop_1(\overline{x}_1); \ldots; x_n \leftarrow dop_n(\overline{x}_n)$, and denotes intuitively the execution in sequence of $dop_1, \ldots, dop_n$; at each step,

x_i is a variable of the result sort of dop_i, which may be used in the sequel for denoting the entity returned as final result of dop_i. We write just $dop_i(\overline{x}_i)$ instead of $x_i \leftarrow dop_i(\overline{x}_i)$ in the case in which dop_i has null result sort. The set $\{x_1, \ldots, x_n\}$ of the binding variables of dt is denoted by $Var(dt)$.

An example of dynamic term is, referring to our running example,

$$c_2 \leftarrow copy(c_1); n \leftarrow delGreen; move(c_2, n, n).$$

denoting intuitively an execution sequence in which first a new circle c_2 is created as a copy of an existing circle c_1, then all the existing green circles (with total area n) are deleted, and finally c_2 is moved horizontally and vertically by n.

Accordingly with the intuition, a non-constant dynamic term dt over $D\Sigma$ and X is evaluated w.r.t. a triple $\rho = \langle A, A, r_A \rangle$, with A d-oid over $D\Sigma$, A in $\mathcal{A}$ initial state, r_A mapping of the free variables X into the Σ-model corresponding to A (this triple is what we have called a dynamic valuation). The result of the evaluation, denoted by $[\![\, dt\,]\!]^\rho_{(D\Sigma, X)}$, is a final state B and a mapping into the Σ-model corresponding to B of the variables $X \cup Var(dt)$. The evaluation of a constant dynamic term dt w.r.t. a d-oid A gives a state B and a mapping of $Var(dt)$ into the Σ-model corresponding to B.

We can now define dynamic sentences. They are of three forms: static sentences interpreted as invariants, *pre-post sentences* $\{\phi_1\} \, dt \, \{\phi_2\}$, generalizing classical Hoare's triples, and *dynamic equations* $dt_1 = dt_2$ stating that dt_1 and dt_2 denote exactly the same state transformation. Note that in pre-post conditions the tracking map allows to relate valuations of variables before and after executing dt, generalizing what is usually achieved in temporal logic by *rigid* variables (see e.g. [9]; see also [3] for some other work concerning relating sentences before and after a state transformation).

These sentences are presented mainly for illustrating the expressive power of our formalism, showing examples of basic formulas which can be actually defined "on top" of a given formalism for static aspects: the definition of an "ideal specification language" for dynamic data-types is outside of the scope of this paper.

Definition 10. For any d-oid signature with variables $\langle D\Sigma, X \rangle$, $D\Sigma$ with static part Σ, the set $\overline{DSen}(D\Sigma)$ (resp. $\overline{DSen}(\langle D\Sigma, X \rangle)$) of the *constant sentences* over $D\Sigma$ (resp. *non-constant sentences* over $\langle D\Sigma, X \rangle$) is defined in Fig. 1, where we write $\vdash d\phi$ for $d\phi \in \overline{DSen}(D\Sigma)$, $X \vdash d\phi$ for $d\phi \in \overline{DSen}(\langle D\Sigma, X \rangle)$.

Definition 11. Let $\langle D\Sigma, X \rangle$ be a d-oid signature with variables, $St(D\Sigma) = \Sigma$. The *satisfaction* of constant dynamic sentences over $D\Sigma$ w.r.t. A, $A \Vdash_{D\Sigma}$, and of non-constant dynamic sentences over $\langle D\Sigma, X \rangle$ w.r.t. ρ, $\rho \Vdash_{(D\Sigma, X)}$, are inductively defined in Fig. 1.

It is straightforward to define the *renaming* of constant dynamic sentences, obtaining altogether a functor $\overline{DSen} : \mathbf{DSig} + \mathbf{DSig}^{Var} \to \mathbf{Set}$.

$$\frac{}{X \vdash \phi} \qquad \phi \in Sen(\langle \Sigma, X \rangle)$$

$$\frac{}{X \vdash \{\phi_1\}\, dt\, \{\phi_2\}} \qquad \begin{array}{c} \phi_1 \in Sen(\langle \Sigma, X \rangle),\, dt \in DT(\langle D\Sigma, X \rangle) \\ \phi_2 \in Sen(\langle \Sigma, X \cup Var(dt) \rangle) \end{array}$$

$$\frac{}{\{dt\}\phi} \qquad dt \in DT(D\Sigma),\, \phi \in Sen(\langle \Sigma, Var(dt) \rangle)$$

$$\frac{}{X \vdash dt_1 = dt_2} \qquad dt_1, dt_2 \in DT(\langle D\Sigma, X \rangle),\, Var(dt_1) = Var(dt_2)$$

$$\frac{}{\vdash dt_1 = dt_2} \qquad dt_1, dt_2 \in DT(D\Sigma),\, Var(dt_1) = Var(dt_2)$$

$$\frac{}{\rho \Vdash_{\langle D\Sigma, X \rangle}\phi} \qquad \phi \in Sen(\langle \Sigma, X \rangle),\, \langle View_A(A), r_A \rangle \Vdash_{\langle \Sigma, X \rangle}\phi$$

$$\frac{}{\rho \Vdash_{\langle D\Sigma, X \rangle}\{\phi_1\}\, dt\, \{\phi_2\}} \qquad \begin{array}{c} (\langle View_A(A), r_A \rangle \Vdash_{\langle \Sigma, X \rangle}\phi_1 \wedge [\![dt]\!]^\rho_{\langle D\Sigma, X \rangle} = \langle B, r_B \rangle) \\ \supset \langle View_A(B), r_B \rangle \Vdash_{\langle \Sigma, X \cup Var(dt) \rangle}\phi_2 \end{array}$$

$$\frac{}{A \Vdash_{D\Sigma}\{dt\}\phi} \qquad ([\![dt]\!]^A_{D\Sigma} = \langle B, r_B \rangle) \supset \langle View_A(B), r_B \rangle \Vdash_{\langle \Sigma, Var(dt) \rangle}\phi$$

$$\frac{}{\rho \Vdash_{\langle D\Sigma, X \rangle} dt_1 = dt_2} \qquad [\![dt_1]\!]^\rho_{\langle D\Sigma, X \rangle} = [\![dt_2]\!]^\rho_{\langle D\Sigma, X \rangle}$$

$$\frac{}{A \Vdash_{D\Sigma} dt_1 = dt_2} \qquad [\![dt_1]\!]^A_{D\Sigma} = [\![dt_2]\!]^A_{D\Sigma}$$

Fig. 1. Dynamic sentences and their satisfaction

The following theorem states our main technical result, i.e. that the canonical construction described until now actually gives a logical formalism with the property that "truth is invariant under change of syntax" .

Theorem 12. *The 6-tuple* $\overline{\mathsf{DF}} = \langle \overline{\mathsf{DSig}}, \overline{St}, \overline{DMod}, |-|, \overline{DSen}, \Vdash \rangle$ *is a dynamic framework.*

5 Conclusion and Further Work

We have presented a canonical construction which associates with any static framework (a logical formalism for specifying static data-types) a corresponding dynamic framework (a logical formalism for specifying dynamic data-types). The relevance of this work is twofold. First, we have shown that d-oids, as already introduced in [1] only as models over a fixed signature, actually form a (pre)institution, moreover parameterized on the underlying (pre)institution chosen for modelling static aspects.

More in general, our work is concerned with the important topic of integrating different formalisms, since we show here how to enrich an existing formalism for expressing static aspects with additional ingredients which allow to handle dynamics. In this paper, the ingredients we choose are, as dynamic models, d-oids (dynamics is modelled by dynamic operations) and, as sentences, pre-post sentences and dynamic equations. Of course different solutions could be adopted, like modelling dynamics by transitions between states instead of operations, and choosing sentences in the style of temporal logic. An interesting question for further work is whether it is still possible with these different choices to get a canonical construction. A final result of this investigation could be a general notion of "sum" between a formalism for static aspects and a formalism for dynamic aspects.

Acknowledgment. I would like to thank all the participants of the informal meeting on FLIRTS (Formalism Logic Institution Relating, Translating and Structuring), held in Genova, October 1995, for many helpful comments and suggestions on a preliminary presentation of this work.

References

1. E. Astesiano and E. Zucca. D-oids: a model for dynamic data-types. *Mathematical Structures in Computer Science*, 5(2):257–282, June 1995.
2. E. Astesiano and E. Zucca. A free construction of dynamic terms. *Journ. of Computer and System Sciences*, 52(1):143–156, February 1996.
3. H. Baumeister. Relations as abstract datatypes: An institution to specify relations between algebras. In *TAPSOFT '95*, LNCS 915, pages 756–771, 1995. Springer Verlag.
4. M. Bidoit and A. Tarlecki. Behavioural satisfaction and equivalence in concrete model categories. In *CAAP '96*, LNCS, 1996. Springer Verlag. To appear.
5. M. Cerioli and J. Meseguer. May I borrow your logic? (transporting logical structures along maps). *Theoretical Computer Science*, 1996. To appear.
6. P. Dauchy and M.C. Gaudel. Implicit state in algebraic specifications. In *IS-CORE'93*, n. 01/93 in Informatik-Berichte, Universitaet Hannover, 1993.
7. H. Ehrig and F. Orejas. Dynamic abstract data types: An informal proposal. *Bull. of EATCS*, 53, June 1994.
8. J.A. Goguen and R.M. Burstall. Institutions: Abstract model theory for computer science. *Journ. ACM*, 39:95–146, 1992.
9. Z. Manna and A. Pnueli. *The Temporal Logics of Reactive and Concurrent Systems*. Springer Verlag, New York, 1992.
10. W. Pawlowski. Context institutions. In *11th ADT Workshop*, LNCS, 1996. Springer Verlag. To appear.
11. M. Große Rhode. *Specification of Transition Categories - An Approach to Dynamic Abstract Data Types*. PhD thesis, TU Berlin, 1995.
12. E. Zucca. From static to dynamic abstract data-types: An institution transformation. Tech. Rep. DISI-TR-1996-1, Univ. Genova, 1996. Submitted for publication.

Author Index

Springer-Verlag
and the Environment

We at Springer-Verlag firmly believe that an international science publisher has a special obligation to the environment, and our corporate policies consistently reflect this conviction.

We also expect our business partners – paper mills, printers, packaging manufacturers, etc. – to commit themselves to using environmentally friendly materials and production processes.

The paper in this book is made from low- or no-chlorine pulp and is acid free, in conformance with international standards for paper permanency.

Lecture Notes in Computer Science

For information about Vols. 1–1045

please contact your bookseller or Springer-Verlag

Vol. 1079: Z.W. Raś, M. Michalewicz (Eds.), Foundations of Intelligent Systems. Proceedings, 1996. XI, 664 pages. 1996. (Subseries LNAI).

Vol. 1080: P. Constantopoulos, J. Mylopoulos, Y. Vassiliou (Eds.), Advanced Information Systems Engineering. Proceedings, 1996. XI, 582 pages. 1996.

Vol. 1081: G. McCalla (Ed.), Advances in Artificial Intelligence. Proceedings, 1996. XII, 459 pages. 1996. (Subseries LNAI).

Vol. 1082: N.R. Adam, B.K. Bhargava, M. Halem, Y. Yesha (Eds.), Digital Libraries. Proceedings, 1995. Approx. 310 pages. 1996.

Vol. 1083: K. Sparck Jones, J.R. Galliers, Evaluating Natural Language Processing Systems. XV, 228 pages. 1996. (Subseries LNAI).

Vol. 1084: W.H. Cunningham, S.T. McCormick, M. Queyranne (Eds.), Integer Programming and Combinatorial Optimization. Proceedings, 1996. X, 505 pages. 1996.

Vol. 1085: D.M. Gabbay, H.J. Ohlbach (Eds.), Practical Reasoning. Proceedings, 1996. XV, 721 pages. 1996. (Subseries LNAI).

Vol. 1086: C. Frasson, G. Gauthier, A. Lesgold (Eds.), Intelligent Tutoring Systems. Proceedings, 1996. XVII, 688 pages. 1996.

Vol. 1087: C. Zhang, D. Lukose (Eds.), Distributed Artificial Intelliegence. Proceedings, 1995. VIII, 232 pages. 1996. (Subseries LNAI).

Vol. 1088: A. Strohmeier (Ed.), Reliable Software Technologies – Ada-Europe '96. Proceedings, 1996. XI, 513 pages. 1996.

Vol. 1089: G. Ramalingam, Bounded Incremental Computation. XI, 190 pages. 1996.

Vol. 1090: J.-Y. Cai, C.K. Wong (Eds.), Computing and Combinatorics. Proceedings, 1996. X, 421 pages. 1996.

Vol. 1091: J. Billington, W. Reisig (Eds.), Application and Theory of Petri Nets 1996. Proceedings, 1996. VIII, 549 pages. 1996.

Vol. 1092: H. Kleine Büning (Ed.), Computer Science Logic. Proceedings, 1995. VIII, 487 pages. 1996.

Vol. 1093: L. Dorst, M. van Lambalgen, F. Voorbraak (Eds.), Reasoning with Uncertainty in Robotics. Proceedings, 1995. VIII, 387 pages. 1996. (Subseries LNAI).

Vol. 1094: R. Morrison, J. Kennedy (Eds.), Advances in Databases. Proceedings, 1996. XI, 234 pages. 1996.

Vol. 1095: W. McCune, R. Padmanabhan, Automated Deduction in Equational Logic and Cubic Curves. X, 231 pages. 1996. (Subseries LNAI).

Vol. 1096: T. Schäl, Workflow Management Systems for Process Organisations. XII, 200 pages. 1996.

Vol. 1097: R. Karlsson, A. Lingas (Eds.), Algorithm Theory – SWAT '96. Proceedings, 1996. IX, 453 pages. 1996.

Vol. 1098: P. Cointe (Ed.), ECOOP '96 – Object-Oriented Programming. Proceedings, 1996. XI, 502 pages. 1996.

Vol. 1099: F. Meyer auf der Heide, B. Monien (Eds.), Automata, Languages and Programming. Proceedings, 1996. XII, 681 pages. 1996.

Vol. 1100: B. Pfitzmann, Digital Signature Schemes. XVI, 396 pages. 1996.

Vol. 1101: M. Wirsing, M. Nivat (Eds.), Algebraic Methodology and Software Technology. Proceedings, 1996. XII, 641 pages. 1996.

Vol. 1102: R. Alur, T.A. Henzinger (Eds.), Computer Aided Verification. Proceedings, 1996. XII, 472 pages. 1996.

Vol. 1103: H. Ganzinger (Ed.), Rewriting Techniques and Applications. Proceedings, 1996. XI, 437 pages. 1996.

Vol. 1104: M.A. McRobbie, J.K. Slaney (Eds.), Automated Deduction – CADE-13. Proceedings, 1996. XV, 764 pages. 1996. (Subseries LNAI).

Vol. 1105: T.I. Ören, G.J. Klir (Eds.), Computer Aided Systems Theory – CAST '94. Proceedings, 1994. IX, 439 pages. 1996.

Vol. 1106: M. Jampel, E. Freuder, M. Maher (Eds.), Over-Constrained Systems. X, 309 pages. 1996.

Vol. 1107: J.-P. Briot, J.-M. Geib, A. Yonezawa (Eds.), Object-Based Parallel and Distributed Computation. Proceedings, 1995. X, 349 pages. 1996.

Vol. 1108: A. Díaz de Ilarraza Sánchez, I. Fernández de Castro (Eds.), Computer Aided Learning and Instruction in Science and Engineering. Proceedings, 1996. XIV, 480 pages. 1996.

Vol. 1109: N. Koblitz (Ed.), Advances in Cryptology – Crypto '96. Proceedings, 1996. XII, 417 pages. 1996.

Vol. 1111: J.J. Alferes, L. Moniz Pereira, Reasoning with Logic Programming. XXI, 326 pages. 1996. (Subseries LNAI).

Vol. 1112: C. von der Malsburg, W. von Seelen, J.C. Vorbrüggen, B. Sendhoff (Eds.), Artificial Neural Networks – ICANN 96. Proceedings, 1996. XXV, 922 pages. 1996.

Vol. 1113: W. Penczek, A. Szałas (Eds.), Mathematical Foundations of Computer Science 1996. Proceedings, 1996. X, 592 pages. 1996.

Vol. 1114: N. Foo, R. Goebel (Eds.), PRICAI'96: Topics in Artificial Intelligence. Proceedings, 1996. XXI, 658 pages. 1996. (Subseries LNAI).

Vol. 1115: P.W. Eklund, G. Ellis, G. Mann (Eds.), Conceptual Structures: Knowledge Representation as Interlingua. Proceedings, 1996. XIII, 321 pages. 1996. (Subseries LNAI).

Vol. 1117: A. Ferreira, J. Rolim, Y. Saad, T. Yang (Eds.), Parallel Algorithms for Irregularly Structured Problems. Proceedings, 1996. IX, 358 pages. 1996.

Vol. 1120: M. Deza. R. Euler, I. Manoussakis (Eds.), Combinatorics and Computer Science. Proceedings, 1995. IX, 415 pages. 1996.

Vol. 1121: P. Perner, P. Wang, A. Rosenfeld (Eds.), Advances in Structural and Syntactical Pattern Recognition. Proceedings, 1996. X, 393 pages. 1996.

Vol. 1122: H. Cohen (Ed.), Algorithmic Number Theory. Proceedings, 1996. IX, 405 pages. 1996.

Vol. 1125: J. von Wright, J. Grundy, J. Harrison (Eds.), Theorem Proving in Higher Order Logics. Proceedings, 1996. VIII, 447 pages. 1996.